TWOSAMPLE test and c.i. [K% confidence] using C, C
   Subcommands: ALTERNATIVE = K
               POOLED
CORRELATION between C, · · · , C
COVARIANCE for C, · · · , C

9. REGRESSION

REGRESS Y in C using K predictors in C, · · · , C
   Subcommands: COEF into C
               RESIDUALS into C
               PREDICT for K, · · · , K
         or PREDICT for C, · · · , C
   Example: REGRESS C1 2 C2 C3 C4 C5 will regress Y in C1 using
            2 predictors in C2 and C3 as well as store the standardized
            residuals in C4 and the estimated Y values ($\hat{Y}$ values) in C5.
BRIEF output at level K
STEPWISE regression of C using predictors C, · · · , C
   Subcommands: FENTER = K
               FREMOVE = K
               FORCE C, · · · , C
               ENTER C, · · · , C
               REMOVE C, · · · , C

10. ANALYSIS OF VARIANCE

ONEWAY using data in C, levels in C
TWOWAY using data in C, levels in C, blocks in C
   Subcommand: ADDITIVE
(or TWOWAY using data in C, factor A levels in C, factor B levels in C)

11. NONPARAMETRIC STATISTICS

CHISQUARE using C, · · · , C
RUNS above and below K for data in C
MANN-WHITNEY [alternative = K] using C and C
WTEST using C
   Subcommand: ALTERNATIVE = K
KRUSKAL-WALLIS test for data in C, levels in C
RANK the values in C, put ranks into C

12. MISCELLANEOUS

IRANDOM K integers between K and K, put into C
ERASE C, · · · , C

13. SUBCOMMANDS

Some MINITAB commands have subcommands to convey additional
information. To use a subcommand, include a semicolon at the end
of the main command; the subcommands follow (one per line). End the
final subcommand with a period.

STATISTICS

STATISTICS

A COMPUTER INTEGRATED APPROACH

Alan H. Kvanli
NORTH TEXAS STATE UNIVERSITY

West Publishing Company
St. Paul   New York   Los Angeles   San Francisco

> **To Ann, who turns a routine day into a unique experience.**

Interior Design: Wendy Calmenson
Cover Design: Delor Erickson, Studio West, Inc.
Composition: Syntax International
Copyeditor: Virginia E. Dunn
Artwork: Rolin Graphics

COPYRIGHT © 1988 By WEST PUBLISHING COMPANY
50 W. Kellogg Boulevard
P.O. Box 64526
St. Paul, MN 55164-1003

Library of Congress Cataloging-in-Publication Data

Kvanli, Alan H.
    Statistics: a computer integrated approach/Alan H. Kvanli.
        p.   cm.
    Includes index.
    ISBN 0-314-60541-X
    1. Statistics—Data processing.   1. Title.
QA276.4.K93 1988
519.5—dc 19

87-31030
CIP

CONTENTS

PREFACE

The question of whether to tackle the seemingly overwhelming job of writing a textbook covering introductory statistics was not an easy one to answer. Certainly the world was not exactly banging on my office door demanding yet another textbook covering a subject that has changed very little over the years. Or has it? As I surveyed and used introductory statistics texts, I saw many that did a fine job of covering statistical concepts, but seemed to intimidate students with an "encyclopedia of statistics" writing style. In addition, none of them allowed the student full use of computers and statistical computer packages. Certainly the ancient statistical theory has not changed over the years but the appreciation and application of statistics has undergone a dramatic change as elaborate calculators and reasonably priced microcomputers have become available.

As a result, I became excited about this project because I believed that a need does exist for such a text. Since computer power within universities is growing at an accelerating (and cheaper) rate, a text integrating statistical software is a necessity in the marketplace. In an effort to appeal to all teaching styles, I decided to design a text which could be used by instructors wanting to fully utilize computer packages, yet one which could still be an excellent text for instructors who prefer a more standard, calculator-based approach.

The text is intended to be an introduction to basic statistics. I assume that the student has a good understanding of basic algebra. Reference is made on a few occasions to calculus applications, but no calculus background is required to read the material. The reading level is interesting and easy to understand without sacrificing any credibility in the descriptive material. It is a derivation free, but not a "black box," approach to teaching the appreciation and application of statistics. I've included a large number of illustrative examples gathered from many disciplines to better guide the student to an understanding of statistical concepts and applications.

To the Instructor

This text can be used for either a one- or two-semester introduction to statistics course. Suggested material to be covered in the first semester would be Chapters 1 through 8, in order, which concludes with an introduction to statistical inference (confidence intervals for the mean). Chapters 9, 10, and 11 could be included in a second-semester course, along with those remaining chapters that you feel are particularly relevant and of interest to your students.

The text has intentionally been written in somewhat of a conversational (user-friendly) style to make it less intimidating to the student. My intent was for the student to read the text; not just use it as a source of homework exercises.

The text fully integrates three popular statistical packages: **MINITAB**, **SAS** and **SPSS**. Both mainframe and microcomputer packages are illustrated and discussed. The featured package throughout all of the chapter examples is **MINITAB**, since these commands are simple English statements and are illustrative of computer capabilities whether or not you use the **MINITAB** package in your course instruction. Corresponding **SAS** and **SPSS** descriptions (mainframe and microcomputer versions) are contained at the ends of chapters—a feature unique to this text. I have fully integrated these packages throughout the text, making it possible for you to include computer usage as part of your course without having to spend a great deal of time explaining the mechanics of a particular package. An introduction to each of these packages is presented at the end of the text. For instructors who wish to avoid computer usage, the text allows for a calculator-based approach—the exercises do not require a computer package and contain reasonably sized data sets.

Other features of the text include:

- a Look Back/Introduction at the start of each chapter to tie the chapter to the relevant material from the preceding chapters. Each chapter closes with a summary section.

- an abundance of exercises using realistic situations. Each chapter also includes a case study containing a larger data set and requiring an in-depth discussion.

- a full treatment of the use of p values to make statistical decisions. These are derived and discussed throughout the entire text.

- three continuous distributions (normal, uniform, and exponential), along with four discrete distributions (uniform, binomial, hypergeometric, and Poisson).

- integration of nonparametric techniques into the chapters containing the corresponding parametric statistical procedures.

- various sampling procedures, along with corresponding sample estimators and confidence intervals, as separate sections in two of the earlier chapters. In this way, the instructor is able to cover this often neglected material without having to spend the time to cover an entire chapter.

□ separate chapters for inference regarding normal parameters (μ, σ) and inference on a binomial parameter (p). Chapters 8, 9, and 10 are strictly devoted to normal inference, both one population (8 and 9) and two populations (10). Binomial inference (one and two populations) is covered in Chapter 11.

□ a large database (1250 observations) containing data from a graduate student population (scores on the GRE, undergraduate and graduate GPA, age, sex, and other variables). Each chapter contains a separate set of exercises that utilize this database. In this way, each student can create, save, and use a subset of the database to better understand the concepts introduced in each chapter.

□ appendixes that provide an introduction to each of the three statistical packages utilized in the text (both mainframe and microcomputer versions).

The following material is also available:

□ an instructor's manual containing chapter objectives and solutions to all exercises.

□ a test bank containing true/false questions, completion exercises, and additional application problems.

□ a student study guide written to put students at ease and guide them through applications of the chapter material.

I certainly hope that this text will meet your classroom needs. If you care to offer comments and suggestions, I would like to hear from you. Address any correspondence to Al Kvanli, P.O. Box 13677, North Texas State University, Denton, Texas 76203.

To the Student

I believe you will find this text to be a readable, easily understood treatment of statistics. My intent is to carefully explain the various statistical concepts and strategies without getting bogged down in unnecessary theory. I have included many examples within each chapter to allow you to see how each procedure works. At the beginning of each chapter you will find a Look Back/Introduction section which will set up the chapter and tie it in with the previous chapters. At the end of each chapter is a summary containing all of the key definitions and concepts introduced within the chapter. At the end of the book you will find introductions to the three computer packages integrated into the text: **MINITAB**, **SPSS**, and **SAS**.

As the old adage goes, "practice makes perfect," and mastering statistics is no exception. To this end, I have included a large number of exercises to help you along the road to perfection. Also, you will find the solutions to the odd-numbered exercises at the end of the text. A study guide, which contains additional examples along with their solutions, has also been prepared. These solutions take you by steps through applications of the various statistical techniques introduced within each chapter.

Acknowledgments

I am extremely grateful to the many people who have helped me along the way. The editorial assistance of Denise Simon was always professional, encouraging, and of immense help. I would especially like to thank Cheryl McQueen who not only typed the entire first draft, but had the patience to hang in there and type the second draft as well. Her skills and enthusiasm are very much appreciated. Many thanks to Anis Kashani, Joanne Killmeyer, Colleen Tarry, Karen Harrell, and Karen Wiley who assisted in the preparation of the solutions manual. Graduate students Viwat Rerksirathia, Ray Mehraban, Abdolreza Rafeh, and Shekar Shetty assisted in providing reliable solutions and cross-checks for the solutions manual. Special appreciation goes to Susan Reiland who assisted in the manuscript preparation.

I'd like to extend a special note of gratitude to Jitendra Sarhad. His creative ability and knowledge of statistics has resulted in his being severely overworked (and underpaid) at times but nevertheless, he found the time to be a real stabilizing force behind this project. His tireless efforts and words of encouragement always came at the right time. His creative contributions (including the case studies and many of the exercises) appear throughout the text and I truly would not have survived this venture without his help.

I certainly want to thank and acknowledge the many reviewers who shared in this project. With special appreciation, I'd like to thank John Brevit, who authored an entertaining and very helpful study guide to accompany the text.

The following list contains the names and affiliations of the many reviewers whose assistance was invaluable in the preparation of this text:

Beyer, William—University of Akron
Boldt, Chris—Eastfield College
Brevit, John—Western Kentucky University
Dowdy, Shirley—West Virginia University
Gervase, Josephine—Manchester Community College
Glaz, Joseph—University of Connecticut
Higgins, James—Moraine Valley Community College
Hilton, James—Grossmont College
Jeffries, Don—Orange Coast College
Lacher, Robert—South Dakota State University
Lentner, Marvin—Virginia Polytechnic Institute and State University
Lund, David—University of Wisconsin, Eau Claire
Morgan, Ron—West Chester University of Pennsylvania
Norton, Julia—California State University, Hayward
Parker, Mary—Austin Community College
Shriner, Don—Frostburg State College
Sinclair, Charles—Portland State University
Turner, David—Utah State University

STATISTICS

A FIRST LOOK
AT STATISTICS

A FIRST LOOK AT STATISTICS

Until recently, many people thought a statistician was someone who helped figure batting averages during a baseball game broadcast. You might wonder how we can devote an entire textbook to compiling numbers and making simple calculations. Surely it cannot be that complicated!

Statistics is the science comprising rules and procedures for collecting, describing, analyzing, and interpreting numerical data. The applications of statistics are evident everywhere. Hardly a day goes by that we are not bombarded by such statements as:

Results show that Crest toothpaste helps prevent tooth decay.

The chance of a NASA rocket failure is higher than was quoted originally.

The state court has ruled that the XYZ Company is guilty of age discrimination in its termination procedure.

Or how about:

The Surgeon General has determined that cigarette smoking is dangerous to your health.

Besides using statistics to inform the public, statisticians help researchers in many areas of decision making.

The use of statistics began as early as the first century A.D., when governments used a census of land and properties for tax purposes. Census taking was gradually extended to include such local events as births, deaths, and marriages. The science of statistics, which uses a sample to predict or estimate some characteristics of a population, began its development during the nineteenth century.

Use of statistical methods has undergone a dramatic change as computers have entered the research environment. Researchers are now able to store and manipulate large collections of data, and once-formidable statistical calculations are reduced to a few key strokes. Sophisticated computer software allows users merely to specify the type of analysis desired and input the necessary data. This textbook will concentrate on three of these statistical packages: MINITAB (a statistical computer package originally designed at Pennsylvania State University specifically for students), SAS (Statistical Analysis System), and SPSS (Statistical Package for the Social Sciences).

Although most statistical functions are performed by professional statisticians, you may have to draw a valid conclusion from a statistical report. Occasionally, however, statistics can obscure the truth or give an erroneous impression. Anyone who has ever changed their plans due to a 90% chance of rainy weather only to sit home on a sunny day can attest to this fact. You often can avoid a bad decision by recognizing statistical errors and bias in the results that you review.

In addition, you may be asked to perform a statistical analysis. Although you may elect to obtain outside assistance, you will need to know when to consult statisticians and how to tell them what you need.

1.1   USES OF STATISTICS IN PRACTICE

Statistics are useful in many disciplines for both describing a collection of data and making a decision based upon these data. Anyone collecting such data, be it a sports announcer or a medical researcher, will quite likely need to use many of the techniques discussed within this text. When the toothpaste commercial states that their brand has been shown to outperform the competing brands, the company had better be able to support the claim with strong statistical evidence. In medical and psychological research, pure conjecture is not a basis for sound decision making, but a properly conducted statistical experiment may provide the necessary support for taking a particular action or medication.

It would be difficult to list all the applied disciplines that use statistical methodology as a basis for investigative research. Examples of the types of questions we will examine in the chapters to follow, gathered from eight different fields of study, are listed below. Illustrations and exercises within the chapter discussion will demonstrate how to approach these and many other research questions.

1 **Psychology** Does a person's memory recall ability improve under the influence of hypnosis?

2 **Education** Does a newly introduced package of geometrical designs and illustrations improve student learning of high school geometry?

3 **Sociology** Is the percentage of adults opposed to legalized abortion higher in the south than in the midwest?

4 **Business/Economics** Is there a difference in the proportion of new cars produced by two manufacturers that requires an engine overhaul before 100,000 miles?

5 **Political Science** Is there a relationship between a voter's age and his or her level of political conservatism?

6 **Medicine** Does a proposed drug used to prevent the common cold reduce one's chances of contracting the disease?

7 **Environmental Research** Based upon numerous water samples, is there sufficient evidence to indicate that the pollution level of the river running alongside the local chemical plant is above the safety level set by the federal government?

8 **Law** Is there evidence to suggest that company XYZ is guilty of age or sex discrimination in their termination of employees?

1.2 SOME BASIC DEFINITIONS

Statistics has specialized definitions for terms crucial to statistical reasoning. In **descriptive statistics**, we collect data and describe them. If we also analyze and interpret the data, we are using **inferential statistics**.

Descriptive statistics are used to describe a large set of data. For example, we can reduce the set of data values to one or more single numbers, such as the average of 150 test scores, or we can construct a graph that represents some feature of the data.

We use inferential statistics to form conclusions about a large group—a population—by collecting a portion of it—a sample. Thus, a **population** is the set of measurements (generally belonging to a group of people or objects) that is of interest. A **sample** is the portion of the population about which information is gathered.

The analyst decides what the population is. Typically, the population is so large that it would be nearly impossible to obtain information about every item in it. Instead, we obtain information about selected population members and attempt to draw a conclusion about all members. In other words, we attempt to infer something about the population using information about only some of the members of this population (that is, the sample). For this procedure to be valid, the sample must be typical of the entire population.

Figure 1.1

Illustration of a
population versus
a sample

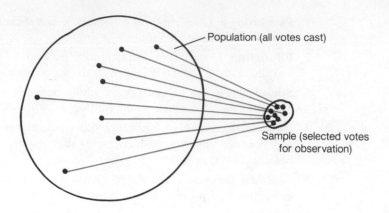

To make an early prediction of the election results for governor of California, analysts could use a sample of voters leaving the voting booths, as illustrated in Figure 1.1. The population consists of all the votes cast in the election. To make a valid statistical inference using a sample, it is crucial that the sample **represent** the population. One way to do this is to collect a sample of size n where each set of n people has the same chance of being selected for the sample. This is a **simple random sample** (Figure 1.1). It is akin to drawing names out of a hat where each name in the hat has the same chance of being pulled out. Thus if our population is all votes cast on the day of the gubernatorial election, a sample of votes cast in only one city would not be representative. We would have to guarantee that these votes would represent the voting of the entire state. A random sample obtained across the entire state would better represent this population.

As another illustration, a researcher in Ohio is interested in the proportion of male athletes attending state universities in Ohio who fail to graduate. Of particular interest is whether or not this value differs from the national rate of, say, 40%. By collecting a simple random sample she can learn (but not prove!) something about this percentage. If, for example, 75% of the individuals in the sample did not graduate, she might be led to believe that the Ohio graduation rate for male athletes is lower than the national average. Note that it is possible for her to conclude such a result when in fact the Ohio graduation rate for male athletes is higher than the national norm, since she is relying on inferential statistics to form this conclusion. Whenever an attempt is made to learn something about a population using sample results, there is always a chance of drawing an incorrect conclusion. The only way of being 100% sure is to list the entire population. Such a sample is called a **census**.

Describing and analyzing samples is basically what statistics is all about. Often we are interested in a particular characteristic of a population (such as the average age), referred to as a **parameter**. Information about this parameter may be obtained by finding the corresponding value in the sample, such as the average age of the 100 (or whatever) people in the sample. This value derived from the sample is defined as a **statistic**. So, a statistic describes a sample and a parameter describes a population. Later chapters will demonstrate how to **infer** something about the large population by using the corresponding statistic obtained from the much smaller sample.

1.3   DISCRETE AND CONTINUOUS NUMERICAL DATA

Proper use of numerical data can be a great aid in making a critical decision. However, using an improper technique or "bad data" can lead us down the wrong path. Generally, the technique we use to analyze data in statistics will depend on the nature of the data. We can distinguish between two types of numerical data.

How do the following two sets of numbers differ?

3, 5, 2, 1, 4, 4, 3, 5, 5, 1, 2, 4
4.31, 11.62, 5.37, 1.55, 3.71, 6.88, 7.23, 9.52, 2.36, 7.42, 6.11, 4.85

The primary difference is that the values in the first data set consist of counting numbers, or integers. Such data are **discrete**. For example, these data may be the coded responses from 12 people who answered a particular question in an opinion survey, where 1 = strongly agree, 2 = agree, 3 = uncertain, 4 = disagree, and 5 = strongly disagree. Note that discrete data may contain a decimal point. Nevertheless, such data have gaps in their possible values. For example, if you throw a single die twice and record the average of the two throws, the possible values are 1, 1.5, 2, 2.5, 3, 3.5, 4, 4.5, 5, 5.5, and 6.

Examples of discrete data that have integer values would include (1) the number of students having a reading deficiency, (2) the number of children in your family, and (3) the total of the two numbers appearing on a throw of two dice. Note that although the second illustration has infinite (theoretically, at least) possible values, the data are discrete. Your family cannot have 2.5 children.

Now consider the second data set in our original example. These data might represent the weight of 12 parcels received at a post office. A list of all the possible values of package weights would be extremely long. In fact, if our scale was completely accurate, the list would be infinite and any value would be possible. Such data are **continuous**: any value over some particular range is possible. There are no gaps in possible continuous data values. For example, although we may say Sandra is 5.5 ft tall, we mean her height is about 5.5 ft; in fact, this value may be 5.50372 ft. Height data are continuous. Or consider the contents of a coffee cup filled by a vending machine. Will the machine release exactly 6 oz every time? Certainly not. In fact, if you were to observe the machine fill five such cups and measure the contents to the nearest .001 oz, you might observe values of 6.031, 5.932, 5.871, 6.353, and 5.612 oz. Here again, any value between, say, 5.5 and 6.5 oz is possible; these are continuous data. Data such as weights, heights, areas, and time are generally continuous data and will be used in many of the examples contained in the chapters to follow.

> Data are classified as
>
> **1** discrete, if there are gaps in the possible values.
> **2** continuous, if any value is possible over some interval.

1.4   LEVEL OF MEASUREMENT FOR NUMERICAL DATA

As well as classifying numerical data as discrete or continuous, we can also label these data as to their **level of measurement**. We will discuss them in order of strength, beginning with the weakest. **Nominal data** are not really numerical at all, but are merely labels or assigned values. Examples include sex (1 = male, 2 = female), manufacturer of automobile (1 = General Motors, 2 = Ford, 3 = Chrysler), or color of eyes (1 = blue, 2 = green, 3 = brown). Assigning a **numerical code** to such data is merely a convenience so that, for example, we can store the information in a computer. Therefore, it makes no sense to perform calculations with such numbers, such as finding their average. What would it mean to claim that, "The average eye color is 2.73"? This is a meaningless statement. Generally, we are interested in the **proportion** of such data in each category. Consider the illustration of the Ohio graduation rate for male athletes in which each individual either graduated or did not graduate. We could assign the code 1 = graduated, 0 = did not graduate. The value of interest here is p, where p = proportion of male athletes who do graduate from college.

Ordinal data can be arranged in order, such as worst to best or F to A (grades on an exam). A classic example of ordinal data is the result of a cross-country race, where 10 people compete and 1 = the fastest (the winner), 2 = the runner-up, and so on, with 10 = the slowest. Here, the order of the values is important (3 finished before 4) but the difference of the values is not. For example, $2 - 1 = 1$ and $10 - 9 = 1$, but this does not imply that 1 and 2 were just as close in the final results as were 9 and 10.

The difference between values of **interval data** does have meaning. It is meaningful to add and average such data. The classic example is temperature, where it is true that the difference in heat between 60 and 61°F is the same as that between 80 and 81°F. Many of the techniques used to analyze data in statistics require data that are at least of this strength.

Figure 1.2

Classifications of numerical data

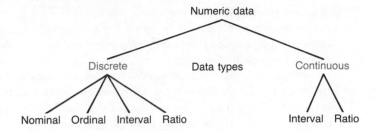

Ratio data differ from interval data in that there is a definite zero point. To decide if your data are interval or ratio, ask yourself whether twice the value is twice the strength. For example, is 100°F twice as hot as 50°F? The answer is no, so these data are interval. Is a 4-acre field twice as large as a 2-acre field? The answer is yes, so these are ratio data. Here the zero point is a field of zero acres. Typically, data consisting of areas, counts, volumes, and weights are ratio data.

The techniques used in statistics generally do not distinguish between interval and ratio data. A summary of the various data classifications is shown in Figure 1.2. Notice that discrete data can have any of the four levels of measurement, whereas continuous data must be interval or ratio.

SUMMARY

Decision making using statistical procedures continues to grow in popularity, since calculators and computers make it easier to avoid "seat of the pants" decisions by analyzing sample results in a scientific manner. Contemporary applications (such as "Does sex education in high school reduce the number of unplanned pregnancies?") covering a variety of disciplines were mentioned.

The science of statistics comprises a set of rules and procedures used to describe numerical data or to make decisions using these data. The group of measurements that are of interest define the **population**. That portion of the population that is selected for observation is a **sample**. Most statistical methods assume that a **simple random sample** of size n has been collected in which each set of n measurements has the same chance of being selected for the sample. A characteristic (such as average) of the population is referred to as a **parameter** and the corresponding sample characteristic is a **statistic**. For example, the average age of a sample of 250 women marrying for the first time, would be a statistic. The average age of *all* women marrying for the first time would be a parameter. A sample that contains the entire population is a **census**.

Descriptive statistics is concerned only with collecting and describing data. **Inferential statistics** is used when tentative conclusions about a population are drawn on the basis of data contained in a representative sample. The question of the effect of sex education upon the number of unplanned pregnancies posed earlier would be an example of inferential statistics if a study was done using a sample of women, some of whom had received sex education and some of whom had not.

Numerical data are either discrete or continuous. **Discrete data** have limited, specific possible values. **Continuous data** can assume any value over some range. A further classification of data is their level of measurement. At the lowest level, **nominal data** are categorical data that are assigned numeric codes. **Ordinal data** are ranked and the order of the data values is meaningful. In **interval data**, both the order of the data and the difference between any two data values have meaning. Finally, **ratio data** have all the properties of interval data and also contain a definite zero point. Most statistical techniques do not distinguish between interval and ratio data, but do require the data to have at least an interval level of measurement.

REVIEW EXERCISES

1.1   A researcher is interested in the amount of time that children spend watching television. She collects data from 100 randomly selected children. Explain what the sample is and what the population of interest might be.

1.2   The manager of Easy Fly Airlines took a sample of 200 people who regularly fly on Easy Fly Airlines and collected information on the ages of these people. The manager then graphed the data to get an idea of the distribution of the ages. After studying the graph, the manager concluded that most of the people who fly Easy Fly Airlines are in the middle age bracket.
(a)   Explain how the manager used descriptive statistics.
(b)   Did the manager use inferential statistics?

1.3   Explain whether the following groups of people or objects would represent a population or a sample.
(a)   A list of 30 students attending a large high school. (*Hint:* Could this be either a sample or a population? Explain.)
(b)   Forty students who were randomly stopped and questioned on a university campus.
(c)   A questionnaire mailed to 200 people randomly selected from the telephone book.
(d)   The list of all possible choices of two cards from a deck of 52 cards.

1.4   Explain whether the following data are continuous or discrete.
(a)   The annual incomes of 20 teachers.
(b)   The number of long distance calls made each month.
(c)   The length of time for each long distance call for a particular month.
(d)   The number of students with a reading deficiency.

1.5   Explain why inferential statistics is not needed if data are collected by taking a census.

1.6   What is the lowest level of measurement for a set of data which would permit the valid calculation of a proportion?

1.7   Your student record contains information including your age, sex, race, current grade point average, and current classification (Freshman, Sophomore, . . .). State whether each of these data is nominal, ordinal, interval, or ratio.

1.8   Do you think nominal data would usually be continuous data or discrete data?

1.9   Give an example of ordinal data that would not be interval data.

1.10 What is the highest level of measurement for each of the following data sets?
      (a)   The chain of command for officers in the army. .
      (b)   The IQs of 50 randomly selected students.
      (c)   The temperatures in Fahrenheit of several classrooms.
      (d)   The Social Security numbers of 12 randomly selected people.
      (e)   The marital status of 15 individuals attending a seminar.

EXERCISES USING THE DATABASE

1. For each of the variables defined in the database (Appendix H), determine if the corresponding data would be classified as discrete or continuous.

2. For each of the variables, what is the highest level of measurement for the corresponding data?

2

DESCRIPTIVE GRAPHS

A LOOK BACK/INTRODUCTION

Chapter 1 introduced some of the basic terms used in statistics. One of the key concepts was that of acquiring data using a sample from a population. It was also emphasized that the proper use of statistics depends on the nature of the data involved. Are the data discrete or continuous? Are the values nominal, ordinal, interval, or ratio?

Once the sample data have been gathered, the problem becomes learning whatever we can from them. One method is to describe the data by means of a graph. A graph allows us to discuss intelligently the "shape" of the data.

Everyone has heard the expression that a picture is worth a thousand words or, more appropriately here, a thousand numbers. This is especially true in statistics, where it may be vital to reduce a large set of numbers to a graph (or picture) that illustrates the structure underlying your data. Very often a quick glance at a graph demonstrates a point much more easily than does a page filled with numbers and words.

To see why you may want to describe a data set by using a descriptive graph, suppose that the vice-president of student affairs at Brookhaven University decided to obtain a sample of 50 students from the population of all sophomores, juniors, and seniors attending Brookhaven. For each student, she records (1) the classification (Sophomore, Junior, Senior), (2) the student's grade point average (GPA), and (3) the number of semester hours that the student has dropped (withdrawn from class) during the past 12 months. The results are shown in Table 2.1.

How can you summarize and present these data in a form that is easily understood? There are many graphical methods to do just that, depending on

Table 2.1

Data from Brookhaven sample

Classification	No. of Students	GPA/No. of Hours Dropped				
Sophomore	22	1.72 6	2.18 0	2.65 9	3.40 11	2.27 17
		1.57 3	3.18 8	1.55 8	1.92 9	3.24 0
		1.08 0	3.96 12	2.63 12	1.69 1	2.31 0
		3.02 3	1.86 13	2.88 6	3.29 12	3.77 11
		2.22 6	1.70 15			
Junior	16	2.47 14	2.11 12	2.40 0	2.76 3	1.44 3
		2.37 10	1.74 0	1.32 1	2.03 4	1.68 5
		1.71 0	3.88 3	2.27 3	1.62 6	2.79 13
		3.36 5				
Senior	12	2.19 5	1.47 0	3.41 6	1.66 0	2.38 17
		2.67 0	2.91 16	3.06 9	2.70 14	2.33 3
		2.83 0	3.75 0			

the nature of the data and what you are trying to demonstrate about them. When presenting data graphically, the first step usually is to combine the data values into a frequency distribution.

2.1   FREQUENCY DISTRIBUTIONS

We need to reduce a large set of data to a much smaller set of numbers that can be more easily comprehended. If you have recorded the population sizes of 500 randomly selected cities, there is no easy way to examine these 500 numbers visually and learn anything. It would be easier to examine a condensed version of this set of data, such as that presented in Table 2.2.

This type of summary, called a **frequency distribution**, consists of classes (such as "10,000 and under 15,000") and frequencies (the number of data values within each class). What do you gain using this procedure? You reduce 500 numbers to 10 classes and frequencies. You can study the frequency distribution in Table 2.2 and learn a great deal about the shape of this data set. For example,

Table 2.2

Frequency
distribution of
populations of
500 cities

Class Number	Size of City	Frequency
1	Under 10,000 people	4
2	10,000 and under 15,000	51
3	15,000 and under 20,000	77
4	20,000 and under 25,000	105
5	25,000 and under 30,000	84
6	30,000 and under 35,000	60
7	35,000 and under 40,000	45
8	40,000 and under 45,000	38
9	45,000 and under 50,000	31
10	50,000 and over	5
		500

approximately 50% of the cities in your sample have a population between 20,000 and 35,000. Also, only 1% of the cities contain 50,000 people or more.

Frequency Distribution for Continuous Data

A frequency distribution is typically condensed from sample data having an interval or ratio level of measurement. When you construct a frequency distribution for continuous data, you need to decide how many classes to use (10 in Table 2.2) and the class width (5000 in Table 2.2).

There is no "correct" **number of classes (K)** to use in a frequency distribution. However, you can best condense a set of data using between 5 and 15 classes. The usual procedure is to choose what you think would be an adequate number of classes and to construct the resulting frequency distribution. A quick look at the resulting distribution will tell you if you have reduced the data too much (not enough classes; K is too small) or not enough (too many classes; K is too large). If you have a very large set of data, you can use a larger number of classes than you would for smaller data sets. Whenever you construct frequency distributions using a computer, select several different values of K and look at the effects of the different choices.

Having chosen a value for K, the next step is to examine

$$\frac{\text{range}}{\text{number of classes}} = \frac{H - L}{K}$$

where H = the highest value in your data and L = the lowest value in your data. Round the result to a value that provides an easy-to-interpret frequency distribution. This value is the **class width (CW)**. For example, a class width of 100 would be easier to use than a class width of 97. The width of each class should be the same. Possible exceptions to this rule are for the first and last classes, which we will discuss later.

Suppose that, for a particular set of data, you have elected to use $K = 10$ classes in your frequency distribution. Suppose also that $H = 106$ and $L = 10$, and so

$$\frac{H - L}{K} = \frac{106 - 10}{10} = 9.6$$

The desirable class width here would be CW = 10.

Now let us use the 50 grade point averages in Table 2.1 to construct a frequency distribution of the GPAs, using six classes. Your first step should be to arrange the data from smallest to largest. This is called an **ordered array**. Both the original data and the ordered data are **raw data** since the values are not grouped into classes. The ordered grade point averages are listed in Table 2.3. Using the ordered data, $H = 3.96$ and $L = 1.08$. Since $K = 6$, you compute CW:

$$\frac{3.96 - 1.08}{6} = .48$$

The best choice here is CW = .5.

There are two rules to remember in selecting the first class: this class must contain L, the lowest data value, and it should begin with a value that makes

Table 2.3

50 grade point averages from the Brookhaven University sample arranged as an ordered array

Raw Data				
1.72	2.18	2.65	3.40	2.27
1.57	3.18	1.55	1.92	3.24
1.08	3.96	2.63	1.69	2.31
3.02	1.86	2.88	3.29	3.77
2.22	1.70	2.47	2.11	2.40
2.76	1.44	2.37	1.74	1.32
2.03	1.68	1.71	3.88	2.27
1.62	2.79	3.36	2.19	1.47
3.41	1.66	2.38	2.67	2.91
3.06	2.70	2.33	2.83	3.75

Ordered Data				
1.08	1.32	1.44	1.47	1.55
1.57	1.62	1.66	1.68	1.69
1.70	1.71	1.72	1.74	1.86
1.92	2.03	2.11	2.18	2.19
2.22	2.27	2.27	2.31	2.33
2.37	2.38	2.40	2.47	2.63
2.65	2.67	2.70	2.76	2.79
2.83	2.88	2.91	3.02	3.06
3.18	3.24	3.29	3.36	3.40
3.41	3.75	3.77	3.88	3.96

Table 2.4

Frequency
distribution of grade
point averages
using six classes

Class Number	Class	Frequency
1	1.0 and under 1.5	4
2	1.5 and under 2.0	12
3	2.0 and under 2.5	13
4	2.5 and under 3.0	9
5	3.0 and under 3.5	8
6	3.5 and under 4.0	4
		50

the frequency distribution easy to interpret. Because $L = 1.08$, the first class should begin with 1.0. The resulting frequency distribution is shown in Table 2.4.

Perhaps you think that six classes are not enough; that is, you have condensed this set of data too much. One indication of this would be that a large portion of your data (say, nearly 50%) lies in one class. Table 2.5 summarizes this set of data using $K = 8$ classes. Here, the class width chosen is CW = .4 because

$$\frac{H - L}{8} = \frac{3.96 - 1.08}{8} = .36$$

As before, the first class begins at 1.0. Notice that the last class contains values between 3.8 and 4.0 (rather than between 3.8 and 4.2), since 4.0 is the largest possible value for this situation.

This table also contains each **relative frequency**, where

$$\text{relative frequency} = \frac{\text{frequency}}{\text{total number of values in data set}}$$

Table 2.5

Frequency
distribution of grade
point averages using
eight classes

Class Number	Class	Frequency	Relative Frequency
1	1.0 and under 1.4	2	.04
2	1.4 and under 1.8	12	.24
3	1.8 and under 2.2	6	.12
4	2.2 and under 2.6	9	.18
5	2.6 and under 3.0	9	.18
6	3.0 and under 3.4	6	.12
7	3.4 and under 3.8	4	.08
8	3.8 and under 4.0	2	.04
		50	1.0

So, for example, in class 2, the relative frequency is .24; this class contains 12 out of the 50 values. The advantage of using relative frequencies is that the reader can tell immediately what percentage of the data values lie in each class.

COMMENTS   The highest and lowest values used to define a class are the **class limits**. For example, in Table 2.4, the lower class limit of class 2 is 1.5 and the upper class limit is 2.0. The **class midpoints*** are those values in the center of the class. Each midpoint in a sense "represents" its class. These values often are used in a statistical graph as well as for calculations performed on the information contained within a frequency distribution. The midpoint of class 2 in Table 2.4 is $(1.5 + 2.0)/2 = 1.75$.

Often, a set of sample data will contain one or two very small or very large numbers quite unlike the remaining data values. Such values are called **outliers**. It is generally better to include these values in one or two **open-ended classes**. The distribution in Table 2.2 contains two open-ended classes: class 1 (under 10,000) and class 10 (50,000 and over). You may need an open-ended class if your data set includes one or more outliers or your present frequency distribution has too many empty classes on the low or high end.

Constructing a Frequency Distribution

1 Gather the sample data.

2 Arrange the data in an ordered array.

3 Select the number of classes to be used.

4 Determine the class width.

5 Determine the class limits for each class by first selecting a lower class limit for the first class which provides a frequency distribution that will be easy to interpret.

6 Count the number of data values in each class (the class frequencies).

7 Summarize the class frequencies in a frequency distribution table.

Frequency Distribution for Discrete Data

When data are discrete, the procedure is almost the same as when they are continuous, except (1) we define the class width (CW) to be the difference between the lower class limits and not the difference between an upper and lower limit (this will also work for continuous data) and (2) the description of each class is slightly different because we no longer use the "and under" definition of each class. Thus, if CW = 5 and the data are continuous, our classes might be 5 and under 10, 10 and under 15, and 15 and under 20. If the data are discrete, they might be 5 to 9, 10 to 14, and 15 to 19. Note that, for the continuous

---

* Class midpoints are often referred to as class marks.

Table 2.6

Frequency distribution of the number of semester hours dropped by 50 students at Brookhaven University

Class Number	Class	Frequency	Relative Frequency
1	0–2 ₅0·5	14	.28
2	3–5	11	.22
3	6–8	7	.14
4	9–11	6	.12
5	12–14	8	.16
6	15–17	4	.08
		50	1.0

data, the class midpoints are 7.5, 12.5, and 17.5. For the discrete data, however, the midpoints are 7, 12, and 17.

Using the sample data in Table 2.1, we can construct a frequency distribution using six classes for the number of semester hours dropped by the 50 students at Brookhaven University. First we develop an ordered array:

0, 0, 0, 0, 0, 0, 0, 0, 0, 0, 0, 0, 1, 1, 3, 3, 3, 3, 3, 3, 3, 4, 5, 5, 5, 6, 6, 6, 6, 6, 8, 8, 9, 9, 9, 10, 11, 11, 12, 12, 12, 12, 13, 13, 14, 14, 15, 16, 17, 17

So, $H = 17$ and $L = 0$. Since

$$\frac{H - L}{K} = \frac{17 - 0}{6} = 2.83$$

we use CW = 3. The resulting frequency and relative frequency distribution is shown in Table 2.6.

EXERCISES

2.1 The following are the scores of the students at a junior college on a statistics exam:

69, 47, 82, 73, 99, 97, 55, 18, 100, 85, 77, 80, 94, 79, 66, 81, 81, 88, 94, 70, 62, 58, 43, 21, 85, 68, 50, 43, 91, 85, 60, 45, 88, 95, 46, 59, 75, 80, 74, 71, 70

(a) Convert the raw data into an ordered array.
(b) If you have to transform the data into a frequency distribution, what value of K would you use? (K = the number of classes).
(c) Calculate the class width for the frequency distribution.
(d) Present the data in the form of a frequency distribution.
(e) What is the difference between frequencies and relative frequencies?
(f) Calculate the frequencies and the relative frequencies of the scores of the junior college students.

2.2 Suppose that the following data are the number of words in the vocabulary of children in a particular age-group:

205, 377, 292, 300, 179, 240, 300, 190, 680, 250, 180, 170, 211, 266, 303, 350, 375, 288, 360, 225

(a)   Are these data discrete or continuous?
(b)   Construct a frequency and relative frequency distribution.
(c)   Identify the class midpoints.
(d)   Is any outlier present in the data?

2.3   A survey was conducted to find out how long homemakers spend food shopping each week. One hundred homemakers were selected randomly in a telephone survey and asked to state the amount of time that they spent food shopping during the past week. The results were:

Hours Shopping	Frequency
0 and under 2	38
2 and under 4	31
4 and under 6	21
6 and under 8	6
8 and under 10	3
10 and over	1

(a)   Construct a relative frequency distribution.
(b)   What are the class limits?
(c)   What are the class midpoints?
(d)   Before the survey, the researchers believed that most homemakers spent no more than 2 hours per week food shopping. Do the data for these 100 homemakers support that opinion?

2.4   Rapid eye movements (REMs) during sleep are associated with periods of dreaming. The length of time for which REM activity was observed was recorded in the case of 18 subjects (time in minutes):

7.00, 7.75, 9.50, 11.60, 10.55, 7.75, 12.00, 10.75, 12.51, 10.91, 8.30, 9.71, 10.50, 11.60, 6.25, 11.75, 9.75, 10.00

(a)   Construct a frequency distribution, using a class width of 1 minute.
(b)   What are the class midpoints?
(c)   What are the lower class limits?

2.5   In evaluating the performance of traffic policemen, one of the measures used was the number of tickets issued for traffic violations. The following data were recorded over a 30-day period:

4, 16, 11, 23, 17, 7, 2, 16, 15, 7, 14, 20, 4, 10, 18, 5, 6, 10, 9, 18, 22, 2, 1, 19, 15, 18, 15, 17, 23, 14

(a)   Are the data discrete or continuous?
(b)   Construct a relative frequency distribution.

2.6   Comment on the "correct" number of classes to be used in a frequency distribution.

2.7   The following data represent the percentage of the labor force estimated to be illegal aliens for certain selected industries in the South:

8.6, 6.2, 11.5, 4.6, 4.5, 8.6, 10.8, 8.9, 1.4, 2.2, 19.2, 32.9, 4.3, 5.5, 8.6, 4.3, 8.5, 6.5, 1.6, 4.0

(a)   Construct a frequency distribution for these data.
(b)   Identify the class width. Is this the only class width that may be used?

2.2 HISTOGRAMS

After you complete a frequency distribution, your next step is to construct a "picture" of these data values using a histogram. A **histogram** is a graphical representation of a frequency distribution. It describes the shape of the data. You can use it to answer quickly such questions as "Are the data symmetric?" and "Where do most of the data values lie?" For the frequency distribution in Table 2.4, the corresponding histogram is illustrated in Figure 2.1. The height of each box represents the frequency of that particular class. The edges of each box represent the class limits.

In a histogram, the boxes must be adjoining (no gaps). For discrete data (such as that in Table 2.6), the right edge of each box is midway between the upper limit of the class contained in this box and the lower limit of the next class. For example, a histogram of Table 2.6 will contain a box between −0.5 and 2.5 (with a height of 14), the next box between 2.5 and 5.5 (height of 11), the next box between 5.5 and 8.5 (height of 7), and so forth. The final box (height of 4) will extend from 14.5 to 17.5.

Avoid constructing a "squashed" histogram by using the vertical axis wisely. The top of this axis (15 in Figure 2.1) should be a value close to the largest class frequency.

A histogram can be constructed using the relative frequency rather than the frequencies. A **relative frequency histogram** of the data in Table 2.4 is shown in Figure 2.2. Notice that the shape of a frequency histogram (Figure 2.1) and a relative frequency histogram (Figure 2.2) are the same. One advantage of using a relative frequency histogram is that the units on the vertical axis are

Figure 2.1

Frequency histogram for the frequency distribution shown in Table 2.4

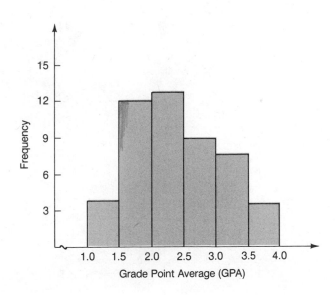

Figure 2.2

Relative frequency
histogram of
the frequency
distribution in
Table 2.4

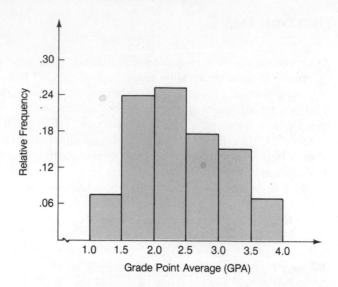

always between zero and one, so the reader can tell at a glance what percentage of the data lies in each class.

Most standard statistical software packages will construct a histogram from your data. Using MINITAB, you can specify the class width and the starting class midpoint, or you can let MINITAB select these values. The output will contain the frequency distribution as well as a graphical representation in the form of a histogram (without the boxes). MINITAB will provide each class

 Figure 2.3

Histogram using
MINITAB where
CW and the first
class midpoint are
not specified

```
MTB > SET INTO C1
DATA> 1.72 2.18 2.65 3.40 2.27 1.57 3.18 1.55 1.92 3.24
DATA> 1.08 3.96 2.63 1.69 2.31 3.02 1.86 2.88 3.29 3.77
DATA> 2.22 1.70 2.47 2.11 2.40 2.76 1.44 2.37 1.74 1.32
DATA> 2.03 1.68 1.71 3.88 2.27 1.62 2.79 3.36 2.19 1.47
DATA> 3.41 1.66 2.38 2.67 2.91 3.06 2.70 2.33 2.83 3.75
DATA> END
MTB > PRINT C1
   C1
    1.72    2.18    2.65    3.40   2.27    1.57   3.18   1.55   1.92   3.24
    1.08    3.96    2.63    1.69   2.31    3.02   1.86   2.88   3.29   3.77
    2.22    1.70    2.47    2.11   2.40    2.76   1.44   2.37   1.74   1.32
    2.03    1.68    1.71    3.88   2.27    1.62   2.79   3.36   2.19   1.47
    3.41    1.66    2.38    2.67   2.91    3.06   2.70   2.33   2.83   3.75

MTB > HISTOGRAM OF C1

   Histogram of C1    N = 50

   Midpoint    Count
      1.2         2    **
      1.6        12    ************
      2.0         6    ******
      2.4         9    *********
      2.8         9    *********
      3.2         6    ******
      3.6         4    ****
      4.0         2    **
```

Figure 2.4

MINITAB histogram
using specified
classes where
CW = .5 and the first
midpoint is 1.25

```
MTB > HISTOGRAM OF C1, FIRST MIDPOINT AT 1.25, WIDTH IS .5

  Histogram of C1    N = 50

  Midpoint    Count
    1.250        4    ****
    1.750       12    ************
    2.250       13    *************
    2.750        9    *********
    3.250        8    ********
    3.750        4    ****
```

frequency next to the corresponding class midpoint (not class limits). Figure
2.3 contains the necessary MINITAB statements and the resulting output, where
the class width and the midpoint of the first class were not specified. Compare
this to the frequency distribution in Table 2.5. Figure 2.4 specified CW = 0.5
and the first midpoint to be 1.25. One can use the output as it appears or use
this information to construct Figure 2.1, which is a graphical representation of
Table 2.4.

2.3   FREQUENCY POLYGONS

Although a histogram does demonstrate the shape of the sample data, perhaps
a clearer method of illustrating this is to use a **frequency polygon**. Here, you
merely connect the midpoints of the top of the histogram boxes (located at the
class midpoints) with a series of straight lines. The resulting multisided figure
is a frequency polygon. Figure 2.5 is an example; once again, the data in Table
2.4 were used.

Figure 2.5

Frequency polygon
using the frequency
distribution in
Table 2.4

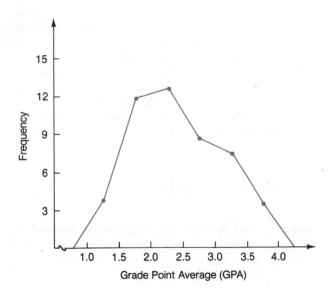

Figure 2.6

Frequency polygon
using footnotes to
handle open-ended
classes. The data
are from Table 2.2.

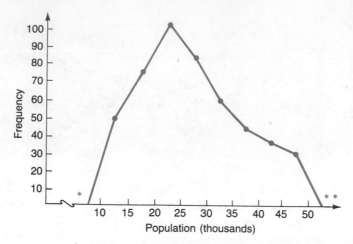

* 4 cities had populations of less than 10,000.
* * 5 cities had populations of 50,000 or greater.

COMMENTS   The polygon can also be constructed from the relative frequency histogram. The shape
will not change, but the units on the vertical axis will now represent relative frequencies.

The polygon must begin and end at zero frequency (as in Figure 2.5). To accomplish this, imagine a class at each end of the corresponding histogram that is empty
(contains no data values). Begin and end the polygon with the class midpoints of these
imaginary classes. Thus, the vertical axis must begin at zero. This need not be true for
the horizontal axis.

How do you handle an open-ended class? The easiest way is to construct a frequency
polygon of the closed classes and place a footnote at each open-ended class location
indicating the frequency of that particular class. Figure 2.6 demonstrates this, using the
data from Table 2.2.

Frequency polygons are usually better than histograms for comparing the shape
of two (or more) different frequency distributions. For example, Figure 2.7 demonstrates
at a glance that women are less satisfied (for the most part) than men in the samples
of males and females selected for the study.

Both histograms and frequency polygons represent the actual number of data values
in each class. Suppose that your annual salary is one of the values contained in a sample
of 250 salaries. One question of interest might be, "What fraction of the people in the
sample have a salary less than mine?" Such information can be displayed using a statistical graph called an ogive.

2.4   CUMULATIVE FREQUENCIES (OGIVES)

Another method of examining a frequency distribution is to list the number of
observations (data values) in the sample that are less than each of the class limits
rather than how many are in each of the classes. You are then determining

Figure 2.7

Frequency polygons
showing scores on a
marriage satisfaction
questionnaire using
samples of males
and females

cumulative frequencies. Take another look at the frequency distribution in
Table 2.4. Table 2.7 shows the cumulative frequencies. Notice that you can
determine cumulative frequencies (column 2) or cumulative relative frequencies
(column 4). The results in Table 2.7 can be summarized more easily in a simple
graph called an **ogive** (pronounced o-jive). The ogive is useful whenever you
want to determine what percentage of data lies below a certain value. Figure 2.8
is constructed by noting that

$$4 \text{ values } (4/50 = .08) \text{ are less than } 1.5$$

$$4 + 12 = 16 \text{ values } (16/50 = .32) \text{ are less than } 2.0$$

$$16 + 13 = 29 \text{ values } (29/50 = .58) \text{ are less than } 2.5, \text{ and so on.}$$

The ogive allows you to make such statements as "Seventy-six percent of the
grade point averages were less than 3.0" and "Fifty percent of the grade point
averages were under 2.35."

Table 2.7

Grade point averages
from Table 2.3
with cumulative
frequencies analyzed

Class Number	Class	(1) Frequency	(2) Cumulative Frequency	(3) Relative Frequency	(4) Cumulative Relative Frequency
1	1.0 and under 1.5	4	4	.08	.08
2	1.5 and under 2.0	12	16	.24	.32
3	2.0 and under 2.5	13	29	.26	.58
4	2.5 and under 3.0	9	38	.18	.76
5	3.0 and under 3.5	8	46	.16	.92
6	3.5 and under 4.0	4	50	.08	1.00
		50		1.0	

Figure 2.8

Ogive for
cumulative relative
frequencies using
data from Table 2.7

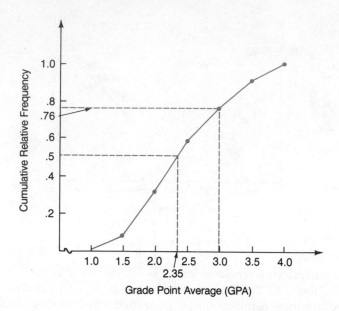

You always begin at the lower limit of the first class (1.0 here). The cumulative relative frequency at that point is always zero because the number of data values less than this number is zero. You always end at the upper limit of the last class (4.0 here). The cumulative relative frequency here is always one because all the data values are less than this upper limit. This ogive value would be n = the number of data values (n = 50 here) if you are constructing a frequency ogive rather than a relative frequency ogive. However, the shape of the ogive is the same for both procedures.

The ogives discussed here are "less than" ogives. Such an ogive always rises from left to right or is flat but never decreases. For example, the number of data values less than 2.0 must be at least as large as the number of values less than 1.5. A "greater than" ogive plots the percentage of data values greater than each class limit. Such an ogive always decreases or remains flat if a particular class is empty.

EXERCISES

2.8   The following were the daily maximum temperatures in Dallas, Texas, for the month of June (in degrees Fahrenheit):

84, 84, 94, 97, 97, 89, 90, 95, 99, 94, 88, 91, 90, 97, 93, 91, 88, 89, 102, 100, 88, 85, 88, 106, 102, 86, 93, 90, 105, 99

  (a)   Convert the data into an ordered array.
  (b)   Present the data in the form of a frequency distribution using 6 classes.
  (c)   Calculate the relative frequencies and the cumulative relative frequencies.

2.9   Using the data in exercise 2.3, construct a relative frequency histogram and an ogive.

2.10   Compute the cumulative relative frequency distribution for the data in exercise 2.2.

2.11 Draw a frequency polygon and an ogive for the data in exercise 2.4.

2.12 Does the shape of an ogive change if the cumulative relative frequencies are used instead of the cumulative frequencies?

2.13 The following is the distribution in a certain town of the population between ages 5 and 39 for the year 1987:

Age	Number
5–9	30,116
10–14	14,633
15–19	29,424
20–24	40,146
25–29	29,424
30–34	44,555
35–39	40,100

(a) Construct a frequency histogram.

(b) What does the shape of the histogram indicate?

(c) If a histogram was constructed from the relative frequency distribution, would the shape of the histogram change? Try it, if you are not sure.

2.14 Draw a frequency histogram that indicates the scores of the students on the statistics exam given in exercise 2.1.

2.15 Price/earnings ratio (P/E) is one of the important ratios for the investor. The following is a list of P/E ratios for some of the major U.S. corporations in the year 1985:

Company	P/E Ratio
Burroughs Corp	11
Gulf & Western	12
Borden	10
Anheuser-Busch Inc.	11
AT & T	17
International Paper	34
Texas Instruments	8
McGraw-Hill	16
General Instruments	18
Polaroid	36
Pennzoil	22
J C Penney	8
Phillips Petroleum	8
Control Data	35
Merrill Lynch	29
Uniroyal	14
Texaco	36
Coca Cola	15
Walmart	24
IBM	12

Source: Wall Street Journal, April 16, 1985.

(a) Construct a frequency distribution.

(b) Construct a histogram for the P/E ratios for the above 20 corporations.

2.16 A frequency distribution of certain discrete data has the following class midpoints:

6, 11, 16, 21, 26

What are the lower and upper class limits for each of the five classes?

2.17 Using the labor force percentages for illegal aliens in exercise 2.7, construct a frequency polygon and an ogive.

2.18 A country library's records show the following information regarding the number of patrons who used the library during the past 30 days:

100, 87, 44, 53, 17, 34, 88, 67, 31, 40, 98, 77, 55, 41, 73, 62, 88, 28, 70, 51, 82, 44, 32, 50, 33, 49, 59, 67, 79, 84

  (a)   Construct a cumulative frequency distribution.
  (b)   Convert the cumulative frequency distribution in (a) into an ogive graph.
  (c)   The number of patrons attending the library was less than or equal to what value 80% of the time?

2.19 The following is a frequency distribution of the number of daily automobile accidents reported for a month in Newark, New Jersey.

Accidents per Day	Frequency
0–3	12
4–7	10
8–11	7
12–15	1
16–19	1

  (a)   Construct a cumulative relative frequency distribution for the data.
  (b)   What percentage of the time do eight or more daily accidents occur?

2.20 Twenty individuals entering a reading speed improvement program were tested to determine their present speed. The following results were obtained (reading rate in words per minute):

250, 185, 225, 348, 403, 195, 380, 409, 500, 175, 400, 278, 515, 618, 267, 354, 400, 315, 376, 288

  (a)   Construct a frequency distribution and histogram.
  (b)   Would a relative frequency histogram have a shape different from the histogram in part (a)?
  (c)   Compute the cumulative relative frequencies and construct an ogive.

2.5   BAR CHARTS

Histograms, frequency polygons, and ogives are used for sample data having an interval or ratio level of measurement. For data having a nominal level, we use a bar chart. For situations producing a sample of ordinal level data with a reasonable set of possible values (such as 1 = strongly agree, 2 = agree, . . . , 5 = strongly disagree) a bar chart can be used to summarize the sample. A bar

Figure 2.9

Bar chart for
number of students
in each classification
using data from
Table 2.1

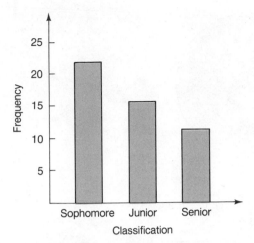

Figure 2.9

Bar chart for
number of students
in each classification
using data from
Table 2.1

chart is similar to a histogram in that the height of each box is proportional to the frequency of that class. Such a graph is most helpful when you have many categories to represent.

Consider the data in Table 2.1. If you are interested in the number of students in each classification, a bar chart will do a good job of summarizing this information (Figure 2.9). Notice that a gap is inserted between each of the boxes in a bar chart. The data here are nominal, so the length of this gap is arbitrary.

Figure 2.10 is an example of a bar chart in which the boxes are constructed horizontally rather than vertically. This enables you to label each category within the box.

Figure 2.10

Bar chart using
horizontal bars
which makes it easy
to place labels
within the boxes

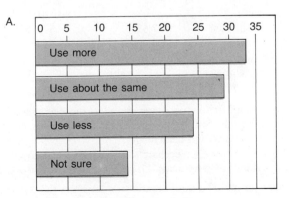

Q. If the price of natural gas goes down by 25% in the next few years, would you and your family use more or less?

Figure 2.11

Pie chart of number
of students in each
classification using
data from Table 2.1

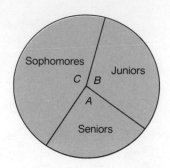

2.6   PIE CHARTS

A **pie chart** is used to split a particular quantity into its component pieces, typically at some specified point in time or over a specified time span. It is a convenient way of representing percentages or relative frequencies rather than frequencies. Figure 2.11 shows a pie chart of the 50 classifications in Table 2.1. To construct a pie chart, draw a line from the center of the circle to the outer edge. Then construct the various pieces of the pie chart by drawing the corresponding angles. For example, the seniors represent 24% of the total number of students (12 out of 50). So angle A in Figure 2.11 is 24% of 360° or 86.4°. Angle B is 32% of 360°, or 115.2°, and angle C is 44% of 360°, or 158.4°.

EXERCISES

2.21 High cholesterol levels in the blood have been associated with an increased risk of heart attack, and diet seems to affect cholesterol levels. Researchers have found that fats with higher percentages of monounsaturated fatty acids reduce the cholesterol level in the blood. The following is the breakdown for some common oils:

Type of Oil	Monounsaturated	Polyunsaturated	Saturated
Olive oil	72.3%	10.6%	17.1%
Peanut oil	48.1%	31.5%	20.4%
Butter	30.0%	4.0%	66.0%
Corn oil	27.6%	57.9%	14.5%
Soybean oil	23.4%	61.0%	15.6%
Safflower oil	13.1%	77.7%	9.2%

Source: University of Texas Health Science Center at Dallas, *Dallas Times Herald,* May 7, 1986.

(a) Construct a bar graph showing the percentage of monounsaturates in the six oils. Similarly, construct two more bar charts for the polyunsaturated and saturated percentages.

(b) Construct a pie chart showing the breakdown of olive oil into the three kinds of fatty acids. Draw pie charts for the other five oils.

2.22 An educational testing organization classifies the test applicants into six categories. Last year, the organization experienced the following applications:

Category	Number of Applicants
Humanities	8,544
Social Sciences	12,304
Life Sciences	8,886
Physical Sciences	14,354
Law	15,721
Other	8,545
Total	68,354

(a) Summarize the information in a bar chart.

(b) Why would a histogram be inappropriate for these data?

(c) Prepare a pie chart for the above data.

2.23 The Consumer Price Index (CPI) is a widely followed measure of inflation. Every 10 years or so, the government revises the CPI to keep it current. Social Security benefits, federal retirement system payments, IRS computations of income tax brackets, and private sector collective bargaining agreements are some of the things affected by the CPI. The last revision of the CPI was in 1978 and was based on 1972–1973 spending patterns. In March 1987, the Bureau of Labor Statistics introduced a revised CPI based on 1982–1984 spending patterns. The following figures provide a quick summary of how the average consumer dollar was spent in the 1970s and 1980s:

Expenditure	Old CPI	New CPI
Housing	38.2%	42.6%
Food and beverage	20.6	17.8
Medical care	5.0	4.8
Apparel and upkeep	7.5	6.5
Transportation	19.1	18.7
Entertainment	4.8	4.4
Other goods and services	4.8	5.2
	100.0	100.0

Source: Bureau of Labor Statistics, Washington, DC, March 1987.

(a) Summarize the information in the form of two pie charts.

(b) What area represents the largest piece of each chart? Is it very much larger than the next piece? How much?

(c) Comment on changes in spending patterns between the old and current CPI.

2.24 The following data indicate the total motor fuel consumption in millions of gallons on the highway for the year 1983 for selected states:

State	Fuel Consumption	State	Fuel Consumption
Wisconsin	1886	Kansas	1147
Arkansas	1082	Alabama	1790
Louisiana	2076	Iowa	1280
Washington	1788	Minnesota	1750
Tennessee	2263	Virginia	2424
Maryland	1849	Missouri	2369
Mississippi	1104	Georgia	2755

Source: Adapted from *The World Almanac and Book of Facts*, 1985.

Construct a cumulative relative frequency distribution and an ogive using these data.

2.25 The number of suicides in hundreds for the past 20 years for a certain metropolitan area are summarized by age in years as follows:

Age Group	Number of Suicides
10–14	.8
15–19	8.0
20–24	16.9
25–29	17.6
30–34	15.7
35–39	15.6
40–44	16.1
45–49	16.7
50–54	17.6
55–59	17.8

(a)   Construct a histogram using the frequency distribution.
(b)   Draw a frequency polygon.
(c)   Comment on the shape of the distribution of the data.

2.26 The following data indicate the percentage of U.S. petroleum imports by source
for the year 1983:

Nation	Percentage of U.S. Petroleum Imports
Algeria	4.8
Indonesia	6.7
Saudi Arabia	6.7
Iran	0.9
Venezuela	8.4
United Arab Emirates	0.6
Canada	10.8
Mexico	16.4
Virgin Islands	5.6
Others	39.1
Total	100.0

Source: The World Almanac and Book of Facts, 1985, p. 176.

Construct a pie chart to illustrate these percentages.

2.27 The following table contains the traffic for the major airports in the United States
in 1983 (total takeoffs and landings, in thousands):

Airport	Traffic
Chicago O'Hare	671.7
Long Beach	422.2
Santa Ana	457.8
Van Nuys	494.3
Atlanta	612.8
Los Angeles	506.1
Dallas/Ft. Worth	435.5
Oakland	356.8
Denver Stapleton	458.1

Source: The World Almanac and Book of Facts, 1985, p. 203.

Present the data in the form of a bar graph.

2.28   The following are various current assets of nonfinancial corporations for the year 1983.

Current Assets	Amount (in billions of $)
Cash	165.8
U.S. government securities	30.6
Notes and accounts receivable	577.8
Inventories	599.3
Other assets	183.7
Total	1557.3

Source: Economic Indicators, January 1985, p. 29.

Summarize the above information using a pie chart.

2.29   "During the 1984–1985 school year, there were 343,113 foreign students enrolled in U.S. colleges and universities. States with the largest number of foreign students were California (47,318), New York (31,064), Texas (29,429), Florida (17,658), Massachusetts (16,357), and Illinois (13,935)." (*Source:* Allen E. Kaye's Immigration column in *INDIA ABROAD*, 43 West 24th St., 7th Floor, New York, NY 10010. March 14, 1986.) The remainder were in other states. Summarize this information in a pie chart.

2.30   Why are gaps inserted between boxes in a bar chart, but not in a histogram using continuous data? How large should the gaps be?

2.7   DECEPTIVE GRAPHS

You might be tempted to be creative in your graphical displays by using, for example, a three-dimensional figure. Such originality is commendable, but does your graph accurately represent the situation? Consider Figure 2.12, which someone drew in an attempt to demonstrate that there are twice as many male teachers as female teachers in a particular high school. The artist constructed a

Figure 2.12

The illustrator wished to show that there are twice as many male teachers as female teachers in a particular high school. However, box B is twice the height and twice the depth of box A, and thus is *four* times the volume.

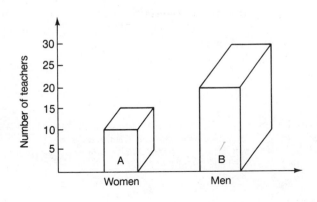

Figure 2.13

Graph of the divorce rate of two cities. The graph is misleading because the vertical axis does not start at zero.

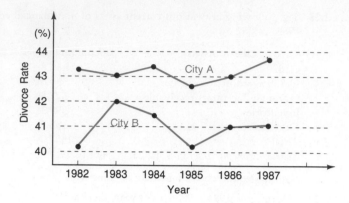

box for the men category twice as high, but also twice as deep as that for the women category. The result is a rectangular solid for men that is, in fact, four times the volume of the one for women. So the illustration is misleading; it appears that there are four times as many male teachers as female teachers.

Another type of deceptive graph is one where the vertical axis is scaled in such a way that small differences appear to be much larger. For example, Figure 2.13 contains a graph of the divorce rate for two cities, plotted over a 6-year period. A glance at this figure might lead you to believe that the divorce rate in City A is nearly twice that for City B. A closer look, however, reveals that *the vertical axis did not start at zero*, which can seriously distort the information contained in such a graph. The 1987 divorce rate for City A appears to be roughly twice that for City B. However, the actual rates are 43.8 for City A and 41.0 for City B. Granted, the rate is higher in City A for the 6 years, but not to the extent that Figure 2.13 seems to indicate.

Such examples, and many others, are contained in an entertaining and enlightening book by Darrell Huff entitled *How to Lie with Statistics*.* Other deceptive graphs mentioned by Huff include bar charts similar to those in Figure 2.14. Here, you may be tempted to conclude that there is a significant difference in the bar heights, either because the vertical axis does not begin at zero (left side) or because the bars are chopped in the middle without a corresponding adjustment of the vertical axis (right side). As an observer, beware of such trickery. As an illustrator do not intentionally mislead your reader by disguising the results through the use of a misleading graph. It tends to give statisticians a bad name!

* Darrell Huff, *How to Lie with Statistics* (New York: Norton, 1954). More recent discussions are included in *Statistics: Concepts and Controversies* by David S. Moore (San Francisco: Freeman, 1979) and *How to Use (and Misuse) Statistics* by Gregory A. Kimble (Englewood Cliffs, NJ: Prentice-Hall, 1978).

Figure 2.14

Two misleading bar charts. The vertical axis of the left chart does not begin at zero, and the bars in the right chart are chopped without a corresponding adjustment in the vertical axis.

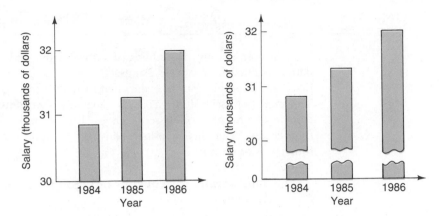

2.8   COMPUTER GRAPHICS ON THE MICROCOMPUTER

Now that you are ready to invest in graph paper, a straight edge, a protractor, and colored pens, you will be happy to learn that there is a much easier method of preparing professional-looking statistical graphs. There are programs available for practically all microcomputers that allow you to construct a variety of multi-colored bar charts, pie charts, and so on. Figure 2.15 was drawn with a micro-computer by using a few simple commands. If you think you will be having to create many graphic summaries, try to obtain access to a computer graphics package and its output. No good report is complete without at least one such graph!

Figure 2.15

An example of microcomputer graphics

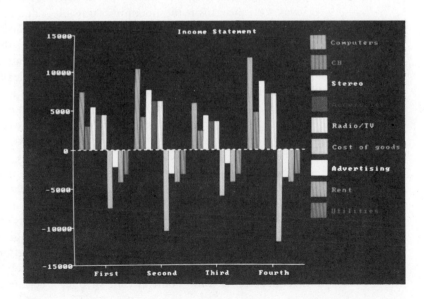

SUMMARY

This chapter examined methods of summarizing and presenting a large set of data contained in a sample using a graph to describe the data. You begin by placing the sample data in order, from smallest to largest (an **ordered array**). The next step is to summarize the data in a **frequency distribution**, which consists of a number of classes (such as "1.5 and under 2.0") and corresponding frequencies. Your final step is to construct an appropriate graph. We discussed four types of graphs:

1 A **histogram** or **frequency polygon** is a graphical view of a frequency distribution.

2 A **bar chart** summarizes frequencies for categorical (nominal) or ordinal data.

3 An **ogive** allows you to illustrate "less than" percentages or frequencies.

4 A **pie chart** presents a percentage breakdown of a particular quantity.

A frequency distribution provides a summary of the sample data by placing them into groups, called **classes**. The number of values in each class is the **class frequency**. For example, there may be 12 values in the class "1.5 and under 2.0." The numbers "1.5" and "2.0" here are the **class limits**, and the difference between consecutive lower class limits is the **class width**. The center of each class is the **class midpoint**. All classes should have the same width, except possibly the first and last class. Such a class may be open ended (as in Table 2.2) if you have a few outliers, or the class limit may be modified to conform to the reality of your situation. Such was the case in Table 2.5 where the upper limit was 4.0, since this is the maximum possible value for this situation.

To construct a frequency distribution, place the data values into an ordered array, from smallest to largest. For comparisons, the same data can be summarized using **relative frequencies**, which indicate the fraction of data values in each class rather than actual counts (**frequencies**).

A **histogram** is a graphical representation of a frequency distribution and generally is used for data having an interval or ratio level of measurement. When the data are nominal or ordinal, a **bar chart** provides a graphical summary. When constructing a bar chart, gaps are inserted between the boxes due to the nature of this data type.

An excellent way to indicate the shape of the data values is to use a **frequency polygon**. This is constructed by replacing the boxes in the histogram with straight lines connecting the midpoint of the top of each box.

An **ogive** allows you to make such statements as, "40% of the data values are less than 2.15." Like a frequency polygon, an ogive consists of many straight lines; in the ogive, the lines increase from zero to one on the vertical axis. A **pie chart** is a circular graph that can be used to represent percentages (relative frequencies) at some point in time or over a certain time period.

What's Next?

A graph, such as a frequency polygon, is an excellent method of describing a set of sample data, but it does have its limitations. For example, examine Figure 2.5 and answer the following question: Where is the middle (center) of the data? One person might argue that it is somewhere around 2.25, whereas someone else might decide that it is some value closer to 2.5. The point is that we need to define what the word middle means and define some method of calculating this value, so that we all get the same result. Such a value is called a numerical measure.

The next chapter will examine a variety of such numerical measures. Rather than reducing a set of sample data to a graph, we will reduce the data to one or more numbers that gives us some information about the data.

REVIEW EXERCISES

2.31 Students at a college are classified as Freshman (Fr), Sophomore (So), Junior (Jr), Senior (Sr), or Graduate (Gr). A survey conducted by the Dean of Student Affairs showed that 30 students interviewed had the following classification:

So, So, Jr, Fr, Gr, Fr, So, Fr, Fr, Gr, Sr, Fr, Gr, So, So, Fr, Sr, Fr, Jr, So, Gr, Fr, Fr, So, Jr, Jr, Sr, Sr, Jr, Jr

(a) The Dean wishes to look at the information presented in a graph showing the frequencies corresponding to the category of classification. What kind of chart would be appropriate, and why?

(b) Prepare a pie chart and a bar chart for the above.

2.32 The following are infant mortality rates (deaths under age one per 1000 live births) for 16 regions of a developing country.

Region	Mortality Rate
1	108
2	37
3	142
4	91
5	69
6	150
7	118
8	91
9	135
10	72
11	86
12	135
13	101
14	100
15	79
16	116

(a) Prepare a frequency distribution for these data.

(b) Draw a frequency histogram and a relative frequency polygon. Are their basic shapes similar?

(c) Compute the cumulative relative frequency distribution.

2.33 As part of a systems review at a space agency, the time taken by astronauts to respond in the case of a particular maneuver was measured by researchers. The following response times were noted for 20 astronauts (time in minutes):

0.70, 0.66, 0.91, 0.77, 1.13, 0.69, 0.80, 1.00, 0.91, 0.86, 1.22, 1.00, 0.72, 0.83, 0.82, 0.79, 0.90, 1.00, 2.13, 0.95

(a) Identify any outliers in the data. How could you adjust for the presence of outliers, when constructing a frequency distribution?

(b) Construct a frequency and relative frequency distribution.

(c) Construct an ogive. What can you say about the majority of astronauts?

2.34 The following is the distribution of the error rate (number of keystrokes in error) as determined by a human resources consultant investigating a pool of computer terminal operators:

No. of Errors	Frequency
0–2	1
3–5	3
6–8	5
9–11	4
12–14	2
	15

(a) Compute the cumulative relative frequencies.

(b) What percentage of the terminal operators makes fewer than nine errors?

(c) What number of errors is most common?

(d) What are the class midpoints?

2.35 The "population explosion" is a serious problem for the poorer nations of the world. The following are the birth rates (birth per 1000 population) for selected countries:

33.6, 28.8, 17.5, 9.8, 11.0, 18.7, 8.2, 14.1, 21.4, 16.3, 11.5, 26.5, 7.9, 10.4, 13.0, 16.6, 19.3, 20.3

(a) Construct a relative frequency distribution for these data.

(b) Construct a cumulative relative frequency distribution (ogive) for the data.

(c) What are the class midpoints for the distribution?

2.36 To measure the attention span of preschool children, a child psychologist conducted a preliminary survey of a group of children in a nursery. She noted the length of time that each child focuses on a specific task (designed as a game) before it was abandoned. The following times were noted (in minutes):

3.00, 1.80, 1.60, 1.60, 2.75, 1.55, 2.50, 1.40, 2.20, 5.10, 3.30, 1.90, 2.70, 5.20, 3.20, 2.80, 1.95, 2.25, 2.65, 4.00

(a) Construct a frequency distribution to summarize the results.

(b) Prepare an ogive. What kind of statement does this enable you to make?

(c) Identify the upper class limits for each class.

2.37 A frequency distribution has 10 classes, with the lowest class being "35–39" and the highest class "80–84". Identify the remaining eight classes.

2.38 The figures below pertain to immigrant visas issued by U.S. foreign service posts under the Immigration and Nationality Act for fiscal years 1981–1985.

Region	1981	1982	1983	1984	1985
Europe	40,080	32,217	31,241	31,757	31,823
Asia	129,029	135,029	142,249	149,095	153,337
Africa	6,539	5,626	5,511	6,204	6,411
N. America	169,858	118,180	129,998	133,325	138,122
S. America	27,211	25,679	27,088	27,568	28,935
Oceania	2,417	2,081	1,938	2,197	2,411
	375,134	318,812	338,025	350,146	361,039

Sources: U.S. Immigration and Naturalization Service; U.S. State Department.

(a) Prepare a bar chart showing the totals only, for 1981–1985.
(b) Prepare a pie chart showing the regional breakdown for 1985.

2.39 The following data represent the scores on a computer graded multiple-choice test. There are 20 questions and each question is worth 5 points. Construct a frequency polygon for the grades of 30 students.

95, 90, 80, 55, 90, 45, 50, 75, 75, 60, 55, 90, 50, 85, 95, 55, 60, 50, 45, 95, 70, 60, 50, 85, 90, 55, 45, 95, 100, 90

2.40 The ages (in years) of the 20 loan officers, 4 vice presidents, and the president of American Bank are

47, 52, 55, 65, 42, 37, 29, 52, 47, 36, 60, 50, 48, 42, 45, 35, 38, 45, 57, 43, 39, 41, 33, 58, 60

(a) Construct a frequency distribution.
(b) Construct a cumulative frequency distribution.
(c) Write a summary statement about the distribution of the ages of the loan officers.

2.41 A psychologist has designed a technique to improve a person's memory. Certain material is given to 30 people to memorize before they learn the technique. Similar material is given to the 30 people after the technique has been taught to them. The difference in the amount of time that it took to memorize the material (before − after) is given below in minutes:

5, 10, 15, 11, 13, 20, 14, 5, 23, 18, 17, 4, 1, 5, 29, 18, 15, 21, 24, 16, 2, 15, 19, 30, 24, 21, 14, 18, 26, 10

(a) Construct a frequency distribution.
(b) Construct a frequency histogram.
(c) Construct a frequency polygon.
(d) Take one class interval and write out, in words, exactly what it tells you.

2.42 As part of an ongoing investigation of stress, various individuals were subjected to increasing levels of noise. There were 10 noise levels, designated 1 to 10, and the first level at which discomfort was felt was noted. The following were the results:

Noise Level	Persons Feeling Discomfort
1	0
2	0
3	1
4	6
5	13
6	25
7	5
8	0
9	0
10	0

(a)   Construct a bar chart using these data.
(b)   What could be said about noise levels 8–10?
(c)   Which noise level formed a threshold of discomfort for 90% of the people?
(d)   What percentage of people did not feel discomfort until this threshold was reached?

2.43   To determine the number of counselors needed by a crisis center hot line, the administrators decided to record the number of calls received each week for 35 weeks. The following data were collected:

23, 22, 38, 43, 24, 35, 27, 25, 23, 22, 52, 31, 30, 41, 29, 28, 37, 25, 29, 31, 24, 49, 33, 25, 27, 25, 34, 32, 21, 23, 24, 18, 48, 23, 16

(a)   Classify the data in a frequency distribution, using eight classes and a class width of 5. The lowest class limit should be 15 and the highest class limit should be 54.
(b)   Also compute the relative frequencies and draw a relative frequency polygon.
(c)   Compute the cumulative relative frequency distribution.
(d)   Interpret the results.

2.44   On an objective true–false test consisting of 30 questions, the following were the number of questions answered correctly by the students:

15, 24, 16, 33, 23, 29, 22, 16, 24, 19, 24, 25, 21, 35, 19, 23, 27, 24, 29, 33, 32, 30, 25, 26, 19, 32, 29, 19, 16, 33

(a)   Prepare a relative frequency distribution.
(b)   Suppose you had to give one of five grades to each student with 5% making an A, 20% B, 50% C, 20% D, 5% F. What cutoff scores would you use in each category? (Percentages are approximate only, to be used as a guideline.)

2.45   The following are the population figures for the 50 states, based on the 1980 U.S. Census:

Alabama	3,890,061	Delaware	595,225
Alaska	400,481	Florida	9,739,992
Arizona	2,717,866	Georgia	5,464,265
Arkansas	2,285,513	Hawaii	965,000
California	23,668,562	Idaho	943,935
Colorado	2,888,834	Illinois	11,418,461
Connecticut	3,107,576	Indiana	5,490,179

Iowa	2,913,387	North Carolina	5,847,429
Kansas	2,363,208	North Dakota	652,695
Kentucky	3,661,433	Ohio	10,797,419
Louisiana	4,203,972	Oklahoma	3,025,266
Maine	1,124,660	Oregon	2,632,663
Maryland	4,216,446	Pennsylvania	11,866,728
Massachusetts	5,737,037	Rhode Island	947,154
Michigan	9,258,344	South Carolina	3,119,208
Minnesota	4,077,148	South Dakota	690,178
Mississippi	2,520,638	Tennessee	4,590,750
Missouri	4,917,444	Texas	14,228,383
Montana	786,690	Utah	1,461,037
Nebraska	1,570,006	Vermont	511,456
Nevada	799,184	Virginia	5,346,279
New Hampshire	920,610	Washington	4,130,163
New Jersey	7,364,158	West Virginia	1,949,644
New Mexico	1,299,968	Wisconsin	4,705,335
New York	17,557,288	Wyoming	470,816

Source: U.S. Bureau of Census, 1980.

(a) Construct a frequency distribution.
(b) Construct a relative frequency distribution.
(c) Construct a relative frequency polygon.
(d) Comment on the way the population is distributed.

2.46 One part of a sociological survey of a group of 40 immigrants in a border state yielded the following data pertaining to the number of years the immigrant had resided in the state:

2, 6, 4, 1, 1, 16, 4, 5, 25, 8, 13, 10, 15, 6, 2, 12, 13, 16, 17, 10, 48, 11, 8, 21, 24, 14, 8, 7, 5, 12, 6, 13, 10, 19, 12, 18, 1, 28, 16, 18

(a) Arrange the data in an ordered array.
(b) Using classes "1 to 5", "6 to 10", and so on, construct a frequency distribution.
(c) Compute the relative and cumulative relative frequencies.
(d) Construct a frequency histogram of the data.
(e) Construct a frequency polygon of the data.
(f) What is the class width, and what are the class midpoints?
(g) Draw an ogive.

2.47 As a part of their studies, anthropologists collect data on the physiological characteristics of a population. The following figures represent the blood cholesterol levels found in 25 members of an African tribe, measured as milligrams of cholesterol per deciliter of blood:

200, 241, 232, 177, 207, 181, 195, 182, 181, 233, 176, 170, 217, 164, 188, 164, 211, 204, 160, 172, 212, 186, 160, 203, 191

(a) Construct a frequency and relative frequency distribution for the data.
(b) Construct a relative frequency histogram.
(c) Construct a frequency polygon.

2.48 In the Presidential election of 1912, won by Woodrow Wilson, the following was the popular vote:

Party	Votes
Democratic (W. Wilson)	6,296,547
Progressive (T. Roosevelt)	4,118,571
Republican (W. M. Taft)	3,486,720
Socialist (E. V. Debs)	900,672
Prohibitionist (E. W. Chafin)	206,275

Source: Blum, J. M., et al., *The National Experience*, 5th ed. (New York: Harcourt Brace Jovanovich, 1981), p. 926.

(a) Why would a bar chart be more appropriate than a histogram for describing the above results?

(b) Prepare a bar chart for the above data.

(c) Describe the results in the form of a pie chart.

CASE STUDY

Suburban School SAT Scores

The Dallas Morning News of December 2, 1986, reported the following average SAT scores for a sample of suburban school districts in its readership area. The chart includes the Verbal, Math, and cumulative Total scores for 1985 and 1986.

School	1985 Verbal	Math	Total	1986 Verbal	Math	Total
Arlington						
Arlington	442	485	927	444	475	919
Lamar	437	477	914	442	481	923
Martin	432	462	894	428	468	896
Sam Houston	421	457	878	402	452	854
Carrollton-Farmers Branch						
Newman Smith	457	511	968	457	494	951
Turner	421	466	887	439	487	926
Cedar Hill	428	455	883	449	481	930
DeSoto	414	452	866	418	461	879
Duncanville	438	460	898	444	478	922
Garland						
Garland	404	476	880	421	454	875
Lakeview Centennial	409	443	852	406	443	849
North Garland	413	468	881	427	468	895
South Garland	434	456	890	442	483	925
Grand Prairie						
Grand Prairie	403	434	837	400	444	844
South Grand Prairie	401	436	837	402	441	843

School	1985 Verbal	1985 Math	1985 Total	1986 Verbal	1986 Math	1986 Total
Grapevine	434	480	914	435	471	906
Highland Park	468	501	969	468	504	972
Irving						
Irving	434	471	905	430	466	896
MacArthur	444	503	947	431	473	904
Nimitz	430	463	893	428	471	899
Lancaster	418	455	873	397	429	826
Lewisville	425	456	881	420	462	882
Mesquite						
Mesquite	442	468	910	442	471	913
North Mesquite	440	478	918	436	476	912
West Mesquite	441	468	909	410	464	874
Plano						
Plano	450	504	954	461	507	968
Plano East	442	481	923	442	479	921
Richardson						
Richardson	442	493	935	436	484	920
Berkner	445	491	936	430	488	918
Lake Highlands	450	494	944	446	477	923
Pearce	462	517	979	456	502	958
Rockwall	438	459	897	432	467	899
Wilmer-Hutchins	340	390	730	410	420	830

Source: "SAT Scores Fall in Suburbs, "*The Dallas Morning News*, December 2, 1986, Used by permission of the *Dallas Morning News*.

Addressing a perception that SAT scores had declined, the newspaper reported that school officials said an 85% increase in the number of students taking the test in 1986 probably contributed to the decline in scores. According to Katherine Oates, a spokeswoman for the Lancaster School District, "with more students taking the test, you have more marginal students taking it."

In Richardson, decreases of 15 to 21 points in average scores at the district's four high schools were reported to be concerning officials. The decline in the district's overall average to 930 was the first decline in six years. Rex Carr, deputy superintendent for planning and personnel, expressed the view that students were "learning the content but lacked test-taking techniques."

Duncanville Assistant Superintendent Art Douglas credited after-school SAT preparation classes and an advanced curriculum for a 24-point jump in the average score at Duncanville High School, where 62% of the senior class took the test. He said, "We've just tried to stay ahead of the game on all counts, I guess you might say."

Case Study Questions

1. As you will learn later in your study of statistics, verifying the validity and accuracy of the above explanations, opinions, and claims is a complicated matter. However, even the limited techniques of descriptive graphs can be very useful in getting a "feel" for the data. A computer package such as SPSS, SAS, or MINITAB might be useful here, if you have one.

   (a)   Put the data into an ordered array and construct two frequency distributions of the total SAT scores, one for 1985 and one for 1986.

   (b)   Print or draw the frequency histograms for 1985 and 1986 total scores. What comments could you make, based on (a) and (b)?

   (c)   Draw the 1985 and 1986 total scores frequency polygons on one and the same chart. (Using different colored inks will improve the presentation.) If indeed SAT scores have declined, what would you expect to see? Would the graph look like Figure 2.7 in your textbook? Could you explain why? Compare Figure 2.7 and your chart for the 1985/1986 scores, and comment on any similarities or differences.

   (d)   Could you make similar graphical comparisons between verbal scores or between math scores for different years, or between math and verbal scores for the same year? Discuss.

2. Suppose you wished to answer questions like, "What percentage of the schools have 1986 scores less than 850, or between 850 and 900?" What kind of graph would you need to construct? Prepare the necessary graph, and answer the questions in the preceding sentence.

3. (a)   In the report, the 1986 average total score for the four Richardson schools was mentioned as being 930. Compute the 1986 average total scores for each of the 17 suburban districts above. (If there is only one high school district, obviously that school's score is also the district's score.) Present the information in the form of a bar chart. Why would a pie chart be inappropriate?

   (b)   The bar chart could be extended to a multiple bar format by showing 1985 and 1986 scores side by side for each district. What would be useful about such a presentation? Compare the kind of information conveyed by such a bar chart, with the kind of information one obtains from the frequency polygons in 1(c).

   (c)   You could get really carried away and compute the verbal, math, and total scores for 1985 and 1986 for each district. Report the information in a multiple bar chart, using different colors for each category.

COMPUTER EXERCISE USING THE DATABASE

Using a sample of 100 observations, complete the following exercises.

1. Construct an ordered array of the verbal GRE scores (variable VGRE). Construct a frequency distribution and histogram using 10 classes.

2. Repeat exercise 1 using variable QGRE (quantitative GRE scores).

3. Repeat exercise 1 using variable GR-GPA (graduate GPA).

4. Repeat exercise 1 using variable AGE.

5. Construct a bar chart summarizing the number of males and females in your sample

using variable SEX. (*Note:* The MINITAB HISTOGRAM command can be used here.)

6. Construct a bar chart summarizing the number of people in each marital category using variable MARITAL. (*Note:* The MINITAB HISTOGRAM command can be used here.)

7. Construct a bar chart consisting of eight bars summarizing the number of never-married males, never-married females, ever-married males, ever-married females, remarried males, remarried females, divorced males, and divorced females in your sample. (*Hint:* Multiply variables SEX and MARITAL using the computer and construct a histogram of the new variable data. What do these frequencies provide? Use exercise 6 to determine the remaining frequencies.)

SPSSX APPENDIX (MAINFRAME VERSION)

Constructing a Histogram

You can use SPSSX to construct a histogram of the grade point averages for the sample of 50 students at Brookhaven University. The SPSSX program listing in Figure 2.16 requests a histogram of the data in Table 2.1. As you can see, it is similar to the procedure in the SPSSX Appendix at the end of the book, which was used to obtain test score averages.

The TITLE command names the SPSSX run.

The DATA LIST command gives each variable a name, and describes the data as being in free form.

The BEGIN DATA command indicates to SPSSX that the input data immediately follow.

Figure 2.16

SPSSX program listing requesting a histogram of data from Table 2.1

```
TITLE    BROOKHAVEN GPA
DATA LIST FREE / GPADATA
BEGIN DATA
1.72    2.18    2.65    3.40    2.27
1.57    3.18    1.55    1.92    3.24
1.08    3.96    2.63    1.69    2.31
3.02    1.86    2.88    3.29    3.77
2.22    1.70    2.47    2.11    2.40
2.76    1.44    2.37    1.74    1.32
2.03    1.68    1.71    3.88    2.27
1.62    2.79    3.36    2.19    1.47
3.41    1.66    2.38    2.67    2.91
3.06    2.70    2.33    2.83    3.75
END DATA
FREQUENCIES VAR=GPADATA/HISTOGRAM
FINISH
```

The next 10 lines contain the data values, which represent the grade point averages of the students in the sample.

The END DATA statement indicates the end of the data.

The FREQUENCIES statement specifies the variable from which we wish to produce a histogram, with the HISTOGRAM statement generating the actual histogram.

The FINISH command indicates the end of the SPSSX program.

Figure 2.17 shows the output obtained by executing the listing in Figure 2.16.

Figure 2.17

SPSSX output obtained by executing the program listing in Figure 2.16

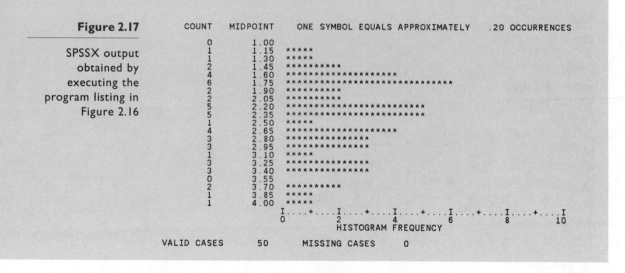

```
COUNT   MIDPOINT      ONE SYMBOL EQUALS APPROXIMATELY     .20 OCCURRENCES
  0      1.00
  1      1.15    *****
  1      1.30    *****
  2      1.45    **********
  4      1.60    *********************
  6      1.75    ********************************
  2      1.90    **********
  2      2.05    **********
  5      2.20    **************************
  5      2.35    **************************
  1      2.50    *****
  4      2.65    *******************
  3      2.80    ***************
  3      2.95    ***************
  1      3.10    *****
  3      3.25    ***************
  3      3.40    ***************
  0      3.55
  2      3.70    **********
  1      3.85    *****
  1      4.00    *****
                 I....+....I....+....I....+....I....+....I....+....I
                 0         2         4         6         8         10
                            HISTOGRAM FREQUENCY

VALID CASES     50      MISSING CASES     0
```

SPSS/PC+ APPENDIX (MICRO VERSION)

Constructing a Histogram

You can use SPSS/PC+ to construct a histogram of the grade point averages for the sample of 50 students at Brookhaven University. The program listing in Figure 2.18 requests a histogram of the data in Table 2.1. As you can see, it is similar to the procedure in the SPSS/PC+ Appendix at the end of the book, which was used to obtain test score averages. All command lines in Figure 2.18 *must end in a period.*

The TITLE command names the SPSS/PC+ run.

The DATA LIST command gives each variable a name, and describes the data as being in free form.

The BEGIN DATA command indicates to SPSS/PC+ that the input data immediately follow.

The next 10 lines contain the data values, which represent the grade point averages of the students in the sample.

The END DATA statement indicates the end of the data.

The FREQUENCIES statement specifies the variable from which we wish to produce a histogram, with the HISTOGRAM statement generating the actual histogram.

Figure 2.18

SPSS/PC+ program listing requesting a histogram of data from Table 2.1

```
TITLE  BROOKHAVEN GPA.
DATA LIST FREE / GPADATA.
BEGIN DATA.
1.72 2.18 2.65 3.40 2.27
1.57 3.18 1.55 1.92 3.24
1.08 3.96 2.63 1.69 2.31
3.02 1.86 2.88 3.29 3.77
2.22 1.70 2.47 2.11 2.40
2.76 1.44 2.37 1.74 1.32
2.03 1.68 1.71 3.88 2.27
1.62 2.79 3.36 2.19 1.47
3.41 1.66 2.38 2.67 2.91
3.06 2.70 2.33 2.83 3.75
END DATA.
FREQUENCIES VAR=GPADATA/HISTOGRAM.
```

Figure 2.19 shows the output obtained from executing the listing in Figure 2.18.

Figure 2.19

SPSS/PC+ output obtained by executing the program listing in Figure 2.18

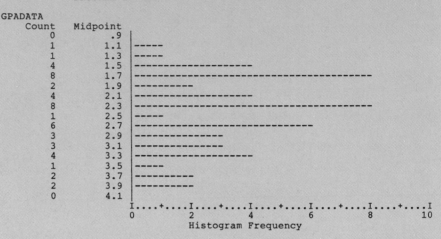

Constructing a
Histogram

You can use SAS to produce a histogram. The SAS program listing in Figure 2.20 was used to request a histogram of the data in Table 2.1. As you can see, it is similar to the procedure in the SAS Appendix at the end of the book, which was used to obtain test score averages.

The TITLE command names the SAS run.

The DATA command gives the data a name.

The INPUT command names and gives the correct order for the different fields on the data lines.

The CARDS command indicates to SAS that the input data immediately follow.

SAS APPENDIX (MAINFRAME VERSION)

Figure 2.20

SAS program listing
requesting a
histogram of data
from Table 2.1

```
TITLE   'BROOKHAVEN GPA';
DATA GPADATA;
INPUT GPA;
CARDS;
1.72
2.18
2.65
3.40
2.27
1.57
3.18
1.55
1.92
3.24
1.08
3.96
2.63
1.69
2.31
3.02
1.86
2.88
3.29
3.77
2.22
1.70
2.47
2.11
2.40
2.76
1.44
2.37
1.74
1.32
2.03
1.68
1.71
3.88
2.27
1.62
2.79
3.36
2.19
1.47
3.41
1.66
2.38
2.67
2.91
3.06
2.70
2.33
2.83
3.75
.
PROC PRINT;
PROC CHART;
   VBAR GPA;
```

The next 50 lines are card images. The first line, for example, represents the grade point average of the first student in the sample. The remaining lines indicate the grade point average of the other 49 students.

The PROC PRINT command directs SAS to list the data that were just read in.

**SAS
APPENDIX
(MAINFRAME
VERSION)**

The PROC CHART command requests a SAS procedure to print a histogram. The VBAR GPA generates a histogram of the variable GPA. The resulting output contains the class midpoints and frequencies.

Figure 2.21 shows the output obtained from executing the program listing in Figure 2.20.

Figure 2.21

SAS output obtained by executing the program listing in Figure 2.20

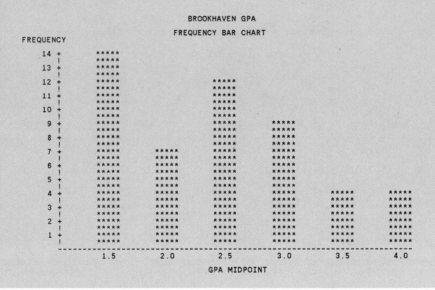

**SAS
APPENDIX
(MICRO
VERSION)**

**Constructing a
Histogram**

You can use SAS to produce a histogram. The SAS program listing in Figure 2.22 was used to request a histogram of the data in Table 2.1. As you can see, it is similar to the procedure in the SAS Appendix at the end of the book, which was used to obtain test score averages.

The TITLE command names the SAS run.
The DATA command gives the data a name.

SAS APPENDIX (MICRO VERSION)

Figure 2.22

SAS program listing requesting a histogram of data from Table 2.1

```
TITLE 'BROOKHAVEN GPA';
DATA GPADATA;
INPUT GPA;
CARDS;
1.72
2.18
2.65
3.40
2.27
1.57
3.18
1.55
1.92
3.24
1.08
3.96
2.63
1.69
2.31
3.02
1.86
2.88
3.29
3.77
2.22
1.70
2.47
2.11
2.40
2.76
1.44
2.37
1.74
1.32
2.03
1.68
1.71
3.88
2.27
1.62
2.79
3.36
2.19
1.47
3.41
1.66
2.38
2.67
2.91
3.06
2.70
2.33
2.83
3.75
PROC PRINT;
PROC CHART;
  VBAR GPA;
RUN;
```

SAS APPENDIX (MICRO VERSION)

The INPUT command names and gives the correct order for the different fields on the data lines.

The CARDS command indicates to SAS that the input data immediately follow.

The next 50 lines are card images. The first line, for example, represents the grade point average of the first student in the sample. The remaining lines indicate the grade point average of the other 49 students.

The PROC PRINT command directs SAS to list the data that were just read in.

The PROC CHART command requests a SAS procedure to print a histogram. The VBAR GPA generates a histogram of the variable GPA. The resulting output contains the class midpoints and frequencies.

The RUN command tells the SAS system to execute the previous SAS statements.

Figure 2.23 shows the output obtained from executing the program listing in Figure 2.22.

```
                              BROOKHAVEN GPA

                           FREQUENCY BAR CHART
FREQUENCY

   14 +        *****
      |        *****           *****
   10 +        *****           *****      *****
      |        *****   *****   *****      *****
    6 +        *****   *****   *****      *****
      |        *****   *****   *****      *****     *****     *****
    2 +        *****   *****   *****      *****     *****     *****
 -----------------------------------------------------------------
                1.5     2.0     2.5        3.0       3.5       4.0

                              GPA MIDPOINT
```

Figure 2.23

SAS output obtained by executing the program listing in Figure 2.22

3

DESCRIPTIVE MEASURES

A LOOK BACK/INTRODUCTION

The first two chapters focused on different types of numerical sample data and methods of summarizing and presenting data. A frequency distribution is used to condense data from a sample into groups (called classes). Different types of statistical graphs can be used to illustrate sample data in different ways. The types of graphs we have discussed so far include the histogram, bar chart, ogive, frequency polygon, and pie chart. The purpose of these graphs is to convey information at a glance about the distribution of the values in your sample.

Every sample data set is a small part of a much larger population. Even if we do not always mention the word "population," it is always there, since the sample values were selected from this group of interest. Every population has properties (called parameters) that describe it. By collecting a set of sample data, we can then estimate these properties by computing statistics and making graphs.

We have seen how to reduce a set of sample data to a graph. It also is helpful to reduce data to one or more numbers (such as an average). Such a number is called a descriptive measure. Because this number is derived from a sample, it also can be called a sample statistic. We will discuss the commonly used descriptive measures and explain what you can expect to learn from each one. In later chapters we will discuss how you can use many of these sample statistics to estimate the corresponding population parameters.

3.1   VARIOUS TYPES OF DESCRIPTIVE MEASURES

A **descriptive measure** is a single number that provides information about a set of sample data. The class of descriptive measures described here consists of four types. Which one you select depends on what you want to measure. These types are:

1 **Measures of central tendency** These answer the questions "Where is the 'middle' of my data?" and "Which data value occurs most often?"

2 **Measures of dispersion** These answer the questions "How spread out are my data values?" and "How much do the data values jump around?"

3 **Measures of position** These answer the questions "How does my value (score on an exam, for example) compare with those of everyone else?" and "Which data value was exceeded by 75% of the data values? by 50%? by 25%?"

4 **Measures of shape** These answer the questions "Are my data values symmetric?" and "If not symmetric, just how nonsymmetric (skewed) are the data?"

3.2   MEASURES OF CENTRAL TENDENCY

The purpose of a **measure of central tendency** is to determine the "center" of your data values or possibly the "most typical" data value. Some measures of central tendency are the mean, median, midrange, and mode. We will illustrate each of these measures using as data the number of family members in 5 randomly selected families.

Family size data: 2, 4, 5, 12, 2

The Mean

The **mean** is the most popular measure of central tendency. It is merely the average of the data. The mean is easy to obtain and explain, and it has several mathematical properties that make it more advantageous to use than the other three measures of central tendency.

The mean is a very popular way of representing a set of values. In this way, you can use a single number as typical of a whole set (sample) of values, such as average salary, average rainfall, or average family size. The income per capita of a certain district, the number of clothes washers per capita, and the number of televisions per capita are all examples of a mean.

The sample mean, $\bar{x}$ (read as "x bar"), is equal to the sum of the data values divided by the number of data values. For the family size data set

$$\bar{x} = \frac{2 + 4 + 5 + 12 + 2}{5} = \frac{25}{5} = 5.0$$

In general, let an arbitrary data set be represented as

$$x_1, x_2, x_3, \ldots, x_n$$

where n is the number of data values. (In the family size data set, $x_1 = 2$, $x_2 = 4$, $x_3 = 5$, $x_4 = 12$, $x_5 = 2$, and n is 5.) Then

$$\bar{x} = \frac{x_1 + x_2 + \cdots + x_n}{n} = \frac{\Sigma x}{n} \qquad (3\text{-}1)$$

The symbol Σ (sigma) means "the sum of." In this case, the sample mean $\bar{x}$ is the sum of the x values divided by n.*

In subsequent chapters, we will be concerned with the mean of the population. The symbol for the population mean is μ (mu). For a population consisting of N elements, denoted by

$$x_1, x_2, x_3, \ldots, x_N$$

the population mean is defined to be

$$\mu = \frac{x_1 + x_2 + \cdots + x_N}{N} = \frac{\Sigma x}{N} \qquad (3\text{-}2)$$

Population: $x_1, x_2, \ldots, x_N$

Population Mean $= \mu = \dfrac{x_1 + x_2 + \cdots + x_N}{N} = \dfrac{\Sigma x}{N}$

* In another application of this symbol, we square each of the sample values and sum these values. For the family size data, this operation would be written as

$$\Sigma x^2 = 2^2 + 4^2 + 5^2 + 12^2 + 2^2$$
$$= 4 + 16 + 25 + 144 + 4$$
$$= 193$$

For these data then $\Sigma x = 25$ and $\Sigma x^2 = 193$.

Sample Values (selected from the population): $x_1, x_2, \ldots, x_n$ where $n \leqslant N$

Sample Mean $= \bar{x} = \dfrac{x_1 + x_2 + \cdots + x_n}{n} = \dfrac{\Sigma x}{n}$

The Median

The **median** of a set of data is the value in the center of the data values when they are arranged from smallest to largest. Consequently, it is in the center of the ordered array.

Using the family size data set, the median **Md** is found by first constructing an ordered array:

$$2, 2, \mathbf{4}, 5, 12$$

The value that has an equal number of items to the right and the left is the median. Thus, Md = 4.

In general, if n is odd, Md is the center data value of the ordered set:

$$\text{Md} = \left(\frac{n+1}{2}\right)\text{st ordered value}$$

Here, the median is the $(5 + 1)/2 = $ 3rd value in the ordered array. Note that for these data, the *position* of the median is 3 (3rd largest value) and the *value* of the median is 4. If n is even, Md is the average of the two center values of the ordered set. Thus, the median of the array 3, 5, 7, 10 is $(5 + 7)/2 = 6.0$.

In our family size data set, one of the five values (12) is much larger than the remaining values—it is an outlier. Notice that the median (Md = 4) was much less affected by this value than was the mean ($\bar{x} = 5.0$). When dealing with data that are likely to contain outliers (for example, personal incomes or prices of residential housing), the median usually is preferred to the mean as a measure of central tendency since the median provides a more "typical" or "representative" value for these situations.

The Midrange

Although less popular than the mean and median, the **midrange (Mr)** provides an easy-to-grasp measure of central tendency. Notice that it is also severely affected (even more than $\bar{x}$) by the presence of an outlier in the data. In general:

$$\text{Mr} = \frac{(\text{smallest value}) + (\text{largest value})}{2} \qquad (3\text{-}3)$$

Using the family size data set,

$$Mr = \frac{2 + 12}{2} = 7.0$$

Compare this to $\bar{x} = 5.0$ and $Md = 4$.

The Mode

The **mode (Mo)** of a data set is the value that occurs the most often. The mode need not occur in the "center" of your data. For the family size data set, the mode is 2, since this value occurs the most often. So, $Mo = 2$, which is clearly not in the "middle" of the data set. One situation in which the mode is the value of interest is the manufacturing of clothing. The most common hat size is what you would like to know, not the average hat size. Can you think of other applications where the mode would provide useful information?

There may be more than one mode if several numbers occur the same (and the largest) number of times, or there may be no mode if all values occur only once.

EXAMPLE 3.1

The budget director of Westview General Hospital obtained a sample of 10 released patients to determine the typical length of stay in the hospital for a particular operation. The length of stay was defined from the time of admission to the time of release. Consequently, a time such as 4 days, 3 hours was recorded as 4.1 days. The 10 lengths of stay (in days) were:

2.5, 4.8, 3.8, 3.2, 16.5, 5.0, 4.6, 4.4, 2.6, 3.1

We find the average length of stay as follows:

$$\bar{x} = \frac{2.5 + 4.8 + \cdots + 3.1}{10} = \frac{50.5}{10} = 5.05 \text{ days}$$

Notice that there is an outlier in the data, namely, 16.5 days. To be safe, you should double-check this figure to make sure that it is, in fact, correct, that is, that there was no mistake in recording or transcribing the value. In the presence of one or two outliers, the median generally provides a more reliable measure of central tendency, so we construct an ordered array:

2.5, 2.6, 3.1, 3.2, **3.8**, **4.4**, 4.6, 4.8, 5.0, 16.5

Consequently,

$$Md = \frac{3.8 + 4.4}{2} = 4.1 \text{ days}$$

The midrange is given by

$$Mr = \frac{2.5 + 16.5}{2} = 9.5 \text{ days}$$

Figure 3.1

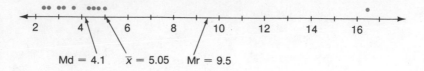

Dot array diagram
of the measures of
central tendency for
a sample of 10
lengths of hospital
stay

This value is severely affected by the presence of the outlier; the midrange of 9.5 days is a poor measure of central tendency for this application.

No mode exists because there are no repeats in the data values. These results are summarized in the graph in Figure 3.1, a **dot array diagram**. Each data value is represented as a dot on the horizontal line. □

EXERCISES

3.1   Compute the mean, median, and mode for the daily maximum temperatures in the Dallas data in exercise 2.8.

3.2   Compute the mean, median, and mode for the traffic ticket data in exercise 2.5.

3.3   Compute the mean, median, and mode for the alien labor force data in exercise 2.7.

3.4   The 10 most frequently run programs during May 1986 at a local university computer center utilized the following times in CPU seconds.

$$317, 277, 209, 197, 122, 72, 54, 32, 18, 15$$

Compute the mean and median for the data.

3.5   The distribution of family income (in dollars) in the United States in 1983 was as follows:

Income Range	Percent Distribution of Families
Under 5000	5.7
5,000 and under 10,000	10.2
10,000 and under 15,000	11.6
15,000 and under 20,000	11.8
20,000 and under 25,000	11.5
25,000 and under 35,000	19.5
35,000 and under 50,000	17.1
50,000 and over	12.6
Total	100.0

Source: Statistical Abstract of the United States, 1985, U.S. Department of Commerce, p. 446.

Estimate the median. Give an interpretation of the value of the median.

3.6   The following data from an experiment on sensory deprivation gives the number of hours that individuals were subject to sensory deprivation in a special room:

$$18, 18, 21, 24, 24, 30, 36, 36, 36, 36, 48, 48, 60, 64, 72$$

Calculate the mean, median, and mode for the number of hours of deprivation.

3.7 A pickup truck has a maximum load capacity of 4500 lb. Determine if the truck is overloaded in each of the following cases:

(a) It carries 10 packages with an average weight of 350 lb.

(b) It carries 5 boxes with a mean weight of 300 lb, 10 packages with a mean weight of 250 lb, and a refrigerator weighing 600 lb.

3.8 The following chart shows the percentage of students by race who scored at or above the 40th percentile on the reading section of the Iowa Test of Basic Skills in 1986, according to the Dallas Independent School District. (Students scoring below the 40th percentile are considered behind their peers and in need of specialized remedial help.)

Grade	White	Black	Hispanic
1st	83	76	76
2nd	90	76	79
3rd	82	60	64
4th	81	60	61
5th	79	57	58
6th	82	60	63
7th	78	52	54
8th	80	56	57
9th	69	29	38
10th	84	44	57
11th	66	20	32

Source: The Dallas Morning News, July 16, 1986.

Compute the mean and median for each racial group. Compare the results and interpret what they suggest.

3.9 Compute the mean, median, and mode for the daily advertising expense of a car dealer using the following data. The daily advertising expense in dollars is given for 20 days.

38, 60, 20, 130, 55, 150, 47, 35, 86, 95, 31, 46, 112, 130, 55, 42, 130, 35, 60, 130

3.10 A student in a social studies class obtained the following results for the semester:

Quiz	80 out of 100
Midterm	128 out of 150
Final	154 out of 200
Term Paper	60 out of 75

Assuming equal weight for the four grades, what is the overall semester average (expressed as a percentage) for this student?

3.11 Ratings measure a percentage of the audience tuned to a particular station, and are used by advertisers when they place orders for commercials. The following data, released by Arbitron Rating Co., pertains to the period April 3, 1986 to June 25, 1986 for radio stations in a major metropolitan area:

Number	Station	Rating
1	KVIL-FM	8.8
2	KKDA-FM	8.5
3	KPLX-FM	7.3
4	WBAP-AM	6.9

Number	Station	Rating
5	KRLD-AM	6.3
6	KMEZ-FM	6.3
7	KSCS-FM	5.7
8	KEGL-FM	5.2
9	KTXQ-FM	5.0
10	KQZY-FM	3.8

Source: *The Dallas Morning News*, July 22, 1986.

Compute the mean, median, and mode for the ratings.

3.3   MEASURES OF DISPERSION

A measure of central tendency, such as the mean, is certainly useful. However, the use of any single value to describe a complete distribution will fail to reveal important facts.

The more homogeneous a set of data is, the better the mean will represent a "typical value." **Dispersion** is the tendency of data values to scatter about the mean $\bar{x}$. If all the data values in a sample are identical, then the mean provides perfect information, and the dispersion is zero. This is rarely the case, however, so we need a measure of dispersion that will increase as the scatter of the data values about $\bar{x}$ increases.

Knowledge of dispersion can sometimes be used to control the variability of your data values in the future. Industrial production operations maintain quality control by observing and measuring the dispersion of the units produced. If there is too much variation in the production process, the causes are determined and corrected using an inspection control procedure.

Some measures of dispersion are the range, mean absolute deviation (MAD), variance, standard deviation, and coefficient of variation. To illustrate the various dispersion measures, we will use the family size data from the previous section: 2, 4, 5, 12, 2.

The Range

The simplest measure of dispersion is the **range** of the data, which is the numerical difference between the largest value and the smallest value. For the family size data

$$\text{Range} = 12 - 2 = 10$$

The range is a rather crude measure of dispersion, but it is an easy number to calculate and contains valuable information for many situations. Stock reports generally give prices in terms of their ranges, citing the high and low prices of the day. When quoting the results obtained from a psychological testing instrument, the range of *possible* scores is of interest as well as the range of the scores

Figure 3.2

A presentation of the family size data showing their variation

obtained from a sample of people using this instrument. The value of the range is strongly influenced by an outlier in the sample data.

Mean Absolute Deviation (MAD)

The purpose of a measure of dispersion is to determine the variability of your data. The more variation there is, the larger this measure should become. Take a look at the family size data illustrated in Figure 3.2. To measure the variation about the sample mean $\bar{x}$ consider the distance from each data value to $\bar{x}$ (that is, $x - \bar{x}$) and its absolute value:

Data Value (x)	$x - \bar{x}$	$|x - \bar{x}|$
2	−3.0	3.0
2	−3.0	3.0
4	−1.0	1.0
5	0.0	0.0
12	7.0	7.0
	$\Sigma(x - \bar{x}) = 0$	$\Sigma|x - \bar{x}| = 14.0$

As a possible measure, consider the average of the $(x - \bar{x})$ values:

$$\frac{\Sigma(x - \bar{x})}{5} = \frac{0}{5} = 0$$

This value is always zero for any set of data because the positive deviations always balance out the negative ones. To overcome this, use the actual distance from each data value to the sample mean, paying no attention to the side of the mean on which it lies, by taking the **absolute value** of each deviation. The **mean absolute deviation (MAD)** is the average of these distances:

$$\text{MAD} = \frac{\Sigma|x - \bar{x}|}{n} \qquad (3\text{-}4)$$

Using the family size data,

$$\text{MAD} = \frac{3.0 + 3.0 + 1.0 + 0.0 + 7.0}{5} = \frac{14.0}{5} = 2.8$$

What is the MAD without the value 12?

$$\bar{x} = \frac{2 + 2 + 4 + 5}{4} = 3.25$$

and so

$$MAD = \frac{|2 - 3.25| + |2 - 3.25| + |4 - 3.25| + |5 - 3.25|}{4}$$

$$= \frac{1.25 + 1.25 + .75 + 1.75}{4} = 1.25$$

Here the MAD is much lower than before. This indicates that the smaller data set has much less variation than does the one containing the outlier.

The Variance and Standard Deviation

By far the most widely used measures of dispersion are the **variance** and **standard deviation**. They resemble the MAD in that they are based on deviations of all the values from the sample mean $\bar{x}$. The problem encountered earlier in examining the sum of each $(x - \bar{x})$ was that the negative deviations balanced out the positive ones. The MAD handled this situation by taking the absolute value of each deviation. Another possibility is to square each of these deviations, thereby removing all the negative signs. We can illustrate this using our example data. Recall that $\bar{x} = 5.0$.

Data Value (x)	$(x - \bar{x})$	$(x - \bar{x})^2$
2	−3.0	9.0
2	−3.0	9.0
4	−1.0	1.0
5	0.0	0.0
12	7.0	49.0
	0	68.0

So $\Sigma(x - \bar{x})^2 = 68.0$.

The obvious thing to do next would be to find the average of these squared deviations:

$$\frac{1}{n} \Sigma(x - \bar{x})^2$$

One use of this particular statistic in subsequent chapters will be as an estimator. In particular, we will need to estimate the variation within an entire population, using sample data collected from the population. However, a better estimator is obtained by dividing the sum of the squared deviations by $n - 1$ rather than by n. This leads to the **sample variance**, s^2. In general

$$s^2 = \frac{\Sigma(x - \bar{x})^2}{n - 1} \tag{3-5}$$

Using the family size data

$$s^2 = \frac{68.0}{5-1} = \frac{68.0}{4} = 17.0$$

The square root of the variance is referred to as the **sample standard deviation**, **s**. In general,

$$s = \sqrt{\frac{\Sigma(x - \bar{x})^2}{n - 1}} \qquad \qquad \textbf{(3-6)}$$

Using the family size data,

$$s = \sqrt{17.0} = 4.123$$

As previously mentioned, the sample variance s^2 is used to estimate the variance of the entire population. The symbol for the **population variance** is σ^2 (read as sigma squared). For a population consisting of N elements

$$x_1, x_2, x_3, \dots, x_N$$

the population variance is defined to be

$$\sigma^2 = \frac{\Sigma(x - \mu)^2}{N} \qquad \qquad \textbf{(3-7)}$$

where μ is the population mean, defined in equation 3-2.

As we saw, the population variance can be obtained by dividing the sum of the squared deviations about μ by the population size N. The sample variance is calculated by dividing the sum of the squared deviations about $\bar{x}$ by the sample size (n) minus one. Had we chosen to divide by n rather than by $n - 1$, the resulting estimator would (on the average) underestimate σ^2. For this reason, we use $n - 1$ in the denominator of s^2.

Population $x_1, x_2, \dots, x_N$

$$\text{Population Variance} = \sigma^2 = \frac{(x_1 - \mu)^2 + \cdots + (x_N - \mu)^2}{N}$$

$$= \frac{\Sigma(x - \mu)^2}{N}$$

$$\text{Population Standard Deviation} = \sigma = \sqrt{\frac{\Sigma(x - \mu)^2}{N}}$$

Sample Values (selected from the population): $x_1, x_2, \ldots, x_n$ where $n \leqslant N$

$$\text{Sample Variance} = s^2 = \frac{(x_1 - \bar{x})^2 + \cdots + (x_n - \bar{x})^2}{n - 1}$$

$$= \frac{\Sigma(x - \bar{x})^2}{n - 1}$$

$$\text{Sample Standard Deviation} = s = \sqrt{\frac{\Sigma(x - \bar{x})^2}{n - 1}}$$

Now consider what the units of measurement are for s and s^2. The units for s are the same as the units for the data. If the data are measured in pounds, then the units for s are pounds. Consequently, the units for the variance s^2 would be (pounds)^2—a rather difficult unit to grasp.

For the family size data,

$$s = 4.123 \text{ members}$$

$$s^2 = 17.0 \text{ (members)}^2$$

For this reason, s (rather than s^2) is typically the preferred measure of dispersion.

There is another way to compute the sample variance. Using equation 3-5 to compute the value of s^2 may have appeared easy enough, but for large data sets, this can be a very tedious procedure that can lead to considerable rounding error. Instead, use

$$s^2 = \frac{\Sigma x^2 - (\Sigma x)^2/n}{n - 1} \qquad \textbf{(3-8)}$$

As before, the standard deviation is the square root of the variance. To illustrate the use of equation 3-8, consider the family size data:

x	x^2
2	4
2	4
4	16
5	25
12	144
25	193

So, $n = 5$, $\Sigma x = 25$, and $\Sigma x^2 = 193$. Consequently, using equation 3-8

$$s^2 = \frac{193 - (25)^2/5}{5 - 1}$$

$$= \frac{193 - 125.0}{4} = 17.0 \qquad \text{(as before)}$$

Also

$$s = \sqrt{17.0} = 4.123 \qquad \text{(as before)}$$

Finally, you may wish to interpret the magnitude of the value of s or s^2, that is, whether your value of s (or s^2) is large or not. This is difficult to determine because the values of s and s^2 depend on the magnitude of the data values. In other words, large data values generally lead to large values of s. For example, which of the following two data sets exhibits more variation?

Data set 1:   2, 2, 4, 5, 12 (family sizes)
Data set 2:   2000, 2000, 4000, 5000, 12,000

As we have already seen, for data set 1, $\bar{x} = 5.0$ and $s = 4.123$. For data set 2, $\bar{x} = 5000$ and $s = 4123$ (we will discuss this later).

Does this mean that data set 2 has a great deal more variation, given that its standard deviation is 1000 times that of data set 1? Another look at the values reveals that the large value of s for data set 2 is due to the large values within this set. In fact, considering the size of the numbers within each data set, the *relative* variation within each group of values is the same. So comparing the standard deviations or variances of two data sets is not a good idea unless you know that their mean values ($\bar{x}$) are approximately equal. The next section deals with another statistical measure that will allow you to compare the relative variation within two data sets.

The Coefficient of Variation

Consider again our two data sets that appear to have the same variation (relative to the size of the data values) yet have vastly different standard deviations. These data sets are

Data set 1:   2, 2, 4, 5, 12   ($\bar{x} = 5.0$, $s = 4.123$)
Data set 2:   2000, 2000, 4000, 5000, 12,000   ($\bar{x} = 5000$, $s = 4123$)

To compare their variation, we need a measure of dispersion that will produce the same value for both of them. The solution here is to measure the standard deviation in terms of the mean; that is, what percentage of $\bar{x}$ is s? This is the **coefficient of variation, CV**. In general,

$$CV = \frac{s}{\bar{x}} \cdot 100 \qquad \textbf{(3-9)}$$

For our example data sets:

$$\text{Data set 1:}\quad CV = \frac{4.123}{5.0} \cdot 100 = 82.46$$

$$\text{Data set 2:}\quad CV = \frac{4123}{5000} \cdot 100 = 82.46$$

So our conclusion here is that both data sets exhibit the same relative variation; s is 82.46% of the mean for both sets.

EXAMPLE 3.2

To review the various measures of dispersion, use the data from Westview Hospital that you used in example 3.1.

Length of stay: 2.5, 4.8, 3.8, 3.2, 16.5, 5.0, 4.6, 4.4, 2.6, 3.1

First, compute the range:

$$\text{(largest value)} - \text{(smallest value)}$$

$$16.5 - 2.5 = 14 \text{ days}$$

To determine the MAD, recall that $\bar{x}$ is 5.05 days:

$$MAD = \tfrac{1}{10}(|2.5 - 5.05| + |4.8 - 5.05| + \cdots + |3.1 - 5.05|)$$
$$= \tfrac{1}{10}(22.9) = 2.29 \text{ days.}$$

Now find the variance and the standard deviation:

$$\Sigma x = 2.5 + 4.8 + \cdots + 3.1 = 50.5$$

and

$$\Sigma x^2 = (2.5)^2 + (4.8)^2 + \cdots + (3.1)^2 = 408.11$$

Then

$$s^2 = \frac{408.11 - (50.5)^2/10}{10 - 1}$$

$$= \frac{408.11 - 255.025}{9} = 17.01 \text{ (days)}^2$$

and

$$s = \sqrt{17.01} = 4.12 \text{ days.}$$

To calculate the coefficient of variation, use the previously obtained values of s and $\bar{x}$, where

$$CV = \frac{4.12}{5.05} \cdot 100 = 81.6$$

The standard deviation is 81.6% of the sample mean.   □

So far, you can reduce a set of sample data to a number that indicates a typical or average value (a measure of central tendency) or one that describes the amount of variation within the data values (a measure of dispersion). The next section will examine yet another set of statistics—measures of position.

EXERCISES

3.12 (a) Compute the standard deviations and coefficients of variation for each of the three racial groups in the Dallas Independent School District data in exercise 3.8.

(b) Is there greater variation in any of the groups?

3.13 The following are the scores made on an aptitude test by a group of job applicants:

53, 55, 43, 14, 64, 39, 65, 22, 17, 74, 36, 24, 13, 28, 40, 96, 92, 32, 92, 36, 18, 100, 84, 65

Calculate the:
(a) Range.
(b) Mean absolute deviation.
(c) Variance.
(d) Standard deviation.
(e) Coefficient of variation.

3.14 Use the data in exercise 2.4 to answer the following:
(a) What is the average length of an REM episode?
(b) What is the variance of the data?
(c) What is the standard deviation of the data?

3.15 The following data give the number of marriages (in thousands) for selected European nations:

Country	Marriages
France	355.0
Italy	325.6
Poland	326.3
Romania	195.9
United Kingdom	416.0
West Germany	328.2

Source: International Encyclopedia of Population (New York: The Free Press, Macmillan, 1982), p. 193.

Compute the:
(a) Mean.
(b) Range.
(c) Mean absolute deviation.
(d) Variance.
(e) Standard deviation.

3.16 The following are some of the largest foreign industrial corporations ranked in the order of sales (in millions of $):

	Company	Sales
1	Royal Dutch	59,417
2	British Petroleum	38,713
3	Unilever	21,749
4	ENI	18,985
5	Fiat	18,300
6	Francaise	17,305
7	Peugeot	17,270
8	Volkswagen	16,766
9	Phillips	16,576
10	Renault	16,117
11	Siemens	15,070
12	Daimler Benz	14,942
13	Hoechst	14,785
14	Bayer	14,196
15	Basf	14,139
16	Toyota	14,012

Source: The World Almanac and Book of Facts, 1981, p. 173.

Compute the mean and median of the sales values.

3.17 Over the years, a number of studies have compared the infant mortality rates (deaths per 1000) of breast-fed versus artificially fed babies. The following table summarizes the results: The mortality rates were selected from various years ranging from 1895 to 1946 with the larger mortality rates occurring in the earlier years.

Study Area	Breast-fed	Artificial
Berlin	57	376
Barmen, Germany	68	379
Hanover, Germany	96	296
Boston	30	212
Paris	140	310
Cologne	73	241
Amsterdam	144	304
8 U.S. cities	76	255
Derby, England	70	198
Chicago	2	84
Liverpool	84	228
Great Britain	9	18

Source: International Encyclopedia of Population (New York: The Free Press, Macmillan, 1982), Table 2, p. 73.

(a) Compute the mean, variance, and standard deviation for each of the two categories.

(b) Do the results seem to indicate that breast-feeding is associated with lower infant mortality, on the average?

(c) Calculate the coefficients of variation. Which set of data has greater variation?

3.18 The following numbers are the age at last birthday for a class of recent college graduates:

19, 21, 21, 22, 23, 24, 22, 23, 25, 24, 30, 26, 25, 24, 26, 32, 35, 36, 24, 28

Calculate the:
(a) Mean.
(b) Range.
(c) Variance.
(d) Standard deviation.

3.19 The game of cricket, which originated in England, is popular in Australia, India, and the West Indies. The following are the runs scored by two players, A and B, in various innings:

	Player A	Player B
	47	66
	0	10
	14	11
	33	22
	101	88
	68	32
	87	40
	14	38
	22	18
	46	41
Total	432	366

(a) Who is the better player? On what basis?
(b) Who is more consistent? Why?

3.20 A spatial orientation test was designed to measure awareness of spatial relationships. The following is a sample of test scores:

28, 15, 16, 20, 24, 1, 16, 35, 36, 39, 21, 18, 40, 12, 31

Compute the:
(a) Mean.
(b) Standard deviation.

3.21 Calculate the variance for the daily advertising expense of the car dealer in exercise 3.9.

3.4   MEASURES OF POSITION

Suppose that you think you are drastically underpaid compared with other people with similar experience and performance. One way to attack the problem is to obtain the salary of these other employees and demonstrate that comparatively you are way down the list. To evaluate your salary compared with those for the entire group, you would use a measure of position. **Measures of position** are indicators of how a particular value fits in with all the other data values. Two commonly used measures of position are (1) a percentile (and quartile), and (2) a Z score.

To illustrate these measures, we will suppose that an employee of PROFILE, which offers psychological testing services, has administered an aptitude test to

Table 3.1

Ordered array of aptitude test scores for 50 applicants ($\bar{x} = 60.36$, $s = 18.61$)				
22	44	56	68	78
25	44	57	68	78
28	46	59	69	80
31	48	60	71	82
34	49	61	72	83
35	51	63	72	85
39	53	63	74	88
39	53	63	75	90
40	55	65	75	92
42	55	66	76	96

50 job applicants. The ordered data are shown in Table 3.1. The mean of the data is $\bar{x} = 60.36$, and the standard deviation is $s = 18.61$. Ms. Jenson received the score of 83. She wishes to measure her performance in relation to all of the applicant scores. We will return to this illustration in example 3.3.

Percentiles

A **percentile** is the most common measure of position. The value of, for example, the 40th percentile is essentially the data value that exceeds 40% of all the data values. In other words, 40% of the data are below and 60% are above the 40th percentile. We will use the job applicant data to determine the 35th percentile. Which data value is 35% of the way between the smallest and largest value? Here the number of data values is $n = 50$ and the percentile is $P = 35$. We define the position of the 35th percentile as follows:

$$n \cdot \frac{P}{100} = 50 \cdot .35 = 17.5$$

Note that whenever $n \cdot P/100$ is *not* a counting number, it should be rounded up to the next counting number. So, 17.5 is rounded up to 18, and the 35th percentile is the 18th value of the ordered values. Referring to Table 3.1, the 35th percentile is 53.

In general, to find the **location** of the Pth percentile, determine $n \cdot P/100$. If this is not a counting number, round it up. If $n \cdot P/100$ is a counting number, the Pth percentile is the average of the number in this location (of the ordered data) and the number in the next largest location.

Now we can use the applicant data to determine the 40th percentile. Here $50 \cdot P/100 = (50)(.4) = 20$. Then

$$40\text{th percentile} = \frac{(20\text{th value}) + (21\text{st value})}{2} = \frac{55 + 56}{2} = 55.5$$

Notice here that the 40th percentile is not one of the data values but an average of two of them. Now work out the 50th percentile yourself. What measure of central tendency uses the same procedure? Based upon the previous discussion, the 50th percentile is the median.

EXAMPLE 3.3   Recall that Ms. Jenson received a score of 83. What is her percentile value?

Solution   Her value is the 45th largest value (out of a total of 50). An initial guess of the percentile here would be

$$P = \frac{45}{50} \cdot 100 = 90$$

However, due to the percentile rules used here, this may be slightly incorrect. Your next step should be to examine this value of P, along with the next two smaller values. The following calculations of $P = 88$, $P = 89$, and $P = 90$ reveal that Ms. Jenson's score is the 89th percentile.

P	$n \cdot P/100$	Pth Percentile
88	$50 \cdot .88 = 44$	$(82 + 83)/2 = 82.5$
89	$50 \cdot .89 = 44.5$	45th value $= 83$
90	$50 \cdot .90 = 45$	$(83 + 85)/2 = 84$

EXAMPLE 3.4   What is the 50th percentile for the applicant data in Table 3.1?

Solution   Here, $n \cdot P/100 = 50 \cdot .5 = 25$. The 50th percentile is an average of the 25th and 26th ordered data values:

$$50\text{th percentile} = \frac{61 + 63}{2} = 62$$

Quartiles

Quartiles are merely particular percentiles that divide the data into quarters, namely:

$$Q_1 = \text{1st quartile} = \text{25th percentile}$$

$$Q_2 = \text{2nd quartile} = \text{50th percentile} = \text{median}$$

$$Q_3 = \text{3rd quartile} = \text{75th percentile}$$

They are used as benchmarks, much like the use of A, B, C, D, and F on examination grades. Using the applicant data in Table 3.1, we can determine

$$n \cdot P/100 = 50 \cdot .25 = 12.5$$

This is rounded up to 13, and $Q_1 = $ 13th ordered value $= 46$.

$$Q_2 = \text{median} = 62$$

from Example 3.4. Finally,

$$n \cdot P/100 = 50 \cdot .75 = 37.5$$

This is rounded up to 38, and $Q_3 = $ 38th ordered value $= 75$.

Z Scores

Another measure of position is a sample Z score, which is based on the mean ($\bar{x}$) and standard deviation of your data set. As with percentiles, a Z score determines the relative position of any particular data value x. The Z score of x is defined as

$$Z = \frac{x - \bar{x}}{s} \qquad\qquad \textbf{(3-10)}$$

Recall from example 3.3 that Ms. Jenson had a score of 83 on the test. For this data set, $\bar{x} = 60.36$ and $s = 18.61$. Her score of 83 is in the 89th percentile. The corresponding Z score is

$$Z = \frac{83 - 60.36}{18.61} = 1.22$$

This means that Ms. Jenson's score of 83 is 1.22 standard deviations to the right of the mean, or above the group's average. Thus, if Z is positive, it indicates how many standard deviations x is to the right of the mean.

A negative value implies that x is to the left of the mean. Again referring to Table 3.1, what is the Z score for the individual who obtained a total of 35 on the aptitude examination?

$$Z = \frac{35 - 60.36}{18.61} = -1.36$$

This individual's score is 1.36 standard deviations to the left of the mean, or below the group's average.

3.5   MEASURES OF SHAPE

A basic question in many applications is whether your data exhibit a **symmetric** pattern. **Measures of shape** determine skewness and kurtosis.

Skewness

The histogram in Figure 3.3 demonstrates a perfectly symmetric distribution. When the data are unimodal and symmetric, the sample mean $\bar{x}$, the sample median Md, and the sample mode Mo, are the same. As the data tend toward a nonsymmetric distribution, referred to as **skewed**, the mean and median drift apart. The easiest method of determining the degree of skewness present in your

Figure 3.3

Example of
symmetric data

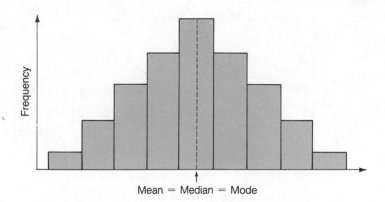

sample data is to calculate a measure referred to as the **Pearsonian coefficient of skewness, Sk**. Its value is given by

$$Sk = \frac{3(\bar{x} - Md)}{s} \qquad \textbf{(3-11)}$$

where s is the standard deviation of the sample data.

The value of Sk ranges from -3 to 3. If the data are perfectly symmetric (a rare event), Sk = 0, because x = Md. For Figure 3.3, Sk is zero. If Sk is positive, then the mean is larger than the median, and we say that the data are skewed right. This merely means the data exhibit a pattern with a right tail, as illustrated in Figure 3.4. We know the mean is affected by extreme values, so we would expect the mean to move toward the right tail, above the median. This results in a positive value of Sk. Similarly, if Sk is negative, then the data are skewed left and the mean is smaller than the median. Figure 3.5 shows a data distribution exhibiting a left tail.

Figure 3.4

An example of right
(positive) skew

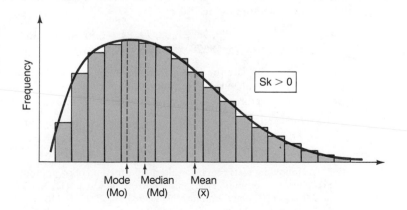

Figure 3.5

Example of left (negative) skew

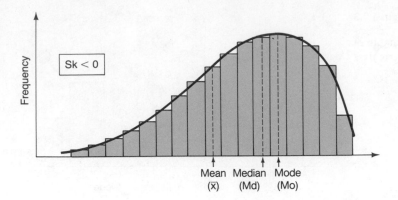

Using the aptitude examination scores in Table 3.1, we have $\bar{x} = 60.36$, $s = 18.61$, and Md $= 62$.

$$Sk = \frac{3(60.36 - 62)}{18.61} = -.26$$

Consequently, a histogram of these data should be just slightly skewed left.

Kurtosis

Sk measures the tendency of a distribution to stretch out in a particular direction. Another measure of shape, referred to as the **kurtosis**, measures the **peakedness** of your distribution. The calculation of this measure is a bit cumbersome, and the kurtosis value is not needed in the remaining text material.* Briefly, this value is small if the frequency of observations close to the mean is high and the frequency of observations far from the mean is low.

EXERCISES

3.22 The following table indicates the scores of a class of students at a junior college on a statistics exam:

18, 58, 71, 83, 89, 96, 21, 62, 74, 84, 90, 97, 43, 66, 75, 86, 92, 98, 47, 66, 77, 86, 94, 100, 55, 68, 78, 88, 95, 100

   (a)   Calculate the 40th percentile.
   (b)   Calculate the 77th percentile.
   (c)   Interpret the meaning of the numbers calculated for the two percentiles.

3.23 In exercise 3.22, how would you evaluate the performance of a student who scored 74 on the exam?

---

* The following texts contain an alternative method of computing sample skewness as well as a procedure for computing the sample kurtosis. L. Ott and D. K. Hildebrand, *Statistical Thinking for Managers* (Boston: Duxbury Press, 1983), p. 27; C. L. Olsen and M. J. Picconi, *Statistics for Business Decision Making* (Glenview, IL: Scott, Foresman, 1983), pp. 127–129.

3.24 Using the data in exercise 3.22, calculate the first and third quartiles. State the results in words.

3.25 Considering the following data:

7, 8, 8, 9, 11, 14, 15, 16, 18, 19, 21, 27, 28, 30, 32, 35

calculate the:
(a)   90th percentile.
(b)   58th percentile.
(c)   The first, second, and third quartiles.

3.26 Gross National Product (GNP) is the market value of all final goods and services produced by an economy in a given time period and it is probably the most important indicator of economic health of a country. The GNP (in billions of constant 1972 $) of the United States for the years 1970–1984 are given below:

Year	GNP
1970	1085.6
1971	1122.4
1972	1185.9
1973	1254.3
1974	1246.3
1975	1231.6
1976	1298.2
1977	1369.7
1978	1438.6
1979	1479.4
1980	1475.0
1981	1512.2
1982	1480.0
1983	1534.7
1984	1639.0

Source: Economic Report of the President, February 1985.

(a)   Calculate the mean, median, variance, and standard deviation of the GNP values.
(b)   Calculate the coefficient of skewness. Are the data skewed to the left?

3.27 Assume a particular set of sample data such that

$$\bar{x} = 49$$

$$s = 18$$

Considering one particular value of $x = 63$:
(a)   Calculate the Z score.
(b)   Interpret the Z score.

3.28 A home economics class determined that the mean time required to cook a thin fish fillet in a microwave oven was 90 seconds with a standard deviation of 15 seconds.
(a)   If you know that the Z score for an individual specimen was 1.50, what was the actual cooking time in seconds for that specimen?
(b)   If you know that a particular specimen took 70 seconds to cook, what was the Z score for this item?

3.29 An elementary school teacher finds that the variance in heights for her class is 36 square inches. If a child with a height of 3 ft 4 in. has a Z score equal to 0.50, what is the average height for the class?

3.30 A county court clerk claimed that a speeding fine of $60 was 2 standard deviations to the right of the mean, which was $50. What was the standard deviation?

3.31 The average distance from home to factory for a group of workers is 8 miles with a standard deviation of 3.5 miles. One worker has a Z score of 1.50. How far is this worker's home from the factory?

3.32 Assume the following about a set of sample data:

$$s = 14$$

$$\bar{x} = 21$$

$$Md = 18.5$$

(a)   What do you observe about the pattern of the data?
(b)   Calculate the coefficient of skewness. What does this value suggest?

3.33 The following data represent the total fertility rates (average number of children that would be borne by a group of women in the child bearing ages if they experienced no mortality) in 15 developing countries for 1980–1985:

Bangladesh	6.15	Thailand	3.52
Fiji	3.50	Colombia	3.93
Indonesia	4.10	Costa Rica	3.50
Malaysia	3.91	Dominican Republic	4.18
Nepal	6.25	Mexico	4.61
Pakistan	5.84	Panama	3.46
South Korea	2.60	Peru	5.00
Sri Lanka	3.38		

Source: Department of International Economic and Social Affairs, United Nations. *World Population Prospects: Estimates and Projections as Assessed in 1984*, Population Studies No. 98 (New York: United Nations Publication, Sales No. E.86.XIII.3, 1986), Table A-12, pp. 102–107.)

(a)   Calculate the mean, median, variance, and standard deviation of the fertility rates.
(b)   Compute the coefficient of skewness. Are the data symmetric or skewed to the right or to the left?

3.6   INTERPRETING $\bar{x}$ AND s

Now that you have gone through several pencils determining the sample mean and standard deviation, what can you learn from these values? The type of question that you can answer is, "How many of the data values are within two standard deviations of the mean?"

Take a look at the data from Table 3.1. Here, $\bar{x} = 60.36$ and $s = 18.61$, and so we obtain

$$\bar{x} - s = 60.36 - 18.61 \qquad \bar{x} + s = 60.36 + 18.61$$
$$= 41.75 \qquad\qquad\quad = 78.97$$

$$\bar{x} - 2s = 60.36 - 37.22 \qquad \bar{x} + 2s = 60.36 + 37.22$$
$$= 23.14 \qquad\qquad\quad = 97.58$$

$$\bar{x} - 3s = 60.36 - 55.83 \qquad \bar{x} + 3s = 60.36 + 55.83$$
$$= 4.53 \qquad\qquad\quad = 116.19$$

Examine these data and observe that (1) 33 out of the 50 values (66%) lie between $\bar{x} - s$ and $\bar{x} + s$; (2) 49 out of the 50 values (98%) lie between $\bar{x} - 2s$ and $\bar{x} + 2s$; and (3) 50 out of the 50 values (100%) lie between $\bar{x} - 3s$ and $\bar{x} + 3s$. Or, put another way: (1) 66% of the data values have a Z score between -1 and 1; (2) 98% have a Z score between -2 and 2, and (3) 100% have a Z score between -3 and 3.

What can we say in general for any data set? There are two types of statements we can make. One of these, **Chebyshev's inequality**, is usually conservative but makes no assumption about the population from which you obtained your data. Following are the components of Chebyshev's inequality:

Chebyshev's Inequality

1 At least 75% of the data values are between $\bar{x} - 2s$ and $\bar{x} + 2s$

2 At least 89% of the data values are between $\bar{x} - 3s$ and $\bar{x} + 3s$

3 In general, at least $(1 - 1/k^2)$ of the data values are between $\bar{x} - ks$ and $\bar{x} + ks$, for $k = 2, 3, 4, \ldots$

Note that if $k = 1$, $1 - 1/k^2 = 0$; so Chebyshev's inequality provides no information on the number of data values to expect between $\bar{x} - s$ and $\bar{x} + s$.

The other type of statement is called the **empirical rule**. We make a key assumption here, namely that the population from which you obtain your sample has a bell-shaped distribution; that is, it is symmetric and tapers off smoothly into each tail. Such a population is called a **normal population** and is illustrated in Figure 3.6. Thus, the data set should have a skewness measure Sk near zero and a histogram similar to that in Figure 3.3. However, the empirical rule is still quite accurate if your distribution is not exactly bell-shaped. Following are the components of the empirical rule.

Figure 3.6

A bell-shaped
(normal) population

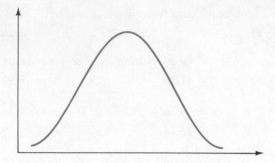

Empirical Rule

Under the assumption of a bell-shaped population:

1 Approximately 68% of the data values will lie between $\bar{x} - s$ and $\bar{x} + s$.

2 Approximately 95% of the data values will lie between $\bar{x} - 2s$ and $\bar{x} + 2s$.

3 Approximately 99.7% of the data values will lie between $\bar{x} - 3s$ and $\bar{x} + 3s$.

Returning to Table 3.1, we can summarize our previous results along with the information provided by Chebyshev's inequality and the empirical rule. The actual percentage of the sample values in each interval, as well as the percentage specified using each of the two rules, is shown in Table 3.2.

As you can see, Chebyshev's inequality is very conservative, but it always works. The empirical rule predicted results close to what was observed. This is not surprising because the skewness measure is only slightly different from zero (Sk = $-.26$).

Table 3.2

Summary of
percentages of
sample values by
interval, using data
from Table 3.1

		PERCENTAGE	
Between	**Actual**	**Chebyshev Inequality**	**Empirical Rule**
$\bar{x} - s$ and $\bar{x} + s$	66% (33 out of 50)	—	≈68%
$\bar{x} - 2s$ and $\bar{x} + 2s$	98% (49 out of 50)	≥75%	≈95%
$\bar{x} - 3s$ and $\bar{x} + 3s$	100% (50 out of 50)	≥89%	≈99.7%

EXAMPLE 3.5   In a random sample of 200 automotive insurance claims obtained from Pearson Insurance Company, $\bar{x} = \$615$ and $s = \$135$. (1) What statement can you make using Chebyshev's inequality? (2) If you have reason to believe that the population of all insurance claims is bell-shaped, what does the empirical rule say about these 200 values?

Solution I   Chebyshev's inequality provides information regarding the number of sample values within a specified number of standard deviations of the mean. For $k = 2$, we have

$$\bar{x} - 2s = 615 - 2(135) = \$345$$

$$\bar{x} + 2s = 615 + 2(135) = \$885$$

We conclude that at least 75% of the sample values lie between $345 and $885. Because $.75 \cdot 200 = 150$, this implies that at least 150 of the claims are between $345 and $885.

For $k = 3$

$$\bar{x} - 3s = 615 - 3(135) = \$210$$

$$\bar{x} + 3s = 615 + 3(135) = \$1020$$

and we conclude that at least 8/9 (89%) of the data values are between $210 and $1020. Here, $\frac{8}{9} \cdot 200 = 177.8$, and so at least 178 of the claims are between $210 and $1020.

Solution 2   If the distribution of the automotive claims at Pearson Insurance Company is believed to be bell-shaped, the empirical rule allows us to draw stronger conclusions. In particular, for $k = 1$, we have

$$\bar{x} - s = 615 - 135 = \$480$$

$$\bar{x} + s = 615 + 135 = \$750$$

and we conclude that approximately 68% of the data values ($.68 \cdot 200 = 136$) are between $480 and $750.

For $k = 2$

$$\bar{x} - 2s = \$345$$

$$\bar{x} + 2s = \$885$$

and we conclude that approximately 95% of the data values ($.95 \cdot 200 = 190$) will lie between $345 and $885.   □

EXERCISES   3.34 (a)   Use Chebyshev's inequality to find the range of fertility rates that will include at least 89% of the data in exercise 3.33.
   (b)   If the empirical rule were applied (assuming a bell-shaped distribution), what percentage of the data will lie within the range determined in part (a) above?

3.35 The mean score of students in a sociology class was 72.5 with a standard deviation of 12. Assuming a bell-shaped distribution, estimate the range of scores within which about 95% of the scores will fall.

3.36 In the annual spaghetti-eating contest of a large city, 100 people participated. The contestants' spaghetti consumption averaged 120 ft with a standard deviation of 14 ft.

    (a) At least how many contestants ate between 92 and 148 feet of spaghetti?

    (b) At least how many contestants ate between 78 and 162 feet of spaghetti?

    (c) Did you have to make any assumptions about the data to answer questions (a) and (b)?

3.37 What do you think will be the value of the coefficient of skewness for a symmetric distribution? Will the mean and median be equal?

3.38 Refer to the data on exam scores in exercise 2.1. Using Chebyshev's inequality, estimate the values that should contain at least 75% of the data.

3.39 A control group of rats is kept on a special diet supplemented with hormone injections. If the average gain in weight after two weeks is 6 oz, with a standard deviation of 2.5 oz, what will be the expected range of weight gain for at least 3 out of 4 rats?

3.40 The mean starting salary of liberal arts–social science majors with a bachelor's degree is $19,250 and the standard deviation is $500. Assuming the population has a bell-shaped (normal) distribution, what will be the range of starting salaries for 95% of the graduates? Also, what percentage of the graduates will fall in the range $18,750 to $19,750 as a starting salary?

3.41 Refer to question 3.29. Between what heights will at least three-fourths of the class fall?

3.42 The mean length of the period of incarceration (actual time served) for violent criminals is 3 years 10 months in a certain state. The standard deviation is 6 months.

    (a) Using Chebyshev's inequality, estimate the range of incarceration periods for at least 89% of the violent criminals.

    (b) Using the empirical rule, determine what percentage of violent criminals will serve between 34 months and 58 months in jail?

3.43 Using Chebyshev's inequality, find the scores on the aptitude test in exercise 3.13 between which at least 89% of the data will fall.

3.44 The average number of visitors to the local museum is 48 per day and the standard deviation is 17. Using the empirical rule, estimate the range within which approximately 99.7% of the visitors will fall.

3.45 A video game arcade operator says that slightly more than two-thirds of his customers spend between $2.00 and $3.00 per visit. The spending pattern has a normal distribution, with a mean amount spent of $2.50. What is the standard deviation? Also, between what amounts will 95% of the customers spend?

3.7   GROUPED DATA

You may have to work with data in the form of a frequency distribution, called **grouped data**, when the raw data are not available. This situation can arise when we find a histogram or frequency distribution in a magazine or newspaper article in which the actual raw data used to construct the histogram are not included. We do not have the data values used to make up this frequency dis-

Table 3.3

Age of 35 teachers
who recently
completed their
masters degrees

Class Number	Class (age in years)	Frequency
1	20 and under 30	5
2	30 and under 40	14
3	40 and under 50	9
4	50 and under 60	5
5	60 and under 70	2
		35

tribution, so we are forced to estimate the sample statistics, in particular the mean, median, and standard deviation.

Estimating the Sample Mean $\bar{x}$

Assume we obtain the frequency distribution shown in Table 3.3, containing the ages of 35 teachers who recently completed their master's degrees. The 35 data values are not available, so we cannot add them up. A procedure that works well for estimating $\bar{x}$ is simply to pretend that the 35 data values are equal to their respective class midpoints. Consequently, there are

$$5 \text{ values at } (20 + 30)/2 = 25$$

$$14 \text{ values at } (30 + 40)/2 = 35$$

$$\vdots$$

$$2 \text{ values at } (60 + 70)/2 = 65$$

The value of $\bar{x}$ is estimated by finding what it is approximately equal to ($\cong$).

$$\bar{x} \cong \frac{(25 + 25 + 25 + 25 + 25) + \cdots + (65 + 65)}{35}$$

$$= \frac{(5)(25) + (14)(35) + (9)(45) + (5)(55) + (2)(65)}{35}$$

$$= \frac{1425}{35} = 40.7$$

Our estimate of the average age of these 35 individuals is

$$\bar{x} \cong 40.7$$

In general,

$$\bar{x} \cong \frac{\Sigma f \cdot m}{n} \qquad \qquad (3\text{-}12)$$

where n = sample size, f = frequency of each class, and m = midpoint of each class.

Estimating the Sample Standard Deviation s

Using the same fictitious data set at the various class midpoints, the variance s^2 can be found in the usual way, using equation 3-8:

$$s^2 = \frac{\Sigma(\text{each data value})^2 - [\Sigma(\text{each data value})]^2/n}{n-1}$$

$$\Sigma(\text{each data value})^2 = \overbrace{(25^2 + 25^2 + \cdots + 25^2)}^{5 \text{ times}}$$

$$+ \overbrace{(35^2 + 35^2 + \cdots + 35^2)}^{14 \text{ times}}$$

$$\vdots$$

$$+ (65^2 + 65^2)$$

$$= (5)(25^2) + (14)(35^2) + (9)(45^2) + (5)(55^2)$$

$$+ (2)(65^2) = 62{,}075$$

Also, $\Sigma(\text{each data value}) = 1425$. This was determined previously when estimating $\bar{x}$.

$$s^2 \cong \frac{62{,}075 - (1425)^2/35}{34} = \frac{4057.14}{34} = 119.33$$

$$s \cong \sqrt{119.33} = 10.92$$

In general,

$$s^2 \cong \frac{\Sigma f \cdot m^2 - (\Sigma f \cdot m)^2/n}{n-1} \tag{3-13}$$

where f, m, and n are as defined in equation 3-12.

The calculations necessary to estimate $\bar{x}$ and s are more easily performed using a table similar to Table 3.4.

Table 3.4

Summary of calculations for grouped data

Class Number	Class	f	m	$f \cdot m$	$f \cdot m^2$
1	20 and under 30	5	25	125	3,125
2	30 and under 40	14	35	490	17,150
3	40 and under 50	9	45	405	18,225
4	50 and under 60	5	55	275	15,125
5	60 and under 70	2	65	130	8,450
				$\Sigma f \cdot m = 1425$	$\Sigma f \cdot m^2 = 62{,}075$

Estimating the Sample Median Md

Using the previous example, we have

$$\text{Md} \cong \left(\frac{35 + 1}{2}\right)\text{th ordered value} = \text{18th ordered value}$$

Where is this value in the frequency distribution? The first class contains the five smallest values, and the first two classes contain the first 19 ordered values $(5 + 14 = 19)$. So the 18th value is in the second class.

We can better approximate the median by assuming that the values in this class (and all classes) are spread evenly between the lower and upper limits. Because the first class contains five values, the median is 13 $(18 - 5)$ values into the second class. This class begins at 30, has a width of 10, and has 14 values in it. So we want to go 13 values into a class of width 10 containing 14 values. The resulting estimate of the median is

$$\text{Md} \cong 30 + \tfrac{13}{14}(10) = 39.3$$

In general,

$$\text{Md} \cong L + \frac{k}{f} \cdot W \qquad\qquad \textbf{(3-14)}$$

where L = lower limit of the class containing the median (called the **median class**), $k = (n + 1)/2 - $ (the number of data values preceding the median class); f = frequency of the median class, and W = class width.

In the previous example, $L = 30$, $f = 14$, $W = 10$, and thus

$$k = \frac{35 + 1}{2} - 5 = 13$$

If n is even (say, $n = 100$), then $(n + 1)/2 = 50.5$, and you need to estimate the 50.5th ordered value. This is halfway between the 50th and the 51st value. The procedure to follow here is exactly the same, except k will not be a counting number.

Remember that these procedures for estimating the sample statistics are used only when the raw data are not available and your only information is a frequency distribution or corresponding histogram. If the actual data values are available, these statistics can be determined exactly, and the estimation procedures described in this section should not be used.

EXERCISES

3.46 The following data are the number of legal abortions (in thousands) by age of mother (in years) in the United States for 1981:

Mother's Age	Number of Legal Abortions
Under 15	15
15 and under 20	433
20 and under 25	555
25 and under 30	316
30 and under 35	167
35 and under 40	70
40 and above	21

Source: U.S. Statistical Abstract, U.S. Department of Commerce, 1985, p. 67.

Compute the median. Is the median an appropriate summary number for these data? Why or why not?

3.47 The following is the summarized information concerning family income (in dollars) for all families living within a specific neighborhood of a major city:

Income Level	Percentage of Families
0 and under 10,000	11
10,000 and under 20,000	19
20,000 and under 30,000	30
30,000 and under 40,000	15
40,000 and under 50,000	10
50,000 and under 60,000	15

Find the mean and median family income.

3.48 In an experiment performed to determine toxicity of a new antidepressant drug, a pharmacologist noted the following results:

Drug Level (milligrams of drug administered)	Number of Fatalities
0.1 and under 0.3	0
0.3 and under 0.5	1
0.5 and under 0.7	2
0.7 and under 0.9	8
0.9 and under 1.1	10
1.1 and under 1.3	10

Compute the mean and standard deviation.

3.49 Calculate the median age using the data in exercise 2.13.

3.50 The following information summarizes a sample of the inmate population of a federal penitentiary according to the number of times the offender has been imprisoned previously:

Number of Incarcerations	Percentage in the Group
0–2	20
3–5	50
6–8	20
9–11	7
12–14	3

Compute the mean, median, and the standard deviation.

3.51 The following is the final distribution of grades in an introductory economics course at a local university:

Grade	Number of Students
90–99	5
80–89	14
70–79	11
60–69	4
50–59	7
	‾‾
	41

$n = 41$

(a)   Calculate the mean score.
(b)   Calculate the standard deviation.
(c)   Interpret the value of the mean and standard deviation.

3.52 The industrial engineer of the Bright Light company is interested in examining the average burning hours (life) of the 100-watt bulbs manufactured. Using the following data, estimate the mean burning hours of a 100-watt bulb:

Burning Hours	Number of Bulbs
0 and under 40	231
40 and under 50	168
50 and under 60	244
60 and under 70	300
70 and under 80	111
80 and under 90	48
90 and under 100	98
	‾‾‾‾
	1200

3.8   CALCULATING DESCRIPTIVE STATISTICS BY CODING

When you use a calculator to determine the sample mean or standard deviation, one problem that can occur is that the data values are too large or too small to "fit" into your calculator. To avoid having the calculator self-destruct in your hands, a procedure referred to as **data adjusting** or **data coding** allows you to derive these statistics using more reasonable data values. You can then work backward to get the desired statistics. To code or adjust the data you subtract (or add) or divide (or multiply) your original data set by a fixed amount. Figure 3.7 demonstrates this procedure using data sets containing several large values. To adjust the data when subtracting (adding) a positive constant to each data value

$$\text{actual } \bar{x} = \text{adjusted } \bar{x} \text{ plus (minus) the constant}$$
$$\text{actual } s = \text{adjusted } s \text{ (no change)}$$

When dividing (multiplying) by a positive constant,

$$\text{actual } \bar{x} = \text{adjusted } \bar{x} \text{ times (divided by) the constant}$$
$$\text{actual } s = \text{adjusted } s \text{ times (divided by) the constant}$$

Figure 3.7

Data coding

Subtracting or Adding Constant

Actual Data   Adjusted Data

1002, 1002, 1004, 1005, 1012 ——————— subtract 1000 ——▸ 2, 2, 4, 5, 12

$\bar{x} = 1000 + 5.0 = 1005.0$ ◂——————— add 1000 ——— $\bar{x} = 5.0$

$s = 4.123$ ◂——————— is the same as——— $s = 4.123$

Dividing or Multiplying by Constant

Actual Data   Adjusted Data

2000, 2000, 4000, 5000, 12,000 ——————— divided by 1000 ——▸ 2, 2, 4, 5, 12

$\bar{x} = 1000 \times 5.0 = 5000$ ◂——————— multiply by 1000——— $\bar{x} = 5.0$

$s = 1000 \times 4.123 = 4123$ ◂——————— multiply by 1000 ——— $s = 4.123$

 Figure 3.8

MINITAB procedure
for describing
sample data

```
MTB > SET INTO C1
DATA> 22 25 28 31 34 35 39 39 40 42 44 44 46 48 49
DATA> 51 53 53 55 55 56 57 59 60 61 63 63 63 65 66
DATA> 68 68 69 71 72 72 74 75 75 76 78 78 80 82 83
DATA> 85 88 90 92 96
DATA> END
MTB > DESCRIBE C1

              N      MEAN    MEDIAN     TRMEAN     STDEV    SEMEAN
     C1      50     60.36     62.00      60.57     18.61      2.63

            MIN       MAX        Q1         Q3
     C1   22.00     96.00     45.50      75.00
```

When you derive sample statistics, you have essentially two options: use a calculator or use a computer. Calculators work well for small data sets but involve too much time (and opportunity for error) for moderate or large sample sizes. Practically all statistical computer packages will provide you with the basic sample statistics (mean, median, variance, and so on) in response to only a few commands once the data have been read in. Figure 3.8 contains the MINITAB commands (along with the output) necessary to derive the basic statistics for the data in Table 3.1. The appendices at the end of the chapter will demonstrate this procedure using SPSS and SAS.

EXERCISES

3.53 Using the subtraction rule, adjust the following data and calculate the mean and standard deviation:

413, 407, 411, 402, 425, 408, 410, 421

3.54 Using the multiplication rule, adjust the following data and calculate the mean and standard deviation:

0.00119, 0.00101, 0.00121, 0.00108, 0.0010, 0.00114, 0.00117, 0.00104, 0.00104, 0.00123, 0.00124

3.55 Using the division rule, calculate the mean and variance for the following data:

200, 600, 800, 1000, 400, 1200, 10,000, 1400, 1600, 400

3.56 Using the division rule, calculate the mean and standard deviation for the following data:

500, 3000, 600, 1000, 400, 300, 100, 2500, 700, 300

3.57 Using the multiplication rule, adjust the following data and calculate the mean and standard deviation:

0.001182, 0.001104, 0.001270, 0.001251, 0.001407, 0.001553, 0.001177, 0.001333, 0.001489, 0.001505

3.58 Using the subtraction rule, adjust the following data and calculate the mean:

1013, 1007, 1011, 1102, 1025, 1008, 1110, 1021, 1111, 1009

3.9   EXPLORATORY DATA ANALYSIS

Exploratory data analysis (EDA) is a recently developed technique that provides easy-to-construct pictures that summarize and describe a set of sample data. Two popular diagrams that fall under this category are **stem and leaf diagrams** and **box and whisker plots**. Stem and leaf diagrams are like histograms since both allow you to see the "shape" of the data. Box and whisker plots are graphical representations of the quartile measures of position, discussed in Section 3.4.

Stem and Leaf Diagrams

Stem and leaf diagrams were originally developed by John Tukey (pronounced Too'-key) of Princeton University. They are extremely useful in summarizing reasonably sized data sets (under 100 values as a general rule), and unlike a histogram, result in no loss of information. By this we mean that it is possible to construct the original data set from a stem and leaf diagram, whereas some of the information in the original data is lost when constructing a histogram. Histograms, however, provide the best alternative when attempting to summarize a large set of sample data.

To illustrate the construction of a stem and leaf diagram, suppose that a local radio station obtained the rainfall at 12 randomly selected weather stations across the state. The recorded rainfall during the past 30 days (in inches) are

3.4, 4.5, 2.3, 2.7, 3.8, 5.9, 3.4, 4.7, 2.4, 4.1, 3.6, 5.1

The stem and leaf diagram for these data is shown in Figure 3.9. Each observation is represented by a **stem** to the left of the vertical line and a **leaf** to the right of the vertical line. For example, the stems and leaves for the first and

Figure 3.9

Stem and leaf
diagram for rainfall
data

```
2 | .3 .7 .4
3 | .4 .8 .4 .6
4 | .5 .7 .1
5 | .9 .1
```

last observation would be

stem	leaf		stem	leaf
3	.4		5	.1

In a stem and leaf diagram, the stems are put *in order* to the left of the vertical line. The leaf for each observation is generally the last digit (or possibly the last two digits) of the data value, with the stem consisting of the remaining first digits. The value 562 could be represented as 5|62 or as 56|2 in a stem and leaf diagram, depending upon the range of the sample data. Whether or not the raw data are ordered, the stem and leaf diagram provides at least a partial ordering of the data. If the diagram is rotated counterclockwise, it has the appearance of a histogram and clearly describes the shape of the sample data.

The aptitude scores of the 50 job applicants contained in Table 3.1 are represented by a stem and leaf diagram in Figure 3.10. From this diagram we observe that the minimum score is 22, the maximum score is 96, and the largest group of scores is between 60 and 69. Also, the five leaves in stem row 3 indicate that five people scored at least 30 but less than 40. The three leaves in stem row 9 tells us at a glance that 3 people scored 90 or better.

For larger data sets, you may want to consider spreading out the stem column by repeating the stem value two or three times. To illustrate, the 50 applicant test scores are put into another stem and leaf diagram in Figure 3.11, where each stem value is repeated twice. The first stem value contains leaves between 0 and 4, the second stem contains leaves between 5 and 9. A MINITAB version of the stem and leaf diagram using these data is shown in Figure 3.12. Notice that MINITAB used the double stems as in Figure 3.11.

In many situations the leaves in your diagram may consist of a *pair* of digits. In Figure 3.13, the verbal score on the graduate record exam (GRE) was

Figure 3.10

Stem and leaf
diagram for aptitude
test scores

```
2 | 2 5 8
3 | 1 4 5 9 9
4 | 0 2 4 4 6 8 9
5 | 1 3 3 5 5 6 7 9
6 | 0 1 3 3 3 5 6 8 8 9
7 | 1 2 2 4 5 5 6 8 8
8 | 0 2 3 5 8
9 | 0 2 6
```

Figure 3.11

Stem and leaf
diagram for aptitude
test scores using
repeated stems

```
2 | 2
2 | 5 8
3 | 1 4
3 | 5 9 9
4 | 0 2 4 4
4 | 6 8 9
5 | 1 3 3
5 | 5 5 6 7 9
6 | 0 1 3 3 3
6 | 5 6 8 8 9
7 | 1 2 2 4
7 | 5 5 6 8 8
8 | 0 2 3
8 | 5 8
9 | 0 2
9 | 6
```

 Figure 3.12

Stem and leaf
diagram of aptitude
test scores
(Table 3.1) using
MINITAB

```
MTB > SET INTO C1
DATA> 22 25 28 31 34 35 39 39 40 42
DATA> 44 44 46 48 49 51 53 53 55 55
DATA> 56 57 59 60 61 63 63 63 65 66
DATA> 68 68 69 71 72 72 74 75 75 76
DATA> 78 78 80 82 83 85 88 90 92 96
DATA> END
MTB > STEM AND LEAF USING C1

Stem-and-leaf of C1        N  = 50
Leaf Unit = 1.0

     1     2  2
     3     2  58
     5     3  14
     8     3  599
    12     4  0244
    15     4  689
    18     5  133
    23     5  55679
    (5)    6  01333
    22     6  56889
    17     7  1224
    13     7  55688
     8     8  023
     5     8  58
     3     9  02
     1     9  6
```

Figure 3.13

Stem and leaf
diagram for verbal
GRE scores

```
3 | 20 70 30 50
4 | 50 10 40 40 20
5 | 60 90 30 50 30 80 50
6 | 20 50 40 70
```

Figure 3.14

Box and whisker
plot for 50 aptitude
test scores (data in
Table 3.1)

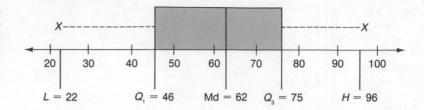

recorded for 20 randomly selected graduate students. Here each stem consists
of a single digit and each leaf represents the last two digits of the student's score,
the last digit of which is always zero.

Box and Whisker Plots

A **box and whisker plot** is a graphical representation of a set of sample data
that illustrates the lowest data value (L), the first quartile (Q_1), the median
(Q_2, Md), the third quartile (Q_3), and the highest data value (H).

In Section 3.4, the following values were determined for the aptitude test
scores in Table 3.1:

$$L = 22 \qquad\qquad Q_3 = 75$$

$$Q_1 = 46 \qquad\qquad H = 96$$

$$Q_2 = \text{Md} = 62$$

A box and whisker plot of these values is shown in Figure 3.14. The ends
of the *box* are located at the first and third quartile with a vertical bar inserted
at the median. The dotted lines are the *whiskers* and connect the highest and
lowest data values to the ends of the box. This means that approximately 25%
of the data values will lie in each whisker and in each portion of the box. If the
data are symmetric, the median bar should be located at the center of the box.
Consequently, the bar location informs you as to the *skewness* of the data; if

 Figure 3.15

Box and whisker
plot of aptitude test
scores (Table 3.1)
using MINITAB

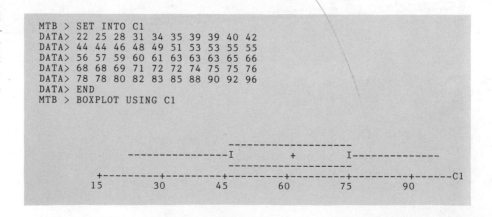

located in the left half of the box, the data are skewed right and if located in the right half, the data are skewed left.

According to Figure 3.14, the data appear to be nearly symmetric (with a slight left skew), which is supported using the stem and leaf diagram in Figure 3.10. A box and whisker plot using MINITAB is shown in Figure 3.15.

EXERCISES

3.59 The following data reflect the number of consecutive hours that a sample of 20 medical interns worked in a local emergency room:

14, 21, 35, 11, 18, 28, 26, 38, 21, 19, 17, 27, 18, 22, 16, 30, 17, 26, 20, 25

Construct a stem and leaf diagram of these data.

3.60 Construct a stem and leaf diagram of the scores in exercise 3.13.

3.61 Construct a stem and leaf diagram of the ages in exercise 3.18.

3.62 The scores on a newly devised instrument for measuring life satisfaction could range from a minimum of 100 to a maximum of 800. A sample of 25 scores contained the following values:

512, 587, 563, 542, 556, 577, 515, 520, 532, 526, 560, 584, 530, 593, 542, 518, 570, 588, 534, 542, 557, 538, 562, 548, 525

The researcher noticed that all of the scores were between 500 and 600, not a good situation due to the lack of variation in these scores. Illustrate this by constructing a stem and leaf diagram of the instrument scores.

3.63 Construct a stem and leaf diagram of the fertility scores in exercise 3.33.

3.64 The ages of 40 individuals working at a telephone crisis center were recorded as follows:

33, 40, 35, 24, 31, 32, 45, 23, 45, 34, 34, 54, 25, 36, 38, 44, 62, 31, 56, 47, 30, 42, 43, 52, 48, 56, 64, 40, 61, 39, 26, 34, 37, 43, 47, 32, 45, 27, 35, 29

(a) Construct a stem and leaf diagram of the data. What conclusions can you make from this picture?

(b) Construct a box and whisker plot of these data. What can you say about the data based upon this plot?

3.65 Construct a box and whisker plot using the scores in exercise 3.22. Does this plot also illustrate the lack of variation in the scores? Discuss.

3.66 What does the box and whisker plot in exercise 3.65 tell you regarding the skewness of these data?

SUMMARY

The purpose of analyzing or describing sample data is to learn more about the population from which it was obtained. Every population has properties that describe it. These properties are referred to as **parameters**. We can estimate these parameters by obtaining a sample and deriving the corresponding sample **statistic**, which is a particular **descriptive measure**.

This chapter has introduced some of the more popular descriptive measures used to describe a set of sample values. **Measures of central tendency** are used to describe a typical value within the sample. They are the **mean** (the average of the sample data), the **median** (the value in the center of the ordered data), the **mode** (that data value occurring the most often), and the **midrange** (an average of the lowest and highest data values). To measure the variation within a set of sample data, we use **measures of dispersion**: the **range** (difference between the highest and lowest data values), the **mean absolute deviation** (an average of the absolute deviations from the sample mean), the **variance** (sum of the squares of the deviations from the sample mean, divided by $n-1$), the **standard deviation** (square root of the variance), and the **coefficient of variation** (standard deviation divided by the mean, times 100).

Percentiles and **quartiles** are **measures of position** and indicate the relative position of a particular value. The first quartile (Q_1) and the third quartile (Q_3) are the 25th and 75th percentiles, respectively. The second quartile (Q_2) is the 50th percentile, which is identical to the sample median. Another measure of position is the **Z score**, which is derived for a particular observation by subtracting the sample mean and dividing by the sample standard deviation.

Finally, the shape of a data set can be described using various **measures of shape**. Two such measures are the sample **skewness** (the degree of symmetry in the data) and **kurtosis** (the tendency of a distribution to stretch out in a particular direction).

The two most commonly used measures are the sample mean and standard deviation. These two statistics can be used together to describe the sample data by applying **Chebyshev's inequality** or the **empirical rule**. The latter procedure draws a stronger conclusion about the concentration of the data values but assumes that the population of interest is bell-shaped (normal).

We examined how to estimate the sample mean and standard deviation when the only information available is a frequency distribution, or **grouped data**. **Data coding** can be used to calculate these two measures more easily when you encounter data sets containing extremely large or small values.

Stem and leaf diagrams and **box and whisker plots** are easy-to-construct pictures that illustrate the shape of the sample data. A stem and leaf diagram is similar to a histogram but results in no loss of information. A box and whisker plot summarizes the lowest and highest data values, along with the three quartiles, in a simple diagram.

REVIEW EXERCISES

3.67 From crime statistics for a large metropolitan area, the following are the number of rapes reported for the 12 months of one year:

120, 98, 104, 87, 90, 125, 114, 99, 100, 88, 105, 107

Calculate the:

    (a) Mean, median, and mode (if any).

    (b) Range.

    (c) Standard deviation.

3.68 The following are the percentage of the population living in urban areas in various regions of the world for the year 1975.

Africa	25.67
Latin America	61.21
North America	71.99
East Asia	30.70
South Asia	22.02
Europe	66.45
Oceania	73.35
Soviet Union	60.90

Source: International Encyclopedia of Population (New York: The Free Press, Macmillan, 1982), Table 1, p. 651.

Compute the:

    (a) Mean and median.

    (b) Mean absolute deviation.

    (c) Variance and standard deviation.

    (d) Third quartile.

    (e) Z score for the data value in part (d).

3.69 The following is a hypothetical distribution of scores on the TOEFL (Test of English as a Foreign Language) conducted by Educational Testing Services, Princeton, New Jersey:

TOEFL Score	Number of Candidates
340 and under 380	3
380 and under 420	7
420 and under 460	14
460 and under 500	22
500 and under 540	24
540 and under 580	16
580 and under 620	10
620 and under 660	4

    (a) Calculate the mean score.

    (b) Calculate the standard deviation.

    (c) Estimate the median.

    (d) Calculate the coefficient of skewness. Is it close to zero? What would that indicate?

    (e) Under the assumption of a bell-shaped curve, estimate the proportion of scores that lies within two standard deviations of the mean. What are the scores that bound this proportion?

3.70 The mean GRE (Graduate Record Examination) Aptitude Test score of the 40 applicants admitted into the Master's degree program in Communication Disorders at a local university was 1150 with a standard deviation of 75. At least how many applicants scored between 1000 and 1300 on the GRE?

3.71 The following frequency distribution was constructed from a sample of 160 items of luggage on a flight from Boston to New York.

Weight (lb)	Number of Luggage Items
20 and under 25	1
25 and under 30	1
30 and under 35	5
35 and under 40	25
40 and under 45	38
45 and under 50	68
50 and under 55	20
55 and under 60	2
60 and under 65	0
	160

(a) What is the average weight of a piece of luggage?
(b) Compute the variance and standard deviation.
(c) Find the median.
(d) Calculate the coefficient of skewness.

3.72 To establish the level of difficulty for a puzzle, the time taken to solve the puzzle was noted for a group of 50 volunteers. The following were the results.

Time to Solve Puzzle (in minutes)	Number of Persons
1 to less than 3	0
3 to less than 5	2
5 to less than 7	11
7 to less than 9	18
9 to less than 11	12
11 to less than 13	5
13 to less than 15	2

(a) Approximate the average time taken to solve the puzzle.
(b) Approximate the standard deviation for this distribution.
(c) Approximate the median.
(d) Approximate the coefficient of skewness. What does it indicate?
(e) Assuming a bell-shaped distribution, between what times would 95% of the people be able to solve the puzzle?
(f) What is the approximate Z score for someone who takes 12 minutes to solve the puzzle?

3.73 The mean is 81, the standard deviation is 9, and one particular x value is 45.
(a) Calculate the Z score.
(b) Interpret the Z score.

3.74 Special-interest Political Action Committees (PACs) have been exerting increasing influence on the outcome of elections at federal, state, and local levels. The following list gives the percentage of total campaign contributions received from PACs by selected state senators in 1985:

District	% from PACs
Austin	44.19
Nacogdoches	50.00
Pasadena	52.93
Lake Jackson	72.30
Bryan	7.83
Duncanville	75.97
Wichita Falls	44.23
Stephenville	73.22
Texarkana	100.00
Abilene	0.00
Rockwell	73.22
Arlington	85.05
Port Arthur	54.43
Lubbock	63.38
Canyon	44.40
Brownsville	54.00

Source: The Dallas Morning News, July 24, 1986.

For parts (a)–(h) calculate the:

(a) Mean.
(b) Median.
(c) Variance.
(d) Standard deviation.
(e) Coefficient of variation.
(f) Coefficient of skewness.
(g) First and third quartile.
(h) Z scores for Arlington and Austin.
(i) Construct a box and whisker plot of these data.
(j) Using Chebyshev's inequality, between what values will be the percentage of total campaign contributions received from PACs by at least three-fourths of the senators?

Note that these statistics are concerned with the percentage of PAC contributions and, since the total contributions differ in each district, these figures cannot be used to estimate corresponding dollar amounts (such as average PAC contributions, in dollars).

3.75 (a) Calculate the mean and standard deviation for these data:

50.2, 53.8, 51.4, 52.2, 50.8, 59.1, 52.8, 57.7, 51.1, 54.3, 55.5, 52.1, 57.6, 55.9, 50.9, 54.7

(b) Construct a stem and leaf diagram for the data in part (a).

3.76 Calculate the mean and standard deviation for the following data:

1000, 700, 400, 100, 800, 20,000, 4000, 300, 900, 600, 200, 500, 2000, 700, 2500, 5500

3.77 The Airport Drug Task Force at a major metropolitan airport reported the following seizures and arrests for the previous year.

Month	Cocaine (kilos)	Currency ($)	Arrests
January	3.8	7,300	8
February	2.8	98,000	17
March	2.6	95,000	7
April	1.8	86,500	18
May	1.0	51,000	4
June	3.5	89,000	12
July	3.2	37,000	5
August	2.7	15,800	10
September	4.5	59,900	8
October	1.3	36,000	2
November	2.9	45,000	6
December	3.0	18,000	2

For parts (a)–(e), calculate in each category (cocaine, currency, and arrests) the:
(a)   Range.
(b)   Mean.
(c)   Median.
(d)   Standard deviation.
(e) Z score for July.
(f)   Construct a stem and leaf diagram for cocaine amounts.
(g)   Construct a box and whisker plot for the number of arrests.

3.78 The comptroller of Public Accounts for a border state reported the following annual percentage change in border jobs for 1985 for 14 counties in his jurisdiction:

4.0, 2.7, 1.8, 0.2, 1.9, 4.6, 5.3, 8.2, 3.5, 3.4, 4.1, −2.5, 1.6, −2.1

Compute the mean, median, variance, and standard deviation.

CASE STUDY

Suburban School SAT Scores— Revisited

In your attempt to get a "feel" for the numbers in the previous case study, (Chapter 2, p. 40), you restricted your efforts to the use of descriptive graphs. While these are useful, they do have their limitations. Suppose you prepared all the different kinds of graphs (bar chart, frequency polygons, etc.) for a meeting of suburban school administrators. At the meeting, you proudly flash the multi-colored charts to those attending. As soon as your presentation is over, you are bombarded with questions like these:

□   What is the amount of decline, if any, in the average total scores from 1985 to 1986?

□   Is there a greater variation in average math scores than in verbal scores?

□   What is the range of total scores for three-fourths of the schools?

□   Do the scores lean more heavily toward low scores or high scores?

As you can see, you will quickly become limited in your presentation if you did not have some useful numbers handy, that is, some descriptive statistics. A combination of charts and numbers is always better. The following questions will help you prepare a better presentation of the data.

Case Study Questions

1. Compute the following, using a computer package if one is available.
    (a) The mean and median math, verbal, and total average scores for 1985 and 1986.
    (b) The variance and standard deviation of the math, verbal, and total scores for the two years.

2. Subtract the 1985 mean scores from the 1986 mean scores to obtain the average change in scores. Do you observe a decline in scores? Enter the mean scores on the respective frequency polygons from Case Study 1 in Chapter 2. Does the chart become more meaningful now?

3. Compare the spread in average math versus average verbal scores for each year. What do you notice?
    (*Comment:* The "proper" way would be to compute the coefficient of variation in each case, but for this particular case, the variances or standard deviations would be acceptable. Can you justify this?)

4. Compare the spread in average math, verbal, and total scores.
    (*Comment:* Why is the use of a variance or standard deviation not satisfactory here? Why do we have to have the coefficient of variation for each in order to make a valid comparison?)

5. Compute the coefficient of skewness for the total average scores for each of the two years, and relate the values you obtain to your histograms and frequency polygons from Case Study 1 in Chapter 2. What do you observe?

6. (Do this part assuming you have a roughly symmetrical distribution, i.e., a bell–shaped (normal) curve.)
    (a) Compute the Z score for a 1986 average total score of 900.
    (b) What are the two total scores that correspond to $Z = -2.0$ and $Z = 2.0$ in 1986?
    (c) What percentage of schools should have total average scores between the two values mentioned in (b)? How many actually do?

7. If you did not have a normal distribution, how would that affect your answer for question 6(c)?

COMPUTER EXERCISES USING THE DATABASE

Using a sample of 100 observations, complete the following exercises.

1. Determine the mean, median, mode, range, variance, standard deviation, and coefficient of variation using the data for variable VGRE (verbal GRE score).

2. Repeat exercise 1 using variable UG-GPA (undergraduate GPA). By comparing the coefficients of variation, discuss which of the two data sets contains more variation.

3. Using variable UG-GPA, how many data values are actually between $\bar{x} - 2s$ and $\bar{x} + 2s$? Between $\bar{x} - 3s$ and $\bar{x} + 3s$? Discuss these values in light of Chebyshev's inequality and the empirical rule.

4. Determine the 30th percentile and the three quartiles using variable AGE.

5. Compute and discuss the Z scores for a student with a verbal GRE score of 610 and an undergraduate GPA of 3.35.

6. Construct a stem and leaf diagram and a box and whisker plot for the age data.

**SPSSX
APPENDIX
(MAINFRAME
VERSION)**

**Solution for
Table 3.1**

You can use SPSSX for the aptitude test scores in Table 3.1. The SPSSX program listing in Figure 3.16 was used to request the calculation of the mean, standard deviation, and other descriptive statistics.

The TITLE command names the SPSSX run.

The DATA LIST command gives each variable a name and describes the data as being in free form.

The BEGIN DATA command indicates to SPSSX that the input data immediately follow.

The next 10 lines contain the data values. Each line represents the test score of five of the 50 applicants.

The END DATA statement indicates the end of the data.

The CONDESCRIPTIVE statement requests an SPSSX procedure to print simple descriptive statistics for the variables in the applicant data set.

The FINISH command indicates the end of the SPSSX program.

Figure 3.16

SPSSX program
listing requesting
descriptive statistics
on data in Table 3.1

```
TITLE    DESCRIPTIVE STATISTICS
DATA LIST FREE / APTEST
BEGIN DATA
22   44   56   68   78
25   44   57   68   78
28   46   59   69   80
31   48   60   71   82
34   49   61   72   83
35   51   63   72   85
39   53   63   74   88
39   53   63   75   90
40   55   65   75   92
42   55   66   76   96
END DATA
CONDESCRIPTIVE APTEST
FINISH
```

SPSSX APPENDIX (MAINFRAME VERSION)

Figure 3.17 gives the output obtained from executing the program listing in Figure 3.16.

Figure 3.17

SPSSX output obtained by executing the program listing in Figure 3.16

```
NUMBER OF VALID OBSERVATIONS (LISTWISE) =         50.00
VARIABLE      MEAN      STD DEV    MINIMUM    MAXIMUM VALID N   LABEL
APTEST       60.360     18.605      22.00      96.00      50
```

SPSS/PC+ APPENDIX (MICRO VERSION)

Solution for Table 3.1

You can use SPSS/PC+ for the aptitude test scores in Table 3.1. The program listing Figure 3.18 was used to request the calculation of the mean, standard deviation, and other descriptive statistics. All command lines in Figure 3.18 *must end in a period.*

The TITLE command names the SPSS/PC+ run.

The DATA LIST command gives each variable a name and describes the data as being in free form.

Figure 3.18

SPSS/PC+ program listing requesting descriptive statistics on data in Table 3.1

```
TITLE DESCRIPTIVE STATISTICS.
DATA LIST FREE / APTEST.
BEGIN DATA.
22 44 56 68 78
25 44 57 68 78
28 46 59 69 80
31 48 60 71 82
34 49 61 72 83
35 51 63 72 85
39 53 63 74 88
39 53 63 75 90
40 55 65 75 92
42 55 66 76 96
END DATA.
CONDESCRIPTIVE APTEST.
```

The BEGIN DATA command indicates to SPSS/PC+ that the input data immediately follow.

The next 10 lines contain the data values. Each line represents the test score of five of the 50 applicants.

The END DATA statement indicates the end of the data.

The CONDESCRIPTIVE statement requests an SPSS/PC+ procedure to print simple descriptive statistics for the variables in the applicant data set.

Figure 3.19 gives the output obtained from executing the program listing in Figure 3.18.

Figure 3.19

SPSS/PC+ output obtained by executing the program listing in Figure 3.18

```
                    DESCRIPTIVE STATISTICS

Number of Valid Observations (Listwise) =       50.00

Variable      Mean     Std Dev   Minimum   Maximum    N   Label

APTEST       60.36      18.61     22.00     96.00     50
```

**SAS
APPENDIX
(MAINFRAME
VERSION)**

**Solution for
Table 3.1**

SAS will compute the descriptive statistics using the aptitude test scores in Table 3.1. The SAS program listing in Figure 3.20 was used to request the calculation of the mean, standard deviation, and other descriptive statistics.

The TITLE command names the SAS run.

The DATA command gives the data a name.

The INPUT command names and gives the correct order for the different fields on the data lines.

The CARDS command indicates to SAS that the input data immediately follow.

SAS APPENDIX (MAINFRAME VERSION)

Figure 3.20

SAS program
listing requesting
descriptive statistics
on data in Table 3.1

```
TITLE   'DESCRIPTIVE STATISTICS';
DATA APTDAT;
INPUT APTEST;
CARDS;
22
44
56
68
78
25
44
57
68
78
28
46
59
69
80
31
48
60
71
82
34
49
61
72
83
35
51
63
72
85
39
53
63
74
88
39
53
63
75
90
40
55
65
75
92
42
55
66
76
96
PROC PRINT;
PROC MEANS;
```

The next 50 lines are card images, with each line representing one applicant's aptitude test score.

The PROC PRINT command directs SAS to list the data that were just read in.

SAS APPENDIX (MAINFRAME VERSION)

The PROC MEANS command requests a SAS procedure to print simple descriptive statistics for the variable, APTEST.

Figure 3.21 is the output obtained from executing the program listing in Figure 3.20.

				DESCRIPTIVE STATISTICS					
VARIABLE	N	MEAN	STANDARD DEVIATION	MINIMUM VALUE	MAXIMUM VALUE	STD ERROR OF MEAN	SUM	VARIANCE	C.V.
APTEST	50	60.36000000	18.60520006	22.00000000	96.00000000	2.63117263	3018.0000000	346.15346939	30.824

Figure 3.21

SAS output obtained by executing the program listing in Figure 3.20

SAS APPENDIX (MICRO VERSION)

Solution for Table 3.1

SAS will compute the descriptive statistics using the aptitude test scores in Table 3.1. The SAS program listing in Figure 3.22 was used to request the calculation of the mean, standard deviation, and other descriptive statistics.

The TITLE command names the SAS run.

The DATA command gives the data a name.

The INPUT command names and gives the correct order for the different fields on the data lines.

The CARDS command indicates to SAS that the input data immediately follow.

The next 50 lines are card images, with each line representing one applicant's aptitude test score.

SAS APPENDIX (MICRO VERSION)

Figure 3.22

SAS program
listing requesting
descriptive statistics
on data in Table 3.1

```
TITLE   'DESCRIPTIVE STATISTICS';
DATA APTDAT;
INPUT APTEST;
CARDS;
22
44
56
68
78
25
44
57
68
78
28
46
59
69
80
31
48
60
71
82
34
49
61
72
83
35
51
63
72
85
39
53
63
74
88
39
53
63
75
90
40
55
65
75
92
42
55
66
76
96
PROC PRINT;
PROC MEANS;
RUN;
```

SAS APPENDIX (MICRO VERSION)

The PROC PRINT command directs SAS to list the data that were just read in.

The PROC MEANS command requests a SAS procedure to print simple descriptive statistics for the variable, APTEST.

The RUN command tells the SAS system to execute the previous SAS statements.

Figure 3.23 is the output obtained from executing the program listing in Figure 3.22.

Figure 3.23

SAS output obtained by executing the program listing in Figure 3.22

```
                      DESCRIPTIVE STATISTICS

Analysis variable : APTEST

N Obs   N        Minimum        Maximum            Mean       Std Dev
----------------------------------------------------------------------
   50  50          22.00          96.00           60.36         18.61
----------------------------------------------------------------------
```

4

BIVARIATE DATA

A LOOK BACK/INTRODUCTION

So far we have discussed ways of obtaining information from a set of numbers (sample). These included reducing the set of data to a numerical measure, such as the mean or the variance, and reducing the set of data to a graph. Both methods are concerned with reducing a set of values of one variable. A **variable** is the characteristic of the population that is being measured or observed. For example, variables of interest might be an individual's height or income. The sample consists of random observations of the variable describing a given population.

We have assumed in these early chapters that someone gives you a sample of measurements of some kind (such as inches or dollars), and your job is to describe these data in some way. In this chapter we discuss the situation in which the population and sample consist of measurements on not *one* variable, but *two*. As a result we can describe not only each variable individually using the results from Chapters 2 and 3, but also how the two variables are related. The following sections will discuss methods of describing the relationship between two variables by using a simple graph or a numerical measure (statistic) to convey the nature and strength of this relationship. As is the usual procedure, we use the sample results to form a conclusion about the population from which the sample was obtained.

4.1   THE NATURE OF BIVARIATE DATA

Bivariate data consist of sample data on two variables, say X and Y, for which the observations are **paired**. This means that the first observation on X is paired (belongs) with the first value of Y, the second on X with the second on Y, and so on.

Suppose that a sample of data on the age (X) and annual income (Y) of eight faculty members at Brookhaven University resulted in the table below:

Person	Age (X)	Income (Y) (in thousands of dollars)
1	40	32
2	31	24
3	50	47
4	53	50
5	36	30
6	55	55
7	37	33
8	45	41

Here ($X = 40$, $Y = 32$), ($X = 31$, $Y = 24$), and so on are **bivariate data**; the two variables being sampled are paired according to the person to whom they belong. An easier way to represent these data is to use **ordered pairs**: (40, 32), (31, 24), and so on. Notice that the X value is the first value within each ordered pair.

As in Chapters 2 and 3, we can describe these data by constructing a picture (called a scatter diagram) and computing a number (numerical measure) that tells us something about the relationship between X and Y. In a **scatter diagram**, each ordered pair in the sample is represented as a point. The X axis is always horizontal and the Y axis is always vertical. A scatter diagram of the eight sample points from the personnel data is shown in Figure 4.1. In this diagram,

Figure 4.1

Scatter diagram for the eight ordered pairs from the faculty salary data

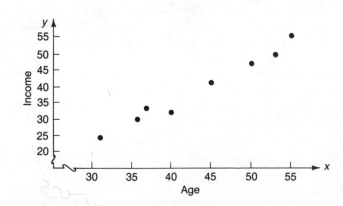

Figure 4.2

A straight line
constructed through
the faculty salary
data

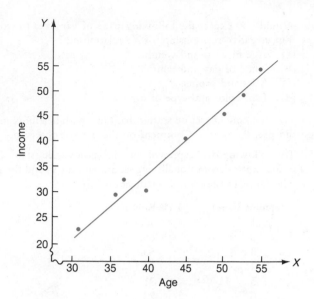

we make better use of the space if we do not begin the X axis and Y axis at zero. If both axes had started at zero, the resulting array of dots would have been very crowded, and not as much information would have been conveyed. So spread the points out as much as possible by labeling the axes properly. Do not forget to label the axes. Here the labels are "Age" and "Income." What type of pattern, if any, do you observe in Figure 4.1?

What we are looking for in a scatter diagram is some underlying pattern. One thing is clear in Figure 4.1: as X increases, so does Y. This means that X and Y have a **positive relationship**. A **negative relationship** exists when Y decreases as X increases. Typically, the gasoline mileage of an automobile (Y) versus its horsepower rating (X) data reveal such a negative relationship.

The pattern in Figure 4.1 is not only positive, it is also nearly **linear**. This means that there is some straight line that "slices through" these points, as shown in Figure 4.2. Remember that a linear pattern does not imply that there is a straight line that connects each of the points—that rarely happens in practice. It does mean that if 10 people tried to draw a straight line through this array of dots, they would all draw the line in nearly the same location with nearly the same direction, or **slope**. Chapter 14 will discuss how to find the "best" line through a set of points that exhibit a strong linear pattern.

So the first step in examining a sample of bivariate data is to plot and examine a scatter diagram of the data. A careful inspection of such a plot can provide much useful information regarding the nature of the data and can help to prevent you from drawing an erroneous conclusion caused by examining computer outputs and skipping a visual inspection of the data.

EXERCISES   4.1   Would you expect the following pairs of variables to exhibit a positive relationship, negative relationship, or no relationship?
(a)   Age in years and weight.
(b)   Time of day and rainfall.
(c)   Sales and profits.
(d)   Hair color and type of car owned.

For exercises 4.2 to 4.8, set up scatter diagrams with headings and appropriately marked
axes, and plot the points. Comment on the pattern observed.

4.2   The following data represent the nine-point-scale (stanine) score obtained by air-
traffic controllers in a qualifying examination (X) and the percentage of candidates
who failed to obtain full certification (Y):

X (Stanine Score)	Y (% Failures)
1	76
2	68
3	55
4	40
5	31
6	22
7	15
8	10
9	5

4.3   Easy Fly Airlines has changed its ticket prices (X) several times during the past
few years. The following results were tabulated for the number of persons (Y)
flying from Los Angeles to Denver at various ticket prices:

X ($)	Y (in 100s)
99	1200
150	1000
130	1050
105	1150
219	700
167	850
180	900

4.4   Test on a nasal decongestant drug resulted in the following figures, where X is the
dosage of decongestant fluid administered and Y is the number of hours of relief
reported by the user:

User	X (ml)	Y (hours)
1	1.0	4.9
2	0.5	2.9
3	2.0	7.7
4	1.5	6.0
5	2.5	9.0
6	3.0	11.2
7	2.2	8.0
8	1.8	7.7
9	3.3	12.5
10	0.7	3.1

4.5   The Texas State Racing Commission is interested in knowing the relationship between attendance (X) and the amount of money wagered (Y). The following are
the data for seven racing days:

X (in hundreds)	Y (in millions)
117	2.07
128	2.09
122	3.14
119	3.33
131	3.76
110	3.00
104	2.20

4.6   A general practitioner in Butte, Montana, was trying to determine if the exodus
of younger families due to poor economic conditions would have an effect on her
practice to the extent that she would need to reduce her staff. She tabulated the
age (X) and number of visits (Y) for a sample of her clients for the preceding year:

X	Y
60	5
80	15
72	9
55	5
44	2
5	1
17	1
12	3
1	6
39	3

4.7   The following data represent the affectivity scores (X = number of words marked
"positive" out of 200) and counseling effectiveness index (Y) for 10 staff members
of a counseling service:

X	Y
184	90
163	85
175	85
150	80
169	76
148	75
120	70
110	65
159	73
135	72

4.8   The very high sound levels that rock-and-roll groups are continuously exposed to
are believed to be associated with loss of hearing. The following data shows the
number of years various rock groups have been performing live concerts (X) and
the percentage of group members diagnosed as having hearing problems (Y):

X (years)	Y (%)
1	30
2	35
2	45
3	60
5	70
4	62
6	75

4.2   CORRELATION COEFFICIENT

Our next objective is to determine a number (numerical measure) that will measure just how linear the pattern in the scatter diagram really is. If this number is "large," the pattern is nearly linear (as in Figure 4.1). If it is "small," X and Y have neither a positive linear relationship nor a negative linear relationship; in this case, X and Y are said to be **uncorrelated**. The numerical measure is called the sample correlation coefficient (r).

Definition

The **correlation coefficient**, **r**, for a sample of bivariate data is a measure of the strength of the linear relationship between the two variables, X and Y, and is computed using

$$r = \frac{\Sigma(x - \bar{x})(y - \bar{y})}{\sqrt{\Sigma(x - \bar{x})^2 \Sigma(y - \bar{y})^2}} \qquad \textbf{(4-1)}$$

The possible values of r range from -1.0 to 1.0.

Here $\bar{x}$ and $\bar{y}$ are the means (averages) of the X and Y values, and the summation is over all n ordered pairs. Notice that the summations in the denominator of equation (4-1) are the numerators for the sample variances of X and Y. An easier expression for computing r using a calculator is

---

$$r = \frac{\Sigma xy - (\Sigma x)(\Sigma y)/n}{\sqrt{\Sigma x^2 - (\Sigma x)^2/n} \sqrt{\Sigma y^2 - (\Sigma y)^2/n}} \qquad \textbf{(4-2)}$$

For large data sets, the only reasonable way to calculate r is to use a computer. At the end of the chapter we will show you how to do this using SAS and SPSSX. A computer-generated scatter diagram along with the corresponding correlation coefficient using MINITAB are contained in Figure 4.3.

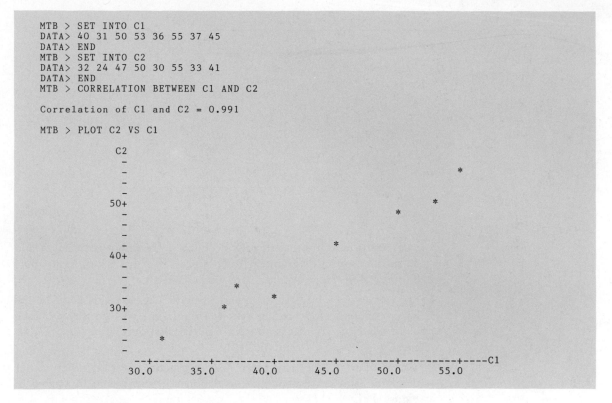

```
MTB > SET INTO C1
DATA> 40 31 50 53 36 55 37 45
DATA> END
MTB > SET INTO C2
DATA> 32 24 47 50 30 55 33 41
DATA> END
MTB > CORRELATION BETWEEN C1 AND C2

Correlation of C1 and C2 = 0.991

MTB > PLOT C2 VS C1
```

Figure 4.3

Computer-generated scatter diagram for faculty salary data using MINITAB

To determine r for the faculty salary data using equation 4-2, format your data and calculations in a table, as shown in Table 4.1.

Using equation 4-2:

$$r = \frac{14{,}195 - (347)(312)/8}{\sqrt{15{,}585 - (347)^2/8}\ \sqrt{13{,}004 - (312)^2/8}}$$

$$= \frac{14{,}195 - 13{,}533}{(23.11)(28.91)} = .991$$

You could have found r using equation 4-1, where $\bar{x} = 347/8 = 43.375$ and $\bar{y} = 312/8 = 39.0$. However, equation 4-2 provides a computationally easier procedure.

The following are some important things to keep in mind regarding the correlation coefficient:

1 r ranges from -1.0 to 1.0.

2 The larger $|r|$ (absolute value of r) is, the stronger the linear relationship.

Table 4.1

Summary of
calculations for
finding r

X	Y	XY	X²	Y²
40	32	1,280	1,600	1,024
31	24	744	961	576
50	47	2,350	2,500	2,209
53	50	2,650	2,809	2,500
36	30	1,080	1,296	900
55	55	3,025	3,025	3,025
37	33	1,221	1,369	1,089
45	41	1,845	2,025	1,681
347	312	14,195	15,585	13,004

3   r near zero indicates that there is no linear relationship between X and Y, and the scatter diagram typically appears to contain a shotgun effect (Figure 4.4A). Here, X and Y are uncorrelated.

4   $r = 1$ or $r = -1$ implies that a perfect linear pattern exists between the two variables in the sample; that is, a single line will go through each point. Here we say that X and Y are **perfectly correlated** (Figure 4.4B and C).

5   Values of $r = 0$, 1, or -1 are rare in practice. Several other scatter diagrams (and corresponding correlation coefficients) are illustrated in Figure 4.4D–F.

6   The sign of r tells you whether the relationship between X and Y is a positive (direct) one or a negative (inverse) one.

7   The value of r tells you nothing about slope of the line through these points (except for the sign of r). If r is positive, the line through these points has positive slope, and, similarly, this line will have negative slope if r is negative. However, a set of data with $r = .9$ will not necessarily have a steeper line passing through it than will a set of data with $r = .4$. All you will observe in the first data set is a set of points that is very close to some straight line with positive slope, but you know nothing (except for the sign) about the slope of this line. See Figure 4.5, where both sets of data provide an r value of .9.

EXAMPLE 4.1

Does alcohol consumption impair a person's recall ability? In a simple pilot study, one individual was given various amounts of alcohol and then given a list of 100 words to memorize (a different list each time). The alcohol doses were given over an extended period of time, so that the effect of one experiment did not affect the results of the next one. The experiment was repeated 10 times using the same individual, the results of which are shown in Figure 4.6 (p. 112).

Figure 4.4

Figure 4.4

Scatter diagrams for various values of the sample correlation coefficient

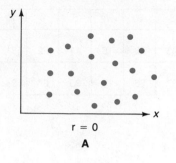

r = 0

A

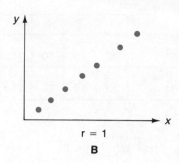

r = 1

B

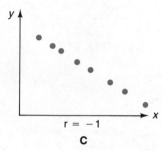

r = −1

C

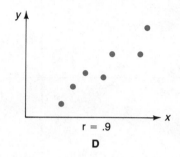

r = .9

D

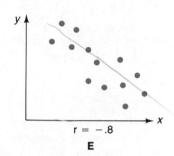

r = −.8

E

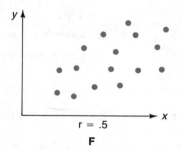

r = .5

F

Figure 4.5

Although (A) has a large slope and (B) has a small slope, both are scatter diagrams for r = .9

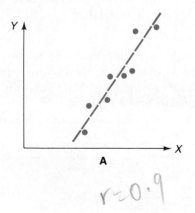

A

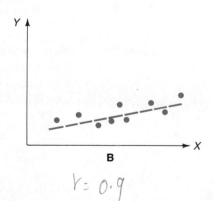

B

r = 0.9   r = 0.9

	1	2	3	4	5	6	7	8	9	10
Level of Intoxication (X)	20	35	10	15	45	28	15	41	22	31
Number of Words Recalled (Y)	30	25	23	26	37	26	18	35	28	28

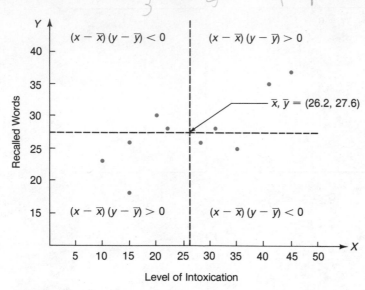

Figure 4.6

Data and scatter diagram for relationship between level of intoxication (X) and number of words recalled (Y)

The variable on the horizontal axis (X) is a measure of the level of intoxication as measured by the alcoholic content contained in the subject's blood sample, and could range from 10 (the minimum possible value) to 50 (the maximum possible value). The Y variable is the number of words (out of 100) that the subject could recall after having 10 minutes to memorize the list.

What is the value of r for these data?

Solution   Here we observe a perhaps surprising result: the relationship between level of intoxication and the number of words recalled is somewhat linear, in a **positive** direction. This means that larger values of Y are generally associated with larger values of X. So we expect to find a value of r that is positive but not as close to 1.0 as before. Table 4.2 shows the calculations we make. Then

$$r = \frac{7677 - (262)(276)/10}{\sqrt{8110 - (262)^2/10}\sqrt{7892 - (276)^2/10}}$$

$$= \frac{445.8}{(35.293)(16.565)} = .763$$

Table 4.2

Summary of
calculations for
example 4.1
(see Figure 4.6)

X	Y	XY	X²	Y²
20	30	600	400	900
35	25	875	1225	625
10	23	230	100	529
15	26	390	225	676
45	37	1665	2025	1369
28	26	728	784	676
15	18	270	225	324
41	35	1435	1681	1225
22	28	616	484	784
31	28	868	961	784
262	276	7677	8110	7892

To better grasp the construction of a correlation coefficient, let's take a look at equation 4-1 and Figure 4.6. Notice that dotted vertical and horizontal lines have been constructed through the point $(\bar{x}, \bar{y})$, dividing the positive X and Y values into four quadrants. In the upper right quadrant, each X value is larger than $\bar{x}$ and each Y value is larger than $\bar{y}$ and so the product $(x - \bar{x})(y - \bar{y})$ is positive. Both $(x - \bar{x})$ and $(y - \bar{y})$ are negative in the lower left quadrant and so again $(x - \bar{x})(y - \bar{y})$ is positive. By a similar argument, $(x - \bar{x})(y - \bar{y})$ is negative in the upper left and lower right quadrants. Consequently, the sum $\Sigma(x - \bar{x})(y - \bar{y})$ will be a large positive number whenever the two variables have an underlying positive relationship since most of the terms in this summation will correspond to points in the upper right and lower left quadrants. A similar argument applies to a negative relationship where the sample values primarily lie in the upper left and lower right quadrants. The two square roots in equation 4-1 are included so that the resulting correlation coefficient will always range from -1 to 1.

In Figure 4.6, we observe six of the points in the "positive" quadrants and four points in the "negative" quadrants. The four points in the negative quadrants are quite close to $(\bar{x}, \bar{y})$ and so the negative contribution to the numerator in equation 4-1 is rather slight. This result is a rather large correlation coefficient, $r = .763$.

EXAMPLE 4.2

In Chapter 2, we summarized a sample of 50 grade point averages (Y) using a frequency distribution and a histogram. The sample (contained in Table 2.1) also included the number of credit hours dropped by each student during the past 12 months (X). These data are listed in Table 4.3. Does there appear to be a linear relationship between these two variables?

Solution

In Figure 4.7, we observe little, if any, pattern to the data, and so it appears that the number of credit hours dropped by a student is not related to the

Table 4.3

Student data for
example 4.2*

Student	1	2	3	4	5	6	7	8	9	10
X	6	0	9	11	17	3	8	8	9	0
Y	1.72	2.18	2.65	3.40	2.27	1.57	3.18	1.55	1.92	3.24
Student	11	12	13	14	15	16	17	18	19	20
X	0	12	12	1	0	3	13	6	12	11
Y	1.08	3.96	2.63	1.69	2.31	3.02	1.86	2.88	3.29	3.77
Student	21	22	23	24	25	26	27	28	29	30
X	6	15	14	12	0	3	3	10	0	1
Y	2.22	1.70	2.47	2.11	2.40	2.76	1.44	2.37	1.74	1.32
Student	31	32	33	34	35	36	37	38	39	40
X	4	5	0	3	3	6	13	5	5	0
Y	2.03	1.68	1.71	3.88	2.27	1.62	2.79	3.36	2.19	1.47
Student	41	42	43	44	45	46	47	48	49	50
X	6	0	17	0	16	9	14	3	0	0
Y	3.41	1.66	2.38	2.67	2.91	3.06	2.70	2.33	2.83	3.75

* Y = GPA, X = number of credit hours dropped during the past 12 months

Figure 4.7

Scatter diagram for
example 4.2 using
data shown in
Table 4.3

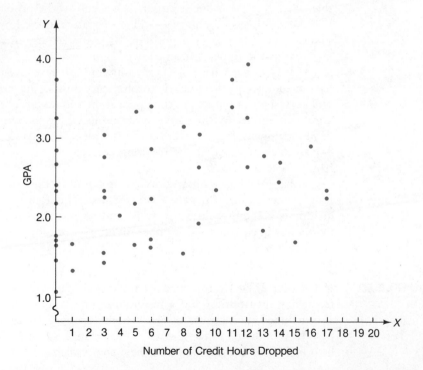

student's grade point average. Consequently, we would expect a value of r close to zero. A summary of the data here is

$$\Sigma x = 314 \qquad \Sigma x^2 = 3414$$

$$\Sigma y = 121.4 \qquad \Sigma y^2 = 321.22$$

$$\Sigma xy = 807.09$$

So

$$r = \frac{807.09 - (314)(121.4)/50}{\sqrt{3414 - (314)^2/50}\sqrt{321.22 - (121.4)^2/50}}$$

$$= \frac{44.698}{(37.975)(5.144)} = .229$$

For this example, r is not close to either $+1$ or -1. Is there a linear relationship or not? At what point do we go from the conclusion that r is "large enough" to indicate that there is an underlying linear relationship between X and Y to the conclusion that no such linear association exists? In other words, would we put $r = .229$ into the "too small" category (no linear pattern exists) or the "large enough" category (there is a linear relationship to these data)?

"Significant" is a commonly used word in statistics. In later chapters, we will provide a more rigorous definition of significance, but for now consider example 4.2. Suppose that these two variables are not related in a linear manner (are uncorrelated). Does this mean that if we selected 50 students, we would obtain a value of r (derived from the sample) exactly equal to zero? The answer is, "No, but it will be close to zero." The reason for this is simply that we are sampling here, and r will be slightly different from zero due to random chance.

Suppose the value of r in example 4.2 was .98 (very large). Is this large sample value due to random chance, or does it suggest that a strong linear relationship exists between these two variables *in the population* from which the sample was selected? We conclude the latter, because to get a value of r that large if X and Y were uncorrelated in the population would be very unlikely. So, we have two situations here:

Situation 1 r is different from zero only by random chance; in fact, X and Y are nearly (or exactly) uncorrelated.

Situation 2 r is significantly different from zero because there is a strong linear association between X and Y in the population.

Where would you classify $r = .229$? Is this value of r large enough to say that there is a linear pattern to these data? We know the linear pattern is not perfect because r is neither 1.0 nor -1.0. How we classify this r value depends on one important piece of information—n, the sample size. A value of, say, $r = .4$ may be considered large enough (far enough away from zero) for us to

Table 4.4

Significant values
of r*

n	Significant Value	n	Significant Value	n	Significant Value
5	.879	15	.514	30	.361
6	.811	16	.497	35	.334
7	.754	17	.482	40	.312
8	.707	18	.468	45	.294
9	.666	19	.456	50	.279
10	.632	20	.444	60	.254
11	.602	22	.423	70	.235
12	.576	24	.404	80	.220
13	.553	26	.388	90	.207
14	.532	28	.374	100	.196

* This table will lead to the conclusion that a linear relationship exists, when in fact it does not, 5% of the time.

conclude that a linear relationship exists (Situation 2) for one sample size but not for another sample size (Situation 1).

One procedure for determining whether r is "large enough" is to use Table 4.4. If the value of $|r|$ exceeds the value in the table for that particular sample size, then we say that the underlying relationship between X and Y is a linear one. However, to apply this table, we need to make an assumption similar to the one that was necessary for applying the empirical rule in Chapter 3. The assumption is that the population of X values and the population of Y values are bell-shaped, or normal. Suppose that you make two histograms: one for the X values and one for the Y values. If either histogram is not approximately symmetrical and bell-shaped in appearance, you cannot use Table 4.4 to reach a valid conclusion. Thus, if $|r|$ exceeds the significant value in Table 4.4 for a specified value of n, then there is evidence of a linear relationship between these two variables in the population; this assumes that both the X and Y populations are bell-shaped (normal).

In our faculty data (Figure 4.1), $r = .991$ (and $n = 8$), which exceeds the value of $r = .707$ from Table 4.4. This situation is illustrated in Figure 4.8. Since $.991 > .707$, we conclude that there is a significant linear association between age and income in the population of *all* faculty members at Brookhaven University—the same conclusion we obtained by simply looking at Figure 4.1. Keep in mind, however, that this conclusion is based on the assumption that both the age and income populations are bell-shaped. Because the sample here consists of only eight values from each population, constructing a histogram to check the shape of these populations is difficult. Devising histograms is more practical for determining the shape of populations using larger sample sizes.

In example 4.1, the computed value of r was .763, which does exceed the value of .632 from Table 4.4 (using $n = 10$). Therefore, it appears that there is a significant linear relationship, although not an overwhelming one, in the

Figure 4.8

Significant values of
r for faculty salary
data (n = 8)

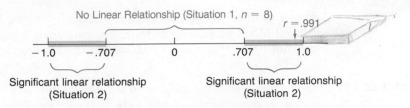

E

population between level of intoxication and the number of words recalled. This is a good example of how just looking at the scatter diagram (Figure 4.6) does not necessarily lead to the correct conclusion.

For the data from example 4.2, as expected, the value of $r = .229$ does not exceed the entry in Table 4.4 for $n = 50$, which is .279. So we conclude that there is not a strong linear association here and that grade point average and the number of credit hours dropped are uncorrelated.

Drawing Conclusions from Table 4.4

There are three points of interest in regard to Table 4.4. First, using this procedure will lead to the conclusion that a linear relationship exists, when in fact there is no such relationship, 5% of the time. There is no way to get around this problem because we are using a sample of (X, Y) values to make a decision concerning the population of all (X, Y) values. However, this table provides only one possible approach. You could construct a table for which the chance of making such an incorrect conclusion is not 5% but is 1%, 10%, or whatever percentage you desire. Be careful, though—there is a tradeoff. The smaller this percentage, the more likely you are to conclude that there is not a linear relationship when in fact there is. We will take a closer look at this situation in Chapter 14, where we examine the population correlation coefficient between X and Y, written as ρ. Since the sample correlation coefficient r is derived using a set of sample data, it is a *statistic* and will later be used to estimate the population *parameter* ρ. The value of ρ is what we would have if we obtained *all* (X, Y) values in the population and used equation (4-1) or (4-2) to derive the corresponding coefficient of correlation.

Second, remember that Table 4.4 can be used whenever both the X and Y populations are bell-shaped, normal populations. When they are not, you should use a much more reliable measure of association between the X and Y variables: the rank correlation, discussed in the next section.

Third, just because there is a strong positive statistical relationship between two variables does not necessarily imply that an increase in X causes an increase in Y. As a simple example, consider $X =$ percentage of gray hairs, and $Y =$ blood pressure. One might expect to observe a high correlation between these two variables, but it is probably absurd to say that an additional gray hair will cause a person's blood pressure to increase. What is actually happening is that

118   CHAPTER 4: BIVARIATE DATA

there is another variable, in this case, age, that is causing both gray hair percentage and blood pressure to increase. So remember, a high statistical correlation does not imply that a *causal* relationship exists between the two variables.

EXAMPLE 4.3   Eight female students auditioned for one of the lead female parts in a high school rendition of *You Can't Take It With You*. Two teachers, Mr. Roberts and Ms. Clauson, were in charge of the auditions. Each teacher ranked the eight students from 1 (best) to 8 (worst). Determine the sample correlation between these two sets of rankings.

Student	Roberts (X)	Clauson (Y)
A	1	2
B	4	3
C	2	1
D	6	6
E	8	7
F	3	5
G	7	8
H	5	4

Here

$$\Sigma x = 36 \qquad \Sigma y^2 = 204$$
$$\Sigma y = 36 \qquad \Sigma xy = 199$$
$$\Sigma x^2 = 204$$

Using equation 4-2:

$$r = \frac{199 - (36)(36)/8}{\sqrt{204 - (36)^2/8}\sqrt{204 - (36)^2/8}} = .881$$

At this point, you may be tempted to use Table 4.4 to decide if this value of r is significantly large. However, the X and Y data here consist of rankings, and so the data are **ordinal**, a weaker form of data than was observed in the previous example. As the next section will demonstrate, when dealing with ordinal data of this type, a better procedure is to use a table especially designed for such data. When we reexamine this set of data in Section 4.3, the conclusion will be that, based upon $r = .881$, there is a significant positive relationship between the rankings of the two teachers. Another way of looking at this result is that the two teachers agree (although not perfectly) with each other. So here r is a **measure of agreement** between the two sets of rankings. For any situation in which you compile multiple rankings, a large value of r is desirable. Whenever the value of r is not significantly large, we are unable to conclude that there is agreement between the two individuals supplying the rankings.   □

EXERCISES 4.9 Are the following values of r for sample size n significant? Use Table 4.4 for determining whether there is a linear relationship.

(a) $r = .473$, $n = 50$.
(b) $r = .223$, $n = 29$.
(c) $r = .175$, $n = 75$.
(d) $r = .819$, $n = 15$.
(e) $r = .750$, $n = 5$.

4.10 Researchers have been trying to identify biological markers of psychiatric disorders. One such candidate has been platelet monoamine oxidase (MAO) activity. Georgotas et al. (1987) conducted a study on 37 depressed patients over the age of 55, who were treated for 5–7 weeks with either nortriptyline, a tricyclic antidepressant, or phenelzine, a monoamine oxidase (MAO) inhibitor. Patients' platelet MAO activity was measured following a drug washout period before treatment. The following data are for 23 patients who responded to the antidepressant treatment:

Patient No.	X = Baseline MAO Activity (nmol product/hr/mg protein)	Y = Final Hamilton Score (rating scale for depression)
1	56.8	3
2	64.0	2
3	47.5	6
4	40.6	0
5	54.4	7
6	44.1	1
7	55.9	10
8	60.9	9
9	41.4	5
10	87.5	3
11	46.3	10
12	35.5	5
13	29.9	7
14	30.8	2
15	29.0	7
16	88.4	6
17	34.7	9
18	56.8	3
19	62.2	0
20	6.9	2
21	41.6	9
22	21.7	6
23	42.4	6

Source: Anastasios Georgotas, Robert E. McCue, Eitan Friedman, and Thomas Cooper. Prediction of response to nortriptyline and phenelzine by platelet MAO activity. *Am. J. Psychiatry* **144** (3): 340 (March 1987), Table I.

(a) Calculate r, the correlation coefficient.
(b) Does there appear to be a linear relationship, using the scatter diagram?
(c) Is the relationship significant, using Table 4.4?
(d) Do your answers to questions (b) and (c) agree?

4.11 Use the data from exercise 4.3, and answer (a)–(d) in exercise 4.10.

4.12 Use the data from exercise 4.5, and answer (a)–(d) in exercise 4.10.

4.13 Use the data from exercise 4.6, and answer (a)–(d) in exercise 4.10.

4.14 Does the movie-going public listen to the movie critics? A student majoring in Mass Media collected the following data, where X represents the critics' rating for a movie on a scale of 1 to 10 (1 being the lowest and 10 the highest rating), and Y represents the film's box-office receipts (gross ticket sales in millions of dollars):

Film	X	Y
Top Gun	6.0	100
Karate Kid II	7.5	70
Aliens	9.0	50
Rambo	6.0	146
Back to the Future	9.0	133
Cocoon	8.5	70
Cobra	6.5	40
Space Camp	4.0	10
Labyrinth	6.5	15
Under the Cherry Moon	2.0	10
Ruthless People	7.0	45
Legal Eagles	7.5	60

(a) Plot a scatter diagram.
(b) Calculate r, the correlation coefficient.
(c) Does there appear to be a positive or negative linear relationship?
(d) Is the relationship significant using Table 4.4?

4.15 The Department of Highway Safety carried out a survey to determine if there was a relationship between usage of seat belts in cars and the number of fatal accidents. In the following data, X represents the percentage of drivers in compliance with the seat belt law and Y represents the number of persons killed in automobile accidents:

X (%)	Y
40	812
44	675
47	720
50	611
52	588
57	600
58	550
60	538
62	386
68	500

(a) Plot the scatter diagram.
(b) Calculate r.
(c) Does there appear to be a positive or negative linear relationship?
(d) Is the relationship significant?

4.16 Use the hearing loss data in exercise 4.8 to answer questions (a)–(d) in exercise 4.15.

4.3   RANK CORRELATION

In example 4.3, Mr. Roberts ranked the students from 1 to 8 without actually assigning each of them a score or number that measured that person's ability. For Mr. Roberts, student A was "better" than student C, maybe only slightly better or maybe a lot—we do not know which. The object here was to obtain a ranking of the eight students, and this type of data is acceptable. In Chapter 1, this type of data was referred to as ordinal data.

Definition

When examining ordinal data consisting of two sets of rankings, the resulting correlation coefficient is referred to as Spearman's **rank correlation**, denoted as r_s. For such data, $r = r_s$. The value of r_s is always between -1.0 and 1.0.

It is sometimes desirable to analyze the relationship between two variables according to the rank of the values of each of the variables in the sample. The variables may be ranked on the basis of ability (as the students were) or any other standard. The use of ranks allows us to measure correlation using characteristics that cannot be expressed quantitatively but that lend themselves to being ranked. For example, it is difficult to measure the ability of an auditioning student quantitatively.

For data consisting of ranks, equation 4-2 can be used to determine r, but an easier expression is

$$r_s = 1 - \frac{6 \cdot \Sigma d^2}{n(n^2 - 1)} \qquad (4-3)$$

where d represents the difference between the individual ranks within each (X, Y) pair. As a check of your calculations, Σd should always equal zero.

Using the student data in example 4.3,

Student	Roberts (X)	Clauson (Y)	d	d^2
A	1	2	-1	1
B	4	3	1	1
C	2	1	1	1
D	6	6	0	0
E	8	7	1	1
F	3	5	-2	4
G	7	8	-1	1
H	5	4	1	1
			0	10

Here, $\Sigma d^2 = 10$, so

$$r_s = 1 - \frac{(6)(10)}{8(64-1)} = .881 \qquad \text{(as in example 4.3)}$$

The interpretation of r_s is similar to that for r, namely:

1 A value of r_s near 1.0 indicates a strong positive relationship.
2 A value of r_s near -1.0 indicates a strong negative relationship.

Ties

When ranking the X values and Y values, you will often encounter a tie whenever two (or more) X values, or Y values, are the same. Equation 4-3 can still be used, provided the number of ties is not excessive, by assigning a rank to these values equal to the **average** rank of the tied positions. For example, if two X values are tied for ranks 3 and 4, both X values are assigned a rank of $(3 + 4)/2 = 3.5$. If three Y values are tied for ranks 6, 7, and 8, all three values are assigned a rank of $(6 + 7 + 8)/3 = 7$. The next largest Y value after these three would be given a rank of 9. This procedure will be illustrated in the example on page 126.

As mentioned earlier, it is sometimes much easier simply to rank a group of people or items without assigning a numerical value to the quality that is being measured. When dealing with rank data, we know that $r = r_s$. We determine the rank correlation for interval data by replacing each of the X values by its rank and each of the Y values by its rank. The resulting r_s value will not be the same as the original r value but will have the same sign and be of roughly the same magnitude.

Let us compute the rank correlation for the sample data in the faculty salary example. Look at the data in Section 4.1. For the X values, 31 is the smallest value, so its rank is 1; 36 is the next largest and receives a rank of 2, and so on. For the Y values, 24 is replaced by its rank of 1, 30 receives a rank of 2, and so on. The ranked data are:

Person	Age (X) Ranks	Income (Y) Ranks	d	d²
1	4	3	1	1
2	1	1	0	0
3	6	6	0	0
4	7	7	0	0
5	2	2	0	0
6	8	8	0	0
7	3	4	−1	1
8	5	5	0	0
			0	2

Figure 4.9

MINITAB calculation
of the rank
correlation of the
age and faculty
salary paired data

```
MTB > SET INTO C1
DATA> 40 31 50 53 36 55 37 45
DATA> END
MTB > SET INTO C2
DATA> 32 24 47 50 30 55 33 41
DATA> END
MTB > RANK C1, PUT INTO C3
MTB > RANK C2, PUT INTO C4
MTB > CORRELATION BETWEEN C3 AND C4

     Correlation of C3 and C4 = 0.976
```

So, $\Sigma d^2 = 2$ and

$$r_s = 1 - \frac{(6)(2)}{8(64 - 1)} = .976$$

For this data set, $r = .991$ and $r_s = .976$, both of which are near 1.0; this indicates a very strong positive relationship within the population. A computer calculation of the rank correlation for these data (using MINITAB) is contained in Figure 4.9.

Another important reason for using the rank correlation (r_s) rather than a correlation of the actual sample data (r) is that we are relieved of one of the key assumptions regarding the type of populations from which you obtained the sample. Using the sample correlation coefficient r and Table 4.4 is appropriate whenever the X and Y values came from symmetrical, bell-shaped (normal) populations. This can be checked by constructing a histogram of the sample X values and a histogram of the sample Y values. If they appear nearly symmetrical, with a bell-shaped appearance, then Table 4.4 offers a reliable procedure for determining if a linear relationship exists or not.

Suppose your histograms appear as in Figure 4.10. Since they are both clearly nonsymmetric and certainly not bell-shaped, using the rank correlation r_s to measure the relationship between X and Y will be much more reliable than using r. Only one of the histograms need be non-bell-shaped to violate

Figure 4.10

Examples of
non-bell-shaped
histograms

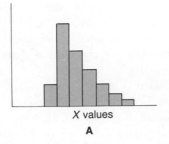

X values

A

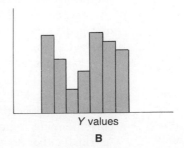

Y values

B

Figure 4.11

MINITAB histograms
for the *X* and *Y*
values from
example 4.2

```
MTB > HISTOGRAM OF C1

  Histogram of C1    N = 50

  Midpoint    Count
     1.2        2    **
     1.6       12    ************
     2.0        6    ******
     2.4        9    *********
     2.8        9    *********
     3.2        6    ******
     3.6        4    ****
     4.0        2    **

MTB > HISTOGRAM OF C2

  Histogram of C2    N = 50

  Midpoint    Count
       0       12    ************
       2        2    **
       4        8    ********
       6        8    ********
       8        2    **
      10        4    ****
      12        6    ******
      14        4    ****
      16        2    **
      18        2    **

MTB > RANK C1, PUT INTO C3
MTB > RANK C2, PUT INTO C4
MTB > CORRELATION BETWEEN C3 AND C4

  Correlation of C3 and C4 = 0.245
```

the assumption necessary to apply Table 4.4. Also, keep in mind that there is nothing wrong with providing both r and r_s if the original data are interval data. In fact, an easy way to spot that the assumptions behind Table 4.4 are not being satisfied is that r and r_s will have substantially different values. If the assumptions are being met, these values should be nearly the same. However, just because these two values are nearly the same does not imply that the assumptions are being satisfied.

In example 4.2, the sample correlation between X = number of credit hours dropped and Y = grade point average was found to be $r = .229$. We used Table 4.4 to determine whether a significant linear relationship existed between these two variables, without examining the assumption that the populations were normal.

In Figure 4.11, MINITAB was used to construct histograms of the X and Y values. Notice that neither histogram is bell-shaped. This suggests that the corresponding populations of all X values and all Y values are not bell-shaped (normal) either. Consequently, using Table 4.4 for this situation was not appropriate. Figure 4.11 also contains a computer solution for the rank correlation,

Table 4.5

Significant values of r_s*	n	Significant Value	n	Significant Value	n	Significant Value
	5	.9000	15	.5179	30	.3620
	6	.8286	16	.5000	35	.3313
	7	.7450	17	.4853	40	.3138
	8	.7143	18	.4716	45	.2955
	9	.6833	19	.4579	50	.2800
	10	.6364	20	.4451	60	.2552
	11	.6091	22	.4241	70	.2360
	12	.5804	24	.4061	80	.2205
	13	.5549	26	.3894	90	.2078
	14	.5341	28	.3749	100	.1970

* This table will lead to a conclusion that a positive or negative relationship exists, when in fact it does not, 5% of the time.

namely, $r_s = .245$. Notice that the r (.229) and r_s values are nearly the same, but the assumption of bell-shaped populations appears to be on shaky ground here.

Significant Values of r_s

Table 4.4 was used to determine if a sample correlation coefficient was large enough to conclude that there was a true underlying linear relationship between the two variables within the population. If $|r|$ exceeded the table value and if r was positive, a significant positive relationship existed; or if r was negative, the two variables had a significant negative relationship. Using Table 4.5 in the same way we use Table 4.4 for r, we can determine if $|r_s|$ is large due to a strong positive or negative association between a pair of variables within the population. As in Table 4.4, there is a 5% chance of concluding that a positive or negative relationship exists when in fact it does not.

In our faculty salary example, $r_s = .976$, which exceeds .7143 using $n = 8$, so we conclude there is a significant positive relationship between age and salary in the population of all faculty members at Brookhaven University. Notice we do *not* say linear here because r_s measures the amount of the positive or negative relationship between X and Y, not necessarily how linear that relationship is. For example, in Figure 4.12, both data sets will provide an r_s value of exactly 1.0 when the data values are converted to ranks. This is because every time X increases, so does Y for both data sets. However, the value of r will be *less* than 1.0 for both data sets because the pattern is not exactly linear.

In example 4.3, $n = 8$ and $r_s = .881$, which exceeds the table value of .7143. This implies that we do have general agreement between the two teachers—a desirable result!

Figure 4.12

Two data sets for
which rank
correlation equals
1.0. Note that the
value of r will be
less than 1.0 because
the pattern is not
exactly linear.

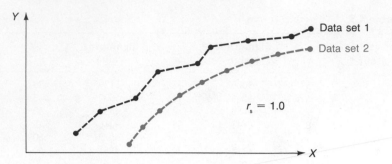

We can rank the data from example 4.1, replacing the X and Y values by their ranks:

Trial	Level of Intoxication Rank	Words Recalled Rank	d	d^2
1	4	8	−4	16
2	8	3	5	25
3	1	2	−1	1
4	2.5	4.5	−2	4
5	10	10	0	0
6	6	4.5	1.5	2.25
7	2.5	1	1.5	2.25
8	9	9	0	0
9	5	6.5	−1.5	2.25
10	7	6.5	.5	.25

Notice that some of the data are tied for assignment to certain ranks. In the X values, there is a tie between second place and third place because both have $X = 15$. This is easily resolved by giving each a rank of $(2 + 3)/2 = 2.5$. There are two ties for the Y values. $Y = 26$ occurs in both the fourth and the fifth ranked positions, so it gets a rank of 4.5 for these two values; $Y = 28$ occurs twice as a tie for the sixth and seventh ranked positions, and both values are assigned a rank of 6.5. One advantage to this procedure is that the sum of the d values is zero, as in situations without ties.

For these data, we have

$$\Sigma d^2 = 16 + 25 + \cdots + 2.25 + .25 = 53$$

and so

$$r_s = 1 - \frac{(6)(53)}{(10)(100 - 1)} = .679$$

Since .679 is larger than .6364 (from Table 4.5), our conclusion (as before) is that there is a positive relationship in the population between level of intoxication and the number of words recalled.

From Figure 4.11, the rank correlation for the 50 observations in Table 4.3 is shown to be $r_s = .245$. It is more appropriate to use rank correlation to determine whether a relationship (positive or negative) exists between these two variables because the X and Y histograms are not bell-shaped. Based on Table 4.5, because $r_s = .245$, which is less than .2800 (corresponding to $n = 50$), we once again conclude that there is not a positive or negative relationship between a student's grade point average and the number of credit hours dropped.

EXERCISES

4.17 For the following set of ranked data

Number	X	Y
1	2	7
2	3	8
3	8	2
4	4	5
5	6	4
6	5	3
7	1	6
8	7	1

(a) Calculate r_s.
(b) Is there a positive or negative relationship?
(c) Is it significant according to Table 4.5?

4.18 Jones and McLeod (1986) report on a study that attempted to identify the structure of executive information systems and evaluate their relationship to decision making. The study centered on the question: "Where and how do senior executives get their decision-making information?" The data were gathered by a variety of means, including personal interviews, questionnaires, and logs of the executives' information transactions. One part of the study revealed the following data:

NUMBER OF INFORMATION TRANSACTIONS AND
AVERAGE TRANSACTION VALUE BY MEDIA

Medium (used to communicate information)	Transaction Quantity		Transaction Value	
	Frequency (%)	Rank by Quantity	Average Value	Rank by Value
Verbal				
Scheduled meeting	5.4	5	7.4	11
Unscheduled meeting	6.3	6	6.2	10
Encounter	3.4	3.5	5.3	9
Social activity	0.1	1	5.0	8
Business meal	1.6	2	3.6	2
Telephone	21.4	11	3.7	3

Medium (used to communicate information)	Transaction Quantity		Transaction Value	
	Frequency (%)	Rank by Quantity	Average Value	Rank by Value
Written				
Memo	19.1 ⁴	9	4.8 ⁵	7
Letter	19.5 ⁵	10	4.2 ²	4
Noncomputer report	9.4 ²	7	4.7 3.5	5.5
Computer report	3.4 ¹	3.5	4.7 3.5	5.5
Periodical	10.4 ³	8	3.1 ₁	1
	100.0 ⁶			

Source: J. W. Jones and R. McLeod, Jr. The structure of executive information systems: An exploratory analysis. *Decision Sci.* **17** (2): 233 (Spring 1986), Table 3.

(a) Compute the correlation coefficient r between transaction frequency and transaction average value. (*Note:* "Value" was determined by the richness of the information obtained from each source.)

(b) Is there a significant correlation between the two variables?

(c) Determine the Spearman rank correlation coefficient r_s.

(d) Compare and comment on the results of questions (a) and (c).

4.19 Use the data in exercise 4.3 and rank the ticket prices and number of passengers.

(a) Calculate r_s.

(b) Is there a positive or negative relationship?

(c) Is it significant?

(d) Does this r_s value lead to the same conclusion as the computations, based on the prices and number of people, in exercise 4.11?

4.20 Elizabeth M. Nuss (1984), Dean for Undergraduate Life at Indiana University in Bloomington, reported on the results of two studies conducted at a major public research university on the issue of academic dishonesty. The first study was a student survey, the second a survey of faculty. She reported the following results:

COMPARISON BETWEEN COLLEGE STUDENTS AND FACULTY ATTITUDES ABOUT THE SERIOUSNESS OF FORMS OF ACADEMIC DISHONESTY

Behavior	Ranking	
	Students	Faculty
Taking an exam for another student	1.5	5
Having another student take an exam for you	1.5	4
Altering or forging an official university document	3	8
Paying someone to write a paper to submit as your own	4	2
Arranging with other students to give or receive answers by use of signals	5	3
Arranging to sit next to someone who will let you copy from his or her exam	6	7
Copying from someone's exam paper without his or her knowledge	7	1

Behavior	Ranking Students	Faculty
Writing a paper for another student	8	6
Allowing another student to copy from you during an exam	9	9
Copying answers from a source without doing work independently	10	11
Getting questions or answers from someone who has already taken the same exam	11	10
Copying a few sentences without footnoting in a paper	12	12
Working on homework with other students when the instructor doesn't allow it	13	13
"Padding" a few items on a bibliography	14	14

The above is based on responses to the question "How serious do you consider these activities to be?" (1 = most serious, 14 = least serious)

Source: E. M. Nuss, Academic integrity: Comparing faculty and student attitudes. *Improving College & University Teaching*, **32**(3): (1984).

(a)   Compute the rank correlation coefficient r_s.

(b)   Is there a significant degree of agreement between students and faculty on the seriousness of various forms of dishonesty?

4.21   Byrnes and Yamamoto (1986) surveyed parents, teachers, and principals in a large southwestern city in the United States to determine their views on grade repetition (i.e., having a low-achieving child repeat a grade in school). One aspect of their survey compared the rankings of various characteristics observed in retained children by teachers and by principals, as shown in the table below.

COMPARISONS BETWEEN TEACHERS AND PRINCIPALS:
CHARACTERISTICS FREQUENTLY OBSERVED IN RETAINED CHILDREN

Characteristics Observed	By Teachers (%)	Rank	By Principals (%)	Rank
Developmentally immature	68	1	57	3
Low motivation	59	2	69	1
Low self-esteem	34	3	60	2
Lack of English proficiency	30	4	26	5.5
Low intelligence	29	5	26	5.5
Learning disabled	28	6	20	7
Discipline problem	21	7	34	4
Shy and nonassertive	12	8	9	9
Emotionally unstable	6	9	14	8

Source: Deborah Byrnes and Kaoru Yamamoto, Views on grade repetition. *J. Res. Dev. Educ.* **20** (1): 14–20 (Fall 1986), Table 3, p. 17.

(a)   Compute the Spearman rank correlation coefficient r_s.

(b)   Is there a significant degree of agreement between the observations of teachers and principals?

(c)   Repeat the above, using the Pearson correlation coefficient r. Do you arrive at the same conclusion as in (b)?

4.22 Refer to exercise 4.14. Rank the movie ratings and box-office receipts, and compute the rank correlation coefficient r_s.

(a)   Compare r_s with r from exercise 4.14.
(b)   Is r_s significant using Table 4.5?
(c)   Does r_s indicate a linear relationship?
(d)   If r and r_s are substantially different what would that indicate? Is there any advantage in using r_s?

4.23 Answer (a)–(d) in exercise 4.19 for the data in exercise 4.5. Compare the result with exercise 4.12.

4.24 Refer back to exercises 4.8 and 4.16. Rank the rock group hearing loss data, and compute the rank correlation coefficient r_s.

(a)   Compare r_s with r from exercise 4.16.
(b)   Is r_s significant using Table 4.5?
(c)   Does r_s indicate a linear relationship?
(d)   If r and r_s are substantially different, what would that indicate?
(e)   What is the advantage in computing r_s?

SUMMARY

This chapter described methods of examining sample data collected on two variables. The population here consists of all pairs of measurements corresponding to these two variables. The sample data are **bivariate data**, since they occur in pairs. A graphical representation of a set of this type of data is a **scatter diagram**, where each pair of data values is represented as a point. Using the scatter diagram, we can determine whether (1) a **positive relationship** exists between the two variables (as one increases, the other increases), or (2) the variables exhibit a **negative relationship** (as one increases, the other decreases), or (3) the variables are **uncorrelated** (the scatter diagram does not exhibit a linear pattern).

To measure the amount of positive or negative association between two variables, we can use the sample correlation coefficient or the sample rank correlation. The **sample correlation coefficient**, r, measures the strength of the linear relationship between the variables. If r is near $+1.0$, then there is a strong positive linear relationship; if r is near -1.0, there is a strong negative linear relationship. If r is near zero, the variables are (nearly) uncorrelated. Table 4.4 is used to determine if $|r|$ is significantly large enough to conclude that a linear relationship exists. This procedure is appropriate whenever both the X and Y populations are normal (bell-shaped) in appearance. The **Spearman rank correlation**, r_s, is found using the ranks of the data and measures the amount of association between the variables. This relationship need not be linear for $|r_s|$ to be large. When dealing with rank data, the values of r and r_s are identical.

Table 4.5 is used to determine whether values of r_s are significantly large. The assumption of normal populations is not necessary here.

A key advantage to using the rank correlation as a measure of association is that it is not necessary to assume (as it is when using r) that the X and Y populations are normal. We illustrated how the rank correlation can be applied to interval type data by replacing each data value by its rank. We thus convert interval data to ordinal data, which require the use of r_s. So, for the same data set, if r is computed using the original interval data, it will not equal r_s computed on the ranks of that same data set because it uses different values. When your original data consist of ranks, you will obtain the rank correlation using either the equation for r or the equation for r_s.

4.25 Industrial psychologists investigated the relationship between the loudness of sounds (X, measured in decibels and recorded as sound level 1 to 10, with 10 being the loudest) and the reaction time for persons exposed to the sound (Y). The following information was collected:

X	Y (seconds)
1	1.90
2	2.00
3	1.50
4	0.90
5	0.50
6	0.70
7	0.05
8	0.02
9	0.01
10	0.01

(a) Draw a scatter diagram for the data.
(b) Compute the correlation coefficient r.
(c) Is there a significant linear relationship?
(d) Compute the rank correlation coefficient r_s.
(e) Compare questions (b) and (d) and comment on the results.

4.26 A survey of the students of Highpoint College gathered the following information with regard to their study time (X, hours per week) and grade point averages (Y):

X	Y
16	4.0
15	3.8
14	3.5
12	3.0
10	2.8
8	2.2
6	1.5
4	1.0
2	0.5
0	0.2

(a)   Plot a scatter diagram of the data.
(b)   Calculate the sample correlation coefficient.
(c)   Interpret the r value.
(d)   Is there a significant relationship between study time and grade point average?

4.27 Law enforcement personnel in one county blamed the parole and probation system for releasing criminals too quickly and contributing to an increase in crime. The following are the relevant statistics for the county:

Year	X (Parolee population)	Y (Major crimes per 100,000 residents)
1981	2803	2360
1982	3513	4177
1983	4562	4300
1984	5711	7063
1985	6278	6160
1986	6935	8224

(a)   Compute the correlation coefficient r.
(b)   Is there a significant linear relationship? Could one conclude that an increase in the number of parolees *causes* an increase in the crime rate?
(c)   Compute the rank correlation coefficient r_s and compare your results with those above.

4.28 Uniformity of judging is important in ice-skating contests. Two judges scored contestants 1 to 6 as follows:

Contestant	X (Judge A)	Y (Judge B)
1	85	80
2	81	82
3	89	87
4	96	92
5	94	94
6	92	90

(a)   Compute r.
(b)   Is it significant?
(c)   Do you think the same pair of judges should be hired for this year's competition?

4.29 The University of Southern North Dakota is one of the top ten party schools. A sample of freshmen taking the English placement examination was obtained which observed the number of fraternity rush parties each person attended during the first week of school.

X (Score on exam)	Y (Number of parties)
80	3
75	4
95	1
100	0
63	4

60	5
86	2
90	2
70	3
88	2

(a) Plot a scatter diagram of the data.
(b) Compute r.
(c) Is it significant?

4.30 (a) A social worker investigating the relationship between income levels and infant mortality rates studied 45 families and found that the correlation coefficient was $-.524$. Does this indicate a significant inverse relationship between income levels and infant mortality rates?

(b) The social worker also found that the rank correlation coefficient was $-.54$. Since the values of r and r_s are close, can we assume that the assumption of both X and Y variables having a bell-shaped distribution is satisfied?

4.31 Using the data in exercise 4.29,
(a) Calculate r_s.
(b) Is it significant?
(c) Compare the value of r_s with the value of r.

4.32 In this chapter, the primary focus has been on the correlation between two variables, X and Y. In actual practice, researchers typically investigate many variables. However, we may still be interested in the correlation between any pair of variables. This information is generally provided in the form of a correlation matrix. For example, the table below was obtained from a study of psychobiomotor assessment of football players. Psychobiomotor assessment refers to the use of psychological, biological, and motor performance capabilities, and the use of such variables to predict successful performance. The study was conducted by Secunda et al. (1986) using 19 male volunteer tryouts for the positions of halfback and fullback on the inaugural football team of a southern university.

CORRELATION MATRIX BETWEEN VARIABLES

	10yd	Prcm	Anrbc	E	H	Q1	Q4	Crit
Fbrc	−.55	−.22	.35	.14	.22	−.34	−.09	.62
10yd		.12	−.20	.10	.14	.03	.22	−.21
Prcm			−.37	.44	.04	.29	.32	−.18
Anrbc				−.10	.26	−.24	−.23	.46
E					.64	.38	−.28	.09
H						−.09	−.56	.34
Q1							−.04	−.62
Q4								−.01

Note: Fbrc = football-receiving ability; 10yd = speed on 10-yd run; Prcm = perceptual-motor run; Anrbc = anaerobic power; E = dominant/aggressive vs. submissive; H = adventurous vs shy; Q1 = experimenting vs. conservative; Q4 = tense vs. relaxed; Crit = criterion (football playing ability)

Source: M. D. Secunda, B. I. Blau, J. M. McGuire, and W. A. Burroughs. Psychobiomotor assessment of football-playing ability. *Int. J. Sports Psychol.* **17** (3): 215–233 (1986), Table IV.

(a)   The correlation between Fbrc and 10yd is $-.55$. Similarly, the r between E and H is .64. What is the correlation between anaerobic power and football playing ability? Comment on the correlation between other variables.

(b)   Identify all the correlation coefficients that are significant in the above matrix.

4.33   The following pairs of observation represent the scores of a test given by a psychologist to a group before an experiment (X) and then to the same group after the experiment (Y).

X	Y	X	Y
2.1	9.4	2.4	12.9
3.4	35.6	1.9	5.2
1.6	3.5	1.3	3.4
2.7	15.4	0.2	0.1
3.2	30.1	1.5	4.6
4.5	52.7	2.1	9.3
1.8	17.4		

(a)   Calculate the correlation coefficient r between X and Y.

(b)   Take the log to the base 10 of Y. Calculate the correlation coefficient between X and log Y.

(c)   Compare the correlations in questions (a) and (b).

4.34   Chemical physics researchers S. G. Lambrakos and N. C. Peterson (1987) reported using a technique called a Firsov inversion procedure to analyze differential cross sections for scattering of high-velocity neutral helium atoms from helium gas. Their paper gave, among other things, the following deflection function obtained from experimentally determined differential cross sections, with beam energy = 10 keV.

X Impact Parameter (Å)	Y Deflection Angle (degrees)
0.8008	2.000
0.7861	2.108
0.7711	2.227
0.7559	2.360
0.7403	2.509
0.7244	2.678
0.7081	2.870
0.6914	3.090
0.6743	3.345
0.6568	3.644
0.6388	4.000
0.6337	4.114
0.6284	4.239
0.6232	4.377
0.6179	4.529
0.6126	4.699
0.6072	4.891
0.6017	5.108
0.5962	5.358
0.5907	5.651
0.5851	6.000

Source: S. G. Lambrakos and N. C. Peterson. Determination of an experimental He potential energy function by inversion of differential cross sections. *J. Chem. Phys.* **86**: 2730 (1987), Table II.

(a) Compute the correlation coefficient r.

(b) Is the relationship between impact parameter (X) and deflection angle (Y) significantly positive or negative, or nonsignificant?

4.35 Researchers in materials science and engineering believe that growth mechanisms may play a fundamental role in the formation of quasicrystalline material. Drehman, Pelton, and Noack (1986) did a study of nucleation and growth kinetics of Pd–U–Si transformation from the metallic glass state to a single-phase quasicrystalline material. The quasicrystalline phase was formed by a process called isothermic annealing at various temperatures from 400 to 455°C, and the corresponding nucleation rate was obtained. (The latter is related to the number of nuclei per unit volume, and is computed by a complicated method that need not concern us here.) The study reported the following results:

X Annealing Temperature (°C)	Y Nucleation Rate (log I in nuclei/cm^3)
400	10.92
410	11.12
410	11.13
420	11.33
430	11.65
440	11.83
440	11.76
445	11.77
450	11.89
455	11.91

Source: A. J. Drehman, A. R, Pelton, and M. A. Noack. Nucleation and growth of quasicrystalline Pd–U–Si from the glassy state *J. Mater. Res.* **1**: 743 (1986).

(a) Draw a scatter plot of the data.

(b) Compute the correlation coefficient r.

(c) For the given range of temperatures, is there a significant linear relationship between annealing temperature and the nucleation rate?

(d) Is the relationship positive or negative?

4.36 A journalist studying a medical report read that in one experiment, the correlation coefficient between age (X) and stroke volume (Y), the quantity of blood pumped through the heart in one heartbeat, for 30 adult males was found to be $-.622$. What does this indicate?

4.37 A suburban city fire department developed a dryness index (based on temperature, rainfall and humidity) and investigated whether there was a relationship between this dryness index (X) and the number of fire alarms per day (Y). Do the following data indicate a significant correlation?

X	Y
88	1
90	1

X	Y
105	2
110	3
98	1
95	4
100	1
86	0
101	1
92	0
100	1

CASE STUDY

Skin Cancer and Sunspot Activity

Malignant melanoma, a type of skin cancer, is a dark-colored tumor of the skin. It has a complex etiology: trauma, heredity, and hormonal activity probably play some role in influencing its outcome. Exposure to solar radiation is believed to have some kind of triggering effect. It is less common in dark-skinned persons; apparently, the dark pigmentation of the skin protects against the sun's ultraviolet rays. (Indeed, a suntan is a result of an increase in melanin pigment, which is the body's way of protecting underlying tissues from injurious solar rays.) Melanoma is more common in fair-skinned individuals. It occurs most frequently in areas of the skin that receive more exposure to the sun, and in those geographic areas closer to the equator, where the intensity of solar radiation is higher.

Houghton, Munster, and Viola (1978) obtained data from the Connecticut Tumor Registry, which has the longest record of state demographic cancer data in the United States. The following table shows the age-adjusted incidence rate for malignant melanoma in the state of Connecticut from 1936 to 1972, for males (Y_1) and for the total population (Y_2). Incidence is measured as number of cases per 100,000 population. Also provided is the sunspot relative number, a measure of solar activity.

Age-adjusted melanoma incidence from the Connecticut Tumor Registry, 1936–1972*

Year	Male (Y_1)	Total (Y_2)	Sunspot Relative Number (X)
1936	1.0	0.9	40
1937	0.8	0.8	115
1938	0.8	0.8	100
1939	1.4	1.3	80
1940	1.2	1.4	60
1941	1.0	1.2	40
1942	1.5	1.7	23
1943	1.9	1.8	10
1944	1.5	1.6	10

Year	Male (Y_1)	Total (Y_2)	Sunspot Relative Number (X)
1945	1.5	1.5	25
1946	1.5	1.5	75
1947	1.6	2.0	145
1948	1.8	2.5	130
1949	2.8	2.7	130
1950	2.5	2.9	80
1951	2.5	2.5	65
1952	2.4	3.1	20
1953	2.1	2.4	10
1954	1.9	2.2	5
1955	2.4	2.9	10
1956	2.4	2.5	60
1957	2.6	2.6	190
1958	2.6	3.2	180
1959	4.4	3.8	175
1960	4.2	4.2	120
1961	3.8	3.9	50
1962	3.4	3.7	35
1963	3.6	3.3	20
1964	4.1	3.7	10
1965	3.7	3.9	15
1966	4.2	4.1	30
1967	4.1	3.8	60
1968	4.1	4.7	105
1969	4.0	4.4	105
1970	5.2	4.8	105
1971	5.3	4.8	80
1972	5.3	4.8	65

* Source: A. Houghton, E. W. Munster, and M. V. Viola. Increased incidence of malignant melanoma after peaks of sunspot activity. *The Lancet* (April 8, 1978), 759–760. Reproduced in D. F. Andrews and A. M. Herzberg, *Data: A Collection of Problems from Many Fields for the Student and Research Worker* (New York: Springer-Verlag, 1985), pp. 199–201. Reprinted with permission.

Case Study Questions

1. **(a)** Compute the sample correlation coefficient r (also known as the Pearson correlation coefficient) between Y_1 (male melanoma incidence) and X (sunspot relative number).

   (b) Is r "large enough" to say that there is a significant linear relationship between incidence of skin cancer and sunspot activity? Use Table 4.4.

   (c) Similarly, compute r using Y_2 (total population melanoma incidence) versus X.

   (d) Answer question (b) again with respect to the Pearson correlation coefficient r obtained in (c). Use Table 4.4.

1. **(e)**   Based on the values of r, are the relationships, if any, between Y_1 and X and between Y_2 and X positive or negative?

   (f)   Plot a scatter diagram to obtain a visual representation of the relationships in the data.

2. **(a)**   In carrying out the above analysis, what was a necessary assumption about the underlying distribution of the variables Y and X?

   (b)   Plot the histograms of the Y and X variables to obtain a crude verification of whether these assumptions are met.

3. **(a)**   Assume that in one or both cases, the requirements of a bivariate normal (bell-shaped) distribution are not satisfied. Is the use of the Pearson correlation coefficient r appropriate under these circumstances? Does the Spearman rank correlation coefficient r_s have the same restrictions?

   (b)   Compute the Spearman rank correlation coefficient r_s for Y_1 versus X, and Y_2 versus X, and address in each case the same question raised in 1(b) above. Use Table 4.5 this time.

   (c)   Does the Spearman rank correlation coefficient provide a better measure of the strength of the linear relationship between Y and X?

   (d)   If there was a substantial difference between r and r_s for a given set of Y and X observations, what might that indicate to you?

COMPUTER EXERCISES USING THE DATABASE

Using a sample of 100 observations, complete the following exercises.

1. Determine the correlation between the verbal GRE scores (variable VGRE) and the corresponding graduate grade point averages (variable GR-GPA). Does there appear to be a significant linear relationship?

2. Determine the correlation between graduate grade point average (variable GR-GPA) and undergraduate grade point average (variable UG-GPA). Is there a significant linear relationship between these two variables?

3. Suppose that someone claims that older people are more satisfied individuals. If we measure life satisfaction using variable LSINDEX, does this claim appear to be justified using the correlation between this variable and the variable AGE?

4. Repeat exercise 1 using the rank correlation to determine if a significant relationship (positive or negative) exists between these two variables.

SPSSX APPENDIX (MAINFRAME VERSION)

Solution in Section 4.1

The Brookhaven University survey of age and salary problem requested a scatter diagram of a set of eight data points and the correlation coefficient between the two variables. You can use SPSSX to solve this problem. The SPSSX program listing in Figure 4.13 was used to request the scatter diagram and correlation coefficient.

The TITLE command names the SPSSX run.

The DATA LIST command gives each variable a name and describes the data as being in free form.

The BEGIN DATA command indicates to SPSSX that the input data immediately follow.

The next eight lines contain the data. The first line, for example, indicates an income of $32,000 and an age of 40 years. The remaining data pairs are interpreted in the same manner.

The END DATA statement indicates the end of the data.

The SCATTERGRAM command requests a plot and specifies that the values for SAL be on the vertical axis and the values for AGE be on the horizontal axis.

The STATISTICS 1 command requests SPSSX to calculate the correlation coefficient.

The FINISH command indicates the end of the SPSSX program.

Figure 4.13

SPSSX program listing requesting scatter diagram and correlation coefficient

```
TITLE   AGE VERSUS SALARY
DATA LIST FREE / SAL, AGE
BEGIN DATA
32 40
24 31
47 50
50 53
30 36
55 55
33 37
41 45
END DATA
SCATTERGRAM  SAL WITH AGE
STATISTICS 1
FINISH
```

SPSSX
APPENDIX
(MAINFRAME
VERSION)

Figure 4.14 shows the output obtained by executing the program listing in Figure 4.13.

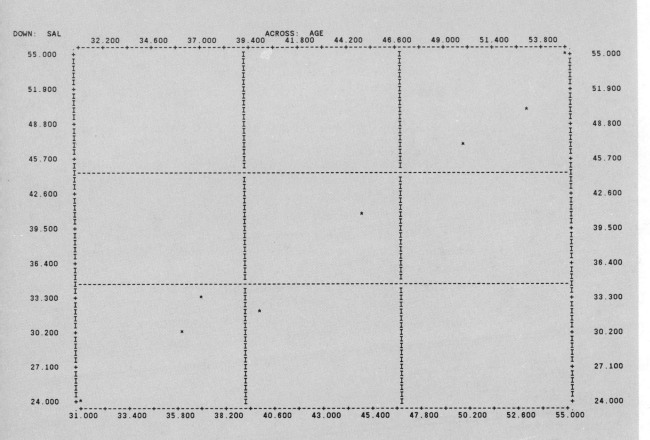

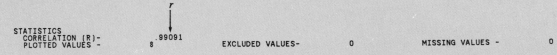

STATISTICS
    CORRELATION (R)-
    PLOTTED VALUES -   8   .99091   EXCLUDED VALUES-   0   MISSING VALUES -   0

Figure 4.14

SPSSX output
obtained by
executing the
program listing in
Figure 4.13

**SPSS/PC+
APPENDIX
(MICRO
VERSION)**

**Solution in
Section 4.1**

The Brookhaven University survey of age and salary problem requested a scatter diagram of a set of eight data points and the correlation coefficient between the two variables. You can use SPSS/PC+ to solve this problem. The program listing in Figure 4.15 was used to request the scatter diagram and correlation coefficient. All command lines in Figure 4.15 *must end in a period.*

The TITLE command names the SPSS/PC+ run.

The DATA LIST command gives each variable a name and describes the data as being in free form.

The BEGIN DATA command indicates to SPSS/PC+ that the input data immediately follow.

The next eight lines contain the data. The first line, for example, indicates an income of $32,000 and an age of 40 years. The remaining data pairs are interpreted in the same manner.

The END DATA statement indicates the end of the data.

The PLOT command requests a plot and specifies that the values for SAL be on the vertical axis and the values for AGE be on the horizontal axis.

The CORRELATION command requests SPSS/PC+ to calculate the correlation coefficient using variables SAL and AGE. The STATISTICS = 1 command will provide this correlation coefficient as part of the output.

Figure 4.15

SPSS/PC+ program
listing requesting
scatter diagram
and correlation
coefficient

```
TITLE   AGE VERSUS SALARY.
DATA LIST FREE / SAL, AGE.
BEGIN DATA.
32 40
24 31
47 50
50 53
30 36
55 55
33 37
41 45
END DATA.
PLOT PLOT=SAL WITH AGE.
CORRELATION VAR=SAL AGE / STATISTICS=1.
```

Figure 4.16 shows the output obtained by executing the program listing in Figure 4.15.

Figure 4.16

SPSS/PC+ output obtained by executing the program listing in Figure 4.15

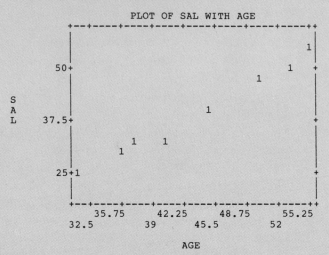

```
                          AGE VERSUS SALARY

                        PLOT OF SAL WITH AGE
        +--+----+----+----+----+----+----+----++
        |                                    1  |
    50+ |                              1     +
        |                          1            |
   S    |                                       |
   A    |                   1                   |
   L 37.5+                                     +
        |          1    1                       |
        |        1                              |
    25+1                                       +
        |                                       |
        +--+----+----+----+----+----+----+----++
           35.75     42.25     48.75     55.25
        32.5      39       45.5      52

                          AGE
```

```
                  AGE VERSUS SALARY

        Variable    Cases       Mean       Std Dev

        SAL           8       39.0000      10.9283
        AGE           8       43.3750       8.7331


                  AGE VERSUS SALARY

        Correlations:  SAL       AGE      r

           SAL       1.0000    .9909**
           AGE        .9909**  1.0000


        N of cases:    8       1-tailed Signif:  * - .01  ** - .001
```

SAS APPENDIX (MAINFRAME VERSION)

Solution in Section 4.1

The Brookhaven University survey of age and salary requested a scatter diagram of a set of eight data points and the correlation coefficient between the two variables. You can use SAS to solve this problem. The SAS program listing in Figure 4.17 was used to request the plot and correlation coefficient.

The TITLE command names the SAS run.

The DATA command gives the data a name.

The INPUT command names and gives the correct order for the different fields on the data lines.

The CARDS command indicates to SAS that the input data immediately follow.

The next eight lines are card images. The first line, for example, indicates an income of $32,000 and an age of 40 years. The remaining input data are interpreted in the same manner.

The PROC PRINT command directs SAS to list the data that were just read in.

The PROC PLOT command requests a SAS procedure to plot. The following line, PLOT SALARY*AGE, requests that the values for SALARY be on the vertical axis of the plot and AGE values be on the horizontal axis.

The PROC CORR command requests a SAS procedure to compute a correlation coefficient. The following statement, VAR SALARY AGE, lists the variables to be used in this computation.

Figure 4.18 shows the output obtained by executing the program listing in Figure 4.17.

Figure 4.17

SAS program listing requesting scatter diagram and correlation coefficient

```
TITLE     'AGE VERSUS SALARY';
DATA      SALAGE;
INPUT     SALARY AGE;
CARDS;
32 40
24 31
47 50
50 53
30 36
55 55
33 37
41 45
PROC PRINT;
PROC PLOT;
   PLOT SALARY*AGE;
PROC CORR;
   VAR SALARY AGE;
```

SAS APPENDIX (MAINFRAME VERSION)

Figure 4.18

SAS output obtained by executing the program listing in Figure 4.17

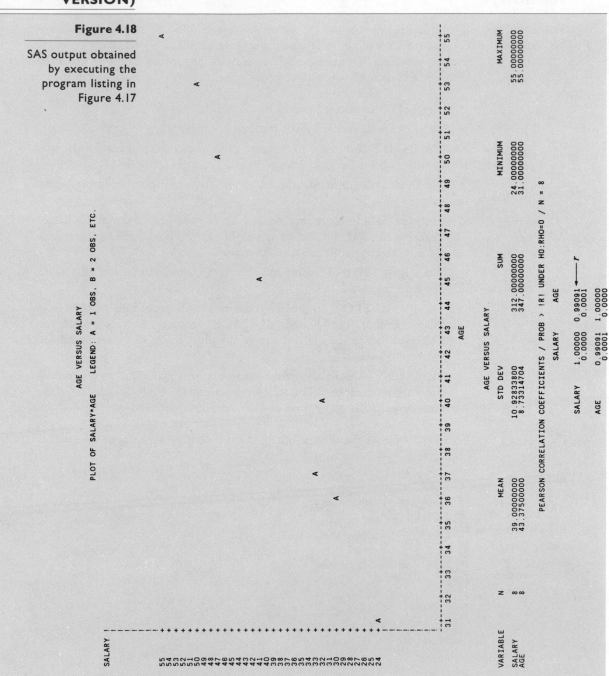

SAS APPENDIX (MICRO VERSION)

Solution in Section 4.1

The Brookhaven University survey of age and salary requested a scatter diagram of a set of eight data points and the correlation coefficient between the two variables. You can use SAS to solve this problem. The SAS program listing in Figure 4.19 was used to request the plot and correlation coefficient.

The TITLE command names the SAS run.

The DATA command gives the data a name.

The INPUT command names and gives the correct order for the different fields on the data lines.

The CARDS command indicates to SAS that the input data immediately follow.

The next eight lines are card images. The first line, for example, indicates an income of $32,000 and an age of 40 years. The remaining input data are interpreted in the same manner.

The PROC PRINT command directs SAS to list the data that were just read in.

The PROC PLOT command requests a SAS procedure to plot. The following line, PLOT SALARY*AGE, requests that the values for SALARY be on the vertical axis of the plot and AGE values be on the horizontal axis.

The PROC CORR command requests a SAS procedure to compute a correlation coefficient. The following statement, VAR SALARY AGE, lists the variables to be used in this computation.

Figure 4.19

SAS program listing requesting scatter diagram and correlation coefficient

```
TITLE   'AGE VERSUS SALARY';
DATA SALAGE;
INPUT SALARY AGE;
CARDS;
32 40
24 31
47 50
50 53
30 36
55 55
33 37
41 45
PROC PRINT;
PROC PLOT;
   PLOT SALARY*AGE;
PROC CORR;
   VAR SALARY AGE;
RUN;
```

**SAS
APPENDIX
(MICRO
VERSION)**

The RUN command tells the SAS system to execute the previous SAS statements.

Figure 4.20 shows the output obtained by executing the program listing in Figure 4.19.

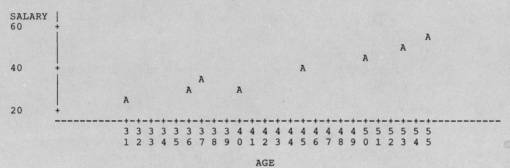

```
                        AGE VERSUS SALARY


         Plot of SALARY*AGE      Legend:  A = 1 obs, B = 2 obs, etc.

SALARY |
60     +
       |                                                        A
       |                                                    A
40     +                                              A
       |                                          A
       |             A                        A
       |          A        A
20     +       A
       ----------+-+-+-+-+-+-+-+-+-+-+-+-+-+-+-+-+-+-+-+-+-+-+-+-+------------
                3 3 3 3 3 3 3 3 3 4 4 4 4 4 4 4 4 4 5 5 5 5 5 5
                1 2 3 4 5 6 7 8 9 0 1 2 3 4 5 6 7 8 9 0 1 2 3 4 5

                                   AGE
```

```
                        AGE VERSUS SALARY


Variable        N       Mean      Std Dev        Sum      Minimum     Maximum

SALARY          8    39.00000    10.92834    312.00000    24.00000    55.00000
AGE             8    43.37500     8.73315    347.00000    31.00000    55.00000
```

```
    Correlation Coefficients / Prob > |R| under Ho: Rho=0 / N = 8
                                SALARY              AGE
              SALARY           1.00000            0.99091 ←— r
                               0.0000             0.0001

                 AGE           0.99091            1.00000
                               0.0001             0.0000
```

Figure 4.20

SAS output obtained
by executing the
program listing in
Figure 4.19

5

PROBABILITY CONCEPTS

A LOOK BACK/INTRODUCTION

We use descriptive statistics to summarize or present data that consist of observations that have already occurred. If these data are drawn from a population, then we describe our sample in some way. If we wish to infer something about the population using the smaller sample, we must deal with uncertainty. To measure the chance that something will occur, we use its **probability**. The concepts of probability form the foundation of all decision making in statistics. By using probabilities, we are able to deal with uncertainty because we are able, at least, to measure it.

To illustrate this idea, suppose that a recent report contained the results of a random sample of 100 homes within a large metropolitan city and stated that the average electric bill was $185. However, the electric company claims that the average bill for all of its customers is $110. Is something wrong here? Do you believe that the *population* mean is $110 based upon the fact that the *sample* mean is $185? Here we need a probability, in particular the probability of observing a sample mean at least this large (that is, $185 or more) assuming that the electric company is correct in their claim. If we decide that this probability is extremely small, we can infer that the *population* claim is incorrect, based upon the *sample* results.

As mentioned in Chapter 1, there is always the possibility of arriving at the wrong decision (maybe the electric company *is* correct) when using sample results to infer something concerning a population. This particular type of question will be addressed in Chapter 9. We begin the journey into probability in this chapter by introducing some basic concepts and discussing various ways of determining probabilities.

5.1 EVENTS AND PROBABILITY

An activity for which the outcome is uncertain is an **experiment**. This need not involve mixing chemicals in the laboratory; it could be as simple as throwing two dice and observing the total. An **event** consists of one or more possible outcomes of the experiment; it is usually denoted by a capital letter.* The following are examples of experiments and some corresponding events:

1 *Experiment:* Rolling two dice. *Events:* A = rolling a total of 7, B = rolling a total greater than 8, C = rolling two 4's.
2 *Experiment:* Taking a teacher competency exam. *Events:* A = pass, B = fail.
3 *Experiment:* Observing the number of people who lose at least 10 lb during a 60-day weight-loss program. *Events:* A_0 = 0 people, A_1 = 1 person, A_2 = 2 people, and so on.

When you estimate a probability, you are estimating the probability of an *event*. For example, when rolling two dice, the probability that you will roll a total of 7 (event A) is the probability that event A occurs. It is written: $P(A)$. The probability of any event is always between zero and one, inclusive.

> **Notation**
>
> $P(A)$ = probability that event A occurs

Classical Definition of Probability

Suppose a particular experiment has n possible outcomes and event A occurs in m of the n outcomes. The **classical definition** of the probability that event A will occur is

$$P(A) = m/n \tag{5-1}$$

This definition assumes that all n possible outcomes have the same chance of occurring. Such outcomes (events) are said to be **equally likely**, and each has probability $1/n$ of occurring. If this is not the case, the classical definition does not apply.

---

* The set of all possible outcomes is often referred to as the sample space.

Consider the experiment of tossing a nickel and a dime into the air and observing how they fall. Event A is observing one head and one tail. The possible outcomes are (H = head, T = tail):

Nickel	Dime
H	H
H	T
T	H
T	T

Thus, there are two ($m = 2$) outcomes that constitute event A of the four possible outcomes ($n = 4$). These four outcomes are equally likely, so each occurs with probability 1/4. Consequently,

$$P(A) = 2/4 = .5$$

Relative Frequency Approach

Another method of estimating a probability is referred to as the **relative frequency** approach. This is based on observing the experiment n times and counting the number of times an event (say, A) occurs. If event A occurs m times, then your estimate of the probability that A will occur in the future is

$$P(A) = m/n \qquad (5\text{-}2)$$

Suppose that a particular busy intersection is observed for 250 days; 220 days have been accident-free. Let A = a randomly chosen day in the future that is free of accidents. Using the relative frequency definition, then

$$P(A) = 220/250 = .88$$

Subjective Probability

Another type of probability is **subjective probability**. This is a measure (between zero and one) of your belief that a particular event will occur. A value of one indicates that you believe this event will occur with complete certainty.

Examples of situations requiring a subjective probability are:

The probability that a new television series will stay on the air at least two seasons.

The probability that last year's NBA champion will repeat this year.

The probability that your recently married cousin, divorced five times already, will once again go down alimony lane.

Although no two people may agree on a particular subjective probability, these probabilities are governed by the same rules of probability, which will be developed later in the chapter.

5.2   BASIC CONCEPTS

One of the services of Parenting Services is to conduct seminars on helping individuals cope with the problems and challenges following divorce. They checked the registration records for a selected sample of 200 people who had recently attended such a seminar. The results obtained are shown in Table 5.1, which is a **contingency** or **cross-tab table**. Such tables are a popular method of summarizing a group by means of two categories, in this case, age and sex. The numbers within the table represent the frequency, or number of individuals, within each pair of subcategories and so the contingency table allows you to determine how these two categories interact.

There are 60 people who are female and under 30; 10 of the people are male and over 45. One person from the total group of 200 is to be selected at random to receive a free set of books. We can define the following events:

M = a male is selected

F = a female is selected

U = the person selected is under 30

B = the person selected is between 30 and 45

O = the person selected is over 45

Because there are 200 people, there are 200 possible outcomes to this experiment. All 200 outcomes are equally likely (the person is randomly selected), so the classical definition provides an easy way of determining probabilities.

Table 5.1

Parenting services survey of seminar attendees

Sex	AGE (YEARS)			
	<30 (U)	30–45 (B)	>45 (O)	Total
Female (F)	60	20	40	120
Male (M)	40	30	10	80
Total	100	50	50	200

The probability of any one single event used to define the contingency table is a **marginal probability**. When you use a contingency table, you can obtain the marginal probabilities by merely counting. For example, of the 200 people, 120 are females. So the probability of selecting a female is

$$P(F) = 120/200 = .6$$

Similarly,

$$P(M) = 80/200 = .4$$
$$P(U) = .5$$
$$P(B) = .25$$
$$P(O) = .25$$

Notice that $P(O) = 50/200 = .25$, which implies that (1) if you repeatedly selected a person at random from this group, 25% of the time this individual would be over 45 years of age, and (2) 25% of the people in this group are over 45 years old. So, a probability here is simply a **proportion**.

The **complement of an event** A is the event that A does not occur. This event is denoted $\bar{A}$. For example, A = it rains tomorrow, $\bar{A}$ = it does not rain tomorrow; or A = you contract a particular disease, $\bar{A}$ = you do not contract this disease.

In our Parenting Services survey, $P(F) = .6$, and so

$$P(\bar{F}) = P(M) = .4$$

Notice that $P(F) + P(\bar{F}) = .6 + .4 = 1.0$. In general, for any event A, either A or $\bar{A}$ must occur. Consequently,

$$P(A) + P(\bar{A}) = 1$$

and so

$$P(\bar{A}) = 1 - P(A)$$

or written another way

$$P(A) = 1 - P(\bar{A})$$

How can we determine what proportion of the people are age 45 or younger?

$$P(\bar{O}) = 1 - P(O) = 1 - .25 = .75$$

Joint Probability

What if we wish to know the probability of selecting a seminar attendee who is male and under age 30? Such a person is selected if events M and U occur.

This probability is written **P(M and U)** and is referred to as a **joint probability**.[*] There are 40 attendees who are male and under 30, so

$$P(M \text{ and } U) = 40/200 = .2$$

What proportion are females between 30 and 45? This is the same as

$$P(F \text{ and } B) = 20/200 = .1$$

because 20 out of 200 satisfy both requirements.

Probability of A or B

In addition to calculating joint probabilities involving two events, we can also determine the probability that either of the two events will occur. In our discussion, "either A or B" will refer to the event that A occurred, B occurred, or both occurred. This probability will be written as

$$P(A \text{ or } B)$$

for any two events A and B.[†]

Now we will calculate the probability of selecting someone who is a female or under 30 years of age. This is $P(F \text{ or } U)$. How many people qualify? There are 120 females and there are 100 people under 30. Is the answer $(120 + 100)/200 = 1.1$? You should realize that this is not correct because a probability is never greater than one. What is the mistake here? The problem is that the 60 females under age 30 were counted twice. How many attendees are female or under 30? The answer is the 120 females plus the 40 males under age 30. So

$$P(F \text{ or } U) = (120 + 40)/200 = .8$$

What is $P(M \text{ or } B)$? The people in the shaded area in Table 5.1 qualify. So,

$$P(M \text{ or } B) = (80 + 20)/200 = .5$$

Conditional Probability

Suppose that someone has some inside information about who has been selected from the group of 200 seminar attendees. This person informs you that the selected individual is under 30 years of age; that is, event U occurred. Armed with this information, we can calculate the probability that the selected person is a female. Given that event U occurred, we have immediately narrowed the number of possible outcomes from 200 to the 100 people under age 30. Each

---

[*] The joint probability of events A and B is often written as $P(A \cap B)$, read as "the probability of A intersect B."

[†] The probability $P(A \text{ or } B)$ can be written as $P(A \cup B)$, read as "the probability of A union B."

of these 100 people is equally likely to be chosen, and 60 of them are female. So the answer is $60/100 = .6$.

Whenever you are given information and are asked to find a probability based on this information, the result is a **conditional probability**. This probability is written as

$$P(A|B)$$

where B is the event that you know occurred, and A is the uncertain event whose probability you need, given that event B has occurred. The vertical line indicates that the occurrence of event B is given, so the expression is read as the "probability of A given B." In the example, $P(F|U) = .6$.

Suppose that you were given no information about U and were asked to find the probability that a female is selected. This is a marginal probability. We earlier determined that $P(F) = .6$. For our example, note that

$$P(F) = P(F|U) = .6$$

This means that being given the information that the person selected is under 30 has no effect on the probability that a female is selected. In other words, whether or not U happens has no effect on whether F occurs. Such events are said to be independent. Thus, events A and B are **independent** if the probability of event A is unaffected by the occurrence or nonoccurrence of event B.

There are a number of ways to demonstrate that any two events A and B are independent.

Events A and B are independent if and only if

$$P(A|B) = P(A) \quad \text{(assuming } P(B) \neq 0\text{)}, \tag{5-3}$$

or

$$P(B|A) = P(B) \quad \text{(assuming } P(A) \neq 0\text{)}, \tag{5-4}$$

or

$$P(A \text{ and } B) = P(A) \cdot P(B). \tag{5-5}$$

You need not demonstrate all three conditions. If one of the equations is true, they are all true; if one is false, they are all false (in which case A and B are not independent). Events that are not independent are **dependent** events.

In our example, are events M and O independent? We previously showed that

$$P(O) = 50/200 = .25$$

Since $P(O|M) = 10/80 = .125$, then $P(O) \neq P(O|M)$, and these events are dependent. Put another way, if someone informs you that event M (a male) has occurred, this does have an effect on whether or not the person selected is over 45 years of age. If you are told that M occurred, the probability that the selected person is over 45 drops from .25 to .125. These events do affect each other and so are dependent events.

We could also approach this by showing that $P(M|O)$ is not the same as $P(M)$:

$$P(M|O) = 10/50 = .2$$

$$P(M) = 80/200 = .4$$

These are not the same values, so events M and O are not independent.

The final option is to show that $P(M \text{ and } O)$ is not the same as $P(M) \cdot P(O)$. This follows since

$$P(M \text{ and } O) = 10/200 = .05$$

$$P(M) \cdot P(O) = (.4)(.25) = .1$$

In our discussion of joint probabilities, we showed that

$$P(M \text{ and } U) = 40/200 = .2$$

Consequently, events M and U can both occur because their joint probability is not zero.

How would you calculate $P(F \text{ and } M)$? One cannot be both a male and a female, so $P(F \text{ and } M) = 0$. Because events M and F cannot both occur, these events are said to be mutually exclusive.

Definition

Events A and B are **mutually exclusive** if A and B cannot both occur simultaneously. To demonstrate that two events, A and B, are mutually exclusive, you must show that their joint probability is zero: $P(A \text{ and } B) = 0$.

EXAMPLE 5.1

Bellmar Laboratories administers a potentially lethal chemical to 10 laboratory mice. Which of these outcomes are mutually exclusive?

A = exactly 9 mice survive

B = fewer than 8 mice survive

C = more than 6 mice survive

Solution

A and B are mutually exclusive events—they cannot both occur.

A and C are not mutually exclusive—if A occurs, so does event C.

B and *C* are not mutually exclusive—if 7 mice survive, both events *B* and *C* will occur.

Note: By "not mutually exclusive" we do not mean that both of these events must occur, only that both could occur. □

EXERCISES

5.1 Assume that there are two red marbles, two blue marbles, and two green marbles in a jar. One marble is picked at random. List the outcomes. Is each outcome equally likely? What is the probability of each outcome?

5.2 If there are 20 sophomores, 10 juniors, and 5 seniors in a classroom, what is the probability of choosing a junior at random? Is this the relative frequency approach to estimating a probability?

5.3 Assume that 20 doctors are chosen at random from the Houston telephone directory. Of these 20, there are 15 who recommend Little's pills and 5 who do not. If a doctor was chosen at random from the city of Houston, estimate the probability that this doctor would recommend Little's pills.

5.4 Suppose a nickel, a dime, and a penny were tossed at the same time. List the possible outcomes by examining whether a head or tail appears on each coin. Is each outcome equally likely? What is the probability of getting a head on each of these three coins?

5.5 Which of the following values cannot be a probability? Why?
   (a) .02.
   (b) 0.
   (c) 5/4.
   (d) 985/1051.

5.6 Four hundred randomly sampled automobile owners were asked whether they selected the particular make and model of their present car mainly because of its appearance or because of its performance. The results were as follows:

Owner	Appearance	Performance	Totals
Male	95	55	150
Female	85	165	250
	180	220	400

   (a) What is the probability that an automobile owner buys a car mainly because of its appearance?
   (b) What is the probability that an automobile owner buys a car mainly because of its appearance and that the automobile owner is a male?
   (c) What is the probability that a female automobile owner purchases the car mainly because of its appearance?

5.7 Craighead and others (1986) conducted a study to examine personality variables of basketball players, starters and nonstarters, at the high school and university level. One of the purposes of the study was to determine if significant differences existed between specific groups of athletes. The following table describes the pool of players investigated:

Players	Starters Black	Starters White	Nonstarters Black	Nonstarters White
High school				
Male	7	6	8	7
Female	3	2	4	1
University				
Male	7	3	9	4

Source: D. J. Craighead, G. Privette, F. Vallianos, and D. Byrkit, Personality characteristics of basketball players, starters and non-starters. *Int. J. Sports Psychol.* **17** (2): 110–119 (1986).

If this group is representative of the population of basketball players, what is the approximate probability that a randomly selected player from this population is:

(a)   A female?

(b)   A university player?

(c)   A nonstarter or a black player?

(d)   A black player?

(e)   A female university player?

(f)   A white starter?

(g)   A white female high school starter?

(h)   A white player, given he or she is a nonstarter?

5.8   The employment center at a university wanted to know the proportion of students who worked and also the proportion of those students who lived in the dorm. The following data were collected:

Living Arrangements	Work Full Time	Work Part Time	Do Not Work	Total
In dorm	19	22	20	61
Not in dorm	25	9	5	39
				100

(a)   What is the probability of selecting a student at random who works either full or part time?

(b)   Given that a student works, what is the probability that the student lives in a dorm?

(c)   What is the probability that a student either works full time or else does not live in the dorm?

5.9   The use of homologous blood transfusions during major surgery still has certain hazards associated with it. Although screening of donated blood and other precautions have virtually eliminated post-transfusion hepatitis of type B, post-transfusion hepatitis of the non-A, non-B type continues to occur. Haugen and Hill (1987) report on a large-scale autologous blood program that has been in effect at Holy Cross Hospital, Fort Lauderdale, Florida, in which patients scheduled for elective orthopedic surgery donated their own blood, which was frozen and stored for subsequent thawing and reuse at the time of their surgery. The report contains the following two tables (which are separate, and not cross–classified):

AGE OF PATIENT DONORS

Age Group	Number
10–19	48
20–29	23
30–39	43
40–49	67
50–59	176
60–69	644
70–79	775
80–91	162
Total	1938

SURGICAL PROCEDURES

Procedure	Number
Total hip replacements	1424
Revision of total hip replacements	197
Total knee replacements	139
Spinal instrumentations	58
Bilateral total knee replacements	37
Spinal fusions	32
Misc. orthopedic procedures	18
Laminectomies	14
Revisions of total knee replacements	8
Revisions of resectional hip arthroplasties	6
Resectional hip arthroplasties	5
Total	1938

Source: R. K. Haugen and G. E. Hill, A large-scale autologous blood program in a community hospital. *J. Am. Med. Assoc.* **257** (9): 1211–1214 (1987).

If an individual patient is randomly selected from the above study, what is the probability that:

(a)   He or she is less than 50 years old?

(b)   He or she is between 50 and 69 years old?

(c)   The patient underwent a total hip replacement?

(d)   The patient had a spinal instrumentation or a laminectomy?

5.10   If events A and B are mutually exclusive, then is the occurrence of event A affected by the occurrence of event B? Can one say that if two events are mutually exclusive, they are not independent?

5.11 A Multinational Peace Force in a Middle Eastern nation was made up of male soldiers from 11 nations. The following table shows the breakdown according to nationality and years of military service:

| Nation | YEARS OF MILITARY SERVICE | | | |
	<1 year	1 to 3 years	>3 years	Total
Britain	9	20	10	39
Canada	30	70	35	135
Colombia	52	300	150	502
Fiji	10	299	190	499
France	1	35	1	37
Italy	2	90	0	92
Netherlands	8	75	25	108
New Zealand	0	15	0	15
Norway	1	2	1	4
Uruguay	5	65	5	75
United States	150	850	187	1187
	268	1821	604	2693

Suppose one soldier is randomly selected.
(a) What is the probability he is French or Italian?
(b) What is the probability he is not from the United States?
(c) Given he is from the United States, what is the probability he has more than 3 years of experience?
(d) Given he has less than 1 year of experience, what is the probability he is British?
(e) What is the probability he is from Fiji and has 1 to 3 years of experience?

5.12 A statistics instructor wishes to find out the relationship between the classification of a student and the student's grade in the course. The following is a breakdown of three sections of an introductory statistics course:

Grade	Freshman	Sophomore	Junior	Senior
A	0	8	9	10
B	0	6	8	11
C	1	7	9	12
D	2	4	1	4
F	0	6	2	1
Total	3	31	29	38

Suppose that one student is randomly selected.
(a) What is the probability that the student is a junior and makes at least a B in the course?
(b) What is the probability that the student does not make an A in the course, given that the student is a senior?
(c) What is the probability that the student makes a D or F in the course?
(d) Let A be the event that a sophomore is taking the course. Let B be the event that the student makes a C in the course. Are events A and B independent? Are events A and B mutually exclusive?

5.13 A group of child prodigies and precocious geniuses were cross-classified by talent and emotional stability, as shown below:

	EMOTIONAL STABILITY		
Talent	Stable (S)	Unstable (U)	Total
Musical genius (M)	11	7	18
Math wizard (W)	15	5	20
Artist (A)	3	5	8
Poet (P)	2	2	4
	31	19	50

Suppose one child is selected at random.
(a) What is the probability the child is a "math wizard"?
(b) What is the probability the child is emotionally stable?
(c) What is the probability the child is an emotionally unstable musical genius?
(d) What is the probability the child is either an artist or a poet?
(e) Given the child is emotionally stable, what is the probability the child is a math wizard?
(f) Given the child is a musical genius, what is the probability the child is emotionally unstable?
(g) Express the above probabilities for (a)–(f) symbolically, such as $P(S)$, $P(M)$ and so on.
(h) Calculate the following probabilities:
   (i) $P(A)$
   (ii) $P(U)$
   (iii) $P(S \text{ and } W)$
   (iv) $P(M \text{ or } W)$
   (v) $P(\bar{M}|U)$
   (vi) $P(S|W)$

5.14 Rofé and Lewin (1986) conducted a study to compare two approaches to the understanding of affiliation under stress: utility theory versus emotional comparison theory. In one part of the study, the affiliative and conversational tendencies of 189 pregnant women who checked into Zahalon Hospital in Tel-Aviv, Israel, to give birth, were cross-classified against an $R–S$ (repressor–sensitizer) variable, as shown below:

	AFFILIATIVE TENDENCY BEFORE DELIVERY		
R–S Variable	Prefer to Be Alone	Prefer to Be with Others	Total
Repressors	33	13	46
Intermediates	49	48	97
Sensitizers	21	25	46
	103	86	189

(Source: Yacov Rofé and Isaac Lewin, Affiliation in an unavoidable stressful situation: An examination of the utility theory. Br. J. Social Psychol. **25** (2): 119–127 (1987).

For an individual selected at random from this group, what is the probability that:

(a) She is a "repressor" or "intermediate"?
(b) Her affiliation is to be alone before delivery?
(c) She is a "sensitizer" and prefers to be with others before delivery?
(d) She is a "sensitizer," or prefers to be alone before delivery?
(e) She prefers to be alone, given that she is a "repressor"?

5.15 A sociologist found that in a certain community, 60% of unmarried mothers ended up on welfare. Within this group on welfare, three-fourths did not finish high school, the rest did finish high school. For those not on welfare, 90% finished high school. Suppose there are 200 unmarried mothers in the community. Complete the contingency table, and calculate the following probabilities:

	On Welfare (W)	Not on Welfare ($\bar{W}$)	Total
Finished high school (H)	30	72	162
Not finished high school ($\bar{H}$)	90	8	98
	120	80	200

(a) $P(\bar{W}) = \dfrac{80}{200} = 0.4$

(b) $P(H \text{ or } W) \rightarrow \dfrac{102}{200} + \dfrac{120}{200} - \dfrac{30}{200} = .96$

(c) $P(H \text{ and } W) \rightarrow \dfrac{30}{200} = 0.15$

(d) $P(\bar{W}|\bar{H}) \rightarrow \dfrac{8}{98} = 0.0816$

(e) $P(H|W) \rightarrow \dfrac{30}{120} = 0.25$

5.3   GOING BEYOND THE CONTINGENCY TABLE

Our Parenting Services example served as an intuitive introduction to probability definitions. The classical approach was used to derive probabilities by dividing the number of outcomes favorable to an event by the total number of (equally likely) outcomes. Not all probability problems, however, are concerned with randomly selecting an individual from a contingency table.

When dealing with two or more events in general, one approach is to illustrate these events by means of a **Venn diagram**. A Venn diagram representing any two events A and B is shown in Figure 5.1.

Figure 5.1

Venn diagram for events A and B. The rectangle represents all possible outcomes of an experiment.

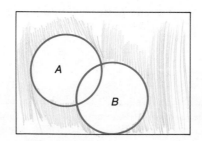

Figure 5.2

Venn diagram for
$P(A) = .4$

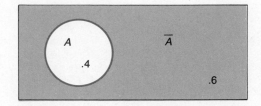

Figure 5.3

$P(A$ and $B)$. The
points in the shaded
area are in A and B.

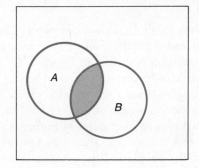

Figure 5.4

$P(A$ or $B)$. The
points in the shaded
area are in A or B.

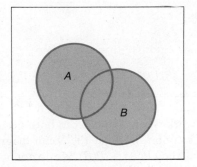

In a Venn diagram, the probability of an event occurring is its corresponding area. This may sound complicated, but it really is not. The Venn diagram for $P(A) = .4$ is shown in Figure 5.2. The area of the rectangle is 1; it represents all possible outcomes. The shaded area is the complement of A, namely $\overline{A}$. Here, $P(\overline{A}) = 1 - P(A) = 1 - .4 = .6$. No effort is made to construct a circle with an area of .4; it is simply labeled .4. The shaded area then represents $\overline{A}$, and the corresponding area must be .6.

Figure 5.3 shows $P(A$ and $B)$, and Figure 5.4 shows $P(A$ or $B)$.

If A and B are mutually exclusive (they cannot both occur), then $P(A$ and $B) = 0$. For example, an auto dealer has data that indicate that 20% of all new cars ordered contain a red interior, whereas 25% have a blue interior. Only

Figure 5.5

Venn diagram of
mutually exclusive
events.
$P(A \text{ and } B) = 0$
$P(A \text{ or } B) =$
$P(A) + P(B) =$
$.2 + .25 = .45.$

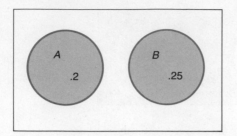

one color is allowed when selecting an interior color. Let A be the event that a red interior is selected and B be the event that a blue interior is selected. A Venn diagram for this situation is shown in Figure 5.5.

Each person can select only one color, so events A and B are mutually exclusive, and the resulting circles do not overlap in the Venn diagram. What is the probability that a person selects red or blue? This is $P(A \text{ or } B)$ and is represented by the shaded area in the circles in Figure 5.5. The Venn diagram allows us to see clearly that this shaded area is $P(A) + P(B) = .2 + .25 = .45$. In other words, 45% of the people will purchase either red or blue interiors. We thus have the following rule.

Rule

If events A and B are mutually exclusive, then

$$P(A \text{ or } B) = P(A) + P(B) \qquad \textbf{(5-6)}$$

This rule does not work when A and B can both occur, but there is an easy way to devise another solution. Look at the Venn diagram for this situation, shown in Figure 5.6. By adding $P(A) + P(B)$, we do not obtain $P(A \text{ or } B)$ because we have counted $P(A \text{ and } B)$ twice. So we need to subtract $P(A \text{ and } B)$ to obtain the actual area corresponding to $P(A \text{ or } B)$. This is the **additive rule of probability**.

Figure 5.6

Venn diagram
illustrating $P(A \text{ or } B)$
and $P(A \text{ and } B)$

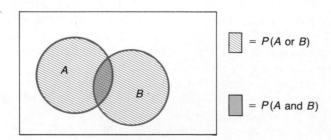

Additive Rule

For any two events, A and B,

$$P(A \text{ or } B) = P(A) + P(B) - P(A \text{ and } B) \qquad (5\text{-}7)$$

Notice that if A and B are mutually exclusive, then $P(A \text{ and } B) = 0$, and we obtain the previous rule; namely, that $P(A \text{ or } B) = P(A) + P(B)$.

EXAMPLE 5.2

Draw a single card from a deck of 52 playing cards. Let S be the event that the card is a seven and H be the event that the card is a heart.

  1 Are these events mutually exclusive?
  2 What is $P(S \text{ and } H)$? $P(S \text{ or } H)$?

Solution 1

There are (1) 52 equally likely outcomes (cards), (2) 4 sevens, and (3) 13 hearts, so

$$P(S) = 4/52$$

$$P(H) = 13/52$$

Both S and H can occur. A seven of hearts is a possible outcome here. S and H are not mutually exclusive.

Solution 2

$P(S \text{ and } H)$ is the probability of selecting a seven of hearts from the deck. There is only one such card, so

$$P(S \text{ and } H) = 1/52$$

A Venn diagram for this situation is shown in Figure 5.7. Using the additive rule, the proportion of draws (probability) that a seven or a heart will be selected from the deck is

$$P(S \text{ or } H) = P(S) + P(H) - P(S \text{ and } H)$$
$$= 4/52 + 13/52 - 1/52$$
$$= 16/52$$

Figure 5.7

$P(S) = 4/52,$
$P(H) = 13/52$

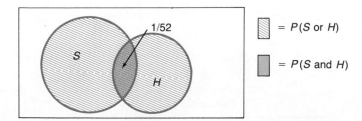

Refer back to Table 5.1. Does the additive rule work here also? It does—this rule works for any two events—but it certainly is a hard way to solve this problem. Suppose we want to find the probability (from our previous example) that the person selected is a female or is under age 30. By inspection, we previously found that

$$P(F \text{ or } U) = 160/200 = .8$$

Using the additive rule, we obtain the same result:

$$
\begin{aligned}
P(F \text{ or } U) &= P(F) + P(U) - P(F \text{ and } U) \\
&= 120/200 + 100/200 - 60/200 \\
&= 160/200 = .8
\end{aligned}
$$

□

Conditional Probabilities

Using Table 5.1, we found that the probability that the person selected is a female (F), given the information that the person selected is under 30 (U), was $P(F|U) = .6$. Our reasoning here was (1) there are 100 people under 30 years of age, (2) 60 of them are female, (3) each of these 100 people is equally likely to be selected, and so (4) the result is $60/100 = .6$. Notice that

$$P(U) = \frac{100}{200} = .5$$

$$P(F \text{ and } U) = \frac{60}{200} = .3$$

$$P(F|U) = \frac{P(F \text{ and } U)}{P(U)} = \frac{.3}{.5} = .6$$

This procedure for finding a conditional probability applies to any two events. Use the Venn diagram in Figure 5.8 to determine $P(A|B)$. Given the information that event B occurred, we are immediately restricted to the lined area (B). What is the probability that a point in B is also in A (that is, event

Figure 5.8

Venn diagram
illustrating a
conditional
probability

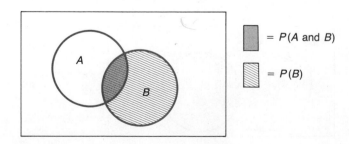

A occurs) given this information? A point is also in A if it lies in the shaded area, and

$$P(A|B) = \frac{\text{shaded area} \; \blacksquare}{\text{striped area} \; \boxslash}$$

$$= \frac{P(A \text{ and } B)}{P(B)}$$

This is the rule for conditional probabilities.

Rule for Conditional Probabilities

For any two events, A and B,

$$P(A|B) = \frac{P(A \text{ and } B)}{P(B)} \qquad \text{(assuming } P(B) \neq 0) \qquad \textbf{(5-8)}$$

and

$$P(B|A) = \frac{P(A \text{ and } B)}{P(A)} \qquad \text{(assuming } P(A) \neq 0) \qquad \textbf{(5-9)}$$

Independent Events

In the discussion of the Parenting Services example, a summary of how to demonstrate that two events are independent was provided in equations 5-3, 5-4, and 5-5. We need demonstrate only that one of these equations holds to verify independence. These three methods of proving independence apply to any two events, not just to contingency table applications.

In many situations, it is unnecessary (or impossible) to prove independence of two events. However, we can often argue convincingly that two events are independent or dependent without resorting to a mathematical proof. Consider these events:

$A =$ You will score above average on the next statistics exam

$B =$ A Republican will become the next President of the United States

Whether event B happens should have no effect on whether event A occurs. So $P(A|B) = P(A)$, and these events are independent. Next, change event A to be: The unemployment rate reaches an all time high. Now whether event B occurs could very well have an effect on whether event A occurs. So it is not safe to assume that $P(A|B) = P(A)$. Notice that we have not discussed the values of $P(A)$ and $P(A|B)$. The probability values are not necessary to show that the events are dependent. The important thing is that $P(A|B) \neq P(A)$, so these events are clearly dependent events.

Joint Probabilities

The rule for conditional probabilities in equations 5-8 and 5-9 can be rewritten as

---

Multiplicative Rule

For any two events, A and B,

$$P(A \text{ and } B) = P(A|B) \cdot P(B) \qquad \textbf{(5-10)}$$
$$= P(B|A) \cdot P(A) \qquad \textbf{(5-11)}$$

This is the **multiplicative rule of probability**. Using equation 5-5, we also have the following rule for two independent events.

Rule

For any two independent events, A and B,

$$P(A \text{ and } B) = P(A) \cdot P(B) \qquad \textbf{(5-12)}$$

You may be wondering how we can use the same equation to define the rule for $P(A|B)$ (equation 5-8) and the rule for $P(A \text{ and } B)$ (equation 5-10). This is not a bad question! It appears that we have used the same rule twice to make two different statements—and in fact we have. However, for any application you encounter, $P(A|B)$ or $P(A \text{ and } B)$ must either be provided or can be determined without resorting to formulas. We can clarify this using our card-drawing example:

S = select a seven

H = select a heart

Here $P(S \text{ and } H)$ (the probability of selecting a seven of hearts) is 1/52. No formulas were necessary to determine this, only a little head scratching.

Now, what is $P(S|H)$? Using equation 5-8,

$$P(S|H) = \frac{P(S \text{ and } H)}{P(H)}$$
$$= \frac{1/52}{13/52}$$
$$= 1/13$$

Assume that you select a card from a deck, examine it, and then discard it. You then select another card. This is called **sampling without replacement**. Let

A = selecting a seven on the first draw

B = selecting a seven on the second draw

What is the probability of drawing two sevens ($P(A$ and $B)$)? If you selected a seven on the first draw, then, of the 51 cards remaining, three are sevens. So $P(B|A) = 3/51$. Again, we used no formulas.

Next, we use the multiplicative rule, equation 5-11:

$$P(A \text{ and } B) = P(B|A) \cdot P(A)$$
$$= 3/51 \cdot 4/52 \cong .0045$$

Notice that $P(A) = 4/52$ because there are four sevens available on the first draw. So you would expect to draw two sevens from a card deck about 45 times out of 10,000, if you are drawing without replacement.

Now suppose you select a card from a deck but replace it before selecting the second card. This is called **sampling with replacement**. What is $P(B|A)$? There are still 52 cards in the deck when you select your second card, and four of these are sevens. So

$$P(B|A) = 4/52 = P(B)$$

If event A occurs, the probability of a seven on the second draw is unaffected. This probability is 4/52 whether or not A occurs; these events are now independent. For this situation,

$$P(A \text{ and } B) = P(A|B) \cdot P(B)$$
$$= P(A) \cdot P(B) \qquad \text{(since they are independent)}$$
$$= 4/52 \cdot 4/52 \cong .0059$$

The probability of getting two sevens is higher when drawing cards with replacement—not a surprising result.

5.4   APPLYING THE CONCEPTS

EXAMPLE 5.3

Among the seniors in a particular high school, 20% have a deficiency in mathematics, 30% have a deficiency in reading ability, and 10% have a deficiency in both. Determine the probability that a senior from this high school has a deficiency in reading, mathematics, or both.

Solution

The most important thing when solving a wordy probability problem is to set up the problem correctly. Your first step when solving any probability application should always be to define the events clearly, using capital letters. Your initial step should be to define

M = senior with a deficiency in mathematics

R = senior with a deficiency in reading

We do not need to define another event for a student deficient in both areas, as we shall see.

We now have

$$P(M) = .2$$
$$P(R) = .3$$

The probability that a selected senior is deficient in mathematics *and* reading is given as .10. This is a joint probability:

$$P(M \text{ and } R) = .1$$

We want to find the probability of M or R. Using the additive rule,

$$P(M \text{ or } R) = P(M) + P(R) - P(M \text{ and } R)$$
$$= .2 + .3 - .1$$
$$= .4$$

So 40% of the seniors in this school are deficient in at least one of the two areas.

Suppose that one-third of the seniors deficient in reading are also deficient in mathematics. How can you translate this statement into a probability? Another way of stating the above sentence is, "Given that the senior is deficient in reading, the probability that the senior is deficient in mathematics is 1/3." In other words, this is a conditional probability:

$$P(M|R) = 1/3 \qquad \qquad \Box$$

EXAMPLE 5.4   Referring to example 5.3, what percentage of the seniors deficient in reading are not deficient in math?

Solution   A Venn diagram for this problem is shown in Figure 5.9. Notice that event M is made up of two components: (1) those people in R and (2) those not in R.

Figure 5.9

Venn diagram for example 5.4

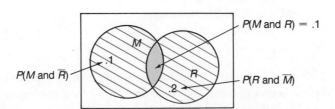

$P(M \text{ and } \overline{R})$ ⟶   .1 M $P(M \text{ and } R) = .1$ R .2 $P(R \text{ and } \overline{M})$

Since $P(M \text{ and } R) = .1$, the area of M that is striped is

$$P(M) - P(M \text{ and } R) = P(M \text{ and } \bar{R})$$
$$= .2 - .1 = .1$$

Similarly, the area of R that is striped is

$$P(R) - P(M \text{ and } R) = P(R \text{ and } \bar{M})$$
$$= .3 - .1 = .2$$

Our question could be stated, "Given that a person is deficient in reading, what is the probability that this person is not deficient in math?" This is a conditional probability: $P(\bar{M}|R)$. Look at the Venn diagram. You know that R occurred, so the outcome is in the R circle. What is the probability that the outcome is not in M? We know that the total area of R is .3 and that the area that is not in M but is in R is .2. So

$$P(\bar{M}|R) = .2/.3 = 2/3 \qquad \square$$

Helpful Hints for Probability Applications

1 Define each event using capital letters.

2 Translate each statement into a probability. Is a particular statement telling you $P(A)$? $P(B)$? $P(A \text{ and } B)$? $P(A \text{ or } B)$? $P(A|B)$? $P(B|A)$?

3 Determine the answer by identifying the probability rule that applies and by using a Venn diagram. Using both allows you to check your logic and your arithmetic.

EXAMPLE 5.5

In a particular downtown industrial complex, 40% of the workers live outside the city limits, 42% take the bus to work, and 75% of the workers living outside the city limits ride the bus. A member of the city council would like to know:

1 What is the probability that, for a randomly chosen worker in this complex, the person lives outside the city limits and will ride the bus?

2 What is the probability that, for a randomly chosen worker in this complex, the person lives outside the city limits or takes the bus to work?

3 What percentage of the bus riders in this complex live outside the city limits?

Solution 1

The first step is to define appropriate events:

A = worker lives outside the city limits
B = worker rides the bus to work

We now translate each of the statements into a probability. We know that

$$P(A) = .4$$

$$P(B) = .42$$

The last statement in the problem can be written as, "Given that a worker lives outside the city limits the probability that this person rides the bus is .75":

$$P(B|A) = .75$$

What is the City Council asking for in question 1? $P(A$ or $B)$? $P(A|B)$? $P(A$ and $B)$? They wish to know the probability that a worker lives outside the city limits and rides the bus. This is $P(A$ and $B)$. Using the multiplicative rule,

$$P(A \text{ and } B) = P(B|A) \cdot P(A)$$
$$= .75 \times .4 = .3$$

Solution 2   For question 2, we wish to know the probability that A or B occurs. By the additive rule,

$$P(A \text{ or } B) = P(A) + P(B) - P(A \text{ and } B)$$
$$= .4 + .42 - .3 = .52$$

Thus, 52% of the workers live outside the city, ride the bus, or both.

Solution 3   Question 3 can be phrased as, "Given that a worker rides the bus, what is the probability that this person lives outside the city?" This is $P(A|B)$:

$$P(A|B) = \frac{P(A \text{ and } B)}{P(B)} = \frac{.30}{.42} \cong .71$$

Therefore, 71% of those workers who ride the bus to work, live outside the city limits.   □

EXERCISES   5.16 In each case below, state whether A and B (1) *must be* mutually exclusive, (2) *are not* mutually exclusive, or (3) *could be* but *need not necessarily* be mutually exclusive.
   (a) $P(A) = .5$, $P(B) = .5$, $P(A \text{ and } B) = 0$.
   (b) $P(A) = .5$, $P(B) = .5$, $P(A \text{ and } B) = .25$.
   (c) $P(A) = .6$, $P(B) = .6$, $P(A \text{ and } B) = .3$.
   (d) $P(A) = .6$, $P(B) = .6$, events other than A and B cannot occur.
   (e) $P(A) = .3$, $P(B) = .6$, $P(A \text{ and } B) = .1$.
   (f) $P(A) = .5$, $P(B) = .5$, events other than A and B can occur.
   (g) $P(A) = .5$, $P(B) = .5$, events other than A and B cannot occur.
   (h)   Try to represent each of the above situations in a Venn diagram.

5.17 Use the additive rule for the data in exercise 5.8 (c) to find the probability that a student either works full time or else does not live in the dorm.

5.18 Suppose one card is randomly picked from a deck of 52 playing cards. Event A is the occurrence of a king. Event B is the occurrence of a spade.
   (a) What is the probability of A and B?
   (b) What is the probability of A or B?
   (c) What is the probability of A given B?

5.19 Using the data in exercise 5.16 (a)–(g), state whether A and B (1) *must be* independent, (2) *are not* independent, or (3) *could be*, but *need not necessarily* be independent.

5.20 If $P(A) = .5$, $P(B) = .2$, and $P(A$ or $B) = .7$, then are the events A and B mutually exclusive? Explain.

5.21 If $P(A) = .5$ and $P(B) = .6$, then are the events A and B mutually exclusive? Explain.

5.22 If the probability that a person orders the morning newspaper is .5 and the probability that a person orders the evening newspaper is .3, and if the probability that a person orders at least one of the newspapers is .7, then what is the probability that a person orders both the morning and evening newspapers?

5.23 If $P(A) = .4$, $P(B) = .3$, and $P(A$ and $B) = .12$, then what is the probability of A given B? Are the events A and B independent?

5.24 The University of Texas Health Science Center at Dallas, a leading center of cholesterol research, recently had a very rare patient: nine-year old Robin Morris of Tyler, Texas. She has a combination of two inherited genetic disorders, a "genetic double whammy," that impair her ability to clear artery clogging fats from her blood. One disorder, which causes high cholesterol, occurs in only one in a million people. (The disorder's most famous victim is Stormie Jones, the world's first patient to undergo a heart and liver transplant.) Robin's other genetic disorder, which causes high levels of fats called triglycerides in her blood, occurs in one in a hundred people. What is the probability of someone inheriting both these genetic disorders, as Robin has? What are you assuming about the two disorders? Does this seem like a reasonable assumption? *Source: The Dallas Morning News*, May 23, 1987, p. 33A.

5.25 Given the following information:

$$P(A) = .5$$
$$P(B) = .5 \qquad (A \text{ and } B \text{ are not mutually exclusive})$$
$$P(A \text{ or } B) = .75$$

   (a) Calculate the following probabilities, representing the situation in a Venn diagram as appropriate:
   (i) $P(A$ and $B)$
   (ii) $P(\overline{A \text{ or } B})$
   (iii) $P(\overline{A} \text{ and } B)$
   (iv) $P(\overline{A} \text{ and } \overline{B})$
   (v) $P(\overline{A} \text{ or } \overline{B})$
   (vi) $P(A \mid B)$
   (vii) $P(\overline{B} \mid \overline{A})$
   (b) Are A and B independent?

5.26 A manufacturer claims that a customer has a 30% chance of noticing a particular flaw in a dress it makes. Two dresses, one without the flaw and one with the flaw, are given to two customers to see whether they can recognize the dress with the flaw.

(a)   What is the probability that a customer will notice the flaw if each of the two customers examines a different dress?

(b)   What is the probability that the flaw will be recognized if the two customers examine both dresses?

(c)   Assume that a customer randomly selects one of the two dresses and then examines the selected dress. What is the probability that the customer will find a flaw?

(d)   If both customers select and examine a dress in the manner described in (c), what is the probability that both customers will recognize the flaw?

5.27 A group of criminals are divided into murderers, rapists, and burglars. The categories of murder and burglary are mutually exclusive (i.e., there are no criminals in the group with both these crimes on their record). Other categories are not mutually exclusive. Half of the group are burglars and half are murderers. One-fifth of the burglars are rapists, and two-fifths of the murderers are rapists. Assume there are 100 criminals in the whole group, and one is randomly selected.

(a)   Draw a Venn diagram for the above categories.

(b)   What is the probability of selecting a rapist?

(c)   What is the probability the criminal selected is a rapist, given this person is a murderer?

(d)   What is the probability the criminal selected is a burglar, given this person is a rapist?

(e)   What is the probability of selecting a criminal guilty of both
    (i)   Murder and burglary?
    (ii)   Murder and rape?
    (iii) Rape and burglary?

(f)   What is the probability of selecting a criminal guilty of:
    (i)   Murder or burglary?
    (ii)   Murder or rape or both?
    (iii) Rape or burglary or both?

5.28 At a certain university, 30% of the students major in mathematics. Of the students majoring in mathematics, 60% are males. Of all the students at the university, 70% are males.

(a)   What is the probability that a student selected at random in the university is a male majoring in mathematics?

(b)   What is the probability that a student selected at random in the university is a male or is majoring in mathematics?

(c)   What proportion of the males are majoring in mathematics?

5.29 Refer to exercise 5.15(b) and (c). Instead of preparing a contingency table as you did earlier, use the given information to draw a Venn diagram and various rules like the Additive Rule and the Multiplicative Rule.

5.30 A supermarket has 40% of its merchandise on sale. Twenty percent of its merchandise consists of nonedible items. Fifty percent of the sale items consists of nonedible items.

(a) What is the probability that an item, selected at random in the supermarket, is nonedible and on sale? *0.2 = 0.50 × 0.40 = 0.9*

(b) What is the probability that an item, selected at random, is either nonedible or on sale? *0.20 + 0.40*

(c) What proportion of nonedible items are on sale? *$\frac{0.2}{0.4} = \frac{1}{2}$*

5.31 A "lucky dip" bag contains 60 calculators, 30 playing card packets, and 10 medallions. All the packages look identical, and each one has an equal likelihood of being picked.

(a) Suppose you have to take one dip, put the package back and take a second dip (i.e., sampling with replacement).
  (i) What is the probability of selecting a calculator both times? *.6 × .6*
  (ii) What is the probability of selecting a calculator and a medallion (in any order)? *$\left(.6 \times .1\right) + \left(\frac{10}{100} \times \frac{60}{100}\right) = .12$*

(b) Suppose you take one dip and *without* putting the package back, take a second dip (i.e., sampling without replacement). Compute the above probabilities for
  (i) Two calculators *$.6 \times \frac{59}{99} = .358$*
  (ii) A calculator and medallion *$\frac{60}{100} \times \frac{10}{99} = .061 + .061$*

5.32 Let $P(A) = .7$, $P(B) = .3$, and $P(A \text{ and } B) = .2$. Find the following probabilities:
(a) $P(\bar{A})$
(b) $P(A \text{ or } B)$
(c) $P(B|A)$
(d) $P(A \text{ and } \bar{B})$
(e) $P(A|\bar{B})$
(f) $P(\bar{A} \text{ and } \bar{B})$
(g) $P(\overline{A \text{ and } B})$

5.33 If a pair of dice is tossed, what is the probability that each of the two dice show the same number? *$\frac{1}{6}$*

5.34 Six men and four women are being considered as finalists for a scholarship award. Two scholarships will be awarded. All the candidates are equally qualified, so the selection will be done randomly.
(a) What is the probability that two men will be selected?
(b) What is the probability that one man and one woman are selected?

5.35 If $P(A|B) = .8$ and $P(A \text{ and } B) = .6$, what is $P(B)$?

$.6 = .8 \times P(B) = \frac{.6}{.8} = .75$

5.5 PROBABILITIES FOR MORE THAN TWO EVENTS

We will illustrate what happens when you encounter more than two events by using three events, A, B, and C. The following rules can easily be extended to any finite number of events. In the applications of probability in the chapters that follow, we typically will be dealing with multiple events that are either mutually exclusive or independent.

Figure 5.10

Three mutually
exclusive events

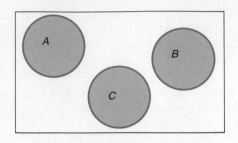

Mutually Exclusive Events

Events A, B, and C are mutually exclusive if no two events can occur simultaneously. A Venn diagram of this situation is shown in Figure 5.10. When dealing with mutually exclusive events, we usually will be interested in the probability that one of these events will occur—that is, $P(A$ or B or $C)$. We can use a simple rule here:

> For mutually exclusive events A, B, and C,
> $$P(A \text{ or } B \text{ or } C) = P(A) + P(B) + P(C) \qquad \textbf{(5-13)}$$

Thus, to determine "or" probabilities when the events are mutually exclusive, you add the respective probabilities.

Independent Events

Events A, B, and C are independent if all of the following are true:

$$P(A \text{ and } B) = P(A) \cdot P(B)$$
$$P(A \text{ and } C) = P(A) \cdot P(C)$$
$$P(B \text{ and } C) = P(B) \cdot P(C)$$
$$P(A \text{ and } B \text{ and } C) = P(A) \cdot P(B) \cdot P(C)$$

Thus the events are independent if the "and" probability for *any* subset of the events (including the set containing all the events) is equal to the corresponding product of marginal probabilities. When dealing with independent events, the probability of interest is usually that all of the events occur, that is, $P(A$ and B and $C)$. Using the fourth condition above, we have for independent events A, B, and C,

> $$P(A \text{ and } B \text{ and } C) = P(A) \cdot P(B) \cdot P(C) \qquad \textbf{(5-14)}$$

Thus, to determine "and" probabilities when the events are independent, you multiply the respective probabilities.

EXAMPLE 5.6

One of the services of Bellmar Laboratories is to test blood samples for the presence of a fatal virus. Their analysis is accurate 98% of the time. One of the requirements of Bellmar is that multiple specimens arriving from the same location be analyzed by different lab technicians using their own individual equipment. If Dr. Blair sends three samples on a particular day, what are the chances that all three will receive correct diagnoses?

Solution

Let

A = first sample is analyzed correctly
B = second sample is analyzed correctly
C = third sample is analyzed correctly

Consequently, $P(A) = P(B) = P(C) = .98$.

We want to find $P(A$ and B and $C)$. Are the events independent? There is no need to use fancy formulas here. The answer is yes, simply because the samples are analyzed by three different individuals, each using different equipment. Therefore,

$$P(A \text{ and } B \text{ and } C) = P(A) \cdot P(B) \cdot P(C)$$
$$= .98 \cdot .98 \cdot .98 \cong .94$$

So 94% of the time, a shipment of three samples will all be analyzed correctly.

EXERCISES

5.36 Given the following information:

$$P(A) = .2$$
$$P(B) = .3$$
$$P(C) = .4$$

where A, B, and C are mutually exclusive events, compute these probabilities:
(a) $P(A$ or B or $C)$
(b) $P(A$ or $C)$
(c) $P(\overline{A \text{ or } C})$
(d) $P(\overline{A}$ or B or $C)$
(e) $P(\overline{A}$ or $\overline{C})$
(f) $P(\overline{A}$ or $\overline{B}$ or $\overline{C})$
(g) $P(A$ and B and $C)$

5.37 Given the following information:

$$P(A) = .4$$
$$P(B) = .3$$
$$P(C) = .2$$

where A, B, and C are independent events, compute these probabilities:

(a) $P(A \text{ and } B \text{ and } C)$
(b) $P(A \text{ and } C)$
(c) $P(\bar{A} \text{ and } C)$
(d) $P(\bar{A} \text{ and } \bar{C})$
(e) $P(\bar{A} \text{ and } \bar{B} \text{ and } \bar{C})$
(f) $P(A \text{ and } B \text{ and } C)$

5.38 In exercise 5.12, are the events "receiving the grade of A," receiving the grade of B," and "receiving a grade of C or better" mutually exclusive? What is the probability that at least one of these events will occur?

5.39 Let $P(A) = .3$, $P(B) = .4$, $P(C) = .5$, $P(A \text{ and } B) = .12$, $P(A \text{ and } C) = .15$, $P(B \text{ and } C) = .2$ and $P(A \text{ and } B \text{ and } C) = .06$. Are the events A, B, and C independent?

5.40 If three events are independent, then are the three events mutually exclusive? If the three events are mutually exclusive, then are the three events independent?

5.41 Three cards are picked with replacement from a deck of 52 playing cards. What is the probability that the first card will be a queen, that the second will be a spade, and that the third will be a king?

5.42 At a local private liberal arts college, 30% of the psychology majors, 25% of the sociology majors, and 20% of the philosophy majors are receiving financial support from the college. If one student is selected at random from each of these three majors:

(a)   What is the probability that all three students are financial support recipients?
(b)   What is the probability that at least two students are financial support recipients?

5.43 A marriage counselor has seen over the years that despite her best efforts, 6 out of 10 couples she sees as clients will end up in divorce court. Suppose she randomly picks three couples from her list of clients.

(a)   What is the probability that all three couples will be divorced eventually?
(b)   What is the probability that only two couples will be divorced eventually?

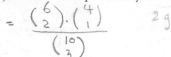

$$= \frac{\binom{6}{2} \cdot \binom{4}{1}}{\binom{10}{3}} \qquad 2\ \text{groups} \begin{cases} 6 \rightarrow \text{divorce} \\ 4 \rightarrow \text{not divorce} \end{cases}$$

5.6   COUNTING RULES

Counting rules determine the number of possible outcomes that exist for a certain broad range of experiments. They can be extremely useful in determining probabilities. For instance, consider an experiment that has 200 possible outcomes, all of which are equally likely to occur. The probability of any one such outcome is $1/200 = .005$.

The question we wish to answer here is, for a particular experiment, how many possible outcomes are there? No set of rules applies to all situations, but we will consider three very popular counting procedures: (1) filling slots, (2) permutations (a special case of filling slots), and (3) combinations.

Filling Slots

We use **counting rule 1** to fill k different slots. Let

n_1 = the number of ways of filling the first slot

n_2 = the number of ways of filling the second slot after the first slot is filled

n_3 = the number of ways of filling the third slot after the first two slots are filled

$$\vdots$$

n_k = the number of ways of filling the kth slot after filling slots 1 through $k - 1$.

The number of ways of filling all k slots is

$$n_1 \cdot n_2 \cdot n_3 \cdot \cdots \cdot n_k$$

EXAMPLE 5.7

When ordering a new car, you have a choice of 8 interior colors, 10 exterior colors, and 4 roof colors. How many possible color schemes are there?

Solution

There are three slots to fill here, interior color (8 ways), exterior color (10 ways), and roof color (4 ways). To answer the question, you simply multiply the number of ways of filling each slot. So the answer is $8 \cdot 10 \cdot 4 = 320$ different color schemes.

The order in which you fill the slots is unimportant. So $n_2 = 10$, regardless of whether or not you have filled the first slot. For some applications, this is not the case. Consider the following example. □

EXAMPLE 5.8

A local PTA group is selecting their officers for the current year. There are 15 individuals in the group, whom we label as $I_1, I_2, \ldots, I_{15}$. They need to select a president, vice president, secretary, and treasurer. How many possible groups of officers are there?

Solution

We have four slots to fill here, president (n_1 ways), vice president (n_2 ways), secretary (n_3 ways), and treasurer (n_4 ways). We know that n_1 is 15. After a president is elected, only 14 people remain, so $n_2 = 14$. By a similar argument, $n_3 = 13$ and $n_4 = 12$. The answer is $15 \cdot 14 \cdot 13 \cdot 12 = 32,760$ different slates of officers. □

Permutations

Example 5.8 is a counting situation in which you select people without replacement. If a particular individual, say I_3, is elected president, then I_3 is not available to fill the remaining slots. Another way of stating the result is that there are 32,760 ways of selecting four people out of 15, where the **order of selection** is important. For example,

I_2 = president

I_6 = vice president

I_{12} = secretary

I_7 = treasurer

is not the same slate of officers as

I_7 = president

I_{12} = vice president

I_2 = secretary

I_6 = treasurer

even though the same people are involved.

The number of ways of selecting k objects (or people) from a group of n distinct objects, where the order of selection is important, is referred to as the number of permutations of n objects using k at a time. This is written

$$_nP_k$$

In example 5.8, $_{15}P_4 = 32,760$. Determining the number of permutations is just a special case of counting rule 1; this is also a slot-filling application.

The symbol $n!$ is read as "n factorial." Its value is determined by multiplying n by all positive integers smaller than n.

$$n! = (n)(n-1)(n-2) \cdots (2)(1) \qquad \textbf{(5-15)}$$

For example,

$$5! = (5)(4)(3)(2)(1) = 120$$

$$1! = 1$$

$$0! = 1 \text{ (by definition)}$$

Notice that $n!$ is the number of ways of filling n slots using n objects. There are n ways of filling the first slot, $(n-1)$ ways of filling the second slot, $(n-2)$ ways for the third slot, and so on.

In example 5.8, the result was obtained by finding $15 \cdot 14 \cdot 13 \cdot 12 = 32,760$. This also can be written as

$$\frac{15 \cdot 14 \cdot 13 \cdot 12 \cdot 11 \cdot 10 \cdot 9 \cdot 8 \cdot 7 \cdot 6 \cdot 5 \cdot 4 \cdot 3 \cdot 2 \cdot 1}{11 \cdot 10 \cdot 9 \cdot 8 \cdot 7 \cdot 6 \cdot 5 \cdot 4 \cdot 3 \cdot 2 \cdot 1} = 32,760$$

This is an application of **counting rule 2**: The number of **permutations** of n objects using k objects at a time is

$$_nP_k = \frac{n!}{(n-k)!} = (n)(n-1)\cdots(n-k+1) \qquad \textbf{(5-16)}$$

EXAMPLE 5.9

How many two-digit numbers can you construct using the digits 1, 2, 3, and 4, without repeating any digit?

Solution

The order of selection is certainly important here—the number 42 is not the same as 24. The answer is $_4P_2$, where

$$_4P_2 = \frac{4!}{(4-2)!} = \frac{(4)(3)(2!)}{2!} = 12$$

These 12 permutations are

12	21	31	41
13	23	32	42
14	24	34	43

Combinations

Take another look at example 5.8, where we selected four people from a group of 15. This time, however, choose a committee of four people from a group of 15 where the order of selection does not matter. Each such committee is one **combination** of the 15 people, using four at a time. For example,

$$I_2 \qquad I_6 \qquad I_{12} \qquad I_7$$

and

$$I_7 \qquad I_{12} \qquad I_2 \qquad I_6$$

are different permutations but are the same combination. These two arrangements are made up of the same individuals; hence they form the same committee or combination.

Clearly, there are not as many combinations (using I_2, I_6, I_{12}, I_7) as there are permutations. The two permutations above form the same combination. There are 24 possible permutations of this combination $(4 \cdot 3 \cdot 2 \cdot 1 = 24)$.

Now we wish to determine how many possible committees (combinations) of four there are for the group of 15. This is written as

$$_{15}C_4$$

Each combination has 24 permutations, so

$$_{15}C_4 = \frac{_{15}P_4}{24} = \frac{32,760}{24} = 1365$$

There are 1365 possible committee combinations. Notice that 24 is the number of permutations of these four numbers (2, 6, 7, and 12); that is, $24 = _4P_4 = 4!$.

Counting rule 3 is used to count the number of possible combinations. The number of combinations of n objects using k at a time is

$$_nC_k = \frac{_nP_k}{k!} = \frac{n!}{k!(n-k)!} \qquad (5\text{-}17)$$

EXAMPLE 5.10   A community support organization must select five members from the total group of 40 people to attend a national conference. How many possible delegations are there?

Solution   The order of selection is not a factor here, so this is a combination problem rather than a permutation problem. The answer is

$$_{40}C_5 = \frac{40!}{5!35!}$$

$$= \frac{(40)(39)(38)(37)(36)(35!)}{(5!)(35!)}$$

$$= \frac{(40)(39)(38)(37)(36)}{(5)(4)(3)(2)(1)}$$

$$= 658,008$$

EXERCISES

5.44   A cafeteria serves four different vegetables, five different main dishes consisting of either fish or meat, and three different desserts. If a customer chooses one serving from each of these three categories, how many different combinations of vegetables, main dishes, and desserts are possible?

5.45   The Political Awareness Club, a new campus organization, has a membership made up of 50 social science majors, 15 mathematics majors, and 35 political science majors. The leadership of the organization is to be selected by random lots, with one person from each major, making a total of three leaders. How many different sets of leaders are possible?

5.46   How many different ways can you select four playing cards from a deck of 52 such that the first is a heart, the second is a diamond, the third is a club, and the fourth is a spade?

$\frac{6}{720} = 6!$

5.47 For a motor-response experiment on rabbits, a professors's assistant has to assign 6 rabbits to 6 different experimental cages. How many different ways can the rabbits be assigned to the cages?

5.48 Seven assistant professors have applied for tenure at a university. However, only two assistant professors can be granted tenure. How many different groups of two assistant professors can be selected to receive tenure?

$_{20}C_5 = \frac{20!}{5! \, 15!} = 15504$

5.49 A large state university owns 5 electron microscopes. A class of 20 biology students has to take turns, 5 students at a time, to use the microscopes. Instead of just dividing the class into four groups of 5 each, the instructor is interested in calculating all the possible groups of 5 students that could be made. How many are there?

5.50 A "psychic" is shown a lineup of five men in an experiment on parapsychology. The psychic is told that two men are convicted thieves, while the other three are average citizens.

$10 \times 9 = 90$

(a) How many different groups of two men could the psychic pick out from the lineup?

(b) What is the probability of the psychic making a perfectly correct choice (picking the two thieves) if she in fact, has no psychic ability and selects two people at random?

$10 \times 9 = 90$

5.51 A committee of 10 people needs to select a committee chairperson and committee secretary. How many different ways can the chairperson and secretary be selected from the committee?

5.52 A builder has five different house styles and three lots on which to build. If each lot must have a different style of house on it, how many configurations of five house styles are possible on these three lots?

$_4P_3 = \frac{4!}{1!} = 4! = 24$

5.53 How many different three-digit numbers can be constructed using the digits 3, 5, 7, and 9 if no digit can be repeated?

5.54 Three immigration officers have to be assigned 12 visa application cases such that the first officer gets 5 cases, the second gets 4 cases, and the third gets 3 cases. How many ways can the cases be assigned?

$\binom{6}{1}\binom{10}{1} = 60$

5.55 A person has 6 different-colored shirts and 10 different-colored trousers. How many color schemes are possible if one shirt and one pair of trousers are chosen?

$\binom{3}{4}\binom{3}{3}\binom{3}{3} = -$

5.56 Try to make up three-letter words, using for the first letter the consonants k, l, t, and z; for the second letter the vowels i, o, u; and for the third letter the consonants p, q, n. The words do not have to make sense. How many different words can you make?

5.57 An objective-type test consists of 20 questions. Part I has 10 questions in a true–false format. Part II has 10 questions in a multiple-choice format with four possible answers (A, B, C, D) to choose from.

(a) In how many ways can the entire test be answered?

(b) What is the probability of working all questions correctly if the answers are chosen randomly?

5.58 A sociologist has classified a population into 5 income groups, 7 regional groups, and 5 racial groups. If an individual is to be classified by income, region, and race, how many classifications are possible?

5.7 SIMPLE RANDOM SAMPLES

Practically all of the applications in later chapters that use probabilities derived from sample results to make a decision concerning the population are based on the assumption that a simple random sample, or more simply, a random sample, was obtained.

We introduced this concept in Chapter 1. A sample of size n, selected from a population of size N, constitutes a **simple random sample** if every possible sample of size n has the same probability of being selected.

In example 5.10, we determined that there were 658,008 possible delegations when selecting 5 people from a group of 40. If we view this group as the population of interest, then $n = 5$ and $N = 40$. Our concern here is to determine the probability that any specified group of 5 individuals will be the designated delegation if simple random sampling is used.

There are $_{40}C_5 = 658,008$ possible delegations, and each has the same probability of being selected. Therefore, the probability that any one combination of people will be picked is $1/658,008 \cong .000002$.

In general, when employing a simple random sample of size n, selected from a population of size N, the total number of possible random samples is

$$_NC_n$$

Also, each of these samples has a probability of being selected of

$$\frac{1}{_NC_n}$$

EXAMPLE 5.11

Your task is to obtain a random sample of two individuals from a group of five employees (E_1, E_2, E_3, E_4, and E_5). What is the probability that you select E_2 and E_5 as your sample?

Solution

There are

$$_5C_2 = \frac{5!}{2!3!} = 10$$

possible random samples. They are:

Sample Number	Sample
1	E_1, E_2
2	E_1, E_3
3	E_1, E_4
4	E_1, E_5
5	E_2, E_3
6	E_2, E_4
7	E_2, E_5
8	E_3, E_4
9	E_3, E_5
10	E_4, E_5

Each random sample (including the one containing E_2 and E_5) has a probability of being selected of $1/10 = .1$.

Obtaining a Random Sample

When N is small, we can put the N names in a hat and pick out n of them for our sample (if you do not have a hat, improvise). This will constitute a random sample. When N is moderately large (for example, 10,000), we need a more practical method of selecting n items from this population. A common procedure is to select n **random numbers** between 1 and 10,000 using a table of random numbers or a computer-generated list of n random numbers. A list of random numbers is provided in Table A-12 at the end of the text. To generate a list of random numbers between 1 and 10,000, one procedure you could use is to

1 Start in any arbitrary position, say, row 5 of column 3.

2 Select a list of random numbers by reading either across or down the table.

3 For each five-digit number selected, place a decimal before the final digit and round this value to the nearest counting number; for example, 24127 would become 2412.7, which is then rounded to 2413.

A computer-generated list of random numbers is easiest to use. Figure 5.11 contains the instructions for generating 100 random numbers between 1 and 10,000 using MINITAB. We will give you the necessary commands in SAS and SPSS at the end of the chapter.

Figure 5.11

Generating random numbers using MINITAB

```
MTB > RANDOM 100 VALUES INTO C1;
SUBC> UNIFORM A=1 AND B=10000.
MTB > PRINT C1

  C1
    4060.03    1256.71    9528.43    6223.37    6046.58
    4103.83    6856.63    8612.84     928.10     193.43
    2557.73    8440.36    6312.44    8538.58    1351.63
    5677.21    9369.35    7003.77    3015.26    2663.19
    3513.16    5353.65    9336.90    5966.16    8866.47
     569.37    1024.59    5734.05    5730.56    4257.37
    2525.70    6125.82    7758.55    9507.50     249.83
    7924.88    9076.84    2033.94    7134.48    1922.90
    7053.21    7943.19    4049.29    9164.25    9939.46
    7990.12    5982.56    1661.29    6109.72    5636.95
    8503.60    4644.28    9101.77    3783.47    7450.04
    8976.78     197.78    2501.04    2318.01    9798.59
    7739.55    7615.11    1322.08    5006.80    9928.16
    7484.27    3596.10    4663.34    5292.32    2811.12
    1499.50    9927.22    2569.22    1837.44    3497.70
    7600.31    9208.42     765.46    5592.59    3878.26
    5219.13    6743.25    6762.23    8409.09    6568.97
    8267.43    1385.38    7249.40    9873.98    5335.84
    9113.31     804.28    7579.61    9202.76     172.12
     355.40     358.71    9080.89    1535.61    4268.78
```

Using either procedure, let us assume that the resulting set of random numbers is 415, 6962, and 4815 (for $n = 3$). Then you must (1) code your population from 1 to 10,000 in some manner and (2) select individuals 415, 4815, and 6962 for your random sample. This topic and extensions of simple random sampling will be further discussed in Chapter 8.

MINITAB, SAS, and SPSS all have options available that allow you to sample randomly from a stored data set and save the results for further analysis. You can find the necessary commands to carry out this procedure in the appropriate user's manual.

To obtain a representative sample when N is extremely large or unknown requires good judgment. Stopping the first 10 people you meet on the street is a very poor way of sampling your population.

You sometimes may be forced to select items for the sample that represent the population as accurately as possible, realizing that a poorly gathered sample can easily lead to an incorrect decision of significant importance. When obtaining a sample of wives of male alcoholics, you will quite likely be forced into obtaining an approximate random sample. However, when such a sample is not a random sample, it is not correct to use probability theory in your analysis.

EXERCISES

5.59 An instructor requires each of 35 students in a class to write a term paper comparing the behavioral school with the classical school of psychology. Two students are selected to present their term papers in class. What is the probability that Georgia and Fred (two students in the class) will be picked?

5.60 A company is trying to encourage women to fill executive positions. For the latest batch of executive trainees, it wishes to fill seven vacancies with 5 women and 2 men. If the company has 7 women and 8 men making a total of 15 finalists for these seven vacancies,

    (a) In how many ways can the seven vacancies be filled if sex is disregarded?

    (b) In how many ways can the seven vacancies be filled if the company insists on 5 women and 2 men?

    (c) What is the probability that the company gets the combination it wants if positions are filled randomly?

5.61 Initial computer screening has given the IRS auditor 12 income tax returns to work on. The auditor intends to randomly select 5 returns from these 12. How many different sets of five returns are possible? If John, Mary, Tom, Ahmed, and Lim are among the 12 originally screened by the computer, what is the probability that all five of them will be selected by this auditor?

5.62 Explain how to select a simple random sample from a population of 100,000 using the computer-generated random numbers in Table A-12.

5.63 The school newspaper photographer takes 10 different pictures of the homecoming queen at the school's football game. All 10 pictures are excellent, so the photographer chooses two at random to place in the school newspaper. What is the probability that the first two pictures taken will be selected?

5.8   BAYES' RULE

Thomas Bayes was an English clergyman who lived in the 1700s. He is credited with developing a procedure that allows you to **revise** probabilities in light of new information. To illustrate this technique, suppose that the information contained in example 5.5 was stated slightly differently, namely, 40% of the workers in this industrial complex live outside the city limits, 75% of the workers living outside the city ride the bus to work, and 20% of the workers living inside the city limits ride the bus to work. Again

A = worker lives outside the city limits

B = worker rides the bus to work

We are then told that

$$P(A) = .4$$
$$P(B \mid A) = .75$$
$$P(B \mid \bar{A}) = .20$$

A particular worker has been randomly selected from this complex and you are told that this person rides the bus to work (event B has occurred). The chances that this worker lives outside the city limits would be written $P(A \mid B)$. The value of this probability is not immediately obvious, but the Bayes procedure will provide an easy method of determining this value.

The value of $P(A) = .4$ was obtained *prior* to receiving any information about the sampled worker and is thus a **prior probability**. The revised probability of A uses the supplied information (namely that event B occurred) and so is a *conditional* probability. Since it is obtained after receiving the new information, it is called a **posterior probability**. Thus for this situation, we have that

$P(A)$ is a *prior* probability

$P(A \mid B)$ is a posterior probability

To structure the Bayes procedure, we start off by constructing a **tree diagram**, as shown in Figure 5.12. The posterior event (event B here) is placed on the far right side along with all possible paths leading to this event. This amounts to two paths for the worker illustration, since either the worker lives outside the city limits (A) or he/she does not ($\bar{A}$). When constructing this diagram, it is crucial that you have the *same* event (B here) running down the far right side of the diagram.

For such a diagram, define

sum of the paths = sum of the product of all probabilities
contained on each path

Figure 5.12

Tree diagram for
worker example

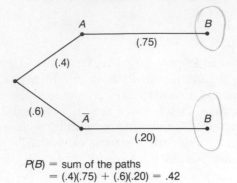

$P(B)$ = sum of the paths
     = $(.4)(.75) + (.6)(.20) = .42$

This probability is always the probability that event B occurs, that is, $P(B)$. Here we have two paths, where the product of the probabilities on the first path is $(.4)(.75)$ and the product of all probabilities on the second path is $(.6)(.20)$. Thus, for Figure 5.12, the sum of the paths is $P(B) = (.4)(.75) + (.6)(.20) = .42$. This is consistent with example 5.5, where it was stated that 42% of the workers rode the bus to work, that is, $P(B) = .42$.

Bayes' rule states that, given the event B, the probability that this event was reached along the ith path corresponding to event E_i, is

$$P(E_i|B) = \frac{P(E_i \text{ and } B)}{P(B)}$$

$$= \frac{i\text{th path}}{\text{sum of the paths}} \tag{5-18}$$

where "ith path" is the product of all probabilities along the ith path.

The original question here, was: given that a particular worker rides the bus to work, what is the probability that this worker lives outside the city limits, that is, $P(A|B)$? Using Figure 5.12, there are two paths here, defined by the events

$E_1 = A$ worker lives outside the city limits
$E_2 = \bar{A}$ worker lives inside the city limits

The probability for each event is also included on the paths, as shown in Figure 5.12. It is important that the probabilities along the initial branches *sum to 1.0.* Here $P(A) + P(\bar{A}) = .4 + .6 = 1.0$.

We are told that event B has occurred and asked to find the probability that we got to event B along the first (top) path. So, according to Bayes' rule, this would be

$$P(A|B) = \frac{\text{1st path}}{\text{sum of the paths}}$$

$$= \frac{(.4)(.75)}{(.4)(.75) + (.6)(.2)}$$

$$= \frac{.30}{.42} \cong .714$$

So 71.4% of the people who ride the bus to work live outside the city limits.

EXAMPLE 5.12

At Truman Elementary School, it has been noted that 60% of the mothers of the children attending the school are married, 35% are divorced, and 5% have never been married. Records also indicate that 6% of the children of married mothers have been identified as having serious emotional problems, 8% of the children of divorced mothers have been so identified, and 12% of the children of never-married mothers have such problems.

(i) What is the probability that a randomly selected child from this school will be identified as having serious emotional problems?

(ii) What percentage of the children identified as having emotional problem have divorced mothers?

Solution to (i)

Define the events

E_1 = child has a married mother

E_2 = child has a divorced mother

E_3 = child has a never-married mother

B = child has been identified as having serious emotional problems

Thus we know that

$$P(E_1) = .60 \qquad P(E_2) = .35 \qquad P(E_3) = .05$$
$$P(B|E_1) = .06 \qquad P(B|E_2) = .08 \qquad P(B|E_3) = .12$$

A tree diagram for this situation is contained in Figure 5.13. According to this diagram, the probability that a randomly selected child from this school will be identified as having serious emotional problems is

$$P(B) = \text{sum of the paths}$$
$$= .07$$

Consequently, 7% of the children attending this school will be in this category.

Figure 5.13

Tree diagram for
example 5.12

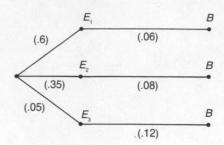

$P(B)$ = sum of the paths
= (.6)(.06) + (.35)(.08) + (.05)(.12)
= .07

Solution to (ii)   According to Figure 5.13 and Bayes' rule, the probability that a child identified as having emotional problems has a divorced mother is

$$P(E_2|B) = \frac{\text{2nd path}}{\text{sum of the paths}}$$

$$= \frac{(.35)(.08)}{.07}$$

$$= \frac{.028}{.07} = .4$$

Thus, 40% of the children identified as having emotional problems have divorced mothers.

EXERCISES   5.64 The Political Action Group, an organization lobbying for tough antipornography legislation, carried out a random telephone survey to determine public attitudes toward pornographic material. The phone interviews showed that 64% were opposed to the unrestricted sales of hardcore pornography, 30% were in favor of unrestricted sale, and 6% had no opinion either way. Of those in opposition, 10% admitted they watched adult movies on cable TV; of those in favor, 70% said they watched adult movies on cable TV; and of those with no opinion, 50% watched adult movies on cable TV.

(a) What is the probability that a randomly selected person will be someone who watches adult movies on cable TV?

(b) What proportion of the persons who watch adult movies on cable TV will be opposed to the unrestricted sale of hard-core pornography?

5.65 A family owns three cars: a Toyota wagon, a Chevy truck, and a Ford sedan. All three are used with equal frequency. Based on owners' experience, the probability

(handwritten at top:)
$\frac{1}{3}$ .15
$\frac{1}{3}$ .25
$\frac{1}{3}$ .25

(handwritten left margin:)

a) $P(B) = (\frac{1}{3} \times 15) + (\frac{1}{3} \times 0.25) + (\frac{1}{3} \times 0.25)$ or → for 1 car

$0.15 + 0.25 + 0.25 = 0.65$ sum

$0.65 \div 3 = 0.2167$ ←

b) $P(T/B) = \frac{0.15}{0.65} = 0.231$ sum

$1 - 0.2167 = 0.7833$

of a car breaking down within 3 years of purchase is as follows: Toyota, 0.15; Chevy, 0.25; Ford, 0.25.

(a) Taking all three cars in account, what is the probability of the family experiencing a car breakdown within 3 years of purchase?

(b) Given that a car breaks down within 3 years of purchase, what is the probability it is a Toyota?

(c) What is the probability the family will not experience any car breakdown within 3 years of purchase? *.783*

(d) If the cars were not used with equal frequency (say, the Toyota was used twice as much as the other two cars), would the probabilities for (a), (b), and (c) remain the same? *NO*

5.66 The Policy Monitor, a weekly newsletter, reports that 52% of the members of Congress are in favor of tougher provisions in an immigration reform bill that would penalize employers who knowingly hire illegal aliens (the so-called "employer sanctions"). Those members opposed to employer sanctions make up 25% of the membership, while the remaining 23% are fence sitters (undecided). Party affiliation of the membership is as follows:

Position on Employer Sanctions
Provision in Immigration

Reform Bill	Democrat	Republican
Favor	60%	40%
Opposed	65%	35%
Undecided	55%	45%

(a) What is the probability that a randomly selected member of Congress, who is a Democrat, will be opposed to employer sanctions?

(b) What percentage of the members of Congress are Democrats?

(c) What is the probability that a randomly selected member of Congress will be in favor of employer sanctions, given this person is a Republican?

5.67 A psychologist investigated the level of life satisfaction of graduate students living in university housing. Half were satisfied, one-third were partly satisfied, and the remainder were dissatisfied. Of those satisfied, 60% were unmarried; of those partly satisfied, 40% were unmarried, and of those dissatisfied, 20% were unmarried. Suppose one graduate student is randomly selected from university housing.

(a) What is the probability the student is unmarried?

(b) What is the probability the student is satisfied if he or she is unmarried?

(c) Are you able to say whether life satisfaction is dependent on marital status, or whether these variables are independent?

5.68 A demographics researcher was analyzing data on immigrants who arrived in America as adults, and their children who grew up here. The following distribution of highest education levels was noted for the immigrants: 40% had a high school education, 35% had a bachelor's degree, and 25% had a master's degree or higher. Of those with a high school education, 90% reported that their children exceeded the parents' educational achievement and income. The percentage of children who surpassed their parents in the other groups were: bachelor's degree, 82%; master's degree or higher, 70%.

(a) What proportion of immigrant children exceed their parents' educational and income levels?

(b)   Given that a randomly selected immigrant's children surpassed the parent in education and income, what is the probability that the immigrant has a bachelor's degree as his highest educational level?

5.69   Thirty percent of university professors surveyed said they subscribed to *Psychology Today*; another 20% read the magazine but did not subscribe themselves. The remainder did not read the magazine at all. Among the subscribers, 60% were psychology professors, while 38% of nonsubscribing readers were psychology professors. Among the nonreaders, only 5% were psychology professors. Suppose one professor is selected at random.

(a)   What is the probability he or she is a psychology professor?

(b)   What is the probability the person selected is a subscriber, given that he or she is a psychology professor?

(c)   What is the probability he or she is *not* a psychology professor?

5.70   In an experiment with a lie-detector machine, a subject has been instructed to answer 3 out of 10 questions falsely, mixing the truthful and false answers at random. The specialist operating the lie detector is not aware of this prearrangement. The specialist has an accuracy rate of 80% (i.e., if a statement is a lie, he detects it 8 out of 10 times). Furthermore, the specialist has an error rate of 10% (i.e., even though a statement is the truth, he wrongly classifies it as a lie 10% of the time). In short, lies are correctly classified 80% of the time and truthful answers are correctly classified 90% of the time.

(a)   What proportion of all answers does the specialist classify correctly?

(b)   If one answer is selected at random, what is the probability that it is actually a lie if the specialist classifies it correctly?

(c)   If one answer is selected at random, what is the probability that it is actually a lie if the specialist classifies it as a lie? (*Hint:* This question is not the same as part (b), where the specialist makes a correct classification. In part (c), the specialist could be wrong. Look carefully at the tree diagram.)

5.71   A suburban town is made up of 10% low-income, 55% medium-income, and 35% high-income households. It is known that 80% of the low-income, 52% of the medium-income, and 8% of the high-income households are regular shoppers at Kwality Discount Mart. A shopper from this community is randomly selected at the mart. What is the probability the shopper is from a:

(a)   Low-income household?

(b)   Medium-income household?

(c)   High-income household?

(d)   What percentage of households in this suburban town shop at Kwality Discount Mart?

SUMMARY

This chapter has examined methods of dealing with uncertainty by applying the concept of probability. An activity that results in an uncertain outcome is called an **experiment**; the possible outcomes are **events**. Uncertainty is measured in terms of the probabilities of events. To determine the value of a particular probability, we used the classical approach, the relative frequency method, and the subjective probability approach. The **classical** definition for the probability of event A occurring assumes that the experiment has n equally likely outcomes and event A occurs in m of the n outcomes with a resulting probability of occurring equal to m/n. When using the **relative frequency** approach, the experiment is observed n times. Letting m represent the number of times event A occurs out of n times, then the resulting probability of event A occurring in the future is m/n. A **subjective probability** is a measure of your belief that a particular event will occur, and like all probabilities, ranges from zero to one, inclusive.

When examining more than one event, say A and B, several types of probabilities can be derived. The probability of A and B occurring is a **joint probability** and is written $P(A$ and $B)$. The **multiplicative rule** is a method of determining a joint probability. The probability of A or B (or both) occurring is written $P(A$ or $B)$ and can be obtained using the **additive rule**. When asked to find a probability given particular information about events, you determine a **conditional probability**. For example, the probability that B occurs given that A has occurred is a conditional probability, written $P(B|A)$. A variation of the multiplicative rule provides a method of determining a conditional probability. The probability of a single event, such as P(the person selected is a female) or P(an individual will die of lung cancer), is a **marginal probability**.

An effective method of determining a probability in complicated situations is to use a **Venn diagram**. When you represent the various events visually, you can often obtain a seemingly complex probability easily.

Two events are said to be **independent** if the occurrence of the one event has no effect on the probability that the other event occurs. Do not confuse "no effect" with "never occur simultaneously." If two events can never occur simultaneously, they are **mutually exclusive**. For example, these two events are certainly independent but are not mutually exclusive (since both events could occur): A, the stock market drops more than two points during a particular week, and B, your car breaks down during the same week.

We discussed various counting rules, including **permutations** and **combinations**. These rules are used to count the number of possible outcomes for experiments that select a certain number of people or objects (k) from a large group of n such objects. When determining the corresponding number of permutations (written $_nP_k$), the order of selection is considered. The number of combinations for this situation (written $_nC_k$) ignores the order of selection and counts only the number of groups that can be obtained.

We also discussed the number of random samples that exist when the population size is known, and we examined methods of obtaining such a sample. In the chapters to follow, any results using a statistical sample assume that the sample is obtained randomly.

A method of revising probabilities was discussed using **Bayes' rule**. This procedure begins with prior probabilities and revises them based upon observed information. The changed probabilities are **posterior probabilities** and can be derived by structuring the problem using a tree diagram.

5.72 The following data were published as part of a study by Winokur and Black (1987) on death rates in former psychiatric patients:

ASSOCIATION OF ORGANIC MENTAL DISORDER WITH
SIGNIFICANT MEDICAL ILLNESS IN PSYCHIATRIC PATIENTS
WHO DIED OF NATURAL CAUSES WITHIN 2 YEARS OF
HOSPITAL DISCHARGE

Group	Organic Mental Disorder		
	Yes	No	Total
Patients 60 years and below			
Significant medical illness	5	15	20
No significant medical illness	0	29	29
Patients over 60 years			
Significant medical illness	22	13	35
No significant medical illness	12	19	31
	39	76	1.15

Source: George Winokur and Donald Black, Psychiatric and medical diagnoses as risk factors for mortality in psychiatric patients: A case-control study. *Am. J. Psychiatr.* **144** (2): 210 (1987).

If this group is representative of the population of psychiatric patients who died of natural causes within 2 years of hospital discharge, what is the approximate probability that a randomly selected member of such a population:

(a)   Is over 60 years old?

(b)   Has an organic mental disorder?

(c)   Is over 60 years old and has an organic mental disorder?

(d)   Is not over 60 years old, or has no organic mental disorder, or both?

(e)   Has an organic mental disorder, given that the individual is over 60 years?

(f)   Died of natural causes?

(g)   Has no significant medical illness and no organic mental disorder?

5.73 A home economics class was developing a menu as part of a project in large-scale food management. A meal was to consist of a main order, a side order, a salad, and a drink. The following options were possible:

Main Order	Side Order	Salad	Drink
Beef burger	French fries	Vegetable	Pepsi
Fish fillet	Taters	Shrimp	Slice
Chicken nuggets	Hush puppies		Mountain Dew
			Iced tea
			Coffee

(a)   How many different meals are possible from this menu?

(b)   Suppose from all these possible meals, one is randomly chosen. What is the probability the meal will contain

(i)   Chicken nuggets?

(ii)   Shrimp salad?

(iii) Chicken nuggets and French fries?

5.74 Consider the following contingency table and calculate the required probabilities.

	A	B	C	Total
X	5	6	4	15
Y	11	18	6	35
Total	16	24	10	50

(a) $P(A)$

(b) $P(\bar{X})$

(c) $P(A \text{ or } C)$

(d) $P(A \text{ or } Y)$

(e) $P(A \text{ and } Y)$

(f) $P(X|B)$

(g) $P(B|X)$

(h) $P(A \text{ or } B \text{ or } C)$

(i) $P(X \text{ and } Y)$

(j) $P(\overline{A \text{ or } Y})$

(k) $P(\overline{A \text{ and } Y})$

(l) $P(\bar{A} \text{ or } \bar{Y})$

(m) $P(\bar{A} \text{ and } \bar{Y})$

5.75 In a particular high school, the probability that a senior decides to go to college after receiving counseling is .75. After the student decides to go to college, the probability that he or she will select Dartmouth College is .10. What is the probability that a high school senior receiving counseling will decide to go to college and select Dartmouth?

5.76 Acquired immune deficiency syndrome (AIDS) is being diagnosed in a variety of population types according to Martha Rogers and Walter Williams (1987), two researchers at the Centers for Disease Control in Atlanta. Consider the following two tables pertaining to women and children, which give the percentage of white, black, and Hispanic AIDS cases in each transmission group, as reported to the Centers for Disease Control for 1981–1986:

ADULT FEMALES WITH AIDS

Transmission Group	White	Black	Hispanic
Intravenous drug abuse	46	62	62
Hemophilia	1	0	0
Blood transfusion	31	4	3
Heterosexual contact	22	34	35
	100	100	100

CHILDREN WITH AIDS

Transmission Group	White	Black	Hispanic
Mother-to-infant	43	91	85
Hemophilia	21	2	2
Blood transfusion	36	7	13
	100	100	100

Source: Martha F. Rogers and Walter W. Williams, AIDS in blacks and Hispanics: Implications for prevention. *Issues in Science and Technology*, pp. 89–94, Spring 1987.

    (a) If a white adult female AIDS patient is randomly selected, what is the probability that:
        (i)   She is an intravenous drug abuser?
        (ii) She got the disease from a blood transfusion?
    (b) If an Hispanic adult female AIDS patient is randomly selected, what is the probability that she got the disease from heterosexual contact?
    (c) If a black child with AIDS is randomly selected, what is the probability that the child got the disease from the mother?
    (d) What is the probability that a child got the disease from a blood transfusion,
        (i)   If the child is white?
        (ii) If the child is black?
        (iii) If the child is Hispanic?

5.77 If $P(A) = .5$ and $P(B) = .8$, are the events A and B mutually exclusive?

5.78 Assume that events A and B are independent. Find the following probabilities where $P(A) = .33$ and $P(B) = .25$.
    (a) $P(A \text{ and } B)$
    (b) $P(A|B)$
    (c) $P(B|A)$
    (d) $P(A \text{ or } B)$
    (e) Are events A and B mutually exclusive?

5.79 Assume events A and B are mutually exclusive. Find the following probabilities where $P(A) = .4$ and $P(B) = .15$.
    (a) $P(A \text{ or } B)$
    (b) $P(A \text{ or } \bar{B})$
    (c) $P(\bar{A} \text{ or } \bar{B})$
    (d) $P(\bar{A} \text{ and } B)$
    (e) $P(A|B)$
    (f) $P(A|\bar{B})$

5.80 Consider the experiment in which a single die is tossed.
    (a) What is the probability that an even number occurs?
    (b) What is the probability that an even number occurs, given that the number is greater than three?
    (c) What is the probability that an even number occurs or that a number greater than three occurs?

5.81 An instructor has 40 questions from which she will draw to make up a 30-question test. How many different tests can the instructor design?

5.82 Using data collected by a telephone survey, a journalist has determined that the primary information source relied upon by adults in the city is distributed as follows: newspapers, one-third of adults; television, one-half of adults and radio, the remainder of adults. Of the newspaper readers, 55% are Republican; of the television viewers, 45% are Republican; and of the radio listeners, 60% are Republican. Suppose that one adult is randomly selected from this community.
    (a) What is the probability he or she is a Republican?

(b)   Given the adult is a Republican, what is the probability he or she relies on television as the primary news source?

(c)   What is the probability the adult is not a Republican?

(d)   Given the adult is a Republican, what is the probability he or she relies on newspapers or radio as the primary news source?

5.83   A sociologist studying the female population of the inner city found that 30% of them were married, 40% were never married, and 30% were divorced. It was also noted that one-fifth of the married women, one-eighth of the single never-married women, and one-third of the divorced women were receiving welfare payments from the government. Suppose one female is selected at random from the inner city.

(a)   What is the probability she is on welfare?

(b)   Given she is on welfare, what is the probability she is divorced?

(c)   What proportion of the inner city female population is not on welfare?

(d)   Given she is on welfare, what is the probability she is not married?

5.84   A student forgot the combination for his bike lock. The combination consists of a sequence of three numbers and each number can range from zero to nine. How many different sequences are possible? (*Note:* In practice, the second and third numbers cannot be the same.)

5.85   If $P(A|B) = .3$, $P(B) = .5$, and $P(A) = .4$, what is $P(A$ or $B)$?

5.86   (a)   A counselor has to meet with 8 clients during the day. She decides to see 5 clients before lunch. In how many different ways could she see these 5 clients?

(b)   For the 3 clients she meets after lunch, the order is also important. In how many ways could she see the three clients, *after* having seen 5 before lunch?

(c)   What is the total number of different ways she can see 8 clients during the day, subject to the above constraints?

(d)   What is the total number of different ways she can see 8 clients during the day if she is not subject to the above constraints?

5.87   At a drug rehabilitation center, 54% of the patients are under treatment for alcoholism, 26% for narcotic drug abuse, and 20% for other kinds of substance abuse. Each client was treated for only one substance abuse. Two-thirds of the alcoholics, half of the drug abusers, and two-fifths of the substance abusers are successful professionals in various fields.

(a)   If one patient is randomly selected from this center, what is the probability he or she is a successful professional?

(b)   What percentage of successful professionals at the center are alcoholics?

5.88   Consider the following set of birthdays: Lee Royal Jenney was born in Warren, Michigan, on January 7, 1882. His daughter, Florence Hume, was born in New York City on January 7, 1925. Her son, Craig Hume, was born in Boston on January 7, 1955. And finally, Craig's son, Chadwick Parker Hume, was born in Dallas, Texas, on—you guessed it—January 7, 1987! What are the odds of that— four generations born on the same day? Many family friends whipped out their pocket calculators to determine the probability of such a thing happening. According to Florence Hume, "It was a little amusing. Everyone came up with a different number." Finally, someone turned to James Andrews, assistant professor of mathematics at the University of Louisville, Kentucky, who got out his calculator and said, "One in 48,727,112, including leap years." He explained his computation as "1 over 365.25 cubed."

Source: The Dallas Morning News, January 30, 1987, p. 21A.

(a) Can you explain why Professor Andrews used the number 365.25?

(b) Do you agree with his answer, or has he left out something?

CASE STUDY

A Case of Consecutive Conception

Perhaps the title should say "cases." Read on and you will understand why. The following news item was reported in *The Dallas Morning News,* September 4, 1983, page 3A. (Used by permission of the Associated Press.)

---

3 Moms Defy Odds—All Have Twins

Johnson City, Tenn.—When three mothers gave birth consecutively early one morning at Johnson City Medical Center Hospital, the results were one in a million: six twins.

Dr. Fred C. Hartley, the obstetrician who supervised the deliveries, said the parents all knew they were having twins.

"There was no surprise about that. The surprise was that they all decided to have them at the same time," Hartley said.

The children—two sets of boys and one set of girls—were born between 2 A.M. and 7:30 A.M. Thursday to three Tennessee women in their 30s.

Specialists at the state Center for Health Statistics said that out of every completed pregnancy, the odds are 1 in 100 that the delivery will result in twins. That means the chances of three consecutive deliveries being twins are 1 in a million.

A fictional detective once said, "When you're on a case you eliminate the impossible, until you are left with the improbable." Now that you have studied probability theory, you should have a better "feel" for the nature of uncertainty, and can certainly appreciate the detective's distinction between the impossible and the improbable. The three consecutive deliveries of twins is a rare event, but not an impossible one, as you can see above.

Case Study Questions

1. Your task is to verify that the probability of one in a million reported above is correct. You may assume that the odds of 1 in 100 for an individual pregnancy are correct.

   (a) Which rule do you have to use to verify the probability?

   (b) The Center for Health Statistics reported the probability of 1/100 for an individual completed pregnancy resulting in twins based on historical data. What kind of probability is this: classical, subjective, or relative frequency?

   (c) Can you represent the 3 consecutive deliveries in a tree diagram, where one branch (T) represents twins and the other branch (N) represents nontwins?

   (d) Is it necessary to assume that each pregnancy is independent of every other pregnancy?

   (e) Try to represent the three deliveries using a Venn diagram, and shade the part that represents the probability of 3 pregnancies resulting in twins.

2. In the above case, the twins were born to different mothers. Would the probability change if we considered the 3 pairs of twins being born to the same mother, that is, the same woman getting pregnant three times in a row and giving birth to twins each time? (The odds of 1 in 100 for any individual pregnancy remain the same.)

3. (a) What is the probability that 5 women in a row would give birth to twins?

   (b) What is the probability that the same woman would give birth to twins 5 times in a row?

   (c) Let n = the number of women. Can you now generalize and state the rule for the probability $P(T_1$ and T_2 and T_3 and ... and $T_n)$, where T_1 stands for the first woman having twins, T_2 stands for the second woman, and so on?

COMPUTER EXERCISES USING THE DATABASE

Using a sample of 100 observations, complete the following exercises.

1. Treating your sample as a random sample from a much larger population, how would you define $P(F)$ where F is the event of selecting a female when selecting a student at random from the graduate student population at Brookhaven University? Use variable SEX. Which of the three probability definitions are you using?

2. Similar to exercise 1, what is $P(O)$, where O is the event that the person selected (male or female) is over 40 years of age? Use variable AGE.

3. Obtain a set of 10 random numbers from a uniform distribution between 1 and 100. Let these 10 values indicate the individuals to use in a random sample of size 10 selected from your set of 100 students. What is the average age of these 10 individuals using variable AGE? Is it equal to the average age of all 100 students? Why or why not?

4. Repeat exercise 3 using variable UG-GPA = undergraduate GPA.

**SPSSX
APPENDIX
(MAINFRAME
VERSION)**

**Random
Number
Generation**

You can use SPSSX to generate random numbers. The SPSSX program listing in Figure 5.14 was used to request the generation of 100 random numbers between 1 and 10,000.

The TITLE command names the SPSSX run.

The INPUT PROGRAM statement allows you to build your own subprograms either to input or to generate data.

The LOOP statement sets up a loop that is terminated by an END LOOP statement. In this example, we are looping 100 times to compute 100 different random numbers.

The RND(UNIFORM(10000)) statement generates a random number from a uniform distribution between 1 and 10,000, and rounds this value to the nearest integer.

The END CASE statement passes control of the loop to the END LOOP statement, which passes control to the loop until 100 random numbers have been generated.

The END FILE statement terminates the loop processing.

The END INPUT PROGRAM statement terminates the INPUT PROGRAM.

The PRINT statement sets up the 100 random numbers that were generated for printing.

The EXECUTE command causes the printing to occur.

Figure 5.14

SPSSX program listing requesting the generation of 100 random numbers between 1 and 10,000

```
TITLE   RANDOM NUMBERS
INPUT PROGRAM
LOOP #1=1 TO 100
COMPUTE RANNUM = RND(UNIFORM(10000))
END CASE
END LOOP
END FILE
END INPUT PROGRAM
PRINT  /RANNUM
EXECUTE
```

SPSSX APPENDIX (MAINFRAME VERSION)

Figure 5.15 shows the output obtained by executing the program listing in Figure 5.14.

Figure 5.15

SPSSX output obtained by executing the program listing in Figure 5.14

```
1396.00        5594.00
4313.00        9826.00
6122.00        3253.00
2908.00        5355.00
1557.00        5077.00
6995.00        2820.00
3463.00        3472.00
4456.00        9211.00
 524.00        4081.00
1032.00         957.00
1412.00        3339.00
 429.00        6208.00
6217.00        6411.00
1536.00        7521.00
7152.00        2050.00
9283.00        1433.00
5781.00        4459.00
2619.00        3691.00
7248.00        4853.00
 371.00         993.00
1006.00        7257.00
7326.00         373.00
2254.00        1747.00
2214.00        2884.00
6101.00        6689.00
9540.00        7612.00
2369.00        8031.00
3385.00        7029.00
 528.00         479.00
7383.00        3943.00
1198.00        2286.00
 589.00        7262.00
1880.00        8825.00
 996.00        1822.00
6759.00        2797.00
6927.00        6955.00
8668.00        9026.00
3512.00        1228.00
4964.00        5077.00
7020.00        1173.00
2060.00        9247.00
7734.00        5173.00
 669.00        2714.00
5271.00        6446.00
6618.00        6438.00
9358.00        8528.00
3157.00         652.00
5227.00        4068.00
7658.00        5667.00
1271.00        8433.00
```

**SPSS/PC+
APPENDIX
(MICRO
VERSION)**

**Random
Number
Generation**

You can use SPSS/PC+ to generate random numbers. The program listing in Figure 5.16 was used to request the generation of 100 random numbers between 1 and 10,000. All command lines in Figure 5.16 *must end in a period.*

The TITLE command names the SPSS/PC+ run.

The DATA LIST FREE / A, BEGIN DATA, the 100 values of 1, and the END DATA statements will provide 100 random numbers when the COMPUTE RANNUM = RND(UNIFORM(10000)) command is executed. These 100 values are from a uniform distribution between 1 and 10,000, and are rounded to the nearest integer. If say, 250 random numbers are desired, the only change necessary is to replace the 100 values of 1 with 250 such values.

The LIST RANNUM command will output the 100 random numbers.

Figure 5.16

SPSS/PC+ program
listing requesting
the generation of
100 random
numbers between
1 and 10,000

```
TITLE RANDOM NUMBERS.
DATA LIST FREE / A.
BEGIN DATA.
1 1 1 1 1 1 1 1 1 1 1 1 1 1 1 1 1 1 1 1 1 1 1 1 1
1 1 1 1 1 1 1 1 1 1 1 1 1 1 1 1 1 1 1 1 1 1 1 1 1
1 1 1 1 1 1 1 1 1 1 1 1 1 1 1 1 1 1 1 1 1 1 1 1 1
1 1 1 1 1 1 1 1 1 1 1 1 1 1 1 1 1 1 1 1 1 1 1 1 1
END DATA.
COMPUTE RANNUM = RND(UNIFORM(10000)).
LIST RANNUM.
```

SPSS/PC+ APPENDIX (MICRO VERSION)

Figure 5.17 shows the output obtained by executing the program listing in Figure 5.16.

RANNUM	RANNUM	RANNUM	RANNUM	RANNUM
2655.00	9872.00	5173.00	197.00	3211.00
3792.00	1948.00	3324.00	7663.00	1452.00
1480.00	4992.00	7247.00	986.00	5776.00
2816.00	6698.00	8443.00	5356.00	7344.00
9208.00	962.00	4572.00	1458.00	5933.00
7057.00	7784.00	4617.00	8588.00	5504.00
3985.00	615.00	1489.00	4465.00	56.00
5374.00	3437.00	8686.00	2012.00	6583.00
7914.00	9681.00	3898.00	7325.00	7162.00
429.00	7596.00	6830.00	5680.00	8975.00
8013.00	1392.00	3577.00	8908.00	6127.00
8931.00	137.00	3916.00	940.00	7413.00
4030.00	1806.00	4650.00	5429.00	425.00
4189.00	880.00	5599.00	7359.00	5208.00
4677.00	592.00	6941.00	2225.00	8593.00
5249.00	9263.00	5780.00	3275.00	7193.00
248.00	8757.00	5613.00	5039.00	
3483.00	4793.00	23.00	9699.00	
6576.00	2909.00	7901.00	5787.00	
7580.00	5387.00	5922.00	5745.00	
6548.00	4127.00	3146.00	3086.00	

Number of cases read =   100   Number of cases listed =   100

Figure 5.17

SPSS/PC+ output obtained by executing the program listing in Figure 5.16

SAS APPENDIX (MAINFRAME VERSION)

Random Number Generation

You can generate random numbers using SAS. The SAS program listing in Figure 5.18 was used to request the generation of 100 random numbers between 1 and 10,000.

The TITLE command names the SAS run.

The PROC MATRIX command allows you to build a table to hold the generated random numbers.

The X = UNIFORM (J(20, 5, 0)) command calls 100 random numbers between 0 and 1 and stores them in a table of 20 rows by 5 columns.

The X = X * 10000 statement scales the numbers up from a decimal number to a value between 1 and 10,000.

The PRINT X FORMAT = F6.0 statement establishes the size of the number to be printed and then prints the 20-by-5 table.

Figure 5.19 shows the output obtained by executing the program listing in Figure 5.18.

Figure 5.18

SAS program listing requesting the generation of 100 random numbers between 1 and 10,000

```
TITLE   'RANDOM NUMBERS';
PROC MATRIX;
X=UNIFORM(J(20,5,0));
X = X * 10000;
PRINT X FORMAT=F6.0;
```

Figure 5.19

SAS output obtained by executing the program listing in Figure 5.18

RANDOM NUMBERS

X	COL1	COL2	COL3	COL4	COL5
ROW1	3970	2905	5822	5538	6562
ROW2	6646	2481	4606	1946	9529
ROW3	5408	8779	5659	5167	8254
ROW4	8065	6826	6648	5708	3155
ROW5	1714	2646	8795	5482	9350
ROW6	8800	5911	3213	6987	1001
ROW7	785	2031	576	3667	5454
ROW8	912	232	3031	8230	2913
ROW9	704	5415	6666	945	6844
ROW10	519	5529	8563	4073	9695
ROW11	9765	5741	3856	9234	2471
ROW12	9653	4164	8642	7860	461
ROW13	4945	3927	9235	2087	8596
ROW14	5875	1951	8002	5427	3335
ROW15	1391	2004	4132	1498	5929
ROW16	3996	7259	5179	4900	776
ROW17	1674	7099	1573	1808	4037
ROW18	1164	1114	12	1264	750
ROW19	3557	7538	6001	1981	1773
ROW20	6554	7086	2010	4441	5020

SAS APPENDIX (MICRO VERSION)

Random Number Generation

You can generate random numbers using SAS. The SAS program listing in Figure 5.20 was used to request the generation of 100 random numbers between 1 and 10,000.

The TITLE command names the SAS run.

The DATA command gives the data a name.

The DO statement sets up a loop that is terminated by the END statement. In this example, we are looping 100 times to compute 100 different random numbers.

The $Y = UNIFORM(0) * 10000$ statement generates a random number from a uniform distribution between 1 and 10,000. The zero in parentheses acts as a seed for the random number generator. This value can be changed to obtain a different set of random numbers.

The $X = ROUND(Y)$ statement rounds the random number to the nearest integer.

The OUTPUT statement stores the X and Y values.

The END statement marks the end of the loop used to generate the random numbers.

The PROC PRINT command prints the (rounded) random numbers as specified by the VAR X statement.

The RUN command tells the SAS system to execute the previous SAS statements.

Figure 5.20

SAS program listing requesting the generation of 100 random numbers between 1 and 10,000

```
TITLE 'RANDOM NUMBERS';
DATA NEW;
DO I = 1 TO 100;
Y = UNIFORM(0) * 10000;
X = ROUND(Y);
OUTPUT;
END;
PROC PRINT;
   VAR X;
RUN;
```

SAS APPENDIX (MICRO VERSION)

Figure 5.21 shows the output obtained by executing the program listing in Figure 5.20.

Figure 5.21

SAS output obtained by executing the program listing in Figure 5.20

OBS	X	OBS	X	OBS	X	OBS	X
1	348	33	5959	65	65	97	4750
2	8338	34	6214	66	5482	98	1006
3	8829	35	8573	67	5466	99	2330
4	9855	36	6154	68	1742	100	7505
5	6751	37	1109	69	8186		
6	2028	38	4997	70	8859		
7	2684	39	1874	71	8833		
8	5214	40	5732	72	6961		
9	2510	41	727	73	1085		
10	7508	42	9298	74	3443		
11	6534	43	2452	75	1329		
12	7289	44	4414	76	5409		
13	1093	45	2130	77	3784		
14	7508	46	8185	78	5200		
15	3199	47	173	79	4596		
16	7742	48	81	80	9158		
17	8184	49	1333	81	4242		
18	2005	50	829	82	3198		
19	2766	51	808	83	5631		
20	315	52	1732	84	85		
21	8895	53	3435	85	2984		
22	7476	54	4829	86	1438		
23	8885	55	8017	87	1396		
24	4459	56	9852	88	6362		
25	3635	57	1327	89	1093		
26	2457	58	541	90	8416		
27	5149	59	5304	91	8687		
28	3710	60	8153	92	6982		
29	6870	61	3806	93	5263		
30	5144	62	5147	94	8921		
31	5304	63	4963	95	3703		
32	8278	64	8594	96	7974		

6

DISCRETE PROBABILITY DISTRIBUTIONS

A LOOK BACK/INTRODUCTION

The early chapters were concerned with describing sample data that had been gathered from a previous experiment, a printed report, or some other source. The data were summarized using one or more numerical measures (for example, a sample mean, variance, or correlation) or using a statistical graph (such as a histogram, bar chart, or scatter diagram).

Chapter 5 introduced you to methods of dealing with uncertainty by using a probability to measure the chance of a particular event occurring. Rules were defined that enable you to compute the various probabilities of interest, such as a conditional or a joint probability. However, so far we have defined only the probability of a certain event happening.

Whenever an experiment results in a numerical outcome, such as the total value of two dice, one can represent the various possible outcomes and their corresponding probabilities much more conveniently by using a **random variable**, the topic of this chapter. Suppose that a recent report indicates that 40% of the people applying for a marriage license in a large metropolitan city have been married previously. An excellent way of describing the chance that exactly 3 couples out of 20 who apply for a marriage license during a particular week have been married before is to use the concept of a random variable.

Random variables can be classified into two categories: discrete and continuous. This chapter will introduce both types but will concentrate on the discrete type. Several commonly used discrete random variables will be discussed, as will methods of describing and applying them.

> **Definition**
>
> A random variable is a function that assigns a numerical value to each outcome of an experiment.

In Chapter 3 we used various statistics (numerical measures) to describe a set of *sample* data. For example, the sample mean and standard deviation provide measures of a "typical" value and variation within the sample, respectively. Similarly, we will use a random variable and its corresponding distribution of probabilities to describe a *population*. Just as a sample has a mean and standard deviation, so does the population from which the sample was obtained. We will use the basic concepts from Chapter 5 to derive probabilities related to a random variable.

6.1   RANDOM VARIABLES

Discrete Random Variables

The probability laws developed in the previous chapter provide a framework for the discussion of random variables. We will still be concerned about the probability of a particular event; often, however, some aspect of the experiment can be easily represented using a random variable. The result of a simple experiment can sometimes be summarized concisely by defining a discrete random variable to describe the possible outcomes. Flip a coin three times. The possible outcomes for each flip are heads (H) and tails (T). According to counting rule 1 from Chapter 5, there are $2 \cdot 2 \cdot 2 = 8$ possible results from three flips. These are TTT, TTH, THT, HTT, HHT, HTH, THH, and HHH. Let

A = event of observing 0 heads in 3 flips (TTT)

B = event of observing 1 head in 3 flips (TTH, THT, HTT)

C = event of observing 2 heads in 3 flips (HHT, HTH, THH)

D = event of observing 3 heads in 3 flips (HHH)

We wish to find $P(A)$, $P(B)$, $P(C)$, and $P(D)$.

Consider one outcome, say, HTH. The coin flips are independent, so we use equation 5-14:

$P(\text{HTH})$ = (probability of H on 1st flip) $\cdot$ (probability of T on 2nd flip)

$\cdot$ (probability of H on 3rd flip)

$= (1/2) \cdot (1/2) \cdot (1/2) = 1/8$

The same argument applies to all eight outcomes. These outcomes are all equally likely, and each occurs with probability 1/8.

Event A occurs only if you observe TTT. This has the probability of occurring one time out of eight:

$$P(A) = 1/8$$

Event B will occur if you observe HTT, TTH, or THT. It would be impossible for HTT and TTH both to occur, so $P(HTT \text{ and } TTH) = 0$. This is true for any combination here, so these three events are all mutually exclusive. Consequently, according to equation 5-13,

$$P(B) = P(HTT \text{ or } TTH \text{ or } THT)$$
$$= P(HTT) + P(TTH) + P(THT)$$
$$= 1/8 + 1/8 + 1/8$$
$$= 3/8$$

By a similar argument,

$$P(C) = 3/8 \qquad \text{(using HHT, HTH, THH)}$$

$$P(D) = 1/8 \qquad \text{(using HHH)}$$

The variable of interest in this example is X, defined as

$$X = \text{number of heads out of three flips}$$

We defined all of the possible outcomes of X by defining the four events A, B, C, and D. This works but is cumbersome. Consider having to do this for 100 flips of a coin! A more convenient way to represent probabilities is to examine the value of X for each possible outcome:

Outcome	Value of X	
TTT	0	1 outcome
THT	1	
TTH	1	3 outcomes
HTT	1	
HHT	2	
HTH	2	3 outcomes
THH	2	
HHH	3	1 outcome

Each outcome has probability 1/8, so the probability that X will be 0 is 1/8, written

$$P(X = 0) = P(0) = 1/8$$

The probability that X will be 1 is 3/8, written

$$P(X = 1) = P(1) = 3/8$$

The probability that X will be 2 is 3/8, written

$$P(X = 2) = P(2) = 3/8$$

The probability that X will be 3 is 1/8, written

$$P(X = 3) = P(3) = 1/8$$

Notice that

$$P(X = 0) + P(X = 1) + P(X = 2) + P(X = 3) = 1/8 + 3/8 + 3/8 + 1/8$$
$$= 1$$

because 0, 1, 2, and 3 represent all the possible values of X.

The values and probabilities for this random variable can be summarized by listing each value and its probability of occurring.

$$X = \begin{cases} 0 & \text{with probability } 1/8 \\ 1 & \text{with probability } 3/8 \\ 2 & \text{with probability } 3/8 \\ 3 & \text{with probability } 1/8 \end{cases}$$

This list of possible values of X and the corresponding probabilities is a **probability distribution**.

In any such formulation of a problem, the variable X is a **random variable**. Its value is not known in advance, but there is a probability associated with each possible value of X. Whenever you have a random variable of the form

$$X = \begin{cases} x_1 & \text{with probability } p_1 \\ x_2 & \text{with probability } p_2 \\ x_3 & \text{with probability } p_3 \\ \vdots \\ x_n & \text{with probability } p_n \end{cases}$$

where $x_1, \ldots, x_n$ is the set of possible values of X, then X is a **discrete random variable**. In the coin-flipping example, $x_1 = 0$ and $p_1 = 1/8$; $x_2 = 1$ and $p_2 = 3/8$; $x_3 = 2$ and $p_3 = 3/8$, and $x_4 = 3$ and $p_4 = 1/8$.

Other examples of a discrete random variable include:

X = the number of space shuttle flights (out of 25) that result in no fatalities.

X = the number of people, out of a group of 50, who will complete a Ph.D. degree.

X = the number of people, out of 200, who make an airline reservation and then fail to show up.

X = the number of students in the Glencoe, Nebraska, freshman high school class who will drop out of school.

Notice that for each example the discrete random variable is a count of the number of people, accidents, and so on that can occur.

EXAMPLE 6.1 You roll two dice, a red die and a blue die. What is a possible random variable X for this situation? What are its possible values and corresponding probabilities? (*Hint:* Roll the dice and observe a particular number. This number is your value of the random variable X. What observations are possible from the roll of two dice?)

Solution There are many possibilities here, including

X = total of the two dice

X = the higher of the two numbers that appear (possible values: 1, 2, 3, 4, 5, 6)

X = the number of dice with 3 appearing (possible values: 0, 1, 2)

Suppose that the random variable X equals the total of the two dice. The next step is to determine the possible values of X and the corresponding probabilities. When you roll the two colored dice, there are $6 \cdot 6 = 36$ possible outcomes, using counting rule 1 from Chapter 5.

Outcome	Red Die	Blue Die	Value of X
1	1	1	2
2	1	2	3
3	1	3	4
4	1	4	5
5	1	5	6
6	1	6	7
7	2	1	3
8	2	2	4
9	2	3	5
⋮	⋮	⋮	⋮
34	6	4	10
35	6	5	11
36	6	6	12

$P(X = 3) = 2/36$

The 36 outcomes are equally likely because the number appearing on each die (1, 2, 3, 4, 5, or 6) has the same chance of appearing. Notice that we are not saying that each value of X is equally likely, as the following discussion will make clear. Each of the above 36 outcomes has probability 1/36 of occurring. If you write down all 36 outcomes and note what can happen to X, your random variable, you will observe:

Value of X	Number of Possible Outcomes
2	1 (rolling a 1, 1)
3	2 (rolling a 1, 2 or 2, 1)
4	3 (rolling a 1, 3 or 3, 1 or 2, 2)
5	4 (and so on)
6	5
7	6
8	5
9	4
10	3
11	2
12	1

Consequently,

$$X = \begin{cases} 2 & \text{with probability } 1/36 \\ 3 & \text{with probability } 2/36 \\ 4 & \text{with probability } 3/36 \\ 5 & \text{with probability } 4/36 \\ 6 & \text{with probability } 5/36 \\ 7 & \text{with probability } 6/36 \\ 8 & \text{with probability } 5/36 \\ 9 & \text{with probability } 4/36 \\ 10 & \text{with probability } 3/36 \\ 11 & \text{with probability } 2/36 \\ 12 & \text{with probability } \underline{1/36} \\ & \phantom{\text{with probability }} 1.0 \end{cases}$$

Because 2 through 12 represent all possible values of X, the total of all probabilities is equal to one.   □

Continuous Random Variables

The previous section introduced you to the discrete random variable where the possible values of X can be listed along with corresponding probabilities. Characteristic of this type of random variable is the presence of gaps in the list of possible values. For example, when throwing two dice, a total of 8.5 cannot occur.

The other type of random variable is the **continuous random variable**, for which any value is possible over some range of values. For a random variable of this type, there are no gaps in the set of possible values. As a simple example, consider two random variables: X is the number of days that it rained in Boston during any particular month, and Y is the amount of rainfall during this month.

Figure 6.1

Example of a
continuous random
variable

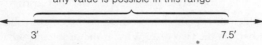

X = height of a randomly selected
adult male in the United States

any value is possible in this range

3'   7.5'

X is a discrete random variable, because it counts the number of days, and consequently there are gaps in the possible values (7.4, for example, is not possible). Y, on the other hand, is a continuous random variable because at least in principle, the amount of rainfall could be any nonnegative value.

Suppose the heights of all adult males in the United States range from 3' to 7.5'. Your task is to describe these heights using such statements as

"15% of the heights are under 5.5'."

"88% of the heights are between 5' and 6'."

We first define the random variable:

X = height of a randomly selected adult male in the United States.

Figure 6.1 shows the range of X.

We are unable to list all of the possible values of X since any height is possible over this range. However, we can still discuss probabilities associated with X. For example, the preceding two statements can be described by using the probability statements

$$P(X < 5.5') = .15$$

$$P(X \text{ is between } 5' \text{ and } 6') = P(5' < X < 6') = .88$$

For this situation, X is a continuous random variable. Probabilities for continuous random variables can be found only for intervals. (Probabilities of exact values are meaningful only for discrete random variables.) Determining probabilities for a continuous random variable will be discussed in Chapter 7.

In Chapter 1, we discussed discrete and continuous data. They are directly related to our present discussion. When you observe a discrete random variable, you obtain discrete data. When you observe a continuous random variable (such as 100 heights), you obtain continuous data.

EXERCISES

6.1 A researcher is investigating the gain in weight (measured as the difference between pre- and post-experimental weights) experienced by rats subject to a special electromechanical stimulus. What random variable(s) would the researcher be interested in? Would the variable(s) be discrete or continuous?

6.2 If a student is taking a multiple-choice test with 10 questions and is interested in the final score, what random variable would be of interest to the student?

6.3 Classify the following random variables as discrete or continuous.
(a) The number of pages in a psychology textbook.
(b) The level of finger dexterity measured on an ordinal scale of 1 to 10.

(c)   The change in body temperature caused by emotional stress.

(d)   The area in square feet of apartments in a ghetto.

(e)   The number of residents in a school district who are opposed to busing.

(f)   The percentage of residents in a school district who are in favor of busing.

6.4   Consider an experiment in which two dice are rolled. Let X be the total of the numbers on the two dice. What is the probability that X is equal to two or four?

6.5   Consider an experiment in which a coin is tossed and a die is rolled. Let X be the number observed from rolling the die. Let Y be the value one if a head appears and zero if a tail appears. List the values that the random variables X and Y can have, along with the corresponding probabilities.

6.6   Raw scores on an achievement test are converted to the stanine scale (levels 1 to 9). Level 5 has 22% of the candidates, levels 4 and 6 have 15% each, and the other levels have 8% each. List all possible outcomes. Let X be the stanine level. What is the value of $P(X = 7)$? of $P(X < 7)$?

6.7   If the random variable X can take on the value of two, three, four, and five with equal probability, what is the probability that X is equal to 3? Assume that X cannot take on any other values.

6.8   Suppose that in answer to the question "Do you approve of the President's performance?" 58% said "Yes," 40% said "No," and 2% had no opinion. Let X take on the values 0, 1, or 2 to represent "Yes," "No," or "No opinion," respectively. What is the value of $P(X = 0)$?

6.2   REPRESENTING PROBABILITY DISTRIBUTIONS FOR DISCRETE RANDOM VARIABLES

There are three popular methods of describing the probabilities associated with a discrete random variable X. They are

List each value of X and its corresponding probability.

Use a bar chart to convey the probabilities corresponding to the various values of X.

Use a function that assigns a probability to each value of X.

Remember our coin-flipping example, in which $X =$ number of heads in three flips of a coin. We can list each value and probability:

$$X = \begin{cases} 0 & \text{with probability } 1/8 \\ 1 & \text{with probability } 3/8 \\ 2 & \text{with probability } 3/8 \\ 3 & \text{with probability } 1/8 \end{cases}$$

This works well when there are only a small number of possible values for X; it would not work well for 100 flips of a coin.

Figure 6.2

A bar chart representation of a discrete random variable

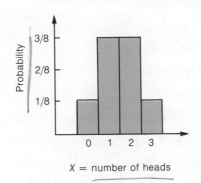

X = number of heads

Using a bar chart is also a convenient way to represent the shape of a discrete distribution having a small number of possible values. For this situation, you construct a bar chart in which the height of each bar is the probability of observing that value of X (Figure 6.2). It is easier to determine the shape of the probability distribution by using such a chart. The distribution in Figure 6.2 is clearly symmetric and concentrated in the middle values.

Using a function (that is, an algebraic formula) to assign probabilities is the most convenient method of describing the probability distribution for a discrete random variable. For any given application of such a random variable, however, this function may or may not be known. Later in the chapter we will identify certain useful discrete random variables, each of which has a corresponding function that assigns these probabilities.

The function that assigns a probability to each value of X is called a **probability mass function (PMF)**. Denoting a particular value of X as x, this function is of the form

$$P(X = x) = \text{some expression (usually containing } x\text{) that produces the}$$
$$\text{probability of observing } x$$
$$= P(x)$$

Not every function can serve as a PMF. The requirements for a PMF are:

1 $P(x)$ is between 0 and 1 (inclusively) for each x
2 $\Sigma P(x) = 1$

EXAMPLE 6.2

Consider a random variable X having possible values of 1, 2, or 3. The corresponding probability for each value is:

$$X = \begin{cases} 1 & \text{with probability } 1/6 \\ 2 & \text{with probability } 1/3 \\ 3 & \text{with probability } 1/2 \end{cases}$$

Determine an expression for the PMF.

Solution   Consider the function

$$P(X = x) = P(x) = x/6 \qquad \text{for } x = 1, 2, 3.$$

This function provides the probabilities

$$P(X = 1) = P(1) = 1/6 \quad \text{(OK)}$$

$$P(X = 2) = P(2) = 2/6 = 1/3 \quad \text{(OK)}$$

$$P(X = 3) = P(3) = 3/6 = 1/2 \quad \text{(OK)}$$

This function satisfies the requirements for a PMF: Each probability is between 0 and 1, and $P(1) + P(2) + P(3) = 1/6 + 1/3 + 1/2 = 6/6 = 1$. Consequently, the function

$$P(x) = x/6 \qquad \text{for } x = 1, 2, 3 \quad \text{(and zero otherwise)}$$

is the PMF for this discrete random variable.   □

EXAMPLE 6.3   Consider Example 6.1, where X is the total of two dice. Determine the PMF for this discrete random variable.

Solution   Consider the expression

$$P(x) = \frac{x - 1}{36} \qquad \text{for } x = 2, 3, 4, \ldots, 12 \quad \text{(and zero otherwise)}$$

If this is the proper PMF, then, for example,

$$P(2) = P(X = 2) = \frac{2 - 1}{36} = 1/36$$

This does appear to be correct, so far. Also,

$$P(5) = P(X = 5) = \frac{5 - 1}{36} = 4/36$$

This also is correct. But now consider

$$P(10) = P(X = 10) = \frac{10 - 1}{36} = 9/36$$

According to our previous solution, we know that $P(10) = 3/36$, not $9/36$. So this particular function is not the PMF for this random variable; the PMF must work for all values of X.

Consider the expression

$$P(x) = \frac{6 - |x - 7|}{36} \qquad \text{for } x = 2, 3, \ldots, 12 \quad \text{(and zero otherwise)}$$

where $|\cdot|$ represents the absolute value of a number. See if you can demonstrate that this function is a bona fide PMF for this example (it is). Do not worry about where this expression came from, but do verify that it works. The truth of the matter is that often PMFs are derived by trial and error.

$$P(a \le x \le b) = P(x \le b) - P(x \le a - 1)$$

Notice that a probability mass function provides a theoretical "model" of the population by describing the chances of observing any particular value of the random variable. You can view the population as what you would obtain if you observed the corresponding random variable indefinitely.   □

EXERCISES

6.9 Is the following function a probability mass function? Why or why not?

$$P(X = x) = x/6 \quad \text{for } x = 0, 2, 4 \quad \text{(and zero otherwise)}$$

6.10 Suppose a test consists of three true–false questions. A student decides to answer each question by tossing a coin, such that heads means true and tails means false. Assume that the classical probability definition will hold for the coin tossing, with $P(\text{Head}) = 0.50$ and $P(\text{Tail}) = 0.50$. Each throw is independent, so the student has a 50% chance of being right on each question.

(a) Draw a tree diagram for the student answering the test questions, labeling the branches right or wrong as appropriate. Compute the probabilities for RRR and RRW and so on (R = right, W = wrong).

(b) Let X be the number of questions answered right. What is the probability mass function of X?

(c) Try to visualize the tree diagram if the test were extended to 20 or more true–false questions. (If that seems a bit much, you will appreciate the binomial distribution in Section 6.5, later.)

6.11 A jar contains four red marbles and three green marbles. Consider an experiment in which two marbles are chosen at random with replacement. Let X be the number of red marbles chosen. What is the probability distribution of X?

6.12 Is the following function a probability mass function? Why or why not?

$$P(X = x) = (x - 2)/6 \quad \text{for } x = 1, 4, 7 \quad \text{(and zero otherwise)}$$

6.13 Is the following function a probability mass function? Why or why not?

$$P(X = x) = x^2/10 \quad \text{for } x = -2, -1, 1, 2 \quad \text{(and zero otherwise)}$$

6.14 Pierson and Shorter (1987) conducted a survey of business schools to determine the hardware and software used by students enrolled in introductory database courses. They found a greater prevalence of relational Data Base Management Systems (DBMS) as opposed to network and hierarchical DBMS. Among other things, they reported the following distribution of DBMS usage:

NUMBER OF DBMS PER SCHOOL USED FOR INSTRUCTIONAL PURPOSES

X = Number of DBMS	Number of Schools	Percentage of Schools
1	34	42.0
2	22	27.1
3	17	21.0
4	5	6.2
5	3	3.7
		100.00

Source: J. K. Pierson and J. D. Shorter, Study reveals emphasis of relational DBMS in introductory data base courses. *Decision Sciences: Theory and Applications;* Proceedings of the Eighteenth Annual Conference of the Southwest Region Decision Sciences Institute, Houston, Texas, March 10–14, 1987.

(a)   What is the probability that any randomly selected school uses more than 2 DBMS packages for instructional purposes?

(b)   Compute the mean and variance of the above discrete distribution. (Refer to the next section, if necessary.)

6.15   A sociologist claims that the distribution of children in Hispanic families is described by the probability function

$$P(X = x) = 0.20 \qquad \text{for } x = 0, 1, 2, 3, 4$$

where X is the number of children in the family.

(a)   What is the probability that a randomly selected Hispanic family will have two children?

(b)   If three Hispanic families are randomly selected, and each selection is independent, what is the probability that two families will have two children each and one family will have one child?

6.16   The same sociologist in exercise 6.15 also claims that the distribution of children in Asian families is given by

$$P(X = x) = 0.25 \qquad \text{for } x = 1, 2, 3, 4 \quad \text{(and zero otherwise)}$$

where X is the number of children in the family. If one Hispanic and one Asian family are randomly selected, each selection being independent, what is the probability that the two families will each have two children?

6.17   Sumpton, Raynes, and Thorp (1987), researchers at the University of Manchester, Manchester, England, conducted a follow-up to an earlier 1971 study by other researchers on patterns of residential care for mentally handicapped people in Great Britain. Among other things, they reported the following distribution of the number of moves, from one place of residence to another, made by people as a function of their type of original residence.

	TYPE OF ORIGINAL RESIDENCE			
X = No. of Moves	Voluntary Home	Hospital	Local Authority Hostel	Total
0	35 (68.6%)	96 (64.4%)	1 (0.7%)	132 (38.2%)
1	13 (25.5%)	35 (23.5%)	107 (73.3%)	155 (44.8%)
2	2 (3.9%)	15 (10.1%)	24 (16.4%)	41 (11.8%)
3	1 (2.0%)	3 (2.0%)	12 (8.2%)	16 (4.6%)
4	0 (0%)	0 (0%)	2 (1.4%)	2 (0.6%)
	51 (100%)	149 (100%)	146 (100%)	346 (100%)

Source: R. C. Sumpton, N. V. Raynes, and D. Thorp. The residential careers of a group of mentally handicapped people: The influence of early residential experience. *Br. J. Mental Subnormality* **23** (Part 1, No. 64): 3–9 (1987).

If we let X = number of moves be a discrete random variable, we have above four discrete distributions, three based on the original residence, and one for the overall group.

(a)   If an individual's original residence was a local authority hostel, what is the probability that the individual made more than 2 moves?

(b)   For an individual originally resident in a voluntary home, what is the probability of making 2 or 3 moves?

(c)   For the group as a whole, what is the probability that an individual made only one move, or did not move at all?

6.18   Suppose a probability mass function is defined to be nonzero at three points, $X = 1, 2$, and 3. If $P(X = 1) = .2$ and $P(X = 2) = .3$, what is $P(X = 3)$?

6.19   There are six marbles in a jar. One marble is white, two are blue, and three are black. A marble is drawn at random. Let the random variable X be equal to one if white is drawn, two if blue is drawn, and three if black is drawn. Verify that the following function is the probability mass function of X.

$$P(X = x) = x/6 \qquad \text{if } x = 1, 2, 3 \quad \text{(and zero otherwise)}$$

6.3   MEAN AND VARIANCE OF DISCRETE RANDOM VARIABLES

Mean of Discrete Random Variables

Chapter 3 introduced you to the mean and variance of a set of sample data consisting of n values. Suppose that these values were obtained by observing a particular random variable n times. The sample mean $\bar{X}$ represents the average value of the *sample data*. In this section, we determine a similar value, the **mean of a discrete random variable**, written as μ. The value of μ represents the average value of the random variable if you were to observe this variable over an indefinite period of time. Thus μ describes the population and so is referred to as a population *parameter*.

Reconsider our coin-flipping example, where X is the number of heads in three flips of a coin. Suppose you flip the coin 3 times, record the value of X, flip the coin three times again, record the value of X, and repeat this process 10 times. Now you have 10 observations of X. Suppose they are

$$2, 1, 1, 0, 2, 3, 2, 1, 1, 3$$

The mean of these data is the *statistic*, $\bar{X}$, defined by

$$\bar{X} = \frac{2 + 1 + 1 + \cdots + 1 + 3}{10}$$

$$= 1.6 \text{ heads}$$

If you observed X indefinitely, what would X be on the average?

$$X = \begin{cases} 0 & \text{with probability } 1/8 \\ 1 & \text{with probability } 3/8 \\ 2 & \text{with probability } 3/8 \\ 3 & \text{with probability } 1/8 \end{cases}$$

So you should observe the value 0, 1/8 of the time, the value 1, 3/8 of the time, the value 2, 3/8 of the time, and the value 3, 1/8 of the time. In a sense, each probability represents the relative frequency for that particular value of X. So the average value of X is

$$\mu = (0)(1/8) + (1)(3/8) + (2)(3/8) + (3)(1/8) = 1.5 \text{ heads}$$

Notice that X cannot be 1.5; this is merely the value of X on the average.

Definition

The average value of the discrete random variable X (if observed indefinitely) is the mean of X. The symbol for this parameter is μ.

We found that $\mu = 1.5$ by multiplying each value of X by its corresponding probability and summing the results:

$$\mu = 1.5 = 0 \cdot P(0) + 1 \cdot P(1) + 2 \cdot P(2) + 3 \cdot P(3)$$

This procedure applies to any discrete random variable, and so we define

$$\mu = \Sigma x P(x) \qquad\qquad \textbf{(6-1)}$$

EXAMPLE 6.4

In example 5.12, mothers of children attending Truman Elementary School were categorized as married, divorced, or never-married. To take another look at the marital background of these mothers, the school principal decided to define a random variable equal to the *number* of marriages for the mother of each child attending the school. He defined this random variable to be

$$X = \text{number of marriages}$$

Based on school records, he has derived the following probability distribution for X:

$$X = \begin{cases} 0 & \text{with probability .05} \\ 1 & \text{with probability .45} \\ 2 & \text{with probability .35} \\ 3 & \text{with probability .10} \\ 4 & \text{with probability .05} \end{cases}$$

Five percent of the mothers have never been married, 45% of the mothers have been married only once, and so on. (Notice that deriving an algebraic expression for the PMF for this distribution would be extremely difficult, if not impossible. This poses no problem, however.)

What is the mean (average value) of X?

Solution   Using equation 6-1,

$$\mu = (0)(.05) + (1)(.45) + (2)(.35) + (3)(.1) + (4)(.05)$$
$$= 1.65$$

The average number of marriages for these mothers is 1.65 marriages.   □

Variance of Discrete Random Variables

We previously considered 10 observations of the random variable that counted the number of heads in three flips of a coin. These data were: 2, 1, 1, 0, 2, 3, 2, 1, 1, 3. We used the notation from Chapter 3 to define the mean of these data, and we obtained $\bar{X} = 1.6$. The variance of these data, using equation 3-8, is $s^2 = .933$. Since s^2 describes a sample, it is a statistic.

Once again, consider observing X indefinitely. For this situation, the average value of X is defined as the mean of X, μ. When we observe X indefinitely, this particular variance is defined to be the variance of the random variable X and is written σ^2 (read as "sigma squared").

$$\sigma^2 = \text{variance of the discrete random variable } X$$

The **variance of a discrete random variable**, X, is a parameter describing the corresponding population and can be obtained by using one of the following expressions, which are mathematically equivalent:

$$\sigma^2 = \Sigma(x - \mu)^2 \cdot P(x) \qquad \textbf{(6-2)}$$
$$\sigma^2 = \Sigma x^2 P(x) - \mu^2 \qquad \textbf{(6-3)}$$

Equation 6-3 generally provides an easier method of determining the variance and will be used in all of the examples to follow. For the coin-flipping example,

$$\sigma^2 = \Sigma x^2 P(x) - \mu^2$$
$$= (0)^2 \cdot 1/8 + (1)^2 \cdot 3/8 + (2)^2 \cdot 3/8 + (3)^2 \cdot 1/8 - (1.5)^2$$
$$= 3 - 2.25 = .75$$

So our final results would be:

Using the Sample of 10 Observations		For the Random Variable X (indefinite number of observations)	
$\bar{X} = 1.6$	$s^2 = .933$	$\mu = 1.5$	$\sigma^2 = .75$
mean	variance	mean	variance
Statistics		Parameters	

In Chapter 3, the square root of the variance, s, was defined to be the standard deviation of the data. The same definition applies to a random variable.

The **standard deviation of a discrete random variable** X is denoted σ, where

$$\sigma = \sqrt{\Sigma(x - \mu)^2 \cdot P(x)} \qquad (6\text{-}4)$$

$$\sigma = \sqrt{\Sigma x^2 P(x) - \mu^2} \qquad (6\text{-}5)$$

EXAMPLE 6.5   Determine the variance and standard deviation of the random variable described in example 6.4 concerning the number of marriages for mothers of children attending Truman Elementary School.

Solution   A convenient method of determining both the mean and variance of a discrete random variable is to summarize the calculations in tabular form:

x	$P(x)$	$x \cdot P(x)$	$x^2 \cdot P(x)$
0	.05	0	0
1	.45	.45	.45
2	.35	.7	1.4
3	.1	.3	.9
4	.05	.2	.8
	1.00	1.65	3.55

So

$$\mu = \Sigma x P(x) = 1.65 \text{ marriages}$$

and

$$\sigma^2 = \Sigma x^2 P(x) - \mu^2 = 3.55 - (1.65)^2$$
$$= .8275$$

Also

$$\sigma = \sqrt{.8275} \cong .91 \text{ marriage} \qquad \square$$

EXERCISES   6.20 From the data in exercise 6.17, calculate the mean and standard deviation for the number of moves made:
(a)   By those originally resident in voluntary homes
(b)   By those originally resident in hospitals
(c)   By those originally resident in local authority hostels
(d)   By the group as a whole

6.21 Several students in a finance class subscribe to the *Wall Street Journal*. If two students are chosen at random, the probability of choosing no students who subscribe is .81. The probability of choosing one student who subscribes and one who does not subscribe is .18, and the probability of choosing two students who subscribe is .01. X is a random variable equal to the number of students who subscribe out of the two chosen at random. Find the mean value of X and the variance of X.

6.22 Find the mean and variance of a random variable X, the probability mass function of which is as follows:

X	P(x)
−2	.12
−1	.3
0	.1
1	.3
2	.18

6.23 Suppose that a coin is flipped three times. Define the random variable X to be equal to twice the number of heads that appear. Determine the mean and variance of X.

6.24 Show that $\Sigma(x - \mu)P(x) = 0$, where the summation is over all outcomes of X, for any discrete random variable.

6.25 Determine the mean and standard deviation of the random variable defined in exercise 6.11.

6.26 Determine the mean and standard deviation of the random variable, the probability mass function of which is defined as follows:

$$P(X = x) = (x - 2)/30 \quad \text{for } x = 3, 12, 21 \quad \text{(and zero otherwise)}$$

6.27 Determine the mean and standard deviation of the random variable, the probability mass function of which is defined as follows:

$$P(X = x) = 1/5 \quad \text{for } x = 1, 2, 3, 4, 5 \quad \text{(and zero otherwise)}$$

6.4   DISCRETE UNIFORM RANDOM VARIABLES

A random variable X follows a discrete uniform distribution if X is discrete and each possible value of X has the same probability of occurring. As a simple illustration, consider the roll of a single die and define X to be the number that appears. Since the six possible values are equally likely to occur, we have a **uniform random variable** defined by

$$X = \begin{cases} 1 & \text{with probability } 1/6 \\ 2 & \text{with probability } 1/6 \\ 3 & \text{with probability } 1/6 \\ 4 & \text{with probability } 1/6 \\ 5 & \text{with probability } 1/6 \\ 6 & \text{with probability } 1/6 \end{cases}$$

What are the chances of rolling a value of 4? This would be

$$P(X = 4) = P(4) = 1/6$$

Consequently, 16.7% of the time you would expect to roll a 4. What are the chances of rolling a value at least 3? This can be written

$$P(X \geqslant 3) = P(3) + P(4) + P(5) + P(6)$$
$$= 1/6 + 1/6 + 1/6 + 1/6$$
$$= 4/6 = 2/3$$

Thus, 66.7% of the time, the value appearing will be 3 or more.

In general, for a discrete uniform random variable,

$$X = \begin{cases} x_1 & \text{with probability } 1/n \\ x_2 & \text{with probability } 1/n \\ \vdots \\ x_n & \text{with probability } 1/n \end{cases}$$

and so

$$P(x) = 1/n \text{ for } x = x_1, x_2, \ldots, x_n.$$

For the die illustration,

$$P(x) = 1/6 \text{ for } x = 1, 2, \ldots, 6.$$

EXAMPLE 6.6

A recent city ordinance has forced local restaurant owners to strictly monitor and regulate the temperature of their kitchens as a safety precaution. The owner of Kozy Korner Restaurant installed a digital thermometer in his kitchen. During the summer months he made over 150 spot checks using the same thermometer setting he has always used during these months. The results indicated that

$X =$ kitchen temperature showing on the thermometer at Kozy Korner during the summer months

follows a discrete uniform distribution defined by (values in degrees Fahrenheit)

$$X = \begin{cases} 68 & \text{with probability } 1/5 \\ 69 & \text{with probability } 1/5 \\ 70 & \text{with probability } 1/5 \\ 71 & \text{with probability } 1/5 \\ 72 & \text{with probability } 1/5 \end{cases}$$

(i)   What is the probability that a spot check in the summer months will reveal a temperature under 70°F?

(ii) What percentage of the time will the restaurant kitchen thermometer show a temperature of at least 70°F during the summer months?

Solution (i) The temperature reading will be under 70°F if it is 68 or 69°F and so

$$P(\text{temperature reading under } 70) = P(68) + P(69)$$
$$= 1/5 + 1/5 = 2/5$$

Consequently, this will occur 40% of the time.

Solution (ii) The kitchen will have a temperature reading of at least 70°F if the temperature is 70, 71, or 72°F. The chance of this occurring is

$$P(X \geqslant 70) = P(70) + P(71) + P(72)$$
$$= 1/5 + 1/5 + 1/5 = 3/5$$

Thus the temperature reading is at least 70°F 60% of the time. □

Mean and Variance of Uniform Random Variables

In example 6.6, we defined a discrete random variable X = the temperature reading (using a digital thermometer providing discrete values) in the kitchen at Kozy Korner Restaurant. What will be the average value of this variable if it is observed indefinitely? This is the mean of X and using equation 6-1, this will be

$$\mu = \Sigma x P(x)$$
$$= (68)(1/5) + (69)(1/5) + (70)(1/5) + (71)(1/5) + (72)(1/5)$$
$$= 70$$

This means that the temperature on the average will be 70°F.

The variance of X can be found using equation 6-3, and is

$$\sigma^2 = \Sigma x^2 P(x) - \mu^2$$
$$= [(68)^2(1/5) + (69)^2(1/5) + (70)^2(1/5)$$
$$+ (71)^2(1/5) + 72^2(1/5)] - (70)^2$$
$$= 4902 - 4900 = 2$$

If possible values of the discrete uniform random variable are **sequential counting numbers** (as in the die example and the Kozy Korner example), it can be shown that

$$\mu = \frac{a + b}{2} \tag{6-6}$$

and

$$\sigma^2 = \frac{n^2 - 1}{12} \tag{6-7}$$

where (1) n is the number of possible values of the random variable X, (2) a is the minimum possible value of X, and (3) b is the maximum possible value. For situations where the possible values of X are not sequential counting numbers (such as 1.5, 2, 2.5, 3 or 3, 6, 9, 12, 15), you can still use equations 6-1 and 6-3 to determine the mean and variance.

To illustrate equations 6-6, and 6-7, for the Kozy Korner Restaurant example there are $n = 5$ possible values of X (namely 68, 69, 70, 71, 72), each having the same probability of occurring. The minimum possible value is $a = 68$ and the maximum value is $b = 72$. Thus

$$\mu = \frac{68 + 72}{2} = 70 \qquad \text{(as before)}$$

and

$$\sigma^2 = \frac{5^2 - 1}{12} = \frac{24}{12} = 2 \qquad \text{(as before)}$$

Finally for this example, the standard deviation of the random variable X is

$$\sigma = \sqrt{2} = 1.414 \; (°F)$$

EXERCISES

6.28 Suppose that the level of finger dexterity in a hypothetical population measured on an ordinal scale of 1 to 10 is a uniformly distributed random variable.
  (a) If one individual is randomly selected from the population, what is the probability that the person's dexterity level will be below 6?
  (b) What percentage of the population have a dexterity level exceeding 7?
  (c) What is the average dexterity level?
  (d) What is the variance of the dexterity level?

←**6.29** A social worker has found that, depending on the complexity of the case, she can handle from 2 to 7 cases of child abuse per week. Over a long period of time, she has also noted that the number of such cases per week follows a uniform distribution between 2 and 7.
  (a) What is the probability that in any given week, the number of cases she has to handle will be more than 3?
  (b) What is the average number of cases per week?

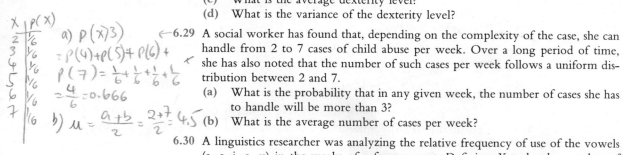

6.30 A linguistics researcher was analyzing the relative frequency of use of the vowels (a, e, i, o, u) in the works of a famous poet. Defining X to be the number of vowels that were used per line, and ignoring duplications of the same vowel, it was noted that whether 1 or 2 or 3 or 4 or all 5 vowels were used in a line was evenly spread out. In other words, X was uniformly distributed.
  (a) What is the probability that in any given line, more than 3 vowels are used?
  (b) What proportion of the lines use 1 or 2 vowels?
  (c) What is the average number of vowels per line?
  (d) What is the standard deviation of the vowel distribution?

6.31 A sociologist with a special interest in the impact of technology on life-style patterns interviewed a group of video cassette recorder (VCR) owners. Among other

Handwritten margin notes:

$$x \quad p(x)$$

x	$p(x)$
3	1/7
4	1/7
5	1/7
6	1/7
7	1/7
8	1/7
9	1/7

a) $P(x \geq 5) = P(5) + P(6) + P(7) + P(8) + P(9) = \left(\frac{1}{7}\right) \times 5 = \frac{5}{7} = 0.714$

b) $P(x=7)$ or $P(x=8)$

$= \frac{1}{7} + \frac{1}{7} = \frac{2}{7} = 0.286$

c) $M = \frac{3+9}{2} = 6$

things, it was observed that the number of movies recorded per month by the VCR owners ranged from 3 to 9 and was a uniformly distributed variable.

(a)   What percentage of VCR owners record at least 5 movies a month?

(b)   What is the probability of selecting (randomly) a VCR owner who records 7 or 8 movies a month?

(c)   What is the average number of movies recorded by VCR owners per month?

6.32   A sports photographer uses a high-speed motorized drive on his 35 mm camera to capture action shots. Only a few photographs are actually used. His editor found from their records that the number of photographs rejected per 36 exposure roll of film was uniformly distributed over a range from 25 to 32.

(a)   What percentage of 36 exposure film rolls used by this photographer will have 30 or more rejects?

(b)   What is the probability that in any one roll of film, 28 or fewer photographs will be rejected?

(c)   What is the average number of photographs rejected per roll of film?

(d)   Compute the variance of the distribution of the number of rejected photographs per roll of film.

6.33   Refer to exercises 6.15 and 6.16.

(a)   Are the probability distributions in these exercises the distributions of discrete uniform random variables?

(b)   Compute the mean, variance, and standard deviation for the distribution of the number of children in the Hispanic families and the Asian families.

6.34   The Bone Breaker Gym is a favorite spot for many weight lifters in training. A recent competition centered on number of push-ups performed. It was noted by a sports statistician that the variable X = number of push-ups was a uniformly distributed random variable in a rather narrow range from 55 to 65 push-ups.

(a)   What is the proportion of weight lifters doing 60 or more push-ups?

(b)   What is the probability of selecting a weight lifter who does fewer than 60 push-ups?

(c)   What is the average number of push-ups for a weight lifter?

(d)   What is the variance in push-ups?

6.35   A textbook copy editor is checking a manuscript for typographical errors. Let the random variable X be the number of errors per page. Assume that X is uniformly distributed over a range of 4 to 15 errors per page.

(a)   What is the average number of errors per page?

(b)   What is the standard deviation of the distribution of errors?

(c)   What percentage of the pages have more than 10 errors?

6.5   BINOMIAL RANDOM VARIABLES

The random variable X representing the number of heads in three flips of a coin is a special type of discrete random variable, a **binomial** random variable.

We now list the conditions for a binomial random variable in general and as applied to our coin flipping example.

A Binomial Situation	Coin Flipping Example
1 Your experiment consists of n repetitions, called **trials**.	**1** n = three flips of a coin.
2 Each trial has two mutually exclusive possible outcomes, referred to as **success** and **failure**	**2** Success = head, failure = tail (this is arbitrary).
3 The probability of a success for each trial is denoted p; the value of p remains the same for each trial.	**3** p = the probability of flipping a head on a particular trial = 1/2.
4 The n trials are independent.	**4** The results on one coin flip do not affect the results on another flip.
5 The random variable X is the number of successes out of n trials.	**5** X = the number of heads out of three flips.

You encounter a binomial random variable when a certain experiment is repeated many times (n trials), the trials are independent, and each experiment results in one of two mutually exclusive outcomes. For example, a randomly selected individual is either male or female, is on welfare or is not, will vote Republican or will not, and so on.

The two outcomes for each experiment are labeled as success or failure. A success need not be considered "good" or "desirable." Instead, it depends on what you are counting at the completion of the n trials. If, for example, the object of the experiment is to determine the probability that three people, out of 20 randomly selected individuals, are on welfare, then a success on each of the $n = 20$ trials will be the event that the person selected on each trial is on welfare.

EXAMPLE 6.7

In example 5.3, it was noted that 30% of the senior students in a particular school have a deficiency in reading ability. Select four seniors from this school. Consider the number of students out of these four that have a reading deficiency. Does this satisfy the requirements of a binomial situation? What is your random variable here?

Solution

Refer to conditions 1 through 5 in our list for a binomial situation.

1 There are $n = 4$ trials, where each trial consists of selecting one senior from this school.

2 There are two outcomes for each trial. We are interested in counting the number of students, out of the four selected, who do have a reading deficiency, so define

success = student has a reading deficiency

failure = student does not have a reading deficiency

3 p = probability of a success on each trial = .3.

4 The trials are independent since the students are selected randomly.

5 The random variable here is X, where

X = number of successes in n trials

= number of seniors (out of four) that have a reading deficiency

All of the requirements are satisfied. Thus, X is a binomial random variable (it is also discrete). □

Counting Successes for a Binomial Situation

How many ways are there of getting two heads out of four flips of a coin? There are six: HHTT, HTHT, HTTH, THHT, THTH, and TTHH. How many ways can you select two people from a group of four people, where the order of selection is unimportant (say you are selecting a two-person committee)? Label the individuals as I_1, I_2, I_3, and I_4. You want to find the number of combinations of four people using two at a time:

$$_4C_2 = \frac{4!}{2!2!} = 6$$

Put these results side by side. The scheme for matching the two results is to select I_1 if H appears on the first flip, select I_2 if H appears on the second flip, and so on.

Two Heads out of Four Flips	Two People from a Group of Four
HHTT	I_1, I_2
HTHT	I_1, I_3
HTTH	I_1, I_4
THHT	I_2, I_3
THTH	I_2, I_4
TTHH	I_3, I_4

You should see a direct correspondence between the two solutions. Our conclusion is that the number of ways of getting two heads out of four flips of a coin is $_4C_2$. Extending this to any number of flips of a coin, the number of ways of getting k heads out of n flips of a coin is $_nC_k$. Finally, for any binomial situation, the number of ways of getting k successes out of n trials is $_nC_k$. We are thus able to determine the probability mass function (PMF) for any binomial random variable.

Once again, let X equal the number of heads out of three flips. Here X is a binomial random variable, with p = .5. Consider any value of X, say, $X = 1$. Then (1) 1/8 is the probability of any one outcome where $X = 1$, such as HTT, and (2) $_3C_1 = 3$ is the number of ways of getting one head (success) out of three flips (trials). Consequently, the probability that X will be one is

$$P(1) = {}_3C_1(1/8) = 3/8$$

The resulting PMF for this situation can be written as

$$P(x) = {}_3C_x \cdot 1/8 \qquad \text{for } x = 0, 1, 2, 3 \quad \text{(and zero otherwise)}$$

Using this function, we obtain the same results as before:

$$P(0) = {}_3C_0(1/8) = 1 \cdot 1/8 = 1/8$$

$$P(1) = {}_3C_1(1/8) = 3 \cdot 1/8 = 3/8$$

$$P(2) = {}_3C_2(1/8) = 3 \cdot 1/8 = 3/8$$

$$P(3) = {}_3C_3(1/8) = 1 \cdot 1/8 = \frac{1/8}{1.0}$$

EXAMPLE 6.8

In example 6.7, the binomial random variable X is the number of students (out of four) who have a reading deficiency. Also, there are $n = 4$ trials (students) with $p = .3$ (30% of the seniors have such a deficiency). Let S denote a success and F a failure. Then define:

$$S = \text{student has a reading deficiency}$$

$$F = \text{student does not have a reading deficiency}$$

What is the probability that exactly two students (out of four) will have this problem?

Solution

This is $P(X = 2)$ or $P(2)$. Consider any one result where $X = 2$, such as $SFSF$. The probability of this result, using equation 5-14, is (probability of S on first trial) $\cdot$ (probability of F on second trial) $\cdot$ (probability of S on third trial) $\cdot$ (probability of F on fourth trial), which is

$$(.3)(.7)(.3)(.7) = (.3)^2(.7)^2$$

Also note that the probability of each result with two S's and two F's ($X = 2$) also is $(.3)^2(.7)^2 = p^2(1 - p)^2$. How many ways can we get two successes out of four trials? This is

$$_4C_2 = \frac{4!}{2!2!} = 6$$

So, the final result here is

$$P(2) = (\text{number of ways of getting } X = 2)(\text{probability of each one})$$

$$= {}_4C_2(.3)^2(.7)^2$$

$$= (6)(.09)(.49) = .265$$

So, 26.5% of the time, exactly two seniors out of four will have a reading deficiency.

Figure 6.3

Binomial probability
mass function for
$n = 4$, $p = .3$

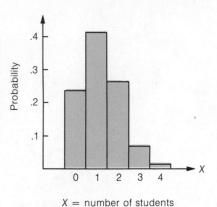

X = number of students

We can extend the results of example 6.8 to obtain the PMF for a binomial random variable:

$$P(x) = {}_nC_x p^x (1 - p)^{n-x} \qquad \text{for } x = 0, 1, 2, \ldots, n \\ \text{(and zero otherwise)} \qquad \textbf{(6-8)}$$

For the previous example, $x = 2$, $n = 4$, and $p = .3$. The complete list of probabilities for this example is

$$X = \begin{cases} 0 & \text{with probability } {}_4C_0(.3)^0(.7)^4 = & .240 \\ 1 & \text{with probability } {}_4C_1(.3)^1(.7)^3 = & .412 \\ 2 & \text{with probability } {}_4C_2(.3)^2(.7)^2 = & .265 \\ 3 & \text{with probability } {}_4C_3(.3)^3(.7)^1 = & .076 \\ 4 & \text{with probability } {}_4C_4(.3)^4(.7)^0 = & \underline{.008} \\ & & 1.001 \end{cases}$$

Note that the total value may be slightly greater or less than 1.0, due to rounding. A graphical representation of this PMF is shown in Figure 6.3. □

Using the Binomial Table

The binomial PMFs have been tabulated in Table A-1 (page A-3) for various values of n and p. The maximum number of trials in this table is $n = 20$. For binomial situations where $n > 20$, one must use an approximation to a binomial probability. This will be considered in Chapter 7.

For the binomial situation in example 6.7, $n = 4$ and $p = .3$. To find $P(2)$, locate $n = 4$ and $x = 2$. Go across the table to $p = .3$ and you will find the corresponding probability (after inserting the decimal in front of the number). This probability is .265. Similarly, $P(0) = .240$, $P(1) = .412$, $P(3) = .076$, and $P(4) = .008$, as before.

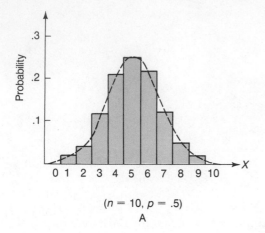

$(n = 10, p = .5)$
A

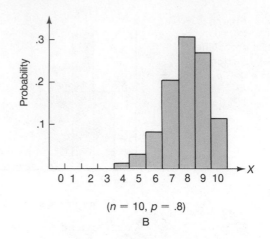

$(n = 10, p = .8)$
B

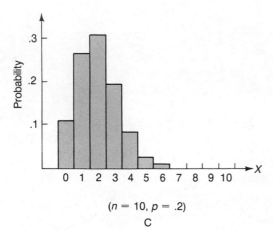

$(n = 10, p = .2)$
C

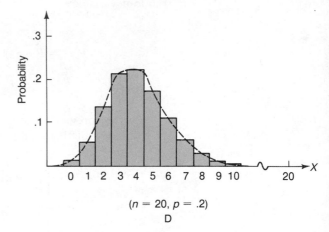

$(n = 20, p = .2)$
D

Figure 6.4

Shape of the
binomial distribution

The probability that no more than two seniors will have a reading deficiency is written $P(X \leqslant 2)$, where

$$P(X \leqslant 2) = P(X = 0) + P(X = 1) + P(X = 2)$$
$$= P(0) + P(1) + P(2)$$
$$= .240 + .412 + .265 = .917$$

This is a **cumulative probability** and is obtained by summing $P(x)$ over the appropriate values of X.

Shape of the Binomial Distribution

Figure 6.4 contains a graphical representation of four binomial distributions. In particular, notice that:

1 When $p = .5$, the shape is perfectly symmetrical and resembles a bell-shaped (normal) curve.

2 When $p = .2$, the distribution is skewed right. This skewness increases as p decreases.

3 For $p = .8$, the distribution is skewed left. As p approaches 1, the amount of skewness increases.

Compare Figure 6.4c and d. Notice that, in both cases, p is .2; however, the number of trials increased from $n = 10$ (in (C)) to $n = 20$ (in (D)). For the larger value of n, the shape of this distribution is nearly bell-shaped, despite the small value of p. This implies that, regardless of the value of p, the shape of a binomial distribution approaches a bell-shaped distribution as the number of trials (n) increases. We will use this fact in the next chapter, when we demonstrate an approximation to the binomial distribution using a bell-shaped (normal) curve.

In summary, the shape of a binomial distribution is:

1 Skewed left for $p > 1/2$ and small n.

2 Skewed right for $p < 1/2$ and small n.

3 Approximately bell-shaped (symmetric) if p is near 1/2 or if the number of trials is large.

Mean and Variance of Binomial Random Variables

In example 6.7, we examined the binomial random variable X representing the number of seniors (out of four) who have a reading deficiency. If you select four seniors, observe X, select four more students, observe X, and repeat this procedure indefinitely, what will X be on the average? This is the mean of X, where, using equation 6-1,

$$\mu = \Sigma x P(x)$$
$$= (0)(.240) + (1)(.412) + (2)(.265) + (3)(.076) + (4)(.008)$$
$$= 1.2 \text{ students}$$

Also, using equation 6-3, the variance of X is

$$\sigma^2 = \Sigma x^2 P(x) - \mu^2$$
$$= [(0)^2(.240) + (1)^2(.412) + (2)^2(.265) + (3)^3(.076)$$
$$+ (4)^2(.008)] - (1.2)^2$$
$$= 2.28 - 1.44 = .84 \text{ (students)}^2$$

and so $\sigma = $ standard deviation of $X = \sqrt{.84} = .92$ student. (Watch the units.)

The good news is that there is a convenient shortcut for finding the mean and variance of a binomial random variable. For this situation, you need not use equations 6-1 and 6-3. Instead, for any binomial random variable,

$$\mu = np \qquad\qquad (6\text{-}9)$$
$$\sigma^2 = np(1 - p) \qquad\qquad (6\text{-}10)$$

How these expressions were derived is certainly not obvious, but let us verify that they work for example 6.7. Here $n = 4$ and $p = .3$, so

$$\mu = (4)(.3) = 1.2 \quad (OK)$$
$$\sigma^2 = (4)(.3)(.7) = .84 \quad (OK)$$

EXAMPLE 6.9

If you repeat example 6.7 using $n = 50$ seniors (rather than $n = 4$), how many reading deficient students will you observe on the average?

Solution

Now X is the number of students (out of 50) who have a reading deficiency. Consequently,

$$\mu = np = (50)(.3) = 15$$

So, on the average, X will be 15 students. For this situation, the variance of X is

$$\sigma^2 = np(1 - p) = (50)(.3)(.7) = 10.5 \text{ (students)}^2$$

Also,

$$\sigma = \sqrt{10.5} = 3.24 \text{ students} \qquad \Box$$

EXAMPLE 6.10

Airline overbooking is a common practice. Many people make reservations on several flights due to uncertain plans and then cancel at the last minute or simply fail to show up. Air Texas is a small commuter airline. Their planes hold only 15 people. Past records indicate that 20% of the people making a reservation do not show up for the flight.

Suppose that Air Texas decides to book 18 people for each flight.

1 Determine the probability that on any given flight, at least one passenger holding a reservation will not have a seat.

2 What is the probability that there will be one or more empty seats for any one flight.

3 Determine the mean and standard deviation for this random variable.

Solution I

The binomial random variable for this situation is $X =$ the number of people (out of 18) who book a flight and actually do appear. For this binomial situation,

$n = 18$ (18 reservations are made) and $p = 1 - .2 = .8$ (the probability that any one person will show up). At least one passenger will have no place to sit if X is 16 or more. Using Table A-1,

$$P(X \geqslant 16) = P(X = 16) + P(X = 17) + P(X = 18)$$
$$= .172 + .081 + .018$$
$$= .271$$

We see that, if the airline follows this policy, 27% of the time one or more passengers will be deprived of a seat—not a good situation.

Solution 2   We want to find the probability that the number of people who actually arrive (X) is 14 or less. Using Table A-1 (where $n = 18$, $p = .8$),

$$P(X \leqslant 14) = .215 + .151 + \cdots + .003 + .001$$
$$= .50$$

(Notice that the four remaining probabilities are nearly zero.) With this booking policy, the airline will have flights with one or more empty seats approximately one-half of the time.

Solution 3   The mean of X is

$$\mu = np = (18)(.8) = 14.4$$

which implies that the average number of people that book a flight and do appear is 14.4.

The standard deviation of X is

$$\sigma = \sqrt{np(1 - p)} = \sqrt{(18)(.8)(.2)} = 1.70 \text{ people.}$$ □

EXAMPLE 6.11   According to the latest crime report statistics, 60% of the people convicted of home burglaries in Riverport will later be convicted of another crime. During the past week, 10 people were convicted of such an offense. What are the chances that (i) none of these individuals will later be convicted of another crime, and (ii) at least 5 of them will be convicted of another crime?

Solution   This is a binomial situation, where

  1 There are $n = 10$ trials (no pun intended!)
  2 Each trial has two outcomes:

    success: person is later convicted of another crime
    failure: person is not later convicted of another crime

(*Note:* Since the questions are concerned with the number of repeat criminals, we defined a success to be that an individual convicted of a home burglary *will* be a repeat offender. This choice is arbitrary—just keep in mind that the binomial random variable counts the number of successes. As mentioned earlier, a "success" need not be a desirable event.)

3   p = probability of success = .6

4   The trials are independent, assuming that the 10 convictions are unrelated.

5   The random variable here is X = number of successes out of n trials = number of repeat offenders out of 10 people convicted of home burglary.

For question (i), we need to determine the probability that X will be zero, $P(0)$. Using Table A-1, you will find that this value is $0+$ (zero to three decimal places). A more accurate value can be determined using equation 6-8, where

$$P(X = 0) = P(0) = {}_{10}C_0(.6)^0(.4)^{10}$$
$$= (1)(1)(.4)^{10}$$
$$= (.4)^{10} = .000105$$

Consequently, this is a very rare event (no repeats out of 10 people convicted) happening approximately one time in ten thousand.

For question (ii), we need to determine the chance that X is at least five, that is, $P(X \geqslant 5)$. At this point you could find $P(5), P(6), \ldots, P(10)$ and add them to obtain the answer. Another procedure is the following:

$$P(X \geqslant 5) = 1 - P(X < 5)$$
$$= 1 - P(X \leqslant 4)$$
$$= 1 - [P(0) + P(1) + P(2) + P(3) + P(4)]$$
$$= 1 - [0 + .002 + .011 + .042 + .111]$$
$$= 1 - .166 = .834$$

As a result, 83.4% of the time you will observe at least 5 repeat offenders out of 10 people convicted of home burglary in Riverport.   □

EXAMPLE 6.12

It is estimated that one out of ten personal computers sold by Videotech will be in for repair before the warranty has expired. During a recent special offer, Videotech sold 20 such computers. Define X to be the number of computers that will be returned to the store for repair before the warranty has expired.

1   What is the probability that at least three computers are returned?

2   What is the probability that no more than one is returned?

3   Determine the mean and standard deviation of X.

Solution 1

The random variable X satisfies the requirements for a binomial random variable with $n = 20$ and $p = .1$. For this situation, a success is defined to be that a computer is returned for repair prior to the expiration of the warranty. The probability of at least three computers being returned is the probability of *three or more* returning. This is

$$P(X \geqslant 3) = 1 - P(X < 3)$$
$$= 1 - P(X \leqslant 2)$$
$$= 1 - [P(0) + P(1) + P(2)]$$
$$= 1 - [.122 + .270 + .285]$$
$$= .323$$

Consequently, the probability of at least three computers being returned is .323.

Solution 2   The chance that no more than one is returned is the probability that X is one or less. This is

$$P(X \leqslant 1) = P(0) + P(1)$$
$$= .122 + .270$$
$$= .392$$

So, this event will occur with probability .392.

Solution 3   The mean of the random variable X is

$$\mu = np = (20)(.1) = 2 \text{ computers}$$

and the standard deviation of X is

$$\sigma = \sqrt{np(1-p)} = \sqrt{(20)(.1)(.9)} = 1.34 \text{ computers.}$$

This implies that on the average, given this same situation, Videotech will have two computers returned before the warranty has expired.   □

EXERCISES   6.36   A journalist randomly selected telephone numbers in a school district and conducted phone interviews to determine residents' position on busing. The school district's Board of Directors already knows that 60% of all the residents are in favor of busing.
(a)   If the journalist telephoned 15 residents, what is the probability that *at least* 10 will be in favor of busing?
(b)   If the journalist telephoned 25 residents, what is the probability that *exactly* 15 will be in favor of busing?

6.37   If four trials are independently conducted and each trial results in a success with probability one-third, what is the probability that exactly two of the four trials result in successes?

6.38   A survey reveals that 60% of the eligible people in a certain county vote during a county election.
(a)   If 20 people in the county who are eligible to vote are chosen at random, what is the probability that exactly 12 vote during the next county election?
(b)   What is the probability that exactly 10 people in the county vote during this election out of the 20 chosen at random?

$n = 4 \, , \, p = \frac{1}{3}$

$P(x=2)$

$C_2 \cdot \left(\frac{1}{3}\right)^2 \left(\frac{2}{3}\right)^2 = .2963$

6.39 According to Blades and Spencer (1987), the use of maps is a complex cognitive skill, yet even very young children can use simple maps to find a location. These researchers conducted a study to find out if young children aged 4–6 years could also use a map to follow a route. A large-scale maze, 25 meters long, was designed in a school playground and 120 children were given maps of the maze. The children had to walk through the maze, making correct route choices at different T-junctions in the maze.

The 120 children were divided into 5 age groups, and each age group was further subdivided into 2 groups. Thus, there were 10 groups, with $n = 12$ in each group. Each child completed six trials in the maze, with three junctions in each trial, so each child had a score of number of correct choices out of a total of 18. If a child ignored the map provided and simply guessed the way at each junction, there was a 0.50 probability of being correct at each junction. *Source:* Mark Blades and Christopher Spencer, The use of maps by 4–6 year old children in a large-scale maze. *Br. J. Dev. Psychol.* **5** (Part 1): 19–24 (1987). If the child guessed at each junction,

(a) What is the probability that an individual child will score 13 or more correct choices out of 18?

(b) In each group of 12 children, what is the probability that 3 or more children score at least 13 (out of 18)?

(c) In each group of 12 children, what is the probability that 5 or more children score at least 13 (out of 18)? (*Hint:* Your answer for part (a) gives the probability of success for parts (b) and (c), where success = a score of 13 or more.

6.40 A lawyer estimates that 40% of the cases in which she represented the defendant were won. If the lawyer is presently representing 10 defendants in different cases, what is the probability that at least 5 of the cases will be won? What are you assuming here?

6.41 A newsstand owner has calculated that 80% of the midday newspapers are sold. If the owner orders 25 midday newspapers daily, what is the probability that on any day 23 or more of the newspapers will be sold?

6.42 A recent study by Yale–Harvard researchers received widespread publicity in the mass media. It made some interesting predictions about "baby boom" women, that is, those born in the mid 1950s. The study suggested that white, college-educated women born in the mid 1950s who were still unmarried at age 30 had a 0.20 probability of ever getting married. If the women were still unmarried at age 35, the probability of their ever getting married dropped to 0.05. *Source:* A 1986 study by Neil Bennett, Yale University sociologist, and others, originally reported in the *Stanford Advocate*, Connecticut, and picked up by the mass media, including cover stories in *People* and *Newsweek* magazines in late 1986. The specific reference used herein is Nancy Shute, syndicated columnist, distributed by *Words by Wire*, January 17, 1987.

(a) Suppose a group of 15 such "baby-boom" women who remain unmarried at age 30 is randomly selected. What is the probability that 3 to 6 women (inclusive) out of these 15 will ever get married?

(b) For the group of 15 in question (a), what is the probability that between 9 and 12 women inclusive will *never* get married?

(c) Compare your results for (a) and (b) and explain any similarity or difference.

(d) Suppose another group of 15 is randomly selected and this time baby-boom women who remain unmarried at age 35 are picked. What is the probability that in this group, fewer than 3 women will ever get married?

6.43 Hot on the heels of the study by Yale University sociologist Neil Bennett and others (see exercise 6.42 above) comes another study by Census Bureau researcher Jeanne E. Moorman that has dramatically different findings. Consider the following table, which shows women's likelihood of marriage, in percent, for various age and education groups:

			WOMEN'S MARRIAGE PROSPECTS				
Age	All Women	Elementary School	Some High School	High School Graduate	Some College	College Graduate	Graduate Degree
25	77.5	74.3	63.6	72.0	81.0	89.1	85.9
30	60.4	68.3	53.8	55.9	59.7	66.3	67.8
35	43.7	60.5	44.4	40.6	38.2	40.9	42.3
40	29.9	50.9	32.7	26.9	23.8	22.8	22.9
45	18.0	39.2	21.0	15.3	12.4	10.8	9.3

Source: Jeanne E. Moorman, U.S. Census Bureau, Washington, DC, as reported by Associated Press, January 14, 1987.

Focusing on 30-year-old women, you will notice that the chances of marriage range from 54% to 68% depending on educational background, and are about 60% overall for women in this group. In other words, the probability is 0.60 according to Census Bureau data, as opposed to 0.20 according to the Yale–Harvard study. Similarly, for 35-year-old women, the chances of marriage are 43.7% (and not 5% as reported by Bennett et al.). Use the Census Bureau (Moorman's) figures to answer the same questions in exercise 6.42 (a)–(d).

(Postscript: Why the enormous difference in results? The Yale–Harvard researchers use the Census Bureau's Current Population Survey of 1982, which is a survey of 60,000 households. After winnowing the numbers, these researchers ended up with 1500 in each age group. Moorman's study was based on 1980 census figures gleaned from one in six households in the United States, obviously a much larger sample. The importance of proper sampling is addressed in Chapter 8.)

6.44 Let the random variable X represent the number of correct responses on a multiple-choice test that has 15 questions. Each question has five multiple-choice answers and there is only one correct answer to each question.

(a) What is the probability that the random variable X is greater than 8 if the person taking the test randomly guesses?

(b) What is the mean value of X if the person randomly guesses?

(c) What is the standard deviation of X if the person randomly guesses?

6.45 A local newspaper report carrying a headline "CRIME WAVE PROPELLED BY COCAINE" stated that a survey of prison inmates in the county revealed that 30% of the prison inmates committed a crime to obtain money for cocaine, and an additional 20% of inmates committed a crime to obtain money for other drugs like marijuana, heroin, etc.

(a)   If a random group of 30 prison inmates is selected from this county, what is the probability that 9 of them committed a crime to get money for cocaine?

(b)   If a random group of 20 prison inmates is selected from this county, what is the probability that 4 to 6 inmates (inclusive) committed a crime to get money for other drugs like marijuana, heroin, etc.?

(c)   If a random group of 20 prison inmates is selected from this county, what is the probability that 4 to 6 inmates (inclusive) committed a crime to get money for cocaine or other drugs?

6.46   Suppose that four out of five women who become pregnant with the aid of fertility drugs experience multiple births (twins, triplets, quadruplets, and so on).

(a)   In a randomly selected group of 16 women who became pregnant with the assistance of fertility drugs, what is the probability that more than 13 will experience multiple births?

(b)   For the group of women in question (a), what is the probability that fewer than 5 will not experience multiple births?

6.47   If X is binomially distributed, with the number of trials equal to 18 and the probability of success equal to 0.12, then what are the mean and standard deviation of X?

6.48   A nursery knows that 90% of its hedges will survive the winter. If 15 hedges are randomly selected and planted, what is the probability that at least 13 hedges will survive the winter?

6.49   A sociology professor surveyed college students' level of "scientific illiteracy" and was surprised to learn that 40% of the students believed that humans and dinosaurs coexisted, even though there is evidence that millions of years separate the two periods of natural history. *Source:* Sociology Department, University of Texas, Arlington, Texas; on-campus survey, Fall 1986. If a group of 8 students is randomly selected, what is the probability of observing the following number of students in this group who believe that humans coexisted with dinosaurs?

(a)   Fewer than 2.

(b)   Between 2 and 4, inclusive.

(c)   More than 5.

HYPERGEOMETRIC AND POISSON
6.6   DISCRETE DISTRIBUTIONS (OPTIONAL)

As mentioned earlier, not all discrete random variables belong to a special category, such as binomial. One example is $X =$ total value of two dice. This is a discrete random variable, but it is not binomial. There are two other widely used discrete distributions worthy of mention: the hypergeometric and Poisson random variables. (The total of two dice does not fall into one of these categories either; not all discrete random variables can be classified as one of these types.)

Hypergeometric Distribution

The conditions for a **hypergeometric random variable** are:

1 Population size $= N$. In this population, k members are S (successes) and $N - k$ are F (failures).

2 Sample size $= n$ trials, obtained without replacement.

3 $X =$ the number of successes out of n trials (a hypergeometric random variable).

The main distinction between a hypergeometric and a binomial situation is that the trials in the former are not independent. As a result, the probability of a success on each trial is affected by the results of the previous trials. This occurs when sampling without replacement from a finite population.

The situation surrounding a hypergeometric random variable is similar to the binomial situation in that you count "successes" in both cases. However, for the hypergeometric situation, you have a finite population (of size N) and you know the number of successes (k) and failures ($N - k$) that make up this population. For example, you might select a random sample of $n = 8$ from a group of $N = 25$ teachers, of which $k = 15$ are female and $N - k = 10$ are male. For this situation, the hypergeometric random variable could be defined as $X =$ the number of teachers (of the 8) who are female.

As in the binomial situation, each trial results in a *success* or *failure*. For this situation define

$$S = \text{success} = \text{person is a female}$$

$$F = \text{failure} = \text{person is a male}$$

In this illustration, out of 25 teachers, 15 are female. So

$$P(S \text{ on first trial}) = 15/25 = .6$$

The conditional probability of S on the second trial is:

$$14/24 = .583 \text{ if first person selected is a female}$$

$$15/24 = .625 \text{ if first person selected is a male}$$

The probability of a success on the second trial is affected by what occurred on the first trial; this is a hypergeometric situation.

The PMF for the hypergeometric random variable is

$$P(x) = \frac{{}_kC_x \cdot {}_{N-k}C_{n-x}}{{}_NC_n} \tag{6-11}$$

for $x = a, a + 1, a + 2, \ldots, b$, where a is the maximum of 0 and $n + k - N$ and b is the minimum of k and n. The value of $P(x)$ is zero for all other values of X.

EXAMPLE 6.13

Determine the probability of observing exactly 3 females and 5 males in the sample of size 8.

Solution

Imagine two containers (the population). One contains 15 S's and the other has 10 F's. The sample consists of eight names, randomly selected from these two containers. If x names are selected from the success container, then $8 - x$ names are selected from the failure container. For this situation, $N = 25$, $k = 15$, and $n = 8$. The possible values for X are from $a =$ maximum of 0 and -2 (0) to $b =$ minimum of 15 and 8 (8). The probability of obtaining three S's and five F's in your sample is

$$P(X = 3) = P(3) = \frac{{}_{15}C_3 \cdot {}_{10}C_5}{{}_{25}C_8}$$

As you will quickly see, the term ${}_nC_n$ gets very large—in fact, it becomes too large for many calculators. The only practical way to evaluate a hyper-geometric probability, short of relying on a computer, is to cancel as many terms as possible in the expression.

The final result here is $P(3) = .106$: 10.6% of the time, you will obtain exactly three females and five males in your sample of size eight. □

EXAMPLE 6.14

The president of Ace Realty Company is investigating a particular neighborhood that has suffered more than most other neighborhoods due to recent layoffs in the local area. She notices that of the 20 houses in this neighborhood, 8 are vacant. If she takes a random sample of 5 houses from this neighborhood, what is the probability that exactly one will be vacant?

Solution

This situation fits the requirements for a hypergeometric random variable, where X is the number of houses (out of 5) that are vacant, $n = 5$, $N = 20$, and $k = 8$. Consequently,

$$P(X = 1) = P(1) = \frac{{}_8C_1 \cdot {}_{12}C_4}{{}_{20}C_5}$$

$$= \frac{\dfrac{8!}{7!1!} \cdot \dfrac{12!}{4!8!}}{\dfrac{20!}{5!15!}}$$

$$= \frac{(8)(495)}{15,504} = .255$$

Approximately 25.5% of the time, in a sample of size 5 from this neighborhood, exactly 1 house will be vacant. □

Mean and Variance of a Hypergeometric Random Variable

As we did with the binomial random variable, we could use the definition of the mean and variance of a discrete random variable contained in equations 6-1 and 6-3. For example,

$$\mu = \Sigma x P(x)$$

where $P(x)$ is the PMF given in equation 6-11.

As in the binomial situation, simpler expressions exist for both the mean and the variance of the hypergeometric random variable. These are

$$\mu = \Sigma x P(x) = \frac{nk}{N} \tag{6-12}$$

$$\sigma^2 = \Sigma x^2 P(x) - \mu^2$$
$$= \frac{k(N-k)n(N-n)}{N^2(N-1)} = \left[n\left(\frac{k}{N}\right)\left(1-\frac{k}{N}\right)\right]\left(\frac{N-n}{N-1}\right) \tag{6-13}$$

For example 6.14, $N = 20$, $k = 8$, $n = 5$. Consequently,

$$\mu = \frac{(5)(8)}{20} = 2.0 \text{ houses}$$

$$\sigma^2 = \frac{(8)(12)(5)(15)}{(20)^2(19)} = .947$$

and so

$$\sigma = \sqrt{.947} = .973 \text{ house}$$

This means that, if we observed this process of sampling these 20 houses indefinitely, we would obtain 2 ($=\mu$) vacant houses on the average. Also, $\sigma = .973$ (or $\sigma^2 = .947$) is our measure of the variation in the observations of this random variable if we observe it over an indefinite period.

Using the Binomial to Approximate the Hypergeometric

Whenever $n/N < .05$, the binomial distribution will provide a good approximation to the hypergeometric distribution. Here, define

$$p = \frac{\text{number of successes in population}}{\text{size of population}} = k/N$$

Then X is the number of successes in the sample. X is approximately a binomial random variable with n trials and probability of success p. Briefly, the binomial approximation works well if your sample size is less than 5% of your population size.

What probability would you obtain had you treated example 6.13 as a binomial situation, where $p = k/N = 15/25 = .6$? Here you have a binomial situation with $n = 8$ and $p = .6$. Using equation 6-8,

$$P(3) = {}_8C_3(.6)^3(.4)^5$$
$$= (56)(.216)(.01024)$$
$$= .124$$

The same result is obtained using Table A-1.

For this example, .106 is the exact probability using the hypergeometric distribution and .124 is the *approximate* probability using the binomial distribution. We did not obtain a very good approximation here. The problem is that the population size is $N = 25$ and the sample size is $n = 8$, which is 32% of the population size.

The Poisson Distribution

The Poisson distribution, named after the French mathematician Simeon Poisson, is useful for counting the number of times a particular event occurs over a specified period of time. It also can be used for counting the number of times an event (such as a manufacturing defect) occurs over a specified area (such as a square yard of sheet metal) or in a specified volume. We will restrict our discussion to counting over time, although any unit of measurement is permissible.

The random variable X for this situation is the number of occurrences of a particular event over a specified period of time. The possible values are 0, 1, 2, 3, ... For X to be a **Poisson random variable**, three conditions must be present:

1 The number of occurrences in one interval of time is unaffected by (statistically independent of) the number of occurrences in any other nonoverlapping time interval. For example, what took place between 3:00 and 3:20 P.M. is unaffected by what took place between 9:00 and 10:00 A.M.

2 The expected (or average) number of occurrences over any time period is proportional to the size of this time interval. For example, we would expect half as many occurrences between 3:00 and 3:30 P.M. as between 3:00 and 4:00 P.M.

This also implies that the probability of an occurrence must be constant over any intervals of the same length. A situation in which this is usually not true is at a restaurant from 12:00 noon to 12:10 P.M. and 2:00 to 2:10 P.M. Due to the differences in traffic flow for these two intervals, we would not expect the arrivals between, say, 11:30 A.M. and 2:30 P.M. to satisfy the requirements of a Poisson situation.

3 Events cannot occur exactly at the same time. More precisely, there is a unit of time sufficiently small (such as one second) during which no more than one occurrence of an event is possible.

Some situations that usually meet these conditions are

The number of arrivals at a local bank over a five-minute interval.

The number of telephone calls arriving at a switchboard over a one-minute interval.

The number of daily accidents reported along a 20-mile stretch of an intercity toll road.

The number of trucks from a fleet that break down over a one-month period.

The number of admissions to a hospital emergency room over a one-hour period.

For each situation, the (discrete) random variable X is the number of occurrences over the time period T. If all the assumptions are satisfied, then X is a Poisson random variable. Define μ to be the expected (or average) number of occurrences over this period of time.* For any application, the value of μ must be specified or estimated in some manner. The Poisson PMF for X follows:

Poisson Probability Mass Function

X = number of occurrences over time period T.

$$P(x) = \frac{\mu^x e^{-\mu}}{x!} \qquad \text{for } x = 0, 1, 2, 3, \ldots \qquad (6\text{-}14)$$

where μ = expected number of occurrences over T.

Equation 6-14 contains the number e. This is an interesting and useful number in mathematics and statistics. To get an idea how this number is derived, consider the following sequence:

$$(1 + 1/2)^2 = 2.25$$

$$(1 + 1/3)^3 = 2.37$$

$$(1 + 1/4)^4 = 2.44$$

$$(1 + 1/5)^5 = 2.49$$

$$\vdots$$

$$(1 + 1/100)^{100} = 2.705$$

$$(1 + 1/1000)^{1000} = 2.717$$

$$\vdots$$

* The symbol λ (lambda) is often used to denote this parameter.

This sequence of numbers is approaching e. The actual value is:

$$e = 2.71828 \ldots$$

One interesting application of the number e occurs when calculating compound interest. For example, if you invest \$100 at 12%, compounded annually, then at the end of the year you will have \$112. However, if your interest is compounded not monthly, not daily, but continuously, the amount in your account will be $(100)(e^{.12}) = (100)(1.1275) = \$112.75$. The difference in these amounts is not as large as you might expect!

We will use e again in Chapter 7.

Mean and Variance of a Poisson Random Variable

Once again, we could use the definition of the mean and variance of a discrete random variable in equations 6-1 and 6-3. However, this is not necessary. It is fairly easy to show that, using equation 6-14,

$$\text{mean of } X = \Sigma x P(x)$$
$$= \mu$$

This is hardly a surprising result, given how μ was originally defined. Also,

$$\text{variance of } X = \sigma^2$$
$$= \Sigma x^2 P(x) - \mu^2$$
$$= \mu$$

So, both the mean and the variance of the Poisson random variable X are equal to μ.

Applications of a Poisson Random Variable

EXAMPLE 6.15

Riverside Memorial Hospital is concerned with staffing its emergency room with enough medical personnel so that patients arriving will have an excellent chance of receiving immediate attention. On the other hand, if a large number of doctors and nurses are assigned to this station but the number of emergency room admissions is not extremely high, this may result in wasted time and expense. Define X to be the number of people who arrive at the emergency room over a one-hour period. Assume that the conditions for a Poisson situation are satisfied with

$$\mu = 4 \text{ people over a one-hour period}$$

A graph of the Poisson probabilities for $\mu = 4$ is contained in Figure 6.5.

1 What is the probability that, over any one-hour interval, exactly four people arrive at the emergency room?

2 What is the probability that more than one person will arrive?

3 What is the probability that you observe exactly six people over a two-hour period?

Figure 6.5

Poisson probabilities for $\mu = 4$

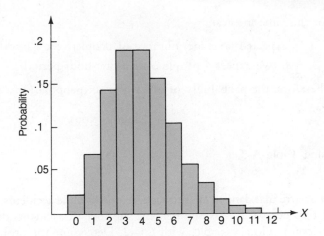

Solution 1

First, this probability is not 1 because $\mu = 4$ is the average number of arrivals over this time period. The actual number of arrivals over some one-hour period may be fewer than four, more than four, or exactly four. The fraction of time that you observe exactly four people is, using Table A-3,

$$P(4) = \frac{4^4 e^{-4}}{4!} = .1954$$

If you observe the emergency room for several one-hour periods, 19.5% of the time you will observe four people arrive.

Solution 2

This is $P(X > 1) = P(X \geqslant 2)$. We could try

$$P(X \geqslant 2) = P(X = 2) + P(X = 3) + \cdots$$
$$= P(2) + P(3) + \cdots$$
$$= .1465 + .1954 + \cdots$$

There is an infinite number of terms here, however, so this is not the way to find this probability. A much better way is to use the fact that the probabilities for all values of X sum to 1. Consequently,

$$P(X \geqslant 2) = 1 - P(X < 2)$$
$$= 1 - P(X \leqslant 1)$$
$$= 1 - [P(0) + P(1)]$$
$$= 1 - \left[\frac{4^0 e^{-4}}{0!} + \frac{4^1 e^{-4}}{1!} \right]$$
$$= 1 - [.0183 + .0733] = .9084$$

Solution 3   For this time interval,

μ = expected (average) number of people over a two-hour time period

= 8 (we expect 4 people over a one-hour period)

Therefore, the probability of observing six people over a two-hour period is

$$\frac{8^6 e^{-8}}{6!} = .1221$$

using Table A-3.   □

EXAMPLE 6.16   At a particular dangerous intersection, automobile accidents occur on the average of two per week. Assume that the number of accidents during any week is a Poisson random variable, with $\mu = 2$. Determine the probability that

  1 No accidents occur during the next week.
  2 There will be more than 10 accidents during the next four weeks.

Solution 1   The Poisson random variable X for this situation is the number of accidents. The average number of accidents over a one-week period is two, so

$$P(X = 0) = \frac{2^0 e^{-2}}{0!} = .135$$

using Table A-3. This means that, 13.5% of the time, no accidents will occur over a one-week period.

Solution 2   The average number of accidents over a four-week period is eight, given that the average is two over a one-week period. Therefore, using Table A-3 with $\mu = 8$:

$$P(X > 10) = 1 - P(X \leqslant 10)$$

$$= 1 - \left[\frac{8^0 e^{-8}}{0!} + \frac{8^1 e^{-8}}{1!} + \cdots + \frac{8^9 e^{-8}}{9!} + \frac{8^{10} e^{-8}}{10!} \right]$$

$$= 1 - [.0003 + .0027 + .0107 + \cdots + .1241 + .0993]$$

$$= .1841$$

Consequently, more than 10 accidents will occur roughly 18% of the time.

□

Poisson Approximation to the Binomial

There will be many times when you are in a binomial situation but n is too large to be tabulated. For such situations, you can use a computer. If that is not convenient, there are methods of approximating these probabilities without

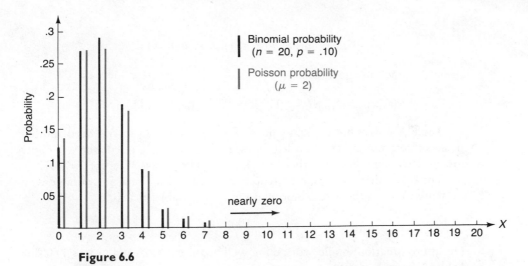

Figure 6.6

Poisson distributions provide a good approximation of binomial probabilities where $n > 20$ and $np \leqslant 7$. Here $n = 20$, $p = .10$.

sacrificing much accuracy. One method is to pretend that your binomial random variable X is a Poisson random variable having the same mean. The corresponding Poisson probability may be much simpler to derive and will serve as an excellent approximation to the binomial probability.

A good approximation to a binomial probability is obtained using the Poisson distribution if n is large and p is small. For most situations, you can trust this approximation if $n > 20$ and $np \leqslant 7$. An illustration using $n = 20$ and $p = .10$ is shown in Figure 6.6. The binomial probabilities are from Table A-1 and the Poisson probabilities are from Table A-3.

EXAMPLE 6.17

In **lot acceptance sampling**, you decide to accept or send back a lot (batch) consisting of many electrical components, machine parts, or whatever, using a randomly selected sample. A shipment of 2500 calculator chips arrive at Cassidy Electronics. The contract specifies that Cassidy will accept this lot if a sample of size 100 from this shipment contains no more than one defective chip. If we assume that 5% of the chips are defective, what is the probability that they will accept the lot?

Solution

We can treat this as a binomial situation (rather than the more complicated hypergeometric situation) because

$$\frac{n}{N} = \frac{100}{2500} = .04 < .05$$

This is approximately a binomial situation with $n = 100$ trials and $p = .05$. The binomial random variable X here is the number of defective chips out of 100.

So

$$P(\text{accept}) = P(X \leqslant 1) = P(0) + P(1)$$

Using the PMF in equation 6-8:

$$P(\text{accept}) = {}_{100}C_0 \cdot (.05)^0(.95)^{100} + {}_{100}C_1 \cdot (.05)^1(.95)^{99}$$

Table A-1 does not contain values of n larger than 20, so you need another means of determining these probabilities. A computer or a calculator will allow you to determine these exactly. The exact answer here is $P(\text{accept}) = .037$.

The other alternative is to pretend that X is a Poisson random variable with the same mean as the actual binomial random variable; that is,

$$\mu = np = (100)(.05) = 5$$

The approximation should work quite well because $n > 20$ and np is $\leqslant 7$. Using Table A-3, with $\mu = 5$,

$$P(0) = .0067$$

and

$$P(1) = .0337$$

Therefore,

$$P(X \leqslant 1) = .0067 + .0337 = .0404$$

Using the Poisson approximation,

$$P(\text{accept}) \cong .0404$$

which is quite close to the binomial value of .037.   □

Summary of the Three Most Commonly Used Discrete Random Variables

Binomial Distribution

1 X denotes the number of successes out of n independent trials. Each trial results in a success (with probability p) or a failure (with probability $1 - p$).

2 PMF is $P(X = x) = P(x) = {}_nC_x p^x(1 - p)^{n-x}$ for $x = 0, 1, \ldots, n$.

3 Mean $= \mu = np$.

4 Variance $= \sigma^2 = np(1 - p)$ and standard deviation $= \sigma = \sqrt{np(1 - p)}$.

5 Probabilities for the binomial random variable are provided in Table A-1.

Hypergeometric Distribution

1 X represents the number of successes in a sample of size n when selecting from a population of size N, containing k successes and $N - k$ failures.

2 PMF is $P(x) = (_kC_x \cdot _{N-k}C_{n-x})/_NC_n)$ for $x = a, a + 1, \ldots, b$, where $a = $ maximum $\{0, n + k - N\}$ and $b = $ minimum $\{k, n\}$.

3 Mean $= \mu = n(k/N)$.

4 Variance $= \sigma^2 = [n(k/N)(1 - k/N)][(N - n)/(N - 1)]$.

Poisson Distribution

1 $X = $ the number of occurrences of a particular event.

2 PMF is $P(x) = (\mu^x e^{-\mu})/(x!)$ for $x = 0, 1, 2 \ldots$.

3 Mean $= \mu$.

4 Variance $= \mu$.

5 Probabilities for the Poisson random variable are provided in Table A-3.

EXERCISES

6.50 Ten people applied for a job as a speech therapist and five of them have a master's degree. Six of the applicants are randomly selected for the job, what is the probability that exactly three have master's degrees?

6.51 Six vegetables are available at a cafeteria. Four vegetables are green and the other two are not green. If three different vegetables are ordered at random, what is the probability that at least two of the vegetables are green?

6.52 A population of 15 objects consists of 8 round objects and 7 square objects. Let the random variable X be equal to the number of round objects selected randomly without replacement in a sample of 9 objects.
(a) Find the $P(2 \leqslant X \leqslant 5)$.
(b) Find the $P(X > 4)$.
(c) Find the mean of the random variable X.
(d) Find the standard deviation of the random variable X.

6.53 A batch of 350 resistors is to be shipped if it is found that a random sample of 15 resistors has two or fewer defective resistors. If it is known that there are 50 defective resistors in the batch, what is the probability that two or fewer of the sample of 15 resistors will be found to be defective?

6.54 In a group of 20 people attending a seminar, 8 are single and 12 are married. If 5 people are randomly selected, answer the following where X is the number of single individuals in the sample.
(a) What is the $P(X \leqslant 2)$?
(b) What is $P(1 \leqslant X \leqslant 3)$?
(c) Find the mean and variance of X.

6.55   In a sample of 10 men, it is found that 6 are physically fit. If 4 men are selected from this sample of 10, what is the probability that no more than three men are physically fit?

6.56   A box contains eight golf balls. Four of these balls are not perfectly round. If three balls are randomly selected without replacement from the eight golf balls, what is the probability that at least one is not perfectly round?

6.57   A factory manufactures rubber grommets to be placed on the stick shift of a car. A sample of 10 grommets is chosen from a box of 200. Let the random variable X be the number of defective rubber grommets. Assume that it is known that there are 10 defective grommets in the box.
   (a)   What is the probability that X is greater than 2?
   (b)   What is the probability that X is equal to zero?
   (c)   What is the standard deviation of X?

6.58   A reviewer of textbooks is editing the text for grammatical errors. Let the random variable X represent the number of grammatical errors made in a particular chapter. Assume that the conditions of a Poisson distribution are satisfied with an average of 10 grammatical errors per chapter.
   (a)   What is the probability that X is less than 7?
   (b)   What is the mean value of X?
   (c)   What is the standard deviation of X?

6.59   Let the random variable X be binomially distributed with $n = 60$ and $p = 0.05$. Use the Poisson distribution to approximate the probability that X is greater than or equal to three.

6.60   A police officer writes an average of two speeding tickets per hour. What is the probability that in one hour the police officer writes no more than one speeding ticket? What assumptions need to be made?

6.61   The principal's office at Lakewood Junior High School sends out an average of 8 parent slips daily. The number of parent slips is assumed to follow a Poisson distribution.
   (a)   What is the probability that for any one day, the number of parent slips sent out will be more than four?
   (b)   What is the standard deviation of the number of parent slips sent out daily?

6.62   A survey indicates that 10% of the people who earn less than 20 thousand dollars a year are homeowners.
   (a)   If a sample of 40 people who earn less than 20 thousand dollars a year is randomly selected, what is the probability that more than 4 people are homeowners?
   (b)   What is the probability that exactly 4 people are homeowners from the sample of 40 people who earn less than 20 thousand dollars a year?

6.63   The owner of a local ticket agency knows that only 6 customers can be handled effectively in a 15-minute interval. If the average number of customers calling during a 15-minute interval is 5, what is the probability that more than 6 customers will call in a 15-minute interval? Assume a Poisson distribution.

SUMMARY

When an experiment results in a numerical outcome, a convenient way of representing the possible values and corresponding probabilities is to use a random variable. A **random variable** takes on a numerical value for each outcome of an experiment. If the possible values of this variable can be listed along with the probability for each value, this variable is said to be a **discrete random variable**. Conversely, if any value of this variable can occur over a specific range, then it is a **continuous random variable**. This chapter concentrated on the discrete type, whereas Chapter 7 will discuss the continuous random variable.

For a discrete random variable, the set of possible values and corresponding probabilities is a **probability distribution**. There are several ways of representing such a distribution, including a list of each value and its probability, a bar chart, or an expression called a **probability mass function** (PMF), which is a numerical function that assigns a probability to each value of the random variable.

If you could observe a random variable indefinitely, you would obtain the corresponding population. A sample then consists of a finite number of random variable observations. In Chapter 3, we introduced ways of describing a set of sample data using various statistics, including the sample mean and variance. Similarly, we can describe a random variable using its mean and variance. Since they describe the population, they are parameters. The mean of a discrete random variable μ is the average value of this variable if observed over an indefinite period. The **mean** is found by summing the product of each value and its probability of occurring. The **variance** of a discrete random variable σ^2 is a measure of the variation for this variable. The **standard deviation** σ also measures this variation and is the square root of the variance.

Four popular discrete random variables discussed in this chapter are the (discrete) uniform, binomial, hypergeometric, and Poisson random variables. A (discrete) **uniform** random variable has a set of outcomes, each of which is equally likely to occur. A **binomial** random variable counts the number of successes out of n independent trials. When the trials are dependent and the population size (N) and the number of successes in the population (k) are known, the **hypergeometric** random variable can be used to describe the number of successes out of the n trials. The **Poisson** random variable is used for situations where you are observing the number of occurrences of a particular event over a specified period of time or space. A table of binomial probabilities is contained at the back of the text in Table A-1. Tables A-2 and A-3 can be used to obtain Poisson probabilities. For these four distributions, shortcut formulas exist for deriving the mean and variance of the random variable. Often the probabilities for one of these discrete distributions are difficult to calculate due to the magnitude of the numbers involved. In many situations, you can use one discrete distribution to approximate the probability for another. Figure 6.7 summarizes how this is done.

Figure 6.7

Summary of how
the three most
common types of
discrete random
variables can be
used to approximate
values for one
another

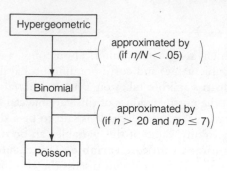

REVIEW EXERCISES

6.64 Is the following function a probability mass function? Why or why not?

$$P(X = x) = x^3/153 \qquad \text{for } x = 1, 3, 5 \quad \text{(and zero otherwise)}$$

6.65 Goodwin and Sanati (1986) describe a new approach to the teaching of a beginning course in Pascal computer programming using a new learning package called PASLAB. There were 322 students taught by the traditional approach in 1984, and 296 students participating under the new conditions in 1985. The following results were reported:

FINAL GRADES UNDER TRADITIONAL (AB84) AND NEW (CA85) CONDITIONS

Definition of Grade	X = Grade	% of Students (AB84)	% of Students (CA85)
Did not pass; completed 1–2 assignments/exams	0	2	4
Did not pass; completed half of assignments/exams	1	6	3
Did not pass; completed almost all assignments/exams	2	15	4
Passed course with grade of "Acceptable"	3	55	68
Passed course with grade of "Distinction"	4	22	21
		100	100

Source: Leonard Goodwin and Mohammad Sanati, Learning computer programming through dynamic representation of computer functioning: Evaluation of a new learning package for Pascal. *Int. J. Man–Machine Studies* **25**: 327–341 (1986).

(a) What is the average grade for the "traditional" students?
(b) What is the average grade for the "new" students?
(c) Compute the variance of the distribution for each of the two groups.

6.66 Assume that a fair die has one blue face, two white faces, and three black faces. Define the random variable X as follows:

$$X = \begin{cases} 1 \text{ if blue} \\ 2 \text{ if white} \\ 3 \text{ if black} \end{cases}$$

Find the probability mass function of X. Construct a bar chart in which the height of each bar is the probability of X.

6.67 Find the mean and variance of the following random variable X with probability mass function $P(x)$.

X	P(x)
-3	.2
0	.1
3	.2
5	.3
10	.2

6.68 A state licensing agency has used historical data to construct the following discrete probability distribution, where the random variable X represents the number of child-care centers that are unlicensed out of every 10 inspected in various counties in the state.

X	P(x)
0	.40
1	.30
2	.15
3	.10
4	.05

(a) Find the mean and standard deviation of this distribution.
(b) Find the probability that at least 2 unlicensed child-care centers are found out of every 10 inspected.

6.69 For a binomially distributed random variable X with 12 trials and with the probability of a success equal to .3, find the following.
(a) $P(X = 7)$.
(b) $P(4 < X \leqslant 6)$.
(c) $P(X > 5)$.
(d) $P(X < 2)$.

6.70 A manager has 10 research projects to assign to either engineer 1 or engineer 2. If each research project is randomly assigned to either one of the two engineers, what is the probability that engineer 1 will be assigned no more than 5 research projects?

6.71 Let the random variable X be equal to the number of apartments vacant in a 20-unit apartment complex at any one time. If the apartment complex has a 20% vacancy rate, what is the probability that more than one but fewer than five apartments are vacant?

6.72 Despite strong health warnings, escalating costs and a variety of sanctions against smoking, three adult Americans in 10 still cling to the habit, according to a Gallup poll based on telephone interviews with 1015 adults, 18 and older, conducted across the nation March 14–18, 1987. Results from polls taken in 1985, 1986, and 1987 are shown below.

GALLUP SURVEY

Percentage of people who responded affirmatively to the question, "Have you, yourself, smoked any cigarettes in the past week?"

Year	Total	Men	Women
1987	30%	33%	28%
1986	31	35	28
1985	35	37	32

Source: Los Angeles Times Syndicate, April 2, 1987.

Assuming that the above percentages are representative of the U.S. population, what is the probability that in a random sample of 10 adults in 1987,

(a) Exactly 3 are smokers?

(b) At least 3 are smokers?

(c) Between 3 and 5 (inclusive) are smokers?

(d) 7 or more are NOT smokers?

6.73 The Family Planning Unit in the Ministry of Health of a developing country calculated that for the nation as a whole, only 28% of married couples used some form of birth control. If a random group of 30 couples is selected from this country, what is the probability that in this group,

(a) Exactly 2 couples used birth control?

(b) At least 2 couples used birth control?

6.74 Consider the following profiles of the freshman class of three universities for the 1986–87 year:

University	Applicants	% Accepted	Enrolled	% Students in State	% Minority Enrolled	% Students on Financial Aid
Duke	12655	26	1543	15	11	36
Stanford	16100	16	1575	38	32	65
Vanderbilt	7001	55	1331	17	5	33

Source: Barron's Profiles of American Colleges, 1986–1987.

(a) In a random sample of 12 applicants to Stanford University's freshman class, what is the probability that

(i)   Between 3 and 4 (inclusive) enroll and are accepted?

(ii) Exactly 4 actually enroll?

(b) In a random sample of 20 students at Duke University, what is the probability that 4 are in-state students?

(c) In a random sample of 15 students at Vanderbilt University, what is the probability that at least one is a minority student?

6.75 Records of an urban library indicate that 5% of all books checked out are never returned. A random sample is taken of 18 books checked out. What is the probability that
(a) Two books are never returned?
(b) Fewer than two books are never returned?
(c) All books are returned?

6.76 An English teacher grading her first test for the semester noticed that the number of grammatical errors per essay ranged from 2 to 10 and seemed to follow a uniform distribution. Let X represent the number of grammatical errors per essay.
(a) State the probability mass function for X.
(b) What is $P(X \geqslant 7)$?
(c) What is $P(4 \leqslant X \leqslant 8)$?
(d) Find the mean and variance of X.

6.77 In the wake of increased terrorism and airplane hijacking incidents in Europe, American tourists canceled their vacation bookings to Europe in large numbers. Seven out of ten cancellations by American tourists cited fear of hijacking as the reason for the cancellation. If a random selection is made of 15 American tourists who canceled their European vacation bookings, what is the probability that in this group of 15, fewer than 10 will cite fear of hijacking as the reason for the cancellation?

6.78 From the fall of 1986, prospective teachers in the state of Texas are required to have passed the Examination for Certification of Educators in Texas (ExCET) before they can be licensed to teach. According to figures released by the Texas Education Agency, in the first statewide test administered, there was an overall passing rate of 80%. However, at the University of Texas at Arlington, 95% of the candidates passed the ExCET.
(a) Do these passing rates of 80 and 95% refer to the same population?
(b) If a group of 20 candidates is randomly selected statewide, what is the probability that at least 16 passed the ExCET?
(c) If a group of 20 candidates is randomly selected from the University of Texas at Arlington, what is the probability that at least 16 passed the ExCET?

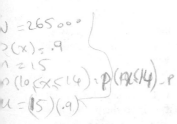

6.79 A report on Asian Americans estimated that by 1985 there were about 265,000 Vietnamese living in the United States. Suppose that 90% of this group are foreign-born. If 15 Vietnamese immigrants are randomly selected, what is the probability that
(a) Between 10 and 14 inclusive are foreign-born?
(b) Fewer than 10 are foreign-born?
(c) All 15 are foreign-born?

6.80 The probability function of a discrete random variable X is given by

$$P(X = x) = \frac{x^3}{100} \quad \text{for } x = 1, 2, 3, 4 \quad \text{(and zero otherwise)}$$

(a) Is this a uniform distribution?
(b) Is this a binomial distribution?

(c)   What is $P(X > 2)$?

(d)   What is $P(1 \leqslant X \leqslant 3)$?

(e)   Compute the mean and standard deviation of this distribution.

6.81 In Greenback County, 20% of the residents are millionaires.

(a)   In a randomly selected group of 10 residents of Greenback County, what is the probability that 5 are millionaires?

(b)   In a randomly selected group of 20 residents of Greenback County, what is the probability that 10 are millionaires?

(c)   Compare your answers to (a) and (b). Are they the same, or are they different? Could you have obtained the answer to (b) by doubling the answer for (a)? Explain the results. (*Hint:* Ask yourself whether 5 successes in 10 trials is identical to 10 successes in 20 trials. Try to imagine the tree diagram for each case.)

6.82 Refer to exercise 6.28. From the hypothetical population mentioned in this exercise, a group of 12 persons is randomly selected. What is the probability that in this group, fewer than 5 persons will have finger dexterity levels below level 4?

6.83 It is believed that 10% of all income tax returns showing a gross income of $200,000 or more are subject to an IRS audit. A random group of 14 tax returns is selected, all showing gross income of $200,000 or higher. What is the probability that

(a)   4 returns are subjected to an IRS audit?

(b)   Fewer than 4 returns are subjected to an IRS audit?

6.84 Over the last several years, the local high school baseball team has won 8 out of 10 games played on the home field. If 4 games played on the home field are randomly selected, what is the probability that the local team won all 4 games? Also, what is the probability that the local team won none of the 4 games?

6.85 An average of five books per week are returned to a book store. Assume that the number of returned books is Poisson distributed.

(a)   What is the probability that fewer than 4 books will be returned in one week?

(b)   What is the standard deviation of the number of books returned in one week?

6.86 A medical research firm finds that 8% of the people who have high blood pressure do not know it. If 35 people who have high blood pressure are randomly selected, what is the probability that at least 3 people do not know they have high blood pressure?

6.87 Ten employees are being reviewed for promotion. Four of the employees are females. If each employee is equally likely to get promoted, what is the probability that two females and three males will be promoted if a total of five promotions are given?

6.88 There are 90 marriage applications on file at a county court house covering the last 30 days. In 50 of these applications, at least one of the individuals had been previously divorced. In the remaining couples, neither person had ever been divorced. If four applications are selected at random, what is the probability that two of the applications will contain at least one divorced individual, and two will not?

6.89 A population consists of 15 employees, 6 of whom have less than two years experience. Let X be equal to the number of employees with less than two years experience in a sample of eight employees drawn from this population.

(a)   Find $P(X = 3)$.

(b)   Find $P(X \leqslant 2)$.

(c)   Find the average value of X.

(d)   Find the standard deviation of X.

6.90 Blair's Moving Company loads an average of 3 boxes of damaged merchandise daily. What is the probability that exactly 3 boxes of damaged merchandise are shipped daily? What is the standard deviation of the number of boxes of damaged merchandise which are shipped daily? Assume a Poisson distribution.

6.91 A songwriter has seven songs, three of which are ballads. If the songwriter chooses two at random, what is the probability that

(a)   Exactly one is a ballad?

(b)   None are ballads?

(c)   Both are ballads?

6.92 Water intoxication is a syndrome that occurs in 3–5% of chronically hospitalized psychiatric patients. It is characterized by profound hyponatremia (serum sodium level < 120 meq/liter) and a variable constellation of neurologic symptoms, including delirium, seizures, coma, and death. Goldman and Luchins (1987) describe a fairly simple approach that they have used to prevent episodes of water intoxication.* It is believed that the frequencies of water intoxication episodes might follow a Poisson distribution. We will assume that a Poisson distribution does apply. For the patients in the Goldman and Luchins study, there were an average of 2 prior episodes per patient over a period of roughly 8 months. (During the study, there were no episodes.) If the preventive measures described by the doctors were not implemented, what is the probability that over a given period of 8 months, a randomly selected patient will experience:

(a)   Two or more episodes of water intoxication?

(b)   Between one and three (inclusive) episodes of water intoxication?

(c)   No episodes of water intoxication?

6.93 Burrus-Bammel and Bammel (1986) in the Division of Forestry at West Virginia University have investigated the effects of population density and crowding at Cranberry Mountain Visitor Center in the Monongahela National Forest in Richwood, West Virginia.† The center is open from 9:00 A.M. to 5:00 P.M. daily. Systematic observations of visitor and vehicle arrivals were made on alternate days for two weeks during the summer of 1983. The frequency of vehicle arrivals was found to be an average of 55.29 vehicles per day, that is, 6.9 vehicles per hour. If we have reason to believe that vehicle arrivals follow a Poisson distribution, what is the probability of observing:

(a)   Less than 10 vehicle arrivals in an hour?

(b)   Between 7 and 10 (inclusive) vehicle arrivals in an hour?

(c)   More than 15 vehicle arrivals in an hour?

* *Source:* M. J. Goldman and D. J. Luchins, Prevention of episodic water intoxication with target weight procedure. *Am. J. Psychiatr.* **144** (3): 365–366 (1987).

† *Source:* Lei Lane Burrus-Bammel and Gene Bammel, Visiting patterns and effects of density at a visitors' center. *J Environ. Educ.* **18** (1): 7–10 (Fall 1986).

CASE STUDY

The Usage of "May" and "Must" by Hamilton and Madison

As you are undoubtedly aware, Alexander Hamilton, John Jay, and James Madison wrote a collection of essays between 1787 and 1788 to persuade the citizens of the State of New York to ratify the Constitution of the United States of America. These essays appeared in newspapers under the pseudonym "Publius" and have come to be collectively referred to as *The Federalist Papers*.

Two researchers (Mosteller and Wallace, 1964) studied the use of various *function* words by Hamilton and Madison. Function words are filler words such as by, to, that, may, must, and so on. The study covered 247 blocks of 200 words each of text known to have been written by Alexander Hamilton and 262 blocks of 200 words each written by James Madison. (The idea behind the study was apparently to discover some usage patterns to assist in case of disputed authorship.)

Although many function words were studied, we give below the distribution of usage of the words "may" and "must" by these illustrious gentlemen, just to give you a flavor of the study, and to obtain a feel for discrete distributions.

DISTRIBUTION OF USAGE OF FUNCTION WORDS

Frequency of Usage	Hamilton		Madison	
	"may"	"must"	"may"	"must"
0	128	173	156	202
1	67	49	63	47
2	32	14	29	10
3	14	9	8	2
4	4	1	4	1
5	1	0	1	0
6	1	1	1	0
	247	247	262	262

Source: F. Mosteller and D. L. Wallace, *Inference and Disputed Authorship: The Federalist* (Reading, MA: Addison-Wesley, 1964). Reprinted with permission from Springer-Verlag New York Inc.

The data in the table are to be read as follows: of the 247 blocks of text by Hamilton, 128 had no occurrences of "may," 67 had one occurrence of "may," and so on.

Case Study Questions

1. If we define the random variable X as the frequency of usage of a particular word by a particular author in a block of text, do any of the four distributions above follow a uniform distribution?

2. If they did follow a uniform distribution, what would be:
   (a) the *probability* of 4 occurrences of the word "must" in Hamilton's block of text?
   (b) the *frequency* of 4 occurrences of the word "must" in Madison's block of text?

3. What is the total number of words of text analyzed by the researchers for
   (a) Hamilton?
   (b) Madison?

4. What is the total number of occurrences of each of the two authors? (For example, how many times in the entire text did the researchers find the use of "may" by Hamilton, and so on.)

5. It had been thought that the distribution of words might follow a *Poisson* distribution. However, this was not the case. It is not clear whether we could fit the data to a binomial distribution. However, let us contrive that just as an exercise, and assume that we have binomial distributions with $p = 0.01$ for the word "may" and $p = 0.02$ for the word "must". We are looking for the word that we are interested to occur 0, 1, 2, . . . times in each block.
   (a) What is the sample size n?
   (b) In any given block of text by Hamilton, what is the probability that:
       (i) "may" occurs exactly one time?
       (ii) "must" occurs less than one time?
   (*Note:* You may want to use the Poisson approximation to the binomial here.)
   (c) In any given block of text by Madison, what is the probability that:
       (i) "may" occurs one time or not at all?
       (ii) "must" occurs between two and three times (inclusive)?
   (*Note:* You may want to use the Poisson approximation to the binomial here.)

COMPUTER EXERCISES USING THE DATABASE

Using a sample of 100 observations, complete the following exercises:

1. Consider the variable SEX. Let p represent the proportion of males in your set of 100 observations. If you were to randomly select 10 observations (with replacement) from the 100 possible, what is $P(X \leq 3)$, where X is the number of males (out of 10)? What type of random variable is X?

2. Repeat exercise 1 where the 10 observations are selected without replacement.

3. Referring to exercise 2, if you were to obtain samples of size 10 indefinitely, what would X be on the average? What is the standard deviation of this random variable?

4. Let p be the proportion of students in your set of 100 observations that have never been married. If you were to randomly select (with replacement) 20 observations from the 100 possible, what is the probability of selecting exactly one student that has never been married? What is the probability that all 20 students are married or have been married at some time?

7

CONTINUOUS PROBABILITY DISTRIBUTIONS

A LOOK BACK/INTRODUCTION

After we described the use of descriptive statistics, we introduced you to the area of uncertainty by using probability concepts and random variables. Random variables offer you a convenient method of describing the various outcomes of an experiment and their corresponding probabilities.

When each value of the random variable as well as its probability of occurring can be listed, the random variable is discrete. The other type, a continuous random variable, can assume any value over a particular range. This includes such variables as X = height, X = weight, and X = time. For this kind of situation, it is impossible to list all values of X, yet you can still make probability statements regarding X if you can make certain assumptions about the type of population.

Making decisions from sample information in statistics is called **statistical inference**. In subsequent chapters, we will develop a formal set of rules to offer you a guide in making statistical decisions. The making of such a decision typically involves one or more assumptions about the population from which the sample was obtained. One such assumption, widely used in statistics, is that the data came from a normal population, which means that you are dealing with a normal random variable.

The concept of a continuous random variable was introduced in Chapter 6. What distinguishes a discrete random variable from one that is continuous is the presence of gaps in the possible values for a discrete random variable. To illustrate, X = total of two dice is a discrete random variable; there are many gaps over the range of possible values and a value of 10.4, for example, is not possible. One can list the possible values of a discrete variable, along with the probability that each value will occur.

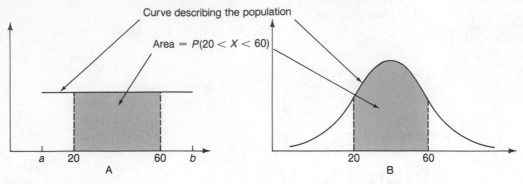

Figure 7.1

Finding a probability for a continuous random variable. (A) a continuous uniform random variable; (B) a normal random variable.

Determining probabilities for a continuous random variable is quite different. For such a variable, any value is possible over a specific range. This means that we are unable to list all of the possible values of this variable. Probability statements for a continuous random variable (say, X) are not concerned with specific values of X (such as the probability that X will equal 50) but rather deal with probabilities over a range of values, such as the probability that X is between 40 and 50, greater than 65, or less than 20, for example.

Such probabilities can be determined by first making an assumption regarding the nature of the population involved. We assume that the population can be described by a curve having a particular shape, such as uniform, normal, or exponential. Once this curve is specified, a probability can be determined by finding the corresponding area under this curve. As an illustration, Figure 7.1 demonstrates two particular curves (called the **uniform curve** and **normal curve**) for which the probability of observing a value of X between 20 and 60 is the area under this curve between these two values. The entire range of probability is covered using such a curve since, for any continuous random variable, the total area under the curve is equal to 1.

The following sections will examine three commonly encountered continuous random variables: the uniform, normal, and exponential random variables. The graphs and descriptive statistics described in the previous chapters can help determine if one of these random variables might be appropriate for a particular situation. If a histogram of the sample data appears nearly flat, the population might be represented by a (continuous) uniform random variable. If the histogram is symmetric with decreasing tails in each end, a normal random variable may be in order. If the sample histogram steadily decreases from left to right, the population of all possible values perhaps can be described using an exponential random variable. In the first two cases, the mean and median should be nearly equal (the population is symmetric) providing a skewness measure near zero. For the exponential case, the median should be less than the mean with a corresponding positive measure of skewness.

Figure 7.2

Relative frequency histogram of a sample of 150 waiting times

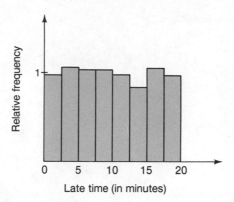

7.1 CONTINUOUS UNIFORM RANDOM VARIABLES

As an example of a simple continuous random variable, consider spinning the minute hand on a clock face. Define a random variable X to be the stopping point of the minute hand. It seems reasonable to assume that, for example, the probability that X is between two and four is twice the probability of observing a value of X between eight and nine. In other words, the probability that X is in any particular interval is proportional to the width of that interval.

A random variable of this nature is a **continuous uniform random variable**. The values of such a variable are evenly distributed over some interval because the random variable occurs randomly over this interval. This random variable bears a distinct resemblance to the discrete uniform random variable in Chapter 6, except that here *any* value is possible over a certain range.

Consider a situation where the citizens of a minority neighborhood are concerned with the length of time that the bus is late during the hours of 6 P.M. to midnight, during which the bus is supposed to arrive every 40 minutes. They contend that the undependable service will make it unsafe for residents to travel alone.

To check out the present waiting time situation, a sample of 150 waiting times was randomly obtained for passengers using the bus at this intersection during these hours. The relative frequency histogram made from these 150 observations is shown in Figure 7.2. The distribution of the sample appears to be a uniform (flat) distribution. The random variable, $X =$ waiting time for a city bus, is a **uniform random variable**. The corresponding smooth curve describing the population is shown in Figure 7.3.

Whenever we describe a continuous random variable, we will use the same strategy, namely

1 A histogram represents the shape of the sample data.
2 A smooth curve represents the assumed shape (distribution) of the population.

Figure 7.3

Uniform distribution
for X = waiting time
(compare with
Figure 7.2)

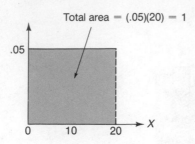

Figure 7.4

Total area for a
uniform distribution

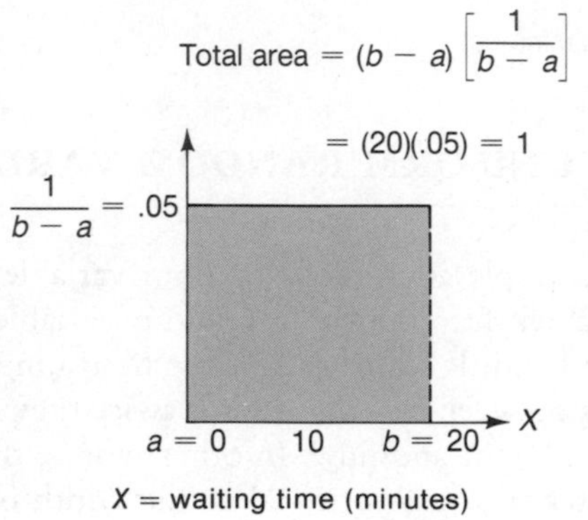

$$\text{Total area} = (b - a)\left[\frac{1}{b - a}\right]$$

$$= (20)(.05) = 1$$

$$\frac{1}{b - a} = .05$$

$a = 0 \quad 10 \quad b = 20$

X = waiting time (minutes)

Notice that the total area here is given by a rectangle, and, as is true of all continuous random variables, this total area must be 1. The area of a rectangle is given by (width) · (height). By making the height of this curve (a straight line, actually) equal to .05, the total area is

$$(20 - 0)(.05) = 1.0$$

In general, the curve defining the probability distribution for a uniform random variable is shown in Figure 7.4. The total area is

$$(b - a)\left[\frac{1}{b - a}\right] = 1.0$$

Mean and Standard Deviation

In Chapter 6, we discussed the mean of (say) 10 observations of the random variable X, written as $\bar{X}$. If you were to observe X indefinitely, then you could obtain the mean of the population, μ. The same concept applies to continuous random variables where, for the city bus example, $\bar{X}$ represents the mean of the 150 observations (the sample) and μ is the mean of all waiting times between 6 P.M. and midnight (the population).

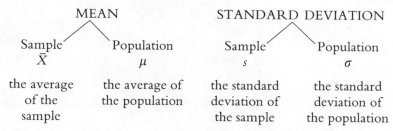

MEAN		STANDARD DEVIATION	
Sample $\overline{X}$	Population μ	Sample s	Population σ
the average of the sample	the average of the population	the standard deviation of the sample	the standard deviation of the population

Refer to Figure 7.4. The mean (μ) of X is the value midway between a and b, namely,

$$\mu = \frac{a + b}{2}$$

The standard deviation of X is a measure of how much variation there would be in X if you were to observe it indefinitely. The standard deviation of the population σ, just like the sample standard deviation s, is a measure of **variability**. The larger σ is, the more variation (jumping around) we would see if X were observed indefinitely. For both the sample and the population, the square of the standard deviation is referred to as the **variance**. It, too, is a measure of the variability of X. The variance of a random variable X is represented by σ^2. The standard deviation of the (continuous) uniform random variable X is given by

$$\sigma = \frac{b - a}{\sqrt{12}}$$

Determining Probabilities

As for all continuous random variables, a probability using a uniform random variable is determined by finding an area under a curve. Suppose, for example, the concerned citizens would like to know what percentage of the time a person will have to wait more than 15 minutes for a city bus between the hours of 6 P.M. and midnight. In Figure 7.5, the shaded area is a rectangle, so its area

Figure 7.5

The probability that X exceeds 15. The shaded area represents the percentage of times that a waiting time is more than 15 minutes.

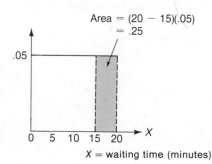

is easy to find:

$$\text{Area} = (\text{length}) \cdot (\text{height})$$
$$= (20 - 15)(.05)$$
$$= .25$$

So 25% of the time a person waits more than 15 minutes.

EXAMPLE 7.1

What is the probability that a person will wait between 5 and 15 minutes? What is the average waiting period?

Solution

Based on Figure 7.6, the chances that a person waits between 5 and 15 minutes is

$$P(5 < X < 15) = .5$$

The average waiting period (mean of X) is

$$\mu = \frac{0 + 20}{2} = 10 \text{ minutes}$$

The standard deviation of X is

$$\sigma = \frac{20 - 0}{\sqrt{12}} = 5.77 \text{ minutes}$$ □

For any continuous random variable, the probability that X is equal to any particular value is zero. So,

$$P(X = 5) = P(X = 15) = 0$$

This follows since there is no area under the curve at $X = 5$ and at $X = 15$. As a result,

$$P(5 \leqslant X \leqslant 15) = P(5 < X < 15) = .5$$

Simulation is an area of statistics that relies heavily on the uniform distribution. In fact, this distribution is the underlying mechanism for this often

Figure 7.6

The probability that X is between 5 and 15

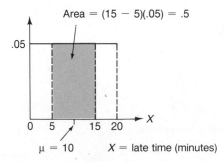

Area = (15 − 5)(.05) = .5

$\mu = 10$ X = late time (minutes)

complex procedure. So, although not many "real world" populations resemble this distribution, the uniform distribution is extremely important in the application of statistics.

EXERCISES

7.1 A random variable X has a uniform distribution between the values 0 and 4.
(a) What is the mean of X? $\frac{4}{2} = 2$
(b) What is the standard deviation of X? $\frac{4}{\sqrt{12}} = 1.155$
(c) What is the height of the probability distribution of X? $\frac{1}{4-0} = .25$
(d) What is the probability that X is greater than 1.23? $(4-1.23)(.25) = 0.6925$

height $\frac{1}{104-100} = .25$
$(101.5-100)(.25)$
$= .375$
$\frac{104+100}{2} = 102.375$

7.2 The temperature of a warming tray is uniformly distributed between the values of 100 and 104°F.
(a) What percentage of the time is the warming tray less than 101.5°F?
(b) What is the mean temperature of the warming tray?
(c) What is the standard deviation of the temperature of the warming tray? $\sigma = \frac{104-100}{\sqrt{12}} = 1.16$

height $\frac{1}{26.3-20} = 0.16$
$P(21.3 < x < 24.6)$
$(24.6-21.3)(.16) = .528$

7.3 The rate at which a swimming pool is filled is uniformly distributed between 20 and 26.3 gallons per minute.
(a) What is the probability that the rate at which the swimming pool is filled at any one time is between 21.3 and 24.6 gallons per minute?
(b) What is the mean rate at which the swimming pool is filled? $\mu = \frac{26.3+20}{2} = 23.15$
(c) What is the standard deviation of the rate at which the swimming pool is filled? $\sigma = \frac{6.3}{\sqrt{12}} = 1.82$

7.4 The response time for a computer system, measured as the delay between query and response in seconds, is a continuous uniformly distributed random variable, ranging from 5 to 15 seconds.
(a) Compute the mean and standard deviation of the distribution.
(b) What percentage of the time is response time less than 8 seconds?
(c) What is the probability that response time is from 8 to 12 seconds, inclusive?

height $\frac{1}{18-7} = 0.091$
$(18-15)(0.091) = 0.273$

7.5 A dance instructor was analyzing a videotape of tap dance routines, using the slow motion option on her video cassette recorder (VCR). She made a preliminary determination that the duration of a particular pattern that was repeated over and over by many performers seemed to follow a uniform distribution over a range of 7 to 18 seconds.
(a) What percentage of the tap dance routines of that particular pattern have a duration greater than 15 seconds?
(b) What is the average duration? $\frac{18+7}{2} = 12.5$
(c) What is the variance of the duration? $\sigma = \frac{18-7}{\sqrt{12}} = 3.175 \Rightarrow \sigma^2 = 10.08$

7.6 The amount of cholesterol contained in the cheeseburgers served by a national fast food chain ranges from 140 to 150 milligrams, and is uniformly distributed over that range.
(a) How much cholesterol does the average cheeseburger contain at this chain?
(b) What is the probability that any one cheeseburger has more than 146 milligrams of cholesterol?

7.7 The quantity of milk in one-gallon cartons of homogenized milk is uniformly distributed with a mean of 128.5 ounces and variance of 2.0834 square ounces. What is the probability that the quantity of milk in a carton is greater than one gallon (i.e., 128 ounces)?

$\mu = 128.5$
$\sigma^2 = 2.0834 \Rightarrow \sigma = 1.44$

$\sigma = \frac{b-a}{\sqrt{12}} \Rightarrow 1.44 = \frac{b-a}{\sqrt{12}} \Rightarrow b-a = 4.988 \Rightarrow b = 4.988+a \quad \rightarrow 4.988+a+a = 257$

$\boxed{a = 126.006}$
$\boxed{b = 130.994}$

$\mu = \frac{b+a}{2} \Rightarrow 128.5 \times 2 = b+a \Rightarrow b+a = 257 \Rightarrow b = 257-a$

height $\frac{1}{b-a} = 0.20$

$\Rightarrow P(X > 128) = (130.994 - 128)(0.20) = 0.5988$

7.8   The automatic cruise control on a car has been set at 55 m.p.h. The actual speed is measured at regular intervals. Suppose the actual speed follows a uniform distribution over a range of 53 to 58 m.p.h.

(a)   What is the average speed?

(b)   What percentage of the time does the car exceed the speed limit of 55 m.p.h.?

7.2   NORMAL RANDOM VARIABLES

The normal distribution is the most important of all the continuous distributions. You will find that this distribution plays a key role in the application of many statistical techniques. When attempting to make an assertion about a population by using sample information, a major assumption often is that the population has a normal distribution.

When discussing measurements such as height, weight, test scores, or time, the resulting population of all measurements often can be assumed to have a probability distribution that is normal.

A histogram constructed from a large sample of such measurements can help determine whether this assumption is realistic. Assume, for example, that data were collected on the scores of 200 students taking a national achievement test. Let X represent an individual score. One thing we are interested in is the shape of the distribution of the 200 scores. Where are they centered? Are they symmetric? The easiest way to approach such questions is to construct a histogram of the 200 values, as illustrated in Figure 7.7. This histogram indicates that the data are nearly symmetric and are centered at approximately 500.

The curve in Figure 7.7 is said to be a **normal curve** because of its shape. A normal curve is characterized by a *symmetric, bell-shaped appearance*, with tails that "die out" rather quickly. Unlike data that are from a uniform distribution, under a normal curve the values are much more concentrated about the mean. We use such curves to represent the *assumed population* of all possible values.

Figure 7.7

Histogram of 200 national achievement test scores. The curve represents all possible values (population). The histogram represents the sample (200 values).

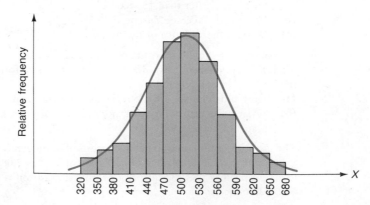

Figure 7.8

Distribution of the
test scores showing
the mean ($\mu = 500$),
the standard
deviation ($\sigma = 75$),
and the inflection
point (P)

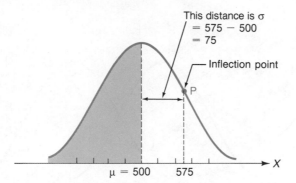

This example contained 200 values observed in a sample. Once again we use a histogram to represent the shape of the sample data and a smooth curve to represent the assumed shape of the population.

If all possible values of a variable X follow an assumed normal curve, then X is said to be a **normal random variable**, and the population is **normally distributed**.

When you assume that a particular population follows a normal distribution, you assume that X, an observation randomly obtained from this population, is a normal random variable. Based on the histogram in Figure 7.7, it appears to be a reasonable assumption that the smooth curve describing the population of all test scores can be approximated using a normal curve centered at 500. Therefore, we will assume that X is a normal random variable, centered at 500.

There are two numbers used to describe a normal curve (distribution); namely, where the curve is centered and how wide it is. The *center* of a normal curve is the mean and is represented by the symbol μ. The *width* of a normal curve can be described using the standard deviation, represented by the symbol σ.

These are illustrated in Figure 7.8, which shows the normal curve representing the population of all test scores. Another way of stating this situation is: X is a normal random variable with $\mu = 500$ and $\sigma = 75$. Notice that the units of μ and σ are the same as the units of the data (exam points).

In Figure 7.8, there is a point P on the normal curve. Above point P, the curve resembles a bowl that is upside down, and below P the curve is "right side up." In calculus, this point is referred to as an **inflection point**. The distance between vertical lines through μ and P is the value of σ. So when using a normal curve, you can represent σ graphically.

Because μ and σ represent the location and spread of the normal distribution, they are called **parameters**. The parameters are used to define the distribution completely. The values of μ and σ of a normal population are all you need to separate it from all other normal populations that have the same bell shape but different location and/or variability. The values of these parameters

Figure 7.9

Two normal curves
with unequal means
and equal standard
deviations

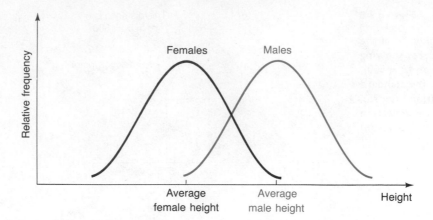

must be specified in order to make probability statements regarding X. As a result, there are infinitely many normal curves (populations), one for pair of values of μ and σ.

In our test score example, the average score of all observations is assumed to be $\mu = 500$. The standard deviation σ is a measure of the variability of the random variable X. For the achievement test scores illustrated in Figure 7.8, the value of this parameter was assumed to be $\sigma = 75$ points.

Consider whether the sample average $\bar{X}$ of the 200 values in our example is the same as μ. It is not. Do not confuse the average of all possible test scores (μ) with the average score of the 200 individuals in the sample ($\bar{X}$). This is an important distinction in statistics. However, if our assumed normal distribution (with $\mu = 500$ and $\sigma = 75$) is correct, then $\bar{X}$ most often will be close to μ in value. We will examine this again in Chapter 8.

The curve in Figure 7.8 is an illustration of a normal random variable with a mean of 500 points and a standard deviation of 75 points. We can compare normal curves that may differ in mean, standard deviation, or both. The normal curves in Figure 7.9 indicate that, on the average, males are taller than females. The mean of the male curve is to the right of the mean of the female curve. The male heights "jump around" about as much as female heights. In other words, there is about the same amount of variation in male and female heights. This is because the standard deviation of each curve is the same; that is, each curve is equally wide.

In Figure 7.10, the two normal curves represent the daily high temperatures for two different cities. It appears that

1   The average maximum temperature for the two cities is the same.

2   The temperatures in City B have more variability. This simply means that there are more cold days and hot days in City B than in City A.

Figure 7.10

Two normal curves
with equal means
and unequal
standard deviations

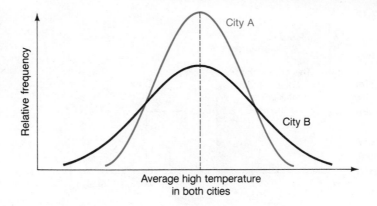

Figure 7.10

Two normal curves with equal means and unequal standard deviations

DETERMINING A PROBABILITY
7.3 FOR A NORMAL RANDOM VARIABLE

You have assumed that the score of an individual taking the national achievement test is a normal random variable with $\mu = 500$ and $\sigma = 75$. Now what? This brings us back to the subject of probability. Before we describe probabilities for a normal random variable, consider one important property of any normal curve (or of any curve representing a continuous random variable, for that matter), namely, that the total area under the curve is 1 (see Figure 7.11). When we described the normal curve as bell-shaped, we also determined that it was symmetrical. If the halves are identical, then the probability above the mean (μ) is equal to .5 and is the same as the probability below the mean. Thus, in Figure 7.8, the shaded area is equal to the nonshaded area under the curve.

Returning to the achievement scores example, what percentage of the time will an individual score X be less than 440? This probability is written as

$$P(X < 440)$$

Figure 7.11

Area under a
normal curve

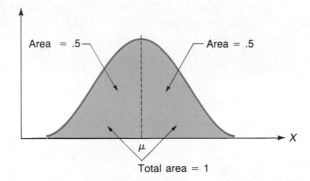

Figure 7.12

Normal curve for achievement test scores showing $P(X < 440)$. The shaded area is the percentage of time that X will be less than 440. (X = score on achievement test.)

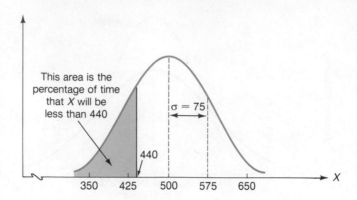

We will discuss how to determine this area (a simple procedure) later in the chapter, but for now just remember that when dealing with a normal random variable, a *probability* is represented by an *area* under the corresponding normal curve. The value of $P(X < 440)$ is illustrated in Figure 7.12. It appears that roughly 20% of the total area has been shaded, so we conclude that (1) roughly 20% of the test scores will be less than 440 points, and (2) the probability that X is less than 440 is approximately .2.

7.4   FINDING AREAS UNDER A NORMAL CURVE

Areas under the Standard Normal Curve

We will begin our discussion by finding the area under a special normal curve, namely, one that is centered at zero ($\mu = 0$) and has a standard deviation of one ($\sigma = 1$). This random variable is typically represented by the letter Z and is referred to as the **standard normal random variable**. As Figure 7.13 demonstrates, Z is as likely to be negative as positive; that is, $P(Z < 0) = P(Z > 0) = .5$. Although you probably never will observe a random variable like Z in practice, it is a useful normal random variable. In fact, an area under any normal curve (as in Figure 7.12) can be determined by finding the corresponding area under the standard normal curve.

Figure 7.13

Standard normal curve

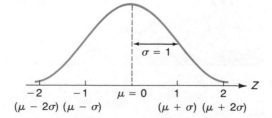

Figure 7.14

Shaded area is .4474
from Table A-4

However, to derive the area under the standard normal curve requires the use of integral calculus. Unfortunately, the integral of the function describing the standard normal curve does not have a simple (closed form) expression. By using excellent approximations of this integral, however, we can tabulate these areas—see Table A-4.

For example, say that we want to determine the probability that a standard normal random variable will be between 0 and 1.62. This is written as

$$P(0 < Z < 1.62)$$

The value of this probability is illustrated in Figure 7.14 and is obtained from Table A-4 by noting that it contains the area under the curve between the mean of zero and the particular value of Z. The far left column of Table A-4 identifies the first decimal place for Z, and you read across the table to obtain the second decimal place.

In our example, we find the intersection between 1.6 on the left and .02 on the top, because $Z = 1.62$. Look at Table A-4; the value .4474 is the area between 0 and 1.62. In other words, the probability that Z will lie between 0 and 1.62 is .4474.

You can begin to see why it is a good idea to sketch the curve and shade in the area when dealing with normal random variables. It gives you a clear picture of what the question is asking and cuts down on mistakes.

EXAMPLE 7.2

What is the probability that Z will be greater than 1.62?

Solution

We wish to find $P(Z > 1.62)$. Examine Figure 7.15. The area under the right half of the Z curve is .5, so, using our value from Table A-4, the desired area here is

$$.5 - .4474 = .0526$$

So the probability that Z will exceed 1.62 is .0526.   □

Figure 7.15

The shaded area represents the probability that Z will be greater than 1.62 ($P(Z > 1.62)$)

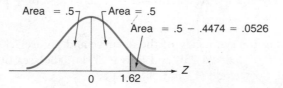

Figure 7.16

Area under the Z curve for $P(Z < 1.62)$

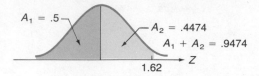

Figure 7.17

Area under the Z curve for $P(1.0 < Z < 2.0)$

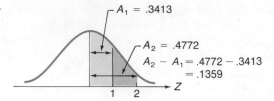

What if we wish to know the probability that Z is equal to a particular value, such as $P(Z = 1.62)$? As in the previous section, there is no area under the curve corresponding to $Z = 1.62$, so

$$P(Z = 1.62) = 0$$

In fact,

$$P(Z = \text{any value}) = 0$$

One nice thing about this fact is that $P(Z \geqslant 1.62)$ is the same as $P(Z > 1.62)$ (that is, .0526). So putting the equal sign on the inequality ($\geqslant$ or $\leqslant$) has no effect on the resulting probability.

By looking at the Z curve in Figure 7.16, you can see that

$$P(Z < 1.62) = .5 + .4474 = .9474$$

As before, this also is $P(Z \leqslant 1.62)$.

Figure 7.17 shows $P(1.0 < Z < 2.0)$ (areas from Table A-4). We see that

$$P(1.0 < Z < 2.0) = P(0 < Z < 2.0) - P(0 < Z < 1.0)$$
$$= .4772 - .3413$$
$$= .1359$$

By subtracting the two areas, we find that the probability that Z will lie between 1.0 and 2.0 is .1359.

We use Figure 7.18 and Table A-4 to determine $P(-1.25 < Z < 1.15)$:

$$P(-1.25 < Z < 1.15) = P(-1.25 < Z < 0) + P(0 < Z < 1.15)$$
$$= A_1 + A_2$$

Using the symmetry of the Z curve and Figure 7.19, the area of A_1 is the same as $P(0 < Z < 1.25)$, and thus is .3944. The area of A_2, from Table A-4, is .3749. So we add A_1 and A_2:

$$.3944 + .3749 = .7693$$

Figure 7.18

Area under the
Z curve for
$P(-1.25 < Z < 1.15)$

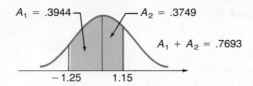

Figure 7.19

Z curve for
$P(0 < Z < 1.25) =$
$P(-1.25 < Z < 0)$

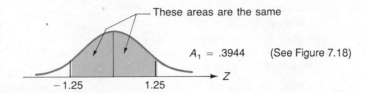

Figure 7.20

Area under the
Z curve for
$P(Z < -1.45)$

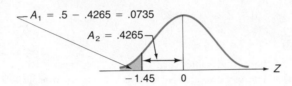

Finally, we can determine $P(Z < -1.45)$ using Figure 7.20. This can be written as $P(Z < 0) - P(-1.45 < Z < 0)$. Using the discussion from Figure 7.19, the area between zero and -1.45 is .4265 (from Table A-4). As a result, Z will be less than (or equal to) -1.45 approximately 7.35% of the time.

EXERCISES

7.9 Explain how the parameters μ and σ determine the graph of a normal distribution.

7.10 Find the area under the standard normal curve banded by the following Z values. Sketch the corresponding area.
  (a) $Z \leqslant 0$
  (b) $Z \leqslant 1.0$
  (c) $Z \geqslant 1.0$
  (d) $Z \leqslant -1.0$

7.11 Find the area under the normal curve between the following Z values. Sketch the corresponding area.
  (a) $Z = 0$ to 1.0
  (b) $Z = 1.0$ to 1.5
  (c) $Z = -1.0$ to 1.0
  (d) $Z = -2.5$ to -1.5

7.12 Find the following probabilities. Sketch the corresponding area.
  (a) $P(Z \leqslant 1.75)$
  (b) $P(Z \geqslant 1.96)$
  (c) $P(-1.0 \leqslant Z \leqslant 2.5)$
  (d) $P(-0.5 \leqslant Z \leqslant 0.5)$

7.13 Find the probability that an observation taken from a standard normal population will be

(a) Between -3 and 1.6
(b) Less than -2.1
(c) Between 0.76 and 1.96
(d) Between -1.65 and 1.65

7.14 Find the value of z for the following probability statements and sketch the corresponding area.

(a) $P(Z \leqslant z) = 0.95$
(b) $P(Z \leqslant z) = 0.10$
(c) $P(Z \geqslant z) = 0.025$
(d) $P(Z \geqslant z) = 0.55$

7.15 Find the value of z for the following probability statements and sketch the corresponding area.

(a) $P(-1.8 \leqslant Z \leqslant z) = 0.6$
(b) $P(0 \leqslant Z \leqslant z) = 0.25$
(c) $P(1.0 \leqslant Z \leqslant z) = 0.1$
(d) $P(-2.8 \leqslant Z \leqslant z) = 0.05$

7.16 Find the two Z values such that

(a) The area bounded by them is equal to the middle 40% of the standard normal distribution.
(b) The area bounded by them is equal to the middle 80% of the standard normal distribution.

7.17 Find the Z values such that the area under the standard normal curve between the Z value and $Z = 1.0$ is equal to 0.10. Find both Z values that make this possible.

7.18 The output from a monitor that measures the amperage of an electronic circuit follows a normal distribution with a mean 0 and variance 1. What proportion of the data would be outside the interval from -2 to 2?

Areas under Any Normal Curve

Take another look at the histogram of the 200 national achievement test scores in Figure 7.7. A normal curve with $\mu = 500$ points and $\sigma = 75$ points was used to describe the population of all test scores. So, $X = $ individual test score is a normal random variable with $\mu = 500$ and $\sigma = 75$.

What happens to the shape of the data if we take each of the 200 scores in this example and subtract 500 (that is, subtract μ)? As you can see in Figure 7.21, the histogram (and corresponding normal curve) is merely "shifted" to the left by 500. It resembles the normal curve for X, except the "new" mean is now zero. The random variable defined by $Y = X - 500$

1 Is a normal random variable.

2 Has a mean equal to zero.

3 Has a standard deviation equal to that of X, that is, 75.

Figure 7.21

Histogram obtained by subtracting $\mu = 500$ (compare with Figure 7.7)

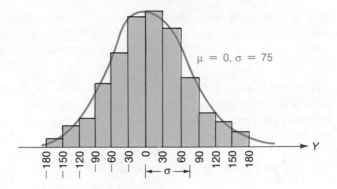

Figure 7.22

Histogram obtained by subtracting $\mu = 500$ and dividing by $\sigma = 75$ (compare with Figures 7.7 and 7.21)

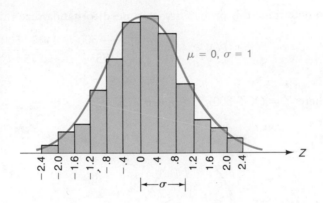

Figure 7.22 shows what happens to the shape of the 200 Y values if each of them is divided by 75 (that is, by σ). Notice the horizontal axis in the histogram and the corresponding normal curve. The resulting normal curve resembles a normal curve with a mean of zero and a standard deviation equal to 1.

Thus, if X is a normal random variable with mean 500 and standard deviation 75, then the random variable defined by $Z = (X - 500)/75$

1 Is a normal random variable.

2 Has a mean equal to zero.

3 Has a standard deviation equal to 1.

This means that, in general, for any normal random variable X,

$$Z = \frac{X - \mu}{\sigma}$$

is a **standard normal random variable**. This procedure of subtracting μ and dividing by σ is referred to as **standardizing** the normal random variable

X. It allows us to determine probabilities for any normal random variable by first standardizing it and then using Table A-4. So the standard normal distribution turns out to be much more important than you might have expected!

EXAMPLE 7.3

The normal curve in Figure 7.12 represented the population of all test scores, with $\mu = 500$ points and $\sigma = 75$ points. What percentage of the scores will be less than (or equal to) 440? Or, put another way, what is the probability that any particular score will be less than 440?

Solution

This is a probability and is written as

$$P(X < 440)$$

To determine this probability, you need to standardize this variable:

$$P(X < 440) = P\left(\frac{X - 500}{75} < \frac{440 - 500}{75}\right)$$
$$= P(Z < -.8)$$

where $Z = (X - 500)/75$ (Figure 7.23).

Figure 7.23

Compare the areas for the X (*top*) and Z (*bottom*) normal curves to find $P(X < 440)$

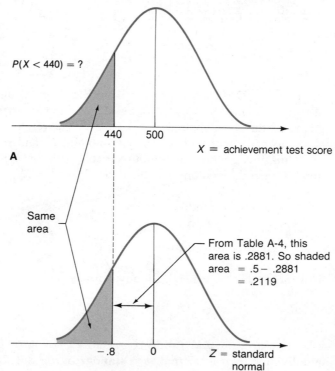

Earlier, by examining Figure 7.12, we estimated this area to be roughly 20%. The actual area, from Figure 7.23, is .2119; that is, it is 21.19% of the total area. The conclusion here is that

$$P(X < 440) = .2119$$

and so 21% of all test scores will be less than 440 points. □

Interpreting Z

What does a Z value of $-.8$ imply in Example 7.3? It simply means that 440 is .8 standard deviation to the left (Z is negative) of the mean. So,

$$\mu - .8(\sigma) = 500 - .8(75)$$
$$= 440$$

Recall that a Z score was defined in exactly the same way in Chapter 3 using a sample mean ($\bar{X}$) and standard deviation (s). In this chapter, we use the population mean (μ) and standard deviation (σ). In general,

1 A positive value of Z designates how many standard deviations (σ) to the right of the mean (μ) X is.

2 A negative value of Z designates how many standard deviations to the left of the mean X is.

EXAMPLE 7.4

The weight of a newborn infant at a large metropolitan hospital is assumed to follow a normal distribution with $\mu = 6.8$ lb and $\sigma = .8$ lb. Determine the probability that an infant will weigh between 5.2 and 6.4 lb.

Solution

This probability can be written as

$$P(5.2 < X < 6.4)$$

Using the standardizing procedure,

$$P(5.2 < X < 6.4) = P\left[\frac{5.2 - 6.8}{.8} < \frac{X - 6.8}{.8} < \frac{6.4 - 6.8}{.8}\right]$$
$$= P(-2.0 < Z < -.5),$$

where Z once again represents the standardized normal random variable, which, for this example, is defined by

$$Z = \frac{X - 6.8}{.8}$$

Refer to Table A-4 and Figure 7.24. Comparing Figures 7.24A and B, the areas are equal:

$$.4772 - .1915 = .2857$$

Figure 7.24

(A) The probability
that X is between
5.2 and 6.4 lb.
(B) The probability
that Z is between
-2.0 and $-.5$.

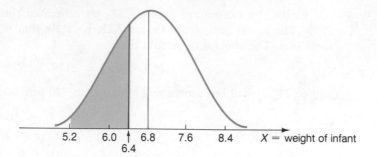

$X =$ weight of infant

A

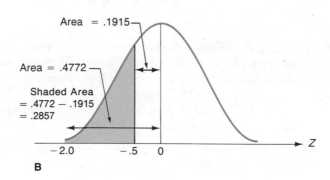

B

So 29% of the infants will weigh between 5.2 and 6.4 lbs. Notice that:

1 5.2 is two standard deviations to the left of the mean: $Z = -2$ and
$5.2 = 6.8 - 2(.8)$.

2 6.4 is .5 standard deviation to the left of the mean: $Z = -.5$ and
$6.4 = 6.8 - .5(.8)$.

3 $P(X = 5.2) = P(X = 6.4) = 0$, so $P(5.2 < X < 6.4) = P(5.2 \leqslant X \leqslant 6.4) =$
$.2857$.

EXAMPLE 7.5   Actuarial scientists in an insurance company formulate insurance policies that
will be both profitable and marketable. For a particular policy, the lifetimes of
the policy holders follow a normal distribution with $\mu = 66.2$ years and $\sigma = 4.4$
years. One of the options with this policy is to receive a payment following
the 65th birthday and a payment every five years thereafter.

1 What percentage of policy holders will receive at least one payment using
this option?

2 What percentage will receive two or more payments?

3 What percentage will receive exactly two payments?

Figure 7.25

The normal curve for policyholder lifetimes. X = age at death in years.

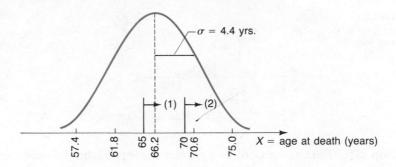

Figure 7.26

Z curve for $P(Z > -.27)$

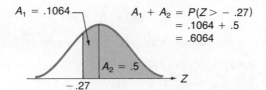

$A_1 = .1064$

$A_1 + A_2 = P(Z > -.27)$
$= .1064 + .5$
$= .6064$

$A_2 = .5$

Solution 1

The normal curve for the policyholder lifetimes is shown in Figure 7.25. To receive at least one payment, the policyholder must live beyond 65 years of age. So we need to determine (see Figure 7.26)

$$P(X > 65) = P[(X - 66.2)/4.4 > (65 - 66.2)/4.4]$$
$$= P(Z > -.27)$$
$$= .1064 + .5$$
$$= .6064$$

So nearly 61% of the policyholders will receive at least one payment.

Solution 2

Because the policyholder receives a payment every five years, he or she will receive two or more payments provided he or she lives to be older than 70 years of age. This means that the probability of two or more payments is determined by (see Figure 7.27)

$$P(X > 70) = P[(X - 66.2)/4.4 > (70 - 66.2)/4.4]$$
$$= P(Z > .86)$$
$$= .5 - .3051$$
$$= .1949$$

So 19.5% of the policyholders will survive long enough to collect two payments.

Figure 7.27

Z curve for $P(Z > .86)$

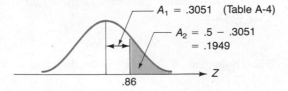

$A_1 = .3051$ (Table A-4)

$A_2 = .5 - .3051$
$= .1949$

Figure 7.28

Z curve for
$P(.86 < Z < 2.00)$

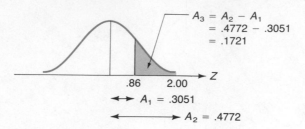

Solution 3   To receive exactly two payments, the policyholder must live longer than 70 years and less than 75 years. This probability is

$$P(70 < X < 75)$$

Using the same standardization procedure (see Figure 7.28):

$$P(70 < X < 75) = P[(70 - 66.2)/4.4 < (X - 66.2)/4.4 < (75 - 66.2)/4.4]$$
$$= P(.86 < Z < 2.00)$$
$$= .4772 - .3051$$
$$= .1721$$

So approximately 17% of the policyholders will receive exactly two payments.   □

7.5   APPLICATIONS WHERE THE AREA UNDER A NORMAL CURVE IS PROVIDED

Another twist to dealing with normal random variables is a situation where you are given the area under the normal curve and asked to determine the corresponding value of the variable. This is a common application of a normal random variable. For example, the manufacturer of a product may want to determine a warranty period during which the product will be replaced if it becomes defective, so that at most 5% of the items are returned during this period. Or, in a testing situation, you may want to set a passing score so that you can expect 30% of the people taking the exam to fail it.

EXAMPLE 7.6   Referring to example 7.3, 80% of the test scores will be less than what value? Recall that $\mu = 500$ and $\sigma = 75$.

Solution   The first step here is to sketch this curve (Figure 7.29A) and estimate the value of X (say X_0) so that

$$P(X < X_0) = .8$$

Because .8 is larger than .5, X_0 must lie to the right of 500.

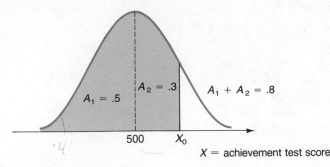

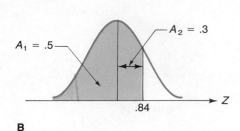

A

B

Figure 7.29

(A) $P(X < X_0) = .8.$
(B) $P(Z < .84) = .8.$

Next, find the point on a standard normal (Z) curve such that the area to the left is also .8 (Figure 7.29B). Using Table A-4, the area between 0 and .84 is .2995. This means that

$$P(Z < .84) = .5 + .2995$$
$$= .7995$$
$$= .8 \quad \text{(approximately)}$$

By standardizing X, we conclude that

$$\frac{X_0 - 500}{75} = .84$$

$$X_0 - 500 = 63$$

$$X_0 = 500 + 63 = 563$$

So 80% of the test scores will be less than 563. □

EXAMPLE 7.7

The city manager of a large northeastern city uses a poverty income index which he defines to be the income level at which 10% of the wage earners fall below and exceeded by the remaining 90%. Assuming that the incomes within this city are normally distributed with a mean of $33,000 and a standard deviation of $14,500, what will be the poverty level index for this situation?

Solution

The normal random variable X here is the income of a worker in this particular city (Figure 7.30A). To find the income under which 10% of the values lie, we need to determine an income level, say X_0, such that

$$P(X \leqslant X_0) = .10$$

Proceeding as before, examine a Z curve having a left tail area of .10 (Figure 7.30B). Using Table A-4,

$$P(-1.28 \leqslant Z \leqslant 0) = P(0 \leqslant Z \leqslant 1.28)$$
$$= .4 \quad \text{(more accurately .3997)}$$

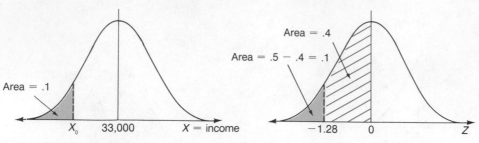

Figure 7.30

(A) $P(X \leqslant X_0) = .10$.
(B) $P(Z \leqslant -1.28) = .10$.

which means that

$$P(Z \leqslant -1.28) = .5 - .4$$
$$= .1$$

So,

$$\frac{X_0 - 33,000}{14,500} = -1.28$$

and

$$X_0 = 33,000 - (1.28)(14,500) \cong 14,400$$

Consequently, the poverty level index here is approximately \$14,400 since 10% of the workers will earn a salary less than or equal to this amount.   □

7.6   ANOTHER LOOK AT THE EMPIRICAL RULE

In Chapter 3, the empirical rule specified that, when sampling from a bell-shaped distribution (which means a normal distribution):

1 Approximately 68% of the data values should lie between $\bar{X} - s$ and $\bar{X} + s$.

2 Approximately 95% of the data should lie between $\bar{X} - 2s$ and $\bar{X} + 2s$.

3 Approximately 99.7% of the data should lie between $\bar{X} - 3s$ and $\bar{X} + 3s$.

Nothing was said at that time about the origin of these numbers. They actually came directly from Table A-4. To see this, consider Figure 7.31, in which

$$P(-1 < Z < 1) \cong .68$$

Figure 7.31

Z curve for
$P(-1 < Z < 1) \cong .68$

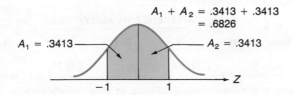

This implies that, for any normal random variable X,

$$P[-1 < (X - \mu)/\sigma < 1] \cong .68$$

That is,

$$P[(\mu - \sigma) < X < (\mu + \sigma)] \cong .68$$

Thus, for a set of data from a normal population, where $\bar{X}$ is the sample mean and s is the sample standard deviation, approximately 68% of the data will be between $\bar{X} - s$ and $\bar{X} + s$.

Similarly, $P(-2 < Z < 2) = .4772 + .4772 = .9544$, so you can expect (approximately) 95% of the data points from a normal (bell-shaped) population to lie between $\bar{X} - 2s$ and $\bar{X} + 2s$.

Finally, $P(-3 < Z < 3) = .4987 + .4987 = .9974$, which leads to the third conclusion of the empirical rule.

EXERCISES

7.19 Let the random variable X be normally distributed with mean 5 and variance 4. Find the following probabilities:
   (a) $P(X \geqslant 5.7)$
   (b) $P(X \leqslant 3.4)$
   (c) $P(2.8 \leqslant X \leqslant 5.1)$
   (d) $P(5.7 \leqslant X \leqslant 6.8)$

7.20 Find the value of x if the random variable X is normally distributed with mean 10 and variance 9.
   (a) $P(X \leqslant x) = 0.51$
   (b) $P(X \geqslant x) = 0.805$
   (c) $P(10 \leqslant X \leqslant x) = 0.05$
   (d) $P(8 \leqslant X \leqslant x) = 0.13$

7.21 Researchers at the World Fertility Survey, London, U.K., and the Population Studies and Training Center, Brown University, Providence, R.I. (Casterline, Williams, and McDonald, 1986), studied the variations in the age difference between spouses. Their comparative analysis covered various countries. One region studied was South Asia, which covered Bangladesh, Pakistan, Sri Lanka, and Nepal. For this region as a whole, they reported the following percentage distribution of the age difference between spouses: once-married women, duration of marriage less than 10 years, female age subtracted from the male age.

$z = \frac{3-6.1}{5.5} = -0.5636$

$z = \frac{5-6.1}{5.5} = -0.2$

z
$0.5636 \longrightarrow 0.2123$

$0.2 \longrightarrow 0.0793$

$0.2123 - 0.0793 = 0.133$

$z = \frac{12-6.1}{5.5} = 1.072$

$1.072 \longrightarrow 0.3577$
$0.5 - 0.3577 = 0.1423$

		DIFFERENCE IN AGE (YEARS)		
Less than 0	**0–4**	**5–9**	**10–14**	**15 or more**
3.6%	32.5%	36.4%	18.8%	8.8%

Source: John B. Casterline, Lindy Williams, and Peter McDonald. The age difference between spouses: Variations among developing countries. *Population Studies* **40**: 353–374 (1986).

Although the actual distribution is slightly skewed, you may assume for this exercise a normal distribution with a mean age difference of 6.1 years and a standard deviation of 5.5 years.

(a) What proportion of the couples in this region have an age difference between 3 and 5 years?

(b) What is the probability for any randomly selected couple the age difference is greater than 12 years?

7.22 The Comptroller of Public Accounts of a state where a lottery bill was pending in the state legislature collected information about the experience of other states. Lottery revenues followed a normal distribution with a mean of 25.4 million dollars and a standard deviation of 2.7 million dollars. What proportion of the states have lottery revenues less than 27 million dollars?

7.23 Harvard University has an extremely large endowment fund ($2.7 billion). However, on a per student basis, Princeton is tops (with $250,530 per student), Harvard is second ($174,080), Caltech is third ($161,080), and Rice University in Houston is fourth ($151,210). Assume that endowment dollars per student is a normally distributed variable with a mean of 80 thousand and a standard deviation of 30 thousand.

$\frac{151,210 - 80000}{30000} = 2.3736$

$z = 0.52$

$-0.52 = \frac{x - 80000}{30000}$

$x = 64400$

$z = \frac{1......-8....}{3....} = 0.666 \to 0.2454$

$.5 - 0.2454 = 0.2546$

(a) Convert the figure for Rice to a *Z* value.

(b) Find the endowment per student such that 30% of universities will have endowments less than this amount.

(c) For any randomly selected university, what is the probability that the endowment per student is greater than 100 thousand dollars?

7.24 To become a member of MENSA, the nationwide organization for people with high IQs, one has to pass the qualifying examination. If the scores on the exam are normally distributed with a mean of 80 and a standard deviation of 25 and if only 20% of the people taking this exam are admitted to the organization, what is the passing score?

7.25 The weight of students in a junior college are normally distributed with a mean of 160 lb and a standard deviation of 18 lb. What is the probability that a student drawn at random will weigh less than 150 lb?

7.26 The mechanics at Quick Brown Fox can tune up a car in an average of 30 minutes with a standard deviation of 5 minutes. If a car arrives for a tune-up 25 minutes before closing, what is the probability that the car will be serviced by closing, assuming the time it takes for a tune-up is normally distributed?

7.27 According to sources at the Texas Education Agency, the education of pregnant school girls has been tied to special education funding since the 1940s, when pregnant teenagers were forced by public opinion to leave the public schools. This con-

Handwritten annotations (left margin):

$$\bar{x} \approx \frac{\Sigma f_i m_i}{n}$$

$$s^2 = \frac{\Sigma f_i m^2 - \frac{(\Sigma f_i m)^2}{n}}{n-1}$$

$$\bar{x} = 16.448$$

$$s^2 =$$

tinued through the 1950s and 1960s. In 1969, Senate Bill 230 listed pregnancy as a temporary handicapping condition qualifying for special education funding. In 1984–85, the following was the distribution of pregnant students receiving special education services in Texas public schools:

Age	Total Number of Pregnant Students		
11 and under 12	1	→ 11.5	
12 and under 13	45	→ 562.5	
13 and under 14	164	→ 2214	
14 and under 15	505	→ 7322.5	
15 and under 16	955	→ 14802.5	
16 and under 17	1337	→ 22060.5	
17 and under 18	1448	→ 25340	
18 and under 19	368	→ 6808	
19 and under 20	52	→ 1014	
20 and under 21	10	→ 205	
21 and under 22	2	→ 43	

Handwritten: column headers $f_i \cdot m$ and $f_i m^2$; totals 4887 and 80383.5

Source: Texas Education Agency, Division of Policy Analysis, *Superintendents Annual Report—Part Three, 1984–85*, issued Nov. 7, 1985.

(a) Compute the mean age of the pregnant students.
(b) Compute the standard deviation of the distribution.
(c) The distribution appears to be normally distributed. Assuming a normal distribution, what is the probability that such a student is under 16 years of age? (Use the mean and standard deviation obtained in parts (a) and (b).)
(d) Assuming a normal distribution, what is the probability that such a student is over 18 years 6 months of age?

Handwritten: 16   16.45

7.28 The yearly cost of dental claims for the employees of D.S. Inc. is normally distributed with a mean of $250 and a standard deviation of $80. At least what yearly cost would be expected for 40% of the employees?

7.29 The Wechsler Deviation Intelligence Quotient is a standardized psychological test scale supposed to measure intelligence. The Stanford-Binet IQ score is another well-known intelligence test scale. The Wechsler IQ scale has a mean of 100 and a standard deviation of 15. The Stanford-Binet IQ scale also has a mean of 100, but has a standard deviation of 16. Both scales follow a normal distribution.

(a) If someone obtains a score of 122.5 on the Wechsler scale, what is the equivalent IQ score on the Stanford-Binet scale?
(b) What percentage of the people tested scored below 90
    (i) On the Wechsler scale?
    (ii) On the Stanford-Binet scale?
(c) What proportion of the population score between 90 and 110 on the Wechsler IQ test?

Handwritten (left margin):

$\mu = 100$ $\mu = 100$
$S = 15$ $\mu = 16$

$$z = \frac{122.5 - 100}{15} = 1.5 \implies 1.5 = \frac{x - 100}{16}$$

$$\implies x = 124$$

7.30 An instructor plans to "curve" the grades for his biology class on the basis of a normal distribution. The overall raw average test score for his class is 72, with a standard deviation of 8 points. If the instructor plans to have 15% of the class obtain a grade of A, 20% B, 30% C, 20% D, and 15% F, what are the score intervals that need to be used?

7.31 If X is normally distributed with a mean of 100, find the standard deviation given that $P(X \geqslant 110) = 0.123$.

7.32 If X is a normally distributed random variable with $P(X \geqslant 2) = 0.1$ and $P(X \leqslant 1) = 0.3$, find both the mean and the standard deviation.

7.33 As part of an experiment conducted for a graduate class in organizational behavior, the time taken to complete an assembly task was measured for two groups of workers. For the first group, the mean time was 10 minutes with a standard deviation of 1.5 minutes. For the second group, the mean time was 11.5 minutes with a standard deviation of 2.0 minutes. Assume that the times for both groups follow normal distributions. A worker from each group is randomly selected. What is the probability that the assembly time for this worker is less than 9 minutes, if the worker is from

(a) The first group?

(b) The second group?

7.34 A group of Air Force trainee pilots was given a newly developed endurance evaluation test. The average score was 71.1 with a variance of 49. Assume that the distribution was normal.

(a) If the Air Force decides that only 40% of the trainees should pass the test, what will be the cutoff (passing) score?

(b) If one trainee is selected at random, what is the probability that this trainee scored 85 or more?

7.35 Monthly seizures of narcotic drugs at large airports follow a normal distribution, with a mean of 3.2 kilos and a standard deviation of 1.1 kilos. What is the probability that, in any given month, the seizures exceed 6 kilos?

7.7　NORMAL APPROXIMATION TO THE BINOMIAL

The binomial random variable was introduced in Chapter 6. It is a discrete random variable used to count the number of successes in a binomial situation.

Characteristics of a Binomial Situation

1 You have n independent identical trials.

2 Each trial is a success (with probability p) or a failure (with probability $1 - p$).

3 The binomial random variable X is the number of successes out of n trials.

4 The mean of X is $\mu = np$, and the standard deviation of X is $\sigma = \sqrt{np(1 - p)}$.

Figure 7.32

Approximating
binomial
probabilities using
a normal curve

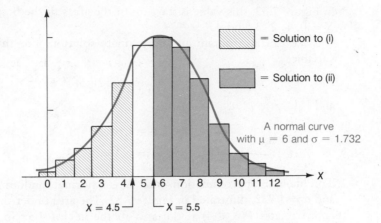

Examples included:

X = the number of heads (successes) out of three flips (trials) of a coin.

X = the number of students that have a reading deficiency (successes) out of a sample of 50 people (trials).

X = the number of airline "no-shows" (successes) out of a sample of 18 people who make airline reservations (trials).

Table A-1 contains values of n (the number of trials) only up to $n = 20$. What can you do for binomial situations where $n > 20$? One method is to use the **normal approximation** to the binomial random variable. Here you pretend that X is a normal random variable having the same mean and standard deviation as the actual binomial random variable. This approximation works well when p is near .5 and in general offers a good estimate when both $np > 5$ and $n(1 - p) > 5$.

Consider 12 flips of a coin. We want to determine (1) the probability of observing no more than four heads and (2) the probability of observing more than five heads. First, notice that a normal approximation is not necessary here. This is a binomial situation with $n = 12$ and $p = .5$, and Table A-1 does contain probabilities for this set of values. We chose this illustration to compare the actual binomial probability to the approximated probability using the normal distribution. Look at Figure 7.32, which demonstrates how we estimate binomial probabilities using a normal curve.

To solve question 1, let X = the number of heads in 12 flips, so X is a binomial random variable. We want to determine $P(X \leqslant 4)$. We can obtain an exact solution using Table A-1:

$$P(X \leqslant 4) = P(0) + P(1) + P(2) + P(3) + P(4)$$
$$= 0 + .003 + .016 + .054 + .121 = .194$$

In Figure 7.32, this value is the sum of the areas of the boxes corresponding to $X = 0, 1, 2, 3$, and 4.

We can also obtain an approximate solution. For this binomial random variable,

$$\mu = np = (12)(.5) = 6$$

and

$$\sigma = \sqrt{np(1 - p)}$$
$$= \sqrt{3} = 1.732$$

To obtain an approximation, treat X as a normal random variable with $\mu = 6$ and $\sigma = 1.732$, illustrated in Figure 7.32. The area under the normal curve that approximates $P(X \leqslant 4)$ is the area to the left of 4.5. So we obtain a better approximation here if we find the area under the normal curve to the left of 4.5, not 4.0. This .5 adjustment is referred to as an **adjustment for continuity**. This adjustment is necessary whenever you approximate a discrete random variable (such as binomial) using a continuous distribution (such as normal). Remember that the discrete distribution has gaps, whereas the continuous does not, so we must assign a portion of the space (probability) between 4 and 5 when we use a continuous distribution to approximate a discrete one. Using Table A-4,

$$
\begin{array}{ccc}
\textbf{Binomial} & & \textbf{Normal} \\
(n = 12, p = .5) & & (\mu = 6, \sigma = 1.732) \\
P(X \leqslant 4) & \cong & P(X \leqslant 4.5) \\
\end{array}
$$

$$= P\left(Z \leqslant \frac{4.5 - 6}{1.732}\right)$$
$$= P(Z \leqslant -.87) = .1922$$

Notice that the approximate solution of .1922 is very close to the actual probability of .194. This is helped in part by the fact that $p = .5$ for this situation, which means that the binomial distribution is perfectly symmetric. As the value of p moves away from .5, larger values of n are necessary to achieve an approximation this good.

Now consider question 2, the probability of observing more than five heads in 12 flips, or $P(X > 5) = P(X \geqslant 6)$. Using Table A-1, we can obtain an exact solution:

$$P(X \geqslant 6) = P(6) + P(7) + \cdots + P(11) + P(12)$$
$$= .226 + .193 + \cdots + .003 + 0 = .613$$

We can also obtain an approximate solution. Using Figure 7.32, the area under the normal curve that corresponds to the darkly shaded area representing the exact solution is the area to the right of 5.5. So, using Table A-4:

$$\begin{array}{cc} \textbf{Binomial} & \textbf{Normal} \\ (n = 12, p = .5) & (\mu = 6, \sigma = 1.732) \\ P(X \geqslant 6) \quad \cong & P(X \geqslant 5.5) \end{array}$$

$$= P\left(Z \geqslant \frac{5.5 - 6}{1.732}\right)$$

$$= P(Z \geqslant -.29) = .6141$$

Again, we obtain a very good approximation, helped by the fact that we are using a perfectly symmetrical binomial distribution.

How to Adjust for Continuity

If X is a binomial random variable with n trials and probability of success $= p$, then

1 $P(X \leqslant b) \cong P\left(Z \leqslant \dfrac{b + .5 - \mu}{\sigma}\right)$

2 $P(X \geqslant a) \cong P\left(Z \geqslant \dfrac{a - .5 - \mu}{\sigma}\right)$

3 $P(a \leqslant X \leqslant b) \cong P\left(\dfrac{a - .5 - \mu}{\sigma} \leqslant Z \leqslant \dfrac{b + .5 - \mu}{\sigma}\right)$

where

$$\mu = np, \qquad \sigma = \sqrt{np(1 - p)},$$

and Z is a standard normal random variable

4 Be sure to convert a $<$ probability to a $\leqslant$, and convert a $>$ probability to a $\geqslant$ before switching to the normal approximation.

EXAMPLE 7.8

In example 6.7, we used the illustration where 30% of the seniors in a certain school had a deficiency in reading ability (a binomial situation). In that example, we considered the number of students having such a deficiency (number of successes) out of a sample of $n = 4$ seniors. No approximation was necessary since we were able to determine all probabilities using Table A-1. Now consider a sample of 100 seniors. Since we have $n = 100$ trials (students) here, we cannot use Table A-1. However, we can obtain a good normal approximation because $np = (100)(.30) = 30$ and $n(1 - p) = (100)(.70) = 70$ are both greater than 5. What are the chances that this sample will reveal no more than 15 of the students having a reading deficiency, that is $P(X \leqslant 15)$?

Solution

X is a binomial random variable with

$$\mu = np = (100)(.30) = 30$$

and

$$\sigma = \sqrt{np(1-p)} = \sqrt{21} = 4.58$$

Therefore, using Table A-4:

Binomial	Normal
($n = 100$, $p = .30$)	($\mu = 30$, $\sigma = 4.58$)
$P(X \leqslant 15)$ $\cong$	$P(X \leqslant 15.5)$

$$= P\left(Z \leqslant \frac{15.5 - 30}{4.58}\right)$$

$$= P(Z \leqslant -3.17) = .0008$$

.4992
⇒ .5 − .4992
= 0.0008

Consequently there is a very small chance of no more than 15 students having this deficiency since you can expect this event to occur only 8 times out of 10,000.   □

EXAMPLE 7.9

In Chapter 6, we discussed a binomial situation in which Air Texas was intentionally overbooking their flights. On a particular flight from Dallas to El Paso, they use a much larger aircraft that holds 200 people. As in our previous example, 20% of the people do not show up for a reserved flight. If Air Texas accepts 235 reservations, what is the probability that at least one passenger will end up without a seat on this flight?

Solution

The binomial random variable X here is the number of people (out of 235) who show up for the flight. For this situation, $n = 235$, and $p = .8$ represents the probability that any one passenger will show up. The mean of this random variable is

$$\mu = (235)(.8) = 188$$

and the standard deviation is

$$\sigma = \sqrt{(235)(.8)(.2)} = 6.13$$

At least one person holding a reservation will be deprived of a seat if $X \geqslant 201$ because the plane holds only 200 people. Once again, we use the normal approximation (Table A-4) to obtain the following probability:

Binomial	Normal
($n = 235$, $p = .8$)	($\mu = 188$, $\sigma = 6.13$)
$P(X \geqslant 201)$ $\cong$	$P(X \geqslant 200.5)$

$$= P\left(Z \geqslant \frac{200.5 - 188}{6.13}\right)$$

$$= P(Z \geqslant 2.04)$$

$$= .5 - .4793$$

$$= .0207$$

So on approximately two flights out of 100, at least one person will be unable to secure a seat.   □

EXERCISES

7.36 A random variable X has a binomial distribution with the probability of a success p equal to 0.25.

  (a) Would it be appropriate to use the normal approximation to the binomial if $n = 30$? If $n = 15$?

  (b) With $n = 40$, use the normal approximation to find $P(2 \leqslant X \leqslant 10)$.

  (c) What is the smallest value that n can be and still have the normal distribution to be appropriate for approximating the binomial distribution?

7.37 Let the random variable X indicate the number of female students chosen (with replacement) in a sample of 20 from a student body with 40% female students.

  (a) Using the binomial table, find the probability that X is greater than 6 and less than 10.

  (b) Use the normal approximation to answer part (a).

  (c) Compare the answers in parts (a) and (b).

Handwritten notes in margin:

$n = 20$
$p = .4$
$\mu = np = 8$
$\sigma = 2.1908$

$P(7 \leqslant X \leqslant 9)$
$= P(X \leqslant 9) - P(X \leqslant 6)$

0.2517

$z = \dfrac{6.5 - 8}{2.1908} = -0.6846$

$z = \dfrac{9.5 - 8}{2.1908} = 0.6846$

7.38 Thirty percent of the computer programmers who are hired to work for Techtronics do not have work experience in programming. If a random sample of 35 computer programmers is selected, what is the probability that fewer than 20 have had experience in computer programming before being hired by Techtronics?

7.39 A travel agency promotes vacation packages by phoning households at random in the evening hours. Historically, only 65% of heads of households are at home when the agency calls. If 30 households are phoned in a given evening, what is the probability that the agency will find between 15 and 25 households, inclusively, with the head of the household at home?

7.40 According to the Census Bureau, 73.2% of the households in the United States are family households and 26.8% are nonfamily households.[*]

  (a) In a random sample of 120 U.S. households, what is the probability that between 50 and 70 are family households?

  (b) In a random sample of 120 U.S. households, what is the probability that less than 30 are nonfamily households?

  (*Hint:* Use the normal approximation to the binomial.)

Handwritten notes in margin:

$\mu = 30 \times .65 = 19.5$
$\sigma = \sqrt{19.5(.35)} = 2.6124$
$P(15 \leqslant X \leqslant 25)$
$P(14.5 \leqslant X \leqslant 25.5)$
$P\left(\dfrac{14.5 - 19.5}{2.6124} \leqslant z \leqslant \dfrac{25.5 - 19.5}{2.6124}\right)$
$-1.9139 \leqslant z \leqslant 2.2967$
$.4719 + .4890 = 0.9609$

7.41 According to annual figures available from the National Center for Health Statistics, 21% of the births in the United States were out of wedlock. Suppose a random sample of 200 births were made. What is the probability that more than 50 births were out of wedlock?

7.42 A fair coin is flipped 50 times. What is the probability that between 20 and 30 heads (inclusive) will be recorded?

7.43 According to Census Bureau figures, the number of children in the United States residing with only one parent is steadily rising. In 1970, about 12% of the nation's children lived in single-parent homes. By 1985, that figure had nearly doubled to 23% of the 62.5 million children under the age of 18. In 90% of those single-parent homes, the parent was the mother. In 1985, approximately 54% of the nation's black children, 29% of Hispanic children and 18% of white children lived with only one parent.[†]

---

[*] *Source:* Bureau of the Census, Household and Family Characteristics: March 1983. *Current Population Reports*, Series p. 20, No. 388, 1984.

[†] *Source: Parade Magazine.* February 1, 1987, p. 11.

294

(handwritten notes in margin: $n = 10$, $p(x=9) =$; $p = 0.90$; $= .387$; $n = 100, p = .18, \mu = 18, \sigma = 14.78$; $p(x \geq 19) \Rightarrow p(z \geq 18.5)$; $= \frac{18.5 - 18}{14.78} = 0.033, p(0.033) = 0.0120$; $0.5 - 0.0120 = .488$)

(a) If a random sample of 10 single-parent homes were taken, what is the probability that in 9 of them the parent is the mother?

(b) In a random sample of 100 white children, what is the probability that more than 18 of them live with only one parent?

(c) In a random sample of 100 black children, what is the probability that more than 18 of them live with only one parent?

(d) If a random sample of 100 children (taken from all children under the age of 18), what is the probability that less than 23 of them live with a single parent who is the mother?

(*Hint:* For part (a), use the binomial tables. For parts (b)–(d) use the normal approximation to the binomial.)

7.44 Vukonic (1986) conducted a study of the expenditures of foreign tourists in Yugoslavia. In a summary of sociodemographic characteristics of foreign tourists, he gave the following distribution by country of origin:

West Germany	54.7%	Italy	5.1%
Austria	12.1%	France	2.3%
U.K.	7.6%	Other	12.4%
Netherlands	5.8%		

Source: Boris Vukonic, Foreign tourist expenditures in Yugoslavia. *Ann. Tourism Res.* **13**: 59–78 (1986).

What is the probability that in a random sample of 250 foreign tourists in Yugoslavia:

(a) At least 150 are from West Germany?

(b) Between 20 and 25 are from Italy or France?

7.45 Suppose that 70% of moviegoers surveyed rated the movie *Top Gun* as most entertaining movie of the summer of 1986. In a sample of 100 moviegoers, what is the probability that 70 or more persons will rate the above movie as "most entertaining"?

7.8 EXPONENTIAL RANDOM VARIABLES

The final continuous distribution we will discuss is the **exponential distribution**. This particular random variable is used in a variety of applications in statistics. Applications include the time between arrivals at a drive-up bank teller, the lifetime of machine components, and the time between reports of a rare disease.

Chapter 6 discussed the Poisson random variable, which often is used to describe the number of arrivals over a specified time period. If the random variable Y, representing the number of arrivals over time period T, follows a Poisson distribution, then X, representing the time between successive arrivals, will be an **exponential random variable**. The exponential random variable has many applications when describing any situation in which people or objects have to wait in line. This line is called a **queue**. People, machines, or telephone calls may wait in a queue.

Figure 7.33

Curve showing the
distribution of an
exponential random
variable

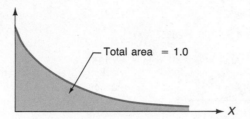

The Exponential Random Variable

The shape of the exponential distribution is represented by a curve that steadily decreases as the value of the random variable X increases. Thus, the larger X is, the probability of observing a value of X at least this large decreases exponentially. This type of curve is illustrated in Figure 7.33.

Determining Probabilities

Determining areas for exponential random variables is not as simple as for uniform ones, but it is easier than for normal random variables because exponential probabilities can be derived on a calculator. Table A-2 can also be used to determine the probability for an exponential random variable.

As Figure 7.34 illustrates, for an exponential random variable X the probability that X exceeds or is equal to a specific value X_0 is

$$P(X \geqslant X_0) = e^{-A \cdot X_0}$$

The parameter A is related to the Poisson random variable we used in discussing arrivals. In fact, the Poisson distribution for arrivals per unit time and the exponential distribution for time between arrivals provide two alternative ways of describing the same thing. For example, if the number of arrivals per unit time follows a Poisson distribution with an average of $A = 6$ arrivals per hour, then an alternative way of describing this situation is to say that the time between arrivals is exponentially distributed with mean time between arrivals equal to $1/A = 1/6$ hour (10 minutes).

Figure 7.34

Curve used for
determining a
probability for an
exponential random
variable

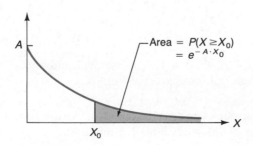

Figure 7.35

Curve showing the probability that X is less than or equal to .5 ($P(X \leqslant .5)$)

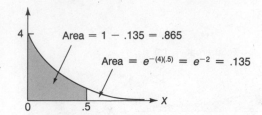

In general, $1/A$ is the average (mean) value of the exponential random variable X. It is also equal to the standard deviation of X. So,

$$\mu = 1/A$$

$$\sigma = 1/A$$

In applications using this distribution, the value of A either will be given or can be estimated in some way.

EXAMPLE 7.10

In example 6.15, the number of people arriving at the Riverside Memorial Hospital emergency room over a one-hour period was assumed to follow a Poisson distribution. The average arrival rate was $A = 4$ people per hour.

1 If you just witnessed an arrival, what is the probability that a new arrival will occur within 30 minutes?
2 If X represents the time between successive arrivals, what are the mean and standard deviation of X?

Solution 1

To determine this probability, we must first convert 30 minutes to .5 hour, since the arrival rate is 4 *per hour*. The desired probability then is $P(X \leqslant .5)$. Values of e^{-x} are contained in Table A-2. Using this table or your calculator (and Figure 7.35) the probability that X *exceeds* .5 is

$$P(X > .5) = P(X \geqslant .5)$$
$$= e^{-A \cdot x_0}$$
$$= e^{-(4)(.5)}$$
$$= e^{-2}$$
$$= .135$$

Consequently, $P(X \leqslant .5) = 1 - .135 = .865$, and so 86.5% of the time, the time between arrivals will not exceed 30 minutes.

Solution 2

Both the mean and standard deviation of X (the time between successive arrivals) are $1/A = 1/4 = .25$ hour (15 minutes).   □

Figure 7.36

Curve showing the probability that X exceeds 15 $(P(X > 15))$

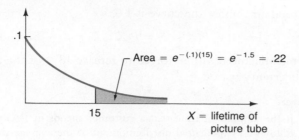

EXAMPLE 7.11

A manufacturer of color televisions has determined that the lifetime of the picture tube follows an exponential distribution with an average lifetime of 10 years. Determine the fraction of picture tubes that

1 Fail after 15 years.

2 Fail before the warranty period of 2 years.

Solution 1

Define X to be the lifetime of a picture tube. Since $\mu = 1/A$, then $A = 1/\mu = .1$. We want to determine $P(X > 15)$, which is illustrated in Figure 7.36.

We see that $X_0 = 15$ and $A = .1$. Using Figure 7.36,

$$P(X > 15) = P(X \geqslant 15)$$
$$= e^{-A \cdot X_0}$$
$$= e^{-(.1)(15)}$$
$$= e^{-1.5}$$
$$= .22$$

So 22% of the television picture tubes will survive longer than 15 years.

Solution 2

Here the problem is to find $P(X < 2) = P(X \leqslant 2)$, which is $1 - P(X > 2)$. Using Table A-2 and Figure 7.37

$$P(X > 2) = e^{-(.1)(2)}$$
$$= e^{-.2}$$
$$= .82$$

Figure 7.37

Curve showing the probability that X is less than 2 $(P(X < 2))$

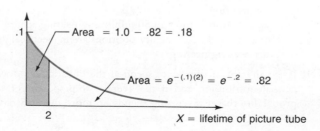

The total area under the curve is 1.0, so

$$P(X < 2) = P(X \leqslant 2) = 1 - .82 = .18$$

The manufacturer will be forced to replace 18% of the tubes during the two-year warranty period.   □

EXERCISES

7.46 If the amount of time that a customer spends in Ricky's Hideaway restaurant follows an exponential distribution and if the average time spent by a customer is 0.75 hour, what is the probability that a customer will spend more than an hour in the restaurant? What is the standard deviation of the amount of time spent by a customer in the restaurant?

7.47 Yellow Rose taxi company estimates that it makes an average of $415 in profits per day. Assuming that the daily profit follows an exponential distribution, what is the probability that on a given day at least $500 in profits will be made?

7.48 The time between reports of a rare disease follows an exponential distribution with a mean of 90 days. What is the probability that the time between successive reports of the disease will exceed 60 days?

7.49 Doigan and Gilkeson (1986), writing for the ASEE Engineering College Faculty and Graduate Student Survey Committee, gave the following distribution of authorized full-time engineering faculty positions in Fall 1985:

Field	Public Schools	Private Schools	All Schools
Aero & Astronautical	4.7%	2.2%	4.1%
Chemical	7.5	9.2	7.9
Civil	15.8	15.4	15.7
Computer Sci/Eng.	7.8	11.0	8.5
Electrical	22.2	23.9	22.6
Industrial	6.5	5.4	6.2
Mechanical	15.3	20.1	18.7
All other	20.2	12.8	16.3
	100.0%	100.0%	100.0%

Source: Paul Doigan and Mack Gilkeson, ASEE survey of engineering faculty and graduate students, fall 1985. *Eng. Educ.* Table 1, p. 53 (October 1986).

(a) In a random sample of 50 public engineering schools, what is the probability that between 10 and 15 faculty positions are in electrical engineering?

(b) In a random sample of 50 private engineering schools, what is the probability that less than 10 faculty positions are in industrial or mechanical engineering?

(c) In a random sample of 50 public engineering schools what is the probability that more than 10 faculty positions are in electrical engineering or computer science/engineering?

7.50 If the amount of time ships spend at the Philadelphia dockyard follows an exponential distribution and if the average ship spends 3.1 days there, what is the probability that a given ship spends no more than 1.5 days at the dockyard?

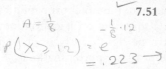

$A = \frac{1}{8}$ $-\frac{1}{8} \cdot 12$

$P(X > 12) = e$

$= .223 \longrightarrow$

7.51 The number of patients arriving at the Lakewood Psychiatric Unit following an unsuccessful suicide attempt is assumed to follow a Poisson distribution at an average rate of 3 per 24 hour period (day). What is the probability that the time between two unsuccessful suicide arrivals of this type is longer than 12 hours?

7.52 The Mylapore County fire department has determined that the amount of time per month spent fighting fires follows an exponential distribution. If the average fire-fighting time per month is 10.4 hours, what is the probability that in a given month no more than 15 hours will be spent fighting fires?

7.53 If the amount of time spent by visitors in a certain zoo follows an exponential distribution and if it is known that the average visitor spends 1.9 hours at the zoo, calculate the probability that a given visitor will spend at least 1.5 hours at the zoo.

3   24

X

SUMMARY

A random variable that can assume any value over a specific range is a **continuous random variable**. This chapter discussed three commonly encountered continuous distributions: the (continuous) uniform, the normal, and the exponential distributions. Each of these distributions has a unique curve that can be used to determine probabilities by finding the corresponding area under this curve. The **uniform** distribution (curve) is flat: values of this random variable are evenly distributed over a specified range. The **normal** distribution is characterized by a bell-shaped curve with values concentrated about the mean. The **exponential** distribution has a shape that steadily decreases as the value of the random variable increases. Table A-2 (or a good calculator) can be used to derive probabilities for the exponential distribution.

We discussed examples illustrating the shape of each distribution. The exact curve for a particular random variable is specified using one or two parameters that describe the corresponding population. As in the case of a discrete random variable, the population consists of what you would obtain if the random variable was observed indefinitely. The resulting average value and standard deviation represent the **mean** and **standard deviation** of the random variable and corresponding population.

There are infinitely many normal random variables, one for each mean (μ) and positive standard deviation (σ). If $\mu = 0$ and $\sigma = 1$, this normal random variable is the **standard normal** random variable, Z. Consequently, there is only *one* normal random variable of this type. Table A-4 gives the probabilities (areas) under the standard normal curve. You can also use this table to determine a probability for any normal random variable if you first **standardize** the variable by defining $Z = (X - \mu)/\sigma$. For this situation, Z represents the number of standard deviations that X is to the right (Z is positive) or left (Z is negative) of the mean.

The normal distribution can be used to approximate binomial probabilities for a large number of trials, n. Because the normal distribution is continuous and the binomial is discrete, the approximation can be significantly improved by **adjusting for continuity** before applying the normal approximation.

REVIEW EXERCISES

7.54 Determine each of the following for a standard normal curve. Sketch the corresponding area.
  (a) $P(0 \leqslant Z \leqslant 1.5)$
  (b) $P(Z \geqslant -3)$
  (c) $P(Z \leqslant -1.88)$
  (d) $P(-2.5 \leqslant Z \leqslant 2.5)$

7.55 In a study conducted by Booth and others (1986), student nurses at the University of Ottawa, Canada, completed the Thurston-Richardson attitude questionnaire and voluntarily took the Canadian Home Fitness Test. They found that the frequency response of heart rates after a second exercise bout ranged from 101 to 190 beats per minute and seemed to follow a normal distribution. The mean heart rate was 145 with a standard deviation of 20.*
  (a)   What proportion of the student nurses had a heart rate of over 160 after the second exercise bout?
  (b)   What is the probability that any given student nurse would have a heart rate of between 150 and 170 after a second exercise bout?

7.56 Some researchers believe there is a genetic predisposition for alcoholism. According to the National Clearing House for Alcoholic Information, 25% of male children of alcoholics ultimately become alcoholics themselves. (For female alcoholics, the figure is 10%.) Suppose a random sample of 160 male children of alcoholics is taken. What is the probability that from 40 to 50 of them (inclusive) ultimately become alcoholics?

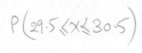

7.57 Nutritionists studying overweight persons found that some could be called "carbohydrate cravers," because consuming high carbohydrate foods like pasta gave them a feeling of well-being. No exact figures are available, but assume that 26% of overweight persons are carbohydrate cravers. If a sample of 100 overweight persons is selected, what is the probability that:
  (a)   30 of them are carbohydrate cravers?
  (b)   More than 30 of them are carbohydrate cravers?

7.58 Suppose that the average length of a telephone call made by teenagers is 45 minutes with a standard deviation of 18 minutes. Assume that the length of a telephone call is a normally distributed variable.
  (a)   What is the probability that a call made by a teenager exceeds one hour in length?
  (b)   What percentage of teenagers' calls are less than 30 minutes long?
  (c)   What is the minimum length of a phone call for at least 75% of teenagers?

7.59 Let X be a normally distributed random variable. Find the values of X that bound the middle 50% of the distribution of X if the mean is 5 and the variance is 9.

* *Source:* Adapted from B. F. Booth, G. S. Richards, D. Sabina, and W. A. R. Orban, Attitudes toward fitness in an undergraduate nursing population. *Int. J. Sports Psychol.* **17** (3): 269–279 (1986).

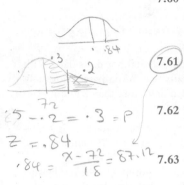

$$25 - .2 = .3 = P$$

$$Z = .84$$

$$.84 = \frac{X - 72}{18} = 87.12$$

7.60   Scores on the English screening exam for international students are distributed normally with a mean of 68 and a standard deviation of 11. Calculate the following:
(a)   The percentage of scores between 70 and 80.
(b)   The percentage of scores that are less than 60.

7.61   The examination committee of the Institute of Chartered Accountants passes only 20% of those who take the examination. If the scores follow a normal distribution with an average of 72 and a standard deviation of 18, what is the passing score?

7.62   During the summer months, the length of a trip made by a taxi driver from a pickup point to destination of passenger is uniformly distributed from 11 to 18 miles. What percentage of the trips exceeds 15 miles?

7.63   The time that a certain drug has an effect on adult persons is considered to be normally distributed, with a mean of 30 hours and a standard deviation of 3.5 hours. What is the probability that any given adult will be affected by the drug for at least 35 hours? What proportion of adults will be affected by the drug for less than 24 hours?

7.64   The weights of the students in a class are normally distributed with a mean of 160 lb and a standard deviation of 15 lb. What proportion of the class weighs
(a)   more than 180 lb?
(b)   less than 130 lb?
(c)   between 140 and 160 lb?

7.65   Based on U.S. Census Bureau figures, people who rent homes are far more likely to move to another part of the country than homeowners. Between 1974 and 1981, the mobility rate for homeowners dropped from 10 to 7%. The mobility rate for renters remained constant at 38%. If a random sample of 80 renters is obtained, what is the probability that fewer than 35 of these renters move to another part of the country?

7.66   Due to a minor misadjustment, a filling machine for 12 fluid ounce cans of soft drink would fill 11.6 fluid ounces, then gradually increase the amount filled on subsequent cans up to 12.5 fluid ounces, at which point the machine would revert to 11.6 fluid ounces. Thus, effectively the quantity of drink dispensed was uniformly distributed over this range (11.6 to 12.5).
(a)   What is the probability that any one can filled by this machine contains between 12.1 and 12.2 ounces?
(b)   What percentage of cans filled by this machine contain less than 12 ounces of soft drink?

7.67   The shelf life of cookies made by a small bakery is considered to be exponentially distributed with a mean equal to 3 days. What percentage of the boxes of cookies placed on the shelf today would still be considered marketable after 2.75 days?

7.68   The time that a certain drug has an effect on a normal human being is considered to be exponentially distributed when a standard dose is taken. If the average length of time that the drug has an effect is 30 hours, what is the probability that any given normal person will be affected by the drug for at least 32 hours? What is the standard deviation for the length of time that the drug affects a person?

7.69   According to a city's budget director, revenue from the city's golf course has been uniformly distributed over the past decade, ranging from $615,000 to $1,115,000 annually.

    (a)   What is the mean of this distribution?

    (b)   What is the standard deviation?

    (c)   What is the probability that in any one year, revenue from the city's golf course exceeds $800,000?

7.70 A hot dog vendor knows that on the average he can sell 200 hot dogs daily with a standard deviation of 28.9. If the number sold daily is approximately normally distributed, what is the probability that on any given day more than 250 hot dogs will be sold?

7.71 At a fast food restaurant, the time spent by a customer in the restaurant is a normally distributed variable. The average amount of time spent is 26 minutes with a standard deviation of 2.8 minutes.

    (a)   What proportion of customers spend more than 30 minutes in the restaurant?

    (b)   If a customer arrives at 11:05 A.M., what is the probability that the customer leaves at 11:30 A.M. or earlier?

7.72 The Big City Police Department feels that it needs to improve its responsiveness to emergency calls. At present, an average of 65 callers a day hang up before dispatchers answer the line, probably getting tired of waiting on hold. Assume that the number of abandoned calls per day is normally distributed and has a standard deviation of 20.

    (a)   What percentage of the days do more than 100 callers hang up before the dispatchers answer the line?

    (b)   On any given day, what is the probability that from 60 to 80 callers (inclusive) abandon the call?

$$P(X \leq 1600) = 1 - e^{-\frac{1}{1800} \times 1600}$$

7.73 Clearvision Company manufactures picture tubes for color television sets and claims that the life spans of their tubes are exponentially distributed with a mean of 1800 hours. What percentage of the picture tubes will last no more than 1600 hours?

7.74 The amount of time each day that the X-ray diagnostic machine is used at a certain hospital is approximately exponentially distributed with a mean of 3.5 hours. What is the probability that the X-ray diagnostic machine will be used at least 2 hours a day?

7.75 A paint sprayer coats a metal surface with a layer of paint between 0.5 and 1.5 millimeters thick. The thickness of the coat of paint is approximately uniformly distributed.

    (a)   What is the mean and standard deviation of the thickness of the coat of paint on the metal surface?

    (b)   What is the probability that paint from this sprayer on any given metal surface will be between 1.0 and 1.3 millimeters thick?

7.76 The Defense Contract Audit Agency (DCAA), which audits defense contractors, has a backlog of unaudited contracts in any given year. Suppose that the dollar amounts of unaudited contracts (in billions of dollars per annum) is normally distributed, with a mean of $74.5 billion and a standard deviation of 11.6 billion. What is the probability that in any given year, the backlog is less than $50 billion?

7.77 If X is a uniform random variable that represents the percentage of time each day that a machine does not work, what is the probability that X is greater than the mean percentage of time that the machine does not work?

7.78 If the random variable X has a uniform distribution between -10 and 10, find the value of x such that $P[X \geqslant x] = 0.25$.

7.79 The marketing division of Goodlife Tires determined the average (mean) life of tires to be 30,000 miles with a standard deviation of 5,000 miles. Given that tire life is a normally distributed random variable, find the following:
(a)   The probability that tires last between 25,000 and 35,000 miles.
(b)   The probability that tires last between 28,000 and 33,000 miles.
(c)   The probability that tires last less than 28,000 miles.
(d)   The probability that tires last more than 35,000 miles.

7.80 Suppose that the national average salary for a public school teacher in the United States is $25,240 with a standard deviation of $2714. If public school teachers' salaries follow a normal distribution, what proportion of public school teachers earn a salary of $22,500 or less?

7.81 The distance traveled by cars with automatic transmissions before they develop transmission problems is, on the average, 48,000 miles, with a standard deviation of 9500 miles. Assume that this variable is normally distributed. What proportion of cars with automatic transmissions develop transmission problems within 30,000 miles?

7.82 A regional supermarket chain knows that 38% of its customer complaints relate to purchases costing less than $10. In a random sample of 50 complaints, what is the probability that more than half of them relate to purchases costing less than $10?

CASE STUDY

What Would Sherlock Holmes Say about Fingerprints and Statistics? It's Elementary, My Dear Watson.

You have undoubtedly at some time or other examined the pattern of swirls and ridges on the tips of your fingertips, the stuff of which detective stories are made, and wondered about them. Fingerprints are formed during the third trimester of pregnancy, and are almost entirely a product of genetic factors. No two fingerprints are alike (even in identical twins) and their pattern remains constant for life.

Despite the uniqueness of individual fingerprints, it is still possible to classify them. Sir Francis Galton, toward the end of the 19th century, was the first to classify all fingerprints into three generic categories: the whorl, the loop, and the arch. Some years later, the pioneering criminologist, Sir Edward R. Henry, extended Galton's work and came up with eight generic types. This came to be known as the Henry system, which revolutionized the identification of criminals.

One characteristic that can be used to distinguish one fingerprint from another is the ridge count. In a loop pattern, there is a point, called the triradius, where three opposing ridge systems come together. A line connecting the triradius to the center of the loop will traverse a certain number of ridges, for each finger. The total number of such ridges crossed for all fingers is the ridge count. The following table shows the distribution of ridge counts for 825 males:

RIDGE COUNTS

Count	Frequency
0–19	10
20–39	12
40–59	24
60–79	40
80–99	73
100–119	100
120–139	90
140–159	117
160–179	139
180–199	100
200–219	67
220–239	36
240–259	10
260–279	4
280–299	3
	825

Sources: C. O. Carter, Multifactorial genetic disease., *Hospital Practice* **5**: 45–59 (1970); H. Cummins and C. Midlo, *Finger Prints, Palms, and Soles*, Blakiston (1943). Adapted from R. J. Larsen and D. F. Stroup, *Statistics in the Real World: A Book of Examples* (New York: Macmillan 1976) p. 21. Adapted and abridged with permission of Macmillan Publishing Company from *Statistics in the Real World: A Book of Examples* by Richard J. Larsen and Donna Fox Stroup. Copyright © 1976 by Macmillan Publishing Company.

Case Study Questions

1. Strictly speaking, the ridge count is not a continuous variable. However, we shall relax this criterion, and treat the above as a normal distribution. Compute the mean and standard deviation of the ridge count distribution.

2. What proportion of males would have ridge counts of 250 or more?

3. Between what values (rounded to the nearest whole number) will the ridge counts for the middle 95% of males lie?

4. If a male criminal suspect is apprehended, what is the probability that his ridge count is 260 or more?

5. Suppose that from a set of fingerprints obtained at the scene of a crime, detectives determined that the ridge count was at least 240. They were unable to be more precise, because the fingerprints were not very clear. What percentage of the male population might have ridge counts that could possibly match the fingerprints found at the scene of the crime?

6. Suppose that the police had three suspects: Mr. A, ridge count 250; Mr. B, ridge count 263, and Mr. C, ridge count 200. Which of the suspects could you eliminate?

7. Using modern laser technology, the investigators subsequently determine that the ridge count exceeds 260. Can you narrow down the search to one suspect? If this suspect is apprehended without other circumstantial or corroborating evidence, is there still a chance of being wrong about the identification of the suspect?

COMPUTER EXERCISES USING THE DATABASE

Using a sample of 100 observations, complete the following exercises:

1. Construct a histogram using variable UG-GPA (undergraduate GPA). What type of distribution does this histogram suggest—Normal? Uniform? Exponential? None of these?

2. Repeat exercise 1 using variable LSINDEX (score on Life Satisfaction Index questionnaire).

3. Repeat exercise 1 using variable QGRE (quantitative score on the graduate record exam).

8

STATISTICAL INFERENCE AND SAMPLING

A LOOK BACK/INTRODUCTION

The previous three chapters laid the foundation for using statistical methods in decision making. Any such decision will have uncertainty associated with it, but we can attempt to measure this uncertain outcome using a probability. Random variables (both discrete and continuous) allow one conveniently to represent certain outcomes of an experiment and their corresponding probabilities. If the experiment fits a particular discrete situation (such as binomial), you can easily determine the probability of certain events or determine the mean (average) value of the related distribution.

If the random variable of interest is continuous, you can make probability statements after assuming the probability distribution involved (such as continuous uniform, normal, exponential, or others not discussed). Both discrete and continuous random variables come into play in all areas of decision making. They allow us to make decisions concerning a large population using the information contained in a much smaller sample.

This is the area of **statistical inference**, which this chapter will introduce by demonstrating how to estimate something about the population (such as the average value μ) by using the corresponding value from a sample (such as the sample average $\bar{X}$). Recall that μ (belonging to the population) is a parameter and $\bar{X}$ (belonging to the sample) is a statistic. When dealing with a normal population, for example, what does one do if the population mean μ is unknown? So far in the text, this value has been specified for you. In this chapter we will discuss methods of estimating population parameters using sample statistics, along with several methods of gathering your sample data.

8.1 RANDOM SAMPLING AND THE DISTRIBUTION OF THE SAMPLE MEAN

In Chapter 3, you learned how to calculate the mean of a sample $\bar{X}$. The sample is drawn from a population having a particular distribution, such as uniform or normal. If you were to obtain another sample (you probably would not, as most decisions are made from just one sample), would you get the same value of $\bar{X}$? Assuming that the new sample was made up of different individuals than was the first sample, then almost certainly the two $\bar{X}$'s would not be the same. So, $\bar{X}$ itself is a random variable. We will demonstrate that, if a sample is large enough, $\bar{X}$ is very nearly normally distributed regardless of the shape of the sampled population. That is, if you were to obtain many samples, calculate the resulting $\bar{X}$'s, and then make a histogram of these $\bar{X}$'s, this histogram would always approximately resemble a bell-shaped (normal) curve.

Simple Random Samples

In Chapter 5, the concept of a simple random sample was introduced. The mechanics of obtaining a random sample range from drawing names out of a hat to using a computer to generate lists of random numbers. For extremely large populations, one is often forced to select individuals (elements) from the population in a nearly random manner.

The underlying assumption behind a random sample of size n is that any sample of size n has the same chance (probability) of being selected. To be completely assured of obtaining a random sample from a finite population, one should number the members of the population from 1 to N (the population size) and, using a set of n random numbers, select the corresponding sample of n population elements for your sample.

This procedure was described in Chapter 5 and is often used in practice, particularly when you have a sampling situation that needs to be legally defensible. However, for situations in which the population is extremely large, this strategy may be impractical, and instead you can use a sampling plan that is nearly random. Several other sampling procedures will be discussed in the last section of this chapter.

The main point of this lengthy discussion is that practically all the procedures presented in subsequent chapters relating to decision making and estimation assume that one is using a random sample. In the chapters that follow, the word sample will mean simple random sample.

Estimation

The idea behind statistical inference has two components:

1 The *population* consists of everyone of interest. By "everyone" we mean all student scores, individual heights, worker stress levels, or whatever else you are interested in measuring or observing. The mean value (for

Figure 8.1

The sample mean $\bar{X}$ is used to estimate the population mean μ. In general, sample statistics are used to estimate population parameters.

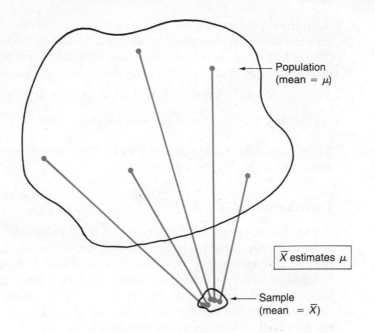

Population (mean $= \mu$)

$\bar{X}$ estimates μ

Sample (mean $= \bar{X}$)

example, average height, average income) of everyone in this population is μ and generally is not known.

2 The *sample* is randomly drawn from this population. Elements of the sample thus are part of the population—but certainly not all of it. The exception to this is a *census*, a sample that consists of the entire population.

The sample values should be selected randomly, one at a time, from the entire population. Figure 8.1 emphasizes our central point, namely, an unknown population **parameter** (such as $\mu =$ the mean value for the entire population) can be **estimated** using the corresponding sample **statistic** (such as $\bar{X} =$ the mean of your sample).

It makes sense, doesn't it? It would be most desirable to know the average value for everyone in the population, but in practice this is nearly always impossible. It may take too much time or money, we may not be able to obtain values for them all even if we want to, or the process of measuring the individual items may destroy them (such as measuring the force required to crack the case of a portable computer). In many instances, estimating the population value using a sample estimate is the best we can do.

EXAMPLE 8.1

In Chapter 7, the weight of a newborn infant was assumed to follow a normal distribution with a mean of $\mu = 6.8$ lb and a standard deviation of $\sigma = .8$ lb. There is no way of knowing that $\mu = 6.8$ lb unless you check the birth record of all babies born at this hospital. Assume that

$$X = \text{weight of a newborn infant}$$

is a normal random variable, but do not assume anything about the mean and standard deviation. Ignoring the standard deviation, estimating μ involves obtaining a random sample of babies and recording their weights. Suppose you obtain a sample of size $n = 10$, with the following results:

$$6.45, \ 7.53, \ 8.11, \ 6.24, \ 5.87, \ 7.03, \ 7.21, \ 6.35, \ 8.42, \ 7.16$$

What is the estimate of μ, based on these values?

Solution   The sample mean would be $\bar{x} = 7.04$ lb. Thus, based on 10 sample values, our best estimate of μ is $\bar{x} = 7.04$ lb.   □

Distribution of $\bar{X}$

Referring to example 8.1, the value of $\bar{X}$ would almost certainly change if you were to obtain another sample. The question of interest here is, if we were to obtain many values of $\bar{X}$, how would they behave? If we observed values of $\bar{X}$ indefinitely, where would they center? That is, what is the **mean** of the distribution for the random variable $\bar{X}$? Is the variation of the $\bar{X}$ values more, less, or the same as the variation of individual observations? This is measured by the **standard deviation** of the distribution for $\bar{X}$.

In example 7.3, it was assumed that the average score on the national achievement test was $\mu = 500$ points, with a population standard deviation of $\sigma = 75$ points. This does not imply that if you obtain a random sample of students taking the test, the resulting sample mean $\bar{X}$ will always be 500. Rather, a little head-scratching should convince you that $\bar{X}$ will not be exactly 500, but $\bar{X}$ should be approximately 500.

Twenty samples of 10 students each and the calculated $\bar{X}$ for each sample are shown in Table 8.1. We will assume that the population parameters are $\mu = 500$ points and $\sigma = 75$ points (Figure 8.2).

The 20 values of $\bar{X}$ are:

493.1, 499.4, 496.1, 497.6, 512.5, 491.5, 496.8, 501.3, 501.1, 502.2, 499.8, 483.1, 506.4, 509.0, 502.8, 506.7, 480.8, 489.9, 526.3, 521.2

Figure 8.2

Assumed distribution of achievement test scores

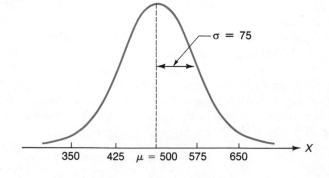

Table 8.1

20 samples of
10 national
achievement
test scores

Sample 1	Sample 2	Sample 3	Sample 4	Sample 5	Sample 6	Sample 7
379	546	524	458	422	507	457
545	572	442	575	469	487	504
501	470	484	392	564	487	638
565	457	645	588	519	582	415
460	481	500	620	615	481	493
370	515	633	424	485	479	560
577	500	509	545	367	478	458
562	539	400	433	576	428	555
480	472	361	513	497	490	439
492	442	463	428	611	496	449
$\bar{X} = 493.1$	499.4	496.1	497.6	512.5	491.5	496.8
$s = 73.90$	42.56	90.39	80.82	81.09	37.97	68.95

Sample 8	Sample 9	Sample 10	Sample 11	Sample 12	Sample 13	Sample 14
569	501	427	391	548	585	564
479	422	564	606	334	493	472
491	744	410	401	478	518	455
538	428	582	431	475	461	624
457	434	501	475	539	433	563
648	515	597	445	487	506	536
443	558	519	525	398	484	458
499	407	406	569	561	455	463
519	436	509	585	539	554	451
370	566	507	570	472	575	504
$\bar{X} = 501.3$	501.1	502.2	499.8	483.1	506.4	509.0
$s = 75.26$	103.19	69.07	80.86	71.67	51.70	59.89

Sample 15	Sample 16	Sample 17	Sample 18	Sample 19	Sample 20
565	630	566	400	504	528
528	424	412	543	510	439
517	457	545	570	389	554
466	698	536	533	605	530
523	564	427	504	540	599
551	516	519	487	582	542
503	377	496	448	484	599
523	419	521	561	433	573
500	573	377	431	686	349
352	409	409	422	530	499
$\bar{X} = 502.8$	506.7	480.8	489.9	526.3	521.2
$s = 59.58$	106.94	67.73	61.71	85.06	76.97

Figure 8.3

Histogram of 20
sample means
generated by
MINITAB. Compare
with Figure 8.2.

```
MTB > SET INTO C1
DATA> 493.1 499.4 496.1 497.6 512.5 491.5 496.8 501.3
DATA> 501.1 502.2 499.8 483.1 506.4 509.0 502.8 506.7
DATA> 480.8 489.9 526.3 521.2
DATA> END
MTB > HISTOGRAM OF C1

   Histogram of C1    N = 20

   Midpoint    Count
        480      1    *
        485      1    *
        490      2    **
        495      3    ***
        500      6    ******
        505      3    ***
        510      1    *
        515      1    *
        520      1    *
        525      1    *
```

Normal

They are not each 500, but they are all close to 500. Using a calculator or computer, you would also find that (1) the average (mean) of these 20 values is 500.88 (this is close to $\mu = 500$), and (2) the standard deviation of these 20 values is 11.17 (this is much smaller than $\sigma = 75$).

The $\bar{X}$ values appear to be centered at $\mu = 500$ points but have much less variation than the individual observations in each of the samples. A histogram of these 20 values generated by MINITAB is contained in Figure 8.3. Based on the shape of this histogram, it seems reasonable to assume that the values of $\bar{X}$ follow a normal distribution, but one that is much narrower than the population of individual scores in Figure 8.2.

8.2   THE CENTRAL LIMIT THEOREM

CLT

Our last example illustrates a useful result, the Central Limit Theorem (CLT).

Central Limit Theorem (CLT)

When using a random sample of size n from a population with mean μ and standard deviation σ, the resulting sample mean $\bar{X}$ has a normal distribution with mean μ and standard deviation $\sigma/\sqrt{n}$. This is true for any sample size n if the underlying population is normally distributed, and it is approximately true for large sample sizes (generally $n > 30$) obtained from any population.

In other words, the distribution of all possible $\bar{X}$ values has an exact or approximate normal distribution, with mean μ and standard deviation $\sigma/\sqrt{n}$.

COMMENT The second part of the CLT is an extremely strong result; it says that you can assume that $\bar{X}$ follows an approximate normal distribution regardless of the shape of the population from which the sample was obtained, if the sample size n is large. For example, if you repeatedly sampled from a population with a uniform distribution, the resulting $\bar{X}$'s would follow a normal (not a uniform) curve.

In Table 8.1, 20 samples of size 10 were obtained, and the corresponding values of $\bar{X}$ were determined. Suppose samples of size 10 were obtained indefinitely and we wished to describe the shape of the resulting $\bar{X}$'s. According to the CLT, $\bar{X}$ will be a normal random variable. We are assuming that the individual scores follow a normal curve (see Figure 8.2), so this will be true for any sample size—in particular, $n = 10$. So the resulting $\bar{X}$'s will describe a normal curve, similar to the curve in Figure 8.3.

Where is the curve centered? According to the CLT, the mean of this normal random variable is the same as that in Figure 8.2; that is, it is the mean of the population from which you are sampling. This is $\mu = 500$, and so, on the average, the value of $\bar{X}$ is $\mu_{\bar{X}} = 500$ points. Notice that the average of the 20 values of $\bar{X}$ that we did observe was 500.88. This value will get closer to, or **tend toward**, 500 as we take more samples of size 10.

What is the standard deviation of the normal curve for $\bar{X}$? As we noted earlier, the 20 values of $\bar{X}$ jump around (vary) much less than do the individual observations in each of the samples. Consequently, the standard deviation of the $\bar{X}$ normal curve will be much less than that of the population curve (describing individual scores) in Figure 8.2. In fact, according to the CLT, this will be

$$\sigma_{\bar{x}} = \frac{\sigma}{\sqrt{n}} \qquad\qquad (8\text{-}1)$$

where σ is the standard deviation of the population ($\sigma = 75$ in Figure 8.2). Consequently,

$$\sigma_{\bar{x}} = \frac{75}{\sqrt{10}} = 23.72$$

Recall that the standard deviation of the 20 observed $\bar{X}$ values was 11.17. This value will tend toward 23.72 if we take more samples of size 10. These results are summarized in Figure 8.4, where $\mu_{\bar{x}} = 500$ and $\sigma_{\bar{x}} = 23.72$.

Basically, the CLT says that the normal curve (distribution) for $\bar{X}$ is centered at the same value as the population distribution but has a much smaller standard deviation. Notice that as the sample size n increases, $\sigma/\sqrt{n}$ values decrease, and so the spread relative to the mean of the $\bar{X}$ curve (that is, the variation in the $\bar{X}$ values) decreases. If we repeatedly obtained samples of size 100 (rather than 10), the corresponding $\bar{X}$ values would lie even closer to $\mu_{\bar{x}} = 500$

Figure 8.4

Normal curves for
population and
sample mean

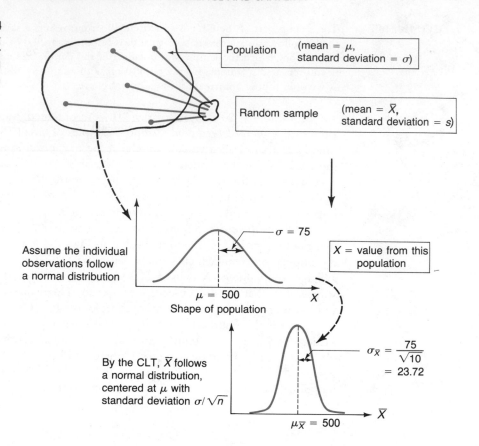

because now $\sigma_{\bar{x}}$ would equal $75/\sqrt{100} = 7.5$. This is illustrated in Figure 8.5.

For the 20 values of $\bar{X}$ in Table 8.1, it was assumed that the population mean was known to be $\mu = 500$, so each of the $\bar{X}$ values estimates μ with a certain amount of error. The more variation in the $\bar{X}$ values, the more error we encounter using $\bar{X}$ as an estimate of μ. Consequently, the standard deviation of $\bar{X}$ also serves as a measure of the error that will be encountered using a sample mean to estimate a population mean. The standard deviation of the $\bar{X}$ distribution is often referred to as the **standard error** of $\bar{X}$.

standard error of $\bar{X}$ = standard deviation of the probability
distribution for $\bar{X}$

$$= \frac{\sigma}{\sqrt{n}}$$

Figure 8.5

Normal curves for
the sample mean
($n = 10, 20, 50, 100$)

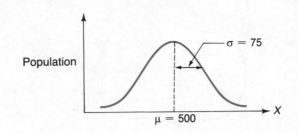

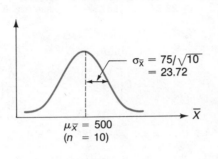

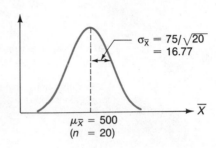

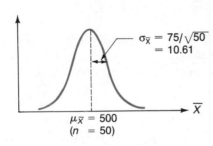

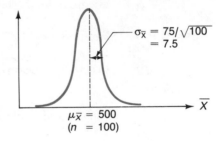

EXAMPLE 8.2

Toddle-Teach has determined that the time required for six-year-olds to put together one of their puzzles is normally distributed with a mean of 20 minutes and a standard deviation of 3 minutes.

1 What is the probability that a six-year-old child takes longer than 22 minutes to assemble the puzzle?

2 What is the probability that the average assembly time for 15 such children exceeds 22 minutes?

3 What is the probability that the average time to assemble the puzzle for 15 children is between 19 and 21 minutes?

Solution 1

The random variable X here is the time required to assemble the puzzle. This was assumed to be a normal random variable, with $\mu = 20$ minutes and $\sigma = 3$ minutes (Figure 8.6). We wish to determine $P(X > 22)$. Standardizing this

Figure 8.6

Curve for puzzle
assembly times

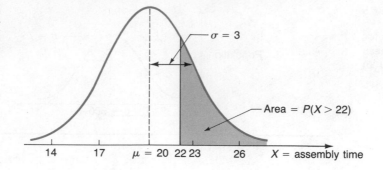

variable and using Table A-4, we obtain

$$P(X > 22) = P\left[\frac{X - 20}{3} > \frac{22 - 20}{3}\right]$$

$$= P(Z > .67)$$

$$= .5 - .2486 = .2514$$

Therefore, a randomly chosen child will require longer than 22 minutes to assemble the puzzle with probability .25.

Solution 2   Figure 8.6 does not apply to this question because we are concerned with the average time for 15 children, not an individual child. Using the CLT, we know that the curve describing $\bar{X}$ (an average of 15 assembly times) is normal with

$$\text{mean} = \mu_{\bar{x}} = \mu = 20 \text{ minutes}$$

$$\text{standard deviation (standard error)} = \sigma_{\bar{x}} = \sigma/\sqrt{n}$$

$$= 3/\sqrt{15} = .77 \text{ minutes} \qquad \text{(Figure 8.7)}.$$

The procedure is the same as in Solution 1, except now the standard deviation of this curve is .77, rather than 3:

$$P(\bar{X} > 22) = P\left[\frac{\bar{X} - 20}{.77} > \frac{22 - 20}{.77}\right]$$

$$= P(Z > 2.60)$$

$$= .5 - .4953 = .0047$$

Figure 8.7

Curve for
$\bar{X}$ = average of 15
puzzle assembly
times

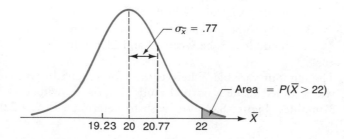

Figure 8.8

Curve for average
assembly time of 15
children. Shaded
area shows
$P(19 < \bar{X} < 21)$.

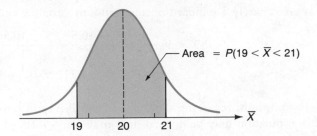

Area $= P(19 < \bar{X} < 21)$

So an average puzzle assembly time for a sample of 15 children will be more than 22 minutes with less than 1% probability; that is, it is very unlikely that the average of 15 assembly times will exceed 22 minutes.

Solution 3 The curve for this solution is shown in Figure 8.8. We wish to find $P(19 < \bar{X} < 21)$.

$$P(19 < \bar{X} < 21) = P\left[\frac{19 - 20}{.77} < \frac{\bar{X} - 20}{.77} < \frac{21 - 20}{.77}\right]$$
$$= P(-1.30 < Z < 1.30)$$
$$= .4032 + .4032 = .8064$$

Thus, a sample of 15 children will produce an average puzzle assembly time between 19 and 21 minutes with probability about .81. □

In example 8.2, it was assumed that the individual puzzle assembly times followed a normal distribution. However, remember that the strength of the CLT is that this assumption is not necessary for large samples. We can answer questions 2 and 3 for any population whose mean is 20 minutes and standard deviation is 3 minutes, provided we take a large sample ($n > 30$). In this case, the normal distribution of $\bar{X}$ is not exact, but it provides a very good approximation.

EXAMPLE 8.3 The price/earnings (P/E) ratio of a stock is usually considered by analysts who put together financial portfolios. Suppose a population of all P/E ratios has a mean of 10.5 and a standard deviation of 4.5.

  1 What is the probability that a sample of 40 stocks will have an average P/E ratio less than nine?

  2 What assumptions are necessary about the population of all P/E ratios in your answer to question 1?

Solution 1 By the CLT, $\bar{X}$ is approximately a normal random variable with mean $= \mu = 10.5$ and standard deviation $= \sigma/\sqrt{n} = 4.5/\sqrt{40} = .71$. So

$$Z = \frac{\bar{X} - 10.5}{.71}$$

is approximately a standard normal random variable, and consequently

$$P(\bar{X} < 9) = P\left[\frac{\bar{X} - 10.5}{.71} < \frac{9 - 10.5}{.71}\right]$$
$$= P(Z < -2.11) = .0174$$

Solution 2   No assumptions regarding the shape of the P/E ratio population are necessary. This population may be normal or it may not—it simply does not matter because we are using a fairly large sample ($n = 40$). The distribution of $\bar{X}$ is approximately normal, regardless of the shape of the population of all P/E ratios. Our only assumptions in Solution 1 were that $\mu = 10.5$ and $\sigma = 4.5$.   □

EXERCISES   8.1   Suppose you roll two dice. Each die is independent of the other and could give a result from 1 to 6 with equal likelihood.
  (a)   List all the possible pairs of results (such as 1, 1 then 1, 2 then 1, 3 and so forth).
  (b)   Calculate the sample mean for each sample (pair) of two dice.
  (c)   Construct the probability distribution of the sample mean, and draw a histogram of the same. What do you notice about the shape of the distribution?

8.2   A southwestern bank issues travelers checks in denominations of $10, $20, $50, $100, and $500. All five amounts have occurred with equal probability.
  (a)   List all possible samples of three from these five denominations. (Denominations may not be repeated.)
  (b)   Calculate the sample mean for each sample of size three.
  (c)   Construct the probability distribution for the sample mean.
  (d)   Draw a histogram showing the distribution of the sample mean ($\bar{X}$).

8.3   A fireworks manufacturer has found that the height reached by his rockets follows a normal distribution, with a mean of 200 feet and a standard deviation of 12 feet.
  (a)   What percentage of the individual rockets achieve a height exceeding 210 feet?
  (b)   What percentage of the time will groups of one dozen rockets have an average height exceeding 210 feet?

8.4   The average length of actual running time (excluding advertisements) for television feature films is 1 hour 40 minutes, with a standard deviation of 15 minutes. If a sample of 49 TV feature films is taken at random, what is the probability that the average running time for this group is 1 hour, 45 minutes or more?

8.5   Let $\bar{X}$ be the average of a sample of size 18 from a normally distributed population with a mean 37 and variance 16. Find the following probabilities.
  (a) $P(\bar{X} \leqslant 35)$
  (b) $P(\bar{X} \geqslant 38)$
  (c) $P(34 \leqslant \bar{X} \leqslant 36.5)$
  (d) $P(36 \leqslant \bar{X} \leqslant 38)$

8.6   A survey of fees charged by marriage counselors revealed that the hourly rate was normally distributed with a mean of $60 and a standard deviation of $20.

(a) What is the probability that any one counselor charges a rate in excess of $70 an hour?

(b) What is the probability that a random group of 25 counselors will charge an average fee in excess of $70 an hour?

8.7 Various researchers have come up with conflicting results in their investigations of the intelligence of the Japanese people. Richard Lynn and Susan Hampson (1986) at the University of Ulster, Coleraine, Northern Ireland, U.K., suggest that the Japanese have a distinctive and idiosyncratic profile of abilities. A standardization sample consisting of between 50 and 60 children in each of 10 age groups was used to obtain the following standardized scores for Japanese children:

JAPANESE MEANS SCORED ON THE AMERICAN McCARTHY SCALES

	Verbal	Perceptual Performance	Quantitative	Memory	Motor	General Cognitive
Means	46.2	56.9	50.5	45.3	54.8	101.3
Std. deviation	8.8	8.3	8.9	7.8	8.6	13.4

Source: Lynn, R. and Hampson, S. Intellectual abilities of Japanese children: An assessment of $2\frac{1}{2}$–$8\frac{1}{2}$-year-olds derived from the McCarthy scales of children's abilities. *Intelligence* **10**: 45 (1986), Table 2.

We may assume the above figures are representative of the Japanese population of $2\frac{1}{2}$–$8\frac{1}{2}$-year-old children.

(a) Assuming a normal distribution, what is the probability that any one Japanese child in this age group has a perceptual performance score less than 50?

(b) What is the probability that a random group of 81 Japanese children in this age category have a mean verbal score between 40 and 50?

(c) What is the probability that a random group of 49 Japanese children in this age category have a mean general cognitive score less than 85?

8.8 Suppose that the average auto fuel consumption in the United States is 6.2 billion gallons per month, with a standard deviation of 2.1 billion gallons. If a random sample of 6 months is taken, what is the probability that the average consumption for these 6 months is less than 5.0 billion gallons per month? What assumptions did you have to make about the population distribution?

Applying the Central Limit Theorem to Normal Populations

The CLT tells us that $\bar{X}$ tends toward a normal distribution as the sample size increases. If you are dealing with a population that has an assumed normal distribution (as in example 8.2), then $\bar{X}$ is normal regardless of the sample size. However, as the sample size increases, the variability of $\bar{X}$ decreases, as is illustrated in Figure 8.5. This means that, for large sample sizes, if you were to get many samples and corresponding values of $\bar{X}$, the values of $\bar{X}$ would be more concentrated around the middle, with very few extremely large or extremely small values.

Look at Figure 8.5, which illustrates the assumed normal distribution of all achievement test scores. We know that (using Table A-4) 95% of a normal

Table 8.2

Sampling from a normal population with $\mu = 500$ and $\sigma = 75$. 95% of the time, the value of $\bar{X}$ will be between $\mu_{\bar{x}} - 1.96\sigma_{\bar{x}}$ and $\mu_{\bar{x}} + 1.96\sigma_{\bar{x}}$

Sample Size	$\sigma_{\bar{x}}$	$\mu_{\bar{x}} - 1.96\sigma_{\bar{x}}$	$\mu_{\bar{x}} + 1.96\sigma_{\bar{x}}$	Conclusion
$n = 10$	23.72	453.5	546.5	95% of the time, the value of $\bar{X}$ will be between 453.5 and 546.5
$n = 20$	16.77	467.1	532.9	95% of the time, the value of $\bar{X}$ will be between 467.1 and 532.9
$n = 50$	10.61	479.2	520.8	95% of the time, the value of $\bar{X}$ will be between 479.2 and 520.8
$n = 100$	7.5	485.3	514.7	95% of the time, the value of $\bar{X}$ will be between 485.3 and 514.7

curve is contained within 1.96 standard deviations of the mean. For a sample size of $n = 10$ from a normal population with $\mu = 500$ and $\sigma = 75$, $\sigma_{\bar{x}} = 23.72$. Now,

$$\mu_{\bar{x}} - 1.96\sigma_{\bar{x}} = 500 - 1.96(23.72) = 453.5$$

and

$$\mu_{\bar{x}} + 1.96\sigma_{\bar{x}} = 500 + 1.96(23.72) = 546.5$$

Thus, if we repeatedly obtain samples of size 10, 95% of the resulting $\bar{X}$ values will lie between 453.5 and 546.5.

This result and the corresponding results using $n = 20$, 50, and 100 are contained in Table 8.2. This reemphasizes that, for larger samples, you are much more likely to get a value of $\bar{X}$ that is close to $\mu = 500$. In practice, you typically do not know the value of μ. However, by using a larger sample size, you are more apt to obtain an $\bar{X}$ that is a good estimate of the unknown μ.

Applying the Central Limit Theorem to Nonnormal Populations

The real strength of the CLT is that $\bar{X}$ will tend toward a normal random variable regardless of the shape of your population. You need a large sample ($n > 30$) to obtain a nearly normal distribution for $\bar{X}$. The CLT also holds when sampling from a discrete population.

Figures 8.9 and 8.10 illustrate the distribution of $\bar{X}$ for two nonnormal populations. Notice that the uniform population (Figure 8.9) is at least symmetric about the mean, so the distribution of the sample mean $\bar{X}$ tends toward a normal distribution for much smaller sample sizes. The U-shaped distribution (Figure 8.10) is another continuous distribution. It is characterized by many

Figure 8.9

Distribution of $\bar{X}$
(uniform population)

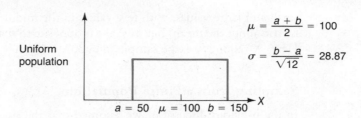

Uniform population

$$\mu = \frac{a + b}{2} = 100$$

$$\sigma = \frac{b - a}{\sqrt{12}} = 28.87$$

$a = 50 \quad \mu = 100 \quad b = 150$

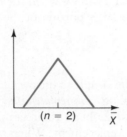

$(n = 2)$

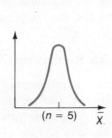

$(n = 5)$

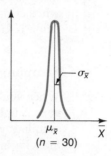

$\mu_{\bar{x}}$
$(n = 30)$

By the CLT, $\mu_{\bar{X}} = \mu = 100$

$$\sigma_{\bar{X}} = \frac{\sigma}{\sqrt{n}} = \frac{28.87}{\sqrt{30}}$$

$$= 5.27$$

Figure 8.10

Distribution of $\bar{X}$
(U-shaped population)

U-shaped population

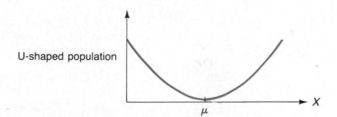

μ

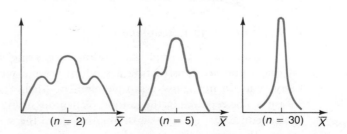

$(n = 2) \qquad (n = 5) \qquad (n = 30)$

small and large values, with few values in the middle. This distribution is symmetric about the mean, but its shape is opposite to that of a normal distribution. Here, $\bar{X}$ requires a large sample ($n > 30$) to attain a normal distribution.

Sampling from a Finite Population

In the previous discussion, we assumed that the population was large enough that the sample was extremely small by comparison. We will now consider whether our results, including the CLT, apply when the exact size of the population is known and the sample is a large portion of the population.

Sampling with Replacement

When you return each element of the sample to the population before taking the next sample element, you are sampling *with replacement*. This is not a common sampling procedure; people generally obtain their sample all at once, which makes it impossible to sample with replacement. When sampling with replacement, it is possible to obtain the same element more than once. For example, the same person could be chosen all three times in a sample of size $n = 3$.

When sampling with replacement, the CLT applies exactly as before, without any adjustments necessary.

Central Limit Theorem: Sampling with Replacement from a Finite Population

When sampling with replacement from a finite population with mean μ and standard deviation σ, the sample mean $\bar{X}$ tends toward a normal distribution with

$$\text{mean} = \mu_{\bar{x}} = \mu$$

$$\text{standard deviation (standard error)} = \sigma_{\bar{x}} = \frac{\sigma}{\sqrt{n}} \qquad \textbf{(8-2)}$$

where n = sample size.

Sampling without Replacement

When you obtain a sample by randomly selecting n different people or items at once, you are constructing a random sample without replacement. In this way, you do not allow a sample element (such as a voter preference or test result) to be repeated in the sample. When discussing a sample mean using such a procedure, we make a slight adjustment to the standard error of the sample mean. Previously the standard error of $\bar{X}$ was $\sigma/\sqrt{n}$. When sampling *without*

$\dfrac{\sigma}{\sqrt{n}} \sqrt{\dfrac{N-n}{N-1}}$

replacement, we multiply $\sigma/\sqrt{n}$ by $\sqrt{(N-n)/(N-1)}$, as illustrated in equation 8-3. Here $(N-n)/(N-1)$ is called the **finite population correction (fpc) factor**. When the sample size n is very small as compared to the population size N the fpc is nearly 1 and can be ignored. We can express this as a rule: the fpc can be ignored whenever $n/N < .05$.

We can also use the CLT in this situation.

Central Limit Theorem: Sampling without Replacement from a Finite Population

When sampling without replacement from a finite population (of size N), with mean μ and standard deviation σ, the sample mean $\bar{X}$ tends toward a normal distribution with

$$\text{mean} = \mu_{\bar{x}} = \mu$$

$$\text{standard deviation (standard error)} = \sigma_{\bar{x}} = \frac{\sigma}{\sqrt{n}} \cdot \sqrt{\frac{N-n}{N-1}} \quad \textbf{(8-3)}$$

where $n =$ sample size.

EXAMPLE 8.4

A group of women managers at Compumart is considering filing a sex-discrimination suit. A recent report stated that the average annual income of all employees in middle-management positions at Compumart is $48,000, and the standard deviation is $8500. A random sample of 45 women in these positions taken from a population of 350 female middle managers at Compumart had an average income of $\bar{X} = \$43,900$. If the population of all female incomes at this level is assumed to have the same mean ($48,000) and standard deviation ($8500) as the distribution of incomes for all employees, what is the probability of observing a value of $\bar{X}$ this low?

Solution

Because we have a large sample, we can assume (using the CLT) that the curve describing $\bar{X}$ is normal, as shown in Figure 8.11. Here, $n = 45$ and $N = 350$. We need to find $P(\bar{X} \leqslant 43,900)$. Standardizing and using Table A-4 we find that

$$P(\bar{X} \leqslant 43,900) = P\left[Z \leqslant \frac{43,900 - 48,000}{1184.75}\right]$$

$$= P(Z \leqslant -3.46) \to .4997 \Rightarrow .5 - .4997 = 0.0$$

$$= .0003$$

So, if the female population has an average salary of $48,000 (and standard deviation of $8500), then the chance of obtaining an $\bar{X}$ as low as $43,900 is extremely small. If we assume that the standard deviation is correct, then, based

Figure 8.11

Distribution of
sample mean of
annual salaries
(assuming
$\mu = \$48{,}000$ and
$\sigma = \$8500$). The
shaded area
represents the
solution to example
8.4, $P(\bar{X} \leqslant 43{,}900)$.

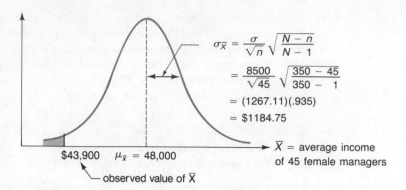

$$\sigma_{\bar{X}} = \frac{\sigma}{\sqrt{n}} \sqrt{\frac{N-n}{N-1}}$$

$$= \frac{8500}{\sqrt{45}} \sqrt{\frac{350 - 45}{350 - 1}}$$

$$= (1267.11)(.935)$$

$$= \$1184.75$$

$\bar{X}$ = average income
of 45 female managers

$\$43{,}900 \qquad \mu_{\bar{x}} = 48{,}000$

observed value of $\bar{X}$

strictly on this set of data, our conclusion would be that the average salary for women at this level is not $48,000 but is less than $48,000. □

With the type of question asked in example 8.4, there is always the chance that we will reach an incorrect decision using the sample data; there is always the chance of error due to sampling. This possible error will be a concern whenever you test a hypothesis. For now, remember that, when dealing with sample data, statistics never *prove* anything. They do, however, support or fail to support a claim (such as $\mu < \$48{,}000$).

EXERCISES

8.9 From a finite population of size 300 that is approximately normally distributed with a mean of 50 and a standard deviation of 10, what is the probability that a random sample of size 30 without replacement will yield a sample mean larger than 55? What is the probability that a random sample of size 30 with replacement will yield a sample mean larger than 55?

8.10 A Washington-based "think tank" has 95 Ph.D.s on its staff. The number of years of experience for these experts is believed to be normally distributed with a mean of 12 years, and a standard deviation of 4.5 years. If a random sample of 15 Ph.D.s is taken from this think tank (without replacement), what is the probability that the average number of years of experience for this sample is less than 15 years?

8.11 The electric bill for 250 households in a small midwestern town was found to have a mean of $120 with a standard deviation of $25 for the month of November. If 10 households are selected at random from the 250 households, what is the probability that the sample mean will be between $110 and $130? What are you assuming about the population?

8.12 Legislators are allowed certain postal privileges to communicate with their constituency. Suppose that for 435 congressmen, the mean amount of annual postal charges used is $1630 with a standard deviation of $170. If a sample of 40 congressmen is obtained, what is the probability that the average annual postal charge for this group of 40 is $1600 or more? If a sample of 20 congressmen is taken, what is the probability that the average annual charge for this group of 20 is $1600 or more? What must you assume to answer the latter question?

(handwritten notes in margin)

$\frac{\sigma}{\bar{x}} = \frac{10}{\sqrt{30}} \cdot \sqrt{\frac{300-30}{299}} = 1.74$

$P(\bar{x} > 55) = ? \quad P\left(z > \frac{55-50}{1.74}\right)$

$P(z > 2.87) = .4979, .5 - .4979$

$= 0.0021 \rightarrow$ without

$\sigma_{\bar{x}} = \frac{10}{\sqrt{30}} = 1.83$

$P\left(z > \frac{55-50}{1.83}\right) = 0.0032$ with

$\frac{n}{N} = \frac{10}{250} = 0.04 < 0.05$

$\sigma_{\bar{x}} = \frac{25}{\sqrt{10}} = 7.91$

$P(\text{$110$} < \bar{x} < 130) = P\left(\frac{110-120}{7.91} < z < \frac{130-120}{7.91}\right)$

$P(z < 1.26)$

$.3962 = 0.7924$

8.13 The mean daily time spent on the telephone by the 60 personnel managers of Retail Products is 1.25 hours; the standard deviation is 0.62 hours. Assuming that the time spent on the telephone is approximately normally distributed, what is the probability that the average amount of time spent on the telephone by 10 different personnel managers selected at random is greater than 1.5 hours?

8.14 As the finite population size gets large for a fixed sample size, explain how the finite population correction factor is affected.

8.15 An organ transplant data base has a list of 600 names of people waiting for suitable organ donors. The average waiting period is 15 months with a standard deviation of 6 months. If a sample of 50 names is randomly pulled out from the waiting list, what is the probability that the average waiting period for this group is 17 months or more?

8.16 An aptitude test on the theory of electronics was given to all of the 275 repair people of A.N.P. Micronics. The mean score on the test was 112 and the standard deviation of the test score was 18.6. Twenty repair people were selected at random. What is the probability that the sample mean would be between 110 and 120? What are you assuming about the distribution of the exam scores?

8.17 One hundred and fifty construction workers drive an average of 11.23 miles to a construction site. The standard deviation of this distance is 4.13 miles. If a random sample of 35 workers is selected, what is the probability that the sample mean of the distance driven to the construction site is greater than 16 miles?

8.18 The Nutrition Disorder Clinic in California has treated 875 patients with obesity problems. The mean weight loss achieved by the clinic for these 875 patients is 37.6 lb with a variance of 64 square pounds.

   (a) If a single patient is chosen at random, what is the probability that the weight loss for this patient is more than 42 lbs assuming the weight losses follow a normal distribution?

   (b) If a group of 25 patients is randomly chosen, what is the probability that the average weight loss for this group is more than 42 lb?

   (c) If a group of 60 patients is randomly chosen, what is the probability that the average weight loss for this group is more than 42 lb?

CONFIDENCE INTERVALS FOR THE
8.3   **MEAN OF A NORMAL POPULATION (σ KNOWN)**

Return to the situation where we have obtained a sample from a normal population with unknown mean μ. We will first consider a case in which we know the variability of the normal random variable, the value of σ (Figure 8.12). (The situation where both μ and σ are unknown will be dealt with in the next section.)

We know that, to estimate μ, the average of the entire population, we obtain a sample from this population and calculate $\bar{X}$, the average of the sample. The sample mean $\bar{X}$ is the estimate of μ and is also called a *point estimate* because it consists of a single number.

In example 8.2, it was assumed that the time for a six-year-old child to assemble a particular puzzle followed a normal distribution, with $\mu = 20$ minutes

Figure 8.12

An example where
the standard
deviation σ is known
but the mean μ is
unknown

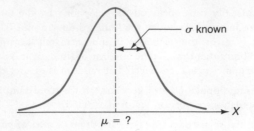

and $\sigma = 3$ minutes. What if μ is not known for all children? A random sample of 25 children's assembly times was obtained with the following results (in minutes):

22.8, 29.3, 27.2, 30.2, 24.0, 23.2, 22.9, 30.3, 27.1, 31.2, 27.0, 32.0, 28.6, 24.1, 28.9, 26.8, 26.6, 23.4, 25.1, 26.6, 25.7, 28.1, 31.5, 24.8, 25.2

Based on these data,

$$\text{estimate of } \mu = \text{sample mean } \bar{X}$$
$$= \frac{22.8 + 29.3 + \cdots + 25.2}{25}$$
$$= 26.9 \text{ minutes}$$

Is this large value of $\bar{X}$ ($= 26.9$) due to random chance? We know that 50% of the samples drawn will have $\bar{X}$ larger than 20, even if $\mu = 20$ (Figure 8.13). Or is this value large because μ is a value larger than 20? In other words, does this value of $\bar{X}$ provide just cause for concluding that μ is larger than 20? We will tackle this type of question in Chapter 9.

How accurate is a derived estimate of the population mean μ? This depends, for one thing, on the sample size. We can measure the precision of this estimate by constructing a **confidence interval (CI)**. By providing the CI, one can make such statements as "I am 95% confident that the average time μ to assemble the puzzle is between 25.7 and 28.1 minutes." For this illustration, (25.7,

Figure 8.13

Distribution of $\bar{X}$ if
$\mu = 20$ minutes

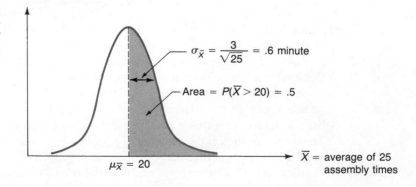

Figure 8.14

$P(-1.96 < Z < 1.96)$
$= .95$

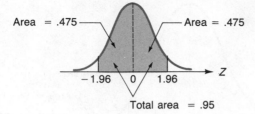

Area = .475   Area = .475

-1.96 0   1.96 Z

Total area = .95

28.1) is called a 95% CI for μ. The following discussion will demonstrate how to construct such a confidence interval.

Using the CLT, we know that $\bar{X}$ is approximately a normal random variable with

$$\mu_{\bar{x}} = \mu$$
$$\sigma_{\bar{x}} = \sigma/\sqrt{n}$$

where μ and σ represent the mean and standard deviation of the population. To standardize $\bar{X}$, you subtract the mean (μ) of $\bar{X}$ and divide by the standard deviation ($\sigma/\sqrt{n}$) of $\bar{X}$. Consequently,

$$Z = \frac{\bar{X} - \mu}{\sigma/\sqrt{n}}$$

is a standard normal random variable. Consider the following statement and refer to Figure 8.14:

$$P(-1.96 \leqslant Z \leqslant 1.96) = .95$$

so

$$P\left(-1.96 \leqslant \frac{\bar{X} - \mu}{\sigma/\sqrt{n}} \leqslant 1.96\right) = .95$$

After some algebra and rearrangement of terms, we get

$$P(\bar{X} - 1.96\sigma/\sqrt{n} \leqslant \mu \leqslant \bar{X} + 1.96\sigma/\sqrt{n}) = .95$$

How does the last statement apply to a particular sample mean $\bar{x}$? Consider the interval

$$(\bar{x} - 1.96\sigma/\sqrt{n}, \bar{x} + 1.96\sigma/\sqrt{n}) \qquad (8\text{-}4)$$

Using the values from our assembly-time example, we have $\bar{x} = 26.9$, $\sigma = 3$, and $n = 25$. The resulting 95% confidence interval is

$$(26.9 - 1.96 \cdot 3/\sqrt{25}, \, 26.9 + 1.96 \cdot 3/\sqrt{25})$$

Figure 8.15

$Z_{.025} = 1.96$,
$Z_{.05} = 1.645$, and
$Z_{.1} = 1.28$

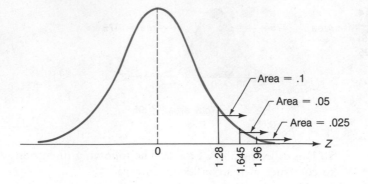

or

$$(25.72, 28.08)$$

Since μ is unknown, then it is unknown whether μ lies between 25.72 and 28.08. However, if you were to obtain random samples repeatedly, calculate $\bar{x}$, and determine the intervals defined in formula (8-4), then 95% of these intervals would contain μ, and 5% would not. For this reason, equation 8-4 is called a **95% confidence interval** for μ. Using our assembly-time illustration, we are 95% confident that the average time to assemble the puzzle, μ, lies between 25.72 and 28.08.

Notation

Let Z_a denote the value of Z with an area to the right of this value equal to a. How can we determine $Z_{.025}$, $Z_{.05}$, and $Z_{.1}$ (Figure 8.15)? Using Table A-4, $Z_{.025} = 1.96$, $Z_{.05} = 1.645$, and $Z_{.1} = 1.28$.

When defining a CI for μ, one can define a 99% CI, a 95% CI, a 90% CI, or whatever. The specific percentage represents the **confidence level**. The higher the confidence level, the wider the CI. The confidence level is written as $(1 - \alpha) \cdot 100\%$, where $\alpha = .01$ for a 99% CI, $\alpha = .05$ for a 95% CI, and so on. Thus, a $(1 - \alpha) \cdot 100\%$ CI for the mean of a normal population μ is

$$[\bar{x} - Z_{\alpha/2}(\sigma/\sqrt{n}), \; \bar{x} + Z_{\alpha/2}(\sigma/\sqrt{n})] \qquad (8\text{-}5)$$

According to the CLT, equation 8-5 provides an approximate CI for the mean of any population, provided the sample size n is large ($n > 30$).

EXAMPLE 8.5

Determine a 90% and a 99% CI for the average puzzle assembly time for all children, using the 25 observations given on page 326.

Solution The sample mean here was $\bar{x} = 26.9$. The population standard deviation is assumed to be 3 minutes. The resulting 90% CI for the population mean μ is

$$26.9 - Z_{.05}(3/\sqrt{25}) \text{ to } 26.9 + Z_{.05}(3/\sqrt{25})$$
$$= 26.9 - 1.645(3/\sqrt{25}) \text{ to } 26.9 + 1.645(3/\sqrt{25})$$
$$= 26.9 - .99 \text{ to } 26.9 + .99$$
$$= 25.91 \text{ to } 27.89 \text{ minutes}$$

The 99% CI for μ is

$$= 26.9 - Z_{.005}(3/\sqrt{25}) \text{ to } 26.9 + Z_{.005}(3/\sqrt{25})$$
$$= 26.9 - 2.575(3/\sqrt{25}) \text{ to } 26.9 + 2.575(3/\sqrt{25})$$
$$= 26.9 - 1.54 \text{ to } 26.9 + 1.54$$
$$= 25.36 \text{ to } 28.44 \text{ minutes}$$

Consequently, we are 90% confident that the mean assembly time for all children is between 25.91 and 27.89 minutes. We are also 99% confident that this parameter is between 25.36 and 28.44 minutes based upon the results of this sample. Notice that the width of the interval increases as the confidence level increases when using the same sample data. □

Discussing a Confidence Interval

The narrower your CI the better, for the same level of confidence. Suppose Toddle-Teach spent $20,000 investigating the average time necessary for six-year-old children to assemble their puzzle. Part of this study included obtaining a CI for the average assembly time μ. Which statement would they prefer to see?

1 I am 95% confident that the average assembly time is between 2 and 50 minutes.

2 I am 95% confident that the average assembly time is between 25 and 27 minutes.

The information contained in the first statement is practically worthless, and that's $20,000 down the drain. The second statement contains useful information; μ is narrowed down to a much smaller range.

Given the second statement, can you tell what the corresponding value of $\bar{X}$ was that produced this CI? For any CI for μ, $\bar{X}$ (the estimate of μ) is always in the center. So $\bar{X}$ must have been 26 minutes.

For the 90% CI in example 8.5, the following conclusions are valid:

1 I am 90% confident that the average assembly time lies between 25.91 and 27.89 minutes.

2 If I repeatedly obtained samples of size 25, then 90% of the resulting CIs would contain μ and 10% would not. (Question from the audience: Does this CI (25.91, 27.89) contain μ? Your response: I don't know. All

I can say is that this procedure leads to an interval containing μ 90% of the time.)

3 I am 90% confident that my estimate of μ (namely, $\bar{x} = 26.9$) is within .99 minute of the actual value of μ.

Here .99 is equal to 1.645 $(\sigma/\sqrt{n})$. This is referred to as the **maximum error, E.**

$$E = \text{maximum error} = Z_{\alpha/2}(\sigma/\sqrt{n}) \qquad \textbf{(8-6)}$$

Be careful! The following statement is not correct: "The probability that μ lies between 25.91 and 27.89 is .90." What is the probability that the number 27 lies in this CI? How about 24? The answer to the first question is 1, and to the second, 0, because 27 lies in the CI and 24 does not. So what is the probability that μ lies in the CI? Remember that μ is a fixed number; we just do not know what its value is. It is not a random variable, unlike its estimator, $\bar{X}$. As a result, this probability is either 0 or 1, not .90. Therefore, remember that, once you have inserted your sample results into equation 8-5 to obtain your CI, the word *probability* can no longer be used to describe the resulting CI.

EXAMPLE 8.6

Refer to the 20 samples of achievement test scores in Table 8.1. Using sample 1, what is the resulting 80% CI for the population mean μ? Assume that σ is 75 points.

Solution

Here, $n = 10$ and $\bar{x} = 493.1$. The confidence level is 80%, so $Z_{\alpha/2} = Z_{.1} = 1.28$ (from Table A-4). Therefore, the resulting 80% confidence interval for μ is

$$493.1 - 1.28(75/\sqrt{10}) \text{ to } 493.1 + 1.28(75/\sqrt{10})$$
$$= 493.1 - 30.4 \text{ to } 493.1 + 30.4$$
$$= 462.7 \text{ to } 523.5$$

So we are 80% confident that μ lies between 462.7 and 523.5. Also, we are 80% confident that our estimate of μ ($\bar{x} = 493.1$) is within 30.4 points of the actual value.   □

EXERCISES

8.19 A random sample of 125 observations is obtained from a normally distributed population with a standard deviation of 5. Given that the sample mean is 20.6, construct a 90% CI for the mean of the population.

8.20 The following data are the values of a random sample from a normally distributed population. Assume that the population variance is 10.2. Construct a 99% CI for the mean of the population.

50.6, 52.3, 48.6, 45.3, 51.8, 50.8, 46.7, 56.1, 47.7, 49.3, 44.9, 57.0, 50.7, 42.6, 49.8, 46.1, 48.7, 51.8, 54.3, 48.4, 50.5

8.21 Based on an analysis of 10 rocket launches, the time from lift-off to maximum thrust is an average of 118 seconds. Assume that the time from lift-off to maximum thrust is a normally distributed variable with a standard deviation of 12 seconds. Construct a 95% confidence interval for the true mean time to maximum thrust.

8.22 The perfectionist owner of Kwik Kar Kare has reduced an oil-change job to a science, and wants to keep it that way. He constantly monitors the performance of his staff. This week, he sampled 15 oil-change jobs performed and found an average time taken of 9.8 minutes per job. From experience, the times follow a normal distribution and the standard deviation of the population is known to be 1.2 minutes. Based on this week's sample, construct a 90% confidence interval for the population mean (average time for an oil-change job).

8.23 The Occupational Safety and Health Administration (OSHA) is investigating the effects of incidental radiation emitted by computer video terminals. One sampling of 25 terminals was found to have an average of 3.50 units of radiation. If the units of radiation are believed to follow a normal distribution and the standard deviation is assumed to be 1.75 units, what is the 90% confidence interval for the mean amount of radiation emitted in the population?

$\frac{\alpha}{2} = 0.05 \rightarrow z_{0.05} = 1.645$

$3.50 + 1.645 \cdot \frac{1.75}{\sqrt{25}} = 4.076$

$= 2.92$

8.24 A real-estate firm takes a random sample of 40 homes from a small suburb of Memphis. The standard deviation of the total square feet of living space per home is 150. Construct a 98% CI for the mean square footage of living space, assuming that the random sample yielded a mean of 1600.

8.25 A manufacturer of ten-speed racing bicycles believes that the average weight of the bicycle is normally distributed with a mean of 22 lb and a standard deviation of 1.5 lb. A random sample of 30 bicycles is selected. If the mean from this sample is 22.8, what is a 96% CI for the mean weight of the bicycle?

8.26 A psychology professor was asssigned to teach a "mass section" of about 380 students enrolled in Introduction to Psychology. To get some idea of the age of the students, the professor randomly chose 15 students and noted their ages, as follows:

18, 23, 24, 20, 21, 19, 27, 24, 19, 20, 25, 20, 18, 26, 20

From previous semesters, ages follow a normal distribution and the standard deviation in ages is known to be 3 years. Construct a 95% confidence interval for the mean age of the class.

$n = 20, \bar{x} = 52.4$

$\sigma = 3.4$

$52.4 \pm 1.96 \frac{3.4}{\sqrt{20}}$

$53.90 \text{ to } 50.91$

8.27 The city traffic manager wants to estimate the average speed of cars on a certain stretch of highway leading to a popular intersection. The speed of a sample of 20 cars was obtained using a radar detector, yielding a mean speed of 52.4 mph. The standard deviation is assumed to be 3.4 mph from prior studies. What is the 95% confidence interval for the true population mean speed of cars approaching this intersection? What are you assuming here?

8.28 As the sample size increases, would a CI given by equation 8-5 get smaller or larger? For a given random sample, would the CI given by equation 8-5 for a 90% CI be larger or smaller than that for an 80% CI?

$\frac{\alpha}{2} = 0.005 \quad z_{0.005} = 2.575$

$8.3 + 2.575 \times \frac{.93}{\sqrt{30}} = 8.74$

$= 7.86$

8.29 A medical researcher would like to obtain a 99% CI for the mean length of time that a particular sedative is effective. Thirty subjects are randomly selected. The mean length of time that the sedative was effective is found to be 8.3 hours for the sample. Find the 99% CI, assuming that the length of time that the sedative is effective is considered to be approximately normally distributed with a standard deviation of .93 hours.

8.30 A safety council is interested in the age at which a person first obtains his or her driver's license. If the ages of people who obtain their driver's licenses is considered to be normally distributed with a standard deviation of 2.5 years, what is an 80% CI for the mean age, given that a random sample of 20 new drivers yields a mean age of 19.3 years?

8.4   CONFIDENCE INTERVALS FOR THE MEAN OF A NORMAL POPULATION (σ UNKNOWN)

If σ is unknown, then it is impossible to determine a confidence interval for μ using equation 8-5 because we are unable to evaluate the standard error $\sigma/\sqrt{n}$. Let us take another look at how we estimate the parameters of a normal population.

When a population mean is unknown, we can estimate it using the sample mean. The logical thing to do if σ is unknown is to replace it by its estimate, the standard deviation of the sample s. But consider what happens when

$$\frac{\bar{X} - \mu}{\sigma/\sqrt{n}}$$

is replaced by

$$\frac{\bar{X} - \mu}{s/\sqrt{n}}$$

This is no longer a standard normal random variable Z. However, it does follow another identifiable distribution, the *t* **distribution**. Its complete name is Student's *t* distribution, named after W. S. Gosset, a statistician in a Guinness brewery who used the pen name Student. The distribution of

$$\frac{\bar{X} - \mu}{s/\sqrt{n}}$$

will follow a *t* distribution, provided the population from which you are obtaining the sample is normally distributed.

The *t* distribution is similar in appearance to the standard normal (Z) distribution in that it is symmetric about zero. Unlike the Z distribution, however, its shape depends on the sample size n. Consequently, when you use the *t* distribution, you must take into account the sample size. This is accomplished by using **degrees of freedom**. For this application using the *t* distribution,

$$\text{degrees of freedom} = df = n - 1$$

Figure 8.16

The t distribution

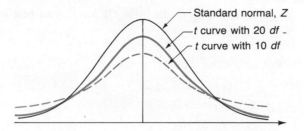

The value of $df = n - 1$ can be explained by observing that, for a given value of $\bar{X}$, only $n - 1$ of the sample values are free to vary. For example, in a sample of size $n = 3$, if $\bar{x} = 5.0$, $x_1 = 2$, and $x_2 = 7$, then x_3 must be 6 because this is the only value providing a sample mean equal to 5.0.

Two t distributions are illustrated in Figure 8.16. Notice that the t distributions are symmetrically distributed about zero but have wider tails than does the standard normal Z. Observe that as n increases, the t distribution tends toward the standard normal Z. In fact, for $n > 30$, there is little difference between these two distributions. Areas under a t curve are provided in Table A-5 on page A-14 for various df. So, for large samples ($n > 30$), it does not matter whether σ is known (Z distribution, Table A-4) or σ is unknown (t distribution, Table A-5) because the t and Z curves are practically the same. For this reason, the t distribution often is referred to as the **small sample distribution** for $\bar{X}$. The Z table can be used as an approximation even if σ is unknown provided n is larger than 30.

Using the t distribution, then, a $(1 - \alpha) \cdot 100\%$ CI for μ is

$$\bar{x} - t_{\alpha/2, n-1}(s/\sqrt{n}) \text{ to } \bar{x} + t_{\alpha/2, n-1}(s/\sqrt{n})$$

where $t_{\alpha/2, n-1}$ denotes the t value from Table A-5 using a t curve with $n - 1$ df and a right tail area of $\alpha/2$.

Do you remember our sample of 25 assembly times that produced a point estimate for μ having a value of $\bar{x} = 26.9$ minutes? This estimate was used in example 8.5, where it was assumed that the population standard deviation was $\sigma = 3$, in constructing a CI for μ. Furthermore, the assembly times were assumed to follow a normal distribution.

Suppose that we do not know σ either. Then the point estimate of the population standard deviation is

$$s = \sqrt{\frac{(22.8^2 + 29.3^2 + \cdots + 25.2^2) - (22.8 + 29.3 + \cdots + 25.2)^2/25}{24}}$$

$$= \sqrt{\frac{18{,}285.14 - (672.6)^2/25}{24}}$$

$$= \sqrt{7.896} = 2.81 \text{ minutes}$$

Using Table A-5 to find a 90% CI for μ, you first determine that

$$t_{\alpha/2, n-1} = t_{.05, 24} = 1.711$$

The resulting 90% CI is

$$26.9 - 1.711(2.81/\sqrt{25}) \text{ to } 26.9 + 1.711(2.81/\sqrt{25})$$
$$= 26.9 - .96 \text{ to } 26.9 + .96$$
$$= 25.94 \text{ to } 27.86$$

Using these data, we are 90% confident that the estimate for the mean of this normal population ($\bar{x} = 26.9$) is within .96 minutes of the actual value. Comparing this result with example 8.5, we notice little difference in the two 90% CIs. This is due mostly to the fact that the estimate of σ ($s = 2.81$) is very close to the assumed value of $\sigma = 3$.

EXAMPLE 8.7

A national research firm recently developed an instrument called the Personality Profile Questionnaire (PPQ) which measures several personality characteristics. One of the subscales of this instrument measures the level of assertiveness of the person completing this 50-item questionnaire. Of primary concern to the firm is the average assertiveness score for males in their senior year of college who are majoring in education.

In a preliminary pilot study, they obtained the scores of 22 randomly selected males who met these criteria, with the following sample scores (possible values range from 50 to 200):

124, 116, 83, 137, 143, 92, 105, 71, 138, 126, 174, 101, 95, 114, 130, 119, 92, 183, 120, 98, 135, 127

What is a 95% CI for the average score on this subscale of the PPQ, assuming that these scores follow a normal distribution?

Solution

Your (point) estimate of σ is the sample standard deviation, $s = 27.19$. Also, your point estimate of μ is $\bar{x} = 119.23$. A 95% CI for the average assertiveness score (if scores were recorded indefinitely) is

$$119.23 - t_{.025, 21}(27.19/\sqrt{22}) \text{ to } 119.23 + t_{.025, 21}(27.19/\sqrt{22})$$
$$= 119.23 - 2.080(27.19/\sqrt{22}) \text{ to } 119.23 + 2.080(27.19/\sqrt{22})$$
$$= 119.23 - 12.06 \text{ to } 119.23 + 12.06$$
$$= 107.17 \text{ to } 131.29$$

We are thus 95% confident that the average assertiveness score is between 107 and 131. Notice here that the maximum error is

$$E = 2.080(27.19/\sqrt{22}) = 12.06$$

which implies that we are 95% confident that the sample mean $\bar{X}$ is within 12 points of the population mean. The MINITAB solution for this example is shown in Figure 8.17.   □

Figure 8.17

MINITAB solution to
example 8.7

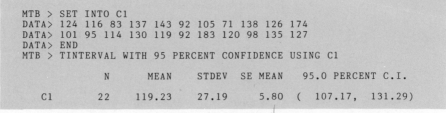

```
MTB > SET INTO C1
DATA> 124 116 83 137 143 92 105 71 138 126 174
DATA> 101 95 114 130 119 92 183 120 98 135 127
DATA> END
MTB > TINTERVAL WITH 95 PERCENT CONFIDENCE USING C1

              N     MEAN    STDEV   SE MEAN    95.0 PERCENT C.I.
      C1     22    119.23   27.19    5.80    ( 107.17,  131.29)
```

EXERCISES

8.31 Find the t values for the following levels and degrees of freedom.

(a) $t_{.10,29}$ (b) $t_{.025,13}$ (c) $t_{.05,18}$

(d) $t_{.90,20}$ (e) $t_{.95,25}$ (f) $t_{.10,40}$

8.32 For a sample of size 21 used to test a hypothesis about a sample mean, what is the t value from a t distribution such that the following are true?

(a) Ninety percent of the area under the t distribution is to the right of the t value.

(b) Ten percent of the area under the t distribution is to the right of the t value.

(c) Five percent of the area under the t distribution is to the left of the t value.

8.33 The eating disorders known as anorexia nervosa and bulimia nervosa have received a fair amount of publicity in the United States. However, the problem is not restricted to the United States. In Japan, Nakai *et al.* (1987) conducted a study of 23 anorexia patients and 13 bulimic patients. Among other things, their paper provided the following data:

SERUM LEVELS IN FEMALE PATIENTS		
	Anorexia Nervosa ($n = 23$)	**Bulimia Nervosa** ($n = 13$)
Zinc (mg/dl)	75.4 ± 14.7	85.4 ± 17.1
Magnesium (mg/dl)	2.88 ± 0.9	2.49 ± 0.2
Calcium (mg/dl)	8.76 ± 0.44	9.1 ± 0.5
Phosphorus (mg/dl)	3.91 ± 0.52	3.9 ± 0.6
Copper (mg/dl)	87.8 ± 13.7	97.3 ± 16.2
Iron (mg/dl)	65.2 ± 23.5	80.5 ± 45.3

Source: Nakai, Y., Kinoshita, F., Koh, T., Tsujii, S., and Tsukada, T., Taste function in patients with anorexia nervosa and bulimia nervosa. *Int. J. Eating Disorders* **6** (2): 257–265 (March 1987), Table 6.

The above style of presentation is quite common in the academic literature. The numbers are in the form: Mean $\pm$ standard deviation. For example, the mean serum level of zinc in the 23 anorexics was 75.4 with a sample standard deviation of 14.7. These are not confidence intervals.

(a) Construct 95% confidence intervals for serum levels of zinc and magnesium for anorexics.

(b) Construct 90% confidence intervals for serum levels of copper and iron for bulimia patients.

8.34 In a study carried out to see how people learn about operating a complex programmable device without a user's guide or other assistance ("instructionless learning"), Shrager and Klahr (1986) provided the following data on gross performance of 7 subjects during a stage called the systematic investigation phase:

Subject	Phase Duration	Key Presses	Words per Minute	Keystroke Rate/min
1	29.3	423	150	14.5
2	22.3	470	70	21.1
3	25.7	362	71	14.1
4	22.3	319	60	14.3
5	23.3	543	125	23.3
6	18.7	391	101	21.0
7	18.8	345	92	17.4
Mean	23.0	408	96	18.0
Std deviation	3.6	78	32	3.8

Source: Jeff Shrager and David Klahr, Instructionless learning about a complex device: The paradigm and observations. *Int. J. Man–Machine Studies* **25**: 160 (1986), Table 2.

Construct the following confidence intervals using the above data:
(a) The 90% CI for the phase duration in minutes.
(b) The 95% CI for the number of key presses.
(c) The 99% CI for the keystroke rate per minute.

8.35 The mean monthly expenditure on gasoline per household in Middletown is determined by selecting a random sample of 36 households. The sample mean is $68 with a sample standard deviation of $17.
(a) What is a 95% CI for the mean monthly expenditure per household on gasoline in Middletown?
(b) What is a 90% CI for the mean monthly expenditure per household on gasoline in Middletown?

8.36 The following figures are the household income (in thousands of dollars) for 18 Hispanic families in a major metropolitan area:

10.8, 20.1, 19.6, 20.8, 27.4, 9.8, 12.6, 17.3, 16.1, 10.7, 26.3, 11.5, 21.7, 30.0, 10.6, 13.1, 16.3, 9.6

(a) Compute the mean and standard deviation for this sample.
(b) Construct an 80% confidence interval for the mean income for Hispanic households in this metropolitan area.

8.37 An apartment-finder service would like to estimate the average cost of a one-bedroom apartment in Kansas City. A random sample of 41 apartment complexes yielded a mean of $310 with a standard deviation of $29. Construct a 90% CI for the mean cost of one-bedroom apartments in Kansas City.

8.38 Based on 12 months' data, an average of 7.6 billion gallons of automobile fuel were consumed in the United States per month, with a standard deviation of 0.4 billion gallons. Use this information to construct a 90% confidence interval for the mean monthly auto fuel consumption in the United States.

8.39 Georgotas and others (1987) studied the response of depressed patients who were treated with either nortriptyline, a tricyclic antidepressant, or phenelzine, a monoamine oxidase (MAO) inhibitor. Fourteen patients did not respond to the treatment. The following data pertain to these nonresponders:

Nonresponder	Baseline MAO Activity (nmol product/hr/mg protein)
1	45.3
2	39.9
3	3.5
4	41.2
5	36.9
6	43.8
7	45.7
8	37.3
9	28.3
10	17.5
11	18.7
12	21.3
13	4.4
14	48.9

Source: Georgotas, A., McCue, R. E., Friedman, E., and Cooper, T., Prediction of response to nortriptyline and phenelzine by platelet MAO activity. *Am. J. Psychiatr.* **144** (3): 338–340 (March 1987).

(a) Compute the mean and standard deviation of this sample.

(b) Construct the 95% confidence interval of baseline MAO activity for the nonresponders.

(c) Comment on the values observed for patient numbers 3 and 13. What would be the effect of eliminating them from the calculations? Why would you wish to remove them, if at all?

8.40 To estimate the true average salary of tenured faculty at American colleges and universities, a graduate student selected a random sample of 40 tenured professors. The following sample statistics were computed:

$$\bar{x} = \$52,700 \quad \text{(annual salary)}$$

$$s = \$10,650$$

Construct a 95% CI for the mean annual salary of tenured faculty.

8.41 The Social Security Administration, in an effort to reduce the burden of more than 22 million routine telephone inquiries on its 1300 district and branch offices, set up 34 telephone service centers. The General Accounting Office (GAO), which audits government agencies, made 4044 random calls in May 1985 to gauge the effectiveness of those telephone centers. The response rate at these centers ranged from 740 calls out of a thousand encountering problems (busy signals, no answer, disconnected, ineffective response) at one center in Jersey City, New Jersey, to only 33 calls per 1000 encountering problems at Grand Prairie, Texas. Taking the sample of 4044 random calls as a whole, the GAO determined that an average of

540 calls out of 1000 experienced easy access, with a standard deviation of 7. Construct the 90% confidence interval for the true population mean (number of calls out of 1000 that experience easy access).

8.42 The Rural Electrification Agency (REA) acts as a banker for about 1000 rural electric cooperatives, which are locally owned, nonprofit companies that generate and distribute power to rural areas that traditionally are more expensive to serve than urban areas. These co-ops have invested heavily in nuclear power plants. The following figures are the wholesale cost of generating electricity using nuclear power, in cents per kilowatt hour, for a random selection of rural cooperatives:

4.20, 3.82, 5.50, 5.18, 3.99, 6.39, 3.08, 7.19, 7.25, 8.74, 8.60, 7.46

Construct the 95% confidence interval for the mean cost per kilowatt hour of nuclear-generated electricity at rural cooperatives.

8.5   SELECTING THE NECESSARY SAMPLE SIZE

Sample Size for Known σ

How large a sample do you need? This is often difficult to determine, although a carefully chosen large sample generally provides a better representation of the population than does a smaller sample. Acquiring large samples can be costly and time-consuming; why obtain a sample of size $n = 1000$ if a sample size of $n = 500$ will provide sufficient accuracy for estimating a population mean? This section will show you how to determine what sample size is necessary when the maximum error E is specified in advance.

In example 8.6, we assumed that the achievement test scores are normally distributed with standard deviation $\sigma = 75$ points but unknown mean μ. Based on the results of sample 1 from Table 8.1, the conclusion was that we were 80% confident that the estimate of μ ($\bar{x} = 493.1$) was within 30.4 points of the actual value of μ for $n = 10$. How large a sample is necessary if we want our point estimate ($\bar{X}$) to be within 15 points of the actual value of μ, with 80% confidence? The value 15 here is the maximum error E defined in equation 8-6. We would like the estimate of μ (that is, $\bar{X}$) to be within 15 of the actual value, so

$$E = 15 = Z_{\alpha/2} \left(75/\sqrt{n}\right)$$

Because the confidence level is 80%, $Z_{\alpha/2} = Z_{.1} = 1.28$. Consequently,

$$15 = (1.28)\frac{75}{\sqrt{n}}$$

$$\sqrt{n} = \frac{(1.28)(75)}{15} = 6.4$$

Squaring both sides of this statement produces

$$n = (6.4)^2 = 40.96$$

Rounding this value up (always), then a sample size of $n = 41$ will produce a CI with $E \leqslant 15$. As a result, $\bar{X}$, your point estimate of μ, will be within 15 points of the actual value, with 80% confidence.

This sequence of steps can be summarized using the following expression:

$$n = \left[\frac{Z_{\alpha/2} \cdot \sigma}{E} \right]^2 \qquad \text{(8-7)}$$

Sample Size for Unknown σ

Equation 8-7 works if σ is known but does not apply to situations where both μ and σ are unknown. There are two approaches to the latter situation.

A Preliminary Sampling

If you have already obtained a small sample, then you have an estimate of σ, namely, the sample standard deviation s. Replacing σ by s in equation 8-7 gives you the desired sample size n. Assuming that the resulting value of n is greater than 30, the $Z_{\alpha/2}$ notation in equation 8-7 is still valid because the actual t distribution here will be closely approximated by the standard normal.

When you do obtain the CI using the larger sample, the resulting maximum error E may not be exactly what you originally specified because the new sample standard deviation will not be the same as that belonging to the smaller original sample.

EXAMPLE 8.8

In example 8.7, a research firm obtained 22 observations consisting of scores measuring assertiveness in a preliminary pilot study. For these data,

$$\bar{x} = 119.23$$

$$s = 27.19$$

How large a sample would they need for $\bar{X}$ to be within 5 points of the population mean, with 95% confidence?

Solution

Based on the results of the original sample, $s = 27.19$, so

$$n = \left[\frac{Z_{\alpha/2} \cdot s}{E} \right]^2 = \left[\frac{(1.96)(27.19)}{5} \right]^2$$
$$= 113.6$$

We round this up to 114. Consequently, a sample size of $n = 114$ would be required to make a statement with this much precision, that is, within 5 points.

□

Obtaining a Rough Approximation of σ

We know from the empirical rule and Table A-4 that 95.4% of the population will lie between $\mu - 2\sigma$ and $\mu + 2\sigma$. Because $(\mu + 2\sigma) - (\mu - 2\sigma) = 4\sigma$, this is a span of four standard deviations. One method of obtaining an estimate of σ is to ask a person who is familiar with the data to be collected these questions:

1   What do you think will be the highest value in the sample (H)?
2   What will be the lowest value (L)?

The approximation of σ is then obtained by assuming that $\mu + 2\sigma = H$ and $\mu - 2\sigma = L$, so

$$H - L = (\mu + 2\sigma) - (\mu - 2\sigma) = 4\sigma$$

Consequently,

$$\sigma \cong \frac{H - L}{4} \qquad\qquad (8\text{-}8)$$

We can use this estimate of σ in equation 8-7 to determine the necessary sample size n.

EXAMPLE 8.9

At the start of this section, we determined the sample size that would be necessary to estimate the population mean for the national achievement test scores to within 15 points with 80% confidence, *assuming that the population standard deviation was $\sigma = 75$ points*. Ms. McQueen is a school counselor who has administered this test many times over the years and would like to estimate this mean to within 15 points with 90% confidence. Unfortunately, there are no school records containing the results of this test, so she has no way of estimating the population standard deviation. However, based upon her past experience, she estimates that the highest observed score is $H = 700$ and the lowest observed score is $L = 350$. Assuming that the achievement test scores follow a normal distribution (centered at μ), how large of a sample would be necessary to estimate μ to within 15 points with 90% confidence?

Solution

Based upon $H = 700$ and $L = 350$,

$$\sigma \cong \frac{700 - 350}{4} = 87.5$$

The sample size necessary to obtain a maximum error $E = 15$ points with 90% confidence is

$$n = \left[\frac{(1.645)(87.5)}{15} \right]^2 = 92.08$$

Thus a sample size of 93 should produce a value of E close to 15. The value of E calculated from the sample of 93 students will not be exactly 15 because the standard deviation s probably will not be exactly 87.5. Estimating σ in this manner, however, produces a value that is "in the neighborhood" of σ. □

EXERCISES

8.43 A 99% CI is to be constructed such that $\bar{X}$ is within 1.5 units of the mean of a normal population. Assuming that the population variance is 30, what sample size would be necessary to achieve this maximum error?

8.44 To be 95% confident that $\bar{X}$ is within .65 of the actual mean of a normal population with a standard deviation of 2.5, what sample size would be necessary?

$$100 = 1.96 \times \frac{200}{\sqrt{n}} \Rightarrow \boxed{16 = n}$$

8.45 The Chamber of Commerce of Tampa, Florida, would like to estimate the mean amount of money spent by a tourist to within $100 with 95% confidence. If the amount of money spent by tourists is considered to be normally distributed with a standard deviation of $200, what sample size would be necessary for the Chamber of Commerce to meet their objective in estimating this mean amount?

8.46 Refer to exercise number 8.41 in which the GAO took a sample of 4044 calls. Suppose that the population, which follows a normal distribution, had a standard deviation of 80. What would be the appropriate sample size that needs to be taken if you desire a 90% CI that is plus or minus 20 calls from the true mean?

$$1.2 = z_{0.05} \times \frac{\sigma}{\sqrt{70}}$$
$$1.645 \quad \frac{\sigma}{\sqrt{70}} = 6.10$$

8.47 If a sample size of 70 was necessary to estimate the mean of a normal population to within 1.2 with 90% confidence, what is the approximate value of the standard deviation of the population?

8.48 As a team project, a group of graduating seniors majoring in elementary education put together a computer-aided instruction (CAI) course for second-graders. Their project supervisor suggested that the CAI course be evaluated for effectiveness by measuring the learning time. Based on other studies, they were told to assume a normally distributed population with a standard deviation of 15 minutes. How large a sample should be taken if it is desired to estimate the true mean learning time to within 5 minutes with 90% confidence?

$$\sigma = 2.3$$
$$3 = 2.33 \times \frac{2.3}{\sqrt{n}}$$
$$= 3.19 \sim 4$$

8.49 A demographic researcher wanted to estimate the number of doctors per 1000 residents in a developing country. Assuming a normal distribution and a standard deviation of 2.3, what is the sample size needed to achieve an accuracy to within 3 doctors at the 98% level of confidence?

8.50 A chemist at International Chemical would like to measure the adhesiveness of a new wood glue. From past experiments, a measure used to indicate adhesiveness has ranged from 7.3 to 11.1 units. To be 98% confident, how large a sample would be necessary to estimate the adhesiveness to within .5 units?

$$\frac{\alpha}{2} = 0.005 \quad Z = 2.575$$
$$= \frac{129-48}{4} = 20.25$$
$$3 = 2.575 \frac{20.25}{\sqrt{n}} \Rightarrow n = 303$$

8.51 A consultant in organizational behavior and group psychology is trying to develop a new instrument for measuring job aptitude for a particular profession. She wishes to establish the mean aptitude test score. A preliminary sample shows that the highest score is 129 and the lowest score is 48. What is the sample size necessary to estimate the true mean aptitude test score to within 3 points, using a 99% confidence level?

8.52 If past data indicate that the maximum value of a random variable X is 155 and the lowest value of X is 120, what is the sample size necessary to be 95% confident in estimating the mean of X to within one? Assume that X is approximately normally distributed.

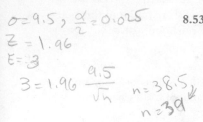

8.53 Suppose you wanted to estimate the average number of yards run by professional NFL running backs per season. One way is to get data on each and every football player that ever existed. Obviously that is not practical, so you would prefer to take a shortcut and obtain an estimate of the true mean by using a sample. If you wanted your estimate to have an accuracy of plus or minus 3 yards, what is the size of the sample that you should take to obtain a 95% confidence interval? You may assume a standard deviation of 9.5 yards.

8.6   OTHER SAMPLING PROCEDURES

To discuss methods of sampling other than simple random sampling, we need to define several terms. These definitions also apply to simple random sampling.

1 Population As before, this refers to the collection of people or objects of interest about which we desire to learn something. It may be as large as the set of all voting adults in the United States or as small or smaller than the set of all students in a particular school. In this section, we will assume that we are sampling from a finite population.

2 Sampling unit This is a collection of elements or an individual element selected from the population. Elements within one sampling unit must not overlap with the elements in other sampling units.

3 Cluster This is a sampling unit that is a group of elements from the population, such as all adults in a particular city block.

4 Sampling frame This is a list of population elements from which the sample is to be selected. Ideally, the sampling frame should be identical to the population. In many situations, however, this is impossible, in which case the frame must be representative of the population.

5 Strata These are nonoverlapping subpopulations. For example, the population of all cigarette smokers can be split into two strata, men and women. You can then use stratified sampling, in which your total sample consists of a sample selected from each individual stratum.

6 Sampling design This is a plan that specifies the manner in which the sampling units are to be selected for your sample. Sampling designs consist of nonprobability sampling (such as convenience, judgment, or quota samples) or probability sampling (including simple random sampling, systematic sampling, stratified sampling, and cluster sampling).

Nonprobability Sampling

There are many situations, such as an exploratory pilot study, where you simply want to gain insight into a population of interest, rather than make a statistical estimate. With nonprobability sampling, you can use a variety of procedures for obtaining a sample, with little or no attention paid to randomization. For example, you may wish to restrict your sample to handpicked individuals who satisfy a certain requirement, such as divorced at least 5 years. When using such a sample, there is no attempt to derive confidence intervals for a population parameter (such as the mean) since a randomization process was not used when selecting individuals for the sample.

Three commonly used nonprobability sampling procedures are:

1 Convenience sampling
2 Judgment sampling
3 Quota sampling

Convenience Sampling A **convenience sample** is the least expensive and easiest to obtain sample, but, as you might expect, the least reliable. A convenience sample is obtained by selecting anyone you wish, such as coworkers and neighbors or by stopping 10 people on the sidewalk. Such a sample can be extremely valuable in gaining insight into a new idea or getting a "feel" for the sentiment of a much larger population. If a convenience sample of 20 neighbors indicate that 19 of the 20 people think your idea is not a good one, it's probably time to go back to the drawing board. Besides, if your friends and acquaintances dislike your idea, chances are the general public will be even less inclined to support it. On the positive side, many a good idea has been born by obtaining information through a convenience sample.

Judgment Sampling For situations where you want to handpick individuals who satisfy certain requirements, a **judgment sample** often works very well. For example, in a study of remarried couples, you might want to restrict your sample to couples who have been successfully remarried for at least 5 years and who have not had marital therapy during that period. With such a judgment sample, you may discover that individuals in this sample will have better insight into the problem you are investigating.

When using a judgment sample, you take the risk that your sample is not representative of the associated population. But if your intent is to gain insight into the population, rather than represent them as a whole, a pilot study conducted using a judgment sample can be very valuable. By interviewing teenagers who have survived a suicide attempt, you may discover issues that trouble most teenagers. The members of your sample perhaps had more severe conflicts that appeared to them to have no solution, except through suicide.

Quota Sampling **Quota sampling** is a type of nonprobability sampling that is a bit more deliberate in an attempt to obtain a sample that is representative of the entire population. To illustrate this technique, if you know that 60% of the registered voters in a particular community are men, you may, in a sample of 100 voters, restrict your sample to 60 males and 40 females. In this way, your sample is a "miniature" of the population.

Very often in a quota sample, the researcher restricts the sample according to more than one criteria, such as age level, religious affiliation, race, or smoke/don't smoke. Such criteria are believed by the researcher to have an effect on the response of the person being interviewed. Certainly one's religious affiliation could have an effect on his or her view of a proposed ordinance forbidding the sale of alcoholic beverages. Consequently, a quota sample would attempt to duplicate the proportions of each religious category present in the community where the ordinance would take effect.

Once the criteria have been determined, typically a judgment sampling technique is used to make the sample selection. This may introduce bias into the results if care is not taken in obtaining a representative cross section of people satisfying the sample criteria. Despite this defect of quota sampling, it is a very popular technique among opinion pollsters and other researchers. The technique does attempt to obtain a representative sample by selecting what amounts to a miniature population, oftentimes saving a great deal of time and money.

Probability Sampling

We introduced you to the concept of simple random sampling in Chapter 5. This is a type of **probability sampling** where the researcher uses a randomization process to improve the precision and reduce the bias of the sampling procedure. Such a process allows you to estimate the sampling error that is present, which then enables you to construct a confidence interval for the parameter under investigation. Such samples need to be carefully controlled and typically are more costly and time-consuming than a nonprobability sample.

Besides simple random sampling, there are other more complex probability sampling procedures. These sampling techniques can be used to improve the precision of the sample estimate by dividing the population into "pieces" or subpopulations, resulting in detailed information describing these subpopulations. Probability sampling procedures include:

1 Simple random sampling
2 Systematic sampling
3 Stratified sampling
4 Cluster sampling

Simple Random Sampling

The results obtained when using a simple random sample were presented earlier and will be summarized here for the usual case of sampling without replacement, where every sample of n elements (from a population of size N) has an equal chance of being selected.

According to the CLT, for large samples the distribution of the sample mean $\bar{X}$ is approximately normal, without any assumptions concerning the shape of the population being sampled. The resulting CI for the population mean μ is an approximate CI for this parameter. If you assume that the population has a normal distribution with mean μ, the CI is exact.

Simple Random Sampling

Population Mean: μ

Estimator: $\bar{X} = \dfrac{\Sigma x}{n}$ (8-9)

Variance of the estimator: $\sigma_{\bar{x}}^2 = \dfrac{\sigma^2}{n} \cdot \dfrac{N-n}{N-1}$ (8-10)

The fpc is $(N-n)/(N-1)$ and can be ignored if $n/N < .05$.

Approximate Confidence Interval: $\bar{X} \pm Z_{\alpha/2}\sigma_{\bar{x}}$

Systematic Sampling

For large populations, obtaining a random sample can be quite cumbersome. Perhaps the head librarian of a large library would like a sample of people using the library over a particular one week period. For them to select people randomly would be nearly impossible. A much easier scheme would be to have them select every kth person entering the library to be included in the sample. This is systematic sampling.

Other situations where systematic sampling is advantageous include:

1 The population consists of N records on a magnetic tape or disk. The sample of n is obtained by sampling every kth record, where k is an integer approximately equal to N/n. For example, if there are $N = 9435$ records and you need a sample of size $n = 100$, then selecting every $9435/100 = 94$th record would result in a systematic sample. Typically, a random starting point (record) is determined, and then every kth record is selected for your sample.

2 The population consists of a collection of files stored consecutively by date of birth. A quick (although not necessarily reliable) method of obtaining a "nearly random" sample would be to select every kth file for your sample. What could cause the sample selected from such a list to be *not* random?

There are many situations in which it is dangerous to use systematic sampling. If there are obvious patterns contained in the sample frame listing, your sample may be far from random. If elements are stored according to days, for example, your sample could consist of data that all belong to Tuesday. If the data are cyclic, your sample might consist of all the peaks or all the valleys of the population. Basically, systematic sampling works best when the order of your population is fairly random with respect to the measurement of interest.

Despite its dangers, systematic sampling can provide an easy method of obtaining a representative sample. If the order of your population is in fact random (no cycles, no obvious patterns of any kind), then a systematic sample can be analyzed as though it were a simple random sample.

Stratified Sampling

A school counselor wants to obtain an estimate of the total family income for mothers of children attending the 6th grade of Truman Elementary School, by collecting a sample of 20 total incomes. One possibility here would be to randomly select 20 children from the school file to obtain a simple random sample. However, the counselor feels that the marital status of the mother will have a large impact, since, for example, divorced mothers do not have the potential of two incomes. Also, there was not an even mix across all marital categories, since she has observed that nearly 40% of the mothers are in the remarried category. The counselor decided to stratify the population of mothers of children attending Truman Elementary using four strata (subpopulations):

Stratum 1: remarried mothers

Stratum 2: ever-married mothers (never-divorced or widowed)

Stratum 3: never-married mothers

Stratum 4: divorced mothers

All 350 children attending Truman had mothers that belonged to one of these four strata since (the counselor observed) all children of divorced or remarried mothers had stayed with this parent.

Stratified sampling is used when the population can be physically or geographically separated into two or more groups (strata), where the variation within the strata is less than the variation within the entire population. The cost of obtaining the stratified sample may be less than that of collecting a random sample of the same size, especially if the sampling units are determined geographically.

The advantages of stratified sampling are:

1 By stratifying, one can obtain more information from the sample because data are more homogeneous within each stratum; consequently, CIs are narrower than those obtained through random sampling.

2 One does obtain a cross section of the entire population.

3 One does obtain an estimate of the mean within each stratum as well as an estimate of μ for the entire population.

We use the following notation:

$$n_i = \text{sample size in stratum } i$$

$$N_i = \text{number of elements in stratum } i$$

$$N = \text{total population size} = \Sigma N_i$$

$$n = \text{total sample size} = \Sigma n_i$$

$$\bar{X}_i = \text{sample mean in stratum } i$$

$$s_i = \text{sample standard deviation in stratum } i$$

Stratified Sampling

Population Mean: μ

Estimator: $\quad \bar{X}_s = \dfrac{\Sigma N_i \bar{X}_i}{N}$ (8-11)

Variance of the estimator: $\quad \sigma_{\bar{x}_s}^2 = \dfrac{\Sigma N_i^2 \left(\dfrac{N_i - n_i}{N_i - 1} \right) \dfrac{s_i^2}{n_i}}{N^2}$ (8-12)

Approximate Confidence Interval: $\quad \bar{X}_s \pm Z_{\alpha/2} \sigma_{\bar{x}_s}$

One method often used to determine the strata sample sizes n_i is to select each sample size **proportional** to stratum size. Consequently,

$$n_i = n \left(\frac{N_i}{N} \right)$$

In this way, you obtain larger samples from the larger strata.

Because you desire an estimator with small variance (that is, one that will not drastically vary from one data set to the next), you should attempt to create strata such that the individual variances s_i^2 are as small as possible.

Assume you would like to obtain a sample of size 20 from the population of size 350 at Truman Elementary School. You want to use a stratified sample with proportional sample sizes because each of the strata has differing sizes and income potential. Here, N_i is the number of mothers belonging to each stratum: $N_1 = 144$, $N_2 = 74$, $N_3 = 47$, and $N_4 = 85$. So

$$N = 144 + 74 + 47 + 85 = 350 \text{ individuals}$$

Your sample sizes are:

$$n_1 = 20\left(\frac{144}{350}\right) \qquad n_3 = 20\left(\frac{47}{350}\right)$$
$$\cong 8 \qquad\qquad\qquad \cong 3$$

$$n_2 = 20\left(\frac{74}{350}\right) \qquad n_4 = 20\left(\frac{85}{350}\right)$$
$$\cong 4 \qquad\qquad\qquad \cong 5$$

Mothers from each of the strata were randomly selected and contacted to obtain the total family income for each sample member. The sample results are (in thousands of dollars):

	Remarried	Ever-Married	Never-Married	Divorced
	$59.7	$42.3	$17.3	$22.7
	66.2	50.7	23.0	28.5
	51.4	41.2	18.7	30.6
	50.3	48.6		24.2
	60.5			20.0
	56.3			
	48.1			
	51.1			
N_i	144	74	47	85
n_i	8	4	3	5
$\bar{X}_i$	$55.45	$45.70	$19.67	$25.20
s_i	$ 6.28	$ 4.66	$ 2.97	$ 4.31

To estimate μ from these data,

$$\bar{X}_s = \frac{(144)(55.45) + (74)(45.70) + (47)(19.67) + (85)(25.20)}{350}$$

$$= 41.24 \text{ (that is, \$41,240)}$$

Also

$$\sigma_{\bar{x}_s}^2 = \left[(144)^2\left(\frac{144-8}{144-1}\right)\frac{(6.28)^2}{8} + (74)^2\left(\frac{74-4}{74-1}\right)\frac{(4.66)^2}{4} \right.$$
$$\left. + (47)^2\left(\frac{47-3}{47-1}\right)\frac{(2.97)^2}{3} + (85)^2\left(\frac{85-5}{85-1}\right)\frac{(4.31)^2}{5} \right] \div (350)^2$$

$$= 157{,}504.25 \div 122{,}500 = 1.286$$

Consequently,

$$\sigma_{\bar{x}_s} = \sqrt{1.286} = 1.13$$

The corresponding approximate 95% CI for the average total family income μ is

$$41.24 - (1.96)(1.13) \text{ to } 41.24 + (1.96)(1.13) = 39.03 \text{ to } 43.45$$

So we are 95% confident that the average total income for the population is between $39,030 and $43,450.

Cluster Sampling

We can sample clusters (groups) within the population rather than collecting individual elements one at a time. For example, to determine the opinions of the members of a statewide PTA organization, you might interview everyone attending several of the local meetings. Of course, the danger here is that possibly (1) the people attending the local meetings that were sampled (clusters) do not represent the population of all PTA members, and (2) the people attending the local meetings do not provide an adequate representation of the local members. As a general rule, it is advisable to select many small clusters rather than a few large clusters to obtain a more accurate representation of your population.

Cluster sampling is preferred to (and less costly than) random and stratified sampling when

1 The only sampling frame that can be constructed consists of clusters (for example, all people in a particular household, school district, or ZIP code area).

2 The population is extremely spread out, or it is impossible to obtain data on all the individual members.

When using cluster sampling, you should randomly select a set of clusters (once they have been clearly defined) for sampling. You can then include all individuals within each cluster selected for the sample (**single-stage cluster sampling**) or randomly select individuals from the sampled clusters to be included in the sample (**two-stage cluster sampling**).

We use the following notation:

M = total number of clusters in the population

m = number of clusters randomly selected for the sample

n_i = number of elements in sample cluster i

$\bar{n}$ = average cluster size of the sampled clusters ($\bar{n} = \Sigma n_i/m$)

N = total population size (N = total of all M cluster sizes that make up the population)

$\bar{N}$ = average cluster size for the population ($\bar{N} = N/M$)

T_i = total of all observations within cluster i (required for the sampled clusters only)

Cluster Sampling (Single Stage)

Population mean: μ

Estimator: $\bar{X}_c = \dfrac{\Sigma T_i}{\Sigma n_i}$ (8-13)

Variance of estimator: $\sigma_{\bar{x}_c}^2 = \left(\dfrac{M-m}{mM\bar{N}^2}\right)\dfrac{\Sigma(T_i - \bar{X}_c n_i)^2}{m-1}$ (8-14)

If $\bar{N}$ is unknown, this can be replaced by its estimate, $\bar{n}$.

Approximate Confidence Interval: $\bar{X}_c \pm Z_{\alpha/2}\sigma_{\bar{x}_c}$

A psychological researcher is interested in studying the leisure time for childless couples living in high-rise condominiums, in particular the number of hours spent each week watching television *together*. He is operating under the suspicion that such couples spend a good deal of time watching television since they are more apt to have a lot of time available for doing things together and are free of child responsibilities and home maintenance. A colleague, however, argues that such couples are typically "on the go" and spend very little time at home "cooped up and watching the tube."

Rather than drawing a sample from all high-rise tenants, you construct a sampling frame consisting of all 18 ($= M$) high-rise apartment complexes. From these, you randomly select a sample of $m = 4$ complexes (clusters). Each couple in these four complexes is then asked how many hours per week they watch television together. You obtain the following results:

	Complex 1	Complex 2	Complex 3	Complex 4
Number of units containing childless couples (n_i)	130	110	155	137
Total number of hours per cluster (complex)	1235	1375	1580	2055

N = the total number of units in the 18 high-rise complexes (population) containing childless couples = 2304, so

$$\bar{N} = 2304/18 = 128.$$

You begin by noting:

$$\Sigma(T_i - \bar{X}_c n_i)^2 = \Sigma T_i^2 - 2\bar{X}_c \Sigma T_i n_i + \bar{X}_c^2 \Sigma n_i^2 \qquad \textbf{(8-15)}$$

Using the sample data,

$$\Sigma T_i = 1235 + \cdots + 2055 = 6245$$

$$\Sigma n_i = 130 + \cdots + 137 = 532$$

$$\Sigma T_i^2 = (1235)^2 + \cdots + (2055)^2 = 10{,}135{,}275$$

$$\Sigma T_i n_i = (1235)(130) + \cdots + (2055)(137) = 838{,}235$$

$$\Sigma n_i^2 = (130)^2 + \cdots + (137)^2 = 71{,}794$$

As a result,

$$\bar{X}_c = \frac{\Sigma T_i}{\Sigma n_i} = \frac{6245}{532} = 11.739$$

Also, using equation 8-15,

$$\frac{\Sigma(T_i - \bar{X}_c n_i)^2}{m - 1}$$
$$= [10{,}135{,}275 - (2)(11.739)(838{,}235) + (11.739)^2(71{,}794)] \div 3$$
$$= 116{,}234.24$$

Consequently,

$$\sigma_{\bar{x}_c}^2 = \frac{18 - 4}{(4)(18)(128)^2} \cdot 116{,}234.24 = 1.379$$

and so

$$\sigma_{\bar{x}_c} = \sqrt{1.379} = 1.174$$

The resulting approximate 95% CI for the average number of television hours (for all 18 complexes) is

$$11.739 - 1.96(1.174) \text{ to } 11.739 + 1.96(1.174)$$
$$= 11.739 - 2.30 \text{ to } 11.739 + 2.30$$
$$= 9.44 \text{ to } 14.04 \text{ hours}$$

Therefore, we are 95% confident that μ lies between 9.44 and 14.04 hours and that we have estimated μ to within 2.3 hours. This example has illustrated single-stage cluster sampling where everyone in the selected clusters (apartment complexes) was used in the sample. Another look at this example indicates that a two-stage cluster procedure might be more practical where each of the four sample clusters is also sampled to obtain the final sample.

EXERCISES

8.54 A realtor would like to estimate the average price of a home in the suburbs of a major metropolitan city. The realtor decides to use stratified random sampling. The population of homes is stratified into the five major suburbs. The results of the stratified sample yield the following statistics. Construct a 95% CI for the mean price of a home in units of one thousand.

Stratum	N_i Number of Houses	n_i Sample Size	$\bar{X}_i$ (in thousands)	s_i^2
Suburb 1	150	22	101.2	64.2
Suburb 2	220	33	80.7	24.3
Suburb 3	140	21	61.4	20.8
Suburb 4	70	11	139.6	53.5
Suburb 5	90	13	76.8	30.1
	670	100		

8.55 A government auditor wanted to estimate the number of days of sick leave used per month by employees of a federal bureau. The employees were stratified by marital status. The random sample for each stratum yielded the following results. Construct a 90% confidence interval for the mean monthly sick days used.

Stratum	No. of Employees	Sample Size	No. of Days of Sick Leave Taken
Divorced	40	7	0, 2, 1, 2, 4, 3, 2
Married	136	23	4, 4, 3, 4, 4, 0, 2, 2, 1, 3, 4, 5, 5, 3, 0, 4, 3, 6, 1, 4, 0, 1, 2
Single	99	17	5, 6, 8, 6, 5, 4, 3, 8, 6, 3, 4, 6, 5, 3, 5, 5, 3
	275	47	

8.56 A sociological researcher specializing in life-style surveys was testing a new instrument to measure post-counseling adjustment levels among adults who had gone through a counseling program following a certain trauma. The adults were stratified by income level. The sample results are shown below. Construct the 95% confidence interval for the mean score obtained on the new instrument.

Income Level	N_i	n_i	$\bar{X}_i$	s_i^2
Lower-middle	864	56	23	16
Upper-middle	1468	95	46	19
High	745	49	48	14
	3077	200		

8.57 A child development expert studying the influence of television on children wanted to estimate the average number of hours a child spends each day watching television. Four neighborhoods were selected from a total of 24 neighborhoods for sampling purposes. Use the following data to find a 90% confidence interval for the mean number of hours spent watching television daily by children.

	Neighborhood		
1	**2**	**3**	**4**
2.3	4.5	1.6	5.6
4.3	2.4	6.0	4.1
1.1	3.6	4.6	2.3
2.0	7.2	5.4	3.7
3.9	4.8	6.5	6.1
4.5	2.1	4.6	4.7
6.0	2.1	1.7	5.8
6.3	1.5	3.2	4.8
3.8	6.8	4.5	5.3
7.0	4.1	3.1	1.9
5.9	3.4	2.6	4.8
3.3			

8.58 In what situation is the use of systematic sampling appropriate? Explain how a systematic sample would be taken from a file of students listed by social security numbers.

8.59 The superintendent of schools for the LaSalle Independent School District wants to estimate the time spent by students on homework. Five schools out of 20 in his jurisdiction were selected for sampling purposes. Find the 95% confidence interval for the mean time spent on homework (data are in hours per day).

School 1: 0.5, 1.5, 3.5, 3.0, 2.0, 4.0, 2.8

School 2: 3.3, 4.5, 2.5, 3.7, 4.6, 5.4, 1.5

School 3: 3.5, 4.6, 5.1, 1.7, 3.6, 5.9, 0.9

School 4: 2.3, 1.4, 1.0, 4.6, 3.8, 6.1, 2.9

School 5: 1.9, 6.1, 4.0, 6.0, 2.6, 4.6, 6.0, 5.0

SUMMARY

This chapter introduced you to **statistical inference**, an extremely important area of statistics. Inference procedures were used to estimate a certain unknown parameter (such as the mean μ or the standard deviation σ) of a population by using the corresponding sample statistic (such as the sample mean $\bar{X}$ or the sample standard deviation s).

The **central limit theorem** (CLT) states that, for large samples, the sample mean $\bar{X}$ always follows an approximate normal distribution. If, in addition, you assume that the population is normally distributed, then $\bar{X}$ will follow an exact normal distribution. The strength of the CLT is that no assumptions need be made concerning the shape of the population, provided the sample is large ($n > 30$). The CLT allows you to make probability statements concerning $\bar{X}$, such as $P(\bar{X} < 150)$. When sampling without replacement from a finite population, the standard deviation of the normal distribution for $\bar{X}$ (the standard

Figure 8.18

The correct table to
use for constructing
a confidence interval
for a population
mean

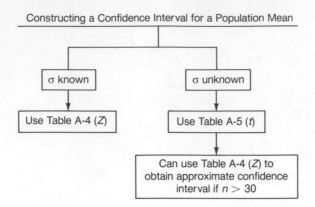

Constructing a Confidence Interval for a Population Mean

error) is obtained by including a **finite population correction** factor (fpc) which adjusts the standard error by including the effect of the known population size N.

The sample mean $\bar{X}$ provides a **point estimate** of μ because it estimates this parameter using a single number. A **confidence interval** (CI) for μ measures the precision of the point estimate. If the population standard deviation σ is known, then the standard normal table (Table A-4) is used to derive the CI. If σ is unknown, it can be replaced by its estimate, the sample standard deviation s. This provided an introduction to the t distribution. The corresponding CI for μ is constructed using the t table (Table A-5) and assumes that the sampled population is normally distributed (that is, that μ is the mean of a normal population). For sample sizes greater than 30, the standard normal table can be used to construct an approximate CI for the population mean when σ is unknown. A summary of this procedure is contained in Figure 8.18.

For many applications, the precision of the point estimate $\bar{X}$ can be determined using the **maximum error** E. When constructing a confidence interval, E is the amount that is added to and subtracted from the point estimate to obtain the desired endpoints of the interval. The sample size n necessary to achieve a desired accuracy can be obtained using a specified value of E. The population standard deviation can be estimated from a preliminary sample, or a rough approximation procedure can be used.

There are many procedures available for obtaining a sample. **Nonprobability sampling** procedures (such as convenience, judgment, or quota samples) provide an inexpensive, easy-to-obtain method of gaining information. They can be extremely useful in a preliminary study but have the disadvantage that they provide no estimate of the sampling error. A **convenience sample** is obtained in a pilot study by selecting anyone you wish for your (perhaps non-representative) sample. A **judgment sample** is used when you want to hand-pick individuals who satisfy certain requirements to gain better insight into a particular problem. A **quota sample** is used to restrict the sample according to one or more criteria, in an attempt to build a sample that is a miniature population.

Probability sampling techniques include simple random, systematic, stratified, and cluster sampling. **Simple random** sampling is employed when every sample of n elements has an equal chance of being selected. **Systematic** sampling selects a random starting point and then selects every kth value for some counting number $k > 0$. This procedure assumes that the population is stored sequentially in some manner, such as on a computer file. **Stratified** sampling is used when the population can be physically or geographically separated into two or more groups (strata), where the variation within the strata is less than the variation within the entire population. When sampling groups of people (clusters) within the population rather than selecting individual elements one at a time, **cluster** sampling is used. When using a particular probability sampling procedure, you can help guard against researcher bias, obtain a more precise sample estimate and obtain a narrower confidence interval for the population mean.

REVIEW EXERCISES

8.60 The mean score for a random group of 45 high school students who took the SAT (Scholastic Aptitude Test) was 965, with a variance of 2500.
   (a) Construct the 95% CI and the 99% CI for the true mean score on the SAT.
   (b) What do you notice about the two intervals, that is, as you increase the confidence level, what happens to the size of the interval?
   (c) Suppose you want to increase the confidence level but do not want the size of the interval to expand. What do you need to do with the sample size in that case?

8.61 An accounting firm has a large pool of secretaries. It is assumed that the time it takes a secretary to type a certain legal document is normally distributed with a mean time of 30 minutes and a standard deviation of 4 minutes.
   (a) What is the probability that a secretary will spend less than 27 minutes typing the legal document?
   (b) In a randomly selected sample of six secretaries, what is the probability that the average time that it takes them to type the legal document is less than 27 minutes?

8.62 The population of an Italian city consume an average of 200 feet of spaghetti per person per week with a standard deviation of 40 feet. Consumption of spaghetti is believed to follow a normal distribution.
   (a) If one Italian is randomly selected from this city, what is the probability that he or she consumes more than 220 feet of spaghetti?
   (b) If a group of 20 Italians is randomly selected, what is the probability that the mean weekly spaghetti consumption per person exceeds 220 feet for this group?
   (c) What percentage of the time will groups of 100 Italians show an average weekly consumption of spaghetti less than 195 feet per person?

8.63 The "Reiss-Epstein-Gursky Anxiety Sensitivity Index" (ASI) is a measure of the fear of anxiety. It is worth noting that this instrument purports to measure fear of anxiety symptoms, rather than the frequency or intensity of these symptoms; that is, anxiety sensitivity and not anxiety itself. McNally and Lorenz (1987) conducted an empirical validation study of the ASI using 48 subjects (44 females and 4 males) who met the DSM-III criteria of the American Psychiatric Association for agoraphobia with panic attacks. Besides the ASI, the subjects also completed two other questionnaires: Body Sensations Questionnaire (BSQ), where the subject rates the degree to which he or she fears certain bodily sensations; and the Agoraphobic Cognitions Questionnaire (ACQ), which estimates the frequency with which different catastrophic thoughts occur during periods of anxiety. The following scores were reported for the 48 agoraphobics:

SCORES ON "FEAR OF ANXIETY" IN AGORAPHOBIC SUBJECTS

Questionnaire	Mean Score	Standard Deviation
ASI	36.25	11.86
BSQ	2.87	0.69
ACQ	2.48	0.63

Source: R. J. McNally and Marleen Lorenz, Anxiety sensitivity in agoraphobics. *J. Behavior Therapy Exp. Psychiatr.* **18** (1): 3–11 (March 1987).

(a)   Construct the 90% confidence interval for the mean ASI score.
(b)   Construct the 95% confidence interval for the mean BSQ score.
(c)   Construct the 99% confidence interval for the mean ACQ score.

8.64   A local merchant would like to estimate the mean amount of money that a family spends at the state fair to within $15. If the amount spent by a family is considered to be normally distributed with a standard deviation of $27, what sample size would be necessary to be 90% confident?

8.65   A random variable is found to range from a high of 50 to a low of 25 from past data. The distribution of the random variable can be approximated by a normal distribution. To estimate the mean of the random variable to within 2.1, what sample size would be necessary in selecting a random sample to achieve a 99% confidence level?

8.66   A group of 18 women using a new contraceptive pill experienced a weight gain of 3.5 lb per year on the average with a standard deviation of 2.8 lb.
(a)   Assuming the population is normally distributed, what is the 95% confidence interval for the true average gain in weight experienced by women using the new pill?
(b)   If the sample size is increased to 61 women, how does that change the procedure you follow to obtain the CI?
(c)   For the sake of comparison, construct the 95% CI for 61 women (using the same mean and standard deviation as part (a)), first with the Z distribution, then with the t distribution. Are the results close? How can you explain this?

8.67   Hyponatremia is the medical term for abnormally low levels of sodium in the blood. It is a syndrome that characterizes episodes of "water intoxication" that sometimes affect chronically hospitalized patients. Goldman and Luchins (1987) reasoned that a decline in mean serum sodium level (MSS) would be associated with an increase in mean body weight (MBW). They set a minimum allowable sodium level (MAS) of 125 meq/liter and computed maximum allowable body weight = MBW × (MSS/MAS). By closely monitoring increases in the patient's weight, it was possible to predict if the patient's sodium level might be dangerously low and close to hyponatremia. During the course of their study involving 8 patients, the researchers found 7 instances when the body weight exceeded the maximum allowable, indicating potentially low sodium levels, as shown below:

Patient	Instances of Excess Weight	Sodium Level (meq/liter)
1	None	Nil
2	1	115
3	None	Nil
4	2	123, 117
5	2	124, 123
6	1	121
7	1	118
8	None	Nil
Total	7 cases, with Mean = 120.2, std. dev. = 3.4	

Source: Goldman, M. B., and Luchins, D. J., Prevention of episodic water intoxication with target weight procedure. Am. J. Psychiatr, **144** (3): 365–366 (March 1987).

(a) Compute the 90% confidence interval for the mean serum sodium level of patients who were detected as having excess weight.

(b) If a 99% CI is desired, with the same level of precision as the 90% CI above, what is the size of the sample that is needed?

(c) The study found the correlation coefficient between change in body weight and serum sodium level to be $r = -0.74$. Does a negative value support the researchers' reasoning in using body weight to detect potential hyponatremia?

8.68 A hospital administrator wants to estimate the number of hours worked per day by nurses in his district. Four (out of 15) hospitals were randomly selected, and the hours worked were recorded. The results are shown below. Find the 95% CI for the mean number of hours worked.

	Hospital 1	Hospital 2	Hospital 3	Hospital 4
Number of nurses	250	220	315	275
Total number of hours per day by cluster (hospital)	2000	2024	3087	3052

8.69 A personnel administrator for Teltronix would like to estimate the amount of term life insurance that an employee carries. Three strata are used for finding a stratified random sample of all employees. From the data, construct a 95% CI for the mean amount of term life insurance that an employee carries (given in units of one thousand dollars).

Stratum	N_i: Total Number in Company	n_i	$\bar{X}_i$	s_i
Employees Paid by the Hour	350	67	28.5	5.7
Engineers and Technicians	112	22	80.6	10.3
Management	57	11	125.2	13.6
	519	100		

8.70 The Controller of Public Accounts is trying to estimate the cost of bringing bridges up to federal standards. Six counties out of 40 counties were randomly selected, and the total of repairs for upgrading to federal standards was obtained, as shown below. Construct the 99% CI for the mean cost per bridge for repairs required to upgrade these bridges to federal standards.

	County					
	1	2	3	4	5	6
Number of Bridges in the Cluster	185	220	320	274	118	370
Total Cost by Cluster (in thousands of dollars)	2465	2750	3144	2680	1765	3685

CASE STUDY

Identifying Learning Disability— Are Subtest Scores Useful?

A growing number of learning disabled (LD) students are applying to colleges. To serve the needs of LD students, many institutions have established support programs, such as tutoring, modified course examinations, waiver of certain university requirements (e.g., foreign language), and special counseling. Due to the extensive services provided, LD support programs attract many students who have learning problems, but are not learning disabled. Because of limited resources, institutions have to verify eligibility and filter out non-LD applicants.

Learning disability is frequently represented as a discrepant pattern of abilities. Hence, in the verification process, a large discrepancy between subtest scores on the ACT or the Scholastic Aptitude Test (SAT) is considered by some to be a significant indicator of LD.

Bennett, Rock, and Chan* conducted a study to assess the validity of SAT Verbal–Math score discrepancies as indicators of college learning disability. Four groups of certified LD students taking special test formats of the SAT were compared to reference groups taking the regular SAT. The four special test formats were Regular Print with extra time, Large Type, Cassette, and Cassette

Source: R. E. Bennett, D. A. Rock, and K. L. Chan, SAT verbal–math discrepancies: Accurate indicators of college learning disability? *J. Learning Disabilities* **20** (3): 189–192 (March 1987). Reprinted with permission.

plus Regular Print. The study compared one pool of 3552 LD students taking the SAT form WSA3 with a reference group of 35,424 students in Texas and California taking the regular SAT in October 1974. A second pool of 2535 LD students taking WSA5 was compared to a national sample of 33,161 high school students taking the test in December 1974. Since the results for both the WSA3 and WSA5 samples were identical, we consider here only the WSA5 sample results, given below.

Table 1

Sample Descriptive Statistics

				WSA5	
Group	N	Median Age (years)	Percentage Male	SAT Verbal Mean (SD)	SAT Mathematical Mean (SD)
Reference	33,161	17.6	52%	424 (107)	459 (113)
LD–Regular	2,050	17.3	74%	365 (92)	401 (104)
LD–Large type	129	17.2	74%	364 (98)	386 (98)
LD–Cassette	110	17.6	73%	318 (77)	354 (82)
LD–Cassette & Regular	246	17.4	82%	344 (87)	369 (89)

Table 2

Means and standard deviations of SAT absolute verbal–mathematical discrepancies in scale–score units

	WSA5		
Group	N	Mean	SD
Reference	33,161	78.8	62.4
LD–Regular	2,050	77.2	59.1
LD–Large type	129	65.0	55.8
LD–Cassette	110	69.8	53.1
LD–Cassette & Regular	246	69.2	57.1

Table 3

SAT absolute verbal–mathematical discrepancies for LD and reference samples matched on SAT total score

	WSA5		
Group	N	Mean	(SD)
Reference	2050	76.3	(61.9)
LD–Regular	2050	77.2	(59.1)
Reference	129	78.8	(59.3)
LD–Large Type	129	65.0	(56.0)
Reference	110	72.9	(57.0)
LD–Cassette	110	69.8	(53.4)
Reference	246	79.8	(62.2)
LD–Cassette & Regular	246	69.2	(57.2)

Figure I

Distributions
of absolute
verbal–mathematical
differences in
SAT scaled
score units

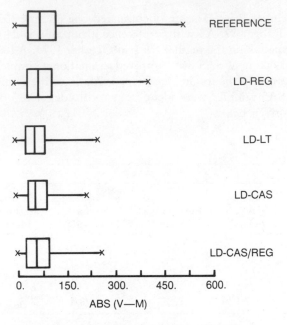

Distributions of absolute verbal-mathematical differences
in SAT scaled score units.

Tables 2 and 3 give the absolute discrepancies in Verbal–Math scores. The difference between the two tables is that in Table 3, the discrepancies are for subgroups of reference and LD students matched on the basis of total SAT scores.

Case Study Questions

1. Refer to Table 1. A reference sample of 33,161 students could be considered to be representative of the nationwide high school student population. Assuming this to be the case, and assuming a normally distributed population of SAT scores,
   (a)   What proportion of high school students nationwide have SAT Verbal scores between 400 and 500?
   (b)   What is the probability that in a random sample of 144 high school students the average SAT Math score is between 400 and 500?

2. Refer to Table 2. Construct 95% confidence intervals for the absolute Verbal–Math discrepancies for:
   (a)   The reference group
   (b)   The LD–Regular group
   (c)   The LD–Large Type group
   (d)   The LD–Cassette group
   (e)   The LD–Cassette + Regular group

3. To further determine the usefulness of discrepancy scores as indicators of LD, the researchers considered the degree of overlap between the LD and reference distributions by a graphical approach. They presented a special kind of graph called

box-and-whisker plots, as shown in Figure 1. The lower and upper boundaries of each distribution are represented by X's, the portion corresponding to the 25th through 75th percentile is represented by a box, and the median is indicated by the vertical line within the box.

Which of the five groups has the largest range of absolute differences?

4. The researchers argued: A high degree of overlap would suggest that the size of the discrepancies found in the LD group is little different from that found in regular students taking the SAT. A low degree of overlap suggests that there is a difference. Interpret this argument and comment on it. Do you agree with the researchers?

5. Finally, consider Table 3, and compare it with Table 2. Do you observe the importance of sampling procedures? Observe how using matched samples of a reference group and LD students brings out very clearly whether there is a difference in discrepancy scores.

6. In future chapters, you will study formal statistical procedures for conducting tests to determine whether significant differences exist between different groups. However, in the light of the above discussion, you should be in a position to give an informal and intuitive answer to the following question. What do you think should be the researchers' conclusion with respect to the issue they were addressing in their paper: Does a discrepancy between Verbal and Math scores on the SAT serve as a useful indicator of college learning disability?

COMPUTER EXERCISES USING THE DATABASE

1. Using the entire 1250 observations, generate 20 random samples of size 5 using the data for variable VGRE (verbal score on the GRE). For each sample determine the sample mean $\bar{X}$. Construct a histogram of these 20 sample means. Also find the mean and variance of these 20 values. Repeat this procedure for samples of size 10, 20, and 30. Does the Central Limit Theorem appear to be operating correctly here? Discuss.

2. Using the first random sample of size 30 in question 1, construct a 90% confidence interval for the mean verbal GRE score. Make a statement concerning the accuracy of using this particular sample mean to estimate the population mean.

3. Repeat exercises 1 and 2 using variable AGE.

9

HYPOTHESIS TESTING FOR
THE MEAN AND VARIANCE
OF A POPULATION

A LOOK BACK/INTRODUCTION

We have seen that statistical inference is used to estimate a population parameter using a sample statistic. For the rest of this book, the mean (μ) and standard deviation (σ) of the parent population will be unknown and will have to be estimated from the sample. Do not forget that even though you have estimated μ or σ, these values still are unknown and will forever remain unknown.

As a measure of how reliable your point estimate of the population mean really is, you can determine a confidence interval for this parameter. For a given confidence level, the narrower your resulting confidence interval is, the more faith you can have in the ability of your sample mean $\bar{X}$ to provide an accurate estimate of the population mean. Also, when the Central Limit Theorem is applicable, you need not worry about the shape of your population (normal, uniform, etc.) when making probability statements regarding $\bar{X}$, provided you have a large sample (generally, $n > 30$). When you do have a large sample, the distribution of $\bar{X}$ closely approximates the normal. This allows you to construct confidence intervals for population means without worrying about the nature of the parent population, simply because it does not matter.

Next, we turn to the situation in which someone makes a claim regarding the value of the population mean μ. For example, when dealing with the scores on the national achievement test in Chapter 7, we assumed that the population average of all scores was $\mu = 500$ points. Where did this value come from? Suppose that the designers of this test claim that the average score on this test is 500 points. By obtaining a sample of scores, can we prove this statement? The answer is an emphatic no; the only way to know the value of μ exactly is to obtain data for all students taking this test; that is, the entire population.

The sample, however, may allow us to reject the claim that μ is 500 points. But since the sample is only a portion of the population, this conclusion may be incorrect. Such is the nature of hypothesis testing.

9.1   HYPOTHESIS TESTING ON THE MEAN OF A POPULATION: LARGE SAMPLE

A newspaper article claims that the average age at which men get married for the first time is not the same as it was 50 years ago; it claims the average age is now 28 years. To investigate the article's claim, you randomly select 75 recently married males and obtain the age at which they were married.* Your results for $n = 75$ are $\bar{x} = 26.4$ years and $s = 5.5$ years.

Let μ represent the population average (mean) age at which U.S. males marry for the first time. We do have a point estimate of μ; namely, $\bar{x} = 26.4$ years is an estimate of μ. Keep in mind that the actual value of μ is unknown (although it does exist) and will remain that way. What we can do is estimate μ using the sample data. This situation can be summarized by considering the following pair of hypotheses:

Null hypothesis:
H_0: $\mu = 28$

Alternative hypothesis:
H_a: $\mu \neq 28$

H_0 asserts that the value of μ that has been claimed to be correct is in fact correct. H_a asserts that μ is some value other than 28 years. The alternative hypothesis typically contains the conclusion that the researcher is attempting to demonstrate using the sample data. In our age example, if you do not believe that the average age is 28, and you expect the data to demonstrate that μ has some other value, H_a is $\mu \neq 28$.

The task of all tests of hypothesis is to **reject** H_0 or **fail to reject (FTR)** H_0. Notice that we do not say reject H_0 or *accept* H_0. This is an important distinction.

In our study of male ages at first marriage, the (point) estimate of μ is $\bar{x} = 26.4$ years. Should we reject H_0, given that it claims that μ is 28? First, we need not worry about the shape of the underlying population of male ages because, by the Central Limit Theorem, $\bar{X}$ is approximately normally distributed for large samples, regardless of the shape of this population. So, $\bar{X}$ is approximately a normal random variable and thus continuous. What is the probability that any continuous random variable is equal to a certain value? In particular,

* The size of this sample is unrealistically small (yet large, statistically).

what is the probability that $\bar{X}$ is exactly equal to 28? The answer to both questions is zero. Thus we see that we cannot reject H_0 simply because $\bar{X}$ is not equal to 28. What we do is to allow H_0 to stand, provided $\bar{X}$ is "close to" 28, and reject H_0 otherwise. To define what "close" means, we need to take an in-depth look at what happens when you test hypotheses.

Type I and Type II Errors

Because the sample does not consist of the entire population, there always is the possibility of drawing an incorrect conclusion when inferring the value of a population parameter using a sample statistic. When testing hypotheses, there are two types of possible error:

Type I error. This occurs if you reject H_0 when in fact it is true. For example, this would occur if you reject the claim (hypothesis) that the population mean is 28 years when in fact it really is true.

Type II error. This occurs if you fail to reject H_0 when in fact H_0 is not true.

	Actual Situation	
Conclusion	H_0 **True**	H_0 **False**
FTR H_0	Correct decision	Type II error
Reject H_0	Type I error	Correct decision

For any test of hypothesis, define

α = the probability of rejecting H_0 when it is true; $\alpha = P(\text{Type I error})$

β = the probability of failing to reject H_0 when it is false; $\beta = P(\text{Type II error})$

For any test of hypothesis, you would like to have control over n (the sample size), α (the probability of a Type I error), and β (the probability of a Type II error). However, in reality, you can control only two of these: n and α, n and β, or α and β. In other words, for a fixed sample size you cannot control both α and β.

Suppose you decide to set $\alpha = .02$. This means that the procedure you use to test H_0 versus H_a will reject H_0 when it is true with a probability of .02. You may wonder why we do not set $\alpha = 0$, so we would never have a Type I error. The thought of never rejecting a correct H_0 sounds appealing, but the bad news is that β (the probability of a Type II error) is then equal to 1; that is, you will always fail to reject H_0 when it is false. If we set $\alpha = 0$, then the resulting test of H_0 versus H_a will automatically fail to reject H_0: $\mu = 28$ whenever μ is, in fact, any value other than 28. If, for example, μ is 50 (hardly the case, but interesting), we would still fail to reject (FTR) H_0—not a good situation at all. We therefore need a value of α that offers a better compromise between the two types of error probabilities. (Note that, for the situation where $\alpha = 0$ and $\beta = 1$, $\alpha + \beta = 1$. As later examples will demonstrate, this is not true in general.)

The value of α you select depends on the relative importance of the two types of error. For example, consider the following hypotheses, and decide if the Type I error or the Type II error is the more serious.

You have just been examined by a physician using a sophisticated medical device, where

H_0: device indicates that you do not have a serious disease

H_a: device indicates that you do have the disease

$\alpha = P(\text{rejecting } H_0 \text{ when it is true}) = P(\text{device indicates that you have the disease when you do not have it})$

$\beta = P(\text{FTR } H_0 \text{ when in fact it is false}) = P(\text{device indicates that you do not have the disease when you do have it})$

For this situation, the Type I error (measured by α) is not nearly as serious as the Type II error (measured by β). Provided the treatment for the disease does you no serious harm if you are well, the Type I error is not serious. But the Type II error means you fail to receive the treatment even though you are ill.

We never set β in advance, only α. This will allow us to carry out a test of H_0 versus H_a. The smaller α is, the larger β is. Consequently, if you want β to be small, then you choose a large value of α. For most situations, the range of acceptable α values is .01 to .1.

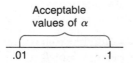

Acceptable
values of α

.01   .1

For the medical-device problem, you could choose a value of α near .1 or possibly larger, due to the seriousness of a Type II error. On the other hand, if you are more worried about Type I errors for a particular test then a small value of α is in order. An illustration of this would be to conclude that a drug has no ill side effects when in fact it does. What if there is no basic difference in the effect of these two errors? If there is no significant difference between the effects of a Type I error versus a Type II error, researchers often choose $\alpha = .05$.

Performing a Statistical Test

The claim that the average age at which men get married is 28 years resulted in the following pair of hypotheses:

$$H_0: \quad \mu = 28$$
$$H_a: \quad \mu \neq 28$$

We decide to use a test that carries a 5% risk of rejecting H_0 when it is correct, that is, $\alpha = .05$. In hypothesis testing, α is referred to as the **significance level** of your test. Using $n = 75$, $\bar{x} = 26.4$ years, and $s = 5.5$ years, we wish to carry out the resulting statistical test of H_0 versus H_a. We decided to let H_0 stand (not reject it) if $\bar{X}$ was "close to" 28. In other words, we will reject H_0 if $\bar{X}$ is "too far away" from 28. We write this as follows:

$$\text{Reject } H_0 \text{ if } |\bar{X} - 28| \text{ is "too large"}$$

or, by standardizing $\bar{X}$, we can

$$\text{Reject } H_0 \text{ if } \left| \frac{\bar{X} - 28}{s/\sqrt{n}} \right| \text{ is "too large"}$$

We rewrite the last statement as

$$\text{Reject } H_0 \text{ if } \left| \frac{\bar{X} - 28}{s/\sqrt{n}} \right| > k, \text{ for some } k$$

What is the value of k? Here is where the value of α has an effect. If H_0 is true, and the sample size is large, then, using the Central Limit Theorem, $\bar{X}$ is approximately a normal random variable with

$$\text{mean} = \mu = 28$$

$$\text{standard deviation} \cong \frac{s}{\sqrt{n}}$$

So, if H_0 is true, $(\bar{X} - 28)/(s/\sqrt{n})$ is approximately a standard normal random variable Z for large samples.* In this case, we reject H_0 if $|Z| > k$, for some k. Suppose $\alpha = .05$. Then,

$$.05 = \alpha = P(\text{rejecting } H_0 \text{ when it is true})$$
$$= P\left(\left| \frac{\bar{X} - 28}{s/\sqrt{n}} \right| > k, \text{ when } \mu = 28 \right)$$
$$= P(|Z| > k)$$

---

* In Chapter 8, we mentioned that $(\bar{X} - 28)/(s/\sqrt{n})$ actually follows a t distribution, but, for large sample sizes, it can be approximated well using the standard normal (Z) distribution. This section deals with large samples, so we will use the Z notation to represent this random variable.

Figure 9.1

The shaded area
represents the
significance level α

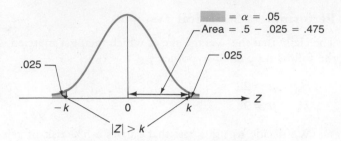

To find the value of k that satisfies this statement, consider Figure 9.1. When $|Z| > k$, either $Z > k$ or $Z < -k$, as illustrated. Since $P(|Z| > k) = .05$, then the total shaded area is .05, with .025 in each tail due to the symmetry of this curve. Consequently, the area between 0 and k is .475, and, using Table A-4, $k = 1.96$. So our test of H_0 versus H_a is

$$\text{Reject } H_0 \text{ if } \left| \frac{\bar{X} - 28}{s/\sqrt{n}} \right| > 1.96$$

and fail to reject H_0 otherwise. So,

$$\text{Reject } H_0 \text{ if } \frac{\bar{X} - 28}{s/\sqrt{n}} > 1.96$$

or

$$\text{if } \frac{\bar{X} - 28}{s/\sqrt{n}} < -1.96$$

This test will reject H_0 when it is true, 5% of the time. This means that there is a 5% risk of making a Type I error.

Using the sample data, we obtained $n = 75$, with $\bar{x} = 26.4$ years and $s = 5.5$ years. Is $\bar{x} = 26.4$ years far enough away from 28 years for us to reject H_0? This was not at all obvious at first glance; it may have seemed that this value of $\bar{X}$ is "close enough to" 28 for us to not reject H_0. Such is not the case, however, because

$$Z = \frac{\bar{X} - 28}{s/\sqrt{n}} = \frac{26.4 - 28}{5.5/\sqrt{75}} = -2.52 = Z^*$$

where Z^* is the **computed value** of Z. Because $-2.52 < -1.96$, we reject H_0. We thus conclude that, based on the sample results and a value of $\alpha = .05$, the average age at which males first get married (μ) is not equal to 28.

Another way of phrasing this result is to say that, if H_0 is true (that is, if $\mu = 28$ years), the value of $\bar{X}$ obtained from the sample (26.4) is 2.52 standard

Figure 9.2

Distribution of $\bar{X}$ if H_0 is true (H_0: $\mu = 28$)

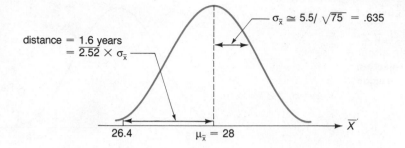

$\sigma_{\bar{x}} \cong 5.5/\sqrt{75} = .635$

distance = 1.6 years = $2.52 \times \sigma_{\bar{x}}$

26.4 $\mu_{\bar{x}} = 28$ $\bar{X}$

deviations to the left of the mean using the normal curve for $\bar{X}$ (Figure 9.2). Because a value of $\bar{X}$ this far away from the mean is very unlikely (that is, with probability less than $\alpha = .05$), our conclusion is that H_0 is not true, and so we reject it.

When testing $\mu =$ (some value) versus $\mu \neq$ (some value), the null hypothesis H_0 always contains the $=$, and the alternative hypothesis H_a always contains the $\neq$. In our example, this resulted in splitting the significance level α in half and including one-half in each tail of the test statistic Z. Consequently, when testing H_0: $\mu =$ (some value) versus H_a: $\mu \neq$ (some value), we refer to this as a **two-tailed test**.

EXAMPLE 9.1

Using the data from our example of male ages at first marriage, what would be the conclusion using a significance level α of .01?

Solution

The only thing that we need to change from our previous solution is the value of k. Now,

$$P(|Z| > k) = \alpha = .01$$

as shown in Figure 9.3. Using Table A-4, $k = 2.575$, and the test is (see Figure 9.4):

$$\text{Reject } H_0 \text{ if } Z > 2.575 \quad \text{or} \quad Z < -2.575$$

Figure 9.3

The shaded area is $\alpha = .01$

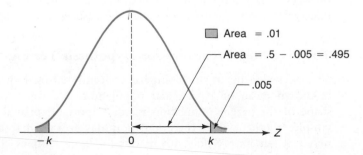

Area = .01

Area = .5 − .005 = .495

.005

−k   0   k   Z

Figure 9.4

We reject H_0 if Z^* falls within either tail—the rejection region

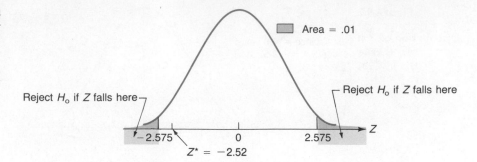

What is the value of $(\bar{X} - 28)/(s/\sqrt{n})$? Our data values have not changed, so the value of this expression is the same: $Z^* = -2.52$.

The region defined by values of Z to the right of 2.575 and to the left of -2.575 in Figure 9.4 is the **rejection region**. The value of k (2.575) defining this region is the **critical value**. Z^* fails to fall in this region, so we fail to reject H_0. In other words, for $\alpha = .01$, the value of $\bar{X}$ is "close enough" to 28 to let H_0 stand; there is insufficient evidence to conclude that μ is different from 28.   □

Accepting H_0 or Failing to Reject H_0?

It may appear that there is no difference between "accepting" and "failing to reject" a null hypothesis, but there *is* a difference between these two statements. When you test a hypothesis, H_0 is presumed innocent until it is demonstrated to be guilty. In example 9.1, using $\alpha = .01$, we failed to reject H_0. Now, how certain are we that μ is exactly 28 years? After all, our estimate of μ is 26.4 years. Clearly, we do not believe that μ is precisely 28. There simply was not enough evidence to reject the claim that $\mu = 28$.

For any hypothesis-testing application, the only hypothesis that can be accepted is the alternative hypothesis H_a. Either there is sufficient evidence to support H_a (we reject H_0) or there is not (we fail to reject H_0). The focus of our attention is whether there is sufficient evidence within the sample data to conclude that H_a is correct. By failing to reject H_0, we are simply saying that the data do not allow us to support the claim made in H_a (such as $\mu \neq 28$), not that we accept the statement made in H_0 (such as $\mu = 28$).

The Five-Step Procedure for Hypothesis Testing

The discussion up to this point has concentrated on hypothesis testing on the unknown mean of a particular population. We want to emphasize that the shape of the parent population is not important, provided you have a large sample. In other words, the population may be a normal (bell-shaped) one or it may not—it simply does not matter for large samples. The steps carried out when attempting to reject or failing to reject a claim regarding the population mean μ are:

Step 1 Set up the null hypothesis H_0 and the alternative hypothesis H_a. If the purpose of the hypothesis test is to test whether the population mean is equal to a particular value (say, μ_0), the "equal hypothesis" always is stated in H_0 and the "unequal hypothesis" always is stated in H_a.

Step 2 Define the test statistic. This is evaluated, using the sample data, to determine if the data are compatible with the null hypothesis. For tests regarding the mean of a population using a large sample, the test statistic is approximately a standard normal random variable given by the equation

$$Z = \frac{\bar{X} - \mu_0}{s/\sqrt{n}} \tag{9-1}$$

where μ_0 is the value of μ specified in H_0.

Step 3 Define a rejection region, having determined a value for α, the significance level. In this region the value of the test statistic will result in rejecting H_0.

Step 4 Calculate the value of the test statistic, and carry out the test. State your decision: to reject H_0 or to FTR H_0.

Step 5 Give a conclusion in the terms of the original problem or question. This statement should be free of statistical jargon and should merely summarize the results of the analysis.

Steps 1 through 5 apply to all tests of hypothesis in this and subsequent chapters. The form of the test statistic and rejection region change for different applications, but the sequence of steps always is the same.

EXAMPLE 9.2   Remember that scores on the national achievement test are reported to be 500 on the average. If the average score is in fact less than 500, you stand the chance of refusing admission to some students who are "better than average." If the average test score is more than 500, then you may be overestimating the ability of certain less able students. The results of this sample are $n = 100$, $\bar{x} = 520$ points, and $s = 78.5$ points. What conclusion would you reach using a significance level of .1?

Solution   **Step 1** *Define the hypotheses.* We will test H_0: $\mu = 500$ versus H_a: $\mu \neq 500$.

Step 2 *Define the test statistic.* The proper test statistic for this problem is

$$Z = \frac{\bar{X} - 500}{s/\sqrt{n}}$$

Figure 9.5

See example 9.2—
the rejection region
is $|Z| > 1.645$

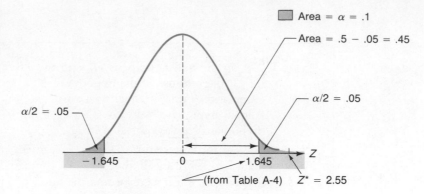

Step 3   *Define the rejection region.* The steps for finding the rejection region are shown in Figure 9.5. We conclude:

$$\text{Reject } H_0 \text{ if } Z > 1.645 \quad \text{or} \quad Z < -1.645.$$

Step 4   *Calculate the value of the test statistic and carry out the test.* The computed value of Z is

$$Z^* = \frac{520 - 500}{78.5/\sqrt{100}} = \frac{20}{7.85} = 2.55$$

Since $2.55 > 1.645$, our decision is to reject H_0. In Figure 9.5, Z^* falls in the rejection region.

Step 5   *State a conclusion.* Based on the sample data, there is sufficient evidence to conclude that the average score on the national achievement test is not 500 points.   □

COMMENTS

In example 9.2, $\bar{X}$ was "far enough away from" 500 for us to reject the claim that the average score is 500 points (H_0). However, remember that you cannot decide what is "far enough away from" without also considering the value of the standard deviation ($s = 78.5$ in example 9.2). This is why the value of s (or σ, if it is known) is a vital part of the test statistic.

Examine the test statistic in example 9.2. Observe that for small s, the "easier" it is to reject H_0. As s becomes smaller, the absolute value of the test statistic Z becomes larger, and the test statistic is more likely to be in the rejection region for a given value of α.

Confidence Intervals and Hypothesis Testing

What is the relationship, if any, between a 95% confidence interval and performing a two–tailed test using $\alpha = .05$? There is a very simple relationship here, namely, when testing $H_0: \mu = \mu_0$ versus $H_a: \mu \neq \mu_0$, using the five-step procedure and a significance level α, H_0 will be rejected if and only if μ_0 does not lie in the $(1 - \alpha) \cdot 100\%$ confidence interval for μ.

The five-step procedure and the confidence interval procedure always lead to the same result. In fact, you can think of a confidence interval as that set of values of μ_0 that would not be rejected by a two-tailed test of hypothesis.

In our age at first marriage for U.S. males example, a sample of 75 ages produced $\bar{x} = 26.4$ and $s = 5.5$. The resulting 95% confidence interval for μ is

$$\bar{X} - k\left[\frac{s}{\sqrt{n}}\right] \quad \text{to} \quad \bar{X} + k\left[\frac{s}{\sqrt{n}}\right]$$

What is the value of k? The population standard deviation (σ) is unknown, so we need to use the t table (Table A-5). We do have a large sample, however, so the t value will be closely approximated by the corresponding Z value (Table A-4). Keep in mind that when dealing with large samples it really does not matter if σ is known or replaced by s. In either case, the standard normal table (Table A-4) gives us the probability points we need.

The value of k that provides a 95% confidence interval here is the same value of k that provides a two-tailed area under the Z curve equal to $1 - .95 = .05$. In other words, we use the same k value that we used in a two-tailed test of H_0 versus H_a—namely, $k = 1.96$. So the 95% confidence interval for μ is

$$\bar{X} - 1.96(s/\sqrt{n}) \quad \text{to} \quad \bar{X} + 1.96(s/\sqrt{n})$$
$$= 26.4 - 1.96(5.5/\sqrt{75}) \quad \text{to} \quad 26.4 + 1.96(5.5/\sqrt{75})$$
$$= 26.4 - 1.24 \quad \text{to} \quad 26.4 + 1.24$$
$$= 25.16 \quad \text{to} \quad 27.64$$

The value of μ we are investigating here is $\mu = 28$, and the corresponding hypotheses were H_0: $\mu = 28$ and H_a: $\mu \neq 28$. Using $\alpha = .05$, our result using the two-tailed test was to reject H_0. Using the confidence interval procedure, we obtain the same result because 28 does not lie in the 95% confidence interval.

Thus, if you already have computed a confidence interval for μ, you can tell at a glance whether to reject H_0 for a two-tailed test, provided the significance level α for the hypothesis test and the confidence level, $(1 - \alpha) \cdot 100\%$, match up.

EXAMPLE 9.3

Repeat the ages at first marriage for U.S. males example using a 99% confidence interval. Is the result the same as in example 9.1, where we failed to reject H_0: $\mu = 28$ using $\alpha = .01$?

Solution

Using $\alpha = .01$, we failed to reject H_0 because the absolute value of the test statistic did not exceed the critical value of $k = 2.575$. The corresponding 99% confidence interval for μ is

$$\bar{X} - 2.575(s/\sqrt{n}) \quad \text{to} \quad \bar{X} + 2.575(s/\sqrt{n})$$
$$= 26.4 - 2.575(5.5/\sqrt{75}) \quad \text{to} \quad 26.4 + 2.575(5.5/\sqrt{75})$$
$$= 26.4 - 1.64 \quad \text{to} \quad 26.4 + 1.64$$
$$= 24.76 \quad \text{to} \quad 28.04$$

Because 28 does (barely) lie in this confidence interval, our decision is to fail to reject H_0—the same conclusion reached in example 9.1.   □

The Power of a Statistical Test

Up to this point, the probability of a Type II error, β, has remained a phantom— we know it is there, but we do not know what it is. One thing we can say is that a wide confidence interval for μ means that the corresponding two-tailed test of H_0 versus H_a has a large chance of failing to reject a false H_0; that is, β is large. Now,

$$\beta = P(\text{FTR } H_0 \text{ if it is false})$$

which means that

$$1 - \beta = P(\text{rejecting } H_0 \text{ if it is false})$$

The value of $1 - \beta$ is referred to as the **power** of the test. Since we like β to be small, we want the power of the test to be large. Notice that $1 - \beta$ represents the probability of making a *correct* decision in the event that H_0 is false, because in this case we *should* reject it. The more powerful your test is, the better.

Determining the power of your test (hence, β) is not difficult. We will illustrate this procedure for the previous two-tailed test of H_0: $\mu = \mu_0$ versus H_a: $\mu \neq \mu_0$, for some μ_0. We will first consider the case where σ is known and then discuss the situation where σ is unknown.

Power of the Test: σ Known

In example 9.2 we looked at the scores on the national achievement test where the hypotheses were H_0: $\mu = 500$ points and H_a: $\mu \neq 500$ points. Assume that the actual population standard deviation is known to be $\sigma = 75$ points. For this situation, our test statistic is (using a sample size of $n = 100$):

$$Z = \frac{\bar{X} - 500}{\sigma/\sqrt{n}} = \frac{\bar{X} - 500}{75/\sqrt{100}} = \frac{\bar{X} - 500}{7.5}$$

Proceeding as in example 9.2, using $\alpha = .10$, we reject H_0 if $Z > 1.645$ or $Z < -1.645$; that is, $|Z| > 1.645$. So, reject H_0 if $(\bar{X} - 500)/7.5 > 1.645$ (same as $\bar{X} > 500 + (1.645)(7.5) = 512.338$) or if $(\bar{X} - 500)/7.5 < -1.645$ (same as $\bar{X} < 500 - (1.645)(7.5) = 487.662$). This way of representing the rejection region is illustrated in Figure 9.6, using the shaded area under curve A. The power of this test is

$$1 - \beta = P(\text{rejecting } H_0 \text{ if } H_0 \text{ is false})$$
$$= P(\text{rejecting } H_0 \text{ if } \mu \neq 500)$$

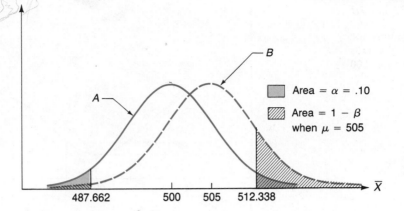

Figure 9.6

The shaded area is the probability of rejecting H_0 if $\mu = 500$ (that is $\alpha = .10$), and the striped area is the probability of rejecting H_0 if $\mu = 505$ (that is, the power of the test $1 - \beta$ when $\mu = 505$).

What is the power of this test if μ is not 500 but is 505? What you have here is a value of $1 - \beta$ for each value of $\mu \neq 500$.

Recall that we reject H_0 if $\bar{X} > 512.338$ or $\bar{X} < 487.662$. The probability of this occurring if $\mu = 505$ is illustrated as the lined area under curve B in Figure 9.6. Now, if $\mu = 505$ and $\sigma = 75$ (assumed), then

$$Z = \frac{\bar{X} - 505}{75/\sqrt{n}} = \frac{\bar{X} - 505}{7.5}$$

is a standard normal random variable. So, in Figure 9.6, the striped area to the right of 512.338 is

$$P(\bar{X} > 512.338) = P\left[\frac{\bar{X} - 505}{7.5} > \frac{512.338 - 505}{7.5}\right]$$

$$= P\left[Z > \frac{7.338}{7.5}\right]$$

$$= P(Z > .98) = .5 - .3365$$

$$= .1635$$

Also, the striped area to the left of 487.662 is

$$P(\bar{X} < 487.662) = P\left[\frac{\bar{X} - 505}{7.5} < \frac{487.662 - 505}{7.5}\right]$$

$$= P(Z < -2.31) = .5 - .4896$$

$$= .0104$$

Adding these two areas, we find that, if $\mu = 505$, the power of the test of $H_0: \mu = 500$ versus $H_a: \mu \neq 500$ is

$$1 - \beta = .1635 + .0104 = .1739$$

Figure 9.7

Power curve for
$H_0: \mu = 500$ versus
$H_a: \mu \neq 500$

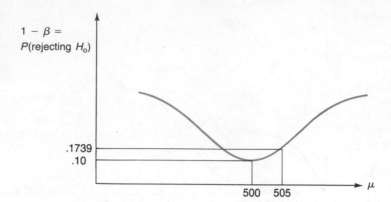

This means that, if $\mu = 505$, the probability of making a Type II error (not rejecting H_0) is $\beta = 1 - .1739 = .8261$ (rather high).

The power of your test increases (it becomes easier to reject H_0) as μ moves away from 500. Using the five-step procedure, which uses the test statistic $Z = (\bar{X} - 500)/(\sigma/\sqrt{n})$, the resulting power curve is illustrated in Figure 9.7. It is symmetric, and its lowest point is located at $\mu = 500$. For this value of μ, H_0 is actually true, so that a Type II error was *not* committed. Nevertheless, the value on the power curve corresponding to $\mu = 500$ is always

$$P(\text{rejecting } H_0 \text{ if } \mu = 500) = \alpha = .10 \quad \text{(for this example)}$$

Power of the Test: σ Unknown

When σ is unknown, we are forced to approximate the power of the test by replacing σ with the sample estimate s. We are dealing with large samples, so we can use Table A-4 (the Z table).

In our discussion of the power of our test for the national achievement test scores, we treated the population standard deviation σ as known. If we make no assumptions about this parameter, we need to approximate the power of the test for $\mu = 505$. We use the sample estimate $s = 78.5$ points (from example 9.2) and use $\alpha = .10$.

We will now reject H_0 if

$$\bar{X} > 500 + 1.645\left[\frac{s}{\sqrt{n}}\right] \quad \text{or} \quad \bar{X} < 500 - 1.645\left[\frac{s}{\sqrt{n}}\right]$$

So, H_0 is rejected, provided

$$\bar{X} > 500 + 1.645\left[\frac{78.5}{\sqrt{100}}\right] = 512.91$$

or

$$\bar{X} < 500 - 1.645 \left[\frac{78.5}{\sqrt{100}} \right] = 487.09$$

The resulting power of the test for $\mu = 505$ is approximately equal to

$$P(\bar{X} > 512.91 \text{ if } \mu = 505) + P(\bar{X} < 487.09 \text{ if } \mu = 505)$$

$$= P\left[Z > \frac{512.91 - 505}{78.5/\sqrt{100}} \right] + P\left[Z < \frac{487.09 - 505}{78.5/\sqrt{100}} \right]$$

$$= P(Z > 1.01) + P(Z < -2.28)$$

$$= (.5 - .3438) + (.5 - .4887) = .1562 + .0113$$

$$= .1675$$

EXERCISES

9.1 Identify the critical value of Z for the following hypothesis tests:
  (a) H_0: $\mu = 50$ H_a: $\mu \neq 50$
      $\alpha = .05$
  (b) H_0: $\mu = 1250$ H_a: $\mu \neq 1250$
      Use a significance level of 10%.
  (c) H_0: $\mu = 8.75$ H_a: $\mu \neq 8.75$
      Probability of Type I error should be $= .01$
  (d) H_0: $\mu = .005$ H_a: $\mu \neq .005$
      $\alpha = .02$

9.2 State what type of error can be made in the following situations:
  (a) The conclusion is to reject the null hypothesis.
  (b) The conclusion is to fail to reject the null hypothesis.
  (c) The calculated value of the test statistic does not fall in the rejection region.

9.3 Explain why the following statements are true or false.
  (a) The probability of the Type I error and the probability of the Type II error always add to one.
  (b) Increasing the value of α increases the value of β.

9.4 According to a sociologist who has studied the buying habits of American consumers, the average American woman purchases 4.7 pairs of shoes a year. To check this claim, one of her students obtained a random sample of 50 women, and discovered that the average number of pairs of shoes purchased per person each year was 3.9 with a standard deviation of 1.4.
  (a) At a 5% significance level, could the student reject the sociologist's claim?
  (b) Suppose that in part (a), the null hypothesis was rejected (assume that even if your decision in part (a) was otherwise). What is the type of error that the student might commit if H_0 is rejected, and what is the probability of committing this type of error?

9.5 The mean of a normally distributed population is believed to be equal to 50.1. A sample of 36 observations is taken and the sample mean is found to be 53.2. The alternative hypothesis is that the population mean is not equal to 50.1. Complete

Handwritten margin notes (top left):

reject 378 if

$|z| > K$, $\alpha = 0.05$

$-1.96 - 1.96 \leftarrow$ rejection region

$z = \dfrac{\bar{x} - \mu}{\frac{\sigma}{\sqrt{n}}} = \dfrac{53.2 - 50.1}{\frac{4}{\sqrt{36}}} = 4.65$

reject H_0

the hypothesis test, assuming that the population standard deviation is equal to four. Use a .05 significance level.

9.6 Given the following statistics, perform the two-tail hypothesis test that the mean of a normally distributed population is equal to 30:

$$n = 75$$
$$\bar{x} = 27.4$$
$$s = 9$$

Use a significance level of .05.

Handwritten margin notes (left):

$\sigma = 1.1$, $\bar{x} = 2.4$, $n = 30$ $\alpha = .1$

$H_0: \mu = 3.1$ $H_a: \mu \neq 3.1$

$\dfrac{2.4 - 3.1}{\frac{1.1}{\sqrt{30}}} = -3.485$, $K = 1.645$

$-3.485 < -1.645$

$\Rightarrow$ reject H_0 since $|z| > K$

9.7 The weights of fish in a certain pond that is stocked regularly are considered to be normally distributed with a standard deviation of 1.1 lb. A random sample of size 30 is selected from the pond and the sample mean is found to be 2.4 lb. Test the null hypothesis that the mean weight of fish in the pond is 3.1 lb. Use a two-tailed test and a 10% significance level.

9.8 A social worker addressing a group of parents said that an average of 30 teenagers attempted to commit suicide every week in the greater New York City area. A sample of 35 weeks showed an average of 18.5 attempted suicides with a standard deviation of 5.4. Do these data support the social worker, or should her claim be rejected?
(a) Use $\alpha = .01$ for the hypothesis test.
(b) Use $\alpha = .10$ for the hypothesis test.

Handwritten margin notes (left):

2.2 – 2.6 } 95% C.I.

$\bar{x} > 2.6$, $\bar{x} < 2.2$ rejection area

$H_0: \mu = 2$ $\alpha = 0.05$

95% C.I. does not contain 2 $\Rightarrow$ reject H_0

9.9 A 95% confidence interval for the mean time that it takes a city bus to complete its route is 2.2 to 2.6 hours. The time that it takes the bus to complete its route is normally distributed. Test the null hypothesis that the mean time to complete the route is 2.0 hours, using a two-tailed test and a 5% significance level.

9.10 A psychologist studying the influence of television on children claims that children between the ages of 2 to 12 watch an average of 25 hours of television a week.
(a) A 95% confidence interval for number of hours of television watched per week is constructed from a sample of 50 children, and is found to be from 19.9 hours to 24.1 hours. Does this provide enough evidence to reject the psychologist's claim? (i.e., reject H_0: $\mu = 25$ hours) at $\alpha = .05$?
(b) The above hypothesis test is to be conducted at a 1% level of significance. A 99% confidence interval is computed, and found to be from 18.8 to 25.2 hours. What is your decision in this case?

Handwritten margin notes (left):

$\mu = 62.5$, $n = 250$, $\bar{x} = 64.2$

$s = 3.8$ $H_a: \mu \neq 62.5$ $\alpha = 0.05$

$\dfrac{64.2 - 62.5}{\frac{3.8}{\sqrt{250}}} = 7.074$

$-1.96 - 1.96$ rejection area

7.074 lies in rejection area

so support H_a.

9.11 The average life expectancy of males in a developing nation was believed to be 62.5 years. However, this belief was based on data that might be considerably out-of-date. A random survey of 250 deaths in that country revealed an average life span of 64.2 years with a standard deviation of 3.8 years. Does this support the hypothesis that average life-expectancy is not 62.5 years? Use $\alpha = .05$.

9.12 A crime reporter was told that, on the average, 3000 burglaries per month occurred in his city. The reporter examined past data, which he used to compute a 95% confidence interval for the number of burglaries per month. The confidence interval was from 2176 to 2784. At a 5% level of significance, do these data tend to support the alternative hypothesis, H_a: $\mu \neq 3000$?

(handwritten left margin annotations for 9.13)

$H_0: \mu = 200$

$H_a: \mu \neq 200$

$n = 35$ $z = \dfrac{115-200}{\frac{35}{\sqrt{35}}} = -14.37$

$\bar{x} = 115$

$s = 35$

$\alpha = .1$

$-1.645 \quad -1.645$

z lies within rejection area

$\Rightarrow$ reject H_0

9.13 The acronym WASP is often used pejoratively to refer to White Anglo-Saxon Protestants. However, WASP also refers to Women Air Service Pilots, who trained at Avenger Field in Sweetwater, Texas, the only all-female Army Air Force base in American history. At a reunion, one veteran WASP recalled that the average logged and certified flight time on joining was 200 hours. However, another veteran said that because of World War II, flight time requirements were modified, so the logged hours for new recruits was not 200 hours. Which of these is correct? To decide this, a sample of 35 WASP's attending the reunion was queried. The result showed an average of 115 flight hours logged by new recruits, with a standard deviation of 35 hours. Using $\alpha = .10$, test the hypothesis $H_0: \mu = 200$ hours against the alternative $H_a: \mu \neq 200$ hours. What is your final conclusion?

9.14 What is the price of a commercial on television? Here are some examples of the cost of a 30 second time slot in the Fall 1986 season: *The Cosby Show* ($380,000), *Family Ties* ($300,000), *Moonlighting* ($215,000), *Dynasty* ($185,000) all the way down to *Sidekicks* ($61,000), *The Colby's* ($63,000), and *Our World* ($30,000). A 95% confidence interval for the cost of a 30-second television spot was computed to have a range from $42,000 to $318,000. Does this support the idea that the cost of the average 30-second television advertisement is $180,000, at $\alpha = .05$?

9.15 Suppose a horticulturist is nurturing a population of 300 transplanted young trees. The horticulturist believes the average growth rate of the trees is 1.2 inches per week. If a sample of 35 trees is observed to have an average growth of 1.35 inches with a standard deviation of 0.40 inch, does this provide enough evidence to reject the horticulturist's claim? Use a significance level of .05.

9.16 The hypotheses for a particular situation are

$$H_0: \quad \mu = 20 \qquad H_a: \quad \mu \neq 20$$

If the population of interest is normally distributed, what is the power of the test for the mean if μ is actually equal to 22? Assume that a sample of size 49 is used and the sample standard deviation is 4.2. Use a significance level of .05.

9.17 Find the power of the test for the mean of a normal population for the following situations if the true population mean is 30 and the population variance is 25. Use a 10% significance level.

(a) $H_0: \quad \mu = 26 \qquad H_a: \quad \mu \neq 26, n = 20$
(b) $H_0: \quad \mu = 36 \qquad H_a: \quad \mu \neq 36, n = 25$
(c) $H_0: \quad \mu = 33 \qquad H_a: \quad \mu \neq 33, n = 25$

(handwritten annotations for 9.17)

a)

$\mu = 30, \sigma = 5 \quad \alpha = .1$

$\bar{x} > 30 + 1.645\left(\dfrac{5}{\sqrt{20}}\right) = 31.839$

$\bar{x} < 30 - 1.645\left(\dfrac{5}{\sqrt{20}}\right) = 28.1608$

$z > \dfrac{31.8-26}{\frac{5}{\sqrt{20}}} = 5.22 \approx .5$

$z < \dfrac{28.1608-26}{\frac{5}{\sqrt{20}}} = 1.9327 = .4732$

$1-\beta = .5 + .4732 = .9732$

9.18 An electro-optical firm currently uses a particular laser component in producing sophisticated graphic designs. The mean time it takes to produce a certain design with the current laser component is 70 seconds with a standard deviation of 8 seconds. A new laser component is bought by the firm and it is believed that the mean time it takes to produce the same design is not equal to 70 seconds, and has a standard deviation of 8 seconds. The research and development department is interested in constructing the power curve for testing the claim that the mean time it takes to produce the same design by the new laser component is not equal to 70 seconds. They have reason to believe that these times are normally distributed. Graph the power function for a sample of size 25 and a significance level of .05.

9.2 ONE-TAILED TEST FOR THE MEAN OF A POPULATION: LARGE SAMPLE

There are many situations in which you are interested in demonstrating that the mean of a population is larger or smaller than some specified value. For example, you may be attempting to demonstrate that the mean pulse rate for male joggers in a particular age group is less than the national norm of 75 beats per minute. Because the situation that you are attempting to demonstrate goes into the alternative hypothesis, the resulting hypotheses would be $H_0: \mu \geq 75$ and $H_a: \mu < 75$. Remember that we said it is standard practice always to put the equal sign in the null hypothesis. In the testing procedure only the **boundary value** is important, so the hypotheses may be written as

$$H_0: \quad \mu = 75$$
$$H_a: \quad \mu < 75$$

In this way, we can identify the distribution of $\bar{X}$ when H_0 is true—namely, $\bar{X}$ is a normal random variable centered at 75 with standard deviation $s/\sqrt{n}$ (or $\sigma/\sqrt{n}$ if σ is known). Because the focus of our attention is on H_a (can we support it or not?), which of the two ways you use to write H_0 is not an important issue. The procedure for testing H_0 versus H_a is the same regardless of how you state H_0.

The resulting test is referred to as a **one-tailed test**, and its uses the same five-step procedure as the two-tailed test. The only change we make is to modify the rejection region: all the error is in a single tail.

EXAMPLE 9.4

A physician suspects that the average diastolic blood pressure for women between 20 and 30 years of age who smoke more than a pack of cigarettes a day is more than the national norm of 80 for women in this age group. A report written by this physician discusses the results of a random sample of 50 female smokers between 20 and 30 years of age, with the following results: average blood pressure is $\bar{x} = 84.4$ with a corresponding standard deviation of $s = 11.1$.

This report failed to offer any statistical conclusion and you have been asked to interpret these results. Do the sample results support the physician's claim, using a significance level of $\alpha = .05$?

Solution

Step 1 An important point to be made here is that H_0 and H_a (as well as α) must be defined before you observe any data. In other words, do not let the data dictate your hypotheses; this would introduce a serious bias into your final outcome. For this application, we want to demonstrate that the population mean μ is more than 80, and so this goes into H_a. The appropriate hypotheses then are $H_0: \mu \leq 80$ and $H_a: \mu > 80$.

Figure 9.8

The one-tailed rejection region is $Z > 1.645$. We reject H_0 if $Z = (\bar{X} - 80)/(s/\sqrt{n}) > 1.645$.

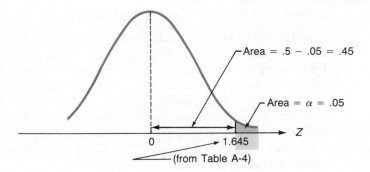

Area = .5 − .05 = .45

Area = α = .05

1.645

(from Table A-4)

Step 2 The test statistic for a one-tailed test is the same as that for a two-tailed test, namely,

$$Z = \frac{\bar{X} - \mu_0}{s/\sqrt{n}} \quad \left(\text{or, if } \sigma \text{ is known, } \frac{\bar{X} - \mu_0}{\sigma/\sqrt{n}} \right)$$

$$= \frac{\bar{X} - 80}{s/\sqrt{n}}$$

Step 3 What happens to Z when H_a is true? Here we would expect $\bar{X}$ to be > 80 (because μ is), so the value of Z should be positive. Consequently, our procedure will be to reject H_0 if Z lies "too far to the right" of zero; that is,

$$\text{Reject } H_0 \text{ if } Z = \frac{\bar{X} - 80}{s/\sqrt{n}} > k \text{ for some } k > 0$$

Since $\alpha = .05$, we will choose a value of k (the critical value) such that the resulting test will reject H_0 (and support the physician's claim) when it is true with a 5% risk of an incorrect decision. This amounts to defining a rejection region in the right tail of the Z curve, the shaded area in Figure 9.8. Using Table A-4, we see that the critical value is $k = 1.645$, and the resulting test of H_0 versus H_a is

$$\text{Reject } H_0 \text{ if } Z = \frac{\bar{X} - 80}{s/\sqrt{n}} > 1.645$$

Step 4 Using the sample results, the value of the test statistic is

$$Z^* = \frac{84.4 - 80}{11.1/\sqrt{50}} = 2.80$$

Because $2.80 > 1.645$, the decision is to reject H_0.

Step 5 The results of this study support the claim that the average blood pressure for women who smoke in this age group is greater than 80. In other words, the sample mean of 84.4 was enough larger than 80 to support the physician's claim. □

One-Tailed Test or Two-Tailed Test?

The decision to use a one-tailed test or a two-tailed test depends on what you are attempting to demonstrate. In example 9.4, the physician had reason to believe that the mean blood pressure for the smokers was *larger* than the population norm (mean) for all women in this age group. Whenever there are medical or theoretical reasons for supporting such a claim, a one-tailed test is in order. A one-tailed test is often referred to as a **directional** test, since the alternative hypothesis specifies the direction of the population mean relative to a specified value (80, in example 9.4). Conversely, a two-tailed test is called a **nondirectional** test.

Keep in mind that the burden of proof is on the alternative hypothesis since the question for any test of hypothesis will be "is there sufficient evidence to support the alternative hypothesis H_a?" If you employ a nondirectional test and you end up rejecting H_0, it is *not correct* to conclude that, say, the population mean exceeds the specified value (such as 80, in example 9.4) simply because the sample mean exceeds this value. All you can conclude is that the population mean is *different* than this specified value. If your intent is to support a directional claim, then use a one-tailed test.

Is it proper to examine the sample mean and then decide whether to use a one-tailed or two-tailed test? As mentioned in the solution to example 9.4, this is *not* a correct procedure since you are allowing the data to dictate your hypotheses. If you elect to use a right-tailed test because the sample mean appears to be very "large," the actual significance level of the test will be much larger than your originally specified value of α. In other words, the chances of rejecting a null hypothesis that is in fact correct will be larger than you think. In deciding between a directional or nondirectional procedure, decide what it is you are attempting to demonstrate *before observing any data*.

The testing of electric fuses is a classic example of a nondirectional testing procedure. A fuse must break when it reaches the prescribed temperature or a fire will result. However, the fuse must not break before it reaches the prescribed temperature or it will shut off the electricity when there is no need to do so. Therefore, the quality-control procedures for testing fuses must be two-tailed.

EXAMPLE 9.5   In the literature issued by Lecrem University, they mention that the average IQ of students entering the university is 125. A local newspaper reporter has some doubts as to whether this claim is correct, since he found several interviews with Lecrem students to be less than overwhelming. He suspects that a sample of students will demonstrate that the population mean (average IQ of all entering Lecrem students) is less than 125. A random sample of 40 students resulted in a sample mean of $\bar{x} = 121.8$ and a standard deviation of $s = 13.2$. Do the sample results support the reporter's claim using a significance level of $\alpha = .01$?

Solution   Step 1   The hypotheses here are H_0: $\mu \geqslant 125$ and H_a: $\mu < 125$.

Figure 9.9

One-tailed rejection region; reject H_0 if $Z < -2.33$

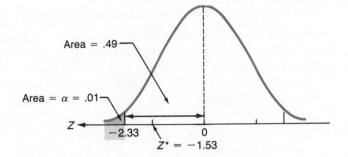

Step 2   The correct test statistic is

$$Z = \frac{\bar{X} - 125}{s/\sqrt{n}}$$

Step 3   For this situation, what happens to Z if H_a is true? The value of $\bar{X}$ should then be smaller than 125 (on the average), resulting in a negative value of Z. So we

$$\text{Reject } H_0 \text{ if } Z = \frac{\bar{X} - 125}{s/\sqrt{n}} < k \text{ for some } k < 0$$

Examine the standard normal curve in Figure 9.9, where the area corresponding to α is the shaded part of the left tail; using Table A-4, the critical value is $k = -2.33$. The test of H_0 versus H_a will be

$$\text{Reject } H_0 \text{ if } Z < -2.33$$

Step 4   The value of your test statistic is

$$Z^* = \frac{121.8 - 125}{13.2/\sqrt{40}} = -1.53$$

Step 5   Consequently, we fail to reject H_0. Using this value of α, there is insufficient evidence to support the reporter's claim that the average IQ of entering freshmen is less than 125.   □

This result is very much tied to the value of α. Using $\alpha = .10$ in example 9.5, we would obtain the opposite conclusion—which you may find somewhat disturbing. You often hear the expression that "statistics lie." This is not true—statistics are merely mistreated, either intentionally or accidentally. One can often obtain the desired conclusion by choosing the value of α that produces a desired conclusion. We therefore reemphasize that you must choose α by weighing the seriousness of a Type I versus a Type II error before seeing the data. A partial remedy for this dilemma will be discussed in Section 9.3.

Large Samples Taken from a Finite Population

For applications in which we take a large sample from a finite population, we make a slight adjustment to the standard error of $\bar{X}$ by including the finite population correction (fpc) factor.

For the finite population case, the standard error (standard deviation) of $\bar{X}$ is not $s/\sqrt{n}$, but instead

$$\text{standard error of } \bar{X} = s_{\bar{x}} = \frac{s}{\sqrt{n}} \cdot \sqrt{\frac{N-n}{N-1}} \qquad \text{(9-2)}$$

Once again using the results of the Central Limit Theorem, the test statistic is an approximate standard normal random variable, given by

$$Z = \frac{\bar{X} - \mu_0}{s_{\bar{x}}}$$

As a result, the five-step procedure can be carried out exactly as before.

EXAMPLE 9.6

In example 8.4, we considered a sample of 45 incomes from a group of female managers at Compumart. The women wished to demonstrate that the average income of the population of 350 female middle managers was less than \$48,000. For this illustration, it was assumed that the population standard deviation was known. Because this assumption is not necessary and perhaps incorrect, a safer procedure would be to use the sample estimate s. The results of the sample were $n = 45$, $\bar{x} = \$43,900$, and $s = \$7140$.

What would be your conclusion from these results, using a significance level of $\alpha = .05$?

Solution

Step 1 The hypotheses are $H_0: \mu \geqslant 48,000$ and $H_a: \mu < 48,000$ where $\mu =$ the average annual income for all females in middle-management positions at Compumart.

Step 2 The corresponding test statistic here is

$$Z = \frac{\bar{X} - 48,000}{s_{\bar{x}}}$$

where

$$s_{\bar{x}} = \frac{s}{\sqrt{n}} \cdot \sqrt{\frac{N-n}{N-1}}$$

Step 3 Using the symmetry of the normal curve and Figure 9.8, the rejection region is: Reject H_0 if $Z < -1.645$.

Step 4 Here,

$$s_{\bar{X}} = \frac{7140}{\sqrt{45}} \cdot \sqrt{\frac{350 - 45}{350 - 1}} = 995.01$$

so our computed test statistic is

$$Z^* = \frac{43{,}900 - 48{,}000}{995.01} = -4.12$$

Since $-4.12 < -1.645$, we (strongly) reject H_0 in favor of H_a.

Step 5 The sample results strongly support the assertion that the female middle managers are underpaid. We reached the same conclusion in example 8.4, where we based this decision on the extremely small probability of observing a value of $\bar{X}$ this small if μ was in fact \$48,000. □

COMMENTS   As mentioned in Chapter 8, the finite population correction factor $(N - n)/(N - 1)$ can be ignored whenever your sample size is less than 5% of the population size; that is, when $n/N < .05$. Such is also the case when using the fpc in hypothesis testing. In the preceding example, $n/N = 45/350 = .129 > .05$. Consequently, ignoring the fpc would have produced a much smaller (and less accurate) value of Z^*.

Large Sample Tests on a Population Mean

Two-Tailed (Nondirectional) Test

H_0: $\mu = \mu_0$

H_a: $\mu \neq \mu_0$

Reject H_0 if $|Z| > Z_{\alpha/2}$ where $Z = (\bar{X} - \mu_0)/s_{\bar{X}}$
($Z_{\alpha/2} = 1.96$ for $\alpha = .05$)

One-Tailed (Directional) Tests

H_0: $\mu \leqslant \mu_0$ H_0: $\mu \geqslant \mu_0$

H_a: $\mu > \mu_0$ H_a: $\mu < \mu_0$

Reject H_0 if $Z > Z_\alpha$ Reject H_0 if $Z < -Z_\alpha$
($Z_\alpha = 1.645$ for $\alpha = .05$)   ($-Z_\alpha = -1.645$ for $\alpha = .05$)

For a finite population with $n/N > .05$,

$$s_{\bar{X}} = \frac{s}{\sqrt{n}} \cdot \sqrt{\frac{N - n}{N - 1}}$$

Otherwise,

$$s_{\bar{X}} = \frac{s}{\sqrt{n}}$$

EXERCISES

9.19 A local political activist complained to the City Council that applicants for rent-subsidized housing provided by the city had to wait an average of 16 months or more. The City Manager replied that while there was a waiting period involved, it was less than 10 months, and not quite as long as the activist claimed. A sample of 40 applicants randomly selected from the city's files revealed an average waiting period of 11 months with a standard deviation of 3 months.

(a)   Use $\alpha = .05$ to test the null hypothesis that the waiting period is 16 months or more against the alternative hypothesis that the waiting period is less than 16 months. Who seems to be correct, the political activist or the City Manager?

(b)   Perform a similar test of the manager's claim that the average waiting period is less than 10 months. Do the data support the City Manager in this respect?

9.20 Find the rejection region of the Z statistic in a hypothesis test of the population mean for the following situations:

(a)   It is believed that the mean monthly advertising expenditure for a company was greater than $2,000. A significance level of .05 is used.

(b)   It is believed that the average length of sick time taken by an employee for firm XYZ is not equal to 5.2 days per year. A significance level of .10 is used.

(c)   It is believed that the mean age of an applicant applying for a particular job is less than 25 years. A significance level of .01 is used.

9.21 A sample of size 20 is drawn from a finite population of size 225. The finite population can be approximated by a normal distribution. The sample size of 20 yields a sample mean of 75.8. The population variance is equal to 16. Test the null hypothesis that the mean of the population is equal to 82.5. Use a two-tailed test and a 10% significance level.

9.22 Carry out the hypothesis test for the mean of the normally distributed population given the following information:

$$H_0: \quad \mu \geqslant 4.5 \qquad \bar{x} = 3.9$$
$$H_a: \quad \mu < 4.5 \qquad \sigma = 1.12$$
$$n = 30$$

9.23 Given the following statistics, perform the hypothesis test that the mean of a normally distributed population is greater than 10.31:

$$n = 48 \qquad \bar{x} = 12.03 \qquad s = 1.8$$

9.24 A delivery company claims that the mean time it takes to deliver frozen food between two particular cities is less than 3.7 hours. A random sample of 50 deliveries yielded a sample mean of 3.3 hours with a sample standard deviation of .2 hours. Test the delivery company's claim using a 10% significance level.

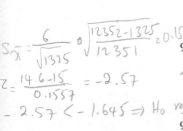

9.25 An auditing firm would like to test the hypothesis that the average customer of a small town's utility service pays the utility bill in less than 15 days after receipt of the bill. The town has only 12,352 customers. A sample size of 1325 yields a sample mean of 14.6 days in which a customer paid the utility bill. The sample standard deviation is 6 days. Test the hypothesis using a 5% significance level.

9.26 Is the average rent in housing estates built by the Liberian National Housing Authority more than $50 per month? Use the following data to address this issue:

HOUSING ESTATES BUILT BY NHA (1970–1980)

Estate	No. of Units	Average Monthly Rent
Cabral	72	$38.61
Goodridge	576	$56.00
New Georgia	226	$43.96

Source: Republic of Liberia 1980, National Housing Authority Annual Report 1980. Adapted from L. Lacey and S. E. Owusu. Self-help shelter and related programs in Liberia. *J. Am. Planning Assoc.* **53** (2): 209 (Spring 1987). Table 1.

(a)   Compute the overall average monthly rent for the 874 units.

(b)   Although this is not strictly a random sample of 874 units, you may assume that for this exercise. Given a standard deviation in monthly rent of $22.00, conduct a hypothesis test at a 5% significance level to decide if the average monthly rent is greater than $50.

(c)   What is your conclusion?

9.27   The average age of first-time drug abusers has been declining alarmingly. It is believed by one agency that the average age of first-time drug abusers is less than 13 years. To check this claim, a sample of 36 first-time drug abusers was obtained. It is found to have a mean of 13.8 years, with a standard deviation of 1.2 years. Does this support the above belief, at a 1% significance level?

9.28   A psychologist claims that the average vocabulary of children in a certain age group is fewer than 450 words. Suppose a sample of 100 children in that age group was obtained, and the average vocabulary is found to be 442 words, with a standard deviation of 15 words. Does this support the psychologist's claim?

(a)   Use $\alpha = .10$ for the test.

(b)   Use $\alpha = .01$ for the test.

9.29   Refer to exercise 2.3. In part (d) of that exercise, you considered whether the data supported the opinion that homemakers spent no more than 2 hours per week food shopping. You probably relied on intuition and rule of thumb when you considered this question. In the light of what you now know about hypothesis testing, answer the following questions.

(a)   What are the possible errors that could be made in taking an intuitive, rule of thumb approach?

(b)   Compute the mean and standard deviation of the sample of data in exercise 2.3.

(c)   Conduct a formal hypothesis test at $\alpha = .05$.

(d)   What are your conclusions?

9.3   REPORTING TESTING RESULTS USING A p VALUE

In example 9.1, we noted that for one value of α we rejected H_0, and for another (seemingly reasonable) value of α we failed to reject H_0. Is there a way of summarizing the results of a test of hypothesis that allows you to determine whether these results are barely significant (or insignificant) or overwhelmingly significant (or insignificant)? Did we barely reject H_0, or did H_0 go down in flames?

A convenient way to summarize your results is to use a p value, often called the *observed* α or *observed significance level*.

> The p **value** is the value of α at which the hypothesis test procedure changes conclusions. It is the smallest value of α for which you can reject H_0 (that is, at which the test is significant).

Consequently, the p value is the point at which the five-step procedure leads us to switch from rejecting H_0 to failing to reject H_0.

Determining the p Value

The p value for any test is determined by replacing the area corresponding to α by the area corresponding to the computed value of the test statistic. In our discussion and example 9.1, using $\alpha = .05$ you reject H_0, and using $\alpha = .01$ you fail to reject H_0. We know that the p value here is between .01 and .05. For this example, the computed value of the test statistic was $Z^* = -2.52$, where the hypotheses are $H_0: \mu = 28$ and $H_a: \mu \neq 28$. The Z curve for this situation is shown in Figure 9.10.

For which value of α does the testing procedure change the conclusions here? In Figure 9.10, if you were using a predetermined significance level α, you would split α in half and put $\alpha/2$ into each tail. So the total tail area represents α. Using Figure 9.11, we reverse this procedure by finding the total tail area corresponding to a two-tailed test with $Z^* = -2.52$; we add the area to the left of -2.52 (.0059) to that to the right of 2.52 (also .0059). This total area is .0118, which is the p value for this application. Thus, if you choose a value of $\alpha > .0118$ (such as .05), you will reject H_0. If you choose a value of $\alpha < .0118$ (such as .01), you will fail to reject H_0.

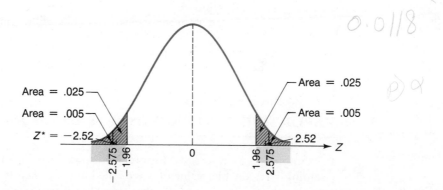

Figure 9.10

Rejection regions for $\alpha = .01$, .05

Figure 9.11

p value is determined by replacing the area corresponding to α (see Figure 9.8) by the area corresponding to Z^*. Here $Z^* = -2.52$, and the p value $=$ 2(.0059) $=$ .0118 (total shaded area).

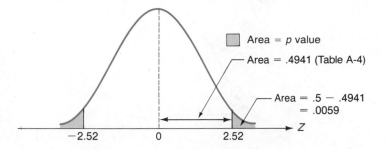

Area $= p$ value

Area $= .4941$ (Table A-4)

Area $= .5 - .4941$
$\quad = .0059$

$-2.52 \qquad 0 \qquad 2.52 \qquad Z$

Procedure for Finding the p Value

1 For H_a: $\mu \neq \mu_0$

$$p = 2 \cdot (\text{area outside of } Z^*)$$

Reason: When using a significance level α, the value of α represents a two-tailed area.

2 For H_a: $\mu > \mu_0$

$$p = \text{area to the right of } Z^*$$

Reason: When using a significance level α, the value of α represents a right-tailed area.

3 For H_a: $\mu < \mu_0$

$$p = \text{area to the left of } Z^*$$

Reason: When using a significance level α, the value of α represents a left-tailed area.

EXAMPLE 9.7

What is the p value for example 9.5?

Solution

The results of the sample were $n = 40$, $\bar{x} = 121.8$, and $s = 13.2$. The corresponding value of the test statistic was

$$Z^* = \frac{121.8 - 125}{13.2/\sqrt{40}} = -1.53$$

The alternative hypothesis is H_a: $\mu < 125$, so the p value will be the area to the left of the computed value, -1.53, as illustrated in Figure 9.12. Notice that the inequality in H_a determines the *direction* of the tail area to be found. The p value here is .063, which is consistent with the results of example 9.5, where we concluded that for $\alpha = .01$, you fail to reject H_0, and for $\alpha = .10$, you reject H_0. That is, the p value is between .01 and .10. □

Figure 9.12

p value for
$Z^* = -1.53$

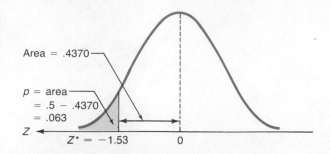

Area = .4370

p = area
= .5 − .4370
= .063

Z $Z^* = -1.53$ 0

  Figure 9.13

MINITAB solution
for example 9.5

```
MTB > SET INTO C1
DATA> 111 141 124 133 128 · · ·
     .
     .  ( the 40 data values )
     .
DATA> END
MTB > TTEST OF MU=125 USING C1;
SUBC> ALTERNATIVE=-1.

   TEST OF MU = 125.00 VS MU L.T. 125.00
```

	N	MEAN	STDEV	SE MEAN	T	P VALUE
C1	40	121.80	13.20	2.09	−1.53	0.067

Most statistical computer packages will provide you with the computed *p* value when testing the mean of a population. The MINITAB solution to example 9.5 is provided in Figure 9.13. This procedure assumes that the population standard deviation (σ) is unknown, and so it uses the command TTEST (as in *t* test). The *p* value in Figure 9.13 is slightly different than the value obtained in example 9.7 since MINITAB uses the *t* distribution to obtain this value. We will discuss this point further in Section 9.4, but for now remember that the *t* random variable is closely approximated by the standard normal *Z* when using a large sample.

Interpreting the *p* Value

We will consider two ways of using the *p* value to arrive at a conclusion. The first is the classical approach that we have used up to this point: We choose a value for α and base our decision on this value. When using a *p* value in this manner, the procedure is:

Reject H_0 if *p* value $< \alpha$

FTR H_0 if *p* value $\geq \alpha$

The second approach is a general rule of thumb that applies to most applications of hypothesis testing on a population mean. We previously stated that typical values of α range from .01 to .10. This implies that for most applications we will not see values of α smaller than .01 or larger than .1. With this in mind, the following rule can be defined:

Reject H_0 if the p value is small $(p < .01)$

FTR H_0 if the p value is large $(p > .1)$

Consequently, if $.01 \leqslant p$ value $\leqslant .1$, the data are inconclusive.

The advantage of this approach is that you avoid having to choose a value of α; the disadvantage is that you may arrive at an inconclusive result.

Now for a brief disclaimer: This rule does not apply to all situations. If a Type I error would be extremely serious, and you prefer a very small value of α when using the classical approach, then you can lower the .01 limit. Similarly, you might raise the .1 limit if the Type II error is extremely critical and you prefer a large value for α. However, this gives a working procedure for most situations you are likely to encounter.

What can you conclude if the p value is $p = .0001$? This value is extremely small as compared with any reasonable value of α. So we would strongly reject H_0. Consequently if, for example, you are making a life threatening decision based on these results, you can breathe a little easier. This data set supports H_a overwhelmingly. On the other hand, if $p = .65$, this value is large as compared with any reasonable value of α. Without question, we would fail to reject H_0.

There is yet one other interpretation of the p value. Essentially, a p value measures how unlikely a particular value of the test statistic is if the null hypothesis is true. This is summarized in the box below.

Another Interpretation of the p Value

1   For a two-tailed test where H_a: $\mu \neq \mu_0$, the p value is the probability that the value of the test statistic, Z^*, will be at least as large (in absolute value) as the observed Z^*, if μ is in fact equal to μ_0.

2   For a one-tailed test where H_a: $\mu > \mu_0$, the p value is the probability that the value of the test statistic, Z^*, will be at least as large as the observed Z^*, if μ is in fact equal to μ_0.

3   For a one-tailed test where H_a: $\mu < \mu_0$, the p value is the probability that the value of the test statistic, Z^*, will be at least as small as the observed Z^*, if μ is in fact equal to μ_0.

In example 9.7, we determined the p value to be .063; the computed value of the test statistic was $Z^* = -1.53$; the hypotheses were H_0: $\mu \geqslant 125$ and H_a: $\mu < 125$. So the probability of observing a value of Z^* as small as -1.53 (that is, $Z^* \leqslant -1.53$) if μ is 125 is $p = .063$.

Based on this description of the p value, if p is small, conclude that H_0 is not true and reject it. We obtain precisely the same result using the classical and rule-of-thumb options of the p value. Small values of p favor H_a, and large values favor H_0.

EXAMPLE 9.8

In example 9.6, we performed a one-tailed test of H_0: $\mu \geqslant 48{,}000$ and H_a: $\mu < 48{,}000$. The sample results were $n = 45$, $\bar{x} = \$43{,}900$, and $s = \$7140$. The calculated value of the test statistic was

$$Z^* = \frac{43{,}900 - 48{,}000}{s_{\bar{X}}}$$

where

$$s_{\bar{X}} = \frac{s}{\sqrt{n}} \cdot \sqrt{\frac{N-n}{N-1}}$$

$$= \frac{7140}{\sqrt{45}} \sqrt{\frac{350-45}{350-1}} = 995.01$$

so $Z^* = -4.12$.

1 What is your conclusion based on the corresponding p value, using $\alpha = .05$?

2 Without specifying a value of α, what would be your conclusion based on the calculated p value?

3 Interpret the p value for this application.

Figure 9.14

Illustration for the p value for example 9.8

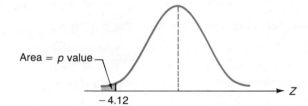

Solution 1 We are unable to determine the p value exactly using Table A-4; however, this area is roughly the same as the area to the left of -4.0 under the Z curve—namely, $.5 - .49997 = .00003$. So $p \cong .00003$. Because p is less than $\alpha = .05$, we reject H_0. Our conclusion is the same as that of example 9.6, which also used $\alpha = .05$, where we concluded that the female managers were underpaid.

Solution 2 We use the general rule of thumb for interpreting the p value. Since $p \cong .00003$; it is extremely small, so we strongly reject H_0 (same conclusion as Solution 1).

Solution 3 We can make the following statements:

1 The significance level at which the conclusion indicated by the testing procedure changes is $\alpha = .00003$.

2 The smallest significance level for which you can reject the null hypothesis is $\alpha = .00003$.

3 The probability of observing a value of the test statistic as small as the one obtained ($\leqslant -4.12$) is .00003, if in fact the population mean is $48,000. □

EXERCISES

9.30 State whether you would reject or fail to reject the null hypothesis in each of the following cases.
   (a) $p = .12$, $\quad \alpha = .05$
   (b) $p = .03$, $\quad \alpha = .05$
   (c) $p = .001$, $\quad \alpha = .01$
   (d) $p = .01$, $\quad \alpha = .001$

9.31 Using the rule of thumb option (not selecting a value of α) in the interpretation of the p value, state whether the test statistic would be statistically significant in the following situations.
   (a) $p = .57$
   (b) $p = .008$
   (c) $p = .12$
   (d) $p = .04$

9.32 State whether the following statements are true or false, and explain your answer where appropriate.
   (a) The p value is determined a priori, whereas the significance level is established a posteriori. (Look up these words in the dictionary if you do not know their meanings; they are useful to have in your vocabulary.)
   (b) In a two-tailed test conducted with a 5% significance level, a p value of .024 indicates that the null hypothesis is rejected.
   (c) In a one-tailed test conducted with 5% significance, a p value of .024 indicates that the null hypothesis is rejected.

9.33 Find p values for the following situations with calculated test statistics given by Z^*.
   (a) H_0: $\mu = 30$, $\quad H_a$: $\mu \neq 30$, $\quad Z^* = 2.38$
   (b) H_0: $\mu \leqslant 20$, $\quad H_a$: $\mu > 20$, $\quad Z^* = 1.645$
   (c) H_0: $\mu \geqslant 15$, $\quad H_a$: $\mu < 15$, $\quad Z^* = -2.54$
   (d) H_0: $\mu = 50$, $\quad H_a$: $\mu \neq 50$, $\quad Z^* = -1.85$

9.34 Test the null hypothesis that the mean of a normally distributed population is less than or equal to 20, assuming that a sample of size 60 yields the statistics

$$\bar{x} = 20.4$$

$$s = 3.0$$

9.35 You will recall that in exercise 9.8, part (a) used $\alpha = .01$ and part (b) used $\alpha = .10$. Do the same exercise using the p value criterion and compare the two approaches.

9.36 Find the p values for the following:
   (a) Exercise 9.7
   (b) Exercise 9.15
   (c) Exercise 9.19, parts (a) and (b)
   (d) Exercise 9.23

9.4   HYPOTHESIS TESTING ON THE MEAN OF A NORMAL POPULATION: SMALL SAMPLE

Our approach to hypothesis testing with small samples when the standard deviation σ is unknown uses the same technique we used for dealing with confidence intervals on the mean of a population: We switch from the standard normal distribution Z to the t distribution. However, we need to examine the distribution of the population when the sample is small—the population distribution determines the procedure that we use. In this section, we have reason to believe that the population has a normal distribution.

Certain variations from a normal population are permissible with the small-sample test. If a test of hypothesis is still reliable when slight departures from the assumptions are encountered, this test is said to be **robust**. If you believe the parent population to be reasonably symmetric, the level of your confidence interval and Type I error (α) will be quite accurate, even if the population has heavy tails (unlike the normal distribution), as shown in Figure 9.15A. However, when using small samples, the small-sample test is not robust for populations that are heavily skewed (see Figure 9.15B). For larger sample sizes, a histogram of your data often can detect whether a population is heavily skewed in one direction.

To reemphasize, the discussion in this section will assume a normal population. In other words, if X is an observation from this population, then X is a normal random variable with unknown mean μ. Also, we will assume that σ is unknown. (If σ is known, your resulting test statistic is $Z = (\bar{X} - \mu_0)/(\sigma/\sqrt{n})$, and the five-step procedure of Section 9.1 allows you to do hypothesis testing on μ.)

The only distinction between using a small and a large sample is the form of the test statistic. Using the discussion from Chapter 8, if we define the test statistic as

$$t = \frac{\bar{X} - \mu_0}{s/\sqrt{n}} \qquad (9\text{-}3)$$

Figure 9.15

(A) Small sample test is valid.
(B) Small sample test is not valid.

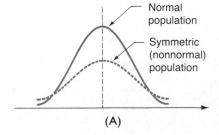

(A)

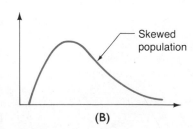

(B)

we now have a t distribution with $n - 1$ degrees of freedom (df). The procedure to use for testing $H_0: \mu = \mu_0$ and $H_a: \mu \neq \mu_0$ is the same five-step procedure, except that the rejection region is defined using the t table (Table A-5) rather than the Z table (Table A-4). This also applies to a one-tailed test. Because we are looking at very small samples (typically $n < 30$), we can ignore the fpc factor.

EXAMPLE 9.9

You may recall from Section 8.2, that a company called Toddle-Teach manufactures an educational toy consisting of a puzzle that children are instructed to assemble as quickly as they can. The national norm (population mean) for 5-year-olds, according to Toddle-Teach, is 20 minutes to assemble the puzzle. The assembly times are believed to follow a normal distribution. A local private school claims that their 5-year-olds are exceptionally bright and, in fact, have a mean assembly time that is less than 20 minutes. You obtain a random sample of 18 five-year-olds, which produces a sample mean of $\bar{x} = 18.73$ minutes and a sample standard deviation of $s = 2.68$ minutes. Are you able to support the claim of the private school using these results?

Solution

Step 1 What you are attempting to demonstrate goes into the alternative hypothesis H_a, so the directional hypotheses are $H_0: \mu \geqslant 20$ and $H_a: \mu < 20$.

Step 2 The test statistic here is

$$t = \frac{\bar{X} - 20}{s/\sqrt{n}}$$

Step 3 The implications of making a Type I error (rejecting a correct H_0) and a Type II error (FTR an incorrect H_0) appear to be the same, so you decide on a significance level of $\alpha = .05$.

As before, we will reject H_0 when the value of the test statistic lies in the left tail (Figure 9.16):

$$\text{Reject } H_0 \text{ if } t < -t_{.05,17} = -1.74$$

because $df = n - 1 = 17$.

Figure 9.16

t distribution; the rejection region is the lightly shaded area to the left of −1.74, for example 9.9

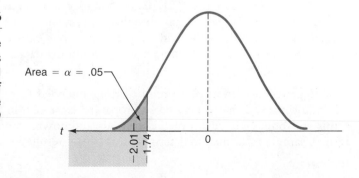

Area = α = .05

Figure 9.17

t curve with 17 df.
The p value is the
area to the left of
$t^* = -2.01$, so we
can say only that it
is between .025
and .05.

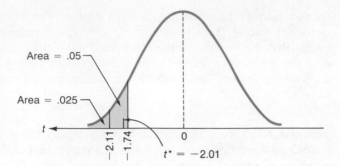

Step 4 For this data set, $n = 18$, $\bar{x} = 18.73$ minutes, and $s = 2.68$ minutes. The
value of your test statistic is

$$t^* = \frac{18.73 - 20}{2.68/\sqrt{18}} = -2.01$$

Because $-2.01 < -1.74$, we reject H_0.

Step 5 The average puzzle assembly time for 5-year-old children at this school
is less than 20 minutes.   □

What is the p value in example 9.9, and what can we conclude based on this
value? We run into a slight snag when dealing with the t distribution because
we are not able to determine precisely the p value. You can see this in Figure
9.17, using Table A-5 (17 df). The p value is the area to the left of $t^* = -2.01$.
The best we can do here is to say that p is between .025 and .05. (*Note:* A
reliable computer package will provide the exact p value.)

Using the classical approach and $\alpha = .05$ we *can* say that p is less than .05,
despite not knowing p exactly. Consequently, we reject H_0. This procedure
always produces the same result as the five-step procedure. Notice that this conclusion does not tell us how strongly we reject H_0, but simply that H_0 *is* rejected
at this significance level.

Suppose we choose not to select a significance level (α) but prefer to base
our conclusion strictly on the calculated p value. We use the rule of thumb and
decide whether p is small ($<.01$), large ($>.1$), or in between. Despite not having
an exact value of p, we can say that this p value falls in the inconclusive range.
These data values do not provide us with any strong conclusion. Your advice to
Toddle–Teach would be to obtain some additional data.

EXAMPLE 9.10 A suggested side effect of a proposed drug intended to reduce blood pressure
in adult males is that it may be accompanied by a gain in body weight. Since
this and other side effects of the drug are unknown, it was decided to use a
small sample of 12 men to determine if the drug resulted in a noticeable weight

gain, defined to be 5 lb or more. Based upon similar experiments in the past, the population of weight gains was believed to be normally distributed. A researcher obtained a sample by randomly selecting men who weighed between 150 and 175 lb. These individuals were made fully aware of the nature and risks of the experiment. The resulting weight gains are (a negative value indicates a weight loss):

$$15.3, \ 12.9, \ -3.2, \ 16.4, \ 4.3, \ 14.6, \ 15.0, \ -2.1, \ 15.5, \ 7.2, \ 9.1, \ 15.2$$

So $n = 12$, $\bar{x} = 10.017$ lb, and $s = 7.024$ lb. Set up the appropriate hypotheses and test them using a significance level of $\alpha = .05$.

Solution

Step 1 Since the purpose of the experiment is to determine if there is a noticeable weight gain, that is, more than 5 lb, the hypotheses are H_0: $\mu \leqslant 5$ and H_a: $\mu > 5$, where μ is the average weight gain for all men in this weight class (between 150 and 175 lb).

Steps 2, 3

$$\text{Reject } H_0 \text{ if } t = \frac{\bar{X} - 5}{s/\sqrt{n}} > t_{.05,11} = 1.796$$

where the *df* are $n - 1 = 11$.

Step 4 The calculated t is

$$t^* = \frac{10.017 - 5}{7.024/\sqrt{12}} = 2.47$$

Because 2.47 exceeds the tabulated value of 1.796, we reject H_0. Also, the *p* value (using Table A-5 and 11 *df*) is the area to the right of 2.47. This is between .01 and .025, which is less than $\alpha = .05$, and so (as before) we reject H_0.

Step 5 These data support the claim that the average gain in weight for men in this weight class is more than 5 lb.   □

A MINITAB solution for this example is contained in Figure 9.18. Note that the calculated (exact) *p* value is .015.

Figure 9.18

MINITAB solution for example 9.10

```
MTB > SET INTO C1
DATA> 15.3 12.9 -3.2 16.4 4.3 14.6 15.0 -2.1 15.5 7.2 9.1 15.2
DATA> END
MTB > TTEST OF MU=5 USING C1;
SUBC> ALTERNATIVE=1.

    TEST OF MU  =   5.00 VS MU G.T.   5.00

                 N       MEAN     STDEV    SE MEAN         T    P VALUE
    C1          12      10.02      7.02       2.03      2.47      0.015
```

Small Sample Tests on a Normal Population Mean

Two-Tailed Test

H_0: $\mu = \mu_0$

H_a: $\mu \neq \mu_0$

Reject H_0 if $|t| > t_{\alpha/2, n-1}$
where n = sample size and

$$t = \frac{\bar{X} - \mu_0}{s/\sqrt{n}}$$

One-Tailed Test

H_0: $\mu \leq \mu_0$ H_0: $\mu \geq \mu_0$

H_a: $\mu > \mu_0$ H_a: $\mu < \mu_0$

Reject H_0 if $t > t_{\alpha, n-1}$ Reject H_0 if $t < -t_{\alpha, n-1}$

EXERCISES

9.37 Summarize the rules as to when to use the Z^* statistic or the t^* statistic to conduct any hypothesis test on the mean of a population.
   (a) When would the use of either Z or t or both be invalid?
   (b) When must you use the finite population correction (fpc) factor, and when could you ignore it?

9.38 Computer tomographic (CT) scanning is a useful, albeit somewhat expensive, diagnostic test for patients with cerebrovascular disease. Hazelton and Earnest (1987) studied two groups of patients: the first group of 93 patients had cerebral infarctions (strokes) treated before CT scanning was available, and the other group of 92 patients had CT scanning as part of a diagnostic evaluation. Among other things, the researchers reported the following clinical data:

Mean Blood Pressure[a] (mm Hg)	Pre-CT Group ($n = 93$)	Post-CT Group ($n = 92$)
Systolic	162 ± 4.0	157 ± 3.6
Diastolic	93 ± 1.7	90 ± 2.0

[a] The blood pressure is given as mean $\pm$ standard error.

Source: A. E. Hazelton and M. P. Earnest. Impact of computed tomography on stroke management and outcome. *Arch. Internol Med.* **147** (2): 218 (February 1987). Table I.

   (a) At $\alpha = .05$, does the above data support the statement: "The average systolic blood pressure of the pre-CT patients is greater than 150"? State the p value for the test.

(b) At $\alpha = .01$, does the above data support the statement: "The average diastolic blood pressure of the post–CT patients is less than 95"? State the p value for the test.

9.39 A school superintendent obtained the following data and explained that they were "school ratings" made by 11 randomly selected teachers in his district.

Teacher	Rating
1	6.21
2	6.53
3	5.31
4	4.19
5	5.17
6	4.69
7	5.09
8	4.12
9	3.84
10	2.95
11	4.75

Handwritten:
$$\bar{x} = \frac{\Sigma x}{n} = \frac{52.85}{11} = 4.805$$
$$S = \sqrt{\frac{\Sigma(x-\bar{x})^2}{n-1}} =$$

The ratings were on a scale of 1 to 8, and were summarized from a questionnaire. Assume that ratings are normally distributed.

(a) Calculate the mean and standard deviation of the above data.

(b) Test the hypothesis that the average rating for the district is 5.0, using a 5% significance level. Also, determine the p value for the test.

(c) Test the hypothesis that the average rating for the district is less than 5.0 using a 5% significance level. Also, find the p value for this test.

(d) If you could not assume a normally distributed population, are the above tests still valid? (*Hint:* Consider robustness.)

Handwritten (left margin):
$$H_0 : \mu = 5$$
$$H_a : \mu \neq 5$$
$$n = 11$$
$$t = \frac{\bar{x} - \mu}{\frac{s}{\sqrt{n}}}$$

9.40 In a study of pituitary adrenocortical response in depressed patients, Holsboer et al. (1987) reported, among other things, the following corticosterone responses in a pituitary adrenocortical response to a 100-μg bolus injection of human CRH in 10 patients with recurrent major depressive episode:

Subject No.	Corticosterone (ng/ml/min)
1	480
2	1,360
3	768
4	711
5	590
6	785
7	324
8	591
9	339
10	790

Mean = 674, Standard deviation = 297

Source: F. Holsboer, A. Gerken, G. K. Stalla, and O. A. Muller. Blunted aldosterone and ACTH release after human CRH administration in depressed patients. *Am. J. Psychiatry*–**144** (2): 230 (February 1987), Table 1.

(a) Do the above data provide sufficient evidence to support a claim that the corticosterone response in depressed patients is less than 750 ng/ml/min? Conduct a hypothesis test at a 10% level of significance.

(b) State the p value for the test.

(c) Based on the p value, would the decision change if the significance level was 5% or 1%?

9.41 Find the p value for the following situations with calculated test statistics given by t^*.

(a) H_0: $\mu = 40$, H_a: $\mu \neq 40$, $t^* = 2.30$, $n = 12$
(b) H_0: $\mu \leqslant 13.6$, H_a: $\mu > 13.6$, $t^* = 2.73$, $n = 19$
(c) H_0: $\mu \geqslant 100.80$, H_a: $\mu < 100.80$, $t^* = 1.25$, $n = 20$
(d) H_0: $\mu = 35.6$, H_a: $\mu \neq 35.6$, $t^* = 1.57$, $n = 11$

9.42 Carry out the hypothesis test for the mean of a normally distributed population given the following information:

(a) H_0: $\mu \leqslant 1.6$
H_a: $\mu > 1.6$
$n = 15$ $\bar{x} = 1.8$ $s^2 = 1.7$ $\alpha = .10$

(b) Find the p value.

9.43 A sample of size 12 is drawn from a finite population of size 300. The finite population can be approximated by a normal distribution. The sample mean and sample standard deviation are 100.6 and 3.7, respectively.

(a) Test the null hypothesis that the mean of the population is equal to 107. Use a 10% significance level.

(b) Find the p value.

9.44 Refer to exercise 9.43. Test the null hypothesis that the mean of the population is equal to 107, if it is known that the population standard deviation is equal to 4. Find the p value.

9.45 The senior executive of a publishing firm would like to train his employees to read faster than 1000 words per minute. A random sample of 21 employees underwent a special speed-reading course. This sample yielded a mean of 1018 words per minute with a standard deviation of 30 words per minute. Using a significance level of .05, test the belief that the speed-reading course will enable the employees to read more than 1000 words per minute. Assume that the reading speeds of persons who have taken the course are normally distributed. Find the p value.

9.46 It is believed that the mean score on an aptitude test of engineers graduating from Safire University is greater than 180. Assuming the scores are normally distributed, test the belief if a random sample of 26 engineers yielded a mean score of 186 with a standard deviation of 10.2. Use the p value to form your conclusion.

9.47 Cyclooxygenase inhibitors are a class of medical drugs. Perlman, Johnson, and Malik (1987) studied pulmonary vascular responses in sheep that were subjected to pulmonary microembolism by means of alpha-thrombin challenge. The study was fairly complicated and involved comparisons between a control group of sheep and two other groups pretreated with two different cyclooxygenase inhibitors, ibuprofen and meclofenamate. Formal comparisons between groups will be studied by you in the next and later chapters, but you can consider the *[handwritten top-left: $t_{.05,2}$ ↓ $\bar{x} \pm 2.920 \frac{s}{\sqrt{n}}$]*

following results for each group:

NEUTROPHIL SUPEROXIDE ANION PRODUCTION INDUCED BY PHORBOL MYRISTATE ACETATE (PMA) IN THE PRESENCE OF IBUPROFEN AND MECLOFENAMATE

Stimulus	O_2 Production[a] $(\text{nmole} \cdot 10^6 \text{ cells}^{-1} \cdot 10 \text{ min}^{-1})$
PMA Control	15.9 ± 3.7
PMA + meclofenamate (0.05 mg/ml)	9.5 ± 2.1
PMA + ibuprofen (0.04 mg/ml)	14.7 ± 6.2

[handwritten right column: 95% C.I. 26.704 – 5.096 ; 15.632 – 3.368 ; 32.804 – –3.404]

[a] Results are given as mean $\pm$ standard error; $n = 3$.

Source: M. B. Perlman. A. Johnson, and A. B. Malik. Ibuprofen prevents thrombin-induced lung vascular injury: Mechanism of effect. *Am. J. Physiol.: Heart Circ. Physiol.* **21** (3): H611 (March 1987), Table 4.

[handwritten left margin:
$H_0 : \mu \geq 15$ reject if
$H_a : \mu < 15$, $t < -2.920$
$t = \dfrac{9.5-15}{2.1} = -2.62$
$t_{.05,2} = 2.920$]

(a) Express the above as 90% confidence intervals.

(b) At $\alpha = .05$, test the hypothesis that neutrophil superoxide anion production is less than 15.0

    (i) For the PMA + meclofenamate group *[handwritten: FTR H_0 since $-2.62 < 2.920$]*

    (ii) For the PMA + ibuprofen group

9.48 The comptroller of National Insurance states that the average claim against the company for an automobile accident is less than $4500. A random sample of 14 claims yielded a mean amount of $4200 with a sample standard deviation of $171. It is believed that the claims are normally distributed. Use the p value criteria to test the comptroller's statement.

9.49 A fifth-grade teacher obtained the following results on the California Test of Mental Maturity (CTMM) from a sample of 12 fifth-graders:

$$\bar{x} = 101.5 \qquad s = 12.7$$

(a) Use the above data to test the hypothesis that the mean score on the CTMM for fifth-graders is greater than 100, at a 5% significance level. Assume a normally distributed population.

(b) State the p value for this test.

9.5 INFERENCE FOR THE VARIANCE AND STANDARD DEVIATION OF A NORMAL POPULATION (OPTIONAL)

Our discussion in Chapters 8 and 9 has been concerned with the mean of a particular random variable or population. In other words, we are trying to decide or estimate what is occurring on the average. Suppose someone involved with a production process that is manufacturing 2-inch bolts, has just been informed that, without a doubt, these bolts are 2 inches long, on the average.

Is there anything else you might like to know about this process? Suppose that one-half of the bolts produced are 1 inch long and the other half are 3 inches. The report was accurate—on the average, the bolts are 2 inches long.* However, such a production process certainly will not satisfy customers, and this company will soon be out of the bolt business.

What was missing in the report was the amount of variation in this production process. If the variation was zero, then every bolt would be exactly 2 inches long—an ideal situation. In practice, there always will be a certain amount of variation in any production process. So we are concerned about not only the mean length μ of the population of bolts but also the variance σ^2 or standard deviation σ of the lengths of these bolts. If the variance is too large, the process is not operating correctly and needs adjustment.

The variance of a population also is of vital interest to someone designing or examining an evaluation instrument. Suppose that a psychological testing device designed to measure marital satisfaction has a possible range of scores from 50 to 500, but nearly everyone using the instrument scores between 300 and 325. Due to the lack of variation in these scores, it is very difficult, if not impossible, to distinguish between those couples who are happily married and those that are not.

In the inference procedures for a population variance (and standard deviation) to follow, we will assume that the population of interest is normally distributed. Unlike the t test, the hypothesis testing procedures and confidence intervals for the variance are very sensitive to departures from the normal population—notably, heavy tails in the distribution or heavy skewness will have a large effect. In other words, the following tests of hypothesis are less robust than are those we discussed earlier.

Confidence Interval for the Variance and Standard Deviation

The point estimate of a population variance is the obvious one—namely, the sample variance. This was discussed in Chapter 8, where we used the variance s^2 of a sample to estimate the variance σ^2 of the much larger population.

When constructing a confidence interval for μ using a small sample, we used the t distribution. Such a distribution is referred to as a **derived distribution** because it was derived to describe the behavior of a particular test statistic. This type of distribution is not used to describe a population, as is the normal distribution in many applications. For example, you will not hear a statement such as "assume that these data follow a t distribution"—normal, uniform, or exponential, maybe, but not a t distribution. The t random variable merely offers us a method of testing and constructing confidence intervals for the mean of a normal population when the standard deviation is unknown, and is replaced by its estimate.

* A statistician is often described as someone who thinks that if one-half of you is in an oven and the other half is in a deep freeze, on the average you are very comfortable.

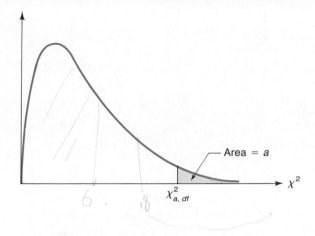

Figure 9.19

Shape of a chi-square
distribution

Another such continuous derived distribution allows us to determine confidence intervals and perform tests of hypothesis on the variance and standard deviation of a normal population. This is the **chi-square** (pronounced ky) distribution, written as χ^2. The shape of this distribution is illustrated in Figure 9.19. Notice that, unlike the Z and t curves, the χ^2 distribution is not symmetric and is definitely skewed right.

For chi-square, as with all continuous distributions, a probability corresponds to an area under a curve. Also, the shape of the chi-square curve, like that of its cousin the t distribution, depends on the sample size n. As before, this will be specified by the corresponding degrees of freedom (df).

When using the χ^2 distribution to construct a confidence interval or perform a test of hypothesis on a population variance or standard deviation, the degrees of freedom are given by

$$df = n - 1$$

Let $\chi^2_{a,df}$ be the χ^2 value whose area to the right is a, using the proper df.

EXAMPLE 9.11   Using a chi-square curve with 12 df, determine $P(\chi^2 > 18.5494)$ and $P(\chi^2 < 6.30380)$.

Solution   Tabulated values for the χ^2 distribution are contained in Table A–6 on page A–15. This table contains right-tail areas (probabilities). Based on this table (see Figure 9.20),

$$P(\chi^2 > 18.5494) = .1$$

This can be written as

$$\chi^2_{.1,12} = 18.5494$$

Figure 9.20

χ^2 curve with
12 *df*. The shaded
area represents
$P(\chi^2 > 18.5494)$.

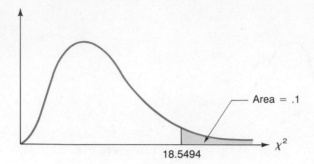

For $\chi^2 = 6.30380$, Table A-6 informs us that the area to the right of 6.30380 is .900. Because the total area is 1, the area to the left of 6.30380 is $1 - .900 = .1$, and so $P(\chi^2 < 6.30380) = .1$. As a result, we can say that

$$P(6.30380 \leqslant \chi^2 \leqslant 18.5494) = 1 - .1 - .1 = .8$$

That is, 80% of the time a χ^2 value (with 12 *df*) will be between 6.30380 and 18.5494.   □

EXAMPLE 9.12   Using example 9.11, determine a and b that satisfy

$$P(a < \chi^2 < b) = .95, \quad \text{with } df = 12.$$

Choose a and b so that an equal area occurs in each tail.

Solution   Figure 9.21 shows the areas for a and b. Using Table A-6,

$a =$ the χ^2 value whose left-tailed area is .025
    $=$ the χ^2 value whose area to the right is .975
    $= 4.40379$

Figure 9.21

χ^2 curve with 12 *df*

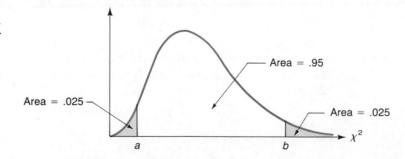

and

$$b = \text{the } \chi^2 \text{ value whose right-tailed area is } .025$$
$$= 23.3367 \qquad \qquad \square$$

To derive a confidence interval for σ^2, we need to examine the sampling distribution of s^2. If we repeatedly obtained a random sample from a normal population with mean μ and variance σ^2, calculated the sample variance s^2, and made a histogram of these s^2 values, what would be the resulting shape of this histogram? It can be shown that the shape will depend on the sample size n and the value of σ^2 but not on the value of the population mean μ. In fact, the values of n and σ^2, along with the random variable s^2, can be combined to define a chi-square random variable, given by

$$\chi^2 = \frac{(n-1)s^2}{\sigma^2} \qquad \qquad \textbf{(9-4)}$$

having a chi-square distribution with $n - 1$ degrees of freedom (df). Therefore, the sampling distribution for s^2 can be defined using the chi-square distribution in equation 9-4.

For example, a sample size of $n = 13$ results in 12 df. From example 9.12, it follows that

$$P(4.40379 < \chi^2 < 23.3367) = .95$$

So,

$$P\left[4.40379 < \frac{12s^2}{\sigma^2} < 23.3367 \right] = .95 \quad \text{using equation 9-4}$$

or

$$P\left[\frac{12s^2}{23.3367} < \sigma^2 < \frac{12s^2}{4.40379} \right] = .95$$

As in all confidence interval constructions, the parameter (σ^2) is bounded between two limits defined by a random variable (s^2). This means that a 95% confidence interval for σ^2 is

$$\frac{12s^2}{23.3367} \quad \text{to} \quad \frac{12s^2}{4.40379}$$

In general, the following procedure can be used to construct a confidence interval for σ^2 or σ. A $(1 - \alpha) \cdot 100\%$ confidence interval for σ^2 is

$$\frac{(n-1)s^2}{\chi^2_{\alpha/2,n-1}} \quad \text{to} \quad \frac{(n-1)s^2}{\chi^2_{1-\alpha/2,n-1}} \tag{9-5}$$

The corresponding confidence interval for σ is

$$\sqrt{\frac{(n-1)s^2}{\chi^2_{\alpha/2,n-1}}} \quad \text{to} \quad \sqrt{\frac{(n-1)s^2}{\chi^2_{1-\alpha/2,n-1}}} \tag{9-6}$$

EXAMPLE 9.13

A practicing therapist is interested in a new instrument designed to measure marital satisfaction. She prefers to use an instrument that has a large amount of variation in the scores of people completing it, so that she can more clearly distinguish between happily and unhappily married couples. She administers the instrument to a random sample of 15 couples. The following scores (believed to come from a normal population) were obtained.

221, 311, 286, 392, 412, 187, 346, 305, 280, 446, 351, 320, 237, 276, 381

For these data, $\bar{x} = 316.73$ and $s = 72.75$. Determine a 90% confidence interval for σ^2 and for σ.

Solution

The corresponding 90% confidence interval for σ^2 is

$$\frac{(15-1)(72.75)^2}{\chi^2_{.05,14}} \quad \text{to} \quad \frac{(15-1)(72.75)^2}{\chi^2_{.95,14}} = \frac{(14)(72.75)^2}{23.6848} \quad \text{to} \quad \frac{(14)(72.75)^2}{6.57063}$$

$$= 3128.4 \quad \text{to} \quad 11{,}276.8$$

The 90% confidence interval for σ would be

$$\sqrt{3128.4} \quad \text{to} \quad \sqrt{11{,}276.8}$$

that is, 55.9 points to 106.2 points.

Hypothesis Testing for the Variance and Standard Deviation

For many applications, we are concerned that the standard deviation or variance of our population may be exceeding some specified value. If this claim is supported, then, for example, you may elect to use a particular psychological or educational instrument, since the scores exhibit enough variation. As you could with the tests of hypothesis examined so far, you can perform a two-tailed test where either too much variation or too little variation is the topic of concern. This is, however, not the usual case.

Hypothesis Testing on σ^2

Two-Tailed Test

H_0: $\sigma^2 = \sigma_0^2$

H_a: $\sigma^2 \neq \sigma_0^2$

Test statistic: $\chi^2 = \dfrac{(n-1)s^2}{\sigma_0^2}$

Reject H_0 if $\chi^2 > \chi^2_{\alpha/2,n-1}$
or if $\chi^2 < \chi^2_{1-\alpha/2,n-1}$

One-Tailed Test

H_0: $\sigma^2 \leqslant \sigma_0^2$ H_0: $\sigma^2 \geqslant \sigma_0^2$

H_a: $\sigma^2 > \sigma_0^2$ H_a: $\sigma^2 < \sigma_0^2$

Reject H_0 if $\chi^2 > \chi^2_{\alpha,n-1}$ Reject H_0 if $\chi^2 < \chi^2_{1-\alpha,n-1}$

EXAMPLE 9.14

Example 9.13 was concerned with the variation in the scores of a marital satisfaction scale. Prior to obtaining the data, the therapist decided that she would use this instrument if the sample results support the hypothesis that the standard deviation of the population is more than 50 points. Using the 15 observations in example 9.13, how would you advise the therapist, using a significance level of .10?

Solution

Step 1 The appropriate hypotheses are H_0: $\sigma \leqslant 50$ and H_a: $\sigma > 50$. (Note that these hypotheses are precisely the same as H_0: $\sigma^2 \leqslant 2500$ and H_a: $\sigma^2 > 2500$. Whether you write H_0 and H_a in terms of σ or σ^2 does not matter; the testing procedure is the same in either case.)

Step 2 The test statistic is

$$\chi^2 = \frac{(15-1)s^2}{(50)^2} = \frac{14s^2}{2500}$$

which has a chi-square distribution with 14 *df*.

Step 3 Using $\alpha = .1$ and Table A-6, the rejection region for this test is

$$\text{Reject } H_0 \text{ if } \chi^2 > 21.0642$$

Step 4   The computed value using the sample data is

$$\chi^{2*} = \frac{(15-1)(72.75)^2}{(50)^2} = 29.64$$

Since $29.64 > 21.0642$, we reject H_0.

Step 5   We conclude that σ is larger than 50 points. There appears to be sufficient variation in the scores of this instrument.

Note that the p value for the test of hypothesis in example 9.14 is the area to the right of 29.64 under the χ^2 curve with 14 df.

Figure 9.22

Illustration for the p value for example 9.14

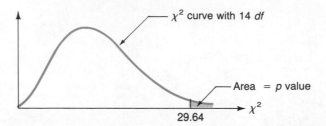

All we are able to determine about this value using Table A-6 is that it is between .005 and .01. Using this information, we arrive at the same decision—namely, reject H_0—because (1) using the classical approach, p is less than $\alpha = .10$, or (2) the p value is extremely small ($<.01$) by the general rule of thumb described in Section 9.3.   □

EXAMPLE 9.15

A company selling decorative wood chips in 30–lb bags is concerned about the variation in the weight of these bags because they have recently acquired a new mechanical packaging device. Based on earlier production tests, management is convinced that the average weight of all bags being produced is, in fact, 30 lb and the bag weights follow a normal distribution. However, the production supervisor has been informed that at least 95% of the bags produced *must* be within 1 lb of the specified weight (30 lb). A random sample of 20 bags was obtained with the following results:

weight of bags (lb):   31.2, 27.5, 30.8, 31.5, 29.5, 31.1, 31.3, 30.7, 26.7, 29.2, 32.1, 28.3, 31.6, 29.2, 31.5, 29.7, 30.4, 31.0, 29.1, 30.5

For these data $\bar{x} = 30.145$ lb and $s = 1.458$ lb. Using a significance level of $\alpha = .05$, what can we conclude?

What is the supervisor being told about σ? Remember that, for a normal population, approximately 95% of the observations will lie within two standard deviations of the mean (empirical rule, Chapter 3). So, if two standard deviations are the same as 1 lb, then the supervisor is being told that σ must be no more than .5 lb. Is there any evidence to conclude that this is not the case—that is, that σ is larger than .5 lb? Let's investigate.

Solution

Step 1 The appropriate hypotheses are $H_0: \sigma \leqslant .5$ and $H_a: \sigma > .5$, where H_a states that production is not meeting required standards.

Step 2 The test statistic is

$$\chi^2 = \frac{(20-1)s^2}{(.5)^2} = \frac{19s^2}{.25}$$

which has a chi-square distribution with 19 *df*.

Step 3 The rejection region for this test (using $\alpha = .05$ and Table A-6) is

$$\text{Reject } H_0 \text{ if } \chi^2 > 30.1435$$

Step 4 The computed value of the test statistic is

$$\chi^{2*} = \frac{(20-1)(1.458)^2}{(.5)^2} = 161.56$$

Since $161.56 > 30.1435$, we reject H_0. This is hardly a surprising result; the point estimate of σ is $s = 1.458$, quite a bit larger than .5.

Step 5 We conclude rather convincingly that σ is larger than .5. The bagging procedure has far too much variation in the weight of the bags produced. □

EXERCISES

9.50 From the tabulated values for the chi-squared distribution, find the following values and indicate graphically where the values fall with respect to other values of the chi-squared distribution.

(a) $\chi^2_{.10,10}$
(b) $\chi^2_{.025,30}$
(c) $\chi^2_{.95,15}$
(d) $\chi^2_{.01,26}$

9.51 A sample of size 25 from a normally distributed population yields a sample standard deviation of 12.8. At the 10% significance level, test the null hypothesis that the population standard deviation is less than or equal to 11.3. Find the p value.

9.52 A sample of size 15 from a normally distributed population yields the sample statistic

$$\Sigma(x - \bar{x})^2 = 180.3$$

(a) Construct a 90% confidence interval for the population variance.
(b) Construct a 90% confidence interval for the population standard deviation.
(c) Test the null hypothesis that the population variance is equal to ten. Use a two-tailed test and a 10% significance level.
(d) Find the p value for (c).

9.53 A sample of size five from a normally distributed population yields the following sample statistics:

$$\Sigma x^2 = 135 \qquad \Sigma x = 23$$

(a)   Construct a 95% confidence interval for the population variance.
(b)   Construct a 95% confidence interval for the population standard deviation.
(c)   Test the null hypothesis that the population standard deviation is equal to 2.8. Use a two-tailed test and a 5% significance level.
(d)   Find the p value for (c).

9.54 A highly sensitive device used to measure electrical activity in the brain is required to have a variance in measurements not exceeding 0.0001 volts^2. If the variance exceeds this tolerance, the device has to be serviced and recalibrated. A sample of 24 measurements taken with the device was found to have a variance of 0.000144 volts^2.

(a)   Conduct a hypothesis test to check whether the variance exceeds 0.0001 volts^2. Use a 5% significance level.
(b)   What assumptions are necessary regarding the population distribution?
(c)   State your conclusion: does the device have to be serviced and recalibrated?
(d)   Find the p value for the test.

9.55 For a random variable χ^2, which has a chi-square distribution with 18 degrees of freedom, determine the values of a and b such that $P(a < \chi^2 < b) = .90$ and such that the areas in each tail are equal, that is, $P(0 < \chi^2 < a) = P(b < \chi^2)$.

9.56 A psychologist has asked a group of experimental subjects to maintain a "satisfaction index." While the average is easily obtained, it is necessary that the total variation should also be kept to a minimum. (Namely, the standard deviation should be less than 5 points.) A random sample of 30 subjects resulted in a standard deviation of 7.1 points. Assuming a normal population,

(a)   Conduct the appropriate hypothesis test at $\alpha = .01$.
(b)   State the p value for the test.
(c)   If the variation is too large, the experiment should be abandoned. What does the above test indicate?

9.57 The salaries for mathematics teachers in secondary schools in Connecticut are believed to be normally distributed with a variance greater than 3000. Test this belief using the following sample statistics:

$$\Sigma(x - \bar{x})^2 = 45{,}130 \qquad n = 14$$

where X represents a math teacher's salary. Use a 1% significance level. Find the p value.

9.58 A medical researcher selects 15 rats at random and gives each the same dose of an antihypertensive drug. The blood pressure of the rats is then recorded. It is believed that the blood pressures are normally distributed. From the sample, the standard deviation of the blood pressures is found to be 3.86. The standard deviation of the blood pressures of the rats without the drug is known to be 4.16.

(a)   Test the hypothesis that the standard deviation of blood pressures for rats given the drug is less than 4.16. Use a significance level of 0.01.
(b)   Find the p value for the test.

9.59 With the widespread use of computer terminals in American society, numerous studies have investigated the interaction between human beings and machines. One study looked at (among other things) the degree of flicker and fuzziness that a human operator can tolerate for characters on the video screen. A scale of measure for flicker and fuzziness was devised by the researchers, who concluded that a variance of .25 was the optimum level. A manufacturer of video terminals claimed that his machines have precisely this level of variance (i.e., .25). A sample of 40 terminals from this manufacturer had a variance of .29. At a one percent significance level, is it possible to reject the manufacturer's claim?

SUMMARY

In Chapter 8 you were introduced to the topic of **statistical inference** by discussing the concept of estimating a population parameter (such as μ or σ) by using a corresponding sample estimate. The reliability of using the sample mean to estimate μ was measured using a confidence interval. This chapter presented the other side of statistical inference—**hypothesis testing** regarding this population parameter, along with a confidence interval for the population standard deviation σ.

For testing against a hypothetical value of the population mean (μ), we introduced a procedure that used the standard normal (Z) distribution for large samples ($n > 30$) and the t distribution for small samples. For small samples, the hypotheses are concerned with the mean of a normal population. However, we are able to discuss the mean of any continuous population when we have large samples by using the Central Limit Theorem.

The two hypotheses under investigation are the **null hypothesis**, H_0, and the **alternative hypothesis**, H_a. Typically, a claim that one is attempting to demonstrate goes into the alternative hypothesis.

Since any test of hypothesis uses a sample to infer something about a population, errors can result. Two specific errors are of great concern when you use the hypothesis-testing procedure. A **Type I error** occurs in the event you reject a null hypothesis when in fact it is true; a **Type II error** occurs when you fail to reject a null hypothesis when in fact it is not true.

The probability of a Type I error is the **significance level** of the test and is written as α. The probability of a Type II error is β; large values of β are associated with small values of α and vice versa. The value of $1 - \beta$ is the **power** of the statistical test and can be determined for each value of the parameter in the alternative hypothesis. To define a test of hypothesis, you select a value of α that considers the cost of rejecting a correct H_0 and failing to reject (FTR) an incorrect H_0. Typical values of α range from .01 to .1.

A five-step procedure was defined for any test of hypothesis. The steps are:

Step 1 Set up H_0 and H_a.

Step 2 Define the **test statistic**, which is evaluated using the sample data.

Step 3 Define a **rejection region**, using the value of α, stating which values of the test statistic will result in rejecting H_0.

Step 4   Calculate the value of the test statistic from the sample data and carry out the test. This will result in rejecting H_0 or failing to reject H_0.

Step 5   Give a conclusion in the language of the problem.

A test such as H_0: $\mu = 100$ versus H_a: $\mu \neq 100$ is called a **two-tailed test** because we reject H_0 whenever the sample estimate of μ ($\bar{X}$) is either too large (test statistic is in the right tail) or too small (test statistic is in the left tail). Similarly, a test on the population variance or standard deviation such as H_0: $\sigma^2 = 25$ versus H_a: $\sigma^2 \neq 25$ also is a two-tailed test.

$$H_0: \ \mu \leqslant 100 \text{ versus } H_a: \ \mu > 100 \quad \text{or} \quad H_0: \ \mu \geqslant 100 \text{ versus } H_a: \ \mu < 100$$

$$\text{or } H_0: \ \sigma^2 \leqslant 25 \text{ versus } H_a: \ \sigma^2 > 25 \quad \text{or} \quad H_0: \ \sigma^2 \geqslant 25 \text{ versus } H_a: \ \sigma^2 < 25$$

are all examples of **one-tailed tests** of hypothesis, since the rejection region lies in either the left tail *or* the right tail.

The tests on a population variance introduced the **chi-square** distribution, χ^2. This distribution was used to construct confidence intervals for σ^2 and σ as well as to define a distribution for the test statistic when performing a test of hypothesis on the variance or standard deviation.

Finally, we discussed why you should always include a **_p_ value** in the results of any hypothesis test. When using a predetermined significance level α, you reject H_0 whenever the p value is less than α and fail to reject H_0 otherwise. Another option you can use is not to select the somewhat arbitrary value of α but simply to reject H_0 whenever the p value is "small" (say, $<.01$), FTR H_0 if it is "large" (say, $>.1$), or decide that the data are inconclusive if the p value lies between these two values. You can also use the p value to measure the enthusiasm (p value very small) with which you reject H_0 or the authority (p value quite large) with which you fail to reject H_0.

REVIEW EXERCISES

9.60   Explain how changes in the significance level affect the following:
   (a)   The rejection region
   (b)   The Type II error

9.61   The federal government provides an enormous amount of funding to the states for research and development, according to a federal official interviewed by a journalist. These funding amounts are believed to follow a normal distribution. The official said that each state, on the average, gets $150 million or more per annum. The journalist thought this figure might be too high, and felt it was less than $150 million. Ten states were selected at random. The average amount of federal funding for R & D received per state was $120 million, with a standard deviation of $45 million. Does this provide enough evidence to reject the federal official's claim, using a 5% significance level? Determine the p value for your test.

9.62   The use of biological materials as metal indicators is a relatively cheap, simple, and reliable method of monitoring atmospheric pollution. Hernandez et al. (1987) collected samples of rosebay leaves (*Nerium oleander*) from February 1985 to January 1986. They reported the following levels of lead concentration (in parts per million) found in rosebay leaves taken from along the roadside at different locations:

LEAD CONCENTRATION IN ROSEBAY LEAVES

Place of Collection	Lead (ppm)
Cuzco Square	33.03
Nuevos Ministerios	58.25
Neptuno Square	42.95
Atocha Square	60.23
Maria Ana de Jesus Square	23.23
Ruiz Gimenez Square	52.67
Cristo Rey Square	48.35
Espana Square	74.25
Casa de Campo Park	14.17
Alonso Martinez Square	38.31
Roma Square	34.83
Conde de Casal Square	54.73
Retiro Park	13.63
Republica del Ecuador Square	66.73
Serrano St./M. de Molina St.	65.63

Source: L. M. Hernandez, Ma. C. Rico, Ma. J. Gonzales, and Ma. A. Hernan. Environmental contamination by lead and cadmium in plants from urban area of Madrid, Spain. *Bull. Environ. Contamination Toxicol.* **38** (2): 206 (February 1987).

(a) Do these data provide sufficient evidence, at $\alpha = .05$, to reject the claim that the mean lead concentration in urban Madrid is at least 50.0 ppm?

(b) State the p value for the test.

(c) Construct a 99% confidence interval for lead concentration.

9.63 Oscillatory afterpotentials (OAP), or late afterdepolarizations, are one mechanism postulated to cause cardiac arrythmias. Terek and January (1987) studied excitability in Purkinje fibers in sheep hearts. They reported, among other things, the following values of current thresholds (CT) with intracellular stimulation at OAP peak, for a basic cycle length (BCL) of 700 ms. The drug used was acetylstrophanthidin.

Fiber	CT at OAP Peak (Drug, nA)
1	290
2	468
3	170
4	266
5	550
6	66
7	320

Source: R. M. Terek and C. T. January. Excitability and oscillatory afterpotentials in isolated sheep cardiac Purkinje fibers. *Am. J. Physiol.: Heart Circ. Physiol.* **21** (3): H649 (March 1987), Table 1.

(a) Compute the mean and standard deviation for the above data.

(b) Based on the above results, and using a 5% significance level, could you support a hypothesis that claimed that CT at OAP peak in sheep cardiac Purkinje fibers was, on the average, greater than 250 nA?

(c) State the p value for your test.

9.64 Although teenage pregnancy is a serious problem, a sociologist feels that, with more and more women following careers of their own and postponing child-bearing, the average age of women at the birth of their first child is greater than 28 years. A sample of 81 females revealed an average age at the birth of the first child of 29.7 years, with a standard deviation of 3 years.

(a) Does this support the sociologist's belief? Use $\alpha = .10$ for your test.

(b) State the p value for the test.

(c) If the sample size were only 25 females, what changes would be necessary in the test? What assumptions are needed?

9.65 The Miser, an economy car, has an EPA gas mileage of 35 miles per gallon on the highway. A local dealer would like to verify that the car's realistic gas mileage is not significantly less than that obtained when the car is driven under "ideal" conditions. A random sample of 55 cars yields a sample mean of 34.4 with a sample standard deviation of 1.8 miles per gallon. Using a significance level of .05, is there sufficient evidence that the mean gas mileage of the Miser is less than 35 miles per gallon on the highway?

9.66 A manufacturer of drugs and medical products claims that a new antiinflammatory drug will be effective for 4 hours after the drug is administered in the prescribed dosage. A random sample of 50 volunteers demonstrated that the average effective time is 3.70 hours with a sample standard deviation of .606 hours. Use the p value criteria to test the null hypothesis that the mean effective time of the drug is 4 hours.

9.67 Indicate what the p values are for the following situations in which the mean of a normally distributed population is being tested.

(a) H_0: $\mu = 31.6$, H_a: $\mu \neq 31.6$ (population variance is known),
    $Z^* = 2.16$

(b) H_0: $\mu = 4.07$, H_a: $\mu \neq 4.07$ (population variance is known),
    $Z^* = -1.35$

(c) H_0: $\mu = 87.6$, H_a: $\mu \neq 87.6$ (population variance is unknown),
    $t^* = 2.51$, $n = 15$

(d) H_0: $\mu = 195.3$, H_a: $\mu \neq 195.3$ (population variance is unknown),
    $t^* = -1.71$, $n = 25$

9.68 The president of a small private liberal arts college stated that the pupil–teacher ratio at his institution had always been less than 7.50. A sample of 8 randomly selected years showed a mean of 6.39 and a standard deviation of 1.2 for the pupil–teacher ratio.

(a) Do these data support the president's belief, at a significance level of .05? Assume a normal distribution.

(b) State the p value for the test.

9.69 The vice-president of academic affairs at a small private college believes that the average full-time student who lives off campus spends about $300 per month for housing. A random sample of 200 full-time students living off campus spent an average of $305 per month with a standard deviation of $70 per month.

(a) Find the p value to determine whether there is sufficient evidence to indicate that a full-time student spends more than \$300 per month on housing.

(b) Would you reject the null hypothesis for the test in question (a) if $\alpha = .01$? if $\alpha = .05$? if $\alpha = .10$?

9.70 The Environmental Protection Fund believes that lead poisoning is far more widespread than is generally supposed, and levels of lead in the blood of children in certain suspected areas exceed the safety limit of 10 micrograms per deciliter. A sample of 45 children drawn from these areas shows a mean lead level of 10.9 micrograms per deciliter in their blood, with a standard deviation of 2.6 micrograms. Assume that lead level is a normally distributed variable.

(a) Does this sample support the claim made by the EPF? Use a 0.01 significance level.

(b) State the p value.

(c) Construct a 99% confidence interval for the mean level of lead in the blood of the children in suspected areas of contamination.

9.71 There are 420 persons attending a conference on the Strategic Defense Initiative (SDI). A reporter randomly selected 35 persons from those attending, to determine the average income of the participants. From this sample, a mean of 38.6 (thousand dollars) and a standard deviation of 6.7 were obtained.

(a) Construct the 95% confidence interval for the mean income of the conference participants.

(b) Test the hypothesis that the average income is less than 40 (thousand dollars). Use a 5% significance level.

(c) Determine the p value.

9.72 Teacher turnover in the Turnsville School District is believed to be higher than the state average, which is 15 per annum. A random sampling of the file records of departed teachers reveals a mean of 17.5 per annum and a variance of 25. The sample covered is eight years and the number of teacher turnovers is believed to follow a normal distribution. Use the p value criteria to test the null hypothesis ($H_0: \mu \leqslant 15$, $H_a: \mu > 15$) with a 5% significance level. What do you conclude?

9.73 There are 210 students in a particular dormitory at City University. The manager would like to know how much time the average student uses the dormitory's recreational facility. Twenty-eight students are randomly selected and questioned. The sample mean is 2.6 hours per week with a standard deviation of .4 hours. The student times are believed to be normally distributed.

(a) Find a 90% confidence interval for the mean time per week that a student spends using the dormitory's recreational facilities.

(b) Test the null hypothesis that the mean time per week that a student spends using the dormitory's recreational facilities is equal to 2.0 hours. Use a 10% significance level.

9.74 A large state university had recently converted to a computerized registration system for enrollment. After the first semester, administrators found that the average time spent registering per student was quite satisfactory, yet there still continued to be substantial complaints and dissatisfaction among students. Further study indicated that although the *average* time might seem satisfactory, there might be too much *variation* in the registration times. It was decided to study the situation during the next semester's registration period. If the standard deviation was greater than 20 minutes, six additional computer terminals would be installed; otherwise, two new computers would be installed. From a random sample, the following data were obtained:

$$\Sigma(x - \bar{x})^2 = 6900, \qquad n = 18$$

Assume the population is normally distributed.

(a) Conduct a hypothesis test to test if the standard deviation exceeds 20 minutes. Use a 5% significance level.

(b) What is the decision indicated by the test: install six new terminals, or two new terminals?

(c) State the p value for the test.

9.75 A nutrition counseling service investigated a behavior pattern that seemed to be fairly common among young females: they would alternate between periods of severe dieting and eating binges. Consequently, their weight showed high volatility, going up and down according to the period. A questionnaire was developed wherein the subjects were asked to rate various menus on a scale of 1 to 100 on different days, and also record their weight. From this a composite index was calculated. As a rule of thumb, if the true variance in the index for a person observed over an extended period was greater than 135, the subject was a likely victim of the dieting–binging syndrome.

Assume that Mary has just completed the questionnaire, but due to lack of time, an extended period of observation was not possible. Instead, just six observations were obtained, with a variance in the index of 146.

(a) Construct a 90% confidence interval for the variance of the index using the observations collected from Mary.

(b) Construct a 99% confidence interval for the variance of the index.

(c) Conduct a hypothesis test of the form

$$H_0: \quad \sigma^2 \leqslant 135$$

$$H_a: \quad \sigma^2 > 135$$

to decide if Mary's sample score of 146 is high enough that it might indicate she is a victim of the above syndrome. Use the p value method.

9.76 Growing use of outpatient services and introduction of the Diagnosis Related Group (DRG) method of payment has led to a decline in the length of hospital stays. Under the DRG program, federal payments to hospitals are set at a predetermined flat fee. The National Center for Health Statistics reports that in 1985, the average hospital stay was 6.5 days. The administrator of a large county hospital decided to compare his own hospital with the national figure. A sample of 100 patient records showed a mean stay of 7.1 days, with a standard deviation of 1.9 days. Does this suggest that the county hospital's average stay is different from the national figure of 6.5 days? Use a 5% significance level.

CASE STUDY

Under the Light of the Silvery Moon

Even in an age when men have flown to the moon and back, folklore and superstition about the sinister and evil influence of the full moon still persist. Poets and writers speak of "moon-struck madness," and movies perpetrate the myths of Dracula, and Dr. Jekyll and Mr. Hyde, and other sundry demonic characters. The possibility that the moon has some influence on human behavior is, of course, part of a broader attitude about planetary influences on the lives of human beings. Millions of men and women even today read the astrologer's column in their daily newspapers.

The negative influence of the moon has been referred to as the "Transylvania effect." Attempts have been made to link it with higher suicide rates, pyromania, epilepsy, and so on. No definite conclusions have been reached. For this case study, we will examine whether the full moon has any effect on the admission rates to the emergency room of a mental hospital.

The following table shows the number of admissions to the emergency room of a Virginia mental health clinic during the 12 full moons from August 1971 to July 1972.

Admissions During Full Moon			
5	12	13	13
13	6	16	14
14	9	25	20

Source: S. Blackman and Don Catalina. The Moon and the emergency room. *Perceptual and Motor Skills* **37**: 624–626 (1973). Adapted from R. J. Larsen and D. F. Stroup. *Statistics in the Real World.* (New York: Macmillan, 1976), pp. 59–64. Adapted and abridged with permission of Macmillan Publishing Company from *Statistics in the Real World: A Book of Examples* by Richard J. Larsen and Donna Fox Stroup. Copyright © 1976 by Macmillan Publishing Company.

For the remaining days of the year, the average number of daily admissions was 11.2. The reasoning behind the study was that if there was a lunar effect, the average number of admissions during the full moon should be significantly higher than for the rest of the year. If this was not so, then one might tend to be skeptical about the Transylvania effect.

Case Study Questions

1. State the null and alternative hypotheses to test the lunar effect.

2. Compute the mean and standard deviation of admissions during the full moon.

3. Which test would you use to decide whether or not the admissions during the full moon support the existence of a lunar effect?

4. Complete the necessary steps of the hypothesis test, at a 5% significance level.

5. State the *p* value for the test.

6. How would you interpret a decision to reject the null hypothesis?

7. How would you interpret a decision not to reject the null hypothesis?

8. In the context of this case, describe in words what a Type II error would be.

9. Is the above sample of 12 full moons a true random sample?

**COMPUTER
EXERCISES
USING THE
DATABASE**

Using a sample of 100 observations, complete the following exercises.

1. A recent report claims that the average verbal score on the GRE (variable VGRE) is greater than 450 for students attending Brookhaven University. Based upon the sample of 100 observations, can this claim be supported using a significance level of .05? Determine the p value.

2. This same report claims that the population standard deviation for the verbal GRE score is 95. Using a two-tailed test, what would be your conclusion, using a significance level of .05? Determine the p value.

3. Construct 95% confidence intervals for the two parameters discussed in exercises 1 and 2.

4. The previous report stated that the average age of students at Brookhaven University is over 30. Can this statement be supported using the sample of 100 observations and a significance level of .05? Determine the p value.

5. Using variables AGE, SEX, and MARITAL, determine a 90% confidence interval for the average age of the remarried males attending Brookhaven University. Be sure to use the small sample approach (using the t distribution) if the resulting sample size is under 30.

**SPSSX
APPENDIX
(MAINFRAME
VERSION)**

**Solution to
Example 9.10**

Example 9.10 was concerned with the computation of means and a t statistic to determine average weight gain. You can use SPSSX to solve this problem by borrowing a technique that actually belongs in Chapter 10 (and will be used again in that chapter). The program listing in Figure 9.23 was used to request the calculation of the t statistic and the corresponding p value. Note that SPSSX automatically assumes a two-tailed test; this was a one-tailed test, so the calculated p value needs to be divided by two.

Figure 9.23

SPSSX program
listing used to
request calculation
of the t statistic and
corresponding
p value

```
TITLE WEIGHTS
DATA LIST FREE / WEIGHT AMT
BEGIN DATA
15.3 5
12.9 5
-3.2 5
16.4 5
4.3 5
14.6 5
15.0 5
-2.1 5
15.5 5
7.2 5
9.1 5
15.2 5
END DATA
T-TEST PAIRS = WEIGHT AMT
```

The TITLE command names the SPSSX run.

The DATA LIST command gives each variable a name and describes the data as being in free form. The variable AMT (abbreviated form of "amount") is defined as the value of the mean contained in the hypotheses. Actually, AMT is a constant (five, here) but is treated as a variable in the analysis.

The BEGIN DATA command indicates to SPSSX that the input data immediately follow.

The next 12 lines contain the data values. Each line contains the weight gain for one individual along with the value for AMT.

The END DATA statement indicates the end of the data.

The T-TEST PAIRS = WEIGHT AMT command computes the t statistic using the variable WEIGHT and the corresponding p value.

Figure 9.24 contains the output obtained by executing the program listing in Figure 9.23.

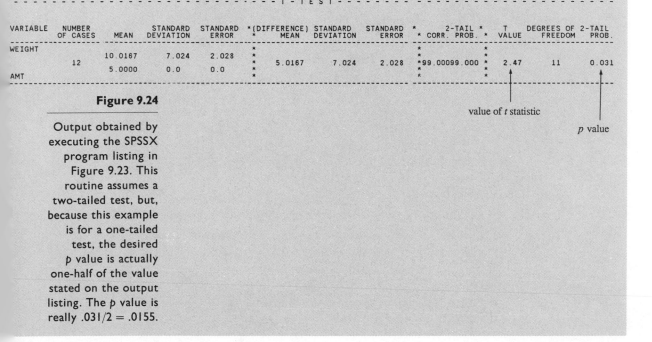

VARIABLE	NUMBER OF CASES	MEAN	STANDARD DEVIATION	STANDARD ERROR	*(DIFFERENCE) MEAN	STANDARD DEVIATION	STANDARD ERROR	2-TAIL CORR. PROB.	T VALUE	DEGREES OF FREEDOM	2-TAIL PROB.
WEIGHT		10.0167	7.024	2.028							
	12				5.0167	7.024	2.028	99.000 99.000	2.47	11	0.031
AMT		5.0000	0.0	0.0							

value of t statistic

p value

Figure 9.24

Output obtained by executing the SPSSX program listing in Figure 9.23. This routine assumes a two-tailed test, but, because this example is for a one-tailed test, the desired p value is actually one-half of the value stated on the output listing. The p value is really $.031/2 = .0155$.

**SPSS/PC+
APPENDIX
(MICRO
VERSION)**

**Solution to
Example 9.10**

Example 9.10 was concerned with the computation of means and a *t* statistic to determine average weight gain. You can use SPSS/PC+ to solve this problem by borrowing a technique that actually belongs in Chapter 10 (and will be used again in that chapter). The program listing in Figure 9.25 was used to request the calculation of the *t* statistic and the corresponding *p* value. Note that SPSS/PC+ automatically assumes a two-tailed test; this was a one-tailed test, so the calculated *p* value needs to be divided by two.

Figure 9.25

SPSS/PC+ program
listing used
to request
calculation of the
t statistic and
corresponding
p value

```
TITLE WEIGHTS.
DATA LIST FREE / WEIGHT , AMT.
BEGIN DATA.
15.3  5
12.9  5
-3.2  5
16.4  5
 4.3  5
14.6  5
15.0  5
-2.1  5
15.5  5
 7.2  5
 9.1  5
15.2  5
END DATA.
T-TEST PAIRS = WEIGHT AMT.
```

The TITLE commands names the SPSS/PC+ run.

The DATA LIST command gives each variable a name and describes the data as being in free form. The variable AMT (abbreviated form of "amount") is defined as the value of the mean contained in the hypotheses. Actually, AMT is a constant (five, here) but is treated as a variable in the analysis.

The BEGIN DATA command indicates to SPSS/PC+ that the input data immediately follow.

The next 12 lines contain the data values. Each line contains the weight gain for one individual along with the value for AMT.

The END DATA statement indicates the end of the data.

The T-TEST PAIRS = WEIGHT AMT command computes the *t* statistic using the variable WEIGHT and the corresponding *p* value.

Figure 9.26 contains the output obtained by executing the program listing in Figure 9.25.

SPSS/PC+ APPENDIX (MICRO VERSION)

```
Paired samples t-test:   WEIGHT
                         AMT

Variable      Number                Standard    Standard
             of Cases     Mean      Deviation    Error

WEIGHT          12       10.0167      7.024       2.028
AMT             12        5.0000       .000        .000
```

(Difference) Mean	Standard Deviation	Standard Error	2-Tail Corr. Prob.	t Value	Degrees of Freedom	2-Tail Prob.
5.0167	7.024	2.028	99.00099.000	2.47	11	.031

value of *t* statistic   *p* value

Figure 9.26

Output obtained by executing the SPSS/PC+ program listing in Figure 9.25. This routine assumes a two-tailed test, but, because this example is for a one-tailed test, the desired *p* value is actually one-half of the value stated on the output listing. The *p* value is really .031/2 = .0155.

SAS APPENDIX (MAINFRAME VERSION)

Solution to Example 9.10

Example 9.10 was concerned with the computation of means and a *t* statistic to determine average weight gain. You can use SAS to solve this problem. The SAS program listing in Figure 9.27 was used to request the calculation of the *t* statistic and the resulting *p* value. Note that SAS automatically assumes a two-tailed test; this was a one-tailed test, so the calculated *p* value needs to be divided by two. Each line represents one card image to be entered.

SAS APPENDIX (MAINFRAME VERSION)

Figure 9.27

SAS program listing used to request calculation of the *t* statistic and corresponding *p* value

```
TITLE 'WEIGHTS';
DATA  WTDATA;
INPUT WEIGHTS;
  HO1=WEIGHTS - 5;
CARDS;
15.3
12.9
-3.2
16.4
4.3
14.6
15.0
-2.1
15.5
7.2
9.1
15.2
PROC PRINT;
PROC MEANS N MEAN T PRT;
  VAR HO1;
  TITLE 'ONE-TAILED TEST';
```

The TITLE command names the SAS run (in single quotes).

The DATA command gives the data a name.

The INPUT command gives the variable a name.

The HO1 = WEIGHTS − 5 statement is used to calculate a new variable, HO1, which is the difference between the variable WEIGHTS and 5.

The CARDS command indicates to SAS that the input data immediately follow.

The next 12 lines are the data values, representing the weight gain for the 12 people.

Figure 9.28

Output obtained by executing the SAS program listing in Figure 9.27. This SAS routine assumes a two-tailed test, but, because this example is for a one-tailed test, the desired *p* value is actually one-half of the value stated on the output listing. The *p* value is really .0309/2 = .01545.

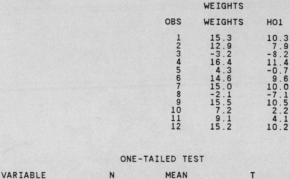

```
                        WEIGHTS

           OBS      WEIGHTS      HO1

            1        15.3       10.3
            2        12.9        7.9
            3        -3.2       -8.2
            4        16.4       11.4
            5         4.3       -0.7
            6        14.6        9.6
            7        15.0       10.0
            8        -2.1       -7.1
            9        15.5       10.5
           10         7.2        2.2
           11         9.1        4.1
           12        15.2       10.2


              ONE-TAILED TEST

VARIABLE       N       MEAN          T       PR>!T!

HO1           12    5.01666667      2.47      0.0309
                                     ↑          ↑
                              value of t statistic   p value
```

The PROC PRINT command directs SAS to list the data that were just entered.

The PROC MEANS command requests a SAS procedure to print the number of observations, the *t* statistic, and the corresponding *p* value.

The VAR statement specifies that the variable HO1 is the variable to be used in computing the statistics.

The TITLE statement specifies the heading for the output.

Figure 9.28 contains the output obtained by executing the program listing in Figure 9.27.

**SAS
APPENDIX
(MICRO
VERSION)**

**Solution to
Example 9.10**

Example 9.10 was concerned with the computation of means and a *t* statistic to determine average weight gain. You can use SAS to solve this problem. The SAS program listing in Figure 9.29 was used to request the calculation

Figure 9.29

SAS program listing
used to request
calculation of the
t statistic and
corresponding
p value

```
TITLE 'WEIGHTS';
DATA WTDATA;
INPUT WEIGHTS;
 HO1 = WEIGHTS - 5;
CARDS;
15.3
12.9
-3.2
16.4
4.3
14.6
15.0
-2.1
15.5
7.2
9.1
15.2
PROC PRINT;
PROC MEANS N MEAN T PRT;
 VAR HO1;
 TITLE 'ONE-TAILED TEST';
RUN;
```

SAS APPENDIX (MICRO VERSION)

```
        WEIGHTS
                                              ONE-TAILED TEST
  OBS    WEIGHTS    HO1
                                    Analysis variable : HO1
    1      15.3     10.3
    2      12.9      7.9
    3      -3.2     -8.2
    4      16.4     11.4          N Obs    N        Mean            T       Prob>|T|
    5       4.3     -0.7          --------------------------------------------------
    6      14.6      9.6            12    12        5.02          2.47       0.0309
    7      15.0     10.0          --------------------------------------------------
    8      -2.1     -7.1
    9      15.5     10.5
   10       7.2      2.2
   11       9.1      4.1                                   Value of t statistic   p value
   12      15.2     10.2
```

Figure 9.30

Output obtained by executing the SAS program listing in Figure 9.29. This SAS routine assumes a two-tailed test, but, because this example is for a one-tailed test, the desired p value is actually one-half of the value stated on the output listing. The p value is really .0309/2 = .01545.

of the t statistic and the resulting p value. Note that SAS automatically assumes a two-tailed test; this was a one-tailed test, so the calculated p value needs to be divided by two. Each line represents one card image to be entered.

The TITLE command names the SAS run (in single quotes).

The DATA command gives the data a name.

The INPUT command gives the variable a name.

The HO1 = WEIGHTS − 5 statement is used to calculate a new variable, HO1, which is the difference between the variable WEIGHTS and 5.

The CARDS command indicates to SAS that the input data immediately follow.

The next 12 lines are the data values, representing the weight gain for the 12 people.

The PROC PRINT command directs SAS to list the data that were just entered.

The PROC MEANS command requests a SAS procedure to print the number of observations, the t statistic, and the corresponding p value.

The VAR statement specifies that the variable HO1 is the variable to be used in computing the statistics.

The TITLE statement specifies the heading for the output.

The RUN command tells the SAS system to execute the previous SAS statements.

Figure 9.30 contains the output obtained by executing the program listing in Figure 9.29.

10

INFERENCE PROCEDURES FOR TWO POPULATIONS

A LOOK BACK/INTRODUCTION

We have learned to describe and summarize data from a single population using a statistic (such as the sample mean, $\bar{X}$), or a graph (such as a histogram). Chapters 8 and 9 introduced you to statistical inference, where we (1) attempted to estimate a parameter (such as the mean, μ) from this population by using the corresponding sample statistic and (2) arrived at a conclusion about this parameter (such as $\mu > 100$) by performing a test of hypothesis. The concept behind hypothesis testing was described, and we paid special attention to the errors (Type I and Type II) that can occur when we use a sample to infer something about a population.

Next, we will learn how to compare two populations. Questions that will be of interest here include:

1 Are the values in population 1 larger, on the average, than those in population 2? (For example, are men taller, on the average, than women?)

2 Do the values in population 1 exhibit more variation than those in population 2? (For example, do male heights vary more than female heights?)

The two populations under observation may or may not be normally distributed. When comparing two population means using large samples, once again using the Central Limit Theorem, it simply does not matter. For small samples, we need to examine the distribution of the populations so we can use the proper procedure to construct confidence intervals and perform tests of hypothesis.

This chapter will discuss two different sampling situations. In the first, random samples from two populations are obtained independently of each other; in the second, corresponding data values from the two samples are matched up, or paired. Paired samples are dependent. The chapter will also introduce you to an area of statistics referred to as **nonparametric statistics** that is extremely useful for certain types of data.

10.1   INDEPENDENT VERSUS DEPENDENT SAMPLES

When making comparisons between the means of two populations, we need to pay particular attention to how we intend to collect sample data. For example, how would you determine the effect of hypnosis on one's ability to recall particular details of an experience or to memorize a list of words? One procedure would be to subject a randomly selected group to an experiment without hypnosis and then repeat the experiment on another group of individuals who are subjected to hypnosis. This is an illustration of *independent* samples.

A safer procedure here would be to use the same individuals for the experiment where each individual goes through the exercise without hypnosis and then is subjected to a similar exercise under the influence of hypnosis. In this way, you can help eliminate the effect of the experimental subject, since some people have better memories than others, hypnosis or not. This is an illustration of using *dependent* samples. For this experiment, it is important that you randomly determine whether each subject first completes the exercise with hypnosis or without it.

Consider another situation. Suppose you are interested in male heights as compared with female heights. You obtain a sample of $n_1 = 50$ male heights and $n_2 = 50$ female heights. You obtain these data:

Observation	Male Heights	Observation	Female Heights
1	5.92'	1	5.36'
2	6.13'	2	5.64'
3	5.78'	3	5.44'
⋮	⋮	⋮	⋮
50	5.81'	50	5.52'

Is there any need to match up 5.92 with 5.36, 6.13 with 5.64, 5.78 with 5.44, and so on? The male heights were randomly selected and the female heights were obtained independently, so there is no reason to match up the first male height with the first female height, the second male height with the second female height, and so on. Nothing relates male 1 with female 1 other than the accident of their being selected first—these are **independent samples**.

What if you wish to know whether husbands are taller than their wives? To collect data, you select 50 married couples. Suppose you obtain the 100 observations from the previous male and female height example. Now, is there a reason to compare the first male height with the first female height, the second with the second, and so on? The answer is a definite yes since each pair of heights belongs to a married couple. The resulting two samples are **dependent** or **paired samples**.

In summary,

1 If there is a definite reason for pairing (matching) corresponding data values, the two samples are *dependent* samples.

2 If the two samples were obtained independently and there is no reason for pairing the data values, the resulting samples are *independent* samples.

Why does this distinction matter? If you are trying to decide whether male heights are, on the average, larger than female heights, the procedure that you use for testing this depends on whether the samples are obtained independently.

Other applications of dependent samples include data from the following situations.

1 Comparisons of *before versus after*. Sample 1: Person's adjustment to a recent divorce *prior to* attending a workshop designed to help the participants deal with this situation. The sample consists of scores on a divorce adjustment scale. Such a test is called a **pretest**. Sample 2: Same person's score on the divorce adjustment scale three months after the workshop. These scores are referred to as **post-test** scores. Why pair the data? Each pair of observations belong to the same person.

2 Comparisons of people with *matching characteristics*. Sample 1: Effect of a vaccination on a male child. Sample 2: Effect of the vaccination on a female child that is identical to the male child in age and body weight. Why pair the data? Pairing allows for comparison between males and females, since the pairs are identical in age and body weight.

3 Comparison of observations *matched by location*. Sample 1: Salary of a starting teacher without a masters degree, each salary coming from a different city. Sample 2: Salary of a starting teacher with a masters degree, collected from the same set of cities. Why pair the data? Each pair of salaries came from the same city.

4 Comparisons of observations *matched by time*. Sample 1: Sales of restaurant A during a particular week. Sample 2: Sales of restaurant B during this week. Why pair the data? Each pair of observations corresponds to the same week of the year.

EXERCISES

10.1 It is desired to test the hypothesis that the income of male television anchors is greater than that for female TV anchors. Fifteen cities are selected, and the incomes for male and female TV anchors are noted for each. Is this a case of independent or dependent samples? If dependent, is it a case of before–after comparisons, or matching by characteristics location or time?

10.2 Two private colleges decided to compare the mean SAT scores of their incoming freshmen. One college gathered 98 scores and the other took a sample of 52 scores. Do the two sets of data represent dependent or independent samples?

10.3 A medical institution is examining the effectiveness of a newly developed drug. The drug was administered to 18 patients whose health condition before and after taking the drug was recorded. Is this a case of dependent or independent samples?

10.4 University professors in a Northeastern state are being asked to evaluate two pension schemes. A rating scale of 1 to 10 is to be used, with 1 being the lowest rating and 10 being the highest.
(a) How could dependent samples be taken?
(b) How could independent samples be taken?

10.5 In a medical experiment, the pulmonary capacity of asthmatic patients is measured before administering a new inhalant drug, and measured again 30 minutes after the drug is administered. Would the data obtained from this experiment be a case of dependent or independent sampling?

10.6 It is desired to compare the weight of newborn male infants with that of newborn female infants. Consider the following situations.
(a) *Sample 1* consists of 15 newly born male infants from 15 randomly selected hospitals.
Sample 2 consists of 15 newly born female infants from 15 randomly selected hospitals.
(b) *Sample 1* consists of 15 newly born male infants from 15 randomly selected hospitals.
Sample 2 consists of 15 newly born female infants from the same 15 hospitals.
Which of these is a case of dependent samples, and which is a case of independent samples?

10.7 In exercise 10.6, part (a), would it be possible for the size of Sample 1 to not be the same as the size of Sample 2? Similarly, answer the same question for 10.6, part (b).

10.8 A school district reported results of a reading test in the following format:

School	White	Nonwhite
1	60	54
2	71	70
3	77	80
⋮	⋮	⋮

(a) If a comparison of Whites and Nonwhites is made, are the data dependent or independent?
(b) Suggest a format for the kind other than the one you identified in part (a) above. In other words, if your answer for (a) was "dependent", then show what "independent" data might possibly look like, and vice versa.

10.2   LARGE INDEPENDENT SAMPLES: COMPARING TWO POPULATION MEANS

When comparing the means of two independent samples from different populations, we can use Figure 10.1 to help visualize the situation. The two populations are shown to be normally distributed, but, because we will be using large samples from these populations, this is not a necessary assumption. For these populations,

$$\mu_1 = \text{mean of population 1}$$

$$\mu_2 = \text{mean of population 2}$$

$$\sigma_1 = \text{standard deviation of population 1}$$

$$\sigma_2 = \text{standard deviation of population 2}$$

For example, in many experiments, you define two populations (groups). Each member of the first group (called the **experimental group**) is given a particular **treatment**, which could be a drug dosage or a particular treatment plan. To determine the effect of this treatment, you construct another group, referred to as the **control group**, which does not receive the treatment.

To evaluate the eyestrain effects of a proposed computer monitor using an amber screen, you would begin by randomly assigning each member of a sample pool of computer users to the experimental group (population 1) or the control group (population 2). The members of the experimental group use the amber monitor, whereas the control group would use the standard monitor with a black and white screen. Each sample member would then complete an exercise and provide a score that measures the eyestrain of this individual. The parameters here would be:

$$\mu_1 = \text{mean of experimental group scores}$$

$$\mu_2 = \text{mean of control group scores}$$

$$\sigma_1 = \text{standard deviation of experimental group scores}$$

$$\sigma_2 = \text{standard deviation of control group scores}$$

Figure 10.1

Example of two populations. Is $\mu_1 = \mu_2$?

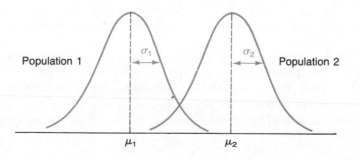

If the amber screen is effective in reducing eyestrain, we would expect that $\mu_1 < \mu_2$, assuming that larger scores indicate more eyestrain. The point estimates discussed in earlier chapters apply here as well—we simply have two of everything because we are dealing with two populations.

The procedure we follow is to obtain a random sample of size n_1 from population 1 and then obtain another sample of size n_2, completely independent of the first sample, from population 2. So, $\bar{X}_1$ is our best (point) estimate of μ_1. Likewise, $\bar{X}_2$ estimates μ_2. The sample standard deviations (s_1 and s_2) provide the best estimates of the population standard deviations (σ_1 and σ_2).

Constructing a Confidence Interval for $\mu_1 - \mu_2$

An experiment was conducted to determine if neurolinguistic programming (NLP) a technique for reducing anxiety, is an effective method of treating speech anxious individuals. A group of 75 subjects believed to have speech anxiety was randomly assigned to an experimental group where each member was subjected to neurolinguistic programming, or to a control group where each member received no treatment. Each member then gave a 4-minute speech on an assigned topic in the presence of a trained observer who scored the individual's speech anxiety. A high score indicated that the subject performed well, that is, with *little* anxiety. The results of the experiment are summarized below.

Population 1 Experimental Group Using NLP	Population 2 Control Group Not Using NLP
$n_1 = 35$	$n_2 = 40$
$\bar{x}_1 = 71.5$	$\bar{x}_2 = 63.1$
$s_1 = 7.1$	$s_2 = 13.2$

Let μ_1 be the average score of the experimental population subjected to NLP (35 of them were sampled) and let μ_2 be the average score of the control population not using NLP (40 of them were sampled).

When dealing with these two populations, the parameter of interest is $\mu_1 - \mu_2$ rather than the individual values of μ_1 and μ_2. Here, $\mu_1 - \mu_2$ represents the difference in average performance between those using neurolinguistic programming and those who are not. If we conclude that $\mu_1 - \mu_2 > 0$, then this means that $\mu_1 > \mu_2$. In this case, the neurolinguistic programming treatment *is* effective.

The point estimator of $\mu_1 - \mu_2$ is the obvious one: $\bar{X}_1 - \bar{X}_2$. For our data, the (point) estimate of $\mu_1 - \mu_2$ is $\bar{x}_1 - \bar{x}_2 = 71.5 - 63.1 = 8.4$. How much more effective is the NLP treatment on all patients suffering from speech anxiety, on the average? We do not know, because this is $\mu_1 - \mu_2$, but we do have an estimate of this value—namely, 8.4 points on the assessing instrument.

What kind of random variable is $\bar{X}_1 - \bar{X}_2$? First, because the samples are moderately large, we know by using the Central Limit Theorem that $\bar{X}_1$ is approximately a normal random variable with mean μ_1 and variance σ_1^2/n_1 and that $\bar{X}_2$ is approximately a normal random variable with mean μ_2 and variance σ_2^2/n_2. Because these are two independent samples, it follows that $\bar{X}_1 - \bar{X}_2$ is also approximately a normal random variable with mean $\mu_1 - \mu_2$ and variance $(\sigma_1^2/n_1) + (\sigma_2^2/n_2)$. Note that the variance of $\bar{X}_1 - \bar{X}_2$ is obtained by adding the variances for $\bar{X}_1$ and $\bar{X}_2$.

By standardizing this normal distribution, we obtain an approximate standard normal random variable defined by

$$Z = \frac{(\bar{X}_1 - \bar{X}_2) - (\mu_1 - \mu_2)}{\sqrt{\dfrac{\sigma_1^2}{n_1} + \dfrac{\sigma_2^2}{n_2}}} \qquad \textbf{(10-1)}$$

We do not need normal populations. The results of equation 10-1 are approximately valid regardless of the shape of the two populations, provided both samples are large (from the Central Limit Theorem). We pointed out that the two populations illustrated in Figure 10.1 need not follow a normal distribution. In fact, they can have any shape, such as uniform, or possibly a discrete distribution of some sort.

This enables us to derive a confidence interval for $\mu_1 - \mu_2$. We know, by using Table A-4, that for the standard normal random variable Z,

$$P(-1.96 < Z < 1.96) = .95$$

Using equation 10-1 (after rearranging the inequalities), we can make the following statement about a random interval prior to obtaining the sample data:

$$P\left[(\bar{X}_1 - \bar{X}_2) - 1.96 \sqrt{\frac{\sigma_1^2}{n_1} + \frac{\sigma_2^2}{n_2}} < \mu_1 - \mu_2 \right.$$
$$\left. < (\bar{X}_1 - \bar{X}_2) + 1.96 \sqrt{\frac{\sigma_1^2}{n_1} + \frac{\sigma_2^2}{n_2}} \right] = .95$$

This produces a $(1 - \alpha) \cdot 100\%$ confidence interval for $\mu_1 - \mu_2$ (large samples), where σ_1 and σ_2 are known, of:

$$(\bar{X}_1 - \bar{X}_2) - Z_{\alpha/2} \sqrt{\frac{\sigma_1^2}{n_1} + \frac{\sigma_2^2}{n_2}} \quad \text{to} \quad (\bar{X}_1 - \bar{X}_2) + Z_{\alpha/2} \sqrt{\frac{\sigma_1^2}{n_1} + \frac{\sigma_2^2}{n_2}} \qquad \textbf{(10-2)}$$

If σ_1 and σ_2 are unknown, we have:

$$(\bar{X}_1 - \bar{X}_2) - Z_{\alpha/2}\sqrt{\frac{s_1^2}{n_1} + \frac{s_2^2}{n_2}} \quad \text{to} \quad (\bar{X}_1 - \bar{X}_2) + Z_{\alpha/2}\sqrt{\frac{s_1^2}{n_1} + \frac{s_2^2}{n_2}} \quad \textbf{(10-3)}$$

Notice that this interval is very similar to the confidence interval for a single population mean using a large sample, namely

$$\text{(point estimate)} \pm Z_{\alpha/2} \text{(standard deviation of point estimator)}$$

To construct the confidence interval if σ_1 and σ_2 are unknown (the usual case), you simply substitute the sample estimates in their place provided you have large samples ($n_1 > 30$ and $n_2 > 30$). Consequently, the confidence interval in equation 10-2 is exact (σ_1, σ_2 known) and the confidence interval in equation 10-3 is approximate (σ_1, σ_2 unknown).

EXAMPLE 10.1

Using the data from the NLP experiment, construct a 90% confidence interval for $\mu_1 - \mu_2$.

Solution

To begin with, the estimate of μ_1 is $\bar{x}_1 = 71.5$, and the estimate of μ_2 is $\bar{x}_2 = 63.1$. We are constructing a 90% confidence interval, so (using Table A-4) we find that $Z_{.05} = 1.645$ (Figure 10.2). The resulting 90% confidence interval for $\mu_1 - \mu_2$ is

$$(\bar{X}_1 - \bar{X}_2) - 1.645\sqrt{\frac{s_1^2}{n_1} + \frac{s_2^2}{n_2}} \quad \text{to} \quad (\bar{X}_1 - \bar{X}_2) + 1.645\sqrt{\frac{s_1^2}{n_1} + \frac{s_2^2}{n_2}}$$

$$= (71.5 - 63.1) - 1.645\sqrt{\frac{(7.1)^2}{35} + \frac{(13.2)^2}{40}}$$

$$\text{to} \quad (71.5 - 63.1) + 1.645\sqrt{\frac{(7.1)^2}{35} + \frac{(13.2)^2}{40}}$$

$$= 8.4 - (1.645)(2.41) \quad \text{to} \quad 8.4 + (1.645)(2.41)$$

$$= 8.4 - 3.96 \quad \text{to} \quad 8.4 + 3.96$$

$$= 4.44 \quad \text{to} \quad 12.36$$

We can summarize this result in several ways:

1 We are 90% confident that $\mu_1 - \mu_2$ lies between 4.44 and 12.36.

2 We are 90% confident that the average score for speech anxious individuals treated using neuro-linguistic programming is between 4.44 and 12.36 higher than for speech anxious subjects not given this treatment.

3 We are 90% confident that our estimate of $\mu_1 - \mu_2$ ($\bar{x}_1 - \bar{x}_2 = 8.4$) is within 3.96 points of the actual value.

Figure 10.2

Finding the pair of
Z values containing
90% of the area
under the curve.
The values are
−1.645 and 1.645.

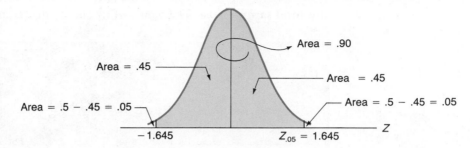

The confidence intervals defined in equations 10-2 and 10-3 will contain $\mu_1 - \mu_2$, 90% of the time. In other words, if you repeatedly obtained independent samples and repeated the procedure in example 10.1, 90% of the corresponding confidence intervals would contain the unknown value of $\mu_1 - \mu_2$, and 10% of them would not. □

Sample Sizes

The amount that you add to and subtract from your point estimate to obtain the confidence interval is the **maximum error**, **E**. For example 10.1, this value is $E = 3.96$ points. If you think that E is too large and you would like it to be smaller, one recourse is to obtain larger samples from your two populations. To determine how large a sample you need, one procedure is to select equal sample sizes. Consider the illustration in example 10.1, where in this case a researcher wants large enough samples so that the difference in sample means is within 2 points (rather than 3.96 in example 10.1) of the difference in population means, with 90% confidence. So, $E = 2$. By insisting on equal sample sizes, where $n_1 = n_2 = n$ (say), then

$$2 = 1.645 \sqrt{\frac{(7.1)^2}{n} + \frac{(13.2)^2}{n}}$$

After some algebraic manipulation, we have

$$n = \frac{(1.645)^2[(7.1)^2 + (13.2)^2]}{2^2} \cong 152.$$

In general, this value is

$$n = \frac{Z_{\alpha/2}^2(s_1^2 + s_2^2)}{E^2}. \tag{10-4}$$

In this illustration, the total sample size is $n_1 + n_2 = 152 + 152 = 304$. A better way to proceed here is to find the values of n_1 and n_2 that **minimize**

the total sample size. The values of n_1 and n_2 that accomplish this are

$$n_1 = \frac{Z_{\alpha/2}^2 s_1 (s_1 + s_2)}{E^2} \qquad \textbf{(10-5)}$$

$$n_2 = \frac{Z_{\alpha/2}^2 s_2 (s_1 + s_2)}{E^2} \qquad \textbf{(10-6)}$$

For this illustration, $Z_{\alpha/2} = Z_{.05} = 1.645$, $s_1 = 7.1$, $s_2 = 13.2$, and $E = 2$. Consequently,

$$n_1 = \frac{(1.645)^2 (7.1)(20.3)}{(2)^2} \cong 98$$

$$n_2 = \frac{(1.645)^2 (13.2)(20.3)}{(2)^2} \cong 182$$

and the total sample size is $98 + 182 = 280$.

A derivation of this result is contained in Appendix B, on page A-39. Keep in mind that, when using these values of n_1 and n_2, the resulting value of E may not be exactly what you previously specified because the values of s_1 and s_2 in the new samples will change. If no prior estimates of s_1 and s_2 are available, each can be roughly estimated using the high/low procedure discussed in Chapter 8, page 340.

Using equations 10-5 and 10-6, observe that if $s_1 = s_2$, your total sample size $(n_1 + n_2)$ will be the smallest when $n_1 = n_2$. If $s_1 > s_2$, you will select $n_1 > n_2$, and if $s_1 < s_2$, you will select $n_1 < n_2$. Finally, note that the ratio of the sample sizes (n_1/n_2) is the same as the ratio of the estimated standard deviations (s_1/s_2).

Hypothesis Testing for μ_1 and μ_2 (Large Samples)

Are men on the average taller than women? How do you answer such a question? We know that we can start by getting a sample of male heights and independently obtain a sample of female heights. Figure 10.3 illustrates this situation.

We proceed as before and put the claim that we are trying to demonstrate into the alternative hypothesis. The resulting hypotheses are

H_0: $\mu_1 \geqslant \mu_2$ (men are not taller, on the average)

H_a: $\mu_1 < \mu_2$ (men are taller, on the average)

We have estimators of μ_1 and μ_2, namely, $\bar{X}_1$ and $\bar{X}_2$. A sensible thing to do would be to reject H_0 if $\bar{X}_2$ is "significantly larger" than $\bar{X}_1$. In this case, the

Figure 10.3

Hypothesis testing
for two populations.
Is $\mu_2 > \mu_1$?

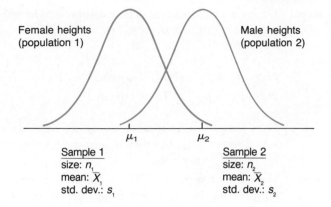

Female heights
(population 1)

Male heights
(population 2)

μ_1 μ_2

Sample 1
size: n_1
mean: $\bar{X}_1$
std. dev.: s_1

Sample 2
size: n_2
mean: $\bar{X}_2$
std. dev.: s_2

obvious conclusion is that μ_2 (the average of all male heights in your population) is larger than μ_1 (for female heights).

To define "significantly larger," we need to know what chance we are willing to take in rejecting H_0 when in fact it is true. This is α (the significance level) and, as before, it is determined prior to seeing any data. Typical values range from .01 to .1, with $\alpha = .05$ generally providing a good trade-off between the Type I and Type II errors. The test statistic here is the same as the one used to derive a confidence interval for $\mu_1 - \mu_2$. We arc dealing with large samples ($n_1 > 30$ and $n_2 > 30$), so this is approximately a standard normal random variable, defined by

$$Z = \frac{\bar{X}_1 - \bar{X}_2}{\sqrt{\dfrac{s_1^2}{n_1} + \dfrac{s_2^2}{n_2}}}$$

$$(10\text{-}7)$$

EXAMPLE 10.2

The researchers using the neurolinguistic programming (NLP) technique suspected that the NLP technique was effective in treating speech anxiety before observing any data. (*Note:* This is important! Do not let the data dictate your hypotheses for you. If you do, you introduce a serious bias into your testing procedure, and the "true" significance level may no longer be the predetermined α.) Here, μ_1 represents the average score for the experimental population subjected to NLP and μ_2 is the average score for those individuals not receiving NLP. Since a large score indicates a *small* level of stress, the question becomes "Is there sufficient evidence to indicate that $\mu_1 > \mu_2$?" Use a significance level of .05.

Solution   Step 1   Define the hypotheses. The question is whether or not the data support the claim that $\mu_1 > \mu_2$, so we put this statement in the alternative hypothesis.

$$H_0: \quad \mu_1 \leqslant \mu_2 \quad \text{(NLP is countereffective)}$$
$$H_a: \quad \mu_1 > \mu_2 \quad \text{(NLP is effective)}$$

Note that, as in Chapter 9, when defining a one-tailed test, the equal sign goes into H_0. In other words, the case where $\mu_1 = \mu_2$ is contained in the null hypothesis.

Step 2   Define the test statistic. This is the statistic that you evaluate using the sample data. Its value will either support the alternative hypothesis or it will not. The test statistic for this situation is equation 10-7:

$$Z = \frac{\bar{X}_1 - \bar{X}_2}{\sqrt{\dfrac{s_1^2}{n_1} + \dfrac{s_2^2}{n_2}}}$$

Step 3   Define the rejection region. In Figure 10.4, where should the null hypothesis H_0 be rejected? We simply ask, what happens to Z when H_a is true? In this case ($\mu_1 > \mu_2$), we should see $\bar{X}_1 > \bar{X}_2$. In other words, Z will be positive. So we reject H_0 if Z is "too large"; that is,

Reject H_0 if $Z > k$ for some $k > 0$

Using $\alpha = .05$, we use Table A–4 to find the corresponding value of Z (that is, k). In Figure 10.4, $k = 1.645$. This is the same value and rejection region we obtained in Chapter 9 when using Z for a one-tailed test in the right tail. The test is

Reject H_0 if $Z > 1.645$

Figure 10.4

Z curve showing rejection region for example 10.2

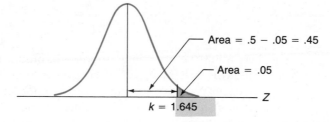

Area = .5 − .05 = .45

Area = .05

$k = 1.645$

Step 4   Evaluate the test statistic and carry out the test. The data collected showed $n_1 = 35$, $\bar{x}_1 = 71.5$, $s_1 = 7.1$ (from the experimental group) and $n_2 = 40$, $\bar{x}_2 = 63.1$, $s_2 = 13.2$ (from the control group). Based on these sample results, can we conclude that $\bar{x}_1 = 71.5$ is significantly larger than $\bar{x}_2 = 63.1$? If we can, the decision will be to reject H_0. The following value of the test statistic will answer our question.

$$Z = \frac{\bar{X}_1 - \bar{X}_2}{\sqrt{\dfrac{s_1^2}{n_1} + \dfrac{s_2^2}{n_2}}} = \frac{71.5 - 63.1}{\sqrt{\dfrac{(7.1)^2}{35} + \dfrac{(13.2)^2}{40}}}$$

$$= \frac{8.4}{2.41} = 3.49 = Z^*$$

Because $3.49 > 1.645$, we reject H_0; $\bar{x}_1$ is significantly larger than $\bar{x}_2$. Therefore we claim that $\mu_1 > \mu_2$.

Step 5 State a conclusion. We conclude that subjects treated with neuro-linguistic programming scored higher (had less stress) than the individuals not subjected to the treatment. □

Using the corresponding p value for the data in example 10.2, what would you conclude using the classical approach (with $\alpha = .05$)? For this example, the p value will be the area under the Z curve (Z is our test statistic) to the right (we reject H_0 in the right tail for this example) of the calculated test statistic, $Z^* = 3.49$. In general,

$$p = p \text{ value} = \begin{cases} \text{area to the right of } Z^* \text{ for } H_a\text{: } \mu_1 > \mu_2 \\ \text{area to the left of } Z^* \text{ for } H_a\text{: } \mu_1 < \mu_2 \qquad \textbf{(10-8)} \\ 2 \text{ (tail area of } Z^*) \text{ for } H_a\text{: } \mu_1 \neq \mu_2 \end{cases}$$

These three alternative hypotheses are your choices for this situation. Once again, H_0: $\mu_1 = \mu_2$ versus H_a: $\mu_1 \neq \mu_2$ is a two-tailed test, and the first two alternative hypotheses in equation 10-8 represent one-tailed tests.

Returning to our example, we can see from Figure 10.5 that the resulting p value is $p = .0002$ (very small). Using the classical approach, because $p <$ the significance level of .05, we reject H_0—the same conclusion as before. In fact, this procedure always leads to the same conclusion as the five-step solution, as we saw in Chapter 9.

Figure 10.5

Z curve showing p value for $Z^* = 3.49$

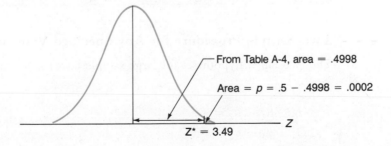

From Table A-4, area = .4998

Area = p = .5 − .4998 = .0002

$Z^* = 3.49$

If we elect not to select a significance level α and instead use only the p value to make a decision, we proceed as before:

Reject H_0 if p is small ($p < .01$).

FTR H_0 if p is large ($p > .1$).

Data are inconclusive if p is neither small nor large ($.01 \leqslant p \leqslant .1$).

For this example, $p = .0002$ is clearly small, and so we again reject H_0. The NLP method of reducing speech anxiety definitely appears to be effective.

COMMENTS   There may well be situations where the severity of the Type I error requires a significance level smaller than .01 on the low end, or the impact of a Type II error dictates a significance level larger than .1 on the upper end. This rule is thus only a general yardstick that applies to most, but certainly not all, situations.

Large Sample Tests for μ_1 and μ_2

Two-Tailed Test

H_0: $\mu_1 = \mu_2$

H_a: $\mu_1 \neq \mu_2$

Reject H_0 if $|Z| > Z_{\alpha/2}$
where

$$Z = \frac{\bar{X}_1 - \bar{X}_2}{\sqrt{\dfrac{s_1^2}{n_1} + \dfrac{s_2^2}{n_2}}}$$

One-Tailed Test

H_0: $\mu_1 \leqslant \mu_2$ H_0: $\mu_1 \geqslant \mu_2$

H_a: $\mu_1 > \mu_2$ H_a: $\mu_1 < \mu_2$

Reject H_0 if $Z > Z_\alpha$ Reject H_0 if $Z < -Z_\alpha$

Two-Sample Procedure for Any Specified Value of $\mu_1 - \mu_2$

The two-tailed hypotheses for large sample tests for μ_1 and μ_2 can be written as

H_0: $\mu_1 - \mu_2 = 0$

H_a: $\mu_1 - \mu_2 \neq 0$

The right-sided one-tailed hypotheses are

$$H_0: \quad \mu_1 - \mu_2 \leqslant 0$$
$$H_a: \quad \mu_1 - \mu_2 > 0$$

The left-tailed hypotheses can be written in a similar manner. The point is that H_0 (so far) claims that $\mu_1 - \mu_2$ is equal to zero or lies to one side of zero (the one-tailed tests).

Suppose the claim is that $\mu_1 - \mu_2$ is more than 10. To demonstrate that this is true, we must make our alternative hypothesis $H_a: \mu_1 - \mu_2 > 10$, and the corresponding null hypothesis is $H_0: \mu_1 - \mu_2 \leqslant 10$.

In general, to test that $\mu_1 - \mu_2 = $ (some specified value, say D_0), the five-step procedure still applies, except the test statistic is now

$$Z = \frac{(\bar{X}_1 - \bar{X}_2) - D_0}{\sqrt{\dfrac{s_1^2}{n_1} + \dfrac{s_2^2}{n_2}}} \qquad (10\text{-}9)$$

Equation 10-9 applies to both one-tailed and two-tailed tests. It can be used to compare two means directly (for example, $H_0: \mu_1 = \mu_2$ versus $H_a: \mu_1 \neq \mu_2$) by setting $D_0 = 0$, as in example 10.2.

EXAMPLE 10.3

In example 10.2, we decided that neurolinguistic programming (NLP) was an effective method of reducing speech anxiety. Another person reviewing these results is interested in whether this technique is better than the method she uses for treating this particular form of anxiety. From past experience, she has found that this treatment results in a score on the speech evaluation that is 5 points higher than when no treatment is used. Consequently, she is curious if the sample in example 10.2 would lead her to conclude that the improvement using NLP is more than 5. What would be her conclusion using a significance level of .05?

Solution

Now the hypotheses are whether the data support the claim that the difference between the two means (experimental and control) is larger than 5. So the hypotheses are $H_0: \mu_1 - \mu_2 \leqslant 5$ and $H_a: \mu_1 - \mu_2 > 5$, where μ_1 is the experimental population mean and μ_2 is the control population mean.

The test statistic is

$$Z = \frac{(\bar{X}_1 - \bar{X}_2) - 5}{\sqrt{\dfrac{s_1^2}{n_1} + \dfrac{s_2^2}{n_2}}}$$

The computed value of Z is

$$Z^* = \frac{(71.5 - 63.1) - 5}{\sqrt{\dfrac{(7.1)^2}{35} + \dfrac{(13.2)^2}{40}}} = \frac{3.4}{2.41}$$

$$= 1.41$$

The testing procedure is exactly as it was previously—reject H_0 if $Z^* > 1.645$. Because $1.41 < 1.645$, we FTR H_0. The difference between the two sample means was not significantly larger than the hypothesized value of 5 points. These data provide insufficient evidence to conclude that the NLP treatment procedure reduces speech anxiety by more than five points beyond the presently used procedure.

EXERCISES

10.9 Determine the value of the test statistic and the p value that would result from the hypothesis test in each of the following cases:

(a) H_0: $\mu_1 - \mu_2 \leqslant 100$, H_a: $\mu_1 - \mu_2 > 100$, $n_1 = 31$, $n_2 = 34$, $\bar{x}_1 = 190$, $\bar{x}_2 = 80$, $s_1 = 25.1$, $s_2 = 20$.

✓ (b) H_0: $\mu_1 - \mu_2 = 0$, H_a: $\mu_1 - \mu_2 \neq 0$, $n_1 = 40$, $n_2 = 64$, $\bar{x}_1 = 4.5$, $\bar{x}_2 = 5.8$, $\sigma_1 = 19.1$, $\sigma_2 = 49.3$.

(c) H_0: $\mu_1 - \mu_2 \geqslant 406$, H_a: $\mu_1 - \mu_2 < 406$, $n_1 = 100$, $n_2 = 100$, $\bar{x}_1 = 1050$, $\bar{x}_2 = 650$, $s_1 = 900$, $s_2 = 330$.

10.10 The Department of Psychology is interested in comparing the utilization of two student laboratories. A crude measure of utilization was used, namely, time between sign-in and sign-out, in hours. Attendance records for the two labs from the previous semester were obtained and 40 students were randomly selected for each lab. The results are summarized below:

	n	$\bar{x}$	s
Lab 1	40	1.5	0.9
Lab 2	40	1.3	1.1

(a) Construct a 90% confidence interval for the true population difference $(\mu_1 - \mu_2)$ in the utilization of the two labs.

(b) If a 99% confidence interval is wanted, but the maximum error E has to remain the same as obtained in part (a), what are the optimum sample sizes n_1 and n_2?

(c) Conduct a hypothesis test to determine if there is any difference in the utilization rate of the two labs. Use $\alpha = .05$.

(d) Indicate the p value for the test.

10.11 Fort Worth and Dallas are two large cities that, despite their geographical closeness, have somewhat different economies. To find out the difference in the amount of unemployment, a statistician randomly interviewed 120 unemployed workers from Fort Worth and 150 unemployed workers from Dallas. These people indicated the number of weeks they had been out of work over the past 52 weeks. The data are as follows. Total number of weeks: Fort Worth, 2031; Dallas, 3713. Standard deviation: Fort Worth, 1.3; Dallas, 2.1. Find a 90% CI for the mean difference in the number of weeks unemployed for the work forces of Fort Worth and Dallas.

10.12 First National Bank and City National Bank are competing for customers who would like to open IRAs (individual retirement accounts). Thirty-two weeks are randomly selected for First National Bank and another 32 weeks are randomly selected for City National. The total amount of deposits from IRAs is noted for each week. A summary of data (deposits in thousands of dollars) from the survey is as follows. First National: $\bar{x} = 4.1$, $s = 1.2$. City National: $\bar{x} = 3.5$, $s = 0.9$. Use a 98% CI to estimate the difference in the mean weekly deposits from IRAs for each bank.

10.13 The role played by cholesterol in the development of "hardening of the arteries" (atherosclerosis) and heart disease has been widely reported. In one experiment, 100 patients considered to be high-risk persons were split into two groups of 50 each. One group, the experimental group, was put on a special diet with a high proportion of fish like salmon, tuna, mackerel, and cod. Fish oil from these deep-sea fish is known to be very rich in Omega-3 fatty acids. The other control group of 50 patients was maintained on a standard diet (high-protein, low-fat, complex carbohydrates and polyunsaturated cooking oil). Both groups experienced a drop in blood cholesterol, measured as milligrams of cholesterol per deciliter of blood. The results are given below:

	Number of Subjects	Average Drop in Cholesterol (mg/dl)	Standard Deviation
Experimental Group (Omega-3 Diet)	50	80	20
Control Group (Polyunsaturated Diet)	50	50	32

(a) Do the above data suggest that the Omega-3 (fish) diet is more effective in reducing blood cholesterol levels than the standard diet? Use $\alpha = 0.05$.

(b) Use the p value criterion to decide the above issue.

10.14 (a) In exercise 10.13 above, construct the 90% and 99% confidence intervals for the difference between the mean drop in cholesterol levels for the two groups.

(b) To estimate the true difference to within 3 mg/dl (that is, $E = 3$), what is the appropriate size of the samples, assuming you want to minimize the total sample size? Use $\alpha = .05$.

10.15 The systems manager of Ace Manufacturing is about to purchase a microcomputer and has narrowed her choices to the Alpha and Gemini models. She is seriously concerned about the cost of maintenance of these two models. After interviewing 40 experts on the Alpha model and a different 40 experts on the Gemini model, she obtained the following information on the cost of maintenance. Average annual cost of maintenance: Alpha, \$46.50; Gemini, \$37.20. Standard deviation: Alpha, \$4.20; Gemini, \$6.10. Do the data indicate a significant difference in the cost of maintenance for the two models? Use the *p* value to justify your answer.

10.16 An education analyst is studying the performance of high-school seniors on the SAT examination. He is specifically testing whether there is any difference between the mean SAT scores of seniors who attended public schools and those who attended private schools. He believes that private schools score more than 50 points higher than public schools. Using the following information, what would be your conclusion at a significance level of 0.05? Do you agree with the analyst? Seniors—public schools: $\bar{x} = 590$, $s = 67$, $n = 40$. Seniors—private schools: $\bar{x} = 680$, $s = 110$, $n = 55$.

10.17 For exercise 10.16, construct a 95% CI for the mean difference in the SAT scores for seniors for the two types of schools. What does this tell you?

10.18 Alan Nakkupuchi, a designer of high-quality stereo systems, is interested in constructing a 95% CI for the difference between the signal/noise ratios for the two models he developed. Data were obtained from each model as follows: Model A: $\bar{x} = 54$, $n = 45$, $s = 8$. Model B: $\bar{x} = 62$, $n = 60$ $s = 17$. Find the 95% CI.

10.19 Test the null hypothesis of no difference in the mean signal/noise ratios for models A and B in exercise 10.18, using a .02 significance level.

10.20 Consider the following two independent random samples: $n_1 = 37$, $s_1 = 210$, $\bar{x}_1 = 1360$, and $n_2 = 64$, $s_2 = 108$, $\bar{x}_2 = 1270$. Test the hypotheses: H_0: $\mu_1 - \mu_2 \leqslant 75$ and H_a: $\mu_1 - \mu_2 > 75$. Use a 10% significance level.

SMALL INDEPENDENT SAMPLES: COMPARING
10.3   TWO POPULATION MEANS USING A *t* TEST

When dealing with small samples from two populations, we need to consider the assumed distribution of the populations because the Central Limit Theorem no longer applies. This section is concerned with comparing two population means when two small independent random samples are used. It differs from the previous section in two respects:

1 We are dealing with small samples.

2 We have reason to believe that the two populations of interest are normal populations. In Figures 10.1 and 10.3, where we had large sample sizes, this was not a necessary assumption. When you use small samples from two populations, one or both of which appear to be not normally distributed, a nonparametric procedure is the proper method for analyzing such data. This will be discussed in Section 10.4.

In Chapter 9, we showed that, when going from large samples to small samples from normal populations, the confidence interval and hypothesis-testing procedures both remained exactly the same, except we use the t distribution, rather than the Z distribution, to describe the test statistic. We will use the same approach for small samples from two populations.

Confidence Interval for $\mu_1 - \mu_2$ (Small Independent Samples)

When using large samples from the two populations to compare μ_1 and μ_2, we used the Z statistic defined by

$$Z = \frac{\bar{X}_1 - \bar{X}_2}{\sqrt{\dfrac{s_1^2}{n_1} + \dfrac{s_2^2}{n_2}}}$$

When using small samples ($n_1 < 30$ or $n_2 < 30$), this statistic no longer approximates the standard normal. To make matters more complicated, it is not a t random variable either. However, this expression is approximately a t random variable with a somewhat complicated expression used to derive the degrees of freedom (df). So we define

$$t' = \frac{\bar{X}_1 - \bar{X}_2}{\sqrt{\dfrac{s_1^2}{n_1} + \dfrac{s_2^2}{n_2}}} \qquad \textbf{(10-10)}$$

This statistic approximately follows a t distribution with degrees of freedom, df, given by

$$df \text{ for } t' = \frac{\left[\dfrac{s_1^2}{n_1} + \dfrac{s_2^2}{n_2}\right]^2}{\dfrac{\left(\dfrac{s_1^2}{n_1}\right)^2}{n_1 - 1} + \dfrac{\left(\dfrac{s_2^2}{n_2}\right)^2}{n_2 - 1}} \qquad \textbf{(10-11)}$$

Admittedly, equation 10-11 is a bit messy, but a good calculator or computer package makes this calculation relatively painless. To be on the conservative side, if df as calculated is not an integer $(1, 2, 3, \ldots)$, it should be rounded *down* to the next integer. As a check of your calculations, the df should be between A and B, where A is the smaller of $(n_1 - 1)$ and $(n_2 - 1)$, and B is $(n_1 - 1) + (n_2 - 1)$.

When finding the *df*, you can scale both s_1 and s_2 any way you wish, provided you scale them both the same way. By scaling, we mean that you can use s_1 and s_2 as is, or you can move the decimal point to the right or left. The resulting *df* will be the same regardless of the scaling used. However, when you evaluate the test statistic *t*, or later perform a test of hypothesis, you must return to the original values of s_1 and s_2.

To derive an approximate confidence interval for $\mu_1 - \mu_2$, we use the same logic as in the previous (large samples) procedure. Thus, a $(1 - \alpha) \cdot 100\%$ confidence interval for $\mu_1 - \mu_2$ (small samples) is:

$$(\bar{X}_1 - \bar{X}_2) - t_{\alpha/2,df}\sqrt{\frac{s_1^2}{n_1} + \frac{s_2^2}{n_2}}$$

$$\text{to}\quad (\bar{X}_1 - \bar{X}_2) + t_{\alpha/2,df}\sqrt{\frac{s_1^2}{n_1} + \frac{s_2^2}{n_2}} \tag{10-12}$$

where *df* is specified in equation 10-11. If *df* is not an integer, round this value down to the next integer.

EXAMPLE 10.4

A local auto dealer wants to know whether single male buyers purchase the same amount of "extras" (such as air conditioning, power steering, exterior trim) as do single females when ordering a new car. A sample of eight males and ten females was obtained. The data consist of the amounts of the ordered extras in hundreds of dollars.

Males (Population 1)	Females (Population 2)
23.00	16.42
23.86	14.20
19.20	21.30
15.78	18.46
30.65	11.70
23.12	12.10
17.90	16.50
30.25	9.20
	21.05
	18.05

Construct a 90% confidence interval for $\mu_1 - \mu_2$, letting μ_1 be the average purchase amount for males and μ_2 be the average purchase amount for females.

Solution   Here is a summary of the data from these two samples.

Sample 1 (Males)	Sample 2 (Females)
$n_1 = 8$	$n_2 = 10$
$\bar{x}_1 = 22.97$ (hundred)	$\bar{x}_2 = 15.90$ (hundred)
$s_1 = 5.40$ (hundred)	$s_2 = 4.05$ (hundred)

Your next step is to get a t value from Table A-5. To do this, you first must calculate the correct df using equation 10-11. This is

$$df = \frac{\left[\dfrac{(5.40)^2}{8} + \dfrac{(4.05)^2}{10}\right]^2}{\dfrac{\left(\dfrac{(5.40)^2}{8}\right)^2}{7} + \dfrac{\left(\dfrac{(4.05)^2}{10}\right)^2}{9}}$$

$$= \frac{27.933}{1.898 + .299} = 12.7$$

Rounding down, we use $df = 12$. Using Table A-5:

$$t_{.10/2,12} = t_{.05,12} = 1.782$$

The resulting 90% confidence interval for $\mu_1 - \mu_2$ is

$$(\bar{X}_1 - \bar{X}_2) - t_{.05,12}\sqrt{\frac{s_1^2}{n_1} + \frac{s_2^2}{n_2}} \quad \text{to} \quad (\bar{X}_1 - \bar{X}_2) + t_{.05,12}\sqrt{\frac{s_1^2}{n_1} + \frac{s_2^2}{n_2}}$$

$$= (22.97 - 15.90) - 1.782\sqrt{\frac{(5.40)^2}{8} + \frac{(4.05)^2}{10}}$$

$$\text{to} \quad (22.97 - 15.90) + 1.782\sqrt{\frac{(5.40)^2}{8} + \frac{(4.05)^2}{10}}$$

$$= 7.07 - 4.10 \text{ to } 7.07 + 4.10$$

$$= 2.97 \text{ to } 11.17$$

So we are 90% confident that the average purchase amount for the male buyers is between \$297 and \$1117 more than the average for female buyers. □

Hypothesis Testing for μ_1 and μ_2 (Small Independent Samples)

The five-step procedure for testing hypotheses concerning μ_1 and μ_2 with large samples also applies to the small-sample situation. The only difference is that Table A-5 is used (rather than Table A-4) to define the rejection region.

EXAMPLE 10.5

In example 10.4, a confidence interval was constructed for the difference between the average purchase amounts for males and females. Can we conclude that these average purchase amounts are in fact not the same? Use a significance level of .10.

Solution   **Step 1**   We are testing for a difference between the two means (not that males purchase more, or vice versa). The corresponding appropriate hypotheses are $H_0: \mu_1 = \mu_2$ and $H_a: \mu_1 \neq \mu_2$:

Step 2   The test statistic is

$$t' = \frac{\bar{X}_1 - \bar{X}_2}{\sqrt{\dfrac{s_1^2}{n_1} + \dfrac{s_2^2}{n_2}}}$$

which approximately follows a t distribution with df given by equation 10-11.

Step 3   You next need the df to determine your rejection region. In example 10.4, we found that $df = 12$. Because $H_a: \mu_1 \neq \mu_2$, we will reject H_0 if t' is too large ($\bar{X}_1$ is significantly larger than $\bar{X}_2$) or if t' is too small ($\bar{X}_1$ is significantly smaller than $\bar{X}_2$). As in previous two-tailed tests using the Z or t statistic, H_0 is rejected if the absolute value of t' exceeds the value from the table corresponding to $\alpha/2$. Using Table A-5, the rejection region for this situation will be

$$\text{Reject } H_0 \text{ if } |t'| > t_{\alpha/2, df} = t_{.05, 12} = 1.782$$

Step 4   The value of the test statistic is

$$t'^* = \frac{22.97 - 15.90}{\sqrt{\dfrac{(5.40)^2}{8} + \dfrac{(4.05)^2}{10}}} = \frac{7.07}{2.299} = 3.08$$

Because $t'^* = 3.08 > 1.782$, we reject H_0. Consequently, the difference between the sample means (7.07) is significantly large, which leads to a rejection of the null hypothesis.

Step 5   There is a difference in the mean purchase amounts for the two sexes.

□

COMMENTS   The hypotheses in example 10.5 could be written as $H_0: \mu_1 - \mu_2 = 0$ and $H_a: \mu_1 - \mu_2 \neq 0$. Having already determined a 90% confidence interval for $\mu_1 - \mu_2$, a much simpler way to perform this two-tailed test (using $\alpha = .10$) would be to reject H_0 if 0 does not lie in the confidence interval for $\mu_1 - \mu_2$, and fail to reject H_0 otherwise. The confidence interval according to example 10.4 is (2.97, 11.17), which does not contain zero, and so we reject H_0 (as before).

This alternative method of testing H_0 versus H_a holds only for a two-tailed test in which the significance level of the test α and the confidence level $[(1 - \alpha) \cdot 100\%]$ of the confidence interval "match up." For example, a significance level of $\alpha = .05$ would correspond to a 95% confidence interval, a value of $\alpha = .10$ would correspond to a 90% confidence interval, and so on.

A MINITAB solution to example 10.5 is provided in Figure 10.6. The calculated p value is $p = .0096$. Based on this extremely small value, we again reject H_0.

Figure 10.6

MINITAB solution to
example 10.5

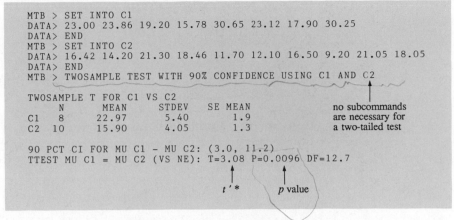

```
MTB > SET INTO C1
DATA> 23.00 23.86 19.20 15.78 30.65 23.12 17.90 30.25
DATA> END
MTB > SET INTO C2
DATA> 16.42 14.20 21.30 18.46 11.70 12.10 16.50 9.20 21.05 18.05
DATA> END
MTB > TWOSAMPLE TEST WITH 90% CONFIDENCE USING C1 AND C2

TWOSAMPLE T FOR C1 VS C2
       N       MEAN      STDEV     SE MEAN
C1     8       22.97     5.40      1.9
C2     10      15.90     4.05      1.3

90 PCT CI FOR MU C1 - MU C2: (3.0, 11.2)
TTEST MU C1 = MU C2 (VS NE): T=3.08 P=0.0096 DF=12.7
```

no subcommands
are necessary for
a two-tailed test

$t'*$ p value

Notice that the procedure in this section for testing μ_1 versus μ_2 and constructing confidence intervals for $\mu_1 - \mu_2$ made no mention as to whether the population variances (or standard deviations) were equal or not. In fact, we can say that this procedure did not assume that $\sigma_1 = \sigma_2$; it also did not assume that $\sigma_1 \neq \sigma_2$. Next, we will examine a special case where we have reason to believe that the standard deviations are equal. For this situation, we will define another t test to detect any difference between the population means.

Special Case of Equal Variances

There are some situations in which we are willing to assume that the population variances (σ_1^2 and σ_2^2) are equal. This is common in many testing situations for which, based on past experience, you are convinced that the variation within population 1 is the same as the variation within population 2.

Another situation in which we may assume $\sigma_1 = \sigma_2$ arises when we obtain two additional samples from the two populations, which we use strictly to determine if the population standard deviations are equal. If there is not sufficient evidence to indicate that $\sigma_1 \neq \sigma_2$, then there is no harm in assuming that $\sigma_1 = \sigma_2$. A procedure for comparing the population standard deviations will be discussed in Section 10.7.

Why make the assumption that $\sigma_1 = \sigma_2$? Remember, we are still interested in the means, μ_1 and μ_2. As before, we would like to obtain a confidence interval for $\mu_1 - \mu_2$ and to perform a test of hypothesis. If, in fact, σ_1 is equal to σ_2, we can construct a slightly stronger test of μ_1 versus μ_2. By "stronger," we mean that we are more likely to reject H_0 when it is actually false. This test is said to be more **powerful**.

For this case, because we believe that $\sigma_1^2 = \sigma_2^2 = \sigma^2$ (say), it makes sense to combine—or **pool**—our estimate of σ_1^2 (s_1^2) with the estimate of σ_2^2 (s_2^2) into one estimate of this common variance (σ^2). The resulting estimate of σ^2 is called the **pooled sample variance** and is written s_p^2. This estimate is merely a weighted average of s_1^2 and s_2^2 defined by

$$s_p^2 = \frac{(n_1 - 1)s_1^2 + (n_2 - 1)s_2^2}{n_1 + n_2 - 2} \qquad \textbf{(10-13)}$$

Constructing Confidence Intervals for $\mu_1 - \mu_2$ To construct the confidence interval, we make two changes in the previous procedure. First, t' is replaced by

$$t = \frac{\bar{X}_1 - \bar{X}_2}{\sqrt{\dfrac{s_p^2}{n_1} + \dfrac{s_p^2}{n_2}}} \qquad \textbf{(10-14)}$$

$$= \frac{\bar{X}_1 - \bar{X}_2}{s_p \sqrt{\dfrac{1}{n_1} + \dfrac{1}{n_2}}} \qquad \textbf{(10-15)}$$

Here (unlike the previous test statistic), t exactly follows a t distribution (assuming the two populations follow normal distributions).

Second, the df for t are much easier to derive:

$$df = n_1 + n_2 - 2$$

So you avoid the difficult df calculation in equation 10-11, but you need to derive the pooled variance s_p^2 using the individual sample variances s_1^2 and s_2^2.

As a check, your resulting pooled value for s_p^2 should be between s_1^2 and s_2^2 since it is a weighted average of these two values.

Hypothesis Testing for μ_1 and μ_2   In hypothesis testing for $\mu_1 - \mu_2$, the previous procedure applies except that t' is replaced by t, where the df used in Table A-5 are $df = n_1 + n_2 - 2$ rather than $df =$ value from equation 10-11.

In examples 10.4 and 10.5, we examined the average amounts spent on "extras" for male and female automobile purchasers. Assume we have determined from previous studies that the variation of the purchase amounts is not affected by sex. Assuming that σ_1^2 (males) $= \sigma_2^2$ (females), how can we construct a 90% confidence interval for $\mu_1 - \mu_2$ and determine whether there is a difference in the mean purchase amounts?

Sample 1 (males)	Sample 2 (females)
$n_1 = 8$	$n_2 = 10$
$\bar{x}_1 = 22.97$ (hundred)	$\bar{x}_2 = 15.90$ (hundred)
$s_1 = 5.40$ (hundred)	$s_2 = 4.05$ (hundred)

Our first step is to pool the sample variances:

$$s_p^2 = \frac{(8-1)(5.40)^2 + (10-1)(4.05)^2}{8 + 10 - 2} = \frac{(7)(29.16) + (9)(16.40)}{16}$$

$$= \frac{351.72}{16} = 21.98$$

$$s_p = \sqrt{21.98} = 4.69 \text{ (hundred)}$$

Is 21.98 between 16.40 and 29.16? Yes. Consequently, $s_p^2 = 21.98$ is our estimate of the common variance (σ^2) of the two populations. To find a 90% confidence interval for $\mu_1 - \mu_2$, we use

$$(\bar{X}_1 - \bar{X}_2) - t_{\alpha/2, df} \sqrt{\frac{s_p^2}{n_1} + \frac{s_p^2}{n_2}}$$

$$\text{to} \quad (\bar{X}_1 - \bar{X}_2) + t_{\alpha/2, df} \sqrt{\frac{s_p^2}{n_1} + \frac{s_p^2}{n_2}}$$

$$(10\text{-}16)$$

where $df = n_1 + n_2 - 2$ and $\alpha = .10$.

Because $n_1 + n_2 - 2 = 16$, we find (from Table A-5) that $t_{.05, 16} = 1.746$. Next,

$$\sqrt{\frac{s_p^2}{n_1} + \frac{s_p^2}{n_2}} = s_p \sqrt{\frac{1}{n_1} + \frac{1}{n_2}}$$

$$= 4.69 \sqrt{\frac{1}{8} + \frac{1}{10}} = 2.225$$

The resulting confidence interval is

$$(22.97 - 15.90) - (1.746)(2.225) \quad \text{to} \quad (22.97 - 15.90) + (1.746)(2.225)$$

$$= 7.07 - 3.88 \text{ to } 7.07 + 3.88$$

$$= 3.19 \text{ to } 10.95$$

Comparing this result to the confidence interval in example 10.4, you see little difference in the two confidence intervals although the interval using the pooled variance is a bit narrower. These intervals often differ considerably, depending on the relative sizes of n_1 and n_2 as well as the relative values of s_1^2 and s_2^2.

Now we wish to test $H_0: \mu_1 = \mu_2$ versus $H_a: \mu_1 \neq \mu_2$. For this particular example, we can, as noted earlier, reject H_0 (using $\alpha = .10$) because zero does not lie in the previously derived confidence interval for $\mu_1 - \mu_2$. When using the five-step procedure, there are only two changes we need to make when using the pooled sample variances. First, when defining our rejection region, we use

Figure 10.7

MINITAB pooled
variances solution
using data from
example 10.4

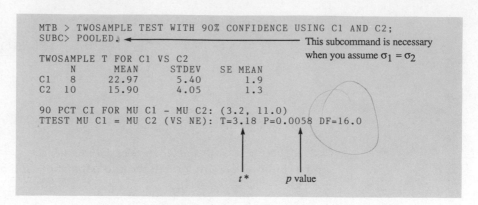

$n_1 + n_2 - 2 = 16$ *df*. From Table A-5, the test procedure is to

$$\text{Reject } H_0 \text{ if } |t| > t_{\alpha/2, df}$$

where $t_{.05,16} = 1.746$.

Second, the value of our test statistic is now

$$t = \frac{\bar{X}_1 - \bar{X}_2}{s_p \sqrt{\dfrac{1}{n_1} + \dfrac{1}{n_2}}} \qquad\qquad \textbf{(10-17)}$$

Hence,

$$t = \frac{22.97 - 15.90}{4.69 \sqrt{\dfrac{1}{8} + \dfrac{1}{10}}} = \frac{7.07}{2.225}$$

$$= 3.18$$

Because $3.18 > 1.746$, we reject H_0; once again the two sample means are significantly different. We conclude that there is a difference in the population mean purchase amounts for males and females.

A MINITAB solution for this example is provided in Figure 10.7. As in example 10.5 (Figure 10.6), we obtain a very small *p* value when pooling the sample variances. For this particular example, we observe little difference in the two solutions.

To Pool or Not to Pool? You might think, based on the previous examples, that it really does not matter whether you assume $\sigma_1 = \sigma_2$ or not. The two confidence intervals were nearly the same and the tests of hypothesis results were very close, with similar *p* values (both .01, rounding to two decimal places). However, this is not always the case. Unless you have strong evidence that the variances are the same, we suggest you not pool the sample variances, and use the test statistic defined in equation 10-10. If you assume that $\sigma_1 = \sigma_2$ and use the t test statistic in equation 10-17, but in fact $\sigma_1 \neq \sigma_2$, your results will be unreliable. This test is quite sensitive to this particular assumption. Also, if σ_1 and σ_2 are the same, then we would expect s_1 and s_2 to be nearly the same. If, in addition, $n_1 = n_2$ (or nearly so), then the computed values of t' and t will be practically identical (including the df). What this means is that you have little to gain by pooling the variances (and using t), but a great deal to lose if your assumption is incorrect.

We will show you in Section 10.7 how to use two samples to test the hypothesis that $\sigma_1 = \sigma_2$. With those results in hand, one possible procedure to use when testing the means would be (1) if you reject $H_0: \sigma_1 = \sigma_2$, then use t' to test $H_0: \mu_1 = \mu_2$, and (2) if you FTR $H_0: \sigma_1 = \sigma_2$, then use t to test $H_0: \mu_1 = \mu_2$.

At first glance this may appear to be statistically sound, but it has some problems. The main one is that these two tests use the same data, and so the tests are not performed independently of one another. Also, your actual significance level may not be the α that you had previously chosen before you saw any data. This can be a valid procedure if you obtain separate samples—one to test the σ values and the other to test the μ values. Again however, caution is in order since the test $H_0: \sigma_1 = \sigma_2$ is very sensitive to the assumption of normal populations and so using small samples to carry out this test can be unreliable. Consequently, if there is reason to believe that the standard deviations might not be equal, a safe procedure is to proceed as if they are not.

By comparing the standard deviations using separate data from the two populations, one can decide whether the pooling procedure should be used when using additional data to test μ_1 versus μ_2. If you reject $H_0: \sigma_1 = \sigma_2$, then the t' statistic in equation 10-10 is the proper test statistic to use on a test for the means because it does not assume that the population standard deviations are equal. On the other hand, if you fail to reject H_0, then the t statistic in equation 10-17, which does assume that $\sigma_1 = \sigma_2$, is the recommended test statistic for testing μ_1 versus μ_2.

Small Sample Tests for μ_1 and μ_2

Two-Tailed Test

$H_0: \quad \mu_1 - \mu_2 = D_0$

$H_a: \quad \mu_1 - \mu_2 \neq D_0$

($D_0 = 0$ for $H_0: \mu_1 = \mu_2$)

Reject H_0 if $|T| > t_{\alpha/2, df}$
where, not assuming $\sigma_1 = \sigma_2$:

$$T = t' = \frac{(\bar{X}_1 - \bar{X}_2) - D_0}{\sqrt{\dfrac{s_1^2}{n_1} + \dfrac{s_2^2}{n_2}}}$$

Or, assuming $\sigma_1 = \sigma_2$:

$$T = t = \frac{(\bar{X}_1 - \bar{X}_2) - D_0}{s_p \sqrt{\dfrac{1}{n_1} + \dfrac{1}{n_2}}}$$

and, for t':

$$df = \frac{\left[\dfrac{s_1^2}{n_1} + \dfrac{s_2^2}{n_2}\right]^2}{\dfrac{\left(\dfrac{s_1^2}{n_1}\right)^2}{n_1 - 1} + \dfrac{\left(\dfrac{s_2^2}{n_2}\right)^2}{n_2 - 1}}$$

and, for t:

$$df = n_1 + n_2 - 2$$

where

$$s_p = \sqrt{\frac{(n_1 - 1)s_1^2 + (n_2 - 1)s_2^2}{n_1 + n_2 - 2}}$$

One-Tailed Test

$H_0: \quad \mu_1 - \mu_2 \leqslant D_0$　　　　　　　$H_0: \quad \mu_1 - \mu_2 \geqslant D_0$

$H_a: \quad \mu_1 - \mu_2 > D_0$　　　　　　　$H_a: \quad \mu_1 - \mu_2 < D_0$

$(D_0 = 0$ for $H_0: \quad \mu_1 \leqslant \mu_2)$　　　$(D_0 = 0$ for $H_0: \quad \mu_1 \geqslant \mu_2)$

Reject H_0 if $T > t_{\alpha, df}$　　　　　　Reject H_0 if $T < -t_{\alpha, df}$

EXERCISES

10.21 Achieving a high score on the LSAT examination is a prerequisite to getting accepted to law school. Scores on the LSAT are considered to be normally distributed. Two law schools decided to compare the mean scores on the LSAT for students enrolled in their schools. Is there sufficient evidence to indicate that the average scores differ between the two schools? Law school 1: $\bar{x} = 680$, $s = 84$, $n = 15$. Law school 2: $\bar{x} = 634$, $s = 92$, $n = 21$. Use a 1% significance level. Assume that the population variances are equal for law school 1 and law school 2.

10.22 Construct a 95% CI for $\mu_1 - \mu_2$ in exercise 10.21. Assume that the population variances are equal for the two law schools.

10.23 Among many phenomenological classifications proposed for schizophrenia, one is based on distinguishing patients with "positive symptoms" (hallucinations, delusions, bizarre behavior, and thinking disorder) from those with "negative symptoms" (blunted affect, emotional withdrawal, poverty of thought, and anhedonia). Different pathological processes have been implicated for each type. Volkow et al. (1987) used positron emission tomography (PET) to measure cerebral metabolism of glucose (which has been shown to reflect brain function) in 18 schizophrenic patients. To create conditions of resting and activation, an eye tracking task was used. The study examined various regions of the brain, but, for this exercise, we shall look at only the results reported for the left frontal region of the brain.

ABSOLUTE REGIONAL GLUCOSE METABOLISM AT BASELINE AND DURING EYE TRACKING TASK FOR PATIENTS WITH CHRONIC SCHIZOPHRENIA

| | 8 Patients with Positive Symptoms | | | | 10 Patients with Negative Symptoms | | | |
| | Baseline | | Task | | Baseline | | Task | |
	Mean	S.D.	Mean	S.D.	Mean	S.D.	Mean	S.D.
Absolute metabolic value (μmol) glucose/100 g. tissue/min) in left frontal brain region	38.25	3.84	39.59	4.05	32.85	5.50	32.67	5.30

Source: N. D. Volkow, A. P. Wolf, P. Von Gelder, J. D. Brodie, J. E. Overall, R. Cancro, and F. Gomez-Mont. Phenomenological correlates of metabolic activity in 18 patients with chronic schizophrenia. Am. J. Psychiatry **144** (2): 154 (February 1987), Table 3.

(a) Assuming equal population variances, is there a significant difference between mean baseline metabolic values in the left frontal region for positive symptom and negative symptom patients? Let $\alpha = .05$.

(b) Not assuming equal population variances, is there a significant difference between mean metabolic values during the eye tracking task for these two groups of patients? Let $\alpha = .05$.

(c) State the p values for the above tests.

10.24 Are you easily distracted when taking a test? It is generally assumed that a quiet, distraction-free situation is necessary for testing. However, some research does not support the general assumption that distractions in the testing situation influence test scores. Many studies have been conducted on this issue. Trentham (1975) randomly selected 72 students from the sixth-grade classes of three elementary schools in Paducah, Kentucky. These 72 students were randomly divided into experimental and control groups. The "Torrance Test of Creative Thinking" (TTCT) was chosen as the instrument for this study. The experimental group was subjected to distractions like an alarm bell going off, blinding lights, books being dropped, and various other interruptions. Conditions were kept as "ideal" as possible for the control group. t scores were calculated for verbal and figural totals on the TTCT using pooled samples, i.e., assuming equal variances. The table below also provides the computed t statistic for comparing two means:

TTCT Subtest	Distraction	Nondistraction	t statistic
Verbal scores	46.80	53.95	2.95
Figural scores	48.68	52.36	1.63

Source: L. L. Trentham. The effect of distractions on sixth-grade students in a testing situation. J. Educ. Measurement **12** (1): 13–17 (Spring 1975).

(a) What are the degrees of freedom for the above tests?

(b) Given the computed t statistic, is there a significant difference between the TTCT Verbal scores of the distraction and nondistraction groups? Let $\alpha = .01$.

(c) Did the distractions imposed during the test have a significant influence on the TTCT Figural scores, at $\alpha = .01$?

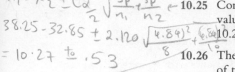

$$\bar{X}_1 - \bar{X}_2 \pm t_{\frac{\alpha}{2}} \sqrt{\frac{s_p^2}{n_1} + \frac{s_p^2}{n_2}}$$

$$38.25 - 32.85 \pm 2.120 \sqrt{\frac{(4.84)^2}{8} + \frac{(6.84)^2}{10}}$$

$$= 10.27 \text{ to } .53$$

10.25 Construct the 95% confidence interval for the true difference between metabolic values at baseline for positive symptom vs. negative symptom patients, in exercise 10.23.

10.26 The production supervisor of Dow Plast is conducting a test of the tensile strength of two types of copper coils. The relevant data are as follows. Coil A: $\bar{x} = 118$, $s = 17$, $n = 9$. Coil B: $\bar{x} = 143$, $s = 24$, $n = 16$. The tensile strengths for the two types of copper coils are approximately normally distributed. Based on the p value, would you reject the hypothesis that $\mu_A - \mu_B = 0$ at a significance level of 7%? Do not assume that the population variances are equal.

10.27 Construct a 99% CI for $\mu_A - \mu_B$ in exercise 10.26. Do not assume that the population variances are equal.

10.28 Using a pooled estimate of the variance in exercise 10.26, test the hypothesis that $\mu_A - \mu_B = 0$. Compare the two answers.

10.29 Repeat exercise 10.13, but this time assume the number of subjects is 20 in each case instead of 50. Other data remain the same. What other additional assumptions might be necessary?

10.30 Do the following sample data support the hypothesis $H_a: \mu_1 - \mu_2 < 10$? The data are sampled from two normally distributed populations. Sample 1: $\bar{x} = 88$, $s = 9$, $n = 12$. Sample 2: $\bar{x} = 80$, $s = 5$, $n = 19$. Use a significance level of .01. Assume that the variances for the two populations are equal.

10.31 Suppose the following hypothesis test is conducted on two normal populations:

$$H_0: \quad \sigma_1^2 = \sigma_2^2, \qquad H_a: \quad \sigma_1^2 \neq \sigma_2^2$$

(a) If a p value of .007 is obtained above, which t test is indicated for a subsequent test of $H_0: \mu_1 = \mu_2$?

(b) If a p value of .140 is obtained for the variance test, which t test is indicated for a subsequent, separate test of $H_0: \mu_1 = \mu_2$?

10.32 A high rate of worker turnover is a major problem in the food-service industry. Although this is widely accepted as an unavoidable fact of life, Mok and Finley (1986) argue that high turnover may be related to factors over which management does have control. They investigated the relationship between job satisfaction and labor turnover at three first-class chain hotels in Hong Kong. A fairly reliable instrument called the Job Descriptive Index (JDI) was used to measure job satisfaction on five components. They first surveyed 373 food-service workers at the 3 hotels. Six months later, turnover data were obtained from the hotels. They found that 39 workers had voluntarily left their jobs. For each of these 39 terminators, they selected 2 stayers who were matched for age, marital status, sex, and length of employment. The JDI component mean scores for the 39 terminators and 78 stayers are given below:

JDI MEAN SATISFACTION SCORES OF TERMINATORS AND STAYERS

JDI Variable	Terminators ($n = 39$)		Stayers ($n = 78$)		t test statistic
	Mean	S.D.	Mean	S.D.	
Work	21.10	9.67	25.88	10.51	2.38
Supervision	28.07	11.83	32.98	11.45	2.16
Pay	11.15	5.73	11.17	5.63	0.02
Promotion	8.64	6.42	10.89	6.55	1.77
Co-Workers	30.20	13.79	35.43	11.21	2.20

Source: C. Mok and D. Finley. Job satisfaction and its relationship to demographics and turnover of hotel food-service workers in Hong Kong. *Int. J. Hospitality Management* **5** (2): 71–78 (1986).

(a) Using the *t* statistic above (based on the pooled variance method) and $\alpha = .05$, compare satisfaction scores of terminators vs. stayers on each of the five JDI components. Is there a significant difference?

(b) The study also reported a correlation of $-.26$ between the overall satisfaction score and turnover, with $p < .01$. Does this corroborate or contradict your findings in (a) based on the means comparisons?

10.4 AN ALTERNATIVE TO THE SMALL SAMPLE t TEST: THE MANN–WHITNEY TEST (OPTIONAL)

A Look at Nonparametric Statistics and the Use of Ranks

So far in Chapters 9 and 10, we have introduced several tests of hypotheses. These tests generally were concerned with things such as the mean or variance of a population. A population mean or variance is referred to as a parameter in statistics, and so these tests are called parametric tests of hypotheses. The common underlying assumption in testing a parameter from a continuous population is that this population has a normal distribution. Any time you use a *t* statistic, you assume a normal population distribution for the data.

What can you do if you have reason to believe that the populations under study are not normally distributed? For example, suppose that data collected previously from these populations have been extremely skewed (not symmetric). One option when dealing with means is to collect large samples. In such a situation, the central limit theorem (CLT) discussed in Section 8.2 assures us that the distribution of the sample estimators is approximately normal regardless of the population distribution. The other alternative, particularly for small or moderate sample sizes, is to use a **nonparametric** statistical procedure that deals with cases in which the assumptions of normality are not necessarily true.

With a nonparametric test the assumptions can be relaxed considerably. A nonparametric test concerning two population means does not assume underlying normal populations—unlike its parametric counterpart, the *t* test discussed in the previous section. Consequently, nonparametric methods are used for situations that violate the assumptions of the parametric procedures.

A common feature for many nonparametric techniques is the use of **ranks**. Given a set of data, we obtain a set of ranks by replacing each data value by its relative position. To illustrate this idea, consider the following eight observations.

$$10.8, \ 6.4, \ 11.7, \ 5.3, \ 9.5, \ 2.5, \ 15.1, \ 10.4$$

Arranged in order, these are

Position	1	2	3	4	5	6	7	8
Value	2.5	5.3	6.4	9.5	10.4	10.8	11.7	15.1

We say that the rank of the value 2.5 is 1, the rank of the value 5.3 is 2, and so forth. Replacing each value by its rank and maintaining the original position produces

$$6, \ 3, \ 7, \ 2, \ 4, \ 1, \ 8, \ 5$$

Many nonparametric procedures use the eight ranks rather than the original data values. By using ranks, we are able to relax the assumptions regarding the underlying populations and develop tests that apply to a wider variety of situations.

You often will encounter an application in which a numeric measurement is extremely difficult to obtain, but a rank value is not. One example of this is a contest; each judge finds it much easier to rank several different people than to assign a numeric value to each one. The data for analysis consist of the rank assigned to each participant. Such data are said to be ordinal because only the relative position of each value has any meaning (see Chapter 1). This is a "weaker" form of data than is the interval form, for which not only the position but also the difference between data values is meaningful. In this text, we have dealt mostly with interval data. Data consisting of temperatures, for example, are interval data; the difference between 60°F and 70°F is the same as that between 65°F and 75°F (10°F). When dealing with ranks, this is not the case; there is no reason to assume that the difference between ranks 1 and 3 is the same as that between ranks 3 and 5.

We will illustrate the use of ranks in this section, which offers another method of testing for equal means (or medians) of two populations using small **independent** samples. This nonparametric technique does not require that the two populations be normally distributed, and is referred to as the Mann–Whitney test.

The Mann–Whitney Test for Small Independent Samples

The previous parametric tests in this chapter were concerned with population means. If two populations have different means, we say that these populations differ in **location**. This implies that population 1 is shifted to the right or left of population 2. When defining the corresponding nonparametric test, the hypotheses will be stated in terms of differing location, rather than differing means.

The **Mann–Whitney test** is named after H. B. Mann and R. Whitney, who developed this test in the 1940s. The purpose of this procedure is to provide a test for differing location that does not require the assumption of normal populations. The test is an alternative to the t tests discussed in the previous section, which do contain this assumption.

The two-tailed hypotheses for the Mann–Whitney test can be written

H_0: the two populations have identical probability distributions

H_a: the two populations differ in location

To use the Mann–Whitney nonparametric technique, we begin by combining (pooling) the two samples into one large sample, and then determining the rank of each observation in the pooled sample. Next, let

$T_1 =$ sum of the ranks of the observations from the first sample in this pooled sample

$T_2 =$ sum of the ranks of the observations from the second sample

The procedure is (if $n_1 = n_2$)

Reject H_0 if T_1 is "too much different" from T_2.

To illustrate this, consider the following pooled sample, where the pooled observations have been arranged in order from smallest to largest. Here A represents a value from population A and B is a value from population B.

Value	A	A	A	A	A	B	B	B	B	B
Rank	1	2	3	4	5	6	7	8	9	10

For this pooled sample, we have

$n_1 = 5$

$T_1 =$ sum of ranks of the five A observations in the pooled sample

$= 1 + 2 + 3 + 4 + 5$

$= 15$

$n_2 = 5$

$T_2 =$ sum of the ranks of the B observations

$= 6 + 7 + 8 + 9 + 10$

$= 40$

Now consider another pooled sample:

Value	A	B	B	A	B	A	A	A	B	B
Rank	1	2	3	4	5	6	7	8	9	10

For this situation, the values are $T_1 = 1 + 4 + 6 + 7 + 8 = 26$ and $T_2 = 2 + 3 + 5 + 9 + 10 = 29$.

In the first pooled sample, there is clear evidence that the second population is shifted to the right of the first population, as indicated by the large difference between $T_1 = 15$ and $T_2 = 40$. This also was evident when you examined this pooled sample, because the values from population A are all less than those from population B. From a parametric point of view, this would imply that $\mu_B > \mu_A$. The Mann–Whitney procedure will result in rejecting H_0: the two populations have identical probability distributions in favor of H_a: the two populations differ in location (or H_a: population 1 is shifted to the left of population 2, had we used a one-tailed test). For the second set of ten pooled observations, there is no indication of a difference in population locations; the A and B values are fairly well mixed in the combined sample. This is evidenced by the values of $T_1 = 26$ and $T_2 = 29$, which are nearly equal here. The Mann–Whitney test will lead to a failure to reject the null hypothesis.

In this section small samples are defined as both n_1 and $n_2 \leqslant 10$. For situations where $n_1 > 10$ or $n_2 > 10$, consult the nonparametric texts in the Extra Readings section at the end of the chapter for a large sample approximation. The Mann–Whitney procedure begins by finding T_1 and T_2 as described previously, and then letting

$$U_1 = n_1 n_2 + \frac{n_1(n_1 + 1)}{2} - T_1 \qquad \textbf{(10-18)}$$

and

$$U_2 = n_1 n_2 + \frac{n_2(n_2 + 1)}{2} - T_2 \qquad \textbf{(10-19)}$$

The Mann–Whitney test is summarized in the accompanying box.

The Mann–Whitney Test for Small Samples

Null hypothesis:

H_0:   the two populations differ in location by an amount equal to D_0 or (if $D_0 = 0$):

H_0:   the two populations have identical probability distributions

Assumptions:

1 Random samples are obtained from each population.
2 The two samples are independent of one another—respective observations are not paired.
3 The sample data are at least ordinal.

Procedure:

1 Assume that $n_1 \leqslant n_2$ (if this is not the case, reverse your populations, so that n_1 is the smaller sample size).
2 Subtract D_0 from each sample 1 value, to obtain "new" sample 1 values. Calculate T_1 and T_2 using these values.
3 Determine U_1 and U_2 from equations 10-18 and 10-19.
4 Use the value from Table A-10 on page A-30 to test H_0 versus H_a where, once again, small p values lead to rejecting H_0.

Two-Sided Test	One-Sided Test	
H_a: the two populations differ in location by an amount $\neq D_0$	H_a: the difference in population locations (population 1 minus population 2) is $> D_0$	H_a: the difference in population locations (population 1 minus population 2) is $< D_0$
or (if $D_0 = 0$):	or (if $D_0 = 0$):	or (if $D_0 = 0$):
H_a: the two populations differ in location	H_a: population 1 is shifted to the right of population 2	H_a: population 1 is shifted to the left of population 2
Reject H_0 if Table A-10 value for U is $< \alpha/2$, where $U =$ minimum of U_1 and U_2.	Reject H_0 if Table A-10 value for U is $< \alpha$, where $U = U_1$.	Reject H_0 if Table A-10 value for U is $< \alpha$, where $U = U_2$.

EXAMPLE 10.6

Example 10.5 was concerned with testing for a difference between males and females in the amount spent on "extras" when purchasing a new automobile. If the individual conducting the study is not willing to assume that the purchase amounts are normally distributed, the same test of hypothesis can be carried out using the Mann–Whitney test which relaxes this assumption.

The sample results (in hundreds of dollars) were:

Male purchase amounts: 23.00, 23.86, 19.20, 15.78, 30.65, 23.12, 17.90, 30.25

Female purchase amounts: 16.42, 14.20, 21.30, 18.46, 11.70, 12.10, 16.50, 9.20, 21.05, 18.05

Use $\alpha = .10$ to test for a difference between male and female purchase amounts.

Solution

The hypotheses are

H_0:   the two populations have identical probability distributions
H_a: the two populations differ in location

The pooled (combined) sample here is

9.20, 11.70, 12.10, 14.20, 15.78, 16.42, 16.50, 17.90, 18.05, 18.46, 19.20, 21.05, 21.30, 23.00, 23.12, 23.86, 30.25, 30.65

Next we indicate which sample each value came from in the pooled sample.

Rank	Male Sample	Female Sample	Ranks for Male Sample	Ranks for Female Sample
1		9.20		1
2		11.70		2
3		12.10		3
4		14.20		4
5	15.78		5	
6		16.42		6
7		16.50		7
8	17.90		8	
9		18.05		9
10		18.46		10
11	19.20		11	
12		21.05		12
13		21.30		13
14	23.00		14	
15	23.12		15	
16	23.86		16	
17	30.25		17	
18	30.65		18	
			$T_1 = 104$	$T_2 = 67$

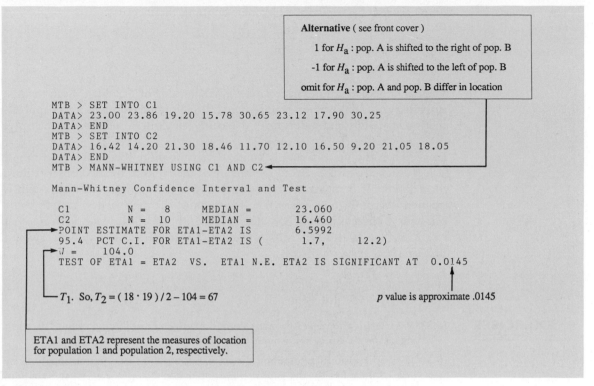

  Figure 10.8

MINITAB procedure
for Mann–Whitney
test (example 10.6)

Using equations 10-18 and 10-19,

$$U_1 = (8)(10) + \frac{(8)(9)}{2} - 104 = 12$$

$$U_2 = (8)(10) + \frac{(10)(11)}{2} - 67 = 68$$

Because this is a nondirectional (two-sided) test, we let $U =$ the minimum of 12 and 68 so $U = 12$.

For $n_1 = 8$, $n_2 = 10$ and $U = 12$, the value in Table A-10 is .0058. Because this is less than $\alpha/2 = .05$, we reject H_0. Based on these data, there is sufficient evidence to indicate a difference between male and female purchase amounts. We obtained the same result (which is not always the case) using the t test procedure in example 10.5. □

The p value for this test is $(2)(.0058) = .0116$. For a one-sided test, the p value would be obtained by finding the value from Table A-10 and not doubling it.

A MINITAB solution for example 10.6 is shown in Figure 10.8. The Mann–Whitney statistic is denoted as W, which is actually the sum of the first sample ranks, T_1. T_2 can be obtained by using the identity

$$T_1 + T_2 = \frac{n(n + 1)}{2}$$

where n = pooled sample size = $n_1 + n_2$. The value of U_1 and U_2 can then be easily derived. The p value obtained here (.0145) is obtained by using an approximation procedure and differs from the exact value (.0116) obtained previously.

Ties When the pooled sample contains two or more identical observations, each is assigned a rank equal to the *average* of the ranks of the tied observations. For example, if there are two observations tied for 6th and 7th place, each is assigned a rank of $(6 + 7)/2 = 6.5$. The rank of the next largest sample value will be 8. If there is a three way tie for 9th, 10th, and 11th place, each is given a rank of $(9 + 10 + 11)/3 = 10$, and the next largest value would be given a rank of 12.

EXERCISES

10.33 What assumptions need be made about the type and distribution of the data when the Mann–Whitney test is used?

10.34 If 15 observations are randomly selected from each of two populations and the sum of the ranks for the sample from population 1 (T_1) is 230, what is the sum of the ranks (T_2) for the sample from population 2?

10.35 Two groups of randomly selected students are given an aptitude test on understanding the financial markets. The first group has eight students, selected from second-semester freshmen, the sum of the ranks (T_1) of the eight students is 65. If the second group has nine students selected from first-semester freshmen, test the null hypothesis that the aptitude of the second-semester freshmen is higher than the aptitude of the first-semester freshmen on understanding the financial markets. Use a 5% significance level.

10.36 A real estate agent claims that the homes in two neighborhoods have the same average value. A random sample from neighborhood A contains the following dollar values: 85,000, 70,000, 74,000, 69,000, 88,000, 89,000. A random sample from neighborhood B consists of the following values: 71,000, 64,000, 68,000, 73,000, 81,000, 69,000, 72,000. Test the alternative hypothesis that the average value of homes in neighborhood B is less than the average value of homes in neighborhood A. Use a 10% significance level.

10.37 The head lawyer of the Brown and Smith firm would like to know whether there is a difference in the number of errors made by the two secretaries employed in the firm. Five randomly selected documents are given to secretary A to type, and five are given to secretary B. The numbers of errors per document is shown in the following table. Using a 5% level of significance, test the hypothesis that there is no difference in the number of errors made by each secretary.

Secretary A	Secretary B
3	2
5	0
4	4
2	3
0	1

10.38 An economist wishes to compare the percent increase in personal income for two suburbs of Chicago. Using the data in the following table, test at the 5% significance level the null hypothesis that there is no difference in the percent increase in personal income.

Suburb A (%)	Suburb B (%)
5.2	2.6
3.1	9.7
10.6	1.2
11.4	1.4
1.2	5.0
0.0	9.8
1.3	11.3
8.4	8.1
9.1	

10.39 A supermarket manager was curious as to which of the two vending machines located at opposite ends of the store was used the most during peak hours. On 10 randomly selected days during the peak hours, the number of users were counted for machine A. On another randomly selected 10 days during the peak hours, the number of users were counted for machine B. From the following data, test at the 5% level of significance the hypothesis that there is no difference in the use of the two vending machines.

Vending Machine A	Vending Machine B
10	9
12	11
13	14
11	10
10	13
15	14
13	9
19	8
11	12
10	13

10.40 Is it possible for the two-sample t test and the Mann-Whitney test to produce different conclusions using the same set of data? Discuss.

10.41 Hypertension (high blood pressure) is acknowledged to be very difficult to actually cure, but it is not too difficult to keep under control. Drugs have generally proven effective, but a team of researchers investigated a biofeedback relaxation technique, which, when combined with drug therapy, gave very promising results. The results given below show the reduction in blood pressure achieved for two independent groups.

Drugs + Biofeedback	Drugs Only
16	15
18	12
25	20
30	18
20	30
50	35
47	40
36	22
60	38
46	

Apply the appropriate nonparametric procedure to decide whether the biofeedback technique combined with drugs is more effective than drugs alone in reducing the blood pressure. Use $\alpha = 0.01$ for the test.

10.5   DEPENDENT SAMPLES: COMPARING TWO POPULATION MEANS USING A t TEST

This section examines the situation in which the two samples are not obtained independently. All discussion up to this point has assumed that the two samples are independent. By not independent, we mean that the corresponding elements from the two samples are paired. Perhaps each pair of observations corresponds to the same city, the same week, the same married couple, or even the same person. This discussion focuses on comparing the two population means for this situation where two dependent samples are obtained from the two populations.

When attempting to estimate or test for the difference between two population means, your first question always should be, "Is there any natural reason to pair the first observation from sample 1 with the first observation from sample 2, the second with the second, and so on?" If there is no reason to pair these data and the samples were obtained independently, the previous methods for finding confidence intervals and testing μ_1 versus μ_2 apply. If the data were gathered such that pairing the values is necessary, then it is extremely important that you recognize this and treat the data in a different manner. We can still determine confidence intervals and perform a test of hypothesis, but the procedure is different.

We will assume here that we have reason to believe the populations follow a normal distribution (as we did in Section 10.3). As a result, we need not worry about large samples versus small samples because we will use the t distribution for our confidence intervals and test of hypothesis, regardless of the sample sizes. Of course, if the samples are large ($n_1 > 30$ and $n_2 > 30$), this distribution is very closely approximated by the standard normal distribution.

If you have reason to suspect that your two populations are not normally distributed, then one alternative is to use a nonparametric procedure—in particular, the Wilcoxon signed rank test (discussed in the next section).

Suppose that Slim-Gym is offering a weight reduction program that they advertise will result in at least a 10-lb weight loss in the first 30 days. Twelve subjects were selected for a study and their weights before and after the weight loss program were recorded. (This is an illustration of the *before vs. after* reason for using dependent samples, previously discussed in Section 10.1.) Define

Population 1: the population of all "before" weights

Population 2: the population of all "after" weights

The following data (in lb) were obtained, where the letter d represents the *difference* of each pair of before and after weights:

Person	1	2	3	4	5	6	7	8	9	10	11	12
Before	132.2	121.6	156.4	167.9	141.7	183.5	137.0	153.1	138.4	145.4	161.0	150.8
After	123.7	111.9	141.1	151.4	124.6	163.8	126.2	145.1	121.9	123.9	156.1	137.9
d	8.5	9.7	15.3	16.5	17.1	19.7	10.8	8.0	16.5	21.5	4.9	12.9
d^2	72.25	94.09	234.09	272.25	292.41	388.09	116.64	64.00	272.25	462.25	24.01	166.41

$$\Sigma d = 8.5 + 9.7 + \cdots + 12.9 = 161.4$$
$$\Sigma d^2 = 72.25 + 94.09 + \cdots + 166.41 = 2458.74$$

Each pair of data values was collected from the same person, so the data values clearly need to be paired—these are dependent samples. It seems reasonable to examine the difference of the two values for each person, so these differences (d), along with the d^2 values, are also shown. We have thus reduced the problem from two sets of values to a single new set.

The parameter of interest here is the **difference** of the population means, μ_d. Put another way, μ_d is the **mean of the population differences**. The following discussion will demonstrate how to use the sample differences to construct a confidence interval for μ_d or perform a test of hypothesis.

Confidence Interval for μ_d Using Paired Samples

The statistic used to derive a confidence interval for μ_d using dependent samples is

$$t_D = \frac{\bar{X}_1 - \bar{X}_2}{s_d/\sqrt{n}} = \frac{\bar{d}}{s_d/\sqrt{n}} \qquad (10\text{-}20)$$

where

$$n = \text{the number of pairs of observations}$$

$$s_d = \text{the standard deviation of the } n \text{ differences}$$

$$= \sqrt{\frac{\Sigma d^2 - (\Sigma d)^2/n}{n-1}}$$

$$df \text{ for } t_D = n - 1$$

This is a t random variable with $n - 1$ df. Notice that the numerator of t_D is the same as before—namely, $\bar{X}_1 - \bar{X}_2$, which is also represented by $\bar{d} = \Sigma d/n$, the mean of the differences. The mean of the differences $\bar{d}$ always is equal to $\bar{X}_1 - \bar{X}_2$. Notice that this can help you in checking your arithmetic when computing the d's.

Using this t statistic, we obtain a $(1 - \alpha) \cdot 100\%$ confidence interval for μ_d:

$$\bar{d} - t_{\alpha/2,n-1} \frac{s_d}{\sqrt{n}} \quad \text{to} \quad \bar{d} + t_{\alpha/2,n-1} \frac{s_d}{\sqrt{n}} \qquad \textbf{(10-21)}$$

EXAMPLE 10.7 Using the Slim-Gym weight reduction data, derive a 95% confidence interval for μ_d, where

$$\mu_d = \text{average weight loss during the weight reduction program}$$

Solution We have

$$\bar{d} = \frac{\Sigma d}{n} = \frac{161.4}{12} = 13.45$$

Notice that

$$\bar{x}_1 = \frac{132.2 + 121.6 + \cdots + 150.8}{12} = 149.08$$

and

$$\bar{x}_2 = \frac{123.7 + 111.9 + \cdots + 137.9}{12} = 135.63$$

so $\bar{d} = \bar{x}_1 - \bar{x}_2 = 13.45$. It checks! Also,

$$s_d = \sqrt{\frac{\Sigma d^2 - (\Sigma d)^2/n}{n-1}} = \sqrt{\frac{2458.74 - (161.4)^2/12}{11}}$$

$$= \sqrt{\frac{287.91}{11}} = \sqrt{26.174} = 5.116$$

The resulting 95% confidence interval for μ_d is

Figure 10.9

MINITAB solution to examples 10.7 and 10.8. This is the correct way to analyze these data.

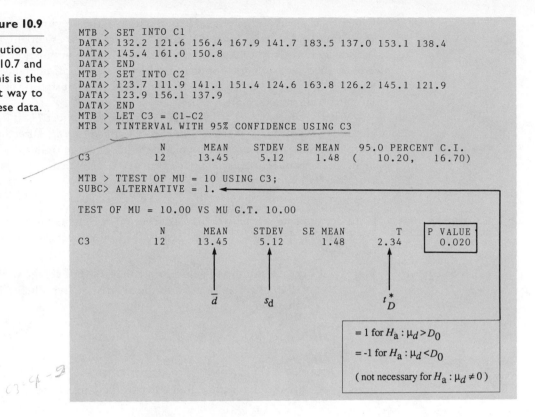

```
MTB > SET INTO C1
DATA> 132.2 121.6 156.4 167.9 141.7 183.5 137.0 153.1 138.4
DATA> 145.4 161.0 150.8
DATA> END
MTB > SET INTO C2
DATA> 123.7 111.9 141.1 151.4 124.6 163.8 126.2 145.1 121.9
DATA> 123.9 156.1 137.9
DATA> END
MTB > LET C3 = C1-C2
MTB > TINTERVAL WITH 95% CONFIDENCE USING C3

               N     MEAN    STDEV   SE MEAN     95.0 PERCENT C.I.
C3            12    13.45     5.12      1.48   (  10.20,   16.70)

MTB > TTEST OF MU = 10 USING C3;
SUBC> ALTERNATIVE = 1.

TEST OF MU = 10.00 VS MU G.T. 10.00

               N     MEAN    STDEV   SE MEAN         T    P VALUE
C3            12    13.45     5.12      1.48      2.34      0.020
```

$\bar{d}$ s_d t_D^*

= 1 for $H_a : \mu_d > D_0$

= -1 for $H_a : \mu_d < D_0$

(not necessary for $H_a : \mu_d \neq 0$)

$$\bar{d} - t_{.025,11}\frac{s_d}{\sqrt{n}} \quad \text{to} \quad \bar{d} + t_{.025,11}\frac{s_d}{\sqrt{n}}$$

$$= 13.45 - 2.201\frac{5.116}{\sqrt{12}} \quad \text{to} \quad 13.45 + 2.201\frac{5.116}{\sqrt{12}}$$

$$= 13.45 - 3.25 \quad \text{to} \quad 13.45 + 3.25$$

$$\doteq 10.20 \quad \text{to} \quad 16.70$$

Based on these data, we are 95% confident that average weight loss using the Slim-Gym weight reduction program is between 10.2 and 16.7 lb. The 95% confidence interval using MINITAB is contained in Figure 10.9 and agrees with the previous result.   □

Hypothesis Testing Using Paired Samples

The test statistic for testing the means using paired samples is

$$t_D = \frac{\bar{d} - D_0}{s_d/\sqrt{n}} \tag{10-22}$$

where D_0 is the hypothesized value of μ_d (10 lb in the previous discussion). If you are testing to determine if $\mu_d \neq 0$, then D_0 is 0. This is essentially the same test statistic used in Chapter 9 to test the mean of a single population, since this test *is* concerned with a single population; the population of differences. In a sense, we have "discarded" the original data, and defined a test statistic using the single sample of differences.

When testing H_0: $\mu_d = D_0$ versus H_a: $\mu_d \neq D_0$, reject H_0 if $|t_D| > t_{\alpha/2, n-1}$. Here $t_{\alpha/2, n-1}$ is obtained from Table A-5 using $n - 1$ *df*. Directional tests are performed in a similar manner by placing α in either the right tail (H_a: $\mu_d > D_0$) or in the left tail (H_a: $\mu_d < D_0$). A summary is provided in the box on paired sample tests for μ_d page 471.

EXAMPLE 10.8   Consider the data on the weight loss program at Slim-Gym. Can you confirm the claim that the weight loss is at least 10 lb? Use a significance level of $\alpha = .05$.

Solution   **Step 1**   We are attempting to demonstrate that μ_d (the average weight loss) is at least 10 lb; this claim goes into the alternative hypothesis. The resulting hypotheses are

$$H_0: \quad \mu_d \leqslant 10$$
$$H_a: \quad \mu_d > 10$$

Step 2   We are dealing with paired data, so the correct test statistic is

$$t_D = \frac{\bar{d} - 10}{s_d/\sqrt{n}}$$

Step 3   What happens to t_D when H_a is true? If $\mu_d > 10$, then we would expect $\bar{d}$ to be larger than 10. So, the test procedure is to

$$\text{Reject } H_0 \text{ if } t_D > k, \quad \text{for some } k > 0$$

What is k? As before, this depends on α and, in the usual manner, we have

$$\text{Reject } H_0 \text{ if } t_D > t_{\alpha, n-1}$$

where $t_{\alpha, n-1}$ is obtained from Table A-5. For this situation, $t_{.05, 11} = 1.796$, and so we

$$\text{Reject } H_0 \text{ if } t_D > 1.796$$

Step 4   Using the sample data,

$$t_D^* = \frac{13.45 - 10}{5.116/\sqrt{12}} = 2.34$$

Because $2.34 > 1.796$, we reject H_0.

Step 5   The average weight loss is at least 10 lb.   □

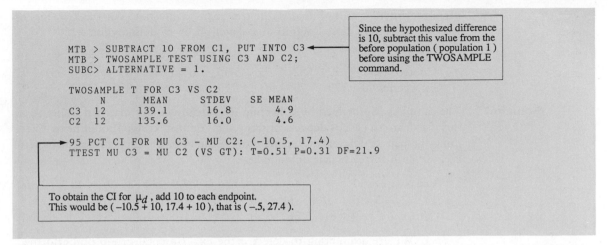

```
MTB > SUBTRACT 10 FROM C1, PUT INTO C3 ◄────
MTB > TWOSAMPLE TEST USING C3 AND C2;
SUBC> ALTERNATIVE = 1.

TWOSAMPLE T FOR C3 VS C2
        N      MEAN     STDEV    SE MEAN
C3     12     139.1     16.8       4.9
C2     12     135.6     16.0       4.6

95 PCT CI FOR MU C3 - MU C2: (-10.5, 17.4)
TTEST MU C3 = MU C2 (VS GT): T=0.51 P=0.31 DF=21.9
```

Since the hypothesized difference is 10, subtract this value from the before population (population 1) before using the TWOSAMPLE command.

To obtain the CI for μ_d , add 10 to each endpoint. This would be (−10.5 + 10, 17.4 + 10), that is (−.5, 27.4).

Figure 10.10

MINITAB solution to example 10.8. This is the INCORRECT way to analyze these data. Compare with Figure 10.9.

A MINITAB solution to example 10.8 is provided in Figure 10.9. Notice that the differences are first derived, and then a standard t test (described in Section 9.4) is used to test that the mean difference (μ_d) is ≤ 10 versus the alternative H_a: $\mu_d > 10$. The resulting p value is $p = .020$, which, using $\alpha = .05$, again results in rejecting H_0 because $p < \alpha$.

What happens if you fail to pair these observations and perform a regular two-sample t test, as we did in Section 10.3 for small independent samples? The results are summarized in Figure 10.10, where we observe an interesting result. The t value (using the test statistic from equation 10-9) now is .51, with a corresponding p value of $p = .31$. This means that, using this test, we now fail to reject H_0. We are unable to demonstrate a weight loss of at least 10 lb, which, according to Figure 10.9, is not a correct conclusion. Figure 10.10 shows convincingly that failing to pair the observations when you should can cause you to obtain an incorrect result. More important, there is nothing to warn you that this has occurred.

EXAMPLE 10.9

A high school geometry teacher is conducting an experiment to determine the effect of using an educational package, consisting of geometric designs and illustrations, to teach geometry. The package is rather expensive, so he wants to determine if using this package results in a significant improvement in the students' comprehension of this material. He randomly assigns students to two groups, a *control* group receiving the standard instruction without these materials and an *experimental* group who do have access to the learning aids. To compensate for student math ability, he selects 10 pairs of students that have equal math ability and assigns one to the control group and the other to the experimental group. In this way, we improve the precision of the statistical test by "masking" the person-to-person variability. This is an example of paired sampling according to *matching characteristics*, previously discussed in Section 10.1.

At the completion of the course, each student is given a test that measures his or her deductive and conceptual ability. Using these data, determine if the mean for population 1 (the control group) is less than the mean of population 2 (the experimental group); that is, the treatment is effective. For this test, higher scores are "better." Use a significance level of .05 to define your test.

Pair	1	2	3	4	5	6	7	8	9	10
Control	72	82	93	65	76	89	81	58	95	91
Experimental	75	79	84	71	82	91	85	68	90	92

Solution   The sample was obtained by collecting data from equally matched students, so this is a clear-cut case of dependent sampling. Your next step should be to determine the paired differences. Define d to be the control score minus the experimental score for each pair of students.

Pair	1	2	3	4	5	6	7	8	9	10	Total
d	-3	3	9	-6	-6	-2	-4	-10	5	-1	-15
d^2	9	9	81	36	36	4	16	100	25	1	317

Step 1   We are attempting to detect if the experimental group outperformed the control group; a one-tailed (directional) test is in order. Let

$$\mu_d = \text{mean for the population of score differences}$$

The correct hypotheses are

$$H_0: \quad \mu_d \geqslant 0$$
$$H_a: \quad \mu_d < 0$$

Steps 2, 3   Using the t_D test statistic, the test will be to

$$\text{Reject } H_0 \text{ if } t_D < -t_{\alpha, n-1}$$

where $t_{\alpha, n-1} = t_{.05, 9} = 1.833$.

Step 4   Using the sample data,

$$\bar{d} = \frac{\Sigma d}{n} = \frac{-15}{10} = -1.50$$

$$s_d = \sqrt{\frac{\Sigma d^2 - (\Sigma d)^2/n}{n-1}}$$

$$= \sqrt{\frac{317 - (-15)^2/10}{9}} = \sqrt{\frac{294.5}{9}}$$

$$= 5.72$$

From these values, we obtain

$$t_D^* = \frac{\bar{d} - 0}{s_d/\sqrt{n}} = \frac{-1.50}{5.72/\sqrt{10}} = -.83$$

Because $-.83 > -1.833$, we fail to reject H_0.

Step 5   Based on these data, there is insufficient evidence to conclude that there is an improvement in the performance of the geometry students as a result of using the education materials.

□

Paired Sample Tests for μ_d

Two-Tailed Test

H_0: $\mu_d = D_0$

H_a: $\mu_d \neq D_0$

Reject H_0 if $|t_D| > t_{\alpha/2, n-1}$

where

1 Each difference d is (sample 1 value) − (sample 2 value)

2 $t_D = \dfrac{\bar{d} - D_0}{s_d/\sqrt{n}}$

3 $\bar{d} = \bar{X}_1 - \bar{X}_2 = \dfrac{\Sigma d}{n}$

4 $s_d = \sqrt{\dfrac{\Sigma d^2 - (\Sigma d)^2/n}{n-1}}$

5 df for $t_D = n - 1$

6 n = number of pairs

One-Tailed Test

H_0: $\mu_d \leq D_0$ H_0: $\mu_d \geq D_0$

H_a: $\mu_d > D_0$ H_a: $\mu_d < D_0$

Reject H_0 if $t_D > t_{\alpha, n-1}$ Reject H_0 if $t_D < -t_{\alpha, n-1}$

EXERCISES

10.42 A hospital is experimenting with the effectiveness of a newly developed drug that controls blood pressure. The blood pressure level is measured using a sphygmomanometer before and after administration of the drug to a sample of hypertensive patients with a history of elevated blood pressure. The question is whether there is a measurable decrease in systolic blood pressure (in mm Hg) after administration of the drug. The difference in blood pressure before and after is believed to be approximately normally distributed.

Patient	Before Drug	After Drug
1	110	94
2	88	81
3	84	82
4	94	88
5	108	97
6	82	85
7	96	77
8	97	89
9	134	110

Handwritten annotations:

$H_0 : \mu_d \leq 0$

$H_a : \mu_d > 0$

$\alpha = .10$

$n = 9$

$d.f. = 8$

$t_{.10, 8} = 2.896$

$t_D = \dfrac{10}{\frac{8.49}{\sqrt{9}}} = 3.53$

$3.53 > 2.896 \Rightarrow$ Reject H_0.

Table columns d and d^2:

d	d^2
16	256
7	49
2	4
6	36
11	121
-3	9
19	361
8	64
24	576
90	1476

$\bar{d} = \dfrac{\Sigma d}{n} = 10$

(a) Using a significance level of .10, can you conclude that the blood–pressure level is lower after the drug is administered?

(b) Should you use an independent or dependent samples statistic to analyze this experiment? → dependent

10.43 Construct a 99% confidence interval for μ_d using the data in exercise 10.42 ($d =$ before $-$ after). $t_{\frac{\alpha}{2}, n-1} = t_{.005, 8} = 3.355$

$$d \pm t_{\frac{\alpha}{2}, n-1} \frac{s_d}{\sqrt{n}}$$

$$10 \pm 3.355 \times \frac{8.49}{\sqrt{9}}$$

$$19.49 \text{ to } .51$$

10.44 Suppose that, in exercise 10.42, the manufacturer of the drug claims that the new drug is effective in reducing the average blood pressure scores by more than 8 mm Hg. Would you support that statement at a significance level of .01?

10.45 The following are No–Pass, No–Play statistics for Dimview Independent School District, listing number of ineligible students barred from extracurricular activities.

School	Boys	Girls
Adams	9	0
Adamson	15	6
Carter	12	15
Hillcrest	15	12
Jefferson	3	0
Kimbell	15	9
Lincoln	12	6
Madison	9	3
Roosevelt	27	15
Smith	9	18
White	18	21
Spruce	15	24

(a) Assuming normally distributed populations, conduct a hypothesis test at $\alpha = .05$ to check the assumption that on the average, there is no difference in the number of boys and girls barred from extracurricular activity.

(b) Find the p value for the test.

(c) Compare the setup of data in this exercise to that in exercise 10.24. Do you observe the difference between dependent samples and independent samples?

10.46 Calculate the 95% confidence interval for the population mean difference, μ_d, in exercise 10.45.

10.47 The following data compare the sales (in billions of dollars) of computer chips, also called semiconductors, from the first quarter of 1985 to the second quarter of 1986.

	Japan	USA
1985 Q1	1.881	2.282
Q2	1.888	2.066
Q3	1.853	1.893
Q4	1.976	1.850
1986 Q1	2.229	1.946
Q2	2.631	2.205

(a) It is commonly said that Japan is outselling and undercutting the United States in the "semiconductor war." If the populations are assumed to be normal, do the above results support that statement at a 10% significance level?

(b) Note that the general trend for Japan's figures is upward, while for the United States, the trend is relatively flat. How could this interfere with the result of the test in (a)?

10.48 Researchers at the State University of New York at Binghamton have said that while older workers were often more productive than their younger counterparts, supervisors tended to rate the older workers lower. Suppose we have the following paired data, giving the actual productivity of older workers and the supervisor's rating:

WORKERS

	1	2	3	4	5	6	7	8
Actual Productivity	82	91	50	82	75	89	90	76
Supervisor's Rating	77	80	65	80	70	83	79	75

Does the above data support the contention that older workers are underrated (i.e., Actual > Supervisor's)? Use $\alpha = 0.05$ and assume normal populations.

10.49 Smart Look, an exercise program developed by Joni Beauty consultants, is claimed to be effective in reducing the weight of a typical overweight woman by more than 17 pounds. To examine the validity of this hypothesis, the program was tried on a group of middle-aged women, and their weights (in pounds) were recorded before and after completion of the exercise program. Assume that the difference in a woman's weight after the program is approximately normally distributed.

Woman	Before	After	Woman	Before	After
1	140	115	5	175	165
2	160	130	6	145	125
3	110	100	7	115	101
4	132	109	8	122	105

Handwritten annotations in table (d, d²):
Woman 1: d=25, d²=625
Woman 2: d=30, d²=900
Woman 3: d=10, d²=100
Woman 4: d=23, d²=529
Woman 5: d=10, d²=100
Woman 6: d=20, d²=400
Woman 7: d=14, d²=196
Woman 8: d=17, d²=289

(a) Using a significance level of .10, what would be your conclusion?

(b) Why did you select the particular test statistic you used to analyze this problem? *dependent*

10.50 Using data from exercise 10.49, test the hypothesis that $H_0: \mu_d = 0$ at a .01 level ($d =$ before − after).

10.51 The effectiveness of an advertisement is usually judged by the extent to which it increases the sales of a certain product. Weekly sales before and after an advertisement are:

Week	Before Campaign	After Campaign
1	3400	4340
2	4100	5270
3	5000	6100
4	3800	4010
5	6200	5750
6	4400	4810
7	3700	3600

Weekly sales are considered to be normally distributed. Construct a 95% CI for μ_d ($d =$ after − before).

Handwritten work in left margin:

$H_0: \mu_d \leq 17$

$H_a: \mu_d > 17$

$\alpha = .10$

$t_D = \dfrac{\bar{d} - D_0}{\frac{S_d}{\sqrt{n}}}, \quad \bar{d} = \dfrac{149}{8} = 18.625$

$S_d = \sqrt{\dfrac{3139 - (149)^2/8}{7}} = 7.21$

$\Sigma d = 149$

$\Sigma d^2 = 31393$

$t_D = \dfrac{18.625 - 17}{\frac{7.21}{\sqrt{8}}} = .638 \sim .64$

$t_{.1,7} = 1.415$

$.64 < 1.415 \Rightarrow$ FTR.

10.52 In exercise 10.51, the advertising department claims that the campaign increased sales by more than \$400 a week. Is there sufficient evidence to support this claim? Base your decision on the resulting p value.

10.6   AN ALTERNATIVE TO THE DEPENDENT SAMPLES t TEST: THE WILCOXON SIGNED RANK TEST (OPTIONAL)

Section 10.4 provided an introduction to an area of statistics called nonparametric statistics. Nonparametric statistical procedures allow you to relax the assumptions regarding the sampled populations and are extremely useful for small samples from populations that quite likely (or knowingly) violate these assumptions. The primary assumption that can be removed is that your populations follow a normal distribution.

Wilcoxon Signed Rank Test for Paired Samples

A method of testing population means using paired sample data was introduced in the previous section, where we used a t test on the sample differences. However, as with all t tests, a key assumption using this method of testing two means is that the differences are normally distributed. When small samples from suspected nonnormal populations are used, or your data consist of ordinal rankings, a nonparametric technique is required. The **Wilcoxon signed rank test** is used for such situations.

The Wilcoxon test begins like its parametric counterpart, the paired-sample t test, by subtracting the data pairs and using the differences to perform the test. As with the paired-sample t test, the hypotheses will be written in terms of the location of the probability distribution for the population differences.

The steps involved in applying the Wilcoxon test are:

1 For each sample pair determine the difference,

$$(\text{sample 1 value}) - (\text{sample 2 value}) - D_0$$

where D_0 is the hypothesized value of the location of the probability distribution for the population differences specified in H_0 and H_a.

2 Arrange the absolute value of these differences in order, assigning a rank to each.

3 Let $T_+ = $ sum of the ranks having a positive value and $T_- = $ sum of the ranks for the negative values.

4 T_+, T_-, or $T = $ the minimum of T_+ and T_- is used to define a test of H_0 versus H_a.

To demonstrate the test, consider the weight loss data provided by a sample of 12 people using the weight loss program at Slim–Gym in example 10.7. The data are repeated below, consisting of the weight before and after the weight loss program.

Weight Before (1)	Weight After (2)	Difference (Before − After − 10) (3)	|Difference| (4)	Rank (5)
132.2	123.7	−1.5	1.5	3 (−)
121.6	111.9	−.3	.3	1 (−)
156.4	141.1	5.3	5.3	7
167.9	151.4	6.5	6.5	8.5
141.7	124.6	7.1	7.1	10
183.5	163.8	9.7	9.7	11
137.0	126.2	.8	.8	2
153.1	145.1	−2.0	2.0	4 (−)
138.4	121.9	6.5	6.5	8.5
145.4	123.9	11.5	11.5	12
161.0	156.1	−5.1	5.1	6 (−)
150.8	137.9	2.9	2.9	5

The hypotheses for this situation are

H_0: the population differences (before − after) are centered at 10

H_a: the population differences are centered at a value > 10.

where $D_0 = 10$.

We proceed by calculating the (sample 1) − (sample 2) − 10 value for each sample pair (column 3). Next, we determine the absolute value of each of these values (column 4) and the rank of each of these absolute values (column 5).

Ties are handled as before by assigning a rank equal to the average of the tied positions. Notice that there is a tie for the 8th and 9th position, so each is assigned a rank of $(8 + 9)/2 = 8.5$. Also, if a pair of observations has a difference equal to zero, then this pair should be deleted from the sample, and n is reduced by one. Other methods exist for handling zero differences, but this is the simplest procedure and works well provided there are not many zero differences.

According to column 3, the negative values are $−1.5$, $−.3$, $−2.0$ and $−5.1$. In column 5, we see their corresponding ranks are 3, 1, 4 and 6. Therefore,

$$T_- = 3 + 1 + 4 + 6 = 14$$

A rule that can simplify the calculations here and serve as a check for arithmetic is that

$$T_+ + T_- = \frac{n(n + 1)}{2}$$

where n = the number of sample pairs. In our example, $n = 12$, so

$$T_+ + T_- = \frac{(12)(13)}{2} = 78$$

which means that $T_+ = 78 - T_- = 78 - 14 = 64$.

Once T_+ and T_- have been obtained, you can use the Wilcoxon signed rank test for testing hypotheses about the location of the population differences.

The Wilcoxon Signed Rank Test for Small Samples (Paired)

Null hypothesis:

H_0:   the population differences are centered at D_0

where "difference" is defined to be population 1 − population 2.
Assumptions:

1 Each data pair is randomly selected.
2 The absolute values of the differences can be ranked.

Procedure:

1 Determine the n differences using each sample pair, where each difference is defined to be sample 1 − sample 2 − D_0.
2 Assign a rank to the absolute value of each difference; define T_+ = sum of the ranks of the positive values and T_- = sum of the ranks of the negative values.

Table A-11 is used to define the rejection region for the following tests:

Two-Sided Test	One-Sided Test	
H_a: the population differences are not centered at D_0.	H_a: the population differences are centered at a value $>D_0$.	H_a: the population differences are centered at a value $<D_0$.
Using the two-sided value from Table A-11, reject H_0 if $T \leqslant$ table value where $T =$ minimum of T_+ and T_-.	Using the one-sided value from Table A-11, reject H_0 if $T_- \leqslant$ table value.	Using the one-sided value from Table A-11, reject H_0 if $T_+ \leqslant$ table value.

EXAMPLE 10.10   Using $\alpha = .05$, are we able to conclude that the weight reduction is at least 10 lbs based upon the Slim-Gym weight loss sample?

Solution The hypotheses here are (1 = before, 2 = after)

H_0: the population differences are centered at 10

H_a: the population differences are centered at a value > 10.

We refer to the "before" population as population 1 and the "after" population as population 2. This agrees with column 3 because our procedure assumes that column 3 is obtained by first subtracting the sample 2 observation *from* the sample 1 observation.

The values of T_+ and T_- were previously derived, where

$$T_- = 14$$

and

$$T_+ = 64$$

The one-sided value in Table A-11 corresponding to $n = 12$, and $\alpha = .05$ is 17.

Consequently, the test is to

$$\text{Reject } H_0 \text{ if } T_- \le 17$$

Because the value of T_- is smaller than 17, we reject H_0 and conclude that there is sufficient evidence to indicate a weight loss of at least 10 lb.

To use MINITAB for the Wilcoxon procedure, you begin by subtracting the two samples and then using the WTEST command on the differences, as illustrated in Figure 10.11. To obtain the value of the test statistic, T_-, use the

Figure 10.11

MINITAB procedure for Wilcoxon signed rank test (example 10.10)

Alternative
1 for H_a : the population differences are centered at a value $> D_0$.
-1 for H_a : the population differences are centered at a value $< D_0$.
omit for H_a : the population differences are not centered at D_0.

```
MTB > SET INTO C1
DATA> 132.2 121.6 156.4 167.9 141.7 183.5 137.0 153.1 138.4
DATA> 145.4 161.0 150.8
DATA> END
MTB > SET INTO C2
DATA> 123.7 111.9 141.1 151.4 124.6 163.8 126.2 145.1 121.9
DATA> 123.9 156.1 137.9
DATA> END
MTB > SUBTRACT C2 FROM C1, PUT INTO C3
MTB > WTEST OF CENTER = 10 USING C3;
SUBC> ALTERNATIVE = 1.

TEST OF MEDIAN = 10.00 VERSUS MEDIAN G.T. 10.00

              N FOR   WILCOXON            ESTIMATED
         N    TEST    STATISTIC  P-VALUE   MEDIAN
C3       12    12       64.0      0.027     13.30
```

T_+. So, $T_- = (12 \cdot 13) / 2 - 64 = 14.$

identity

$$T_- + T_+ = \frac{n(n + 1)}{2}$$

where n is the number of paired observations (12 here). The approximate p value for this example is .027.   □

EXAMPLE 10.11

Example 10.9 was concerned with the effect of using additional learning materials to teach high school geometry. One group (the control group) received the standard method of instruction and another group (the experimental group) was taught using the learning materials, consisting of plastic designs and illustrations. Students were paired according to their prior ability in mathematics and each was given a test at the completion of the course to measure his or her mastery of the subject.

The teacher conducting the experiment is reluctant to assume that the differences in the scores for each pair follow a normal distribution, and due to the small sample sizes (10 in each group), decides to use the nonparametric Wilcoxon procedure. What would be his conclusion using a significance level of $\alpha = .05$?

Solution

The hypotheses now are

H_0:   the population differences are centered at 0

H_a: the population differences are centered at a value < 0

where a difference is defined by subtracting the experimental group score (population 2) from the control group score (population 1).

Pair	(1) Control Group Score	(2) Experimental Group Score	(3) Difference	(4) |Difference|	(5) Rank
1	72	75	−3	3	3.5
2	82	79	3	3	3.5 (+)
3	93	84	9	9	9 (+)
4	65	71	−6	6	7.5
5	76	82	−6	6	7.5
6	89	91	−2	2	2
7	81	85	−4	4	5
8	58	68	−10	10	10
9	95	90	5	5	6 (+)
10	91	92	−1	1	1

Using the one-sided value from Table A-11, the procedure will be to reject H_0 if $T_+ \leq 11$. Notice that there are two ties in column 4, with a tie for 3rd and 4th place (each receiving a rank of 3.5) and a tie for 7th and 8th place (with a corresponding assigned rank of 7.5). The positive differences here are 3, 9, and 5, with corresponding ranks of 3.5, 9, and 6. As a result,

$$T_+ = 3.5 + 9 + 6 = 18.5$$

Since $18.5 > 11$, we fail to reject H_0 and conclude that there is insufficient evidence to indicate that the learning materials improve student performance.

□

EXERCISES

10.53 What assumption needs to made about the sample data used in the Wilcoxon signed rank test? What assumption needs to be made about the distribution of the population?

10.54 From 12 paired observations, it is found that by ranking the magnitude of the differences of the observations in each pair, T_+ (the sum of the ranks of the positive differences) is 27. Using a .05 significance level, can it be concluded that the population differences are centered at a value unequal to zero?

10.55 Assessment of phobic behavior has included self-report, physiological, and *in vivo* behavioral measures. All these measures have certain shortcomings. Milby, Mizes, and Giles (1986) investigated the validity of five convenient measures of phobic behavior. Responses of 14 clinically phobic patients were compared on phobic and neutral scenes presented within the context of systematic desensitization therapy. The researchers used the signed rank test (one-tailed) to compare subjects' scores on phobic and neutral scenes for the five measures being investigated. To minimize calculations for you the corresponding p values for the comparison tests are provided:

Phobic Behavior Measure	Phobic Scene		Neutral Scene		p value
	Mean	S.D.	Mean	S.D.	
1. Latency to clear scene visualization	12.7	8.90	8.85	6.04	< .029
2. Clarity of scene visualization ratings	7.52	1.82	8.89	2.38	< .006
3. Subjective units of discomfort (SUD) at clear scene visualization	26.66	12.80	7.29	7.76	< .001
4. Postscene latency to relaxation	26.44	14.38	14.74	12.89	< .001
5. SUDs at relaxation	9.48	8.48	6.02	8.04	< .01

Source: J. B. Milby, J. S. Mizes, and T. R. Giles. Assessing the process of desensitization therapy: Five practical measures. *J. Psychopathol. Behavioral Assessment* **8** (3): 241–251 (September 1986).

The researchers considered five hypotheses, related to the 5 measures.

(a) At $\alpha = .05$, is the latency to clear scene visualization significantly greater for phobic scenes?

(b) At $\alpha = .01$, is the clarity of scene visualization rating lower for phobic scenes?

(c) At $\alpha = .01$, do phobic scenes have higher SUDs at clear scene visualization than neutral scenes?

(d)   At $\alpha = .01$, is there a longer postscene latency to relaxation for phobic scenes?

(e)   At $\alpha = .01$, is the SUDs score at relaxation higher for phobic scenes?

(f)   Which of the five phoboic measures were found to be statistically useful?

(g)   Why would the Mann–Whitney test be inappropriate for the above data?

10.56 An insurance company believes that employees who have a college degree when hired progress faster in the company than those who do not. Pairs of employees are randomly selected; each pair consists of two people hired at the same time, one person with a college degree and the other without a college degree. The percent increase in pay for the employees after 3 years is recorded below. At a 5% level of significance, can you conclude that employees who have a college degree when hired progress faster than those who do not?

Without College Degree (%)	With College Degree (%)
10	13
9	10
8	6
13	13
14	18
7	10
12	11
11	15
16	20
12	13
9	8
18	16
9	12
15	17
10	9
11	13
10	9

10.57 Seven randomly selected faculty members were asked to evaluate two research project proposals on a scale from 1 to 10, with a higher score indicating a more acceptable proposal. The scores follow. Using a 5% significance level, can you conclude that the proposal for research project 2 is more acceptable than the proposal for research project 1?

Research Project 1	Research Project 2
5	7
3	5
6	9
7	6
8	9
4	6
7	10

10.58 For exercise 10.42, test at the 5% significance level that the blood pressure is less after the drug is administered. Use the Wilcoxon signed rank test. Compare the results using the paired t test and a 5% significance level.

10.59 Do exercise 10.50 without the assumption of normally distributed populations.

10.60 Use the data from exercise 10.48 and apply the Wilcoxon signed rank test to establish or reject the contention that older workers are underrated by their supervisors.

10.61 Refer to exercise 3.17 in which infant mortality rates for breast-fed versus artificially fed babies were provided for 12 locations. The data are reproduced below for easy reference.

Area	Breast-Fed	Artificial
Berlin	57	376
Barmen, Germany	68	379
Hanover, Germany	96	296
Boston	30	212
Paris	140	310
Cologne	73	241
Amsterdam	144	304
Liverpool	84	228
8 U.S. Cities	76	255
Derby, England	70	198
Chicago	2	84
Great Britain	9	18

In exercise 3.17 (b), you addressed the question of whether breast-feeding is associated with lower infant mortality by using an intuitive approach. Now, use a formal approach and apply the Wilcoxon procedure to answer the same question: Is infant mortality significantly lower for breast-fed babies? Use $\alpha = .05$.

COMPARING THE VARIANCES
OF TWO NORMAL POPULATIONS
10.7   USING INDEPENDENT SAMPLES (OPTIONAL)

Once again we concentrate on independent samples from two normal populations, only this time we focus our attention on the variation of these populations rather than on their averages. This is illustrated in Figure 10.12. When estimating and testing σ_1 versus σ_2, we will not be concerned about μ_1 and μ_2. They may be equal, or they may not—it simply does not matter for this test procedure.

In psychological testing applications, you may want to compare the variation of two different testing instruments. You generally prefer a wide spread (large variation) in scores designed to measure aptitude, stability, contentment, or whatever. When comparing two instruments having the same *possible* range of values (such as 0 to 100), you might elect to use that instrument having a larger variation in scores since it better distinguishes between stable and unstable, content and not content, and so on.

When using the *t* test to compare population means based upon the results of small independent samples, you must pay attention to the population standard

Figure 10.12

Comparing two
standard deviations.
Is $\sigma_1 = \sigma_2$?

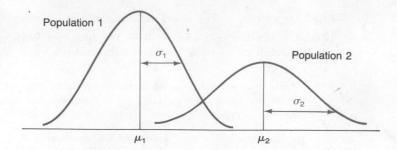

deviations (variances). Based on your belief that σ_1 does or does not equal σ_2, you select your corresponding test statistic for testing the means, μ_1 and μ_2.

In Sections 10.3 and 10.5, when trying to decide if $\mu_1 = \mu_2$, we examined the difference between the point estimators $\bar{X}_1 - \bar{X}_2$. If $\bar{X}_1 - \bar{X}_2$ was large enough (in absolute value), we rejected H_0: $\mu_1 = \mu_2$. When looking at the variances, we use the **ratio of the sample variances**, s_1^2 and s_2^2, to derive a test of hypothesis and construct confidence intervals. We do this because the distribution of $s_1^2 - s_2^2$ is difficult to describe mathematically, but s_1^2/s_2^2 does have a recognizable distribution when in fact σ_1^2 and σ_2^2 are equal. So we define

$$F = \frac{s_1^2}{s_2^2} \qquad\qquad (10\text{-}23)$$

If you were to obtain sets of two samples repeatedly, calculate s_1^2/s_2^2 for each set, and make a histogram of these ratios, the shape of this histogram would resemble the curve in Figure 10.13. This is the **F distribution**. Its shape resembles the chi-square curve—it is nonsymmetric, skewed right (right-tailed), and the corresponding random variable is never negative. There are many F curves, depending on the sample sizes n_1 and n_2. The shape of the F curve becomes more symmetric as the sample sizes n_1 and n_2 increase. As later chapters will demonstrate, the F distribution has a large variety of applications in statistics. Right-tail areas for this random variable have been tabulated in Table A-7.

Figure 10.13

Shape of the
F distribution

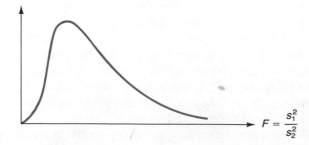

As a final note here, the F statistic in equation 10-23 is highly sensitive to the assumption of normal populations. For larger data sets, it is recommended that you examine the shape of the sample data when using this particular F statistic.

When using the t and χ^2 statistics, we needed a way to specify the sample size(s) because the shape of these curves changes as the sample size changes. The same applies to the F distribution. There are two samples here, one from each population, and we need to specify both sample sizes. As before, we use the degrees of freedom (df) to accomplish this, where

$$v_1 = df \text{ for numerator} = n_1 - 1$$

$$v_2 = df \text{ for denominator} = n_2 - 1$$

So, the F statistic shown in Figure 10.13 follows an F distribution with v_1 and v_2 df provided $\sigma_1^2 = \sigma_2^2$ ($\sigma_1 = \sigma_2$). What happens to F when $\sigma_1 \neq \sigma_2$? Suppose that $\sigma_1 > \sigma_2$. Then we would expect s_1 (the estimate of σ_1) to be larger than s_2 (the estimate of σ_2); we should see

$$s_1^2 > s_2^2$$

or

$$F = \frac{s_1^2}{s_2^2} > 1$$

Similarly, if $\sigma_1 < \sigma_2$, then we expect an F value <1. We will use this reasoning to define a test of hypothesis for σ_1 versus σ_2.

Hypothesis Testing for $\sigma_1 = \sigma_2$

Is $\sigma_1 = \sigma_2$? We use the same five-step procedure for testing a hypothesis concerning the two standard deviations. Your choice of hypotheses is (as usual) a two-tailed test or a one-tailed test. For the two-tailed test, the hypotheses are $H_0: \sigma_1 = \sigma_2$ ($\sigma_1^2 = \sigma_2^2$), and $H_a: \sigma_1 \neq \sigma_2$ ($\sigma_1^2 \neq \sigma_2^2$). For the one-tailed test, the hypotheses are $H_0: \sigma_1 \leq \sigma_2$, and $H_a: \sigma_1 > \sigma_2$ (Figure 10.14A) or $H_0: \sigma_1 \geq \sigma_2$, and $H_a: \sigma_1 < \sigma_2$ (Figure 10.14B).

Notice that the hypotheses can be written in terms of the standard deviations (σ_1 and σ_2) or the variances (σ_1^2 and σ_2^2); if $\sigma_1 = \sigma_2$, then $\sigma_1^2 = \sigma_2^2$.

Figure 10.14

Unequal population variances

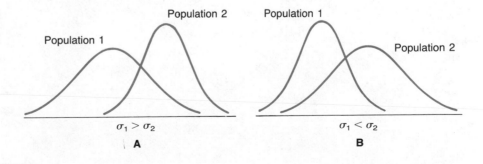

Population 1   Population 2   Population 1   Population 2

$\sigma_1 > \sigma_2$ $\sigma_1 < \sigma_2$

A   B

Figure 10.15

F curve with 10 and
12 *df* for probability
that *F* exceeds 2.75
(2.75 is from Table
A-7b)

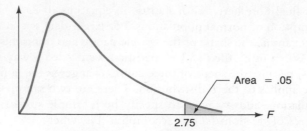

Area = .05

2.75

F

Right-tail areas under an *F* curve are provided in Table A-7. Notice that we have a table for areas of .1 (Table A-7a), .05 (Table A-7b), .025 (Table A-7c), and .01 (Table A-7d). These are the most commonly used values. For each table, the *df* for the numerator (v_1) run across the top, and the *df* for the denominator (v_2) run down the left margin.

Suppose we want to know which *F* value has a right-tail area of .05, using 10 and 12 *df*. Let the *F* value whose right-tail area is *a*, where the *df* are v_1 and v_2, be

$$F_{a, v_1, v_2}$$

For example, $F_{.05, 10, 12} = 2.75$ (Figure 10.15).

Notice that Table A-7 contains right-tail areas only. Later, we will show you how to find left-tail areas. We can, however, define each of our tests of hypothesis as a right-tailed test by simply and arbitrarily putting the larger sample variance in the numerator for a two-tailed test. Then *F* will always be $\geqslant 1$. This procedure is summarized in the accompanying box.

Hypothesis Tests for σ_1 and σ_2

Two-Tailed Test

H_0: $\sigma_1 = \sigma_2$

H_a: $\sigma_1 \neq \sigma_2$

$$F = \frac{\text{larger of } s_1^2 \text{ and } s_2^2}{\text{smaller of } s_1^2 \text{ and } s_2^2}$$

Reject H_0 if $F > F_{\alpha/2, v_1, v_2}$

where

$$v_1 = \begin{cases} n_1 - 1 & \text{if } s_1^2 \geqslant s_2^2 \\ n_2 - 1 & \text{if } s_1^2 < s_2^2 \end{cases}$$

and

$$v_2 = \begin{cases} n_2 - 1 & \text{if } s_1^2 \geqslant s_2^2 \\ n_1 - 1 & \text{if } s_1^2 < s_2^2 \end{cases}$$

One-Tailed Test

H_0: $\sigma_1 \leqslant \sigma_2$ H_0: $\sigma_1 \geqslant \sigma_2$

H_a: $\sigma_1 > \sigma_2$ H_a: $\sigma_1 < \sigma_2$

$$F = \frac{s_1^2}{s_2^2}$$ $$F = \frac{s_2^2}{s_1^2}$$

Reject H_0 if $F > F_{\alpha, v_1, v_2}$ Reject H_0 if $F > F_{\alpha, v_2, v_1}$
where   (be careful about order) where

$$v_1 = n_1 - 1$$ $$v_2 = n_2 - 1$$

and   and

$$v_2 = n_2 - 1$$ $$v_1 = n_1 - 1$$

EXAMPLE 10.12

A study is being conducted to compare the results of two instruments designed to measure the scholastic aptitude of graduating high school seniors. The standard instrument used by most schools, called the ATEST, has a possible range of scores from 0 to 500. There is some concern however, that the variation of these scores is not very large. The other instrument, called the QTEST, has the same possible range but is suspected to have a larger variation in scores.

To investigate this, 25 students, ranging from poor to superior in high school performance, were selected to take the ATEST. Twenty-five students having similar academic performance backgrounds, were selected for the QTEST. Everyone showed up to take the ATEST but 4 people were ill the day of the QTEST and so the final sample sizes were $n_1 = 25$ and $n_2 = 21$.

Using the sample results given below, can you conclude that the variance of the QTEST population is larger than the variance of the ATEST population? Use $\alpha = .05$.

ATEST	QTEST
$n_1 = 25$	$n_2 = 21$
$\bar{x}_1 = 315.8$	$\bar{x}_2 = 328.1$
$s_1 = 28.1$	$s_2 = 47.2$

Solution

Step 1 The purpose of the test is to determine if one standard deviation (or variance) is larger than the other; this calls for a one-tailed test. The suspicion is that σ_2 is larger than σ_1, so this statement is put in the alternative hypothesis. The resulting hypotheses are

H_0: $\sigma_1 \geqslant \sigma_2$

H_a: $\sigma_1 < \sigma_2$

Step 2   The appropriate test statistic is

$$F = \frac{s_2^2}{s_1^2}$$

Step 3   Because the *df* are $v_1 = 25 - 1 = 24$ and $v_2 = 21 - 1 = 20$, we find $F_{.05,20,24} = 2.03$. The test of H_0 versus H_a will be to

Reject H_0 if $F > 2.03$

Step 4   The computed F value is

$$F^* = \frac{(47.2)^2}{(28.1)^2} = 2.82$$

Because $2.82 > 2.03$, we reject H_0.

Step 5   On the basis of these data and this significance level, the QTEST scores do exhibit more variation than the ATEST scores.   □

EXAMPLE 10.13

Using the data from example 10.4, examine the variances only. Can you conclude that there is a difference in the variances of the male and female populations, using a significance level of .05?

Solution

From these data, we determined that $s_1 = 5.40$ (hundred dollars) and $s_2 = 4.05$ (hundred dollars), with $n_1 = 8$ and $n_2 = 10$.

Step 1   We are trying to detect a difference in the two variances (not whether one exceeds the other); a two-tailed test should be used. We define

H_0: $\sigma_1 = \sigma_2$ (or $\sigma_1^2 = \sigma_2^2$)
H_a: $\sigma_1 \neq \sigma_2$ (or $\sigma_1^2 \neq \sigma_2^2$)

Step 2   The test statistic here is

$$F = \frac{\text{larger of } s_1^2 \text{ and } s_2^2}{\text{smaller of } s_1^2 \text{ and } s_2^2}$$

Step 3   The value $F_{.025,7,9}$ from Table A-7c is 4.20. So our test of H_0 versus H_a is to

Reject H_0 if $F > 4.20$

Step 4   Since $s_1^2 = (5.40)^2$ is larger than $s_2^2 = (4.05)^2$, the computed value of the test statistic is

$$F^* = \frac{(5.40)^2}{(4.05)^2} = 1.78 < 4.20$$

Therefore, the decision is to fail to reject H_0.

Step 5 There is not sufficient evidence to conclude that the two variances are unequal. If additional data are obtained to test the population means (such as $H_0: \mu_1 = \mu_2$ versus $H_a: \mu_1 \neq \mu_2$) using small samples, the correct procedure would be to use the t statistic described earlier, which assumes that σ_1 and σ_2 are equal, provided both populations are believed to be normally distributed.   □

Confidence Interval for σ_1^2/σ_2^2

Consider an F curve with v_1 and v_2 df. To construct a 95% confidence interval for σ_1^2/σ_2^2, you first need to find the values of F_L and F_U, where (Figure 10.16)

F_L has a left-tail area $= .025$

F_U has a right-tail area $= .025$

F_U can be found directly from Table A-7c. It is F_L that poses a problem, however, because Table A-7 contains only right-tail areas and the F distribution is not symmetric. However,

$$F_L = \frac{1}{F_{.025, v_2, v_1}} \qquad\qquad (10\text{-}24)$$

Notice that we switched the df used when finding F_U because F_U can be written

$$F_U = F_{.025, v_1, v_2}$$

The confidence interval for σ_1^2/σ_2^2 is then

$$\frac{s_1^2/s_2^2}{F_U} \quad \text{to} \quad \frac{s_1^2/s_2^2}{F_L}$$

Figure 10.16

F curve with v_1 and v_2 df showing F values used for a 95% confidence interval

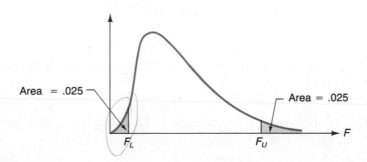

In general, we have a $(1 - \alpha) \cdot 100\%$ confidence interval for σ_1^2/σ_2^2 (independent samples):

$$\frac{s_1^2/s_2^2}{F_U} \quad \text{to} \quad \frac{s_1^2/s_2^2}{F_L} \qquad\qquad \textbf{(10-25)}$$

where

$$F_U = F_{\alpha/2,v_1,v_2}$$
$$F_L = 1/F_{\alpha/2,v_2,v_1}$$
$$v_1 = n_1 - 1$$
$$v_2 = n_2 - 1$$

EXAMPLE 10.14

Using the data from example 10.12, determine a 95% confidence interval for σ_1^2/σ_2^2.

Solution

Here, $n_1 = 25$, $s_1 = 28.1$ and $n_2 = 21$, $s_2 = 47.2$. So we need

$$F_U = F_{.025,24,20} = 2.41$$
$$F_L = 1/F_{.025,20,24}$$
$$= 1/2.33$$
$$= .43$$

The 95% confidence interval for σ_1^2/σ_2^2 is

$$\frac{(28.1)^2/(47.2)^2}{2.41} \quad \text{to} \quad \frac{(28.1)^2/(47.2)^2}{.43} \qquad \text{or} \qquad .147 \quad \text{to} \quad .824$$

As a result, we are 95% confident that σ_1^2/σ_2^2 is between .147 and .824. A confidence interval for σ_2^2/σ_1^2 can be obtained by *inverting* the confidence limits for σ_1^2/σ_2^2 and *reversing* the order. Consequently, a 95% confidence interval for σ_2^2/σ_1^2 is

$$1/.824 \quad \text{to} \quad 1/.147$$

that is

$$1.21 \quad \text{to} \quad 6.80.$$

Thus we are 95% confident that σ_2^2 is between 1.21 and 6.80 times as large as σ_1^2.   □

EXAMPLE 10.15

For the data from example 10.12, determine a 95% confidence interval for σ_2/σ_1. Use the results of example 10.14.

Solution This is obtained simply by finding the square root of each endpoint of the confidence interval for σ_2^2/σ_1^2. Your 95% confidence interval for σ_2/σ_1 will be

$$\sqrt{1.21} \quad \text{to} \quad \sqrt{6.80} \qquad \text{or} \qquad 1.10 \quad \text{to} \quad 2.61 \qquad \qquad \square$$

EXERCISES

10.62 Using the sample statistics in exercise 10.18, test the following hypothesis with a significance level of .10: $H_0: \sigma_1^2 \geqslant \sigma_2^2$ and $H_a: \sigma_1^2 < \sigma_2^2$.

10.63 Construct a 95% CI for σ_1^2/σ_2^2 in exercise 10.18.

10.64 A computer program generates data that are approximately normal. Two independent samples are generated:

Sample 1		Sample 2	
17	26	24	29
25	34	32	34
30	19	29	26
28	25	25	30

Using a significance level of .05, would you reject the null hypothesis that the variance of the population from which Sample 1 was taken is less than or equal to the variance of the population from which Sample 2 was taken?

10.65 Construct a 95% CI for the ratio of the population standard deviations in exercise 10.64.

10.66 Determine which of the following sets of hypotheses are equivalent:
(a) $H_0: \sigma_1^2/\sigma_2^2 \leqslant 1$ and $H_a: \sigma_1^2/\sigma_2^2 > 1$.
(b) $H_0: \sigma_2^2/\sigma_1^2 \geqslant 1$ and $H_a: \sigma_2^2/\sigma_1^2 < 1$.
(c) $H_0: \sigma_2^2 \geqslant \sigma_1^2$ and $H_a: \sigma_2^2 < \sigma_1^2$.
(d) $H_0: \sigma_1 \leqslant \sigma_2$ and $H_a: \sigma_1 > \sigma_2$.

10.67 Suppose that a sample of size 16 is chosen from population 1 and a sample of size 26 is drawn from population 2. Assume that both populations are normally distributed. If a 90% CI for the ratio of the variance of population 1 to the variance of population 2 is .367 to 1.753, what is the point estimate of the ratio of the two population variances?

10.68 The transportation of frozen food by trucks requires that the temperature should be maintained within a narrow range. If the temperature is kept too low, extra fuel is consumed and unnecessary costs are incurred. If the temperature is too high, health standards would be violated, and there is the danger of food spoilage and bacterial contamination. Two models of refrigeration units were being compared for their variation from a given temperature setting. A special sensor attached to the units took readings every hour. The variance of temperatures for Model 1 was 1.44 degrees Centigrade^2. The variance for Model 2 was 2.15 degrees Centigrade^2. In both cases, the sample consisted of 16 readings for each model.
(a) Construct the 95% confidence interval for the ratio of the two variances (Model 2/Model 1).
(b) At a 5% significance level, is the variation of temperature in Model 2 significantly greater than that for Model 1?
(c) State the p value for your test.

10.69 The Water Pollution Prevention Council (WPPC) had recommended that the discharge of industrial waste and effluents into rivers in the district should be done at a slow and steady rate of 100 lb/hour (i.e., 2400 lb/day). Industrial plants tended to concentrate their effluent discharge activity in the night shift. The WPPC found that although companies might technically achieve an average discharge rate of 2400 lb/day, the rivers could not cope with the erratic rate of discharge. The effluents needed to be released throughout the day, rather than all at night. The following was recommended: the true variance of the discharge rate should not exceed 600. The following results were obtained from samples of 21 observations each.

Factory A: variance 585
Factory B: variance 618

(a)   At a 5% significance level, is the variance for Factory A less than that for Factory B?
(b)   State the p value for the test.
(c)   Construct the 90% confidence interval for the ratio of the two variances (Factory B/Factory A).

10.70 Using the data in exercise 10.69, answer the following questions:
(a)   Is Factory A exceeding the recommended standard? Use $\alpha = 0.05$.
(b)   Is Factory B exceeding the recommended standard? Use $\alpha = 0.01$.
(*Hint:* You are not comparing the two variances. Consider the performance of Factory A and Factory B separately.)

SUMMARY

This chapter has presented an introduction to statistical inference for two populations. We examined tests of hypothesis and confidence intervals for the means and variances (for example, whether they are equal) of the two populations, using both independent and dependent samples.

When we used large independent samples to test the population means, we defined a test statistic having an approximate standard normal distribution, which we also used to define a confidence interval for $\mu_1 - \mu_2$. When testing for a difference in two population means using small independent samples ($n_1 < 30$ or $n_2 < 30$), we must first decide if we are willing to assume that the two populations are normally distributed. For the case of two normal populations, although we are concerned with the means, we must pay special attention to whether we also assume that the population standard deviations (σ_1 and σ_2) are equal.

If we do not assume that the σ values are equal, we use a test statistic for μ_1 versus μ_2 having an approximate t distribution. This statistic also results in an approximate confidence interval for $\mu_1 - \mu_2$. If we assume that the σ values are equal, then we use a procedure that pools the sample variances and results in a test statistic having an exact t distribution. We also derived confidence intervals for $\mu_1 - \mu_2$ for this situation.

For the case in which we are not willing to assume that both populations are normally distributed, a nonparametric test can be used which does not require this assumption. This test is the **Mann–Whitney** test, which used the ranks of the sample data to derive a value of the test statistic.

When two samples are obtained such that corresponding observations are paired (matched), the resulting samples are dependent or paired. When we assume that the population differences are normally distributed a t statistic can be used to test the mean of the population differences μ_d and to construct a confidence interval for μ_d. We need not be concerned about whether the population standard deviations are equal for this situation because the test statistic uses the differences between the paired observations, a new variable. For the situation where the population differences are not assumed to be normally distributed, another nonparametric procedure, called the **Wilcoxon signed rank test**, can be used to conduct a test for the location of the population differences when dealing with dependent (paired) samples. The ranks of the absolute values of the sample differences are used to derive a value of the test statistic.

To determine whether two population variances (or standard deviations) are the same, we introduced the **F distribution**. This distribution is nonsymmetric (right skew) and assumes that two independent samples were obtained from normal populations. Probabilities (areas under the curve) for the F random variable are contained in Table A-7. Using this distribution, we can perform two-tailed tests (such as $H_a: \sigma_1 \neq \sigma_2$) or one-tailed tests (such as $H_a: \sigma_1 > \sigma_2$) on the two standard deviations. We also use it to construct a confidence interval for σ_1^2/σ_2^2 or σ_1/σ_2.

CASE STUDY

A Look at the Prescription Drug Market

The retail market for prescription medicines has a number of anomalies. As pointed out by Balsmeier (1977) the purchasing consumer is not the primary demander of drugs; a physician selects the drug an individual is to purchase. Consumers have very little experience in purchasing any given prescription medicine or its specific quantities or dosages. They would be hard put to say what would constitute a normal price for a given prescription. Furthermore, substitution of a brand-name prescription by a generic drug is not permitted in many states. These characteristics might tend to inhibit price competition among retail pharmacies.

The Balsmeier study examined the retail prescription drug market in a metropolitan and a rural market on price levels and pricing practices. The hypotheses tested were:

H_1: There is no difference in the price charged for prescription drugs between the chain drug store and the privately owned drug store in the urban market.

H_2: There is no difference in the price charged for prescription drugs between the rural drug store and the metropolitan drug store.

H_3: There is no difference in the markup on an average prescription size and a prescription written for one hundred units.

Statistical data were gathered from both a metropolitan area and five rural counties. Seventeen drugs were selected and price data were obtained for them. The results pertaining to the three hypotheses being tested are given in the three tables below:

Table I

Relative prices—
Chain vs. private
drug stores in the
urban market

Type	Mean	S.D.	p value for t test
Chain	1.21	0.69	$p > .05$
Private	1.35	0.82	

Table II

Relative prices—
rural vs.
metropolitan drug
stores

Location	Mean	S.D.	p value for t test
Rural	1.66	1.73	$p < .05$
Metropolitan	1.31	0.69	

Table III

Unit prices—
Average prescription
vs. 100-units
prescription

Drug	Average Prescription Mean Price per Unit	100-Units Mean Price per Unit
Aldomet 250	$.103	$.097
Hydrodiuril 50	.100	.087
Indocin 25	.128	.118
Triavil 2-25	.174	.163
Valium 5	.158	.104
Premarin 1.25 mg	.107	.094
Ampicillin 250	.177	.173
Omnipen 250	.203	.199
Vibramycin 100	.973	.938
Keflex 250	.463	.459
Ovulin	.025	.025
Dimetapp Extentabs	.144	.138
Inderal 10 mg	.060	.059
Darvon 65 comp.	.114	.101
Lasix 40 mg	.141	.123
Sinequan 25	.187	.191
Polycillin 250	.154	.268

Source: P. W. Balsmeier. Pricing practices in a retail market for prescription drugs. *Proceedings of the Southwest Region American Institute for the Decision Sciences Conference,* Fuller, 1977, pp. 87–91. Proceedings are published by the Southwest Region of the Decision Sciences Institute (formerly the American Institute for Decision Sciences). Reprinted with permission.

Case Study Questions

1. The p values in Tables I and II pertain to the t test for pooled samples. What assumption was necessary about the variances of chain vs. private and rural vs. metropolitan?

2. What is the formal statistical style of stating the three hypotheses to be tested?

3. At $\alpha = .05$, what do you conclude about hypothesis H_1?

4. At $\alpha = .05$, what do you conclude about hypothesis H_2?

5. Are the figures in Table III an example of independent samples, or paired dependent data?

6. If you assumed normal populations, which test procedure could you use for hypothesis H_3?

7. If you could not assume normal populations, which test procedure could you use for hypothesis H_3?

8. Assume normal populations and complete all the steps to test hypothesis H_3. What is your conclusion, at $\alpha = .05$? State the p value for the test.

Extra Readings

Conover, W. J., *Practical Nonparametric Statistics*, 2d ed. (New York: Wiley, 1980).

Siegel, Sidney, *Nonparametric Statistics for the Behavioral Sciences* (New York: McGraw-Hill, 1956).

REVIEW EXERCISES

10.71 A sociologist was interested in studying the question of whether home ownership versus renting had anything to do with the mobility rate and overall stability of family life. As an initial step, the investigation focused on two groups. The first group were all home owners, the second group were all renters. In considering mobility, any moves from one place to another within a radius of 50 miles were ignored. The length of stay in the community for each family was noted. The following results were obtained.

Home owners: Sample size 49, average stay 4.2 years, standard deviation 2.3 years.

Renters: Sample size 36, average stay 2.8 years, standard deviation 2.7 years.

Do the above results indicate that home owners are more "stable" and "less mobile" (as measured by longer stays in a community, on the average) than renters? Use $\alpha = .05$. Find the p value.

10.72 The sociologist mentioned in exercise 10.71 extended the study to another aspect: how long the primary earner in the family had held his or her current job. This time, the following results were noted:

	Sample Size	Avg Number of Years on Current Job	Standard Deviation
Homeowners	49	2.5	1.2
Renters	36	2.7	1.4

(a) Do the above results indicate that home owners are more "stable" and "less mobile" (as measured by longer periods on the current job, on the average) than renters? Use $\alpha = 0.05$. Find the p value.

(b) Compare your results with those of exercise 10.71, especially the p values in each case. What lesson can one draw from this? (*Hint:* How you define "stable" or "mobile" and how these qualities are measured—do these matter?)

10.73 Holsboer et al. (1987) measured pituitary adrenocortical responses to human corticotropin-releasing hormone (CRH) in 10 patients with a major depressive episode according to DSM-III criteria, and compared these with responses of 10 normal control subjects. The following results were reported:

RESPONSE TO INJECTION OF HUMAN CRH IN 10 DEPRESSED PATIENTS AND 10 NORMAL CONTROL SUBJECTS

Response Measured	Depressed Group		Control Group	
	Mean	**S.D.**	**Mean**	**S.D.**
Aldosterone (pg/ml/min)	1711	357	2717	547
Cortisol (ng/ml/min)	8324	3675	9186	2342

Source: F. Holsboer, A. Gerken, G. K. Stalla, and O. A. Muller. Blunted aldosterone and ACTH release after human CRH administration in depressed patients. *Am. J. Psychiatry* **144** (2): 230 (February 1980), Table 1.

(a) Is there a significant difference between the two groups in the aldosterone response? Conduct a hypothesis test at $\alpha = .05$, assuming equal variances for the two groups.

(b) State the p value for the above test. Would your decision change if $\alpha = .01$?

(c) Repeat (a) and (b), testing for a significant difference in the cortisol response, not assuming equal variances for the two groups.

10.74 Use the data from exercise 10.73 to construct 90%, 95%, and 99% confidence intervals for $\mu_C - \mu_D$, where μ_C is the average for the control group's aldosterone response and μ_D is the average for the depressed group's aldosterone response.

10.75 An important new area of mathematics made possible by the tremendous speed and computational power of the so-called "supercomputers" is *nonlinear mathematics*, which permits the modeling of complex systems. Indeed, it has given birth to the "science of complexity" (with apologies to the experts and lawyers in Washington)! This field is currently on the leading edge of theoretical physics. One researcher remarked that the average number of years of experience in advanced mathematics needed in this field was more than 12 years, as opposed to 6 years in other fields. Test the hypothesis:

$H_0: \quad \mu_1 - \mu_2 \leqslant 6$

$H_a: \quad \mu_1 - \mu_2 > 6$

where μ is the average experience in years (1 = those in nonlinear mathematics, 2 = those in other fields). Use the following sample results, and a significance level of 5%. Assume the populations are normally distributed, with equal variances.

Handwritten annotations:

(a) $\alpha = .05$, $\sigma_1^2 = \sigma_2^2$, $n = 10$

$H_0: \mu_1 = \mu_2$ (independent)

$H_a: \mu_1 \neq \mu_2$

$t = \dfrac{\bar{x}_1 - \bar{x}_2}{sp\sqrt{\dfrac{1}{n_1} + \dfrac{1}{n_2}}}$

$d.f. = n_1 + n_2 - 2 = 10 + 10 - 2 = 18$

$sp = \sqrt{\dfrac{(n_1-1)s_1^2 + (n_2-1)s_2^2}{n_1 + n_2 - 2}}$

$= \sqrt{\dfrac{9(357)^2 + 9(547)^2}{18}}$

$SP = 461.88$

$t = \dfrac{1711 - 2717}{461.88\sqrt{\dfrac{1}{10} + \dfrac{1}{10}}}$

$= \dfrac{-1006}{206.56} = -4.87$

$\dfrac{\alpha}{2} = .025$, $t_{.025,18} = 2.101$

Reject if $t < -2.101$

(a) $-4.87 < -2.101 \Rightarrow$ Reject

(b) p value $< .01$

$\bar{x}_1 - \bar{x}_2 \pm t_{\frac{\alpha}{2}, d.f.} \sqrt{\dfrac{s_1^2}{n_1} + \dfrac{s_2^2}{n_2}}$

		$\bar{x}$ (years)	s
Nonlinear Math (1)	7	14.2	1.5
Other Fields (2)	7	7.0	2.0

10.76 (a) Construct a 95% confidence interval for $\mu_1 - \mu_2$ in exercise 10.75.

(b) Repeat exercise 10.75, but do not assume equal population variances. Indicate the p value for the test.

10.77 The Hamilton Rating Scale for Depression is an instrument commonly used by professionals in the mental health fields. Loosen, Marciniak, and Thadani (1987) have studied the thyroid-stimulating hormone (TSH) response and its relationship to psychiatric history. It is believed that a blunted TSH response may have utility as a marker of past episodes of depression or alcoholism, or as a true marker of trait. The study covered 32 healthy volunteers who had never sought or received psychiatric treatment. Nine of them had a family history of depression or alcoholism. From a preliminary analysis of the subjects, the researchers obtained the following scores on the Hamilton scale for depression:

		Hamilton Score	
Group	n	**Mean**	**S.D.**
Without any history	23	0.9	1.5
With some history	9	4.0	3.6

Source: P. T. Loosen, R. Marciniak, and K. Thadani. TRH-induced TSH response in healthy volunteers: Relationship to psychiatric history. *Am. J. Psychiatry* **144** (4): 456 (April 1987).

(a) Assuming equal variances, use the pooled sample method to test the hypothesis that the Hamilton scores are significantly higher in the group with a personal or family history of depression or alcoholism. Let $\alpha = .01$, and state the p value.

(b) The researchers actually used the Mann–Whitney test and obtained $p < .005$. Does this agree with your results in (a)?

(c) Give the two major reasons why the researchers would prefer to use the Mann–Whitney test, rather than other tests studied in this chapter. (*Hint:* Normality and independence.)

10.78 An economist believes that the cost of a typical basket of goods bought by a family of four costs more in Atlanta, Georgia, than it does in Houston, Texas. Seven grocery stores were randomly selected in each of the two cities to collect the following data (cost of basket of goods, in dollars). Using a significance level of .10, do the data support the economist's belief?

Atlanta	Houston
30	37
33	36
41	39
43	40
37	32
39	36
41	32

(a) $\alpha = .05,$

$H_0: \mu_1 = \mu_2$

$H_a: \mu_1 \neq \mu_2$

reject if $|t| > t_{.025, n_1 + n_2 - 2}$

for electrical eng.

d.f. $= 36 + 19 - 2 = 53$

$t_{.025, 53} \approx 2.01$

$1.94 < 2.01 \Rightarrow FTR !!$

For industrial:

d.f. $= 13 + 18 - 2 = 29$

$t_{.025, 29} = 2.045$

$0.16 < 2.045 \Rightarrow FTR$

For civil:

d.f. $= 10 + 12 - 2 = 20$

$t_{.025, 20} = 2.086$

$0.19 < 2.086 \Rightarrow FTR$

material

d.f. $= 14 + 17 - 2 = 29$

$t_{.025, 29} = 2.045$

$1.12 < 2.045 \Rightarrow FTR.$

10.79 A number of institutions of higher education around the country televise engineering education. In 1983, the Virginia legislature created a cooperative graduate engineering program, wherein engineering classes were broadcast live from special classrooms at the University of Virginia and other institutions via the public television network. The on-campus professor had regular campus students in front of him, but also televised lectures to students at remote locations. The remote classroom was equipped with TV receivers and audio facilities, which permitted remote students to ask questions and participate in discussions. The same admission standards and examinations were applied to on-campus and off-campus (TV) students. An evaluation committee, representing participating institutions and the State Department of Information Technology, was formed to assess the program. One basic question raised was: Do the TV students perform academically on a par with their on-campus counterparts? Wergin et al. report the following results for the 1983–84 academic year:

GRADE POINT AVERAGE, ON-CAMPUS VS. TV STUDENTS, BY COURSE AND TERM[a]

Course	Fall			Spring		
	n	$\bar{x}$	t	n	$\bar{x}$	t
Electrical Engg.						
On-campus	36	3.40	1.94	13	3.41	2.30
TV	19	3.07		9	2.59	
Industrial Engg.						
On-campus	13	3.49	0.16	11	3.60	3.10
TV	18	3.45		12	2.79	
Civil Engg.						
On-campus	10	3.20	0.19	14	3.36	0.17
TV	12	3.11		9	3.39	
Material Science						
On-campus	14	3.59	1.12	11	3.39	6.00
TV	17	3.40		18	2.95	

[a] n = number of students, $\bar{x}$ = mean GPA, t = t-test results, pooled sample.

Source: J. F. Wergin, D. Boland, and T. W. Haas. Televising graduate engineering courses: Results of an instructional experiment. *Eng. Educ.* 110 (November 1986).

(a) In the fall term, was there a significant difference in course grades between on-campus and TV students, for any of the four engineering courses offered? Let $\alpha = .05$ for your hypothesis tests. no sig. diff for all 4.

(b) In the spring term, did on-campus students perform better than TV students in all courses except civil engineering? Let $\alpha = .05$ for your hypothesis tests.

(c) Without conducting any formal tests, but just by examining the figures in the above table, and taking into account (a) and (b), comment on whether you would agree with the statement: "The trend toward increasing differences in GPA between on-campus and TV students during the year was due primarily to the declining performance of TV students (except in civil engineering) compared to the relatively steady performance of on-campus students throughout the year."

10.80 A microbiologist wants to know whether there is any difference in the time in hours it takes to make yogurt from two different starters. Seven batches of yogurt were made with each of the starters. The following data give the times it took to make each batch. At the .05 significance level, do the data indicate a difference in the time it takes to make yogurt from the two starters?

Lactobacillus Acidophilus	Bulgarius
6.8	6.1
6.3	6.4
7.4	5.7
6.1	6.5
8.2	6.9
7.3	6.3
6.9	6.7

10.81 Repeat exercise 10.78, assuming normal population and equal population variances. State the p value for the test.

10.82 (a) Is it possible for the two-sample t test (dependent samples) and the Wilcoxon signed rank test to produce different conclusions using the same set of data? Discuss.

(b) Why is it not necessary to make assumptions about equal or unequal population variances in this case?

10.83 (a) In exercise 10.80, assume normally distributed populations and construct a 95% confidence interval for $\mu_L - \mu_B$ (L = Lactobacillus, B = Bulgarius).

(b) Use the CI to test the hypothesis that there is no difference in the mean times, i.e., $\mu_L - \mu_B = 0$, at a 5% significance level.

(c) If it is desired to construct a 99% CI with a maximum tolerance for error of ± 0.5 hour, what are the optimum sample sizes?

10.84 A pharmacologist was interested in comparing two drugs used for insomnia. Since the drugs had potentially toxic side effects, the drugs were administered to two groups of mice. The first group of 50 mice was given drug A, resulting in an average time to fall asleep of 24 minutes and a standard deviation of 8 minutes. Identical dosages of drug B were given to another independent group of 50 mice, resulting in an average time to fall asleep of 26 minutes and a standard deviation of 12 minutes. Test the hypothesis that there is no difference between the effectiveness of the two drugs at $\alpha = 0.05$. What do your results indicate?

10.85 (a) Use the data from exercise 10.84 to construct a 95% confidence interval for $\mu_A - \mu_B$. Does this CI indicate the same conclusion that was reached in 10.84?

(b) If the CI for $\mu_A - \mu_B$ should have a maximum error of 2 minutes, what are the optimum sample sizes?

10.86 A specialist in criminal psychology was testing two instruments designed to measure aggressiveness and hostility. A large variation was desirable. A group of 31 hard-core criminals with multiple convictions was administered Instrument A. A variance of 78 was observed. A similar group of 41 hard-core criminals who had Instrument B administered had a variance in scores of 120. Is the variation in B significantly higher than in A? Use $\alpha = .01$, and find the p value.

**COMPUTER
EXERCISES
USING THE
DATABASE**

Using a sample of 100 observations, complete the following exercises:

1. Using variables SEX and MARITAL, determine if the average age (variable AGE) for the remarried males is the same as for the remarried females, if the significance level is .05. Be sure to use the appropriate small samples or large samples procedure. What are you assuming here?

2. Using variables LSINDEX (score on Life Satisfaction Index) and SEX, can you conclude that males attending Brookhaven are more satisfied in life on the average than females, at a significance level of .05?

3. Construct a 95% confidence interval for $\mu_M - \mu_F$ where μ_M is the average graduate grade point average for males and μ_F is the average graduate grade point average for females attending Brookhaven. Use variables GR–GPA and SEX. What can you learn from this confidence interval?

4. Someone has reported that the variation of quantitative scores on the GRE (variable QGRE) is more for males than for females. At a significance level of .05, can this statement be supported?

**SPSSX
APPENDIX
(MAINFRAME
VERSION)**

**Solution to
Example 10.4**

Example 10.4 was concerned with a *t* test using two independent samples to compare two population means. The problem was to determine if the purchase amount of automobile "extras" is the same for males and females. The SPSSX program listing in Figure 10.17 was used to test the variances and means using two independent samples. For the test on the means, the value of the test statistic and the *p* value are provided for the case where the

Figure 10.17

SPSSX program listing requesting a two-sample test for the variances and means

```
TITLE PURCHASE AMOUNTS
DATA LIST FREE/AMOUNT SEX
BEGIN DATA
23.00 0
23.86 0
19.20 0
15.78 0
30.65 0
23.12 0
17.90 0
30.25 0
16.42 1
14.20 1
21.30 1
18.46 1
11.70 1
12.10 1
16.50 1
9.20 1
21.05 1
18.05 1
END DATA
PRINT / AMOUNT SEX
T-TEST GROUPS=SEX(0,1) / VARIABLES=AMOUNT
```

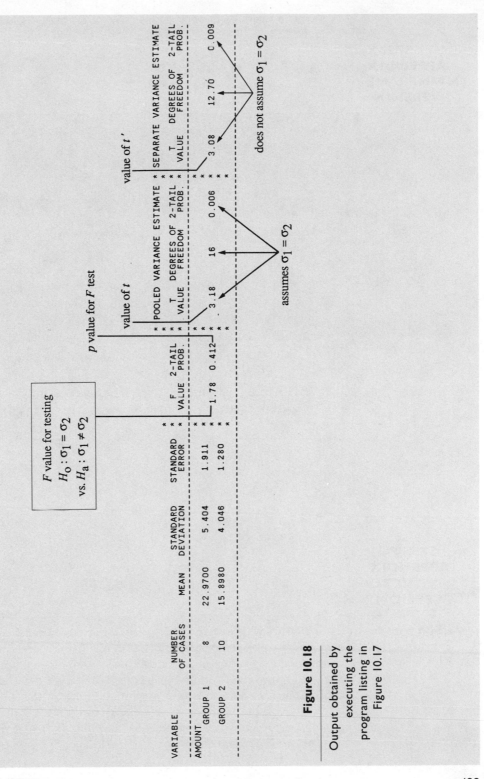

Figure 10.18

Output obtained by executing the program listing in Figure 10.17

**SPSSX
APPENDIX
(MAINFRAME
VERSION)**

population variances are assumed equal and the case where it is not assumed that the variances are equal.

The TITLE command names the SPSSX run.

The DATA LIST command gives each variable a name and describes the data as being in free form.

The BEGIN DATA command indicates to SPSSX that the input data immediately follow.

The next 18 lines contain the data values, with each line representing an individual's purchase amount and sex (0 = male, 1 = female). The first line, for example, represents a male with a purchase amount of 23 (hundred).

The END DATA statement indicates the end of the data.

The PRINT command requests a printout of the input data.

The T-TEST command compares two sample means. The GROUPS and VARIABLES subcommands divide the cases into two groups for a comparison of the two sample means.

Figure 10.18 contains the output obtained by executing the program listing in Figure 10.17.

**SPSS/PC+
APPENDIX
(MICRO
VERSION)**

**Solution to
Example 10.4**

Example 10.4 was concerned with a *t* test using two independent samples to compare two population means. The problem was to determine if the purchase amount of automobile "extras" is the same for males and females. The SPSS/PC+ program listing in Figure 10.19 was used to test the variances and means using two independent samples. For the test on the means, the value of the test statistic and the *p* value are provided for the case where the population variances are assumed equal and the case where it is not assumed that the variances are equal.

SPSS/PC+ APPENDIX (MICRO VERSION)

Figure 10.19

SPSS/PC+ program listing requesting a two-sample test for the variances and means

```
TITLE PURCHASE AMOUNTS.
DATA LIST FREE/AMOUNT SEX.
BEGIN DATA.
23.00 0
23.86 0
19.20 0
15.78 0
30.65 0
23.12 0
17.90 0
30.25 0
16.42 1
14.20 1
21.30 1
18.46 1
11.70 1
12.10 1
16.50 1
9.20 1
21.05 1
18.05 1
END DATA.
LIST VARIABLES=AMOUNT SEX.
T-TEST GROUPS=SEX(0,1) / VARIABLES=AMOUNT.
```

The TITLE command names the SPSS/PC+ run.

The DATA LIST command gives each variable a name and describes the data as being in free form.

The BEGIN DATA command indicates to SPSS/PC+ that the input data immediately follow.

The next 18 lines contain the data values, with each line representing an individual's purchase amount and sex (0 = male, 1 = female). The first line, for example, represents a male with a purchase amount of 23 (hundred).

The END DATA statement indicates the end of the data.

The LIST command requests a printout of the input data.

The T-TEST command compares two sample means. The GROUPS and VARIABLES subcommands divide the cases into two groups for a comparison of the two sample means.

Figure 10.20 contains the output obtained by executing the program listing in Figure 10.19.

SPSS/PC+ APPENDIX (MICRO VERSION)

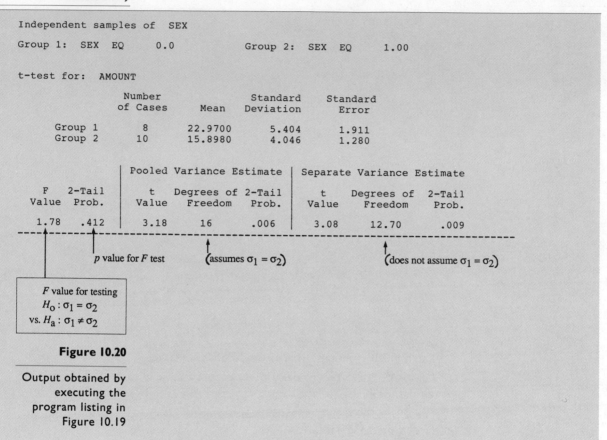

Independent samples of SEX

Group 1: SEX EQ   0.0   Group 2: SEX EQ   1.00

t-test for: AMOUNT

	Number of Cases	Mean	Standard Deviation	Standard Error
Group 1	8	22.9700	5.404	1.911
Group 2	10	15.8980	4.046	1.280

		Pooled Variance Estimate			Separate Variance Estimate		
F Value	2-Tail Prob.	t Value	Degrees of Freedom	2-Tail Prob.	t Value	Degrees of Freedom	2-Tail Prob.
1.78	.412	3.18	16	.006	3.08	12.70	.009

p value for F test $\left(\text{assumes } \sigma_1 = \sigma_2\right)$ $\left(\text{does not assume } \sigma_1 = \sigma_2\right)$

F value for testing
$H_0 : \sigma_1 = \sigma_2$
vs. $H_a : \sigma_1 \neq \sigma_2$

Figure 10.20

Output obtained by executing the program listing in Figure 10.19

**SPSSX
APPENDIX
(MAINFRAME
VERSION)**

**Solution to
Example 10.8**

Example 10.8 was concerned with the computation of the t statistic comparing the means of two populations using paired (dependent) samples. The problem was to determine if the average weight loss was more than 10 lb at the completion of the weight loss program. The SPSSX program listing in Figure 10.21 was used to request a mean, t value, and p value. Note that SPSSX assumes a two-tailed test. This was a one-tailed test, so the calculated p value must be divided by two.

Figure 10.21

SPSSX program
listing used to
request a
two-sample t test
for two population
means using paired
(dependent)
samples

```
TITLE WEIGHT REDUCTION
DATA LIST FREE/BEFORE AFTER
BEGIN DATA
132.2 123.7
121.6 111.9
156.4 141.1
167.9 151.4
141.7 124.6
183.5 163.8
137.0 126.2
153.1 145.1
138.4 121.9
145.4 123.9
161.0 156.1
150.8 137.9
END DATA
COMPUTE BEF10 = BEFORE - 10
PRINT / BEFORE AFTER
T-TEST PAIRS = BEF10 AFTER
```

The TITLE command names the SPSSX run.

The DATA LIST command gives each variable a name and describes the data as being in free form.

The BEGIN DATA command indicates to SPSSX that the input data immediately follow.

The next 12 lines contain the data values, which represent the before and after weight, respectively, of each individual. The first value, for example, represents a person who weighed 132.2 before and 123.7 after.

The END DATA statement indicates the end of the input data.

The COMPUTE command defines the variable BEF10 to be the before weight minus 10.

The PRINT command requests the input data to be printed.

The T-TEST command compares two sample means. The PAIRS subcommand names the variables being compared.

Figure 10.22 contains the output obtained by executing the program listing in Figure 10.21.

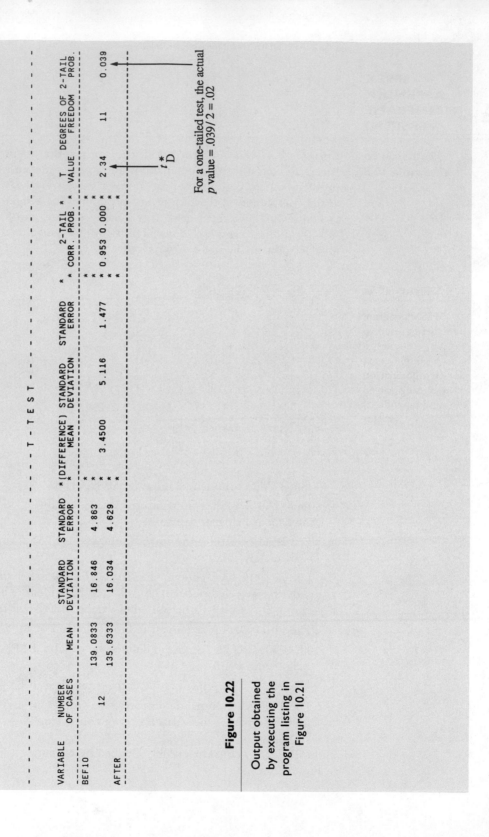

Figure 10.22

Output obtained by executing the program listing in Figure 10.21

SPSS/PC+ APPENDIX (MICRO VERSION)

Solution to Example 10.8

Example 10.8 was concerned with the computation of the *t* statistic comparing the means of two populations using paired (dependent) samples. The problem was to determine if the average weight loss was more than 10 lb at the completion of the weight loss program. The SPSS/PC+ program listing in Figure 10.23 was used to request a mean, *t* value, and *p* value. Note that SPSS/PC+ assumes a two-tailed test. This was a one-tailed test, so the calculated *p* value must be divided by two.

Figure 10.23

SPSS/PC+ program listing used to request a two-sample *t* test for two population means using paired (dependent) samples

```
TITLE WEIGHT REDUCTION.
DATA LIST FREE/BEFORE AFTER.
BEGIN DATA.
132.2 123.7
121.6 111.9
156.4 141.1
167.9 151.4
141.7 124.6
183.5 163.8
137.0 126.2
153.1 145.1
138.4 121.9
145.4 123.9
161.0 156.1
150.8 137.9
END DATA.
COMPUTE BEF10 = BEFORE - 10.
LIST VARIABLES=BEFORE AFTER.
T-TEST PAIRS = BEF10 AFTER.
```

The TITLE command names the SPSS/PC+ run.

The DATA LIST command gives each variable a name and describes the data as being in free form.

The BEGIN DATA command indicates to SPSS/PC+ that the input data immediately follow.

The next 12 lines contain the data values, which represent the before and after weight, respectively, of each individual. The first value, for example, represents a person who weighed 132.2 before and 123.7 after.

The END DATA statement indicates the end of the input data.

The COMPUTE command defines the variable BEF10 to be the before weight minus 10.

The LIST command requests the input data to be printed.

The T-TEST command compares two sample means. The PAIRS subcommand names the variables being compared.

**SPSS/PC+
APPENDIX
(MICRO
VERSION)**

```
        WEIGHT REDUCTION

Paired samples t-test:  BEF10
                        AFTER

Variable   Number                Standard   Standard
           of Cases   Mean       Deviation  Error

BEF10        12       139.0833   16.846     4.863
AFTER        12       135.6333   16.034     4.629
```

(Difference) Mean	Standard Deviation	Standard Error	2-Tail Corr.	Prob.	t Value	Degrees of Freedom	2-Tail Prob.
3.4500	5.116	1.477	.953	.000	2.34	11	.039

t^*_D

For a one-tailed test, the actual
p value $= .039 / 2 = .02$

Figure 10.24

Output obtained
by executing the
program listing in
Figure 10.23

Figure 10.24 contains the output obtained by executing the program listing in Figure 10.23.

**SPSSX
APPENDIX
(MAINFRAME
VERSION)**

**Solution to
Example 10.6**

Example 10.6 used the Mann–Whitney U test to determine if the purchase amounts of automobile "extras" is the same for males and females. The SPSSX program listing in Figure 10.25 was used to compute the Mann–Whitney test statistic.

SPSSX APPENDIX (MAINFRAME VERSION)

Figure 10.25

SPSSX program listing used to determine the Mann–Whitney test statistic

```
TITLE PURCHASE AMOUNT
DATA LIST FREE/AMOUNT SEX
BEGIN DATA
23.00 0
23.86 0
19.20 0
15.78 0
30.65 0
23.12 0
17.90 0
30.25 0
16.42 1
14.20 1
21.30 1
18.46 1
11.70 1
12.10 1
16.50 1
9.20 1
21.05 1
18.05 1
END DATA
PRINT / AMOUNT SEX
NPAR TESTS M-W = AMOUNT BY SEX (0,1)
```

The TITLE command names the SPSSX run.

The DATA LIST command gives each variable a name and describes the data as being in free form.

The BEGIN DATA command indicates to SPSSX that the input data immediately follow.

The next 18 lines contain the data values, with each line representing each individual's purchase amount and sex (0 = male, 1 = female). The first line, for example, represents a male with a purchase amount of 23 (hundred).

The END DATA statement indicates the end of the input data.

The PRINT command requests a printout of the input data.

The NPAR TESTS command indicates that we wish to perform a nonparametric Mann–Whitney test, comparing the purchase amounts (variable AMOUNT) by sex (variable SEX).

Figure 10.26 contains the output obtained by executing the program listing in Figure 10.25.

SPSSX APPENDIX (MAINFRAME VERSION)

Figure 10.26

Output obtained by executing the program listing in Figure 10.25

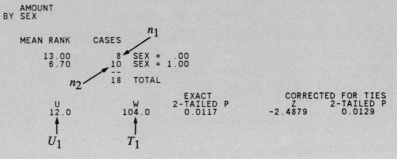

$$\text{So, } T_2 = \frac{n(n+1)}{2} - T_1$$

$$= \frac{(18)(19)}{2} - 104$$

$$= 67$$

Also,

$$U_2 = \frac{n_1 n_2 + n_2(n_2+1)}{2} - T_2$$

$$= \frac{(8)(10) + (10)(11)}{2} - 67$$

$$= 68$$

```
- - - - - MANN-WHITNEY U - WILCOXON RANK SUM W TEST

        AMOUNT
     BY SEX

     MEAN RANK      CASES           n1
        13.00           8   SEX = .00
         6.70          10   SEX = 1.00
                       --
              n2       18   TOTAL

                                     EXACT              CORRECTED FOR TIES
         U             W          2-TAILED P          Z          2-TAILED P
        12.0         104.0         0.0117          -2.4879        0.0129

        U1            T1
```

SPSS/PC+ APPENDIX (MICRO VERSION)

Solution to Example 10.6

Example 10.6 used the Mann–Whitney U test to determine if the purchase amounts of automobile "extras" is the same for males and females. The SPSS/PC+ program listing in Figure 10.27 was used to compute the Mann–Whitney test statistic.

SPSS/PC+ APPENDIX (MICRO VERSION)

Figure 10.27

SPSS/PC+ program listing used to determine the Mann–Whitney test statistic

```
TITLE PURCHASE AMOUNT.
DATA LIST FREE/AMOUNT SEX.
BEGIN DATA.
23.00 0
23.86 0
19.20 0
15.78 0
30.65 0
23.12 0
17.90 0
30.25 0
16.42 1
14.20 1
21.30 1
18.46 1
11.70 1
12.10 1
16.50 1
9.20 1
21.05 1
18.05 1
END DATA.
LIST VARIABLES=AMOUNT SEX.
NPAR TESTS M-W = AMOUNT BY SEX(0,1).
```

The TITLE command names the SPSS/PC+ run.

The DATA LIST command gives each variable a name and describes the data as being in free form.

The BEGIN DATA command indicates to SPSS/PC+ that the input data immediately follow.

The next 18 lines contain the data values, with each line representing each individual's purchase amount and sex (0 = male, 1 = female). The first line, for example, represents a male with a purchase amount of 23 (hundred).

The END DATA statement indicates the end of the input data.

The LIST command requests a printout of the input data.

The NPAR TESTS command indicates that we wish to perform a nonparametric Mann–Whitney test, comparing the purchase amounts (variable AMOUNT) by sex (variable SEX).

Figure 10.28 contains the output obtained by executing the program listing in Figure 10.27.

**SPSS/PC+
APPENDIX
(MICRO
VERSION)**

Figure 10.28

Output obtained
by executing the
program listing in
Figure 10.27

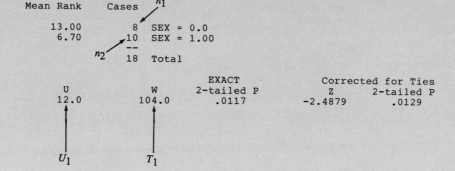

$$\text{So, } T_2 = \frac{n(n+1)}{2} - T_1$$

$$= \frac{(18)(19)}{2} - 104$$

$$= 67$$

Also,

$$U_2 = n_1 n_2 + \frac{n_2(n_2+1)}{2} - T_2$$

$$= (8)(10) + \frac{(10)(11)}{2} - 67$$

$$= 68$$

```
                    PURCHASE AMOUNT

  - - - - - Mann-Whitney U - Wilcoxon Rank Sum W Test

        AMOUNT
      by SEX

     Mean Rank    Cases    n₁

        13.00        8    SEX = 0.0
         6.70       10    SEX = 1.00
                     --
              n₂     18    Total


                                    EXACT           Corrected for Ties
        U             W         2-tailed P        Z        2-tailed P
       12.0         104.0        .0117         -2.4879       .0129



        U₁            T₁
```

**SPSSX
APPENDIX
(MAINFRAME
VERSION)**

**Solution to
Example 10.11**

Example 10.11 used the nonparametric signed ranks test to determine whether
there was a difference between an experimental group and a control group
in geometry performance. The SPSSX program listing in Figure 10.29 was
used to compute the Wilcoxon signed ranks test statistic.

Figure 10.29

SPSSX program
listing used to
determine the
Wilcoxon test
statistic

```
TITLE GEOMETRY
DATA LIST FREE/ EXPER CONTROL
BEGIN DATA
75 72
79 82
84 93
71 65
82 76
91 89
85 81
68 58
90 95
92 91
END DATA
PRINT / EXPER CONTROL
NPAR TESTS  WILCOXON=ALL
```

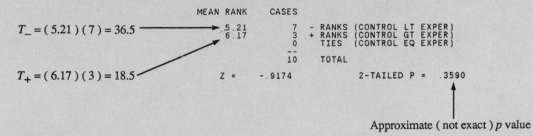

```
      - - - - - WILCOXON MATCHED-PAIRS SIGNED-RANKS TEST
                    EXPER
                WITH CONTROL

                    MEAN RANK      CASES
                                     7    - RANKS (CONTROL LT EXPER)
                       5.21          3    + RANKS (CONTROL GT EXPER)
                       6.17          0      TIES (CONTROL EQ EXPER)
                                    --
                                    10      TOTAL

                    Z =    -.9174           2-TAILED P =  .3590
```

$$T_- = (5.21)(7) = 36.5$$

$$T_+ = (6.17)(3) = 18.5$$

Approximate (not exact) p value

Figure 10.30

Output obtained
by executing the
program listing in
Figure 10.29

The TITLE command names the SPSSX run.

The DATA LIST command gives each variable a name and describes the data as being in free form.

The BEGIN DATA command indicates to SPSSX that the input data immediately follow.

The next 10 lines contain the data values; each line represents the performance for a pair of individuals.

The END DATA statement indicates the end of the input data.

The PRINT command prints the values of EXPER and CONTROL.

The NPAR TESTS command indicates that we wish to perform the nonparametric Wilcoxon signed ranks test and output ALL of the results.

Figure 10.30 contains the output obtained by executing the program listing in Figure 10.29.

**SPSS/PC+
APPENDIX
(MICRO
VERSION)**

**Solution to
Example 10.11**

Example 10.11 used the nonparametric signed ranks test to determine whether there was a difference between an experimental group and a control group in geometry performance. The SPSS/PC+ program listing in Figure 10.31 was used to compute the Wilcoxon signed ranks test statistic.

Figure 10.31

SPSS/PC+ program
listing used to
determine the
Wilcoxon test
statistic

```
TITLE GEOMETRY.
DATA LIST FREE/ EXPER CONTROL.
BEGIN DATA.
75  72
79  82
84  93
71  65
82  76
91  89
85  81
68  58
90  95
92  91
END DATA.
LIST VARIABLES=EXPER CONTROL.
NPAR TESTS   WILCOXON=ALL.
```

The TITLE command names the SPSS/PC+ run.

The DATA LIST command gives each variable a name and describes the data as being in free form.

The BEGIN DATA command indicates to SPSS/PC+ that the input data immediately follow.

The next 10 lines contain the data values; each line represents the performance for a pair of individuals.

The END DATA statement indicates the end of the input data.

The LIST command prints the values of EXPER and CONTROL.

The NPAR TESTS command indicates that we wish to perform the nonparametric Wilcoxon signed ranks test and output ALL of the results.

Figure 10.32 contains the output obtained by executing the program listing in Figure 10.31.

SPSS/PC+ APPENDIX (MICRO VERSION)

Figure 10.32

Output obtained by executing the program listing in Figure 10.31

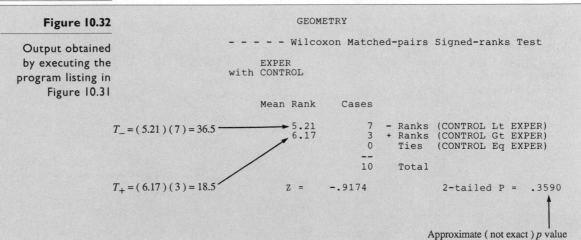

GEOMETRY

- - - - - Wilcoxon Matched-pairs Signed-ranks Test

EXPER
with CONTROL

Mean Rank	Cases		
5.21	7	- Ranks	(CONTROL Lt EXPER)
6.17	3	+ Ranks	(CONTROL Gt EXPER)
	0	Ties	(CONTROL Eq EXPER)
	--		
	10	Total	

$T_- = (5.21)(7) = 36.5$

$T_+ = (6.17)(3) = 18.5$

Z =   -.9174   2-tailed P =   .3590

Approximate (not exact) p value

SAS APPENDIX (MAINFRAME VERSION)

Solution to Example 10.4

Example 10.4 was concerned with a t test using two independent samples to compare two population means. The problem was to determine if the purchase amount of automobile "extras" is the same for males and females. The SAS program listing in Figure 10.33 was used to test the variances and means using two independent samples. For the test on the means, the value of the test statistic and the p value are provided for the case where the population variances are assumed equal and the case where it is not assumed that the variances are equal.

The TITLE command names the SAS run (in single quotes).

The DATA command gives the data a name.

The INPUT command names and gives the correct order for the variables in the data to follow.

**SAS
APPENDIX
(MAINFRAME
VERSION)**

Figure 10.33

SAS program listing
requesting a
two-sample test for
the variances and
means

```
TITLE 'PURCHASE AMOUNT';
DATA EXTRAS;
INPUT AMOUNT SEX;
CARDS;
23.00 0
23.86 0
19.20 0
15.78 0
30.65 0
23.12 0
17.90 0
30.25 0
16.42 1
14.20 1
21.30 1
18.46 1
11.70 1
12.10 1
16.50 1
9.20 1
21.05 1
18.05 1
PROC PRINT;
PROC TTEST;
   CLASS SEX;
   TITLE 'INDEPENDENT SAMPLES TTEST';
```

The CARDS command indicates to SAS that the input data immediately follow.

The next 18 lines contain the data values, with each line representing each individual's purchase amount and sex (0 = male, 1 = female). The first line, for example, represents a male with a purchase amount of 23 (hundred).

The PROC PRINT command directs SAS to list the input data.

The PROC TTEST command compares the means of two groups of observations. The subcommand CLASS identifies SEX as the variable to be classified in this example. The subcommand TITLE provides a report heading for the output.

Figure 10.34 shows the output obtained by executing the program listing in Figure 10.33.

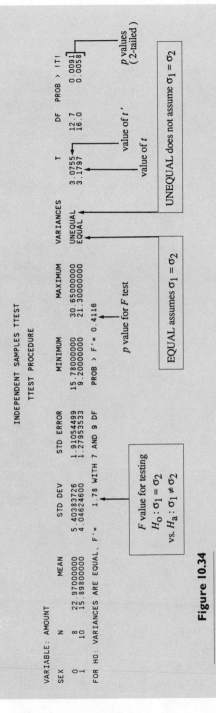

Figure 10.34

Output obtained
by executing the
program listing in
Figure 10.33

SAS APPENDIX (MICRO VERSION)

Solution to Example 10.4

Example 10.4 was concerned with a *t* test using two independent samples to compare two population means. The problem was to determine if the purchase amount of automobile "extras" is the same for males and females. The SAS program listing in Figure 10.35 was used to test the variances and means using two independent samples. For the test on the means, the value of the test statistic and the *p* value are provided for the case where the population variances are assumed equal and the case where it is not assumed that the variances are equal.

Figure 10.35

SAS program listing requesting a two-sample test for the variances and means

```
TITLE 'PURCHASE AMOUNT';
DATA EXTRAS;
INPUT AMOUNT SEX;
CARDS;
23.00  0
23.86  0
19.20  0
15.78  0
30.65  0
23.12  0
17.90  0
30.25  0
16.42  1
14.20  1
21.30  1
18.46  1
11.70  1
12.10  1
16.50  1
9.20   1
21.05  1
18.05  1
PROC PRINT;
PROC TTEST;
 CLASS SEX;
 TITLE 'INDEPENDENT SAMPLES TTEST';
RUN;
```

The TITLE command names the SAS run (in single quotes).

The DATA command gives the data a name.

The INPUT command names and gives the correct order for the variables in the data to follow.

The CARDS command indicates to SAS that the input data immediately follow.

The next 18 lines contain the data values, with each line representing each individual's purchase amount and sex (0 = male, 1 = female). The first line, for example, represents a male with a purchase amount of 23 (hundred).

The PROC PRINT command directs SAS to list the input data.

The PROC TTEST command compares the means of two groups of observations. The subcommand CLASS identifies SEX as the variable to be classified in this example. The subcommand TITLE provides a report heading for the output.

The RUN statement tells the SAS system to execute the previous SAS statements.

Figure 10.36 shows the output obtained by executing the program listing in Figure 10.35.

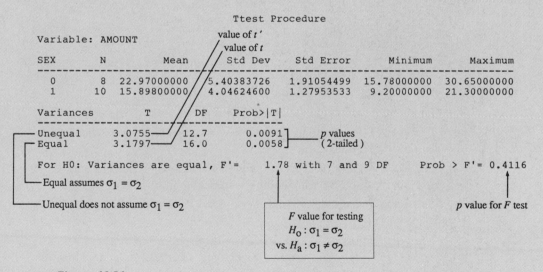

Figure 10.36

Output obtained by executing the program listing in Figure 10.35

SAS APPENDIX (MAINFRAME VERSION)

Solution to Example 10.8

Example 10.8 was concerned with the computation of the t statistic comparing the means of two populations using paired (dependent) samples. The problem was to determine if the average weight loss was more than 10 lb at the completion of the weight loss program. The SAS program listing in Figure 10.37 was used to request a mean, t value, and p value. Note that SAS assumes a two-tailed test. This was a one-tailed test, so the calculated p value must be divided by two.

Figure 10.37

SAS program listing used to request a two-sample t test for two population means using paired (dependent) samples

```
TITLE 'WEIGHT REDUCTION';
DATA WEIGHT;
INPUT BEFORE AFTER;
DIFF=BEFORE-AFTER-10;
CARDS;
132.2 123.7
121.6 111.9
156.4 141.1
167.9 151.4
141.7 124.6
183.5 163.8
137.0 126.2
153.1 145.1
138.4 121.9
145.4 123.9
161.0 156.1
150.8 137.9
PROC PRINT;
PROC MEANS N MEAN T PRT;
 VAR DIFF;
  TITLE 'TWO DEPENDENT SAMPLES';
```

The TITLE command names the SAS run.

The DATA command gives the data a name.

The INPUT command names and gives the correct order for the variables in the data to follow.

The DIFF = BEFORE − AFTER − 10 statement is used to compute a new variable, DIFF, which is the weight loss minus 10 lb.

The CARDS command indicates to SAS that the input data immediately follow.

The next 12 lines contain the data values. Each line represents the before weight and the after weight, respectively. For example, the first line represents a person who weighed 132.2 lb before and 123.7 lb after.

The PROC PRINT command directs SAS to list the input data.

The PROC MEANS command requests an SAS procedure to print simple statistics for the variable in the following subcommand, VAR DIFF. The TITLE subcommand provides a report heading for the output.

Figure 10.38 contains the output obtained by executing the program listing in Figure 10.37.

Figure 10.38

Output obtained by
executing the SAS
program listing in
Figure 10.37

WEIGHT REDUCTION

OBS	BEFORE	AFTER	DIFF
1	132.2	123.7	-1.5
2	121.6	111.9	-0.3
3	156.4	141.1	5.3
4	167.9	151.4	6.5
5	141.7	124.6	7.1
6	183.5	163.8	9.7
7	137.0	126.2	0.8
8	153.1	145.1	-2.0
9	138.4	121.9	6.5
10	145.4	123.9	11.5
11	161.0	156.1	-5.1
12	150.8	137.9	2.9

TWO DEPENDENT SAMPLES

VARIABLE	N	MEAN	T	PR>!T!
DIFF	12	3.45000000	2.34	0.0394

For a one-tailed test, the actual p value = .0394 / 2 = .02

t^*_D

**Solution to
Example 10.8**

Example 10.8 was concerned with the computation of the t statistic comparing the means of two populations using paired (dependent) samples. The problem was to determine if the average weight loss was more than 10 lb at the completion of the weight loss program. The SAS program listing in Figure 10.39 was used to request a mean, t value, and p value. Note that SAS assumes a two-tailed test. This was a one-tailed test, so the calculated p value must be divided by two.

The TITLE command names the SAS run.

The DATA command gives the data a name.

The INPUT command names and gives the correct order for the variables in the data to follow.

SAS APPENDIX (MICRO VERSION)

Figure 10.39

SAS program listing used to request a two-sample *t* test for two population means using paired (dependent) samples

```
TITLE 'WEIGHT REDUCTION';
DATA WEIGHT;
INPUT BEFORE AFTER;
DIFF=BEFORE-AFTER-10;
CARDS;
132.2 123.7
121.6 111.9
156.4 141.1
167.9 151.4
141.7 124.6
183.5 163.8
137.0 126.2
153.1 145.1
138.4 121.9
145.4 123.9
161.0 156.1
150.8 137.9
PROC PRINT;
PROC MEANS N MEAN T PRT;
 VAR DIFF;
 TITLE 'TWO DEPENDENT SAMPLES';
RUN;
```

Figure 10.40

Output obtained by executing the SAS program listing in Figure 10.39

WEIGHT REDUCTION

OBS	BEFORE	AFTER	DIFF
1	132.2	123.7	-1.5
2	121.6	111.9	-0.3
3	156.4	141.1	5.3
4	167.9	151.4	6.5
5	141.7	124.6	7.1
6	183.5	163.8	9.7
7	137.0	126.2	0.8
8	153.1	145.1	-2.0
9	138.4	121.9	6.5
10	145.4	123.9	11.5
11	161.0	156.1	-5.1
12	150.8	137.9	2.9

TWO DEPENDENT SAMPLES

Analysis variable : DIFF

N Obs	N	Mean	T	Prob>|T|
12	12	3.45	2.34	0.0394

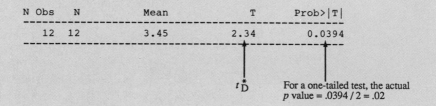

t_D^* For a one-tailed test, the actual *p* value = .0394 / 2 = .02

SAS APPENDIX (MICRO VERSION)

The DIFF = BEFORE − AFTER − 10 statement is used to compute a new variable, DIFF, which is the weight loss minus 10 lb.

The CARDS command indicates to SAS that the input data immediately follow.

The next 12 lines contain the data values. Each line represents the before weight and the after weight, respectively. For example, the first line represents a person who weighed 132.2 lb before and 123.7 lb after.

The PROC PRINT command directs SAS to list the input data.

The PROC MEANS command requests an SAS procedure to print simple statistics for the variable in the following subcommand, VAR DIFF. The TITLE subcommand provides a report heading for the output.

The RUN statement tells the SAS system to execute the previous SAS statements.

Figure 10.40 contains the output obtained by executing the program listing in Figure 10.39.

SAS APPENDIX (MAINFRAME VERSION)

Solution to Example 10.6

Example 10.6 used the Mann–Whitney U test to determine if the purchase amounts of automobile "extras" is the same for males and females. The SAS program listing in Figure 10.41 was used to compute the Mann–Whitney test statistic.

The TITLE command names the SAS run.

The DATA command gives the data a name.

The INPUT command names and gives the correct order for the variables in the data to follow.

The CARDS command indicates to SAS that the input data immediately follow.

The next 18 lines contain the data values; each line represents an individual's purchase amount and sex (0 = male and 1 = female).

SAS APPENDIX (MAINFRAME VERSION)

Figure 10.41

SAS program listing used to determine the Mann–Whitney test statistic

```
TITLE 'PURCHASE AMOUNT';
DATA EXTRAS;
 INPUT AMOUNT SEX;
CARDS;
23.00 0
23.86 0
19.20 0
15.78 0
30.65 0
23.12 0
17.90 0
30.25 0
16.42 1
14.20 1
21.30 1
18.46 1
11.70 1
12.10 1
16.50 1
9.20 1
21.05 1
18.05 1
PROC PRINT;
PROC NPAR1WAY WILCOXON;
 VAR AMOUNT;
 CLASS SEX;
```

The PROC PRINT command directs SAS to list the input data. The Mann–Whitney test is one version of a Wilcoxon test. The following command set was used to compute the sample statistic:

PROC NPAR1WAY WILCOXON
VAR AMOUNT
CLASS SEX

Using this command set, we are able to compare the purchase amounts (variable AMOUNT) by sex category (variable SEX).

Figure 10.42 contains the output obtained by executing the program listing in Figure 10.41.

Figure 10.42

Output obtained by executing the SAS program listing in Figure 10.41

```
                        PURCHASE AMOUNT
ANALYSIS FOR VARIABLE AMOUNT CLASSIFIED BY VARIABLE  SEX

              WILCOXON SCORES (RANK SUMS)

                         SUM OF   EXPECTED   STD DEV      MEAN
LEVEL              N     SCORES   UNDER HO   UNDER HO     SCORE

              0    8     104.00    76.00      11.25      13.00
              1   10      67.00    95.00      11.25       6.70

        WILCOXON 2-SAMPLE TEST (NORMAL APPROXIMATION)
        (WITH CONTINUITY CORRECTION OF .5)
        S= 104.00    Z= 2.4434    PROB >!Z!=0.0145

        T-TEST APPROX. SIGNIFICANCE=0.0258

        KRUSKAL-WALLIS TEST (CHI-SQUARE APPROXIMATION)
        CHISQ=  6.19    DF= 1    PROB > CHISQ=0.0129
```

Approximate (not exact) p value

**SAS
APPENDIX
(MICRO
VERSION)**

**Solution to
Example 10.6**

Example 10.6 used the Mann–Whitney U test to determine if the purchase amounts of automobile "extras" is the same for males and females. The SAS program listing in Figure 10.43 was used to compute the Mann–Whitney test statistic.

Figure 10.43

SAS program listing
used to determine
the Mann–Whitney
test statistic

```
TITLE 'PURCHASE AMOUNTS';
DATA EXTRAS;
 INPUT AMOUNT SEX;
CARDS;
23.00 0
23.86 0
19.20 0
15.78 0
30.65 0
23.12 0
17.90 0
30.25 0
16.42 1
14.20 1
21.30 1
18.46 1
11.70 1
12.10 1
16.50 1
9.20 1
21.05 1
18.05 1
PROC PRINT;
PROC NPAR1WAY WILCOXON;
 VAR AMOUNT;
 CLASS SEX;
RUN;
```

The TITLE command names the SAS run.

The DATA command gives the data a name.

The INPUT command names and gives the correct order for the variables in the data to follow.

The CARDS command indicates to SAS that the input data immediately follow.

The next 18 lines contain the data values; each line represents an individual's purchase amount and sex (0 = male and 1 = female).

The PROC PRINT command directs SAS to list the input data.

SAS APPENDIX (MICRO VERSION)

The Mann–Whitney test is one version of a Wilcoxon test. The following command set was used to compute the sample statistic:

PROC NPAR1WAY WILCOXON
VAR AMOUNT
CLASS SEX

Using this command set, we are able to compare the purchase amount (variable AMOUNT) by sex category (variable SEX).

The RUN statement tells the SAS system to execute the previous SAS statements.

Figure 10.44 contains the output obtained by executing the program listing in Figure 10.43.

Figure 10.44

Output obtained by executing the SAS program listing in Figure 10.43

```
                          PURCHASE AMOUNTS


                        NPAR1WAY PROCEDURE

Wilcoxon Scores (Rank Sums) for Variable AMOUNT
             Classified by Variable SEX


                        Sum of      Expected      Std Dev        Mean
     SEX        N       Scores      Under H0      Under H0       Score

      0         8       104.0         76.0      11.2546287    13.0000000
      1        10        67.0         95.0      11.2546287     6.7000000

     Wilcoxon 2-Sample Test (Normal Approximation)
     (with Continuity Correction of .5)

     S=  104.000      Z=  2.44343912       Prob >|Z| =      0.0145

     T-Test approx. Significance =    0.0258
```

Approximate (not exact) p value

```
     Kruskal-Wallis Test (Chi-Square Approximation)
     CHISQ=  6.1895       DF=  1       Prob > CHISQ=      0.0129
```

11

ESTIMATION AND TESTING FOR POPULATION PROPORTIONS

A LOOK BACK/INTRODUCTION

By now you should be comfortable with the concepts of estimation and hypothesis testing. If, for example, you have reason to believe that the population is normally distributed, you can then estimate the necessary population parameters (such as the mean and standard deviation) using the corresponding sample statistics. You should be well aware that there is always the risk of arriving at an incorrect conclusion when using sample information to infer something about an entire population. Due to the Central Limit Theorem, the assumptions regarding normality can be relaxed when making inferences about population means if large samples are used.

Chapters 8, 9, and 10 have mostly concentrated on normal populations. We provided you with confidence intervals for the mean and the variance of a single normal population. We examined how to check a statement regarding one of these parameters (such as $\mu < 100$ or $\sigma > .50$) using a test of hypothesis. Then this concept was extended to comparing the means or variances of two populations.

Now we return to the binomial situation, in which we are interested in the proportion of your population that has a certain attribute. This would include personal attributes, such as the proportion of a population that is remarried or the proportion of a population that has an allergic reaction to a proposed drug. We can also examine proportions as they relate to a particular physical attribute, such as the proportion of houses in a particular city that are appraised at over $200,000.

We are interested in a single parameter, referred to as **p**, which is the **proportion** of the population having this attribute. For example, suppose that a recent report claims that only 10% of all registered voters in a certain area are in favor of forced busing for school children ($p = .10$). Or suppose it has been reported that a lower proportion of families with children favor busing than do those without children. How can we estimate the actual proportions here and test these claims?

11.1 ESTIMATION AND CONFIDENCE INTERVALS FOR A POPULATION PROPORTION

A test for a population proportion is dealing with a binomial situation. Using the definitions from Chapter 6, each member of your population is either a success or a failure. These words can be misleading; it is necessary only that each person (or object) in your population either have a certain attribute (a success) or not have it (a failure). So we define p to be the proportion of successes in the population; that is, the proportion that have a certain attribute.

Do not confuse the notation $p =$ a population proportion with the previously used shorthand for a p value. They do not mean the same thing. We hope that the context will make it clear which of the two p's is being described.

In Chapter 6, we assumed that p is known. For any binomial situation, perhaps p is known, or more likely it was estimated in some way. This chapter will examine how you can estimate p by using a sample from the population. Also, we can support (or fail to support) claims concerning the value of p. The final section in this chapter will compare two samples from two separate populations.

Point Estimate for a Population Proportion

Suppose that a politician in Riverdale is interested in the proportion of Riverdale registered voters who favor a proposal to allow the sale of alcoholic beverages in restaurants. Since this is a fairly controversial campaign issue, he decided to investigate the situation by obtaining a random sample of 200 registered voters. Of these, 117 said they would be in favor of this proposal. What can you say of the proportion p of all voters who favor this proposal?

We view this problem as a binomial situation and define success as a person who favors the proposal and failure as a person who is opposed to the proposal. Consequently, p is the proportion of successes in the population (proportion of all voters who favor the proposal). Remember that p, like μ and σ previously, will remain unknown forever. To estimate p, we obtain a random sample and observe the proportion of successes in our sample. We use $\hat{p}$ (read as "p hat") to denote the estimate of p, which is the proportion of successes in the sample. Here, $\hat{p} =$ proportion of voters in the sample who are in favor of the liquor proposal, so $\hat{p} = 117/200 = .585$.

In general,

$$\hat{p} = \text{estimate of } p$$
$$= \text{proportion of sample having a specified attribute} \quad \textbf{(11-1)}$$
$$= x/n$$

where n = sample size and x = the number of sample observations having this attribute.

The symbol $\char"005E$ is used to denote an estimate. Distinguish between $\hat{p}$, which is obtained from sample information, and p, the population proportion being estimated by $\hat{p}$. This is the same type of difference that we previously recognized between a sample mean $\bar{X}$ and a population mean μ.

Confidence Intervals for a Population Proportion (Using a Small Sample)

The calculations involved in determining a confidence interval for p using a small sample are fairly complex. To make them easier, we have listed 90% and 95% confidence intervals for sample sizes of $n = 5, 6, \ldots, 20$ in Table A-8. For sample sizes other than these, you can (1) use the large sample confidence interval (described next) or (2) extend Table A-8 by consulting your local statistician. Or you can use a computer subroutine to derive additional values for this table. An explanation of the method used to generate these confidence limits is included at the end of the table.

Using Table A-8 is much like using Table A-1, the table of binomial probabilities. Let n = sample size and x = the observed number of successes in your sample. Based on these values, the confidence interval (p_L, p_U) can be obtained directly from the table.

EXAMPLE 11.1

A private company is considering the purchase of 200 Beagle microcomputers to monitor seismic activity. These computers will be placed in outdoor stations where they must be able to operate in extremely cold weather. If the computers will operate in temperatures as low as $-10°F$, the company will purchase them. Beagle, anxious to demonstrate the reliability of their system, has agreed to subject 15 computers to a "cold test." Let p = proportion of all Beagle computers that will function at $-10°F$.

Of the 15 sample computers, three of them stopped operating at or above $-10°F$. What can you say about p? Construct a 95% confidence interval for p.

Solution

Let a success be that a computer survives the cold test (still functions at $-10°F$). We observe 12 successes out of 15 in the sample. So,

$$\hat{p} = 12/15 = .8$$

Using Table A-8 for $n = 15$, $x = 12$, and $\alpha = .05$, we find $p_L = .519$ and $p_U = .957$. The corresponding 95% confidence interval for p is

$$p_L \text{ to } p_U = .519 \text{ to } .957$$

So we are 95% confident that the actual (population) percentage of Beagle computers that can function at $-10°F$ is between 51.9 and 95.7%.   □

Confidence Intervals for a Population Proportion (Large Sample Approximation)

When dealing with large samples, the Central Limit Theorem once again provides us with a reliable method of determining approximate confidence intervals for a population proportion. For each element in your sample, assign a value of one if this observation is a success (has the attribute) or zero if this observation is a failure (does not have the attribute). Using the Riverdale voters example to illustrate, for each person in the sample, we assign 1 if this person favors the proposal to allow the sale of alcohol and 0 if this person is opposed to the proposal. So what is $\hat{p}$? We can write this as

$$\hat{p} = \frac{\overbrace{1 + 1 + \cdots + 1}^{117 \text{ times}} + \overbrace{0 + 0 + \cdots + 0}^{83 \text{ times}}}{200} = \frac{117}{200} = .585$$

In this sense, then, $\hat{p}$ is a **sample average**: it is an average of zeros and ones. As a result, we can apply the Central Limit Theorem to $\hat{p}$ and conclude that $\hat{p}$ is (approximately) a normal random variable for large samples. This works reasonably well provided np and $n(1 - p)$ are both greater than 5. So the distribution of $\hat{p}$ (large sample; $np > 5$ and $n(1 - p) > 5$) can be summarized: $\hat{p}$ is (approximately) a normal random variable with

$$\text{mean} = p$$

$$\text{standard deviation (standard error)} = \sqrt{\frac{p(1 - p)}{n}}$$

By standardizing this result, we have

$$Z = \frac{\hat{p} - p}{\sqrt{\dfrac{p(1 - p)}{n}}} \tag{11-2}$$

which is approximately a standard normal random variable. This allows us to use Table A-4 to construct a confidence interval for p. This confidence interval is obtained in an identical manner used to construct previous confidence intervals with the standard normal distribution, namely

$$\text{(point estimate)} \pm Z_{\alpha/2} \cdot \text{(standard deviation of point estimator)} \tag{11-3}$$

Thus, a $(1 - \alpha) \cdot 100\%$ confidence interval for p (large sample; np and $n(1 - p) > 5$) is

$$\hat{p} - Z_{\alpha/2} \sqrt{\frac{\hat{p}(1 - \hat{p})}{n}} \quad \text{to} \quad \hat{p} + Z_{\alpha/2} \sqrt{\frac{\hat{p}(1 - \hat{p})}{n}} \tag{11-4}$$

where $\hat{p}$ = sample proportion. Notice that we have used $\hat{p}$ and $1 - \hat{p}$ under the square root in expression 11-4 rather than p and $1 - p$. This was necessary because p is unknown and must be replaced by its estimate $\hat{p}$. As we observed in previous chapters, replacing an unknown parameter by its estimate works well provided our sample is large enough. For this situation, both np and $n(1 - p)$ should be greater than five.

The mean of the random variable $\hat{p}$ is the (unknown) value of p. In other words, the average value of $\hat{p}$ is the parameter it is estimating. Such an estimator is said to be **unbiased**. If we obtained random samples indefinitely, the resulting $\hat{p}$'s on the average will equal p. This is a desirable property for a sample estimator to have. We have actually discussed two other unbiased estimators previously: $\bar{X}$ is an unbiased estimator of the population mean (μ) and s^2 is an unbiased estimator of the population variance (σ^2).

EXAMPLE 11.2

Using the data from the Riverdale voters, what is a 90% confidence interval for the proportion of all voters who favor the liquor proposal?

Solution

Using Table A-4, $Z_{\alpha/2} = Z_{.05} = 1.645$. Also, $\hat{p} = 117/200 = .585$. So the 90% confidence interval for p is

$$.585 - 1.645 \sqrt{\frac{(.585)(.415)}{200}} \quad \text{to} \quad .585 + 1.645 \sqrt{\frac{(.585)(.415)}{200}}$$

$$= .585 - .057 \text{ to } .585 + .057$$

$$= .528 \text{ to } .642$$

Based on the sample data, we are 90% confident that the percentage of voters who favor the liquor proposal is between 52.8 and 64.2%. □

EXAMPLE 11.3

Suppose that in 1986, The Stepfamily Institute became interested in the proportion p of remarried couples who divorce within the first three years of remarriage. A recent report speculated that this proportion (percentage) was 30%, but they had some doubts as to the accuracy of this figure. They obtained a sample of 150 remarried couples from across the United States who were remarried between 1982 and 1984. The sample of 150 contained 61 couples who had divorced within the first three years following the remarriage. Determine a 95% confidence interval for the population proportion, p, of all remarried couples divorcing within the first three years.

Solution

Let p = proportion of remarried couples who divorce within the first three years. Based on the sample of 150 couples, we have

$$\hat{p} = 61/150 = .4067$$

Because $Z_{.025} = 1.96$, the 95% confidence interval for p is

$$.4067 - 1.96 \sqrt{\frac{(.4067)(.5933)}{150}} \quad \text{to} \quad .4067 + 1.96 \sqrt{\frac{(.4067)(.5933)}{150}}$$

$$= .4067 - .0786 \text{ to } .4067 + .0786$$

$$= .328 \text{ to } .485$$

Consequently, we are 95% confident that our estimate of p ($\hat{p} = .4067$) is within .0786 of the actual value of p. In other words, this sample estimates the actual percentage of remarried couples divorcing within the first three years to within 7.86%, with 95% confidence.   □

Choosing the Sample Size (One Population)

Suppose that you want your point estimate $\hat{p}$ to be within a certain amount of the actual proportion p. In example 11.3, the maximum error E was $E = .0786$, that is, 7.86%. What if the Stepfamily Institute wants to estimate the parameter p to within 3%, with 95% confidence? Now,

$$E = 1.96 \sqrt{\frac{p(1 - p)}{n}} \tag{11-5}$$

We have an earlier estimate of p ($\hat{p} = .4067$) using the sample of size 150; this can be used in equation 11-5. The purpose is to extend this sample in order to obtain this specific maximum error E. The specified value of E is .03, so,

$$E = .03 = 1.96 \sqrt{\frac{(.4067)(.5933)}{n}}$$

Therefore,

$$\sqrt{\frac{(.4067)(.5933)}{n}} = \frac{.03}{1.96}$$

Squaring both sides and rearranging leads to

$$n = \frac{(1.96)^2(.4067)(.5933)}{(.03)^2} = 1029.95$$

Rounding up (always), we come to the conclusion that a sample of size $n = 1030$ couples will be necessary to estimate p to within 3%. Notice that the sample size necessary to obtain this much accuracy is considerably larger than the original sample size of 150. Also, in deriving this value, nothing was mentioned concerning the size of the *population*. The large sample size that resulted here was a consequence of the desired accuracy—not the size of the population from which the sample is to be obtained.

In general, the following equation provides the necessary sample size to estimate p with a specified maximum error E and confidence level $(1 - \alpha) \cdot 100\%$:

$$n = \frac{Z_{\alpha/2}^2 \hat{p}(1 - \hat{p})}{E^2} \qquad \textbf{(11-6)}$$

In this illustration, we used an estimate of p from a prior sample to determine the necessary sample size using equation 11-6. If the sample of size n, based on this equation, is our first and only sample, then we have no estimate $\hat{p}$. There is a conservative procedure we can follow here that will guarantee the accuracy (E) that we require. Look at the curve of different values of $\hat{p}(1 - \hat{p})$ in Figure 11.1. Consider these values:

$\hat{p}$	$\hat{p}(1 - \hat{p})$
.2	.16
.4	.24
.5	.25
.7	.21
.9	.09

Note that the largest value of $\hat{p}(1 - \hat{p})$ is .25.

If we make $\hat{p}(1 - \hat{p})$ in equation 11-6 as large as possible, this will provide a value of n that will result in a maximum error that is sure to be less than the specified value. So we can formulate this rule: If no prior estimate of p is available, a conservative procedure to determine the necessary sample size from equation 11-6 is to use $\hat{p} = .5$.

Figure 11.1

Curve of values of
$\hat{p}(1 - \hat{p})$

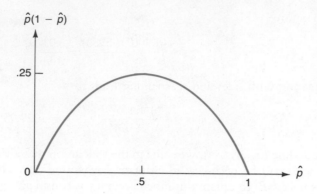

EXAMPLE 11.4

Suppose that the Stepfamily Institute wants to estimate the proportion p of remarried couples who divorce within the first three years to within 3%, with 95% confidence. However, they have *no* prior knowledge of this proportion. Their intent is to obtain a large enough sample the first time, so that they can estimate the population proportion with this much accuracy. How large a sample is required?

Solution

We have no prior knowledge of p, so we use $\hat{p} = .5$ to obtain a sample size of

$$n = \frac{(1.96)^2(.5)(.5)}{(.03)^2} = 1067.1$$

To obtain an estimate of p with a maximum error of $E = .03$, we will need a sample size of $n = 1068$ couples. With a sample of this size, we can safely say that the point estimate $\hat{p}$ will be within 3% of the actual value of p, with 95% confidence (however, this is a very large sample). ☐

Randomized Response Technique

What proportion of people cheat on their taxes? have extramarital affairs? use drugs? These are sensitive questions and it is difficult to obtain an honest response to such questions since people are often hesitant to confess to such activities.

By using the randomized response technique, you offer each subject two questions, one the sensitive question (say, question A) and the other a nonsensitive question (question B). The subject randomly selects either question A or question B using a random outcome technique, such as flipping a coin or using a table of random numbers. Question B is chosen so that the probability a person answering the question will provide a Yes response *is known*. The randomized response technique will then provide an estimate of the proportion of people answering Yes to the sensitive question.

Suppose that each member of a group of 100 people is instructed to select a bingo chip upon entering the room. Twenty percent of the chips begin with a "B," twenty percent begin with "I," and so on. The chips are to be returned to the container at the completion of the session. Each member is given a sheet containing two questions:

Question A: Did you cheat on last year's tax return?

Question B: Does your bingo chip begin with B?

Next, each person is instructed to flip a coin and if a head appears, answer question A and if a tail appears, answer question B. In this way, you remove the threat of answering a sensitive question, since each person can claim that a Yes response was to question B.

We have the following probabilities at the start of the session:

P(question A) = .5 (assuming fair coins)

P(question B) = .5

P(Yes|question B) = .2 (20% of the chips begin with B)

P(Yes|question A) = p (unknown).

This leads to the tree diagram in Figure 11.2.

According to the tree diagram procedure described in Chapter 5, it follows that

$$P(\text{answer Yes}) = (.5)(p) + (.5)(.2)$$

Suppose that at the completion of the session, 38 people answered Yes and 62 answered No. This information provides an estimate of P(answer Yes); namely

Figure 11.2

Tree diagram for randomized response technique

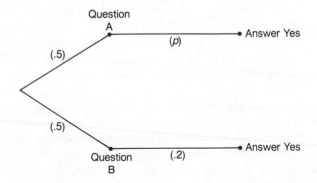

$38/100 = .38$. Thus

$$.38 = (.5)(\hat{p}) + (.5)(.2)$$

$$(.5)(\hat{p}) = .28 \text{ and so } \hat{p} = .56$$

Consequently, the estimate of the proportion of people cheating on last year's tax return is .56 (56%). The randomized response technique can be used for a variety of sensitive questions and allows you to be creative in the construction of the nonsensitive question (question B). All that is required is that P(answer Yes|question B) is known.

EXERCISES

11.1   According to the United States Bureau of Labor Statistics, in 1950, about 27% of women ages 55 to 64 were in the United States labor force. However, in 1985, the percentage is expected to be much higher. Based on trends in the past few decades it should be close to 40%. Suppose you wish to estimate the true proportion of women in that age group who are working, at a 95% confidence level and allowing a maximum error of plus or minus 4%.

(a)   What is the appropriate size of the sample that should be taken to achieve this?

(b)   If you relax the tolerance to $\pm 5\%$ accuracy, what difference does this make in the necessary sample size?

11.2   The Center of Health Data wants to estimate the proportion of American couples ages 20 to 24 who are infertile. A random sample of 18 couples was taken, and 2 were found to be infertile.

(a)   What is the 90% confidence interval for the population proportion of infertile couples?

(b)   What is the sample size that will be needed if the Center of Health Data wishes to obtain an estimate that is within 2.5% of the actual (unknown) proportion at the same level of confidence i.e., 90%?

11.3   The favorite bedtime ritual to lull a child to sleep is probably the telling of a story. In a survey of 700 parents, 52% said that they still read or tell their children a story before tucking them in.

(a)   Construct the 95% confidence interval for the proportion of parents who practice the bedtime story ritual.

(b)   What is the maximum error of your estimate?

(c)   How would you reduce the error?

11.4   A company wishes to estimate the proportion p of its employees who went on sick leave during the past six months. A random sample of 18 employees was taken; of them, 10 went on sick leave. Construct a 90% confidence interval for p. Use Table A-8.

11.5   Using the information in 11.4, construct an approximate 90% confidence interval for the proportion of all employees who went on sick leave during the past six months. Use the normal approximation to the binomial. How does it compare with the exact confidence interval in exercise 11.4?

11.6   A math workshop will be offered only if the student demand is sufficiently high. What is the required sample size necessary to estimate with 90% confidence the

$$.03 = Z_{.05} \sqrt{\frac{.5(1-.5)}{n}}$$

proportion of students who would register for the workshop if we specify the value of .03 for the maximum error E?

11.7 The National Assessment of Educational Progress published a report titled "Literacy: A Profile of America's Young Adults." Of 3600 Americans between the ages 21 and 25 who were tested, 38.5% were unable to read at the 11th grade level, 20% could not read at the 8th grade level, and 6% were unable to read at the 4th grade level. Assuming these 3600 persons tested are representative of the whole population, construct the following confidence intervals:

(a) The 90% confidence interval for the percentage unable to read at the 11th grade level.

(b) The 95% confidence interval for the percentage unable to read at the 8th grade level.

(c) The 99% confidence interval for the percentage unable to read at the 4th grade level.

11.8 Law enforcement officers say that the National Maximum Speed Law of 1974, which set the speed limit at 55 m.p.h., is violated regularly by 70% of motorists. Milton Copulos of the Heritage Foundation in Washington, DC, calls it "the most widely violated statute since prohibition." An enterprising journalist decided to verify this information. Armed with a radar speed detector, he measured the speed of 260 cars on the highway and found that 154 cars exceeded the speed limit.

(a) What is the 95% confidence interval for the proportion of motorists in violation of the speed limit?

(b) To get the estimate within 5% of the true population proportion, what is the size of the sample that should be drawn?

11.9 At the Parkside Medical Facility, out of 143 patients who underwent coronary bypass surgery, 18 patients died.

(a) What is the point estimate for the mortality rate?

(b) What is the 99% interval estimate for the mortality rate?

(c) What is the sample size needed to achieve a maximum error of 3% at this level of confidence?

11.10 While many people dream of hitting some lottery jackpot or some other prize and retiring to a tropical island, a survey of 18 lottery winners found that 13 of them made no changes in their employment situation. Use Table A-8 to construct the 90% and 95% confidence intervals for the percentage of lottery winners who make no changes in their employment situation.

11.11 A manufacturer of microcomputers purchases electronic chips from a supplier which claims its chips are defective only 5% of the time. Determine the sample size that would be required to estimate the true proportion of defective chips if we wanted our estimate $\hat{p}$ to be within 1.25% of the true proportion, with 95% confidence.

11.12 Each member of a group of 200 teenagers selected a chip from a container containing 1000 chips, uniformly ranging from 1 to 10 (integers only). They were then given a sheet containing two questions—question A: Have you ever experimented with cocaine?, and question B: Does your chip contain the number 1? The instructions were to flip a coin and answer question A if heads appeared and question B if tails appeared. At the completion of the exercise, there were

38 yes responses and 162 no responses. Determine $\hat{p}$, where p = population proportion of teenagers who have experimented with cocaine.

11.13 Harry Truman once said, "The buck stops here." However, presidents are not the only ones who are passed the buck. The following interesting data was published in an article by Jill Kiecolt (1987) of Louisiana State University:

WHO IS TO BLAME FOR NATIONAL
ECONOMIC PROBLEMS

Who is to Blame?	Number	%
Congress	181	18.7
President	110	11.3
Labor unions	313	32.3
Big business	328	33.8
Big business and labor	10	1.0
Other combinations	28	2.9
Total sample:	970	100.0

Source: K. J. Kiecolt. Group consciousness and the attribution of blame for national economic problems. Am. Politics Quart. **15** (2): 214 (April 1987), Table 1.

(a) Construct a 90% confidence interval for the proportion of people that blame Congress for national economic problems.

(b) Construct a 95% confidence interval for the proportion of people that did not blame the President for national economic problems.

(c) Construct a 99% confidence interval for the proportion of people that blame Big Business for national economic problems.

11.14 Japan is considered to be one of the major competitors of the United States in the world of electronics. Japanese companies generally hire a worker for life, and Japanese employees are extremely loyal to their employer. However, according to *Japan Electronics Update*, a publication of the American Electronics Association, this attitude may be changing. A survey of 830 workers by a Japanese newspaper indicated that workers may not be so reluctant to switch jobs. In response to the question "Have you ever considered changing your job?", 30% of those surveyed said they had changed jobs, and another 31% reported that they had been contacted by an executive search firm (a relatively new trend in Japan) but had not changed jobs.

(a) Construct the 90% confidence interval for the percentage of Japanese workers who have changed jobs.

(b) Construct the 95% confidence interval for the percentage of Japanese workers who have neither changed jobs nor been approached by an executive search firm.

11.15 Increased health consciousness, higher social pressures, and greater awareness of smoking hazards have led to a decline of smoking in the workplace. According to one survey, 156 out of 200 top executives interviewed were nonsmokers, although out of these nonsmokers, 61.5% used to smoke. Construct the 95% confidence interval for the percentages of the top executives who are ex-smokers (used to smoke but no longer do).

11.2 HYPOTHESIS TESTING FOR A POPULATION PROPORTION

How can you statistically reject a statement such as, "At least 60% of all heavy smokers will contract a serious lung or heart ailment before age 65"? Perhaps someone merely took a wild guess at the value of 60%, and it is your job to gather evidence that will either shoot down this claim or let it stand if there is insufficient evidence to conclude that this percentage actually is less than 60%. We set up hypotheses and test them much like before, only now we are concerned about a population proportion p rather than the mean or standard deviation of a particular population.

Hypothesis Testing Using a Small Sample

Because confidence intervals can be used to perform a test of hypothesis, we will use Table A-8 to conduct such a test. Table A-8 contains sample sizes of $n = 5$ to 20 and $\alpha = .05$ and .10. If $np > 5$ and $n(1 - p) > 5$, the large-sample approximation will provide an accurate test. For sample sizes contained in Table A-8, use the procedure outlined in the accompanying box.

Hypothesis Testing (Small Sample; n Is between 5 and 20)

Two-Tailed Test

H_0: $p = p_0$

H_a: $p \neq p_0$

1 Obtain the $(1 - \alpha) \cdot 100\%$ confidence interval from Table A-8; that is, (p_L, p_U), using $x = $ the observed number of successes.

2 Reject H_0 if p_0 does not lie between p_L and p_U.

3 Fail to reject H_0 if $p_L \leqslant p_0 \leqslant p_U$.

One-Tailed Test

H_0: $p \leqslant p_0$ H_a: $p > p_0$	H_0: $p \geqslant p_0$ H_a: $p < p_0$
1 Obtain the $(1 - 2\alpha) \cdot 100\%$ confidence interval from Table A-8; that is, (p_L, p_U), using $x = $ the observed number of successes.	**1** Obtain the $(1 - 2\alpha) \cdot 100\%$ confidence interval from Table A-8; that is, (p_L, p_U), using $x = $ the observed number of successes.
2 Reject H_0 if $p_0 < p_L$.	**2** Reject H_0 if $p_0 > p_U$.
3 Fail to reject H_0 if $p_0 \geqslant p_L$.	**3** Fail to reject H_0 if $p_0 \leqslant p_U$.

Notice that, for a one-tailed test, we double α when finding the confidence interval for p from Table A-8. For example, if $\alpha = .05$, then $2\alpha = .10$, and so we retrieve a 90% confidence interval from the table. As a result, this particular binomial table can be used only when $\alpha = .025$ or $.05$ for a one-tailed test.

EXAMPLE 11.5

In example 11.1, suppose that the company interested in the Beagle microcomputers will purchase them if Beagle's claim that the proportion p of all Beagle computers that can survive these cold temperatures is greater than .75 (75%) can be shown to be true. Do the data support this claim using $\alpha = .05$?

Solution

The claim under investigation goes into the alternative hypothesis. The appropriate hypotheses are

$$H_0: p \leqslant .75 \quad \text{and} \quad H_a: p > .75$$

We observed, in the sample of 15 computers, $x = 12$ successes (computers that survived). Because $\alpha = .05$, we double this ($2\alpha = .10$) and refer to Table A-8 for a 90% confidence interval for p when $n = 15$, $x = 12$. This is

$$(p_L, p_U) = (.560, .943)$$

We will reject H_0 provided p_L lies to the right of $p_0 = .75$. Because $p_L = .560$ is less than $p_0 = .75$, we fail to reject H_0.

Based on the evidence gathered from this sample, we cannot demonstrate that p is greater than the required 75%. Notice that we are not accepting H_0—we simply fail to reject it. This means that the point estimate $\hat{p} = 12/15 = .8$ is not enough larger than .75 to justify the claim made in H_a. The fact that $\hat{p}$ exceeds .75 may be due to the sampling error that is possible when using a sample statistic $\hat{p}$ to infer something about the population parameter p.

□

Hypothesis Testing: Large Sample Approximation

The standard five-step procedure will be outlined for testing H_0 versus H_a when attempting to support a claim regarding a binomial parameter p using a large sample. To define a test statistic for this situation, the approximate standard normal random variable contained in equation 11-2 is used.

The rejection region for this test is defined by determining the distribution of the test statistic, given that H_0 is true. This means that the unknown value of p in equation 11-2 is replaced by the value of p specified in H_0 (say, p_0). For a one-tailed test, the boundary value of p in H_0 is used. This procedure is summarized in the next box.

Hypothesis Testing [Large Sample; np_0 and $n(1 = p_0)$ both greater than 5]

Two-Tailed Test

H_0: $p = p_0$

H_a: $p \neq p_0$

Reject H_0 if $|Z| > Z_{\alpha/2}$ where

$$Z = \frac{\hat{p} - p_0}{\sqrt{\dfrac{p_0(1 - p_0)}{n}}}$$

One-Tailed Test

H_0: $p \leq p_0$ H_0: $p \geq p_0$

H_a: $p > p_0$ H_a: $p < p_0$

Reject H_0 if $Z > Z_\alpha$ Reject H_0 if $Z < -Z_\alpha$

Notice that the form of the test statistic is that used in previous large sample statistics, namely

$$Z = \frac{\text{(point estimate)} - \text{(hypothesized value)}}{\text{(standard deviation of point estimator)}} \qquad (11\text{-}7)$$

EXAMPLE 11.6

In example 11.2, we estimated the proportion (p) of voters in Riverdale who are in favor of the new liquor proposal. Using the sample of 200 voters, can you conclude that this percentage is greater than the required 50% for this proposal to be approved? Use a significance level of $\alpha = .10$.

Solution   **Step 1** Your hypotheses should be

$$H_0: \quad p \leq .5$$

$$H_a: \quad p > .5$$

Step 2 Since $np_0 = (200)(.5) = 100$ and $n(1 - p_0) = (200)(.5) = 100$ are both > 5, the large-sample test statistic can be used, namely,

$$Z = \frac{\hat{p} - p_0}{\sqrt{\dfrac{p_0(1 - p_0)}{n}}} = \frac{\hat{p} - .5}{\sqrt{\dfrac{(.5)(.5)}{200}}}$$

Step 3 The testing procedure, using $\alpha = .10$, will be to

$$\text{Reject } H_0 \text{ if } Z > Z_{.10} = 1.28$$

Step 4 Using the sample data, $\hat{p} = 117/200 = .585$, so

$$Z^* = \frac{.585 - .5}{\sqrt{\dfrac{(.5)(.5)}{200}}} = \frac{.085}{.0354} = 2.40$$

Because $2.40 > 1.28$, we reject H_0 in favor of H_a.

Step 5 This sample indicates that the population proportion of voters who favor this proposal is greater than 50%.   □

In example 11.6, the computed test statistic was $Z^* = 2.40$. Figure 11.3 shows the Z curve and the calculated p value, which is .0082. Using the classical approach, because $.0082 < \alpha = .10$, we reject H_0. If we choose to base our conclusion strictly on the p value (without choosing a significance level α), this value would be classified as small—it is less than .01. Consequently, using this procedure, we once again reject H_0.

Figure 11.3

Z curve showing p value for example 11.6

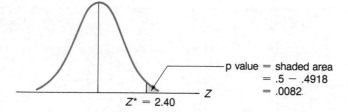

EXAMPLE 11.7 In example 11.3, we estimated the proportion of remarried couples who divorce within the first three years of the remarriage. An earlier report quoted this percentage to be 30%. Do the data gathered by the Stepfamily Institute in example 11.3 lead you to the conclusion that this percentage is different from 30%? Use $\alpha = .05$.

Solution

Step 1 We wish to see if p is different from 30%, so we should use a two-tailed test with hypotheses

$$H_0: \quad p = .30$$
$$H_a: \quad p \neq .30$$

Figure 11.4

Z curve showing p value (twice the shaded area) for example 11.7

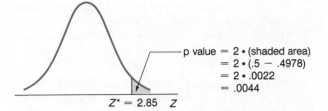

p value = 2 • (shaded area)
= 2 • (.5 − .4978)
= 2 • .0022
= .0044

$Z^* = 2.85$ Z

Step 2 Here $np_0 = (150)(.30) = 45$ and $n(1 - p_0) = (150)(.70) = 105$. Both are >5, so the appropriate test statistic is

$$Z = \frac{\hat{p} - p_0}{\sqrt{\dfrac{p_0(1 - p_0)}{n}}} = \frac{\hat{p} - .3}{\sqrt{\dfrac{(.3)(.7)}{150}}}$$

Step 3 With $\alpha = .05$, the test procedure of H_0 versus H_a will be to

Reject H_0 if $|Z| > 1.96$

Step 4 Using $\hat{p} = 61/150 = .4067$,

$$Z^* = \frac{.4067 - .3}{\sqrt{\dfrac{(.3)(.7)}{150}}} = \frac{.1067}{.0374} = 2.85$$

Because $2.85 > 1.96$, we reject H_0.

Step 5 The percentage of remarrieds divorcing within three years is different than 30% and so the earlier report appears to be in error assuming that both studies are dealing with the same population. As a reminder, because $\alpha = .05$, this particular test will reject H_0 when in fact it is true 5% of the time.

In example 11.7, $Z^* = 2.85$. What is the p value? This is a two-tailed test, so we need to double the right-tail area, as illustrated in Figure 11.4. So $p = 2 \cdot .0022 = .0044$. Thus, using either the classical procedure (comparing the p value to $\alpha = .05$) or basing our decision strictly on the p value, we reject H_0 because of this extremely small p value. □

EXERCISES

11.16 It is believed that more than 10% of the United States population suffers from some form of allergy. To test this belief, a random sample of 250 persons was selected. It was found that 32 persons suffered from some kind of allergy.
(a) Does this sample support the above belief, at a significance level of .05?
(b) Calculate the p value for this test.

11.17 The Institute for Social Research recommended the earliest implementation of a drug education program in schools, saying that drug abuse was reaching epidemic proportions. The Institute claimed that 18% of high school seniors have used cocaine. A random survey of 60 high school seniors found that 12 of them admitted that they had tried cocaine.

(a) Does this survey support the Institute's survey? Conduct a hypothesis test using a significance level of .10.

(b) State the p value for this test.

11.18 An adoption agency officer said that in 1986, 9.7% of all American couples in the age group 20–24 were infertile. A quick random survey of 15 couples in that age group found 2 infertile couples.

(a) Construct a 95% confidence interval (use Table A-8) for the proportion of infertile couples.

(b) Use the confidence interval to test the validity of the adoption agency officer's claim at a significance level of 0.05.

11.19 A politician well known for his rhetoric was making a speech criticizing the government's handling of education of minorities. Among other things he said, "25% of minorities are unable to write a brief description of a job they want." Suppose you wished to verify this claim. From a sample of 50 minority individuals, you find 6 persons who are unable to write a simple job description. Conduct a hypothesis test using a significance level of .05. Do you think the politician was exaggerating (i.e., the percentage is actually lower)?

11.20 There are about 87 million households in the United States. In 1970, about 71% of the households were occupied by married couples. In the last 15 years, the trend toward living alone has accelerated. One demographer believes that at present the proportion of households occupied by married couples is 58%. A random sample of 20 households reveal that 12 are occupied by married couples. At a significance level of 10%, are you in a position to contradict the demographer and say that this person is probably wrong?

11.21 For $np > 5$ and $n(1 - p) > 5$, how large must n be if $p = .03$?

11.22 A recent item in *Working Women* magazine stated that U.S. businesses make about 350 billion photocopies a year, of which 130 billion end up in the trash. Also, 29% of the "good" copies generally are filed and not used again, while 8% of the copies are made for employees' personal use. A female executive decided to investigate her own office. She obtained a random sample of 200 photocopies made, and determined that 30 ended up in the trash, 80 "good" copies were filed, 70 good copies were mailed out, and 20 were made for the personal use of employees. Conduct three hypothesis tests, using $\alpha = 0.05$, to decide if her office is different from the national figures given in *Working Women* magazine, in the following categories:

(a) The proportion of photocopies trashed

(b) The proportion of photocopies filed

(c) The proportion of photocopies made for personal use

(d) State the p value in each case.

11.23 The editor of a famous weekly magazine is concerned about typographical errors and believes that about 1.5% of the number of lines printed have at least one error. An examination of 650 different lines revealed 11 lines that had at least one error. Do the data support the editor's belief? Use a significance level of .10.

11.24 In a debate on ethical standards of therapists, one side said that 10% of the therapists were exploiting their patients and taking advantage of them sexually.

The other side said that in a national survey undertaken by Kenneth Pope, a clinical psychologist in California, of the 585 psychologists polled, 87% (95% of the men and 76% of the women) admitted that they might have been attracted to their clients, at least on occasion, but did not reveal this to their clients. Overall, 93.5% of the therapists had put their clients' interest first, and only 6.5% had made sexual advances. Leaving aside the issue of ethics, are the data from Pope's survey significant enough to statistically reject the claim that 10% of the therapists exploit their patients sexually? Use $\alpha = 0.05$.

11.25 A significant trend in higher education has been the increasing enrollments in business programs. Accompanying this increase has been a shortage of doctorally qualified faculty. Dolecheck (1987) sent a questionnaire to 369 MBA graduates of Northeast Louisiana University (NLU) who had received their MBA degree between May 1965 and May 1984. Research of this sort generally does not obtain a 100% reply rate. Of the 175 usable replies to the questionnaire, 33 respondents had opted to pursue a doctorate and teach, while 147 had chosen to work in business or government. Those who had opted to pursue a doctorate and teach were asked at what point in their lives they made their decision. The percentage distribution of their responses is given below:

POINT IN TIME DECISION TO PURSUE A DOCTORATE
WAS MADE ($n = 33$)

Point in Time	Percent
Before entering undergraduate program	0.0
While pursuing undergraduate degree	3.0
After undergrad. degree but before entering MBA	9.1
While pursuing MBA degree	36.4
After receiving MBA degree	51.5
	100.0

Source: M. Dolecheck, "University teaching: Perceptions of MBAs who earned the doctorate and those who did not. *Proceedings of the 18th Annual Conference of the Southwest Region, Decision Sciences Institute*, Houston, Texas, March 10–14, 1987, Table IV.

(a) Construct a 95% confidence interval for the proportion of persons who decide to pursue a doctorate while pursuing the MBA.

(b) Does the above data support the contention that more than half of those who decide to pursue a doctorate do so after receiving their MBA degree? Use a 5% significance level for the test, and state the p value.

11.26 An instructor believes that, of the students who take a certain course, there are more students who have not taken the prerequisites for the course than students who have taken the prerequisites. The instructor randomly selected 70 students and found that only 30 students had taken the prerequisites for the course. Do these data support the instructor's beliefs? Use a significance level of .10. In addition, state the p value for the above test. Based on the p value, would you reject the null hypothesis at the .01 level?

11.27 An Oklahoma police officer complained that fewer than 1 out of every 5 motorists were observing the seat-belt law. The Oklahoma Highway Safety Officer recently staked out 129 busy street intersections across the state and found that only 16% of the observed drivers and front seat passengers were wearing seat belts. Does this support the officer's complaint? Use a significance level of 5%, and state the p value for your test.

11.28 According to the American Academy of Pediatrics (AAP), the average child in America watches about 20,000 TV commercials each year, more than 60% of them promoting cereal, candy, and toys. An educator was so shocked by these figures that she conducted her own survey. From a random sample of 80 TV advertisements directed at children, she found that 52 promoted cereal, candy, or toys.
   (a) Do these data support the statement made by the AAP, at a significance level of 10%?
   (b) State the p value for this test.
   (c) Based on the p value, would you reject the null hypothesis at a 1% level of significance?

11.29 KNNN, a television news channel, claimed that more than 65% of its subscribers had an annual income of $40,000 or more. A random sample of 160 subscribers was interviewed; 71% of them had incomes of $40,000 or more.
   (a) Does this information support KNNN's claim? Use a significance level of .05.
   (b) Calculate and interpret the p value for part (a).

11.30 A political scientist believed that the influence of PACs (Political Action Committees) had reached a point where more than half of the politicians received a majority of their campaign contributions from PACs. A sample of 50 politicians' records revealed 22 cases where a majority of campaign funds came from PACs. At a significance level of 5%, is the political scientist supported in his belief about the influence of PACs?

11.3   COMPARING TWO POPULATION PROPORTIONS (LARGE INDEPENDENT SAMPLES)

Consider the following questions:

Is the divorce rate higher in California than it is in New York?

Is there a difference in the proportion of high school boys who would attend a drug-awareness program and the proportion of high school girls who would attend?

Is there a higher rate of lung cancer among cigarette smokers than there is among nonsmokers?

These questions are concerned with proportions from two populations. Our method of estimating these proportions will be exactly as it was for one population. We simply have two of everything—two populations, two samples, two estimates, and so on. In this section, it will be assumed that the two samples are obtained independently.

For example, consider the question concerning attendance at a drug-awareness program. A researcher in a large midwestern city is curious to see if the proportion of high school boys who would attend differs from the proportion for high school girls. Population 1 consists of all high school boys in the city, n_1 = number of boys in the sample, and x_1 = number of boys in the sample who agree to attend the program. Similarly population 2 is the group of all high school girls in the city, n_2 = number of girls in the sample, and x_2 = number of girls in the sample who agree to attend the program.

Define a success to be that a student commits to attending the program. (Keep in mind that "success" is merely a label for the trait you are interested in. It need not be a desirable trait.) Our point estimator of p_1 will be as before

$$\hat{p}_1 = \frac{\text{observed number of successes in the sample}}{\text{sample size}}$$

$$= \frac{\text{number of boys in the sample that agree to attend}}{n_1}$$

That is, the point estimator of p_1 is

$$\hat{p}_1 = x_1/n_1 \qquad \text{(11-8)}$$

Similarly, our point estimator of p_2, obtained from the second sample, is

$$\hat{p}_2 = x_2/n_2 \qquad \text{(11-9)}$$

For the two-population case, the parameter of interest will be the difference between the two population proportions, $p_1 - p_2$. The next section will discuss a method of estimating $p_1 - p_2$ by using a point estimate along with a corresponding confidence interval.

Confidence Interval for $p_1 = p_2$ (Large Independent Samples)

The logical estimator of $p_1 - p_2$ is $\hat{p}_1 - \hat{p}_2$, the difference between the sample estimates. What kind of random variable is $\hat{p}_1 - \hat{p}_2$? Since we are dealing with large independent samples (where $n_1\hat{p}_1$, $n_1(1 - \hat{p}_1)$, $n_2\hat{p}_2$, and $n_2(1 - \hat{p}_2)$ are each larger than five), it follows that $\hat{p}_1 - \hat{p}_2$ is (approximately) a normal random variable with

$$\text{mean} = p_1 - p_2$$

and

$$\text{standard deviation} = \sqrt{\frac{p_1(1-p_1)}{n_1} + \frac{p_2(1-p_2)}{n_2}}$$

In a previous section, we observed that $\hat{p}_1$ is a sample mean, where the sample consists of observations that are either a 1 (a particular event occurred) or a 0 (this event did not occur). Because the two samples are obtained independently, the results extend to this situation, leading to the approximate normal distribution for $\hat{p}_1 - \hat{p}_2$. Since $\hat{p}_1$ and $\hat{p}_2$ are unbiased estimators of p_1 and p_2, respectively, the mean of the estimator $\hat{p}_1 - \hat{p}_2$ is $p_1 - p_2$; that is, $\hat{p}_1 - \hat{p}_2$ is an unbiased estimator of $p_1 - p_2$. Notice that the variance of $\hat{p}_1 - \hat{p}_2$ is obtained by adding the variance of $\hat{p}_1$, or $p_1(1-p_1)/n_1$ to the variance of $\hat{p}_2$, or $p_2(1-p_2)/n_2$.

To evaluate the confidence interval, we are forced to approximate the confidence limits by replacing p_1 by $\hat{p}_1$ and p_2 by $\hat{p}_2$ under the square root sign. This approximation works well provided both sample sizes are large. So we derive a $(1 - \alpha) \cdot 100\%$ confidence interval for $p_1 - p_2$ (large independent samples; $n_1\hat{p}_1$, $n_1(1 - \hat{p}_1)$, $n_2\hat{p}_2$, and $n_2(1 - \hat{p}_2)$ are each > 5):

$$(\hat{p}_1 - \hat{p}_2) - Z_{\alpha/2} \sqrt{\frac{\hat{p}_1(1-\hat{p}_1)}{n_1} + \frac{\hat{p}_2(1-\hat{p}_2)}{n_2}} \tag{11-10}$$

$$\text{to} \quad (\hat{p}_1 - \hat{p}_2) + Z_{\alpha/2} \sqrt{\frac{\hat{p}_1(1-\hat{p}_1)}{n_1} + \frac{\hat{p}_2(1-\hat{p}_2)}{n_2}}$$

where $\hat{p}_1 = x_1/n_1$ and $\hat{p}_2 = x_2/n_2$ are the sample proportions. Observe that the construction of this confidence interval uses the "usual" procedure employing Table A-4, described in expression 11-3.

EXAMPLE 11.8   Of a random sample of $n_1 = 250$ high school age boys, 85 agreed to attend a drug-awareness program. A second sample, obtained independently of the first, consisted of $n_2 = 250$ high school girls and 98 of them agreed to attend the program. Construct a 99% confidence interval for $p_1 - p_2$.

Solution   We have $\hat{p}_1 = 85/250 = .34$ and $\hat{p}_2 = 98/250 = .392$. Also, $Z_{\alpha/2} = Z_{.005} = 2.575$, using Table A-4. The resulting confidence interval for $p_1 - p_2$ is

$$(.34 - .392) - 2.575 \sqrt{\frac{(.34)(.66)}{250} + \frac{(.392)(.608)}{250}}$$

$$\text{to} \quad (.34 - .392) + 2.575 \sqrt{\frac{(.34)(.66)}{250} + \frac{(.392)(.608)}{250}}$$

$$= -.052 - .111 \text{ to } -.052 + .111$$

$$= -.163 \text{ to } .059$$

This confidence interval leaves us unable to conclude that either group has a higher attendance percentage. We are 99% confident that the percentage of boys that will attend is between 16.3% lower to 5.9% higher than for the female percentage. □

EXAMPLE 11.9

The Stepfamily Institute is conducting another research project to determine if the divorce rate (p_1) for six states in the New England area is higher than the divorce rate (p_2) for five southern states. A random sample of $n_1 = 400$ marriages in the New England states revealed 180 divorces. Another sample of $n_2 = 300$ marriages in the southern states resulted in 102 divorces. Determine a 95% confidence interval for $p_1 - p_2$.

Solution

Our proportion estimates are

$$\hat{p}_1 = 180/400 = .45$$

$$\hat{p}_2 = 102/300 = .34$$

The 95% confidence interval for $p_1 - p_2$ is

$$(.45 - .34) - 1.96\sqrt{\frac{(.45)(.55)}{400} + \frac{(.34)(.66)}{300}}$$

$$\text{to} \quad (.45 - .34) + 1.96\sqrt{\frac{(.45)(.55)}{400} + \frac{(.34)(.66)}{300}}$$

$$= .11 - .072 \text{ to } .11 + .072$$

$$= .038 \text{ to } .182$$

So we are 95% confident that (1) our estimate of the difference in the divorce rates, namely $\hat{p}_1 - \hat{p}_2 = .11$, is within 7.2% of the actual difference, and (2) the divorce rate is between 3.8 and 18.2% higher in the New England states than it is in the Southern states. □

Choosing the Sample Sizes (Two Populations)

In Chapter 10, we discussed how to select samples from two populations when the desired accuracy of the point estimate of the difference between two population means is specified—this is the maximum error E. If E is, say, 10 lb, then what sample sizes (n_1 and n_2) are necessary for the point estimate of $\mu_1 - \mu_2$ (namely, $\bar{X}_1 - \bar{X}_2$) to be within 10 lb of the actual value, with 95% (or whatever) confidence? Using the results contained in Appendix B, values of n_1 and n_2 were provided in Chapter 10 that minimized the total sample size, $n_1 + n_2$, for this specific value of E.

We encounter a similar situation when dealing with two population proportions, p_1 and p_2. If a maximum error of, say, $E = .10$ is specified, then the question of interest is: What sample sizes (n_1 and n_2) are necessary for the point estimate of $p_1 - p_2$ (namely, $\hat{p}_1 - \hat{p}_2$) to be within .10 of the actual value, with 95% (or whatever) confidence?

The maximum error E always is the amount that you add to and subtract from the point estimate when determining a confidence interval. When dealing with two proportions, this is

$$E = Z_{\alpha/2} \sqrt{\frac{p_1(1 - p_1)}{n_1} + \frac{p_2(1 - p_2)}{n_2}} \qquad \textbf{(11-11)}$$

To evaluate this expression, you will need estimates of p_1 and p_2. You have two options. If you have previously obtained small samples from these two populations, then you can use the resulting sample estimates $\hat{p}_1$ and $\hat{p}_2$. The purpose then will be to extend these samples to obtain better accuracy in the point estimate, $\hat{p}_1 - \hat{p}_2$. If no information regarding p_1 and p_2 is available, then you can use the conservative approach discussed in Section 11.1 by letting $\hat{p}_1 = \hat{p}_2 = .5$.

By applying the results of Appendix B to this situation, the sample sizes n_1 and n_2 that minimize the total sample size $n_1 + n_2$ are given by

$$n_1 = \frac{Z_{\alpha/2}^2(A + B)}{E^2} \qquad \textbf{(11-12)}$$

$$n_2 = \frac{Z_{\alpha/2}^2(C + B)}{E^2} \qquad \textbf{(11-13)}$$

where

$$A = p_1(1 - p_1)$$
$$B = \sqrt{p_1 p_2(1 - p_1)(1 - p_2)}$$
$$C = p_2(1 - p_2)$$

To determine A, B, and C, estimates of p_1 and p_2 should be substituted for p_1 and p_2 by using one of the two options described.

EXAMPLE 11.10   Using the data from example 11.9, determine what sample sizes are necessary for the estimate of the difference between the two divorce rates to be within .03 of the actual value, with 95% confidence, if (1) the results from example 11.9 are available and (2) no sample information is available.

Solution 1 The specified maximum error is $E = .03$. Sample data have been collected regarding these proportions, so we use the corresponding estimates to determine the sample sizes necessary to obtain this degree of accuracy. Using Table A-4, $Z_{\alpha/2} = Z_{.025} = 1.96$. Here, $\hat{p}_1 = .45$ and $\hat{p}_2 = .34$. Consequently,

$$A = \hat{p}_1(1 - \hat{p}_1)$$
$$= .2475$$

$$B = \sqrt{\hat{p}_1\hat{p}_2(1 - \hat{p}_1)(1 - \hat{p}_2)}$$
$$= .2357$$

$$C = \hat{p}_2(1 - \hat{p}_2)$$
$$= .2244$$

To obtain the smallest possible total sample size for the required accuracy, the two sample sizes should be

$$n_1 = \frac{(1.96)^2(.2475 + .2357)}{(.03)^2} \cong 2063$$

(remember—always round up), and

$$n_2 = \frac{(1.96)^2(.2244 + .2357)}{(.03)^2} \cong 1964$$

providing a total sample size of $n_1 + n_2 = 4027$ couples. Notice that there is a huge "price to pay" for this increased accuracy—the total sample size has increased from 700 in example 11.9 to over 4000.

Solution 2 If no prior estimates of p_1 and p_2 are available, using $\hat{p}_1 = \hat{p}_2 = .5$ will result in sample sizes n_1 and n_2 that, when obtained, will provide a maximum error no larger than the specified value of $E = .03$. Here, $A = (.5)(.5) = .25$. Similarly, $B = C = .25$, so

$$n_1 = n_2 = \frac{(1.96)^2(.25 + .25)}{(.03)^2} \cong 2135$$

Consequently, a total sample size of $n_1 + n_2 = 4270$ couples will be necessary for $\hat{p}_1 - \hat{p}_2$ to be within .03 of the actual value of $p_1 - p_2$, with 95% confidence. □

Hypothesis Testing for p_1 and p_2 (Large Independent Samples)

Suppose that a recent report stated that, based on a sample of 500 people, 35% of all cigarette smokers had at some time in their life developed a particular fatal disease. On the other hand, 25% of the nonsmokers in the sample acquired the disease. Can we conclude from this sample that, because $\hat{p}_1 = .35 > \hat{p}_2 = .25$, the proportion ($p_1$) of all smokers who will acquire the disease exceeds the proportion (p_2) for nonsmokers? In other words, is $\hat{p}_1$ significantly larger than $\hat{p}_2$?

After all, even if $p_1 = p_2$, there is a 50/50 chance that $\hat{p}_1$ will be larger than $\hat{p}_2$ because for large samples, the distribution of $\hat{p}_1 - \hat{p}_2$ is approximately a bell-shaped (normal) curve centered at $p_1 - p_2$—which, if $p_1 = p_2$, would be zero.

Are the results of the sample significant, or are they due simply to the sampling error that is always possible when estimating from a sample? Your alternative hypothesis can be that two proportions are different (a two-tailed test) or that one exceeds the other (a one-tailed test). As before, we will assume that the two random samples are obtained independently. The possible hypotheses are these: for a two-tailed test,

$$H_0: \quad p_1 = p_2$$
$$H_a: \quad p_1 \neq p_2$$

and for a one-tailed test,

$$H_0: \quad p_1 \leqslant p_2$$
$$H_a: \quad p_1 > p_2$$

or

$$H_0: \quad p_1 \geqslant p_2$$
$$H_a: \quad p_1 < p_2$$

One possible test statistic to use here would be the standard normal (Z) statistic that was used to derive a confidence interval for $p_1 - p_2$, namely,

$$Z = \frac{\hat{p}_1 - \hat{p}_2}{\sqrt{\dfrac{\hat{p}_1(1 - \hat{p}_1)}{n_1} + \dfrac{\hat{p}_2(1 - \hat{p}_2)}{n_2}}} \qquad (11\text{-}14)$$

In all previous tests of hypothesis, we always examined the distribution of the test statistic when H_0 was true. For a one-tailed test, we assumed the boundary condition of H_0, which in this case would be $p_1 = p_2$. Because of this, whenever we obtained a value of the test statistic in one of the tails, our decision was to reject H_0 because this value would be very unusual if H_0 were true. This reasoning was used for test statistics that followed a Z, t, χ^2, or F distribution.

We use the same approach here. If $p_1 = p_2 = p$ (say), we can improve the test statistic in equation 11-14. For this situation, p is the proportion of successes in the combined population. Our best estimate of p is the proportion of successes in the combined sample. So define

$$\bar{p} = \frac{x_1 + x_2}{n_1 + n_2}$$

Thus, assuming $p_1 = p_2$, $\hat{p}_1 - \hat{p}_2$ is approximately a normal random variable with

$$\text{mean} = p_1 - p_2 = 0$$

and

$$\text{standard deviation} = \sqrt{\frac{p_1(1 - p_1)}{n_1} + \frac{p_2(1 - p_2)}{n_2}}$$

$$\cong \sqrt{\frac{\bar{p}(1 - \bar{p})}{n_1} + \frac{\bar{p}(1 - \bar{p})}{n_2}}$$

The resulting test statistic for p_1 versus p_2 (large independent samples; $n_1\hat{p}_1$, $n_1(1 - \hat{p}_1)$, $n_2\hat{p}_2$, and $n_2(1 - \hat{p}_2)$ are each greater than 5) is

$$Z = \frac{\hat{p}_1 - \hat{p}_2}{\sqrt{\dfrac{\bar{p}(1 - \bar{p})}{n_1} + \dfrac{\bar{p}(1 - \bar{p})}{n_2}}} \qquad \text{(11-15)}$$

where

$$\hat{p}_1 = x_1/n_1$$

$$\hat{p}_2 = x_2/n_2$$

$$\bar{p} = \frac{x_1 + x_2}{n_1 + n_2}$$

Observe that the form of this test statistic is the same as for the single population case, described in equation 11-7. The test procedure is the standard routine when using the Z distribution. For a two-tailed test,

$$H_0: \quad p_1 = p_2$$

$$H_a: \quad p_1 \neq p_2$$

Reject H_0 if $|Z| > Z_{\alpha/2}$

where Z is defined in equation 11-15. For a one-tailed test,

$$H_0: \quad p_1 \leq p_2$$

$$H_a: \quad p_1 > p_2$$

Reject H_0 if $Z > Z_\alpha$

or

$$H_0: \quad p_1 \geq p_2$$

$$H_a: \quad p_1 < p_2$$

Reject H_0 if $Z < -Z_\alpha$

EXAMPLE 11.11

Use the data from example 11.8 and determine whether there is any difference between the proportions of male and female students who would attend a drug awareness program. Let $\alpha = .01$.

Solution

The five-step procedure is the correct one. The confidence interval derived in example 11-8 would produce the same result as the five-step procedure if the test statistic were the one defined in equation 11-14. The correct procedure here is to use the Z statistic in equation 11-15 as your test statistic.

Step 1 Since we are looking for a difference between p_1 and p_2, define

$$H_0: \quad p_1 = p_2$$

$$H_a: \quad p_1 \neq p_2$$

Step 2 The test statistic is

$$Z = \frac{\hat{p}_1 - \hat{p}_2}{\sqrt{\dfrac{\bar{p}(1 - \bar{p})}{n_1} + \dfrac{\bar{p}(1 - \bar{p})}{n_2}}}$$

Step 3 Using $\alpha = .01$, then $Z_{\alpha/2} = Z_{.005} = 2.575$. The test procedure will be to

$$\text{Reject } H_0 \text{ if } |Z| > 2.575$$

Step 4 Since $n_1 = 250$, $x_1 = 85$, and $n_2 = 250$, $x_2 = 98$, then

$$\bar{p} = \frac{x_1 + x_2}{n_1 + n_2} = \frac{183}{500} = .366$$

Therefore, our estimate of the proportion of all students that would attend the program (if $p_1 = p_2$) is $\bar{p} = .366$ (36.6%). Also, $\hat{p}_1 = 85/250 = .34$, and $\hat{p}_2 = 98/250 = .392$. The value of the test statistic is

$$Z^* = \frac{.34 - .392}{\sqrt{\dfrac{(.366)(.634)}{250} + \dfrac{(.366)(.634)}{250}}} = \frac{-0.052}{0.0431} = -1.21$$

Because $|Z^*| = 1.21 < 2.575$, we fail to reject H_0.

Step 5 There is insufficient evidence to conclude that a difference exists between the percentage of boys who would attend and the percentage of girls who would attend.

The Z curve and calculated p value for example 11.11 are shown in Figure 11.5. The p value is twice the shaded area (this was a two-tailed test) and is .2262, which is extremely large. Using the classical approach, because .2262 > $\alpha = .01$, we fail to reject H_0—there is insufficient evidence to indicate a difference in the population proportions. As a reminder, this reasoning always leads to the same conclusion as the five-step procedure. Because .2262 exceeds any reasonable value of α, we fail to reject H_0 quite strongly for this application.

□

Figure 11.5

Z curve showing p value (twice the shaded area) for example 11.11

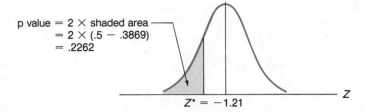

$$p \text{ value} = 2 \times \text{shaded area}$$
$$= 2 \times (.5 - .3869)$$
$$= .2262$$

$Z^* = -1.21$

EXAMPLE 11.12

In example 11.9, the Stepfamily Institute suspected that the divorce rate in the New England area (p_1) was higher than the divorce rate in the southern states (p_2). Based on these data, what would they conclude, using a significance level of .05?

Solution

Step 1 We wish to know whether the data warrant the conclusion that p_1 is larger than p_2. Placing this in the alternative hypothesis leads to

$$H_0: \quad p_1 \leqslant p_2$$
$$H_a: \quad p_1 > p_2$$

Steps 2, 3 Using the test statistic in equation 11-15, the resulting one-tailed test procedure would be to

$$\text{Reject } H_0 \text{ if } Z > Z_{.05} = 1.645$$

Step 4 We have

$$\hat{p}_1 = 180/400 = .45$$

and

$$\hat{p}_2 = 102/300 = .34$$

Also,

$$\bar{p} = \frac{180 + 102}{(400 + 300)} = \frac{282}{700} = .403$$

Consequently,

$$Z^* = \frac{.45 - .34}{\sqrt{\dfrac{(.403)(.597)}{400} + \dfrac{(.403)(.597)}{300}}} = \frac{.11}{.0375} = 2.93$$

Because $2.93 > 1.645$, we reject H_0.

Step 5 There is evidence of a higher divorce rate in the New England area.

The Z curve and calculated p value for example 11.12 (a one-tailed test) are shown in Figure 11.6. The p value is .0017. This is definitely a very small p value and (as before) leads to rejecting H_0 using a significance level of .05.

Figure 11.6

Z curve showing the calculated p value for example 11.12

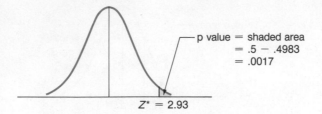

Based on this p value alone, we arrive at the same conclusion—namely, that there is evidence of a higher rate of divorce in the New England states.   □

EXERCISES   11.31   The U.S. Supreme Court's decision in *Illinois Brick Co v. Illinois* (431 U.S. 720, 1977) is believed to be a significant decision which ended nearly a decade of uncertainty with regard to plaintiff's standing to sue for antitrust damages; direct purchasers, with two exceptions, were given exclusive rights to sue. Joyce and McGuckin (1986) used data relating Department of Justice (DOJ) criminal price-fixing cases and the defendants in those cases to private follow-on cases, in order to assess the impact of the *Illinois Brick* decision. Their major focus was whether the decision has had a positive or negative effect on convicted price-fixers' exposure to private damage claims, as measured by number of follow-on cases. The following data shows the number of follow-on cases before and after the *Illinois Brick* decision:

DISTRIBUTION OF U.S. CRIMINAL ANTITRUST CASES AND FOLLOW-ON CASES

	Before Illinois Brick 1968–1977	After Illinois Brick 1977–1980
DOJ cases with one or more follow-ons	100	46
DOJ cases with no follow-ons	87	106
Total DOJ cases	187	152

Source: J. M. Joyce and R. H. McGuckin. Assignment of rights to sue under *Illinois Brick:* An empirical assessment. *The Antitrust Bulletin* 247 (Spring 1986), Table 2.

(a)   Define p_1 and p_2 and show the hypotheses that you would set up to test the impact of the *Illinois Brick* decision on exposure to private follow-on damage claims,

(i)   Using the proportion of DOJ cases with follow-on.

(ii)   Using the proportion of DOJ cases without follow-on.

(b)   At a 5% significance level, has there been a significant increase in the percentage of DOJ criminal cases with no follow-on? What does this suggest about the impact of *Illinois Brick*?

11.32 Use the data in exercise 11.31 to construct a 90% confidence interval for the difference between the two proportions (i.e., the proportion of DOJ cases with no follow-ons before Illinois Brick and the proportion of DOJ cases with no follow-ons after Illinois Brick).

11.33 How does the value of the test statistic given in equation 11-15 change if $\hat{p}_1$, the proportion of successes for sample 1, is replaced by the proportion of failures for sample 1, and if $\hat{p}_2$, the proportion of successes for sample 2, is replaced by the proportion of failures for sample 2?

11.34 An educational analyst was interested in examining the proportions of foreign students in two colleges, Xavier and Safire. A random sample of 68 students was taken from Safire College; nine of them were foreign students. A similar random sample of 110 students was taken at Xavier College; 14 of them were foreign students. Can we conclude from the data that the proportions of foreign students in the two colleges differ? Use a significance level of .05.

11.35 Construct a 95% confidence interval for the difference of the two proportions in exercise 11.34.

11.36 Preliminary results from a 5-year study at Drexel University of the relationship between computer usage and social values suggest that "people who use computers are more tolerant of a white-collar crime than society in general." Suppose you wished to investigate this claim by sampling two independent groups, one from the general population without any special computer expertise and one from persons with a computer background. You are interested in comparing the difference in the proportions of persons tolerant of white-collar crime in the two groups. Allowing for a maximum error 0.06 and a desired confidence level of 95%, what are the sample sizes that need to be drawn?

11.37 A psychologist wished to investigate whether managers with MBA degrees were less sensitive to human relations than those without MBA degrees. Suppose she wishes to take two independent samples, one from each group, to compare the proportion in each group who pass a sensitivity test. The maximum error that can be tolerated is .10, using a confidence level of 99%. What are the sizes of the samples, n_1 and n_2, that need to be taken to achieve this?

11.38 The legislature of a Southern state passed a rule commonly called "no-pass, no play," which prohibits a student failing in any subject from participating in any extracurricular activity for six weeks. Data have been collected for football, volleyball, cross country, and band for the first six-week grading period. A sample taken in 1985 found that 26.91% of 275 students sampled were sidelined for six weeks for failing one or more courses. A sample of the same size taken in 1986 found 17.8% of the students sidelined for failing one or more courses. The assistant superintendent for secondary schools said, "We are very pleased with the improvement. It shows coaches and students are taking the rule seriously." Is the superintendent justified in making this conclusion (i.e., is the proportion sidelined in 1986 significantly lower than the proportion sidelined in 1985)? Conduct a hypothesis test at $\alpha = .05$ to decide this, and also state the p value of your test.

11.39 Refer to exercise 11.38. The data also showed that 28 out of 100 freshmen, and 22 out of 100 junior varsity players were ineligible in 1986 due to the no-pass, no-play rule. Does this provide sufficient evidence to indicate that the failure rate for freshmen is significantly higher than the failure rate for juniors? Use $\alpha = .01$, and also state the p value for your test.

11.40   Storper and Christopherson (1987), in a case study of the U.S. Motion Picture Industry, have argued that for much of its history, the industry was under the control of major studios, and film production was a routinized, mass-production factory-like process. However, in the 1950s a new trend evolved, and in the contemporary motion picture industry, most films are made by independent production companies, which subcontract work to small specialized firms. Dividing firms into independent, major, and mini-major categories, they provide the following data for 1960 and 1980:

NUMBER OF PRODUCTIONS BY ORGANIZATION TYPE

Type of Organization	1960	1980
Independent	42	129
Major	100	69
Mini-major	9	24
Total	151	222

Source: M. Storper and S. Christopherson. Flexible specialization and regional industrial agglomerations: The case of the U.S. motion picture industry. *Ann. Assoc. Am. Geographers* **77** (1): 104–117 (1987).

(a)   Defining p_1 = proportion of independent productions in 1960 and p_2 = proportion of independent productions in 1980, conduct a hypothesis test at $\alpha = .10$ to determine if there was a significant increase in independent productions, as suggested by the authors. What is the p value for the test?

(b)   Conversely, you could also test for a decline in the proportion of major productions. Define p_1 and p_2 in terms of majors, and show how you would conduct the hypothesis test.

(c)   Without doing any more calculations, use the p value obtained in part (a) to decide if the increase in independent productions was significant at $\alpha = .01$.

SUMMARY

You will often encounter a situation in which you are concerned with a population **proportion**, rather than the mean or variance of a continuous random variable. For example, the parameter of interest might be the proportion p of adults earning less than \$15,000 a year—rather than the average salary μ or the standard deviation σ of the salaries. The usual procedure of estimating a population parameter using the sample estimator allows us to derive a point estimate and construct a confidence interval for p. When the sample is small, Table A-8 provides an exact confidence interval for p. For large samples, the Central Limit Theorem can be applied to determine an approximate confidence interval, provided that both $n\hat{p}$ and $n(1 - \hat{p})$ are greater than 5. A **randomized response technique** can be used to estimate population proportions of a sensitive nature. This procedure allows each subject to randomly select one of the two questions to answer, one the sensitive question of interest and the other a nonsensitive question with a known probability of a Yes response.

When the desired accuracy of the point estimate p is specified in advance, you can determine the sample size necessary to obtain this degree of accuracy for a certain confidence level. To derive this sample size, an estimate of p is necessary. You can calculate this value using a previous sample estimate, or, if no information is available, using a conservative procedure and making $\hat{p} = .5$.

When you investigate a statement concerning a population proportion, you can use a statistical test of hypothesis. For small samples, the confidence interval from Table A-8 provides an exact procedure for either a one- or two-tailed test. For tests of hypothesis when a large sample is used, a test statistic having an approximate standard normal distribution can be used.

To compare two population proportions (p_1 and p_2), two independent random samples are obtained, one from each population. Procedures for large independent samples generally provide an accurate confidence interval or test of hypothesis whenever $n_1\hat{p}_1$, $n_1(1 - \hat{p}_1)$, $n_2\hat{p}_2$, and $n_2(1 - \hat{p}_2)$ each exceeds five. Using a standard normal approximation, we can construct a confidence interval for $p_1 - p_2$. If the accuracy of this estimate is specified, the sample sizes necessary to obtain this level of accuracy, as well as to minimize the total sample size $n_1 + n_2$, can be obtained.

Two population proportions can be compared by using two large independent samples to evaluate a test statistic having an approximate standard normal distribution. We examined procedures for a one-tailed test (for example, H_a: $p_1 > p_2$) or a two-tailed test (H_a: $p_1 \neq p_2$). The rejection region for these tests are defined using the areas from Table A-4.

REVIEW EXERCISES

11.41 The argument has been put forward that racial inequalities in the United States have become less dependent on the prejudices of individual white persons and more a consequence of "normal" institutional arrangements that allegedly serve to disadvantage blacks. Tuch and Taylor (1986) obtained a national sample of 372 white adults to analyze their beliefs and attitudes about the racial implications of institutional practices, such as: college admissions criteria (are standardized test scores biased against blacks?); legal services (are blacks less able than whites to afford high quality legal services?); seniority systems (do such systems work against blacks in industries where blacks were excluded in the past?); and other institutional practices which we will not consider in this exercise. The relevant proportions are given below:

WHITE ADULTS' BELIEFS ABOUT INSTITUTIONAL IMPACT

	Belief	
Institutional Practice	Impedes Racial Equality	Does Not Impede Racial Equality
College admissions ($n = 299$)	62.2%	37.8%
Legal services ($n = 350$)	91.7%	8.3%
Seniority systems ($n = 308$)	63.3%	36.7%

Source: S. A. Tuch and M. C. Taylor. Whites' opinions about institutional constraints on racial equality. *Sociol. Social Res.* **70** (4): 268–271 (July 1986).

(a) You will observe that the sample sizes specified for each institutional practice ($n = 299$, etc.) are not equal to the total national sample of 372 white adults. Can you suggest why this would be the case?

(b) Construct the following confidence intervals:

  (i) The 90% confidence interval for the proportion of white adults who believe that college admissions criteria impede racial equality.

  (ii) The 95% confidence interval for the proportion of white adults who believe that legal services affordability impedes racial equality.

  (iii) The 99% confidence interval for the proportion of white adults who believe that seniority systems do not impede racial equality.

11.42 ABC's Monday Night Football had a rating of 19.7 and a share of 32 in 1985. "Rating" is the percentage of United States homes with television; "share" is the percentage of sets in use that are tuned to a particular program. Although the actual process of determining these figures is fairly sophisticated, let us assume for the sake of simplicity that the "share" is computed from a sample of 1000 homes.

(a) What is the 95% confidence interval for ABC's 1985 share for Monday Night Football?

(b) What is the maximum error of your estimate?

11.43 Refer to exercise 11.42. Suppose you decide to conduct an experiment of your own. You select 20 homes randomly from the phone book, and call them. In 5 homes, the TV is off. In the remaining 15 homes, 5 are tuned in to ABC's Monday Night Football.

(a) What is the 95% confidence interval for ABC's share?

(b) What is the maximum error of your estimate?

(c) To achieve a maximum error of plus or minus 4% for a 95% confidence interval, what is the size of the sample you would need to take?

11.44 Green et al. (1987) reported that numerous patients had been referred to their psychiatric consultation service because of crying, under the assumption that some mental illness was the underlying cause. However, they found that, in many cases, the crying was due to neurological disease and could be referred to as being "pathological." Out of 296 consultations, 46 patients were classified as "prominent criers" and 233 as "noncriers." (We will ignore a third group of 17 classified as "nonprominent criers.") Their study examined a number of hypotheses, one of which was concerned with whether the proportions of patients with psychiatric and neurological diagnoses were similar for criers and noncriers. The relevant data are given below:

DISTRIBUTION OF DIAGNOSES

Diagnosis	Noncriers	Prominent Criers
Psychiatric disorder only	148	9
Neurological disorder only	50	15
Other categories	35	22
Total	233	46

Source: R. L. Green, T. W. McAllister, and J. L. Bernat. A study of crying in medically and surgically hospitalized patients. Am. J. Psychiatry **144** (4): 442–447 (April 1987).

(a) Test the hypothesis that prominent criers are less likely than noncriers to have psychiatric disorders only. Let $\alpha = .01$. (*Hint:* Define p_1 = proportion of noncriers diagnosed as having psychiatric disorder only, and so on.)

(b) Test the hypothesis that prominent criers are more likely than noncriers to have neurological disorders only. Let $\alpha = .01$.

11.45 Brittin and Nossaman (1986) conducted a survey to obtain information concerning the use of iron cooking utensils. The survey actually consisted of two samples. In the first sample (the "fair sample"), 182 persons were interviewed at the 1982 Panhandle South Plains Fair in Lubbock, Texas. For the second sample (the "convention sample"), 184 subjects completed a slightly modified questionnaire at the 1982 American Dietetic Association Convention in San Antonio, Texas. While the study actually reported the results for as many as 37 different variables, we will focus on just one variable in this exercise, the proportion that prepared most of their own meals:

Variable	Fair Sample	Convention Sample
Prepare most of own meals	154	139
Sample size	182	184

Source: H. C. Brittin and C. E. Nossaman. Use of iron cookware. Home Econ. Res. J. **15** (1): 43–51. (September 1986), Table 2.

(a) Is there a significant difference between the proportions that prepare most of their own meals for the fair sample (p_1) versus the convention sample (p_2)? Let $\alpha = .05$.

(b)   Combine the two samples and treat them as one sample of 366 persons, of whom 293 persons prepare most of their own meals. Construct a 99% confidence interval for the proportion of persons in the population who prepare their own meals.

11.46   In a newspaper survey of 1770 teenagers, 88% said their parents have done a good job of rearing them, 67% said their parents' house rules are fair, and 64% said their discipline is "about right." Construct the following confidence intervals:

(a)   The 90% CI for the proportion who say their parents have done a good job of rearing them.

(b)   The 95% CI for the proportion who *do not* say their parents' house rules are fair.

(c)   The 99% CI for the proportion who say their discipline is "about right."

11.47   (a)   In exercise 11.46, what is the maximum error of the estimate for parts (a), (b), and (c)?

(b)   If you maintain the same level of confidence in each case, what is the sample size needed for parts (a), (b), and (c) above if you will tolerate a margin of error of plus or minus 5%?

11.48   A foreign student heard an evangelist on TV describing the general decline in teenage morality and claiming that 1 out of 5 births in the United States is out of wedlock. This was unheard of in the student's native country. To verify this claim, the student took a random sample of 75 birth records and found that in 13 of them the mother was unmarried. At a significance level of .05, is the student in a position to reject the evangelist's claim about 1 in 5 births being out of wedlock?

11.49   A statistician reported to a car insurance company a CI for the proportion of convertible cars that had been involved in major accidents during the past year. The 95% CI for p was reported to be the interval from .10 to .36.

(a)   What is the statistician's estimate of p?

(b)   What is the maximum error of estimate (E) of the proportion for this CI?

(c)   Approximately what sample size did the statistician use?

11.50   A sociologist believed that the proportion of families living below the poverty line was higher for Hispanics than for the general population. A sample of 60 families from the general population contained 9 families below the poverty line. A similar, independent sample of 60 Hispanic families contained 11 families below the poverty line.

(a)   At a significance level of 5%, does this support the sociologist's belief?

(b)   What is the p value for the test?

(c)   What would be your decision using a significance level of .01?

11.51   Must a CI for a proportion contain the true proportion of the population? Explain what the "level of confidence" means for a CI.

11.52   A market-research firm believed that the proportion of households with more than four family members in county 1 was greater than the proportion of households with more than four family members in county 2. The firm gathered random samples of 180 and 155 from counties 1 and 2, respectively. The number of households with more than four members was 74 from county 1 and 61 from county 2.

(a) From these data, can we conclude that the proportion of households with more than four members is higher in county 1 than in county 2? Use a significance level of .01.

(b) Calculate the *p* value. Using the *p* value, would you reject the null hypothesis at the .01 level?

11.53 *People's Choice*, a monthly magazine, claimed that more than 40% of its subscribers had an annual income of $50,000 or more. In a random sample of 62 subscribers, 30 had incomes of $50,000 or more. Does this information substantiate the magazine's claim? Use a significance level of .01.

11.54 In a survey of 1985 U.S. home buyers, 86% of first-time buyers were two-income families, whereas only 72% of repeat buyers were two-income families. It is desired to estimate the difference between these proportions for the year 1986. An error of 10% will be tolerated, and the desired confidence level is 99%. What is the required sample size?

11.55 A market research firm is interested in testing the hypothesis that the proportion of students who own a car is the same for the local state university campus and a local private college. They interviewed 240 students from the state university and 270 from the private college. The number of students who did not own a car was 78 at the state university and 82 at the private college. Using a .02 significance level, test the hypothesis.

11.56 For exercise 11.55, construct a 95% confidence interval for the difference of the true proportions of students who own cars at the two campuses.

11.57 In a study of illegitimacy in the parish of Lanheses in the northwestern province of Minho, Portugal, ethnologist Caroline Brettell (1985) provided the following data obtained from the Parish Registers of Lanheses:

ILLEGITIMACY RATIOS IN LANHESES

Period	A Illegitimate Baptisms	C Total Baptisms	A/C Illegitimacy Ratio
1900–1919	77	573	13.4
1920–1939	65	763	8.5
1940–1959	30	922	3.2
1960–1969	6	386	1.5

Source: C. B. Brettell. Male migrants and unwed mothers: Illegitimacy in a northwestern Portuguese town. *Anthropology,* **9** (1–2): 95 (1985), Table 4.

(a) What kind of trend do you observe in the illegitimacy ratio?

(b) How would you compute the overall illegitimacy ratio for the entire period 1900–1969?

(c) Compute a 95% confidence interval for the illegitimacy ratio for 1960–1969.

(d) Construct a 95% confidence interval for the difference in illegitimacy ratios between the periods 1900–1939 (p_1) and 1940–1969 (p_2).

(e) Use the confidence interval from (d) above to answer the question: Is there a significant difference between p_1 and p_2?

11.58 A social commentator was severely critical of the so-called "revolving door" hiring practices, in which defense officials leave government service and immediately go to work for defense contractors, often the same ones they dealt with as officials. He claimed more than one-third of defense officials go through the revolving door. A Government Accounting Office (GAO) study of 5000 former top officials found that 2100 oversee the same programs as civilians that they administered as government officials (i.e., 2100 went through the revolving door). Does this GAO report lend statistical support to the social commentator's claim? Use $\alpha = .01$. Also, state the p value.

11.59 A researcher wanted to compare the proportion of registered voters in the general population with the proportion of registered voters among minorities. A sample of 120 eligible adults from the general population contained 82 registered voters, whereas a sample of 100 eligible minority adults contained 31 registered voters. Construct the 95% confidence interval for the difference between the two proportions.

11.60 The Bureau of Crime Data believes that "women who report abusive husbands to the police are less likely to be attacked again in the next six months." Consider the following sample results:

Group A	Sample Size	No. Abused Again in 6 Months
Women attacked by husband or ex-husband and reported the matter to the police	1280	192
Group B		
Women abused, but did not report it to the police	1280	525

(a) Do the above data indicate that the proportion in group A is significantly lower than the proportion in group B, at $\alpha = .01$?

(b) Interpret the null and alternative hypotheses, and relate their meaning to what the Bureau of Crime Data said above.

CASE STUDY

A Look at American and Japanese Management Styles

Japanese management practices and quality control techniques have been the subject of much discussion and research during the past two decades. Sullivan and Nonaka addressed the question: Just how different is Japanese management from American/European practices? They theorized that a greater Japanese commitment to an organizational learning perspective contributes to differences in the strategy-formulating behavior of Japanese and American executives.

The purpose of their research was to explore the organizational learning practices of Japanese managers as compared to a similar group of American managers. Senior American managers were considered to be company presidents since they are the individuals who carry out the theory of action in American firms. However, in Japanese firms, company presidents typically are not active managers. Consequently, the Japanese sample mostly focused on *bucho*, those managers who are above section chiefs, since they carry out the same senior functions as American company presidents.

Responses to a questionnaire were received from 75 American company presidents, and 75 responses were randomly selected from 422 Japanese managers who participated in the study. The English questionnaire was translated into Japanese and then translated back into English by an American who did not have access to the original. Minor discrepancies were identified and resolved. A portion of the results are summarized in the table below.

Information Management Approaches of Senior American ($n_1 = 75$) and Japanese ($n_2 = 75$) Managers		Senior American Managers Prefer	Senior Japanese Managers Prefer
	1. To communicate realistic rather than idealistic goals.	73%	65%
	2. To inspire employees by talking about values rather than rewards.	52%	95%
	3. To assign tasks which challenge employees to use greater effort and ability than they have ever done before rather than tasks which require routine or average effort and ability.	58%	73%
	4. To stress employee task rotation rather than specialization.	60%	75%
	5. To do some employee training themselves.	68%	62%
	6. To stress learning from mistakes rather than avoiding mistakes.	81%	89%
	7. To base employee knowledge in experience rather than in rules, manuals or information systems.	61%	78%
	8. To encourage overlapping, loosely defined projects to foster information sharing.	70%	51%
	9. To give general, rather than specific directions to project teams to foster freedom in their deciding how to proceed.	72%	89%

Source: Sullivan, J. J. and Nonaka, I., *J. of International Business Studies,* **XVII**, (3), 129–142 (Fall 1986). Reprinted with permission of Hong Kong Baptist College, Hong Kong.

A study of the table reveals some apparently significant differences between the proportions of American and Japanese managers who prefer a particular management style. For example, 52% of the American managers prefer to inspire employees by talking about values rather than rewards, as compared to 95% for the Japanese managers. Are such sample percentages statistically significant? The following questions will take a closer look at these percentages.

Case Study Questions

1. Do you agree with the authors' procedure of randomly selecting the 75 Japanese responses from the 422 managers who participated in the study? Why not simply select the first 75 responses?

2. Give the point estimate for the number of American managers who stress task rotation rather than specialization.

3. Conduct a 95% confidence interval for the parameter in item (2) above.

4. Give the point estimate for the difference between the proportions of American managers (p_1) and Japanese managers (p_2) who prefer to do some employee training themselves. (Use $p_1 - p_2$).

5. Construct a 99% confidence interval for item (4) above.

6. Do the data support the hypothesis that more than half of the American managers inspire employees by talking about values rather than rewards? Conduct a hypothesis test at a 5% significance level, and state the p value.

7. Is there evidence to indicate that a higher percentage of American managers (p_1) prefer to communicate realistic rather than idealistic goals than do the Japanese managers (p_2)? Conduct a hypothesis test at the 1% level of significance, and state the p value.

8. Is there evidence to indicate that a higher percentage of American managers prefer to encourage overlapping, loosely defined projects to foster information sharing? Conduct a test of hypothesis at the 5% level of significance, and state the p value.

COMPUTER EXERCISES USING THE DATABASE

Using a sample of 100 observations, complete the following exercises.

1. Let p represent the proportion of males attending Brookhaven. Derive an estimate of p and construct a 90% confidence interval for p. Test the claim that p is .4 using a two-tailed test and a significance level of .10.

2. Using the estimate of p in exercise 1, how large of a sample would be necessary if you wanted to estimate this parameter to within 3%, with 90% confidence?

3. Let p_1 represent the proportion of ever-married males and p_2 the proportion of ever-married females attending Brookhaven. Construct a 95% confidence interval for $p_1 - p_2$. Test the claim that $p_1 - p_2 > .04$, using a significance level of .05.

4. Let p_1 represent the proportion of males attending Brookhaven with an undergraduate GPA over 3.00 and p_2 be the corresponding proportion for females. At a significance level of .05, can you conclude that $p_1 < p_2$?

5. What is an estimate of the proportion of all students attending Brookhaven that have a graduate GPA of 4.0? Construct a 95% confidence interval for this parameter.

12

ANALYSIS OF VARIANCE

A LOOK BACK/INTRODUCTION

In Chapter 10, we considered a question of the type, "Do men have the same height as women?" By this we mean, is the average height of males equal to the average height of females? We were interested in the means of two populations and performed a test of hypothesis, using, for example, H_0: $\mu_M = \mu_F$ and H_a: $\mu_M \neq \mu_F$. This works well when dealing with two populations, but how can we compare the means of more than two populations? For example, we might wish to examine the average yards gained per game for four different football running backs to see whether they are the same. Our hypotheses become

$$H_0: \quad \mu_1 = \mu_2 = \mu_3 = \mu_4$$

$$H_a: \quad \text{not all } \mu\text{'s are equal}$$

We test such a hypothesis by first collecting four samples, one from each of the four players. We will see that to compare these four means one pair at a time is not the correct approach. This results in six different pairwise tests, and what was intended to be a testing procedure with, say, a 5% significance level, results in a much higher significance level. In other words, the overall significance level α is larger than the predetermined value. The correct procedure for this situation is to examine the variation of the yards gained per game, both (1) within each of the samples (examining the variability of each sample alone) and (2) among the four samples (for example, are the values in sample 1 larger or smaller, on the average, than the values in the other samples?).

In Chapter 10, we saw that, when trying to decide if $\bar{X}_1$ is "significantly different" from $\bar{X}_2$, a key part of the answer rested on the values of s_1 and s_2, the variation within the two samples. Both s_1 and s_2 affect the width of the confidence interval for $\mu_1 - \mu_2$. Consequently, we infer something about the means of several populations by utilizing the variation of the resulting samples. Hence the term, analysis of variance—our next topic.

12.1   COMPARING TWO MEANS: ANOTHER LOOK

We will begin with an example. Checker Cab Company is trying to decide which brand of tires to use for the coming year. Based on current price and prior experience, they have narrowed their choice to two brands, say, brands 1 and 2. A recent study examined the durability of these tires by using a machine with a metallic device that wore down the tires. The time it took (in hours) for five tires of each brand to blow out are shown below.

Brand 1	Brand 2
6.8	2.9
8.1	1.7
5.3	5.0
6.2	3.2
9.2	4.1

Let μ_1 and μ_2 be the average blowout times for *all* brand 1 and brand 2 tires, respectively. We wish to determine whether the data allow us to conclude that $\mu_1 \neq \mu_2$, using $\alpha = .10$.

We examined the same type of question in Chapter 10; we are dealing with two small independent samples. In Chapter 10, we advised against assuming that σ_1 was equal to σ_2. As a result, we generally used a t test that did not pool the sample variances. However, when examining more than two normal populations (the main concern of this chapter), the following testing procedure for detecting a difference in the population means requires that the populations have the same distribution if, in fact, the population means are equal. Consequently, it can be used only when we are willing to assume that the population variances are equal (or approximately equal). The analysis of variance procedure is not extremely sensitive to departures from this assumption if equal-sized samples are obtained from each population. A procedure for verifying this assumption (similar to the F test used to compare two variances in Chapter 10) will be discussed in this chapter.

As a result, we will assume that we have reason to believe that the variation of brand 1 blowout times is the same as for brand 2; that is, $\sigma_1 = \sigma_2$. Using the approach discussed in Chapter 10, we first find

$$s_p^2 = \text{pooled variance} = \frac{(n_1 - 1)s_1^2 + (n_2 - 1)s_2^2}{n_1 + n_2 - 2}$$

where n_1 = sample size for brand 1, n_2 = sample size for brand 2, and s_1^2, s_2^2 = sample variances for brand 1, brand 2. Using the sample data,

Brand 1	Brand 2
$n_1 = 5$	$n_2 = 5$
$\bar{x}_1 = 7.12$	$\bar{x}_2 = 3.38$
$s_1 = 1.5450$	$s_2 = 1.2478$

Consequently,

$$s_p^2 = \frac{(4)(1.5450)^2 + (4)(1.2478)^2}{8}$$

$$= 15.766/8$$

$$= 1.972$$

and so

$$s_p = \sqrt{1.972} = 1.4043$$

The appropriate hypotheses are $H_0: \mu_1 = \mu_2$ and $H_a: \mu_1 \neq \mu_2$. The resulting test statistic is

$$t = \frac{\bar{X}_1 - \bar{X}_2}{s_p \sqrt{\dfrac{1}{n_1} + \dfrac{1}{n_2}}}$$

$$= \frac{7.12 - 3.38}{1.4043 \sqrt{\dfrac{1}{5} + \dfrac{1}{5}}} = \frac{3.74}{.8882}$$

$$= 4.211$$

That is, $t^* = 4.211$.

We are dealing with a two-tailed test using a t statistic with $(n_1 - 1) + (n_2 - 1) = 4 + 4 = 8$ df, so the test procedure is to

$$\text{Reject } H_0 \text{ if } |t^*| > t_{\alpha/2, df} = t_{.05,8} = 1.86$$

Comparing $t^* = 4.211$ to 1.86, we reject H_0 and conclude that the mean blow-out times for the two brands are not the same. Looking at the sample data, we can say that the mean score for brand 1 in the sample ($\bar{x}_1 = 7.12$) is significantly different from the mean score for brand 2 in the sample ($\bar{x}_2 = 3.38$).

The Analysis of Variance Approach

We need to introduce two new terms. The previous example examined the effect of one **factor** (brand), consisting of two **levels** (brand 1 and brand 2). If you want to apply this to four different brands then you still have one factor but you now have four levels.

The purpose of **analysis of variance (ANOVA)** is to determine if this factor has a significant effect on the variable being measured (blowout times, in our example). If, say, the brand factor is significant, then the mean blowout times for the populations will not be equal. Consequently, testing for equal means among the populations is the same as attempting to answer the question, "Is there a significant effect on blowout time due to this factor?"

This section examines the effect of a single factor on the variable being measured, **one-factor ANOVA**. Extensions of this technique include ANOVA procedures that determine the effect of two or more factors operating simultaneously. These factors may be qualitative (such as sex or brand) or quantitative (such as a person's income level).

All 10 values in the Checker Cab Company example are different, and we observe a variation in these values. We will look at two sources of variation: (1) variation within the samples (levels), and (2) variation between the samples.

Within-Sample Variation   When you obtain a sample, you usually obtain different values for each observation. The five sample values for brand 1 vary about the mean $\bar{x}_1 = 7.12$, as measured by $s_1 = 1.545$. Likewise, the five values in the second sample also exhibit some variation ($s_2 = 1.2478$) about $\bar{x}_2 = 3.38$. These are the **within-sample variations**. They are used when estimating the common population variance, say σ^2. This procedure provides an accurate estimate of σ^2, whether or not the sample means are equal.

Between-Sample Variation   When you compare the two samples, you observe that the values for brand 1 are larger, on the average, than are those for brand 2. This is summarized in the sample means, where $\bar{x}_1 = 7.12$ appears to be considerably larger than $\bar{x}_2 = 3.38$. So there is a variation in the 10 values due to the brand; that is, due to the factor. This is **between-sample variation**. In general, if this variation is large, we expect considerable variation among the sample means. The between-sample variation is also used in another estimate of the common variance, σ^2, provided the population means are equal. In other words, if the means are equal, the between-sample and within-sample estimates of σ^2 should be nearly the same. As we will see later in this section, we can derive a test of hypothesis procedure for determining whether the means are equal by comparing these two estimates.

Measuring Variation   When using the ANOVA approach, we measure these two sources of variation by calculating various **sums of squares, SS**. We determine

SS(factor) measures: between-sample variation (also called SS(between))

SS(error) measures: within-sample variation (also called SS(within))

SS(total) = SS(between) + SS(within)

$\qquad\qquad$ = SS(factor) + SS(error)

In addition, each of the first two sums of squares will have corresponding degrees of freedom *df*, which are determined from the number of terms that make up this particular SS. The *df* are given by

$$df \text{ for factor} = (\text{number of levels}) - 1$$
$$= (\text{number of brands}) - 1$$
$$= 2 - 1 = 1$$
$$df \text{ for error} = (n_1 - 1) + (n_2 - 1)$$
$$= n_1 + n_2 - 2$$
$$= 5 + 5 - 2 = 8$$

We will show how to determine these sums of squares and how we combine them, and their *df*, into another test statistic for testing $H_0: \mu_1 = \mu_2$ against $H_a: \mu_1 \neq \mu_2$. The beauty of this approach is that it extends nicely to the situation in which you wish to compare more than two means using a single test.

Determining SS(factor) SS(factor) is the sum of squares that determines whether the values in one sample are larger or smaller on the average than the values in the second sample.

$$SS(\text{factor}) = n_1(\bar{x}_1 - \bar{x})^2 + n_2(\bar{x}_2 - \bar{x})^2 \qquad \textbf{(12-1)}$$

where $\bar{x}_1, \bar{x}_2$ are the two sample means and

$$\bar{x} = \frac{\Sigma(\text{all } x \text{ values})}{n} = \frac{n_1\bar{x}_1 + n_2\bar{x}_2}{n_1 + n_2}$$

and $n = n_1 + n_2 = $ total sample size.

A method of determining this sum of squares that is much easier using a calculator is

$$SS(\text{factor}) = \left[\frac{T_1^2}{n_1} + \frac{T_2^2}{n_2}\right] - \frac{T^2}{n} \qquad \textbf{(12-2)}$$

where $T_1 = $ total of the sample 1 observations, $T_2 = $ total of the sample 2 observations, $T = $ grand total $= T_1 + T_2$.

Determining SS(total) SS(total) is a measure of the variation in all $n = n_1 + n_2$ data values. You obtain its value as though you were finding the variance of these n values, except that you do not divide by $n - 1$. So,

$$SS(\text{total}) = \Sigma(x - \bar{x})^2 \qquad \textbf{(12-3)}$$

or (after some algebra similar to that used in Chapter 3),

$$SS(\text{total}) = \Sigma x^2 - \frac{(\Sigma x)^2}{n}$$

$$= \Sigma x^2 - \frac{T^2}{n} \tag{12-4}$$

Determining SS(error) SS(error) is the measure of the variation within each of the samples. Its value simply is the numerator of the pooled variance s_p^2, obtained using the previous t test. Thus,

$$SS(\text{error}) = \underset{\text{first sample}}{\Sigma(x - \bar{x}_1)^2} + \underset{\text{second sample}}{\Sigma(x - \bar{x}_2)^2} \tag{12-5}$$

or written another way,

$$SS(\text{error}) = \Sigma x^2 - \left[\frac{T_1^2}{n_1} + \frac{T_2^2}{n_2} \right] \tag{12-6}$$

Given that

$$SS(\text{total}) = SS(\text{factor}) + SS(\text{error})$$

a much easier way to find this value is

$$SS(\text{error}) = SS(\text{total}) - SS(\text{factor}) \tag{12-7}$$

Let us return to the Checker Cab Company example. To find the SS(factor) here, we first determine

$$T_1 = 6.8 + 8.1 + 5.3 + 6.2 + 9.2$$
$$= 35.6$$
$$T_2 = 2.9 + 1.7 + 5.0 + 3.2 + 4.1$$
$$= 16.9$$
$$T = T_1 + T_2 = 35.6 + 16.9 = 52.5$$

So, using equation 12-2,

$$SS(factor) = \frac{35.6^2}{5} + \frac{16.9^2}{5} - \frac{52.5^2}{10}$$

$$= 310.594 - 275.625$$

$$= 34.969$$

To find SS(total), the only new term we need to evaluate is

$$\Sigma x^2 = \text{sum of each data value squared}$$

$$= 6.8^2 + 8.1^2 + \cdots + 3.2^2 + 4.1^2$$

$$= 326.37$$

So, using equation 12-4 (the value 275.625 was obtained in SS(factor)),

$$SS(total) = \Sigma x^2 - \frac{T^2}{n}$$

$$= 326.37 - 275.625$$

$$= 50.745$$

Finally, we find SS(error) by subtraction:

$$SS(error) = SS(total) - SS(factor)$$

$$= 50.745 - 34.969$$

$$= 15.776$$

ANOVA Test for H_0: $\mu_1 = \mu_2$ versus H_a: $\mu_1 \neq \mu_2$ To begin with, the procedure we are about to define is valid for a two-tailed test only. In other words, the alternative hypothesis must be that the two means differ, not that one is larger than the other (a one-tailed test). (When examining more than two means the alternative hypothesis will be that at least two of the means are unequal and H_0 will be that all the means are equal.) The next step, when using the ANOVA procedure, is to determine something resembling an "average" sum of squares, referred to as a **mean square**. We compute a mean square for only SS(factor) and SS(error), not for SS(total).

$$MS(factor) = SS(factor)/df \text{ for factor}$$

$$= SS(factor)/1$$

(12-8)

Note that the *df* for this term always is (number of levels) $- 1$. In this section, we are dealing with two levels (populations), and so here *df* is 1.

$$MS(error) = SS(error)/df \text{ for error}$$

$$= SS(error)/(n_1 + n_2 - 2)$$

(12-9)

We denote the common variance of the two normal populations as σ^2. So, $\sigma^2 = \sigma_1^2 = \sigma_2^2$. If the null hypothesis—H_0: the means are equal—is true, then, because the populations have identical means and variances, this implies that under H_0 we are dealing with a single population. The ANOVA procedure is based on a comparison between two separate estimates of the variance σ^2. The first estimate is derived using the variation among the sample means (only two in the previous example). The other estimate is determined using the variation within each of the samples.

The ANOVA procedure is based on a comparison of these two estimates of σ^2 because they should be approximately equal provided H_0 is true. We have derived these two estimates:

MS(factor) = estimate of σ^2 based on the variation among the
sample means

MS(error) = estimate of σ^2 based on the variation within each
of the samples

Our new test statistic for testing H_0: $\mu_1 = \mu_2$ versus H_a: $\mu_1 \neq \mu_2$ is the ratio of these two estimates:

$$F = \frac{\left(\begin{array}{c}\text{estimated population variance based on the}\\ \text{variation among the sample means}\end{array}\right)}{\left(\begin{array}{c}\text{estimated population variance based on the}\\ \text{variance within each of the samples}\end{array}\right)} \quad \text{(12-10)}$$

$$= \frac{\text{MS(factor)}}{\text{MS(error)}}$$

This test statistic follows an F distribution, which was first introduced in Chapter 10 as a ratio of two variance estimates. The degrees of freedom (df) for the F statistic in equation 12-10 are the df for factor and the df for error; that is, in our present examples the df for F are 1 and $(n_1 + n_2 - 2)$. Because the F statistic is based on a comparison of two variance estimates, this technique is called analysis of variance.

This is our second encounter with the F distribution. In Chapter 10, we used this distribution to compare two population variances (σ_1^2 and σ_2^2). The shape of this distribution is illustrated in Figure 12.1 and is tabulated in Table A-7. Remember that the shape of the F curve is affected by both the df for the numerator (= 1 here) and the df for the denominator (= $n_1 + n_2 - 2$ here).

Defining the Rejection Region What happens to the F statistic when H_a is true, that is, when $\mu_1 \neq \mu_2$? In this case, we would expect $\bar{X}_1$ and $\bar{X}_2$ to be "far apart." As a result, the estimate of the variance σ^2 using the between-sample

Figure 12.1

Shape of the F distribution shown by F curve with 1 and $n_1 + n_2 - 2$ df

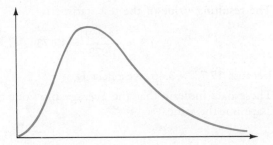

variation (measured by MS(factor)) will be larger than the estimate of σ^2 based on the within-sample variation (measured by MS(error)). This implies that we should reject H_0 in favor of H_a whenever the ratio of these two estimates is large—in which case the computed F value is in the right tail. Consequently, the test procedure will be to

$$\text{Reject } H_0 \text{ if } F^* > F_{\alpha, v_1, v_2}$$

where $v_1 = df$ for numerator $=$ (number of levels) $- 1 = 1$, $v_2 = df$ for denominator $= n_1 + n_2 - 2$, and F_{α, v_1, v_2} is obtained from Table A-7 with a right-tail area $= \alpha$.

EXAMPLE 12.1

Using the data from the Checker Cab Company example and the previously calculated sums of squares, test $H_0: \mu_1 = \mu_2$ versus $H_a: \mu_1 \neq \mu_2$, where $\mu_1 =$ average blowout time for brand 1, if observed indefinitely, and $\mu_2 =$ average blowout time for brand 2. Use a significance level of $\alpha = .10$.

Solution

Step 1 The hypotheses are as defined—$H_0: \mu_1 = \mu_2$ and $H_a: \mu_1 \neq \mu_2$.

Step 2 The test statistic is

$$F = \frac{\text{MS(factor)}}{\text{MS(error)}}$$

Step 3 The rejection region (using Table A-7a) is

$$\text{Reject } H_0 \text{ if } F > F_{.10, 1, 8} = 3.46$$

Step 4 From the previous calculations, SS(factor) $= 34.969$ and SS(error) $= 15.776$. So,

$$\text{MS(factor)} = \text{SS(factor)}/1$$
$$= 34.969$$

and

$$\text{MS(error)} = \text{SS(error)}/(n_1 + n_2 - 2)$$
$$= 15.776/8$$
$$= 1.972$$

The resulting value of the test statistic is

$$F^* = \frac{34.969}{1.972} = 17.73$$

Because $17.73 > 3.46$, we reject H_0.

Step 5 These data indicate that the average blowout times for brands 1 and 2 are not the same.

COMMENTS

Compare our first treatment of the Checker Cab Company problem with example 12.1. Both solutions led to the same conclusion, namely, that the two average blowout times are not the same. In fact, both solutions always lead to the same conclusion when comparing two means. Furthermore, the p values for both solutions are the same, as illustrated in Figure 12.2. The values were obtained using a computer program (available in many statistical packages) that provides an exact p value for a t or F statistic, given the computed value and corresponding degrees of freedom.

The computed value of the F statistic is equal to (the computed value of the t statistic)^2 because $17.73 = (4.211)^2$. This is true whenever you have an F statistic or table value with 1 df in the numerator. So,

$$F_{\alpha,1,v} = [t_{\alpha/2,v}]^2$$

for any $v \geqslant 1$. Furthermore, the table values satisfy the same relationship, namely,

$$F_{.10,1,8} = 3.46 = (1.86)^2 = [t_{.05,8}]^2$$

We see that the two tests are identical. Both tests contain the same hypotheses and so produce the same conclusion and p value. Furthermore, the computed value and

Figure 12.2

p values for the solution to the Checker Cab Company example. (A) Solution using pooled variance t test. (B) Solution using ANOVA (see example 12.1).

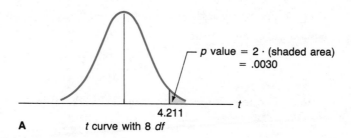

p value = 2 · (shaded area) = .0030

4.211

A t curve with 8 df t

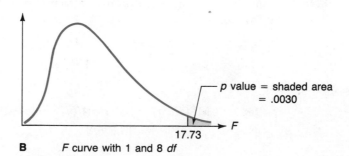

p value = shaded area = .0030

17.73

B F curve with 1 and 8 df F

the table value for the F statistic are the square of the corresponding values using the t statistic. This comparison applies only when the F statistic has 1 df in the numerator—that is, when there are two factor levels (as in this illustration). As mentioned previously, the advantage of the ANOVA approach is that it extends very easily to the situation of comparing means for more than two populations (covered in the next section).

The ANOVA Table Rather than carrying out the five-step procedure using the F statistic, an easier method is to use an **ANOVA table** of the various sums of squares. The format of this table is as follows*:

Source	df	SS	MS	F
Factor	1	SS(factor)	$MS(factor) = \dfrac{SS(factor)}{1}$	MS(factor)/MS(error)
Error	$n - 2$	SS(error)	$MS(error) = \dfrac{SS(error)}{n - 2}$	
Total	$n - 1$	SS(total)		

To fill in this table, you compute the necessary sums of squares along with the mean squares and insert them. Notice that $n = n_1 + n_2 =$ total sample size and that column 4 (MS) = column 3 (SS) divided by column 2 (df).

The ANOVA table for example 12.1 follows:

Source	df	SS	MS	F
Factor	1	34.969	34.969	17.73
Error	8	15.776	1.972	
Total	9	50.745		

Summary of the ANOVA Approach for One-Factor Tests In example 12.1, we concluded that a difference existed between the two means because the variation between the two samples (measured by MS(factor)) was much greater than the variation within the samples (measured by MS(error)). Thus, the ratio of these values was very large and F^* fell in the rejection region. Consequently, we rejected H_0: $\mu_1 = \mu_2$. What this means in the language of ANOVA is that there is a significant effect on tire blowout time due to the brand factor.

To carry out the F test, we first randomly obtain observations, called **replicates**, from each population. Example 12.1 used five replicates (blowout

---

* The headings under the "Source" column will vary, depending on the computer package. SS(factor) often is labeled "between groups" (SAS) or "among groups"; SS(error) often is labeled "within groups" (SAS) or "residual" (SPSS) or "error" (MINITAB).

Figure 12.3

Dot array diagram
of replicates in
example 12.1

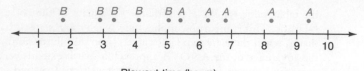

Blowout time (hours)

Figure 12.4

Dot array diagram
where between-
sample and within-
sample variations
are nearly the same.
The *F* statistic would
not lie in the
rejection region.

Blowout time (hours)

times) from each of the two populations. It is not necessary to obtain the same number of replicates from each population.

Figure 12.3 is a dot-array diagram of the data in example 12.1, where the symbol *A* represents a value from the brand 1 sample and *B* represents the brand 2 sample. You do not need to be an expert statistician to observe that a clear difference exists between the scores of the two populations. The variation within the *A*'s alone and the *B*'s alone is your within-sample variation. Because the distances from the *A* values to the *B* values are much larger than the distances among the *A* values alone, the between-sample variation is quite large, as we have already observed.

Suppose instead that your dot-array diagram looks like Figure 12.4. Now the two sources of variation appear to be nearly the same and there is no obvious difference between the two brands. The resulting *F* statistic here would not lie within the rejection region and we would not be able to demonstrate, using the ANOVA approach, a difference between the two population means.

ONE-FACTOR ANOVA
12.2   COMPARING MORE THAN TWO MEANS

In the previous section, we examined a single factor with two levels. Our concern was whether there was any difference between the two levels of this factor. This amounted to performing a test of hypothesis on the means of two populations. Because we were dealing with the effect of a single factor, this was a one-factor (or one-way) ANOVA.

In general, one-factor ANOVA techniques can be used to study the effect of any single factor on questionnaire scores, exam performance, and the like. This factor can consist of any number of levels—say, *k* levels. To determine if the levels of this factor affect our measured observations, we examine the hypotheses

$$H_0: \quad \mu_1 = \mu_2 = \cdots = \mu_k$$

$$H_a: \quad \text{not all } \mu\text{'s are equal}$$

Suppose we are interested in whether there is a difference in general life satisfaction among four different marital groups of females living in Riverport. Life satisfaction will be measured using a questionnaire consisting of 50 items concerning a person's level of satisfaction at work, at home, and within personal relationships. The score on this instrument is called the Life Satisfaction Index. Is there any difference in the average life satisfaction for these four groups? To answer this question, we test

$$H_0: \quad \mu_1 = \mu_2 = \mu_3 = \mu_4$$

$$H_a: \quad \text{not all } \mu\text{'s are equal}$$

We have a single factor (marital status). One possibility is to examine these samples one pair at a time using the t statistic discussed in the previous section. This appears to be a safe way to proceed here, although there are $_4C_2 = 6$ such pairs of tests to perform this way. The main problem with this approach when performing many tests of this nature is determining the probability of making an incorrect decision. In particular, what value does α have, where α is the probability of rejecting H_0: all μ's are equal, when in fact it is true? You set α in advance but, after performing, say, six of these pairwise tests ($\mu_1 = \mu_2, \mu_1 = \mu_3, \ldots$), what is your overall probability of concluding that at least one pair of means are not equal when they actually are? This is a difficult question. The overall probability is not the significance level α with which you started for just one pair. So we need an approach that will test for the equality of these four means using a *single* test. This is what the ANOVA approach does.

Assumptions Behind the ANOVA Analysis

When using the ANOVA procedure, there are three key assumptions that must be satisfied. They are basically the same assumptions that were necessary when testing two means using small independent samples with the pooled variance approach. These assumptions are:

1 The replicates are obtained independently and randomly from each of the populations. The value of one observation has no effect on any other replicates within the same sample or within the other samples.

2 The observations (replicates) from each population follow (approximately) a normal distribution.

3 The normal populations all have a common variance σ^2. We expect the values in each sample to vary about the same amount. The ANOVA procedure will be much less sensitive to violations of this assumption when we obtain samples of equal size from each population.

Deriving the Sums of Squares

When examining, say, k populations, the data will be configured somewhat like this:

	Level 1	**Level 2**	$\cdots$	**Level k**
	$\vdots$	$\vdots$		$\vdots$
	n_1 replicates	n_2 replicates		n_k replicates
	$\vdots$	$\vdots$		$\vdots$
Totals	T_1	T_2		T_k

This resembles the data from example 12.1, where $k = 2$ and $n_1 = n_2 = 5$ replicates. To derive the sums of squares for this situation, we will extend the results in equations 12-2, 12-4, and 12-6 to

$$SS(\text{factor}) = \left[\frac{T_1^2}{n_1} + \frac{T_2^2}{n_2} + \cdots + \frac{T_k^2}{n_k}\right] - \frac{T^2}{n} \qquad \textbf{(12-11)}$$

$$SS(\text{total}) = \Sigma x^2 - \frac{T^2}{n} \qquad \textbf{(12-12)}$$

$$SS(\text{error}) = \Sigma x^2 - \left[\frac{T_1^2}{n_1} + \frac{T_2^2}{n_2} + \cdots + \frac{T_k^2}{n_k}\right] \qquad \textbf{(12-13)}$$

$$= SS(\text{total}) - SS(\text{factor}) \qquad \textbf{(12-14)}$$

Here, n = the total number of observations = $n_1 + n_2 + \cdots + n_k$, and $T = \Sigma x$ = the sum of all n observations = $T_1 + T_2 + \cdots + T_k$. Also, to find Σx^2, you square each of the n observations and sum the results.

The ANOVA Table

The good news is that the format of the ANOVA table is the same regardless of the number of populations (levels), k. The only change from the two-population case is that

$$df \text{ for factor} = k - 1$$
$$df \text{ for error} = n - k$$

As before, the total *df* are $n - 1$. The resulting ANOVA table follows.

Source	df	SS	MS	F
Factor	$k - 1$	SS(factor)	$\text{MS(factor)} = \dfrac{\text{SS(factor)}}{k - 1}$	MS(factor)/MS(error)
Error	$n - k$	SS(error)	$\text{MS(error)} = \dfrac{\text{SS(error)}}{n - k}$	
Total	$n - 1$	SS(total)		

$$MS(factor) = SS(factor)/df \text{ for factor}$$
$$= \frac{SS(factor)}{k - 1} \qquad \text{(12–15)}$$

$$MS(error) = SS(error)/df \text{ for error}$$
$$= \frac{SS(error)}{n - k} \qquad \text{(12–16)}$$

The test statistic for testing $H_0: \mu_1 = \mu_2 = \cdots = \mu_k$ versus H_a: not all μ's are equal, is

$$F = \frac{MS(factor)}{MS(error)}$$

which has an F distribution with $k - 1$ and $n - k$ *df*.

As in the two-sample case, the procedure is to reject H_0 when the variation among the sample means (measured by MS(factor)) is large compared to the variation within the samples (measured by MS(error)). Consequently, the test will be to reject H_0 whenever F lies in the right-tailed rejection region, defined by the significance level α.

EXAMPLE 12.2

A researcher in Riverport decided to take a look at the female population to determine if there is a difference in life satisfaction among women in various marital categories. She considered four subpopulations (groups), namely:

Population 1: never-married females

Population 2: ever-married females (married only once to the same man)

Population 3: remarried

Population 4: divorced

The Life Satisfaction Index questionnaire was given to random samples of six women selected from each of the four groups. The researcher wants to know if there is any difference in general life satisfaction among these four groups of women. Past experience has indicated that scores on this instrument follow a normal distribution.

Population:	1 Never Married	2 Ever Married	3 Remarried	4 Divorced
	41	66	43	47
	54	49	28	36
	77	82	51	20
	95	62	48	25
	64	75	30	31
	69	93	23	38
Total (T)	400	427	223	197
Average ($\bar{X}$)	66.67	71.17	37.17	32.83
Variance (s^2)	348.27	242.17	135.77	93.37

The four sample averages are $\bar{x}_1 = 66.67$, $\bar{x}_2 = 71.17$, $\bar{x}_3 = 37.17$, $\bar{x}_4 = 32.83$. The never marrieds and the ever marrieds appear to have significantly larger scores on the instrument. In other words, it appears that there is a significant between-group variation. But do these sample means provide sufficient evidence to reject $H_0: \mu_1 = \mu_2 = \mu_3 = \mu_4$ where each μ represents the average of all scores for Riverport women in this marital category? Use the ANOVA procedure to answer this question with $\alpha = .05$.

Solution   The assumptions behind this analysis are that (1) the samples were obtained randomly and independently from each of the four populations and (2) the scores within each marital category follow a normal distribution, with a common variance, say, σ^2. So

$$n = n_1 + n_2 + n_3 + n_4 = 24$$

and

$$T = \Sigma x = T_1 + T_2 + T_3 + T_4$$
$$= 400 + 427 + 223 + 197$$
$$= 1247$$

$$SS(\text{factor}) = \left[\frac{T_1^2}{n_1} + \frac{T_2^2}{n_2} + \frac{T_3^2}{n_3} + \frac{T_4^2}{n_4} \right] - \frac{T^2}{n}$$

Therefore,

$$SS(\text{factor}) = \frac{400^2}{6} + \frac{427^2}{6} + \frac{223^2}{6} + \frac{197^2}{6} - \frac{1247^2}{24}$$
$$= 71,811.17 - 64,792.04$$
$$= 7019.13$$

$$SS(total) = \Sigma x^2 - \frac{T^2}{n}$$

$$= [41^2 + 54^2 + \cdots + 31^2 + 38^2] - \frac{1247^2}{24}$$

$$= 75{,}909 - 64{,}792.04$$

$$= 11{,}116.96$$

$$SS(error) = SS(total) - SS(factor)$$

$$= 11{,}116.96 - 7019.13$$

$$= 4097.83$$

This is the ANOVA table for this analysis.

Source	df	SS	MS	F
Factor	$k - 1 = 3$	7019.13	7019.13/3 = 2339.71	2339.71/204.89 = 11.42
Error	$n - k = 20$	4097.83	4097.83/20 = 204.89	
Total	23	11116.96		

The computed F value using the ANOVA table is $F^* = 11.42$. Since $\alpha = .05$, we use Table A-7 (b) to find that $F_{.05,3,20} = 3.10$. Comparing these two values, $F^* = 11.42 > 3.10$, so we reject H_0.

We conclude that the average life satisfactions for the four groups are not the same. This confirms our earlier suspicion based upon the variation among the four sample means. Our results indicate that the marital status factor *does* have a significant effect on general life satisfaction (as measured by the Life Satisfaction Index).   □

The Assumptions behind ANOVA and a Test for Equal Variances

Use of independent random samples is of extreme importance when using the ANOVA procedure. The F test used for comparing the population means in the ANOVA table is very sensitive to departures from this assumption, so the safest way to guard against incorrect conclusions is to use random sampling techniques. In many situations, however, this may be difficult or impossible, such as when using the same set of people for before–after experiments. One solution to this problem is to modify your study design, such as by using a randomized block design, discussed later in this chapter.

Lack of **normality** within the populations is not a critical matter provided the departure is not too extreme. The F test used to test the means is not severely affected by populations that are somewhat nonnormal in nature. One way of making the ANOVA procedure even less sensitive to this assumption is to use large samples.

If the variances of the population are not equal, the F test used in the ANOVA procedure for testing the means is only slightly affected, provided the sample sizes are equal (or nearly so). Another alternative is to use a non-parametric test for the population means that is less sensitive than the ANOVA procedure to the assumption of equal variances. This is the Kruskal–Wallis test and will be discussed in Section 12.6. There is, however, a very simple test of hypothesis for verifying the assumption of equal variances.

In Chapter 10, an F test was defined for determining whether two normal population variances (or standard deviations) are equal. A similar test is used when you are comparing more than two normal population variances, provided the sample sizes are equal.* Here the hypotheses are

H_0: $\sigma_1^2 = \sigma_2^2 = \cdots = \sigma_k^2$

H_a: at least two variances are unequal (the k variances are not the same)

We warned you in Chapter 10 about the dangers of using the same data to test both the variances and the means. This argument applies to tests of more than two populations. A better procedure is to use a different data set for testing H_0: the variances are equal. This requires a much larger data set than is necessary if you use the same data for both tests. The test for equal variances is the **Hartley test**; the test statistic is defined to be

$$H = \frac{\text{maximum } s^2}{\text{minimum } s^2} \qquad (12\text{-}17)$$

which is simply the ratio of the largest sample variance divided by the smallest of these k variances. For the case where $k = 2$, Hartley's test is identical to the F test discussed in Section 10.7.

If H_0 is false, then the test statistic will be "large," so the testing procedure is to reject H_0 if the computed value of H lies in the right tail. The rejection region for a 5% level of significance can be obtained from Table A-13. This region depends on the number k of populations or levels and the number of observations in each sample.

Suppose we use the data from example 12.2 only for testing the hypotheses

H_0: $\sigma_1^2 = \sigma_2^2 = \sigma_3^2 = \sigma_4^2$

H_a: at least two variances are unequal

Using $\alpha = .05$ and Table A-13, because $k = 4$ and there are six observations in each sample, the test is to

Reject H_0 if $H > 13.7$

* When the sample sizes are unequal, a computationally more difficult test for equal variances can be performed, derived by M. S. Bartlett. For details, see J. Neter, W. Wasserman, and M. Kutner, *Applied Linear Statistical Models*, 2d ed. (Homewood, IL: Richard D. Irwin, 1985), pp. 618–622.

Using the data summary in example 12.2, the minimum s^2 is 93.37 and the maximum s^2 is 348.27. Consequently,

$$H = \frac{348.27}{93.37} = 3.73$$

which is less than 13.7, and so the conclusion is that we have no reason to suspect unequal variances for this situation.

If other data are available for testing the means, the assumption of equal variances behind the ANOVA procedure appears to be safe.

Confidence Intervals in One-Factor ANOVA

When we deal with normal populations, as we are here, we can supply

1 A point estimate of each mean, μ_i; for example, an estimate of μ_2 is $\bar{X}_2$.

2 A point estimate of each mean difference, $\mu_i - \mu_j$; for example, an estimate of $\mu_1 - \mu_3$ is $\bar{X}_1 - \bar{X}_3$.

When using the ANOVA procedure, the populations are believed to have a common variance, say, σ^2. To estimate this variance, we use an estimate of σ^2 that does not depend on whether the population means are equal. This is the within-sample variation, measured by MS(error). The point estimate of σ^2 is

$$s_p^2 = \text{pooled variance}$$
$$= \text{MS(error)}$$

where MS(error) is defined in equation 12-16.

In previous chapters, we always supplied a confidence interval along with a point estimate to provide a measure of how reliable this estimate really is. The narrower the confidence interval, the more faith you have in your point estimate. A $(1 - \alpha) \cdot 100\%$ confidence interval for μ_i is

$$\bar{X}_i - t_{\alpha/2,\, n-k}\frac{s_p}{\sqrt{n_i}} \quad \text{to} \quad \bar{X}_i + t_{\alpha/2,\, n-k}\frac{s_p}{\sqrt{n_i}} \qquad \textbf{(12-18)}$$

where

k = number of populations (levels)

n_i = number of replicates in the ith sample

n = total number of observations

$s_p = \sqrt{\text{MS(error)}}$

and $t_{\alpha/2,df}$ is the value from Table A-5 with $df = df$ for error $= n - k$, and right-tail area $= \alpha/2$.

A $(1 - \alpha) \cdot 100\%$ confidence interval for $\mu_i - \mu_j$ is

$$
(\bar{X}_i - \bar{X}_j) - t_{\alpha/2,\, n-k}s_p \sqrt{\frac{1}{n_i} + \frac{1}{n_j}}
$$

$$
\text{to} \quad (\bar{X}_i - \bar{X}_j) + t_{\alpha/2,\, n-k}s_p \sqrt{\frac{1}{n_i} + \frac{1}{n_j}}
$$

(12-19)

EXAMPLE 12.3

Using the data from example 12.2, construct a 95% confidence interval for the average score for the never-married population. Also determine a 95% confidence interval for the difference between the average scores for the ever married and the remarried populations.

Solution

First, your point estimate of the never married population mean is $\bar{x}_1 = 66.67$. Using the ANOVA table from example 12.2,

$$
s_p^2 = \text{MS(error)} = 204.89
$$

and so

$$
s_p = \sqrt{204.89} = 14.314
$$

Because $n = 24$ and $k = 4$, the resulting 95% confidence interval for the never married population mean is

$$
66.67 - t_{.025,20}(14.314)\sqrt{\tfrac{1}{6}} \quad \text{to} \quad 66.67 + t_{.025,20}(14.314)\sqrt{\tfrac{1}{6}}
$$

$$
= 66.67 - (2.086)(14.314)(.408) \quad \text{to} \quad 66.67 + (2.086)(14.314)(.408)
$$

$$
= 66.67 - 12.19 \quad \text{to} \quad 66.67 + 12.19
$$

$$
= 54.48 \quad \text{to} \quad 78.86
$$

As a result, we are 95% confident that the average score on the Life Satisfaction Index for this group of women is between 54.48 and 78.86.

The 95% confidence interval for $\mu_2 - \mu_3$ is

$$
(\bar{X}_2 - \bar{X}_3) - t_{.025,20}s_p \sqrt{\frac{1}{n_2} + \frac{1}{n_3}}
$$

$$
\text{to} \quad (\bar{X}_2 - \bar{X}_3) + t_{.025,20}s_p \sqrt{\frac{1}{n_2} + \frac{1}{n_3}}
$$

$$
= (71.17 - 37.17) - (2.086)(14.314)\sqrt{\tfrac{1}{6} + \tfrac{1}{6}}
$$

$$
\text{to} \quad (71.17 - 37.17) + (2.086)(14.314)\sqrt{\tfrac{1}{6} + \tfrac{1}{6}}
$$

$$= 34 - (2.086)(14.314)(.577) \quad \text{to} \quad 34 + (2.086)(14.314)(.577)$$
$$= 34 - 17.23 \quad \text{to} \quad 34 + 17.23$$
$$= 16.77 \quad \text{to} \quad 51.23$$

This means that we are 95% confident that the ever married group will score between 17 and 51 points higher, on the average than the remarried group.

□

A Word of Warning The procedure we used in example 12.3 for determining confidence intervals is reliable, providing you decide which intervals you want computed *before* you observe your data. For example, constructing a confidence interval for the difference of two population means having the corresponding largest and smallest sample means is not an accurate procedure. If you do this, you let the data dictate which confidence interval you determine.

When using the procedure in example 12.3 to construct confidence intervals for the difference of two population means, it is important to keep the number of such intervals as small as possible. This is because the probability of any one interval containing the true population difference is $1 - \alpha$, but the probability that *all* the intervals contain their respective population differences is not $1 - \alpha$. In other words, if $\alpha = .05$, the overall confidence interval of this procedure is not 95%; it is something much less than 95%. To effectively compare all possible pairs of means, you will need to use a technique that will allow you to make all possible comparisons between population means, while maintaining the Type I error rate at α. This is called a **multiple comparison** procedure and one such procedure is discussed following example 12.5.

EXAMPLE 12.4

A researcher hired by Comptek, a computer software development firm, is interested in the effect of educational level on the job knowledge of the company's employees. He administers an exam to a randomly selected group of people having various educational backgrounds.

In the sample of 15 employees, 6 have only high school diplomas, 5 have only bachelor's degrees, and 4 have master's degrees. The exam scores are:

	High School Diploma	Bachelor's Degree	Master's Degree
	81	94	88
	84	83	89
	69	86	78
	85	81	85
	84	78	
	95		
Total (T)	498	422	340
Average ($\bar{X}$)	83.0	84.4	85.0

What would be your conclusion using a significance level of .10?

Solution   Examining the sample means, you might be tempted to conclude that the higher a person's level of education, the higher their score on the exam. But is there a significant difference among these three means? An ANOVA analysis will clarify this.

The assumptions necessary here are:

1 The scores were obtained randomly and independently from each of the three populations.

2 The exam scores for each of the 3 populations follow a normal distribution, with means μ_1, μ_2, and μ_3. The scores in each of the samples are assumed to have the same amount of variation. Because the sample sizes are not the same, the Hartley test for equal variances cannot be used here. As discussed earlier, we prefer not to use the same data for testing both the means and variances, and so a better procedure would be to obtain additional data (with equal sample sizes) for testing the equality of these three variances.

We begin by calculating the necessary sum of squares.

$$SS(\text{factor}) = \left[\frac{T_1^2}{n_1} + \frac{T_2^2}{n_2} + \frac{T_3^2}{n_3}\right] - \frac{T^2}{n}$$

where

$$T = \Sigma x = T_1 + T_2 + T_3$$
$$= 498 + 422 + 340 = 1260$$

$$n = n_1 + n_2 + n_3$$
$$= 6 + 5 + 4 = 15$$

So,

$$SS(\text{factor}) = \frac{498^2}{6} + \frac{422^2}{5} + \frac{340^2}{4} - \frac{1260^2}{15}$$
$$= 105{,}850.8 - 105{,}840.0$$
$$= 10.8$$

$$SS(\text{total}) = \Sigma x^2 - \frac{T^2}{n}$$

$$= [81^2 + 84^2 + \cdots + 78^2 + 85^2] - \frac{1260^2}{15}$$

$$= 106{,}424 - 105{,}840$$

$$= 584$$

$$SS(\text{error}) = 584 - 10.8 = 573.2$$

Finally, because $k = 3$ and $n = 15$,

$$df \text{ for factor} = k - 1 = 2$$

$$df \text{ for error} = n - k = 12$$

$$df \text{ for total} = n - 1 = 14$$

The resulting ANOVA table is

Source	df	SS	MS	F
Factor	2	10.8	5.4	.11
Error	12	573.2	47.8	
Total	14	584		

The hypotheses are

$$H_0: \quad \mu_1 = \mu_2 = \mu_3$$

$$H_a: \quad \text{not all } \mu\text{'s are equal}$$

where each μ_i represents the average score of all employees having this particular educational level at Comptek.

We will reject H_0 if

$$F^* > F_{.10, 2, 12} = 2.81$$

Because $.11 < 2.81$, we fail to reject H_0.

We conclude that there is not sufficient evidence to indicate that the average performance on the exam is different among the three groups. As usual, we do not accept H_0; that is, we do not conclude that these three means are equal. There is simply not enough evidence to support the claim that employees with a higher educational level are better performers at this particular company.

The factor in example 12.4 was the educational level of the employee; it had three levels. The results show that we are unable to demonstrate that this factor has a significant effect on exam performance.   □

EXAMPLE 12.5   In example 12.4, before the exam was given, the researcher decided to construct a 95% confidence interval for the average exam score of all people holding a master's degree and the difference between the average exam scores for personnel with a master's degree and those with only a high school diploma. What are the confidence intervals?

Solution   The point estimates are

for μ_3: $\bar{x}_3 = 85.0$

for $\mu_3 - \mu_1$: $\bar{x}_3 - \bar{x}_1 = 85.0 - 83.0 = 2.0$

To construct the confidence intervals, you first need an estimate of the common variance of these three populations. Based on the results of example 12.4, this is

$$s_p^2 = \text{MS(error)} = 47.8$$

so

$$s_p = \sqrt{47.8} = 6.91$$

Because $n = 15$, $k = 3$, and $n_3 = 4$, the 95% confidence interval for μ_3 is

$$\bar{X}_3 - t_{.025,12}s_p \sqrt{\frac{1}{n_3}} \quad \text{to} \quad \bar{X}_3 + t_{.025,12}s_p \sqrt{\frac{1}{n_3}}$$

$$= 85.0 - (2.179)(6.91)(.5) \quad \text{to} \quad 85.0 + (2.179)(6.91)(.5)$$

$$= 85.0 - 7.53 \quad \text{to} \quad 85.0 + 7.53$$

$$= 77.47 \quad \text{to} \quad 92.53$$

Figure 12.5

MINITAB solution
for examples 12.4
and 12.5

```
MTB > READ INTO C1 C2
DATA> 81 1  ◄──────────── factor level:    1 = H.S. diploma
DATA> 84 1                                  2 = Bachelors degree
DATA> 69 1                                  3 = Masters degree
DATA> 85 1
DATA> 84 1
DATA> 95 1
DATA> 94 2
DATA> 83 2
DATA> 86 2
DATA> 81 2
DATA> 78 2
DATA> 88 3
DATA> 89 3
DATA> 78 3
DATA> 85 3
DATA> END
     15 ROWS READ
MTB > ONEWAY USING DATA IN C1, LEVELS IN C2

ANALYSIS OF VARIANCE ON C1
SOURCE      DF        SS        MS         F
C2           2      10.8       5.4      0.11
ERROR       12     573.2      47.8
TOTAL       14     584.0
                              INDIVIDUAL 95 PCT CI'S FOR MEAN
                              BASED ON POOLED STDEV
LEVEL       N      MEAN     STDEV  -------+---------+---------+---------
    1       6    83.000     8.367  (------------*-----------)
    2       5    84.400     6.107   (-------------*------------)
    3       4    85.000     4.967    (-------------*--------------)
                                    -------+---------+---------+---------
POOLED STDEV =    6.911              80.0      85.0      90.0
```

The 95% confidence interval for $\mu_3 - \mu_1$ is

$$(\bar{X}_3 - \bar{X}_1) - t_{.025,12}s_p \sqrt{\frac{1}{n_3} + \frac{1}{n_1}} \quad \text{to} \quad (\bar{X}_3 - \bar{X}_1) + t_{.025,12}s_p \sqrt{\frac{1}{n_3} + \frac{1}{n_1}}$$

$$= (85.0 - 83.0) - (2.179)(6.91)(.645)$$
$$\quad \text{to} \quad (85.0 - 83.0) + (2.179)(6.91)(.645)$$

$$= 2.0 - 9.71 \quad \text{to} \quad 2.0 + 9.71$$

$$= -7.71 \quad \text{to} \quad 11.71$$

Consequently, we are 95% confident that the average exam score of all employees with master's degrees is between 7.71 lower to 11.71 higher than those with a high school diploma only. This implies that the data do not allow us to say that the employees with master's degrees performed better on the exam than those with high school degrees.

A MINITAB solution to this example is shown in Figure 12.5. The output contains summary information for each sample, the ANOVA table, and a graphical representation of the confidence interval for each population mean.

□

Multiple Comparisons: A Followup to the One-Factor ANOVA Procedure

If the one-factor ANOVA procedure leads to a rejection of H_0: all population means are equal, a logical question would be "which means do differ?" In other words, rejecting the ANOVA null hypothesis informs us that the means are not all the same, but provides no clue as to which of the population means are different. As was discussed prior to example 12.4, performing a series of t tests to compare all possible pairs of means is not a good idea, since the chances of making at least one Type I error (concluding that a difference exists between two population means when in fact they are the same) using such a procedure is much larger than the predetermined α used for each of the t tests.

What is needed is a technique that compares all possible pairs of means in such a way that the probability of making *one or more* Type I errors is α. This is a **multiple comparison** procedure. There are several methods available for making multiple comparisons; the one presented here is **Tukey's** test for multiple comparisons. (Tukey is pronounced Too'-key).

Tukey's procedure is based upon a statistic that uses the largest and smallest sample means. The form of this statistic is

$$Q = \frac{\text{maximum } (\bar{X}_i) - \text{minimum } (\bar{X}_i)}{\sqrt{\text{MS(error)}/n_r}} \qquad (12\text{-}20)$$

where (1) maximum $(\bar{X}_i)$ and minimum $(\bar{X}_i)$ are the largest and smallest sample means, respectively; (2) MS(error) is the sample variance; and (3) n_r is the number of replicates in each sample.

Notice that Tukey's procedure assumes that each sample contains the same number (n_r) of replicates. Critical values of the Q statistic are contained in Table A-9. Define

$Q_{\alpha, k, v}$ = critical value of the Q statistic from Table A-9, using a significance level of α; k is the number of sample means (groups), and v is the degrees of freedom associated with MS(error).

Multiple Comparison Procedure

1 Find $Q_{\alpha, k, v}$ using Table A-9.

2 Determine $D = Q_{\alpha, k, v} \cdot \sqrt{\dfrac{\text{MS(error)}}{n_r}}$

where MS(error) is the sample variance and n_r is the number of replicates in each sample. For one-factor ANOVA, MS(error) is the same as s_p^2.

3 Place the sample means in order, from smallest to largest.

4 If two sample means differ by more than D, the conclusion is that the corresponding population means are unequal. In other words, if $|\bar{X}_i - \bar{X}_j| > D$, this implies that $\mu_i \neq \mu_j$.

To illustrate this procedure, reconsider example 12.2. Here we concluded that the average life satisfaction scores were not the same for the four groups of females. The four groups were

Group 1: never-married ($\bar{x}_1 = 66.67$)
Group 2: ever-married ($\bar{x}_2 = 71.17$)
Group 3: remarried ($\bar{x}_3 = 37.17$)
Group 4: divorced ($\bar{x}_4 = 32.83$)

For this study, there were $n_r = 6$ replicates in each sample with a resulting pooled sample variance of $s_p^2 = \text{MS(error)} = 204.89$. The study contained $k = 4$ groups and the degrees of freedom for the error sum of squares was $v = n - k = 24 - 4 = 20$. Using a significance level of .05, we begin by finding $Q_{.05, 4, 20}$ in Table A-9. This value is 3.96. Next we determine

$$D = Q_{.05, 4, 20} \cdot \sqrt{\frac{\text{MS(error)}}{n_r}}$$
$$= (3.96) \sqrt{204.89/6}$$
$$= 23.14$$

The sample means, in order, are

$$\overline{32.83, \, 37.17,} \, 66.67, \, 71.17$$

Two sample means are significantly different using the Tukey procedure if they differ by an amount greater than $D = 23.14$. Here there are four significant differences, namely

$$\bar{x}_1 - \bar{x}_4 = 66.67 - 32.83 = 33.84 > 23.14$$

$$\bar{x}_1 - \bar{x}_3 = 66.67 - 37.17 = 29.50 > 23.14$$

$$\bar{x}_2 - \bar{x}_4 = 71.17 - 32.83 = 38.34 > 23.14$$

$$\bar{x}_2 - \bar{x}_3 = 71.17 - 37.17 = 34.00 > 23.14$$

The conclusion from the multiple comparison analysis is that $\mu_1 \neq \mu_4$, $\mu_1 \neq \mu_3$, $\mu_2 \neq \mu_4$, and $\mu_2 \neq \mu_3$. There is no evidence of a difference between the never-married (group 1) and the ever-married (group 2) populations or between the remarried (group 3) and divorced (group 4) populations. This is indicated by the two overbars connecting these two pairs of sample means. In general, there is no evidence to indicate a difference in the population means for any group of sample means under such a bar.

One-Factor ANOVA Procedure

Assumptions

1 The replicates are obtained independently and randomly from each of the populations. The value of one observation has no effect on any other replicates within the same sample or within the other samples.

2 The observations (replicates) from each population follow (approximately) a normal distribution.

3 The normal populations all have a common variance, σ^2. We expect the values in each sample to vary about the same amount. The ANOVA procedure will be much less sensitive to this assumption when we obtain samples of equal size from each population.

Hypotheses

H_0: $\mu_1 = \mu_2 = \cdots = \mu_k$

H_a: not all μ's are equal

Note that H_a is not the same as H_a': all μ's are unequal; H_a states that at least two of the μ's are different.

Sums of Squares

$$\text{SS(factor)} = \left[\frac{T_1^2}{n_1} + \frac{T_2^2}{n_2} + \cdots + \frac{T_k^2}{n_k} \right] - \frac{T^2}{n}$$

where $n = n_1 + n_2 + \cdots + n_k$ and $T = \Sigma x = T_1 + T_2 + \cdots + T_k$.

$$SS(\text{total}) = \Sigma x^2 - \frac{T^2}{n}$$

$$SS(\text{error}) = SS(\text{total}) - SS(\text{factor})$$

$$= \Sigma x^2 - \left[\frac{T_1^2}{n_1} + \frac{T_2^2}{n_2} + \cdots + \frac{T_k^2}{n_k} \right]$$

Degrees of Freedom

df for factor $= k - 1$

df for error $= n - k$

df for total $= n - 1$

Note that $(k - 1) + (n - k) = n - 1$.

ANOVA Table

Source	df	SS	MS	F
Factor	$k - 1$	SS(factor)	$MS(\text{factor}) = \dfrac{SS(\text{factor})}{k - 1}$	MS(factor)/MS(error)
Error	$n - k$	SS(error)	$MS(\text{error}) = \dfrac{SS(\text{error})}{n - k}$	
Total	$n - 1$	SS(total)		

where $MS = $ mean square $= SS/df$.

Testing Procedure

$$\text{Reject } H_0 \text{ if } F^* > F_{\alpha, k-1, n-k}$$

where $F_{\alpha, k-1, n-k}$ is obtained from Table A-7.

EXERCISES

12.1 An instructor wanted to test whether there was a difference in effectiveness of four different teaching techniques. Four groups of students were taught using the four methods, one technique for each group.

(a) If the instructor examined the groups for mean differences one pair at a time, how many t tests would have to be performed?

(b) Would a series of t tests for every possible pair be an appropriate procedure? What could be a problem with that approach?

(c) What is the advantage of using an ANOVA procedure instead of repeated t tests?

(handwritten in margin, top left) 108.

(handwritten equations, left margin)
$$SS_{Factor} = \frac{(16.4)^2}{7} + \frac{(21.4)^2}{7} - \frac{(37.8)^2}{14}$$
$$= 1.786$$
$$SS_T = 110.58 - \frac{(37.8)^2}{14}$$
$$= 8.52$$
$$SS_E = 6.731$$

12.2 A team of psychologists and social workers is investigating the impact that parents' divorce has on the academic performance of their children. A pilot survey used the one-factor ANOVA approach to compare the GPA for high school students in the year following the divorce of their parents, with a control group of students who had not experienced any divorce of their parents. Consider the following data:

S	$\bar{x}$	Family Status	Student's GPA							T
	2.34	Parents divorced within last year (Experimental group	2.4	2.8	3.5	3.0	1.5	1.2	2.0	16.4
	3.06	Parents not divorced (Control group)	3.2	2.9	4.0	3.3	2.5	2.0	3.5	21.4

(handwritten: 37.8)

(handwritten ANOVA table, left margin)

Source	d.f.	SS	MS	F
Factor	1	1.78	1.789	3.189
Error	12	6.73	0.561	
Total	13	8.52		

(a) How many levels of the factor are there? *2*
(b) How many replicates are there at each level? *7*
(c) Compute the "within-sample" and "between-sample" sources of variation, and set up an ANOVA table.
(d) Assuming the underlying ANOVA design assumptions are satisfied, use the ANOVA table to test the average GPA for the two groups. At a 5% significance level, is there sufficient evidence to indicate that the family status (divorced vs. nondivorced) does seem to affect the student's academic performance?
(e) Find the *p* value for the test.
(f) Since the one-factor ANOVA technique depends on comparing two sources of variation ("within-sample" vs. "between-sample"), additional extraneous sources of variation can be considered to be "noise" causing "interference" in the experiment. Suggest possible sources of "noise" in the above setup; for example, the student's level of maturity, the effect of different grading policies in different schools, the socio-economic status of the family. (Later in the chapter, different experimental designs will be considered to help control some of these sources of variation.)

(handwritten, left margin)
$$\alpha = .05$$
$$F_{.05, 1, 12} = 4.75$$
$$F^* < 4.75 \Rightarrow FTR.$$
$$p \geq .1$$

12.3 In exercise 12.2, obtain the following confidence intervals:
(a) The 99% CI for the mean GPA of students in the experimental group (whose parents divorced within the last year).
(b) The 95% CI for the mean difference between the GPA for the two groups. (Use Control group minus Experimental, i.e., $\mu_C - \mu_E$.)

(handwritten, left margin)
$$SP = \sqrt{0.561}$$
$$\alpha = .01$$
$$\bar{x}_1 \pm t_{.005, 12} \cdot \frac{SP}{\sqrt{7}}$$
$$2.34 \pm 3.055 \cdot \frac{\sqrt{0.561}}{\sqrt{7}}$$
$$3.205 - 1.48$$

$$\alpha = .05$$
$$.06 \to 2.34 = 0.72$$
$$0.72 \pm 2.179 \cdot \sqrt{.561} \cdot \sqrt{\frac{1}{7} + \frac{1}{7}}$$
$$1.77 - (-0.35)$$

12.4 A graduate student in educational psychology obtained the cooperation of a mathematics teacher in the third grade to conduct an experiment to compare drill in number operations versus concrete applications of numbers. The dependent variable was the arithmetic scores of the children on a standardized test. The class was divided into two groups at random. One group was taught by the drill method, the other by emphasizing concrete applications. The children were then tested for their proficiency on the topics covered. Their scores are given below (the highest possible score being 100).

Drill Method	Concrete Applications
64	82
81	75
77	90
65	69
72	78
80	88
86	67
75	92
66	79

(a)   Set up an ANOVA table for the above data.

(b)   Conduct a hypothesis test at $\alpha = 0.05$ to determine if there is a difference in the drill method versus concrete applications in the effect these have on the children's learning (as measured by the arithmetic scores). State your conclusion.

(c)   Find the p value for the test.

12.5   Research conducted by social scientists in the Northeast suggests that although older workers are often more productive than their younger counterparts, supervisors tend to rate the older workers lower. Consider the following experimental setup, where the rating is shown on a scale of 1 to 10. The observations are random and independent.

Age Group	Ratings by Supervisors							
<30 years	7.2	5.5	8.0	7.5	6.3	9.0	6.6	7.1
31–45 years	6.1	7.9	5.8	8.0	6.8	7.3	8.2	7.7
>45 years	5.6	6.0	4.9	6.8	5.3	7.0	5.9	5.8

(a)   State the number of factor levels, and the number of replicates at each level.

(b)   If the assumptions of an ANOVA design are satisfied, what is the 95% confidence interval for:

　　(i)   The average rating for those less than 30 years of age?

　　(ii)   The average difference between ratings for the "less than 30 years" and the "more than 45 years" groups?

(c)   Construct an ANOVA table and test the hypothesis that there is no significant difference among the three age groups. At a 10% significance level, is there sufficient evidence to say that age level seems to affect the worker's rating?

(d)   State the p value for the test.

(e)   Refer to the remarks in exercise 12.2, part (f). Could you suggest a source of extraneous variation ("noise")?

(f)   Using a significance level of .05, perform a multiple comparisons procedure (if appropriate).

12.6   A dietician wanted to know whether there was any difference in three types of diet therapy for overweight men. Ten men were chosen for method 1, 12 for method 2, and 13 for method 3. The method 3 group received no treatment;

it was a control group. The percentage of body weight lost using each method follows. A negative value indicates weight was gained.

Method 1	Method 2	Method 3
5	1	1
7	2	−2
8	5	3
4	−2	−1
10	3	0
11	4	1
6	6	−1
8	1	4
−3	4	3
1	−3	2
	−1	1
	1	−2
		0

(a) What is the between-sample variation?

(b) What is the within-sample variation?

(c) Test the null hypothesis that all three methods work equally well using a .10 significance level.

(d) Suppose you were not sure whether the populations had equal variances. Would the Hartley test be appropriate in this case? Explain your answer.

12.7 The science of ergonomics studies the influence of "human factors" in technology, i.e., how human beings relate to and work with machines. With the widespread use of computers for data processing, computer scientists and psychologists are getting together to study human factors. One typical study investigated the productivity of secretaries with different word processing programs. An identical task was given to 18 secretaries, randomly allocated to three groups. Group 1 used a primarily menu-driven program, Group 2 used a command-driven program, and Group 3 was a mixture of both approaches. The secretaries all had about the same level of experience, typing speed, and computer skills. The time (in minutes) taken to complete the task was observed. The results are shown below.

Group 1 (Menu-Driven)	Group 2 (Command-Driven)	Group 3 (Mixed)
12	14	10
15	11	8
11	13	9
12	12	10
10	11	7
13	14	8

(a) Do the necessary calculations to construct an ANOVA table, and test the hypothesis that there is no difference between the three types of word-processing programs (i.e., on the average, the time taken to complete the task is about the same). Use $\alpha = .05$.

(b) State the p value for the test.

(c) Does the type of word-processing software used affect the performance of the secretaries?

(d) If the secretaries had different levels of experience, typing speed, and computer skills, how would it affect the data? (Would it be an extraneous source of variation, or "noise"? Would it tend to increase the "within-sample" variation, or the "between-sample" variation, or both, or neither?)

(e) Using a significance level of .05, perform a multiple comparisons procedure (if appropriate).

12.8 In exercise 12.7, obtain the following confidence intervals.

(a) The 90% CI for the mean time taken to complete the task by the secretaries using the menu-driven software.

(b) The 99% CI for the difference between the means for Group 1 and Group 2.

12.9 Astral Airlines recently introduced a nonstop flight between Houston and Chicago. The vice president of marketing for Astral decided to run a test to see whether Astral's passenger load was similar to that of its two major competitors. Ten daytime flights were picked at random from each of the three airlines and the percent of unfilled seats on each flight was as follows:

Astral	10	14	12	10	8	13	11	8	12	9
Competitor 1	12	9	8	9	9	10	12	7	11	10
Competitor 2	15	10	15	8	14	9	8	11	10	12

Use a significance level of .05 and perform an ANOVA procedure. Find the p value.

12.10 Would a multiple comparisons analysis be appropriate for exercise 12.9? Discuss why or why not.

12.11 Catalytic converters are required on automobiles in the United States to reduce air pollution from exhaust gases. A government agency was comparing three different catalytic converters for their effectiveness. The converters had been installed randomly on 15 automobiles, divided into three groups. The following data give the measured residue pollutants in milligrams after exhaust gases passed through the catalytic converter for a specified period.

Converter "C"	Converter "F"	Converter "M"
1.4	1.7	2.1
2.0	1.5	1.6
1.8	1.2	1.9
0.9	1.6	0.9
1.6	1.6	1.4

(a) Use an ANOVA table to obtain the within-sample and between-sample variations.

(b) At a 5% significance level, do the above data suggest that there is a difference in effectiveness among the three catalytic converters?

(c) State the p value for the test. Would the conclusion be different at the 1% level of significance?

(d) In the above experiment, the researcher made sure that all 15 automobiles were the same make and year models with roughly the same mileage; the same fuel was used for all cars; and the cars were operated under similar conditions of weather and temperature. Why would such precautions be necessary?

(e) Using a significance level of .05, perform a multiple comparisons procedure, if appropriate.

12.12 The workers at a calculator assembly plant wish to bargain for more breaks during the work day. The manager believes that increasing the number of 15-minute breaks will affect productivity. The workers currently receive three breaks during the 8-hour work day. The manager decides to run a test by choosing four groups of five workers each and giving one group three breaks, the next group four breaks, and so on. The number of calculators assembled per day is recorded for five days. Test the manager's claim using an ANOVA procedure with a .10 significance level. Find the p value.

3 Breaks	200	205	197	210	205
4 Breaks	210	203	201	197	199
5 Breaks	198	190	185	188	180
6 Breaks	197	180	190	192	175

12.13 In exercise 12.12, divide the original data by 100, and then construct the ANOVA table on the coded data. What do you notice? Is the ANOVA table identical to that in exercise 12.12? Do you reach the same conclusions regarding the hypothesis tested?

12.14 Independent samples of size 16 are drawn from each of four normally distributed populations. The resulting sample standard deviations are: $s_1 = 2.0$, $s_2 = 2.5$, $s_3 = 2.2$, $s_4 = 2.5$. Do the data provide sufficient evidence that a significant difference exists in the population standard deviations of the four populations? Use a .10 significance level.

12.15 Social psychologists developed a questionnaire designed to measure alienation. The questionnaire was administered to a random group of women, who were then classified into three groups: low, medium, and high alienation scores. The researchers were interested in whether the degree of alienation in women had anything to do with the ages of the women. Consider the following data, listing the ages for the women in the three categories. Use the ANOVA method to test the hypothesis that age, on the average, is the same for all the three groups. Use $\alpha = .05$, and determine the p value.

Low Alienation	Medium Alienation	High Alienation
24	25	34
31	32	45
42	27	38
28	35	51
26	41	40
30	28	37
	25	48
	30	

12.16 The task of air traffic controllers is crucial in ensuring the safety of airline passengers. A manufacturer of radar screens has hired an ergonomics expert to evaluate three different types of radar screen displays. In one aspect of the study, 21 air traffic controllers, with similar age and experience, were subjected to simulated emergency conditions and their reaction times were noted (in tenths of a second),

$$SS_{Factor} = \frac{80^2}{7} + \frac{(125)^2}{7} + \frac{76^2}{7} - \frac{(281)^2}{21}$$

$$= 211.52$$

$$SS_T = 4163 - \frac{(281)^2}{21}$$

$$= 402.95$$

$$SS_E = 191.43$$

d.f	SS	MS	F
Fact	2	211.52	105.76	9.94
E	18	191.43	10.64	
T	20	402.95		

$$F_{.10, 2, 18} = 2.62$$
$$P < 0.01$$

$$Q_{.05, k, r} = Q_{.05, 3, 18} = 3.61$$

$$D = 3.61 \cdot \sqrt{\frac{10.64}{7}} = 4.45$$

10.86, 11.43, 17.86

$\bar{x}_3, \bar{x}_1, \bar{x}_2$

$|\bar{x}_3 - \bar{x}_2| = 7 > 4.45$

$|\bar{x}_1 - \bar{x}_2| = 6.43 > 4.45$

$\mu_3 \neq \mu_2$

$\mu_1 \neq \mu_2$

$F^* 2.62$

$F.10, 2, 18 = 2.62$

as shown below:

$T = 281$

Display 1	Display 2	Display 3
8	15	15
16	19	8
12	18	12
15	16	10
11	24	7
7	13	11
11	20	13

$\bar{x}_1 = 11.43$ $\bar{x}_2 = 17.86$ $\bar{x}_3 = 10.86$

$T_1 = 80$ $\Sigma x^2 = 4163$

$T_2 = 125$

$t_3 = 76$

(a)   If the assumptions of an ANOVA design are applicable, do the above data suggest that there is a significant difference in average reaction times for the three displays? Use $\alpha = .10$.

(b)   Determine the p value. Would your conclusion change if $\alpha = .05$ or $\alpha = .01$?

(c)   What would happen if the air traffic controllers had different backgrounds (e.g., some young, some older, some relatively inexperienced, some quite experienced)?

(d)   Using a significance level of .05, perform a multiple comparisons procedure, if appropriate.

12.17   A target-shooting club performed an experiment on a randomly selected group of 21 beginning shooters to determine whether shooting accuracy is affected by the method of sighting: right eye only open, left eye only open, or both eyes open. The 21 beginners were randomly divided into three groups of seven each. Each group went through the same training and practicing procedures with one exception—the use of eyes for sighting. The scores from each shooter were coded and are:

Right eye	2	0	8	2	5	3	1
Left eye	3	6	0	9	0	4	2
Both eyes	1	7	2	7	1	5	2

(a)   Do the data provide sufficient evidence that a difference exists among the mean scores of the three methods? Use a .10 significance level. Find the p value.

(b)   Find a 95% CI to estimate the mean score of the coded data for each method.

12.3   DESIGNING AN EXPERIMENT

The previous section introduced you to one-factor (or one-way) ANOVA. In this type of analysis, you randomly obtain samples from each of your k populations (levels) describing a single factor. The variable that is being measured (such as the score on the Life Satisfaction Index) is referred to as the **dependent** variable. Since replicates (repeat observations) are obtained in a completely random manner from each population, this type of sampling plan is called a **completely randomized design**. In general, the scheme used to collect the sample data reflects a type of **experimental design**. This section will discuss other experimental designs, including the randomized block design and the two-way factorial design.

Let us consider using the Life Satisfaction Index to examine various group differences and what type of design would be appropriate for each situation.

Situation 1: The Completely Randomized Design

We have already examined the completely randomized design in the previous section, where we obtained independent samples from four marital groups comprising the female population. Replicates are obtained independently within each population and the four samples are not related in any way. This particular example contained one factor (marital status) consisting of four levels. The research question when using this design was: Is there a difference in life satisfaction among the four groups of women? The corresponding null hypothesis was

$$H_0: \quad \mu_1 = \mu_2 = \mu_3 = \mu_4$$

Essentially, this type of analysis (called one-way ANOVA) will fail to reject H_0 if the sample means are "close together" and reject this hypothesis otherwise.

Situation 2: The Randomized Block Design

Suppose that you wanted to investigate whether husbands and wives differ in their level of life satisfaction. For your study, you select 15 (or however many) married couples and have each spouse fill out the questionnaire. This is an example of a **randomized block design**. The configuration of your sample results would resemble the scheme below, where each x represents a score on the Life Satisfaction Index.

	Husbands	**Wives**	
Couple 1	x	x	(1st block)
Couple 2	x	x	(2nd block)
Couple 3	x	x	(3rd block)
⋮			
Couple 15	x	x	(15th block)

This design consists of one *factor* (sex of spouse) with two *levels* (husband/wife) and 15 *blocks* (couple 1, couple 2, . . . , couple 15). Unlike the completely randomized design (Situation 1), the husband and wife samples *are not independent*, since the data are paired by couple. For example, the first male value is not independent of the first female value, since they both belong to the same married couple. We encountered this very same design in Chapter 10, where we compared two population means using paired (that is, blocked) samples. When using the randomized block design, you can also compare the means of more than two populations using a blocking strategy to gather your data.

The question of interest here is: Is there a difference between husband and wife scores on this instrument? Similar to the completely randomized design, the question is whether the factor of interest (sex of spouse, here) has a significant effect on the Life Satisfaction score. The difference between this and the completely randomized situation is that here we use a blocking strategy, rather than independent samples, to obtain a more precise test for examining differences in the factor level means. The null hypothesis for this illustration is that the group (factor level) means are identical, that is

$$H_0: \quad \mu_H = \mu_W$$

where μ_H represents the average Life Satisfaction Index score for husbands and μ_W is the average score for wives.

The analysis for the randomized block design will be discussed in the next section, but essentially this procedure removes the effect of the blocks (couples, here) before testing for a difference between the factor level means. Consequently, this design removes the block effect from the error sum of squares in the completely randomized design. Several examples in the next section will illustrate this technique.

Situation 3: The Two-Way Factorial Design

The two-way factorial design is very similar to situation 1, the one-way analysis of variance, except now two factors are of interest to the researcher. Using the Life Satisfaction Index, you may want to investigate the effect of sex (factor A) and marital status (factor B) on the level of life satisfaction for all individuals in the population. The previous one-way ANOVA illustration examined the effect of marital status for the female population and concluded that this factor was indeed significant. The inclusion of the sex factor accomplishes two things: first, you can determine if the sex of the individual has an effect on this person's life satisfaction and second, you can investigate whether the relationship between marital status and the life satisfaction score is different for males and females. In other words, whether factor B relates to the dependent variable (Life Satisfaction Index score) depends upon the level of factor A. This type of effect is called **interaction** between factors A and B. This differs from the randomized block design where it is assumed that no interaction is present between the factor of interest and the blocks.

Consequently, there are three sets of hypotheses that can be tested using the two-way factorial design. The corresponding null hypotheses are:

$H_{0,A}$: factor A (sex) is not significant

$H_{0,B}$: factor B (marital status) is not significant

$H_{0,AB}$: there is no interaction between factor A and factor B

The first two hypotheses are similar to those tested in one-way ANOVA, the third hypothesis is unique to the two-way (or higher) factorial design.

To collect data for this design, you would obtain a score for *every* combination of a factor A level and a factor B level. This particular illustration consists of two levels for factor A (male/female) and four levels for factor B (never married/ever married/remarried/divorced), and is called a **2 × 4 factorial**

Figure 12.6

Illustration of a
2 × 4 factorial design
using 3 replicates

Factor B

		Never Married	Ever Married	Remarried	Divorced
Factor A	Male	x,x,x	x,x,x	x,x,x	x,x,x
	Female	x,x,x	x,x,x	x,x,x	x,x,x

design. Here, "ever married" means that a person has been continually married to the same spouse. Consequently, data are collected for the eight possible factor level combinations, referred to as **treatments**. Data values within the same treatment are termed **replicates**. Furthermore, it is necessary when using this type of design *to obtain more than one replicate for each treatment*. An illustration using three replicates (two would be sufficient) is shown in Figure 12.6. Each x represents a score on the Life Satisfaction Index. The actual two-way analysis of variance for this illustration will be demonstrated in Section 12.5.

EXERCISES

12.18 (a) Name the three types of experimental design discussed in a preceding section.
    (b) Which design or designs does not involve the necessity of having replicates?
    (c) What is "interaction" between factors and which design permits testing for interaction?
    (d) In each of the three designs, how many dependent variables are there?
    (e) If a one-way (one-factor) ANOVA design has four levels of the factor and six replicates at each level, how many treatments are being considered and how many total number of observations are made?
    (f) If a 4 × 6 two-way factorial design is chosen for an ANOVA-based experiment, what are the number of treatments being considered and the *minimum* number of total observations necessary?
    (g) What is the advantage of "blocking" and when is it necessary?
    (h) What are some potential problems that can arise with the randomized block design?

12.19 In exercise 12.2, the team of researchers were able to control for extraneous influences such as student's level of maturity and socioeconomic status of the family. Indeed, a fairly homogeneous group of students were available. However, they were unable to control the effect of the students coming from different schools.
    (a) Which design would be appropriate under these circumstances?
    (b) Construct a diagram showing how the data would look.
    Let us go one step further. The researchers also consider the possibility of taking all observations only from one school. They are now interested in looking at two factors (family status = divorced or nondivorced and socioeconomic status = low, medium, high) and at possible interactions between these factors.
    (c) Which is the appropriate experimental design?
    (d) Construct a diagram to show the setup of the data.
    (e) If the researchers decide on three replicates for each treatment, what is the total number of students they will need?

12.20 Take exercise 12.5 and change the format to a randomized block design, using the same supervisor for one member each from each of the three age groups.

  (a) How many supervisors will you need, if the total observations are kept the same?

  (b) Construct a diagram of the data.

  (c) Other than reduction in number of supervisors, what would be an advantage of the randomized block design?

  (d) How would you ensure that observations within each block are randomized?

12.21 Consider issues such as those raised in exercises 12.18 and 12.19, and try to apply them to exercise 12.7, exercise 12.11, exercise 12.15 and exercise 12.16. In each case, try to imagine what a completely randomized block design and a two-way factorial design might look like, if the design were extended beyond a simple one-factor format. Try to identify a potential second factor (or block effect) that might be appropriate. Construct a diagram of the data in each case.

12.4   RANDOMIZED BLOCK DESIGN

The previous section described the difference between the randomized block and completely randomized designs. Rather than obtaining independent samples from the k populations, the data for a randomized block design are organized into homogeneous units, referred to as **blocks**. Within each block, any predictable difference in the observations is due to the effect of the factor of interest, such as sex or marital status.

Consider a situation where two workers at Ace Manufacturing are given different tasks to perform. The question of interest is whether the time required to successfully perform these tasks is different for the two workers. The times (in minutes) are shown below.

Task	Worker 1	Worker 2
1	72	77
2	56	54
3	83	85
4	42	47
5	35	30
6	61	52
7	50	61
8	77	84
9	28	35
10	55	48
11	65	60
12	42	48

Here the factor of interest is the worker (2 levels) and the dependent variable is the time required to complete a task. Each task represents a block and so this is a randomized block design consisting of 12 blocks. This situation also fits the **paired sample** design, discussed in Section 10.5, provided we assume the populations are normally distributed. The randomized block design is an extension of Section 10.5, where we can now consider more than two populations.

To illustrate a randomized block design consisting of more than two populations, suppose we are interested in comparing the quality of life for three California cities, using an instrument that considers various factors affecting the quality of living, such as climate and economy. Each city is scored from 0 to 100, where a larger score indicates more quality. To control for the effect of the evaluator, each evaluator scores all three cities under consideration. The following results were obtained.

Evaluator	City 1	City 2	City 3
1	68	72	65
2	40	43	42
3	82	89	84
4	56	60	50
5	70	75	68
6	80	91	86
7	47	58	50
8	55	68	52
9	78	77	75
10	53	65	60

To determine if there is a factor (city) effect in these 30 ratings, we must first account for the block (evaluator) effect. If we ignore this effect and analyze these data as though the three samples are independent, we may come to the conclusion that the three cities do not differ significantly in the quality of life when in fact, such is not the case. In other words, by using the randomized block design, we have a more precise test for detecting a difference among the three cities. This example illustrates a randomized block design with a single factor consisting of 3 levels and 10 blocks. The general appearance of such a design is shown in Table 12.1.

When using the randomized block design, the various levels should be applied in a random manner within each block. In our California cities example, each person should not always examine City 1 first, City 2 second, and the City 3 last. Instead these three cities should be assigned in a randomized order for each person—hence the name "randomized block design."

The assumptions for the randomized block design are:

1 The observations within each factor level/block combination are obtained from a normal population.

2 These normal populations have a common variance, σ^2.

Furthermore, we assume that the factor effects are the same within each block; that is, there is no interaction effect between the factor and the blocks.

Table 12.1

The randomized
block design

	FACTOR LEVEL (POPULATION)					
Block	**1**	**2**	**3**	$\cdots$	**k**	**Total**
1	x	x	x	$\cdots$	x	S_1
2	x	x	x	$\cdots$	x	S_2
3	x	x	x	$\cdots$	x	S_3
$\vdots$						
b	x	x	x	$\cdots$	x	S_b
Total	T_1	T_2	T_3	$\cdots$	T_k	T
Sample mean	$\bar{X}_1$	$\bar{X}_2$	$\bar{X}_3$	$\cdots$	$\bar{X}_k$	

The analysis using the randomized block design is similar to that for the one-factor ANOVA, except that the total sum of squares (SS(total)) has an additional component. Now,

$$SS(total) = SS(factor) + SS(block) + SS(error)$$

where SS(block) measures the variation due to the blocks. Consequently, this design extracts the block effect, as measured by SS(block), from the error sum of squares in the completely randomized design.

If you use the randomized block design when blocking is not necessary, SS(block) will be very small in comparison to the other sums of squares. Referring to Table 12.1, this will occur when $S_1, S_2, \ldots, S_b$ are nearly the same. If all the S_i's are equal, then SS(block) = 0. The effect of the blocks will be significant whenever you observe a lot of variation in these block totals.

The factor sum of squares for the randomized block design is thus

$$SS(factor) = \frac{1}{b}[T_1^2 + T_2^2 + \cdots + T_k^2] - \frac{T^2}{bk}$$

where

$n = $ number of observations $= bk$

$T_1, T_2, \ldots, T_k$ represent the totals for the k factor levels

$S_1, S_2, \ldots, S_b$ are the totals for the b blocks

$T = T_1 + T_2 + \cdots + T_k$
$ = S_1 + S_2 + \cdots + S_b$
$ = $ total of all observations

$$SS(blocks) = \frac{1}{k}[S_1^2 + S_2^2 + \cdots + S_b^2] - \frac{T^2}{bk}$$

$$SS(total) = \Sigma x^2 - \frac{T^2}{bk}$$

where Σx^2 = sum of the squares for each of the n (= bk) observations.

$$SS(error) = SS(total) - SS(factor) - SS(blocks)$$

The degrees of freedom are

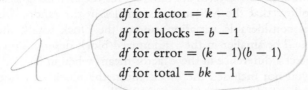

df for factor = $k - 1$
df for blocks = $b - 1$
df for error = $(k - 1)(b - 1)$
df for total = $bk - 1$

The ANOVA table for a blocked design is very similar to the one-factor ANOVA table. There is one additional row because you now include the effect of the various blocks in your design:

Source	df	SS	MS	F
Factor	$k - 1$	SS(factor)	MS(factor) = $\dfrac{SS(factor)}{k - 1}$	F_1 = MS(factor)/MS(error)
Blocks	$b - 1$	SS(blocks)	MS(blocks) = $\dfrac{SS(blocks)}{b - 1}$	F_2 = MS(blocks)/MS(error)
Error	$(k - 1)(b - 1)$	SS(error)	MS(error) = $\dfrac{SS(error)}{(k - 1)(b - 1)}$	
Total	$bk - 1$	SS(total)		

where

$$MS(factor) = SS(factor)/(k - 1)$$
$$MS(blocks) = SS(blocks)/(b - 1)$$
$$MS(error) = SS(error)/[(k - 1)(b - 1)]$$

Hypothesis Testing

Is there a difference in the average rating of the three California cities in our illustration? In other words, is there a significant city effect on the quality of life ratings? The hypotheses for this situation are H_0: $\mu_1 = \mu_2 = \mu_3$ and H_a: not all the means are equal. We determine the test statistic exactly as we did for the one-factor ANOVA:

$$F_1 = \frac{MS(factor)}{MS(error)} \qquad \text{(12-21)}$$

where the mean square values are obtained from the ANOVA table. Notice that the MS(factor) value is the same regardless of whether or not you block. However, when you use the block effect in the design, the SS(error) is smaller than the SS(error) value obtained using the completely randomized design. The degrees of freedom in the error term are different for the two designs, so it does not necessarily follow that the MS(error) is smaller in the randomized block design. If there is considerable variation among the block totals, then quite likely MS(error) *will* be smaller in the randomized block design. You are thus more likely to detect a difference in the k factor means when in fact a difference does exist. Had you not included the block effect, the block variation would have been included in SS(error), resulting in a smaller F value. This value often becomes small enough not to fall in the rejection region, leading you to conclude that no difference exists. But perhaps there is a difference among these means (the factor does have a significant effect) that will go undetected if the appropriate experimental design is not used.

For the randomized block design, the test procedure will be to

$$\text{Reject } H_0 \text{ if } F_1 > F_{\alpha, v_1, v_2}$$

where $v_1 = k - 1$ and $v_2 = (k - 1)(b - 1)$. So, once again, we reject H_0 if the F statistic falls in the right-tail rejection region, this time using Table A-7 with $v_1 = k - 1$, the *df* for factor, and $v_2 = (k - 1)(b - 1)$, the *df* for error.

Now suppose we wish to determine whether the effect of the person evaluating the cities is significant. We are attempting to determine whether there is a block effect, so the hypotheses are

H_0':   there is no effect due to the evaluators (blocks) (the block means are equal)

H_a':   there is an effect due to the evaluators (the block means are not all equal)

The corresponding test uses the "other" F statistic from the randomized block ANOVA table, namely,

$$F_2 = \frac{\text{MS(blocks)}}{\text{MS(error)}} \qquad \textbf{(12-22)}$$

and the test procedure is to

$$\text{Reject } H_0' \text{ if } F_2 > F_{\alpha, v_1', v_2'}$$

where $v_1' = b - 1$ and $v_2' = (k - 1)(b - 1)$.

Let us re-examine our data. We will use $\alpha = .05$. Here, $k = 3$ levels (cities), $b = 10$ blocks (evaluators), and $n = bk = 30$ observations.

Evaluator	City 1	City 2	City 3	Totals
1	68	72	65	205
2	40	43	42	125
3	82	89	84	255
4	56	60	50	166
5	70	75	68	213
6	80	91	86	257
7	47	58	50	155
8	55	68	52	175
9	78	77	75	230
10	53	65	60	178
Total	629	698	632	1959
$\bar{X}$	62.9	69.8	63.2	

$$SS(factor) = \frac{1}{10}[629^2 + 698^2 + 632^2] - \frac{1959^2}{30}$$
$$= 128{,}226.9 - 127{,}922.7$$
$$= 304.2$$

$$SS(blocks) = \frac{1}{3}[205^2 + 125^2 + \cdots + 178^2] - \frac{1959^2}{30}$$
$$= 133{,}627.7 - 127{,}922.7$$
$$= 5705.0$$

$$SS(total) = [68^2 + 40^2 + \cdots + 75^2 + 60^2] - \frac{1959^2}{30}$$
$$= 134{,}107 - 127{,}922.7$$
$$= 6184.3$$

$$SS(error) = SS(total) - SS(factor) - SS(blocks)$$
$$= 6184.3 - 304.2 - 5705.0$$
$$= 175.1$$

So

$$MS(factor) = SS(factor)/(k - 1)$$
$$= 304.2/2 = 152.1$$

$$MS(blocks) = SS(blocks)/(b - 1)$$
$$= 5705.0/9 = 633.9$$

$$MS(error) = SS(error)/[(k - 1)(b - 1)]$$
$$= 175.1/18 = 9.73$$

The resulting ANOVA table is

Source	df	SS	MS	F
Factor (City)	2	304.2	152.1	$152.1/9.73 = 15.63$ (F_1)
Blocks	9	5705.0	633.9	$633.9/9.73 = 65.15$ (F_2)
Error	18	175.1	9.73	
Total	29	6184.3		

We first consider the hypotheses

H_0: $\mu_1 = \mu_2 = \mu_3$

H_a: not all μ's are equal

where

$$\mu_1 = \text{average rating for City 1 (estimate is } \bar{x}_1 = 62.9)$$

$$\mu_2 = \text{average rating for City 2 (estimate is } \bar{x}_2 = 69.8)$$

$$\mu_3 = \text{average rating for City 3 (estimate is } \bar{x}_3 = 63.2)$$

Because $F_1 = 15.63 > F_{.05,2,18} = 3.55$, we reject H_0 and conclude that there is a difference in the perceived quality of the three cities. This is not a surprising result because it appears that City 2 scored much higher than City 1 and City 3. This means that the factor (city) does have a significant effect on the ratings.

We also wish to test

H_0': there is no block effect

H_a': there is a block effect

$S_1, S_2, \ldots, S_{10}$ appear to contain considerable variation, so our initial guess is that there is a block effect. Carrying out the statistical test, we see that

$$F_2 = 65.15 > F_{.05,9,18} = 2.46$$

Consequently, we strongly reject H_0', in favor of H_a'. The effect of the person doing the three evaluations is significant.

A MINITAB solution for this problem is shown in Figure 12.7.

What would the result have been had we treated these 30 observations as replicates, 10 from each of the 3 cities? In other words, what would happen if we failed to recognize that blocking was necessary and we incorrectly used the one-factor ANOVA? Because both SS(factor) and SS(total) do not change, the only difference is a new SS(error). Therefore,

$$\text{SS(error)} = \text{SS(total)} - \text{SS(factor)}$$
$$= 6184.3 - 304.2$$
$$= 5880.1$$

Figure 12.7

MINITAB solution for quality of life data

```
MTB > READ INTO C1-C3
DATA> 68  1  1
DATA> 40  1  2
DATA> 82  1  3
DATA> 56  1  4
DATA> 70  1  5
DATA> 80  1  6
DATA> 47  1  7
DATA> 55  1  8
DATA> 78  1  9
DATA> 53  1 10
DATA> 72  2  1
DATA> 43  2  2
DATA> 89  2  3
DATA> 60  2  4
DATA> 75  2  5
DATA> 91  2  6
DATA> 58  2  7
DATA> 68  2  8
DATA> 77  2  9
DATA> 65  2 10
DATA> 65  3  1
DATA> 42  3  2
DATA> 84  3  3
DATA> 50  3  4
DATA> 68  3  5
DATA> 86  3  6
DATA> 50  3  7
DATA> 52  3  8
DATA> 75  3  9
DATA> 60  3 10
DATA> END
     30 ROWS READ
MTB > TWOWAY USING DATA IN C1, LEVELS IN C2, BLOCKS IN C3;
SUBC> ADDITIVE.

ANALYSIS OF VARIANCE  C1

SOURCE        DF        SS        MS
C2             2     304.20    152.10
C3             9    5704.97    633.89
ERROR         18     175.13      9.73
TOTAL         29    6184.30
```

C_3 contains block values: 1 = Person 1
2 = Person 2, etc.

C_2 contains factor (city) levels: 1 = City 1
2 = City 2
3 = City 3

F value not provided

$F_1 = 152.10 / 9.73 = 15.63$

factor

$F_2 = 633.89 / 9.73 = 65.15$

block

Also, in the one–factor ANOVA design,

$$df \text{ for total} = (df \text{ for factor}) + (df \text{ for error})$$

So

$$df \text{ for error} = (df \text{ for total}) - (df \text{ for factor})$$
$$= 29 - 2 = 27$$

The resulting F value will be

$$F = \frac{MS(\text{factor})}{MS(\text{error})} = \frac{304.2/2}{5880.1/27} = .70$$

Because $F^* = .70$ is much less than $F_{.05,2,27} = 3.35$, we fail to detect a difference in the three means, μ_1, μ_2, and μ_3. This is the effect of assuming independence among the samples when it does not exist. This emphasizes that failing to recognize the need for a randomized block design can have serious consequences!

EXAMPLE 12.6

Consider the data gathered using the 12 tasks and 2 workers at Ace Manufacturing. Can we conclude that there is a difference in the speed of the two workers? Use a significance level of .05.

Solution

The various block and factor totals can be obtained by summing across and down the array of data.

Task	Worker 1	Worker 2	Total
1	72	77	149
2	56	54	110
3	83	85	168
4	42	47	89
5	35	30	65
6	61	52	113
7	50	61	111
8	77	84	161
9	28	35	63
10	55	48	103
11	65	60	125
12	42	48	90
Total	666	681	1347
$\bar{X}$	55.5	56.75	

$$\text{SS(factor)} = \frac{1}{12}[666^2 + 681^2] - \frac{1347^2}{24}$$
$$= 75,609.75 - 75,600.375 = 9.375$$

$$\text{SS(blocks)} = \frac{1}{2}[149^2 + 110^2 + \cdots + 90^2] - \frac{1347^2}{24}$$
$$= 81,992.5 - 75,600.375 = 6392.125$$

$$\text{SS(total)} = [72^2 + 56^2 + \cdots + 60^2 + 48^2] - \frac{1347^2}{24}$$
$$= 82,239 - 75,600.375 = 6638.625$$

$$\text{SS(error)} = \text{SS(total)} - \text{SS(factor)} - \text{SS(blocks)}$$
$$= 237.125$$

This leads to the following ANOVA table. Note that the *df* for the factor are $2 - 1 = 1$, for blocks are $12 - 1 = 11$, for total are $24 - 1 = 23$, leaving $23 - 1 - 11 = 11$ *df* for error.

Source	df	SS	MS	F
Factor	1	9.375	9.375	$F_1 = 9.375/21.56 = 0.43$
Blocks	11	6392.125	581.10	$F_2 = 581.10/21.56 = 26.95$
Error	11	237.125	21.56	
Total	23	6638.625		

The factor here represents the worker, with the corresponding hypotheses:

$$H_0: \quad \mu_1 = \mu_2$$

$$H_a: \quad \mu_1 \neq \mu_2$$

where μ_1 is the average completion time for worker 1, and μ_2 is the average completion time for worker 2.

Since $F_1 = 0.43 < F_{.05,1,11} = 4.84$, we fail to reject H_0 and conclude that there is not sufficient evidence to indicate a difference in the two workers; that is, the effect of the factor is not significant. We would have obtained an identical result had we used the paired sample two-tailed *t* test discussed in Section 10.5; the *p* values for the *t* test and for F_1 are identical.

As a final note, the block (task) effect is highly significant since $F_2 = 26.95 > F_{.05,11,11} \cong 2.85$ (using 10 and 11 degrees of freedom). The decision to use a randomized block design was certainly a wise one. □

Constructing a Confidence Interval for the Difference between Two Population Means

We can construct a confidence interval for the difference between any pair of means, $\mu_i - \mu_j$. Remember, however, that we must determine which confidence intervals we will construct before observing the data. Do not fall into the trap of letting the data dictate which confidence intervals you construct.

When using the randomized block design, our estimate of the common variance σ^2 is now

$$s^2 = \text{estimate of } \sigma^2$$

$$= \text{MS(error)} = \frac{\text{SS(error)}}{(k-1)(b-1)} \tag{12-23}$$

Thus, a $(1 - \alpha) \cdot 100\%$ confidence interval for $\mu_i - \mu_j$ is

$$(\bar{X}_i - \bar{X}_j) - t_{\alpha/2, df} \cdot s \cdot \sqrt{\frac{1}{b} + \frac{1}{b}}$$

$$\text{to} \quad (\bar{X}_i - \bar{X}_j) + t_{\alpha/2, df} \cdot s \cdot \sqrt{\frac{1}{b} + \frac{1}{b}} \tag{12-24}$$

where df = degrees of freedom for the t statistic (Table A-5) = $(k - 1)(b - 1)$; b = number of blocks; k = number of factor levels; and s is determined from equation 12-23.

EXAMPLE 12.7

Assume you have not yet observed the data from the California cities example, and you decided to construct a 95% confidence interval for the difference between the average ratings for Cities 2 and 3. What does it tell you?

Solution

Using the ANOVA table for these data,

$$s^2 = MS(error) = 9.73$$

and so $s = 3.12$. Also, $t_{.025, 18} = 2.101$ using Table A-5. The resulting 95% confidence interval for $\mu_2 - \mu_3$ is

$$(\bar{X}_2 - \bar{X}_3) - t_{.025, 18}s \sqrt{\frac{1}{10} + \frac{1}{10}} \quad \text{to} \quad (\bar{X}_2 - \bar{X}_3) + t_{.025, 18}s \sqrt{\frac{1}{10} + \frac{1}{10}}$$

$$= (69.8 - 63.2) - (2.101)(3.12)(.447)$$
$$\text{to} \quad (69.8 - 63.2) + (2.101)(3.12)(.447)$$
$$= 6.6 - 2.93 \quad \text{to} \quad 6.6 + 2.93$$
$$= 3.67 \quad \text{to} \quad 9.53$$

We are thus 95% confident that the average for City 2 is between 3.67 and 9.53 points higher than that for City 3. City 2 appears to be superior to the City 3 in quality of living. □

EXERCISES

12.22 A particular application contains four blocks and four levels for the factor of interest. The totals for each of the four blocks are given: $S_1 = 170$, $S_2 = 184$, $S_3 = 182$, $S_4 = 240$, and the totals for each of the four levels of the factor are given as $T_1 = 120$, $T_2 = 240$, $T_3 = 210$, $T_4 = 206$. Construct the ANOVA table for the randomized block design and assume that the total sum of squares is 2836.

12.23 Explain the differences between a completely randomized design and a randomized block design. What is the purpose of using blocks in a design?

12.24 Complete the following ANOVA table for a randomized block design. Find the p value.

Source	df	SS	MS	F
Factor	7			
Blocks	3	105.6		
Error		90.8		
Total		336.5		

12.25 The investigation of parental divorce and academic performance discussed in exercise 12.2 was modified to control for the differences in academic standards. It was felt that the latter was an extraneous variable whose influence needed to be eliminated by the technique of blocking. A fresh set of data was collected, one pair for each school. The table below contains the GPA for the students, obtained from six schools:

				SCHOOL			
Family Status	1	2	3	4	5	6
Parents divorced within last year	2.9	1.5	3.5	2.8	2.5	3.0
Parents not divorced	3.3	2.0	3.0	3.1	2.9	3.1

(a) Since the observations for the divorced vs. nondivorced groups are obtained as one pair per school, are the data independent?

(b) Identify the factor and the block in this design.

(c) What does this design assume about interaction between "family status" and "school"?

(d) Compute the ANOVA table and conduct a hypothesis test at a 5% significance level to determine if there is a difference in the academic performance of the two groups (divorced parents vs. nondivorced).

12.26 A computer firm compared the performance of four of its compilers. Five different programs were tested. The time required to compile each of the five programs was recorded. The observations are given in seconds. Do the data provide sufficient evidence to indicate that there is a difference in the performance of the compilers at the .05 significance level? Is there a significant difference due to blocks?

	Compiler			
	1	2	3	4
Program 1	31.103	24.315	33.058	22.013
Program 2	30.111	25.216	34.698	21.001
Program 3	29.903	25.347	33.872	20.314
Program 4	30.013	24.136	35.671	24.316
Program 5	31.981	24.977	34.751	22.591

12.27 In a randomized block design, if the degrees of freedom for the error sum of squares is given as 12 and the degrees of freedom for the factor sum of squares is 3, can you find the number of blocks used in the experiment? If yes, how many were used? If the total sum of squares is given as 520, the error sum of squares as 110, and the sum of squares due to blocks as 280, can you find the F test for this experiment? If yes, what is it?

12.28 The Green Thumb lawn-care company is testing three different formulas for a fertilizer especially designed for lawns in Denton County. To adjust for variation in the soil, the formulas are tested in 13 locations. The growth rate of the grass is recorded. Do the coded results indicate a difference in the lawn growth due to the formulas at the .05 level of significance?

Location	Formula 1	Formula 2	Formula 3
1	3.1	3.0	2.7
2	2.6	2.4	2.5
3	2.9	2.1	2.3
4	3.5	3.4	3.1
5	3.8	3.7	3.2
6	2.9	2.5	2.6
7	3.1	3.3	3.2
8	3.4	2.9	3.1
9	3.1	3.2	3.2
10	3.3	3.0	2.9
11	2.7	2.4	2.2
12	2.8	2.3	2.1
13	3.4	3.0	3.1

12.29 The following table shows the reading scores obtained by students on a standardized test of verbal ability. The scores are in blocks from the first to eleventh grades and are classified by race.

Grade	White	Hispanic	Black	Asian	TOTAL
1st	82	76	75	69	302
2nd	88	80	76	81	325
3rd	81	64	60	80	285
4th	82	58	62	70	272
5th	77	63	47	85	272
6th	82	55	69	67	273
7th	75	67	72	75	289
8th	69	59	56	77	261
9th	78	70	39	80	267
10th	84	60	44	71	259
11th	68	55	50	70	243
	866	707	650	825	3048

Assume the requirements of a randomized block design have been met. Compute the ANOVA table for the above data, and test the hypothesis that there is no difference between reading scores for the four racial groups. Use $\alpha = .05$. State the p value for the test.

12.30 The study in exercise 12.7 was modified such that only 6 secretaries were used. Each secretary had a different typing speed. Each secretary tested all the three word-processing software packages. The same task could not be used for testing all 3 packages, since the "learning-effect" would come into play, so each secretary performed three separate tasks. However, the tasks were of essentially the same length and difficulty level. Furthermore, which task was assigned to which word processor was randomly determined, and the order in which the three word processors were tested was also randomly decided. The secretaries relaxed between tasks to avoid "fatigue effects." Thus, a randomized block design was achieved, with secretaries constituting blocks and the 3 observations (levels of the factor) within each block being randomized. The following data were obtained (the secretary's typing speed in words per minute is given in parentheses for reference purposes, and the body of the table contains time taken to complete the tasks):

Secretary	Group 1 (Menu)	Group 2 (Command)	Group 3 (Mixed)
1 (75 wpm)	9	10	7
2 (65 wpm)	12	11	9
3 (55 wpm)	12	14	11
4 (50 wpm)	13	13	11
5 (45 wpm)	16	15	13
6 (30 wpm)	18	16	15

(a) Compute the ANOVA table for the above data.

(b) Conduct a hypothesis test to address the question: Is there a significant difference between the three word processors (as measured by the performance of the secretaries)? Use $\alpha = .10$.

(c) Determine the p value. Does the conclusion change at $\alpha = .05$ and at $\alpha = .01$?

(d) Is the block (secretary's) effect significant at $\alpha = .01$?

12.31 Suppose the following results were given from a computer printout: SS(factor) = 293.1, SS(blocks) = 5160.2, and SS(error) = 170.2. Assume four factor levels and ten blocks were used. Develop the ANOVA table and test for the effect of the factor and also for the effect of blocks. Use a significance level of .05.

12.32 Although there is no universal agreement in sports academia as to what constitutes a "good athlete," the literature seems to indicate that psychobiomotor characteristics (a combination of psychological, biophysical and motor skills) may help in identifying athletic abilities. A study by Secunda *et al.* (1986) attempted to discover useful psychobiomotor predictors of football-playing ability among 19 college football players trying out for halfback and fullback positions on the inaugural football team of the University of Central Florida. Psychological variables were measured by Catell's 16 Personality Factors Test (16PF). Anaerobic power, which is the maximum amount of energy that the body can provide in a very short period of time, was the major biological factor considered. Motor-skill factors included in the investigation were running speed, perceptual-motor speed, and football pass-receiving ability. Our focus in this exercise is not on these variables. The purpose of all these complicated variables was to predict football-playing ability, so the question remains, how was this to be measured? From a task analysis of the offensive backfield position, 15 sub-variables were rated on their relative importance for this position, for example, ability to follow plays, rushing ability, stamina, blocking ability, overall speed, and so on. These formed the basis of a football skills checksheet, which was used by three different coaches to rate each player. The idea in this aspect of the investigation was to assess the reliability of the coaches' ratings. With 19 players and 3 coaches, a randomized block ANOVA design was obtained, wherein coaches were the factor, and players were the block variable. The study provided the following

ANOVA table, some parts of which have been intentionally left out for you to fill in:

Source	SS	df	MS	F	p value
Raters (Coaches)	103.37		51.685		
Subjects (Players)	151.75		8.431		
Residual (Error)	35.24		.979		
Total	290.36	56			

Source: M. D. Secunda, B. I. Blau, J. M. McGuire, and W. A. Burroughs, Psychobiomotor assessment of football-playing ability. *International Journal of Sports Psychology.* **17** (3): 215–233, (1986) Table III.

(a) Complete the above ANOVA table by calculating the relevant degrees of freedom, the F statistics for the factor and the block variables, and also reporting the p value associated with each F ratio.

(b) Is there a significant difference between the coaches' ratings? Let $\alpha = .01$ for the test.

(c) Would a significant difference in coaches' ratings suggest some difficulties with measuring the criterion (dependent) variable, namely, football-playing ability?

12.33 Anorexia nervosa and bulimia nervosa are eating disorders characterized by an abnormal attitude toward eating behavior. There is some suspicion that normal taste function is impaired in these disorders. Nakai *et al.* (1987) studied the taste function in patients with these disorders. Using a procedure they called the "filter paper disc method," the researchers obtained taste recognition scores for the patients. Filter paper discs, 0.5 cm in size, were immersed in five different concentrations of solutions of four taste substances: sucrose, for a sweet taste; sodium hydrochloride, for a salty taste; tartaric acid, for a sour taste; and quinine, for a bitter taste. The discs were placed on the subject's tongue with a forceps, after rinsing the mouth with distilled water. Each correctly identified solution earned one point on a recognition score scale, with a maximum of 20 points. Confusing one taste for another was counted toward a dysgeusia score; a score higher than 3 in this respect was classified as dysgeusia. Of the patients studied, seven anorexia patients were tested again after behavioral and family therapy. These seven anorexics had shown a substantial improvement in weight gain after treatment. The recognition and dysgeusia scores of these seven anorexics before and after the therapy are summarized below.

ANOREXIA PATIENTS	RECOGNITION SCORES		DYSGEUSIA SCORES	
I.D. NUMBER	BEFORE	AFTER	BEFORE	AFTER
2	12	13	5	4
8	8	13	4	1
9	11	12	2	8
12	6	16	5	0
16	8	19	4	0
19	11	10	1	0
21	6	15	7	0

Source: Y. Nakai, F. Kinoshita, T. Koh, S. Tsujii and T. Tsukada, Taste function in patients with anorexia nervosa and bulimia nervosa. *International Journal of Eating Disorders.* **6** (2): 257–265 (March 1987) Tables 3 and 5.

(a) The weight gain experienced by these anorexics signified an improvement in their condition after treatment. If it was shown that these patients also experienced an improvement in taste function, would you say that the study has been successful in establishing a significant role played by taste function impairment in anorexia? Could you generalize these results to bulimia? (Note: You may assume the assumptions underlying ANOVA are satisfied. Actually, they may not be satisfied, but the above set-up is nevertheless fairly illustrative of a randomized block design, sometimes called a "repeated measures" design.)

(b) Use the ANOVA technique at 5% significance to decide if there was a significant change in the taste recognition scores after treatment. Interpret the results.

(c) What is the 95% confidence interval for the difference in taste recognition scores after and before treatment?

(d) Use the ANOVA technique at 1% significant to decide if there was a significant change in the dysgeusia scores after treatment. Does this corroborate and support the conclusions in (a) and (b)?

12.34 According to Hernandez *et al.* (1987), "plants are important in the biogeochemical cycle of heavy metals: lead associated with leaves or other deciduous tissue is recycled relatively fast." As part of a study designed to determine the environmental contamination by lead and cadmium in certain plants, the researchers obtained measurements of lead in rose-bay leaves (*Nerium oleander*) from 15 locations in the city of Madrid, Spain. After collecting, washing, drying, grinding, reducing, dissolving and filtering the leaf material, the researchers determined the residues of metals in $\mu g/g$, i.e., parts per million, by means of a Perkin–Elmer Model 2280 atomic absorption spectrophotometer. Measurements were taken in winter, spring, summer and autumn to observe seasonal changes, if any, as shown below:

THE EFFECT OF THE SEASONS ON THE LEAD CONTENT OF ROSE-BAY LEAVES

Place of collection	Winter	Spring	Summer	Autumn
Cuzco Square	52.3	30.4	18.6	41.2
Nuevos Ministerios	81.0	78.2	28.9	56.9
Neptuno Square	79.5	40.2	26.0	41.7
Atocha Square	74.1	62.4	46.0	61.0
María Ana de Jesus Square	36.3	21.5	16.6	23.1
Ruiz Giménez Square	81.4	76.2	22.7	48.2
Cristo Rey Square	77.4	39.9	32.0	58.9
España Square	115.0	100.5	31.6	75.0
Casa de Campo Park	19.1	12.2	12.2	14.8
Alonso Martínez Square	49.0	30.5	30.7	50.6
Roma Square	60.9	34.1	20.1	35.4
Conde de Casal Square	94.8	52.0	30.3	61.0
Retiro Park	20.0	11.7	10.5	14.7
República del Ecuador Square	80.8	73.9	44.9	71.3
Serrano St./María de Molina St.	93.6	48.2	62.7	72.6

Source: L. M. Hernandez, Ma. J. Gonzales and Ma. A. Hernan, Environmental contamination by lead and cadmium in plants from urban area of Madrid, Spain. *Bulletin of Environmental Contamination and Toxicology*. **38** (2): 203–208, (February 1987) Table 1.

(a)   Explain how you would recognize the above as a block design ANOVA. Identify the block and the treatment.

(b)   Assume the requirements of the block design are met. Compute the ANOVA table for the above data, and perform a hypothesis test to decide if the change in the seasons has a significant effect on the lead content of rose-bay leaves. Use $\alpha = .05$. State the p value.

(c)   Find the 95% confidence interval for:

(i)   the average lead content of the leaves in winter.

(ii)   the difference between the average lead content of the leaves in winter and summer.

12.35   An old issue in the social and behavioral sciences is the relationship between a sense of power and control within individuals and their position in a social hierarchy. The individualistic point of view would explain success in terms of personal feelings, attitudes and values. The socialistic point of view would explain personal values in terms of social position. Julian Rotter developed an instrument, commonly called Rotter's "locus of control" scale, that purports to measure whether an individual is internally directed (i.e., the individual feels he or she determines the course of his or her life) or externally directed (i.e., the course of life is externally determined by forces outside the individual's control). Researchers had reason to suspect that there was some difference between the scores of black males and white males on Rotter's locus of control personality scale. Thus, in an experiment dealing with locus of control and income levels of males, the investigators decided to block the scores by race, as shown below. The scores have a theoretical range from 11 to 44, with lower scores suggesting internal direction and high scores suggesting external direction.

| | Locus of Control Scores | |
Income Level	Black Males	White Males
< $20,000	30.5	26.1
$20,000 to < $35,000	28.0	23.4
$35,000 to < $50,000	24.8	22.5
$50,000 to < $65,000	22.3	22.0
⩾ $65,000	21.5	20.2

(a)   Do the above data suggest a significant difference between scores on the Rotter scale for the various income levels? Use $\alpha = .05$.

(b)   Find the p value for the test.

(c)   Is there a significant difference between black and white male locus of control scores, at $\alpha = .05$?

12.5   THE TWO-WAY FACTORIAL DESIGN

The two-way factorial design was introduced in Section 12.3. For this type of experiment, the researcher is considering two factors of interest, say factor A and factor B. Of concern will be whether the individual factors have a significant effect on the observed variable (called the dependent variable) as well as the combined effect of the two factors.

Figure 12.8

Scores on
assertiveness/
managerial potential
exam

	Single	Married
Male	Low	High
Female	High	Low

Consider a simple example where the dependent variable is the score on a test designed to measure assertiveness and managerial potential. The factors are sex and marital status (single or married). Each of these two factors consists of two *levels*. Suppose we observed a significant difference between the male and female scores. Thus we would conclude that factor A (sex) is significant. The analysis procedure to investigate this hypothesis will be described in this section. If a significant difference between the scores of the single and married subjects is observed, then we would conclude that factor B (marital status) is also significant.

Suppose that a closer look at the scores revealed that the married males and single females scored high, but the single males and married females scored low on the test, as illustrated in Figure 12.8.

Consequently the relationship between sex and the dependent variable (exam score) *depends upon the marital status*, since this relationship is different for the single and married groups. Similarly, the relationship between marital status and the dependent variable depends upon the particular level of factor A (sex). This is an illustration of **interaction** between factors A and B. A method of detecting interaction using a simple graph, along with a statistical test of hypothesis, will be explained.

Degrees of Freedom

In a two-way factorial design, each level of factor A is combined with each level of factor B when obtaining the sample data. Suppose that factor A has a levels and factor B has b levels, as shown in Figure 12.9. Each x represents a test score.

If we record one observation for each factor A and factor B combination (referred to as a **treatment**), then we have $n = ab$ total observations. The degrees of freedom (df) for each factor is one less than the number of levels and the degrees of freedom for the interaction term is the product of the factor A df and the factor B df. Consequently,

df for factor A $= a - 1$

df for factor B $= b - 1$

df for interaction $= (a - 1)(b - 1)$

df for total $= n - 1 = ab - 1$

We have a bit of a problem here. This design, like all experimental designs, must contain a source of variation due to error, that is, the unexplained

Figure 12.9

Layout for two-way
factorial design

Factor B

		1	2	...	b
Factor A	1	x	x		x
	2	x	x		x
	⋮				
	a	x	x		x

variation. Suppose $a = 4$ and $b = 3$. Then the remaining df for error is (df for total) − (df for factor A) − (df for factor B) − (df for interaction), which in this case is $11 - 3 - 2 - 6 = 0$. It can be shown that the error df is zero regardless of the values of a and b. Since it will be necessary to measure this unexplained variation, this design requires that you obtain repeat observations (**replicates**) for each treatment. An illustration (using two replicates) including the various totals needed to carry out the analysis is shown in Figure 12.10. In general you will need two or more replicates at each treatment. The number of replicates at each treatment need not be the same, but we consider here only the case where there are r replicates at each treatment.

In the replicated design, the degrees of freedom are

df for factor A $= a - 1$

df for factor B $= b - 1$

df for interaction $= (a - 1)(b - 1)$

df for total $=$ (number of observations $- 1$)
$\qquad = abr - 1$

Now the df for error $= (abr - 1) - (a - 1) - (b - 1) - (a - 1)(b - 1)$
$\qquad = ab(r - 1).$

Figure 12.10

Illustration of two
replicates in a
two-way factorial
design ($r = 2$)

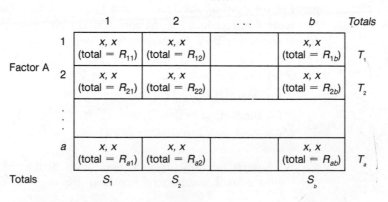

Sums of Squares and Mean Squares

The necessary sums of squares can be computed in a manner similar to that used in the previous designs. Using Figure 12.10, the following expressions can be used to find the corresponding sums of squares.

$$\text{factor A: } SSA = \frac{1}{br}\left[T_1^2 + T_2^2 + \cdots + T_a^2\right] - \frac{T^2}{abr} \quad \textbf{(12-25)}$$

where T = total of all n observations (that is, $T = T_1 + T_2 + \cdots + T_a$).

$$\text{factor B: } SSB = \frac{1}{ar}\left[S_1^2 + S_2^2 + \cdots + S_b^2\right] - \frac{T^2}{abr} \quad \textbf{(12-26)}$$

$$\text{interaction: } SSAB = \frac{1}{r}\left[\Sigma R^2\right] - SSA - SSB - \frac{T^2}{abr} \quad \textbf{(12-27)}$$

where the sum in the brackets is the sum of all the squares of the replicate totals, illustrated in Figure 12.10.

$$\text{total: } SS(\text{total}) = \Sigma x^2 - \frac{T^2}{abr} \quad \textbf{(12-28)}$$

where Σx^2 is the sum of the squares for each of the $n = abr$ observations. By subtraction,

$$SS(\text{error}) = SS(\text{total}) - SSA - SSB - SSAB \quad \textbf{(12-29)}$$

The corresponding mean squares can be obtained by dividing each sum of squares by the corresponding degrees of freedom. Thus we have

$$MSA = SSA/(a-1) \quad \textbf{(12-30)}$$

$$MSB = SSB/(b-1) \quad \textbf{(12-31)}$$

$$MSAB = SSAB/[(a-1)(b-1)] \quad \textbf{(12-32)}$$

$$MS(\text{error}) = SS(\text{error})/[ab(r-1)] \quad \textbf{(12-33)}$$

This analysis can be summarized in the following ANOVA table:

Source	df	SS	MS	F
Factor A	$a - 1$	SSA	$MSA = \dfrac{SSA}{a - 1}$	$F_1 = MSA/MS(error)$
Factor B	$b - 1$	SSB	$MSB = \dfrac{SSB}{b - 1}$	$F_2 = MSB/MS(error)$
Interaction	$(a - 1)(b - 1)$	SSAB	$MSAB = \dfrac{SSAB}{(a - 1)(b - 1)}$	$F_3 = MSAB/MS(error)$
Error	$ab(r - 1)$	SS(error)	$MS(error) = \dfrac{SS(error)}{ab(r - 1)}$	
Total	$abr - 1$	SS(total)		

In Section 12.2, we examined the effect of marital status (using four levels) on general life satisfaction within the female population. Suppose that we want to expand this study by including the male population, thus introducing sex as another factor in the design. By including this factor, we can study the effect of sex (factor A) and marital status (factor B), as well as the interaction effect between sex and marital status, on the scores of the Life Satisfaction Index. This results in a 2 × 4 factorial design, since factor A consists of two levels and factor B has four levels. We decide to use three replicates for each of the eight treatment combinations, requiring scores from 24 different subjects. The sample results are shown below, where the values in parentheses are the replicate totals for each of the treatments.

Marital Status (factor B)

Sex (factor A)		Never Married	Ever Married	Remarried	Divorced	Total	Average
	Male	58, 65, 60 (183)	47, 56, 40 (143)	72, 86, 90 (248)	81, 75, 68 (224)	798	66.50
	Female	67, 58, 74 (199)	75, 81, 77 (233)	52, 61, 48 (161)	51, 42, 35 (128)	721	60.08
	Total	382	376	409	352	1519	
	Average	63.67	62.67	68.17	58.67		

Using the previous discussion, we can derive the necessary sums of squares.

$$SSA = \frac{1}{(4)(3)} [798^2 + 721^2] - \frac{1519^2}{24} = 247.04$$

$$SSB = \frac{1}{(2)(3)} [382^2 + 376^2 + 409^2 + 352^2] - \frac{1519^2}{24} = 274.125$$

$$SSAB = \frac{1}{3}[183^2 + 143^2 + 248^2 + 224^2 + 199^2 + 233^2 + 161^2 + 128^2]$$

$$- 247.04 - 274.125 - \frac{1519^2}{24}$$

$$= 3943.127$$

$$SS(total) = [58^2 + 65^2 + 60^2 + \cdots + 51^2 + 42^2 + 35^2] - 1519^2/24$$

$$= 5246.96$$

Consequently,

$$SS(error) = 5246.96 - 247.04 - 274.125 - 3943.127$$

$$= 782.67$$

The degrees of freedom here will be:

Sex factor: $a - 1 = 2 - 1 = 1$
Marital status factor: $b - 1 = 4 - 1 = 3$
Interaction: $(a - 1)(b - 1) = (1)(3) = 3$
Error: $ab(r - 1) = (2)(4)(3 - 1) = 16$
Total: $abr - 1 = (2)(4)(3) - 1 = 24 - 1 = 23$

These calculations and the resulting mean squares can be summarized in the following ANOVA table.

Source	df	SS	MS	F
Sex	1	247.04	247.04	$F_1 = 247.04/48.92 = 5.05$
Marital Status	3	274.125	91.375	$F_2 = 91.375/48.92 = 1.87$
Interaction	3	3943.127	1314.38	$F_3 = 1314.38/48.92 = 26.87$
Error	16	782.67	48.92	
Total	23	5246.96		

Hypothesis Testing

When using a two-way factorial design, you can test for the significance of factor A, factor B, and the interaction of the two factors. For factor A, the null hypothesis is that the means are equal across the factor A levels. Written another way, we can define the following hypotheses:

$H_{0,A}$:   factor A is not significant ($\mu_M = \mu_F$)
$H_{a,A}$:   factor A is significant ($\mu_M \neq \mu_F$)

The corresponding test statistic is

$$F_1 = \text{MSA/MS(error)} \qquad \textbf{(12-34)}$$

and the testing procedure is to reject $H_{0,\text{A}}$ if

$$F_1 > F_{\alpha, v_1, v_2}$$

where F_{α, v_1, v_2} is from Table A-7, $v_1 = df$ for factor A $= a - 1$, and $v_2 = df$ for error $= ab(r - 1)$.

Similarly, to test for equal means of the factor B levels, we can define the hypotheses:

$H_{0,\text{B}}$:   factor B is not significant ($\mu_1 = \mu_2 = \mu_3 = \mu_4$)

$H_{a,\text{B}}$:   factor B is significant (not all μ_i's are equal)

The test statistic for determining the factor B effect is

$$F_2 = \text{MSB/MS(error)} \qquad \textbf{(12-35)}$$

and factor B is significant ($H_{0,\text{B}}$ is rejected) if

$$F_2 > F_{\alpha, v_1, v_2}$$

where $v_1 = df$ for factor B $= b - 1$, and $v_2 = df$ for error $= ab(r - 1)$.

The final set of hypotheses is concerned with the interaction effect between the two factors. The hypotheses for this procedure can be stated

$H_{0,\text{AB}}$:   there is no significant interaction between factor A and factor B

$H_{a,\text{AB}}$:   there is significant interaction between factor A and factor B

The test statistic is the remaining F statistic in the ANOVA table, namely

$$F_3 = \text{MSAB/MS(error)} \qquad \textbf{(12-36)}$$

and the test procedure is to reject $H_{0,\text{AB}}$ if

$$F_3 > F_{\alpha, v_1, v_2}$$

where $v_1 = df$ for interaction $= (a - 1)(b - 1)$ and $v_2 = df$ for error $= ab(r - 1)$.

Multiple Comparisons

The method of multiple comparisons discussed in Section 12.2 for the one-factor ANOVA procedure (Tukey's method) can be used to examine pairwise differences between the various treatment means in two-way factorial designs. Since factor A has a levels and factor B has b levels, there are ab such means that can be compared, one pair at a time.

For the two-way factorial design, we use Table A-9, to find $Q_{\alpha,k,v}$, where α is the desired experimentwise significance level, $k = ab$ is the number of treatment means, and v is the degrees of freedom associated with MS(error). Any two (sample) treatment means differing by more than

$$D = Q_{\alpha,k,v} \cdot \sqrt{\frac{MS(error)}{r}}$$

will imply that the corresponding population means are unequal. This procedure will be illustrated in the next example.

EXAMPLE 12.8

Using the previous ANOVA table constructed using the Life Satisfaction Index and the two factors, sex (factor A) and marital status (factor B), determine whether (i) factor A is significant, (ii) factor B is significant, (iii) there is significant interaction between sex and marital status, and (iv) which pairs of the eight (population) treatment means are unequal. Use a significance level of .05.

Solution to (i)

The df for the F statistic are $v_1 = 1$ and $v_2 = 16$. Using Table A-7, $F_{.05,1,16} = 4.49$, and the test is to reject $H_{0,A}$ if $F_1 > 4.49$. Since $F_1 = 5.05 > 4.49$, we conclude that the sex factor is significant. Examining the raw data, we observe that the sample mean for the males is $798/12 = 66.5$, and the female average is $721/12 = 60.08$. Thus we conclude that the difference between these sample means *is* significant with higher life satisfaction occurring in the male population.

Solution to (ii)

For the marital status factor, the df for the F statistic are $v_1 = 3$ and $v_2 = 16$, with a corresponding table value of $F_{.05,3,16} = 3.24$. Since $F_2 = 1.87 < 3.24$, the marital status factor is *not significant*. This is in contrast to the conclusion reached in example 12.2, where this factor *was* significant for the female population. Actually, a significant result would be obtained here also if we performed a one-way ANOVA on the female scores only. However, in the combined male/female sample, we observe that the high (low) female scores were balanced by the low (high) male scores within each marital category. Consequently, there is insignificant variation in the means for the four marital groups, leading to the "fail to reject $H_{0,B}$" conclusion.

Solution to (iii)

The discussion in the solution to part (ii) indicates the presence of interaction between the two factors. This means that the relationship between marital status and life satisfaction is not the same for males and females. The four male means are $183/3 = 61.00$ (never-married), $143/3 = 47.67$ (ever-married), $248/3 = 82.67$

Figure 12.11

Illustration of interaction effect. (A) interaction effect in example 12.8. (B) hypothetical situation containing no significant interaction between sex and marital status.

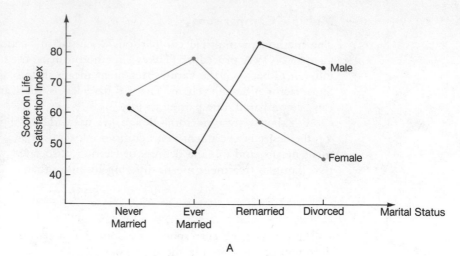

A

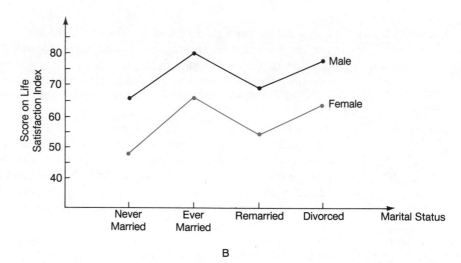

B

(remarried), and $224/3 = 74.67$ (divorced). The corresponding means for the female sample are 66.33, 77.67, 53.67, and 42.67. These means are shown in Figure 12.11A, where interaction effect is very apparent since the male and female lines *are not parallel*. When no interaction exists between the two factors, this graph should contain lines that are *nearly parallel*, as illustrated in Figure 12.11B.

The statistical test here supports this conclusion, since there is significant interaction provided F_3 is larger than $F_{.05,3,16} = 3.24$. Here, $F_3 = 26.87$ and so we once again conclude that there is significant interaction between sex and marital status for this population.

Figure 12.12

MINITAB solution for two-way factorial design in example 12.8

```
MTB > READ INTO C1-C3
DATA> 58 1 1  ◄─────────
DATA> 65 1 1              ┌─────────────────────────────────┐
DATA> 60 1 1              │ data are in column C1            │
DATA> 47 1 2              │ level of factor A in column C2   │
DATA> 56 1 2              │ level of factor B in column C3   │
DATA> 40 1 2              └─────────────────────────────────┘
DATA> 72 1 3
DATA> 86 1 3
DATA> 90 1 3
DATA> 81 1 4
DATA> 75 1 4
DATA> 68 1 4
DATA> 67 2 1
DATA> 58 2 1
DATA> 74 2 1
DATA> 75 2 2
DATA> 81 2 2
DATA> 77 2 2
DATA> 52 2 3
DATA> 61 2 3
DATA> 48 2 3
DATA> 51 2 4
DATA> 42 2 4
DATA> 35 2 4
DATA> END
     24 ROWS READ
MTB > TWOWAY USING DATA IN C1, A LEVELS IN C2, B LEVELS IN C3

    ANALYSIS OF VARIANCE  C1

    SOURCE          DF        SS         MS
    C2               1      247.0      247.0
    C3               3      274.1       91.4
    INTERACTION      3     3943.1     1314.4
    ERROR           16      782.7       48.9
    TOTAL           23     5247.0
```

The MINITAB solution for this example is contained in Figure 12.12. Notice that the same TWOWAY command used for the randomized block design is used for the two-way factorial design. Due to the presence of replications, MINITAB assumes a possible interaction effect and includes this line in the ANOVA table.

Solution to (iv)

A multiple comparisons analysis will determine if the average life satisfaction for never-married males is the same as for divorced males, the average for ever-married males is the same for divorced females, and so forth. There are $2 \cdot 4 = 8$ means here, providing $_8C_2 = 28$ possible pairwise comparisons. In general for a two-way factorial design there are $_{ab}C_2$ possible pairs of means that can be compared using the multiple comparison procedure.

The critical value corresponding to $\alpha = .05$, $k = 8$, and $\nu = 16$ (the df associated with the error sum of squares) from Table A-9 is $Q_{.05,8,16} = 4.90$.

Since MS(error) = 48.92 and there are $r = 3$ replicates at each treatment level, we next determine

$$D = Q_{\alpha,k,v} \cdot \sqrt{\frac{MS(error)}{r}}$$

$$= (4.90) \sqrt{\frac{48.92}{3}}$$

$$= 19.79$$

Consequently any pair of sample treatment means differing by more than 19.79 will imply that the corresponding population means are unequal.

The eight sample means are obtained by dividing the corresponding replicate totals (R) by $r = 3$. Placing them in order, we obtain

42.67	47.67	53.67	61.00	66.33	74.67	77.67	82.67
(F, D)	(M, EM)	(F, R)	(M, NM)	(F, NM)	(M, D)	(F, EM)	(M, R)

Here, M and F represent the sex (factor A) and EM, NM, R, and D represent the marital category (factor B).

Since $61.00 - 47.67 = 13.33 < 19.79$, we *cannot* conclude that $\mu_{M,NM} \neq \mu_{M,EM}$. This is represented by the overbar connecting these two sample means. Consider the divorced males and divorced females. Here $74.67 - 42.67 = 32.00 > 19.79$, and so we conclude that there *is* a difference in average life satisfaction for these two groups; that is, $\mu_{M,D} \neq \mu_{F,D}$. This can be observed in the above sample means, since there is no overbar connecting these two means. Continuing this procedure, we arrive at the following summary for the multiple comparisons analysis:

Males only: $\mu_{M,EM} \neq \mu_{M,D}$, $\mu_{M,EM} \neq \mu_{M,R}$, $\mu_{M,NM} \neq \mu_{M,R}$

Females only: $\mu_{F,D} \neq \mu_{F,NM}$, $\mu_{F,D} \neq \mu_{F,EM}$, $\mu_{F,R} \neq \mu_{F,EM}$

Males and females: $\mu_{F,D} \neq \mu_{M,D}$, $\mu_{F,D} \neq \mu_{M,R}$, $\mu_{M,EM} \neq \mu_{F,EM}$, $\mu_{F,R} \neq \mu_{M,D}$, $\mu_{F,R} \neq \mu_{M,R}$

Consequently, we observe three significant differences in each of the male and female populations, and five significant differences in life satisfaction when comparing marital categories across both sexes.   □

EXERCISES

12.36 In exercise 12.25, the one-factor design of exercise 12.2 was extended to a randomized block design, with schools acting as the block variable. Suppose now that the investigators wish to consider whether parental divorce might affect girls and boys differently. In order to do this, they implemented the two-way factorial design, with sex as one factor and family status as another factor. Three students' GPAs were observed for each sex and family status combination. The data are summarized below.

Family Status	Sex Boys	Girls
Parents divorced within last year	2.2, 1.5, 3.0	2.8, 2.0, 2.7
Parents not divorced within last year	2.9, 3.3, 3.1	3.2, 3.0, 2.8

(a) How many treatments are there?

(b) How many replicates are there for each treatment?

(c) Compute the ANOVA table for the above design.

(d) At a 5% significance level is the family status factor significant? State the p value.

(e) At a 5% significance level, is the sex factor significant? State the p value.

(f) At a 5% significance level, is there a significant interaction effect, i.e., are the GPAs different for boys and girls dependent on whether or not the parents were divorced? State the p value.

12.37 The comparative study of word-processing software in exercise 12.30 was modified to take into account different types of keyboards: enhanced keyboard, modified keyboard, and standard keyboard. Keyboard layout and type of software could not be assumed to be independent, because it was possible that a certain type of software might actually be enhanced by a certain type of keyboard (e.g., one with special function keys). In other words, interaction between factors was possible. Therefore, a 3 × 3 factorial design was implemented. The table below gives the completion time in minutes, with three observations for each treatment "cell."

Keyboard Type	Software Type Group 1 (Menu)	Group 2 (Command)	Group 3 (Mixed)
Enhanced	9, 8, 10	8, 7, 7	8, 10, 10
Modified	14, 14, 13	10, 14, 12	12, 10, 14
Standard	15, 18, 17	18, 16, 15	15, 15, 14

(a) Calculate the ANOVA table for the above experiment.

(b) Assume that the assumptions of an ANOVA have been satisfied. Is there a significant difference in the three word processors, as measured by the productivity of the secretaries? Use $\alpha = .05$.

(c) Do the different keyboards seem to affect productivity, as measured by completion times? Use $\alpha = .05$.

(d) Is there a significant interaction between software type and keyboard type, at $\alpha = .05$?

(e) For parts (b), (c), and (d), find the corresponding p value for each test.

12.38 The government agency in exercise 12.11 also considered an alternative experimental design, which would take into account weather conditions. They were interested not so much in the weather effects alone, since it was reasonable to expect that lower operating temperatures would affect the efficiency of all the catalytic converters; rather, they were interested in the interaction between the two factors. For example, at normal temperatures, all three converters might be equally efficient, but at low temperatures, there might be a difference in pollution readings for different converters. Three temperature levels were chosen as a second factor, with the three converters as the first factor, in a 3 × 3 two-way factorial design. Two cars were observed for each of the nine treatments, for a total of 18 observations. It was assumed that the populations were normally distributed, and that other assumptions of an ANOVA design were satisfied for the following data on residual pollutants:

Temperature	Catalytic Converter		
	"C"	"F"	"M"
Cold	2.5, 2.8	2.2, 2.3	1.9, 2.0
Normal	1.8, 2.0	1.8, 1.7	1.9, 1.7
Hot	1.2, 1.5	1.4, 1.6	1.6, 1.5

(a) At $\alpha = 0.10$, is there a difference between the efficiency of the converters?

(b) At $\alpha = 0.10$, is there a significant interaction between temperature levels and converter efficiency?

(c) Find the p values for the above tests, and determine if the conclusions change at $\alpha = .01$ or $\alpha = .05$.

(d) Is it still necessary to ensure that cars are the same make and year models, and use the same fuel? If these variables were not controlled, how would it affect the experiment?

12.39 Consider exercise 12.15 again. It is known that there are some religious taboos against birth control and abortion. Suppose that the researchers decide to introduce another factor, religion, to be considered at three levels, namely, Catholic, Protestant, and non-Christians.

(a) If it is possible to assume that alienation and religion are independent, would the randomized block design be acceptable? Show in a block diagram what the data would look like.

(b) Consider the possibility that religious faith might reinforce alienation, or conversely, that it might mitigate alienation. Thus, the two factors could interact, and might not be independent. Show with a block diagram how a two-way factorial design might be set up in this case.

(c) In the randomized block design, you would have a total of 9 observations only. Why would a complete factorial design have at least 18 observations?

(d) Why would observations in a complete factorial design be increased in multiples of 9, that is, 18, 27, 36, and so on?

12.40 The traditional approach to teaching has been didactic, that is, to have the teacher assume responsibility for imparting knowledge. In recent years, the idea of children working cooperatively in groups of two or more for instruction and problem-solving, the idea of students teaching students, has enjoyed some popularity. Buckholdt and Wodarski (1978) studied the effects of different systems of reinforcement on such cooperative behavior by means of a 2 × 2 factorial ANOVA design. One factor used as an independent variable was reinforcement method, which was either "non-contingent" (children received reinforcement merely for participating in the experiment) or "group-contingent" (children received points based on the average score of the group). A second factor in the experiment was training in tutoring. One group of children received training in how to recognize need for help, how to ask for help, how to search for others who may need assistance, and how to provide feedback and support for other students. Another group did not receive any training in tutoring. The dependent variable used was a measure of reading comprehension. A portion of the ANOVA table for the experiment is given below.

Source	df	MS
Factor A (Reinforcement)	1	276.14
Factor B (Tutoring)	1	85.55
A × B Interaction	1	28.75
Error	36	18.59

(*Source:* D. R. Buckholdt and J. S. Wodarski, The effects of different reinforcement systems on cooperative behaviors exhibited by children in classroom contexts. *J. of Research and Development in Education.* **12** (1): 50–67 (1978) Table 1.

(a) What is the sum of squares for the reinforcement factor?

(b) What is the total sum of squares in the experiment?

(c) What is the total number of observations?

(d) How many treatments are there? Identify each of them.

(e) How many replications are there in each treatment cell?

(f) Complete the ANOVA table by obtaining the F values for the two factors and the interaction, and the associated p values.

(g) At 1% significance, is the effect of reinforcement significant?

(h) At 5% significance, is the effect of training in tutoring significant?

(i) At 5% significance, is there a significant interaction between reinforcement and training in tutoring?

12.41 Numerous studies have been conducted on the link between blood cholesterol levels and the risk of heart attacks. The following data represent a simplified version of one large study. We will consider two age groups (35–39 and 55–57) and two cholesterol level (181 or lower, and 245 or higher), although the actual study was more complex. For each of the four treatments in the simplified 2 × 2 factorial design, there were three groups under observation. Over a period of 6 years, the death rates were observed for each treatment group. The death rates per thousand persons are given below, for deaths caused by heart disease.

Age	Cholesterol Level (mg/dl of blood)	
	≤181	≥245
35–39 years	0.59, 0.64, 0.60	4.57, 5.0, 4.75
55–57 years	8.31, 9.0, 8.73	15.78, 16.2, 15.6

(a) Is age a significant factor in determining the risk of death by heart disease? Use $\alpha = .05$, and find the p value.

(b)   Is there a significant difference in death rates from heart disease due to the high and low cholesterol levels in the blood? Use $\alpha = .05$, and find the p value.

12.42   An older view of "interpersonal distance" (IPD) was that an individual's personal space consisted of a series of concentric circles that remained more or less constant. Later research has moved toward a dynamic view which postulates that IPD can shrink or expand depending on a number of factors. Tennis and Dabbs (1975) studied the influence of age, sex and setting (corner vs. center) on IPD. We shall examine here just the effect of two factors, age and sex, on IPD preferences among first, fifth, ninth and twelfth grade public school students, and college sophomores. The researchers used a factorial ANOVA design. The ANOVA table is shown below:

Source	SS	df	MS	F	p value
Age	1342276	4			
Sex	83232	1			
Age × Sex	533020	4			
Error	1802250	90			

Source: G. H. Tennis and J. M. Dabbs, Jr., Sex, setting and personal space: First grade through college. Sociometry. **38** (2): 385–394 (1975) Table 1.

(a)   How many levels of the factor Age were used, and what were they?

(b)   What is the total number of observations taken?

(c)   How many treatments were there, and how many replicates were used for each treatment?

(d)   Complete the ANOVA table by calculating the mean squares, F ratios and the p values.

(e)   Is the effect of Age on IPD significant, at $\alpha = 0.05$?

(f)   Is the effect of Sex on IPD significant, at $\alpha = 0.05$?

(g)   Is there a significant interaction between Age and Sex, at $\alpha = 0.05$?

12.6   COMPARING MORE THAN TWO POPULATIONS: THE KRUSKAL–WALLIS TEST (OPTIONAL)

When comparing the means of more than two populations, a popular technique is the ANOVA procedure discussed in Section 12.2. The assumptions behind this technique include that you are dealing with normally distributed populations; the F test used in the ANOVA table is invalid unless all of the populations are nearly normally distributed with equal variances.

The nonparametric counterpart to the one-way ANOVA method is the **Kruskal–Wallis test**. It is named after W. H. Kruskal and W. A. Wallis, who published their results in 1952. This test, like many other nonparametric procedures, is relatively new, unlike most of the parametric hypothesis tests, which were developed much earlier. The assumption of normal populations is not necessary for the Kruskal–Wallis test, which makes it an ideal technique for samples exhibiting a nonsymmetric (skewed) pattern. It is also less sensitive than the ANOVA procedure to the assumption of equal variances. The test also is useful when the data consist of rankings (ordinal data) within each sample.

The assumption of normal populations becomes quite critical when dealing with small samples. As we've seen in many of the earlier tests of hypothesis, this assumption can be relaxed when larger samples are used due to the Central Limit Theorem. However, many experiments in the medical and behavioral sciences use human subjects in studies that are potentially dangerous, very time-consuming, or perhaps of a delicate nature. Many experiments of a business nature dealing with product comparisons result in the destruction of the product being tested. Consequently, small samples are often a necessity for such experiments and nonparametric techniques are widely used to analyze the resulting data.

The Kruskal–Wallis test is actually an extension of the Mann–Whitney U test for two independent samples discussed in Section 10.4. Both procedures require that the sample values have a measurement scale that is at least ordinal, which means that each sample can be ranked from smallest to largest.

The hypotheses for this situation are similar to the Mann–Whitney hypotheses in that they are stated in terms of differing population location. The Kruskal–Wallis hypotheses are

H_0: the k populations have identical probability distributions

H_a: at least two of the populations differ in location.

Procedure

You first obtain random samples of size $n_1, n_2, \ldots, n_k$ from each of the k populations. The total sample size is $n = n_1 + n_2 + \cdots + n_k$. As with the Mann–Whitney procedure, you next pool the samples and arrange them in order, assigning a rank to each. For ties, you assign the average rank to the tied positions.

Let $T_i =$ the total of the ranks from the ith sample. The Kruskal–Wallis test statistic (KW) is

$$\text{KW} = \frac{12}{n(n+1)} \sum_{i=1}^{k} \frac{T_i^2}{n_i} - 3(n+1) \qquad (12\text{-}37)$$

The distribution of the KW statistic approximately follows a chi-square distribution with $k-1$ degrees of freedom (df). This approximation is good even if the sample sizes are small. The chi-square distribution was first introduced in

Chapter 9 when testing the variance of a normal population. Chi-square probabilities (areas) are contained in Table A-6 and, as the figure in this table illustrates, the chi-square random variable is nonnegative, with a shape that is skewed right, much like the F distribution.

To test H_0 versus H_a, the procedure will be to

Reject H_0 if KW is "large";

that is, if KW is in the right tail of the chi-square curve. This right-tail critical value is obtained from Table A-6 using a significance level $= \alpha$ and $df = k - 1$.

The Kruskal–Wallis Test

Hypotheses

H_0: the k populations have identical probability distributions

H_a: at least two of the populations differ in location.

Assumptions

1 Random samples are obtained from each of the k populations.

2 The individual samples are obtained independently.

3 Values within each sample can be ranked.

Procedure

The individual samples are pooled and then ranked from smallest to largest. Letting $T_i =$ the sum of the ranks of the ith sample items, the KW statistic is determined using equation 12-37.

The null hypothesis, H_0, is rejected if

$$ \text{KW} > \chi^2_{\alpha, df} $$

where $\chi^2_{\alpha, df}$ is the value from Table A-6 corresponding to $df = k - 1$, with a right-tail area $= \alpha$.

Table 12.2

Life Satisfaction Index scores and ranks for example 12.9

Never Married	Rank	Ever Married	Rank	Remarried	Rank	Divorced	Rank
41	9	66	18	43	10	47	11
54	15	49	13	28	4	36	7
77	21	82	22	51	14	20	1
95	24	62	16	48	12	25	3
64	17	75	20	30	5	31	6
69	19	93	23	23	2	38	8
	$T_1 = 105$		$T_2 = 112$		$T_3 = 47$		$T_4 = 36$

EXAMPLE 12.9

In example 12.2, we examined the difference among the Life Satisfaction Index scores for four groups of women, using marital status as the factor of interest. Suppose we had little experience using this instrument, and are reluctant to assume that the four populations were normally distributed. Since small samples (six in each) were used, we decide to test for group differences using the Kruskal–Wallis test. The data are repeated in Table 12.2.

Do these data indicate a difference in the level of life satisfaction for the four groups of women using a significance level of .05?

Solution

There are $k = 4$ populations here, so we need the $\chi^2_{.05,3}$ value from Table A-6. Based on this value, the testing procedure is to

$$\text{Reject } H_0 \text{ if KW} > \chi^2_{.05,3} = 7.81$$

From Table 12.2, we are able to compute the value of the KW statistic using the ranks of the observations in the pooled sample

$$\text{KW} = \frac{12}{(24)(25)}\left[\frac{105^2}{6} + \frac{112^2}{6} + \frac{47^2}{6} + \frac{36^2}{6}\right] - 3(25) = 15.247$$

As a check of your calculations at this point, make sure the ranks sum to $n(n + 1)/2$. For this example, $n = $ total number of observations $= 24$, and so the total of the ranks should be $(24)(25)/2 = 300$; here, $105 + 112 + 47 + 36 = 300$.

The calculated KW value exceeds 7.81, and so our conclusion is the same as that arrived at using the one-way ANOVA procedure, namely there *is* a difference in life satisfaction among the four groups of women. From the small values of T_3 and T_4, it appears that there is greater life satisfaction among the never-married and ever-married women within this particular population.

Finally, the p value here is $<.005$, indicating a very strong conclusion. In other words, these data indicate a clear difference among the four groups. This p value is illustrated in Figure 12.13.   □

The Kruskal–Wallis test statistic is computed using MINITAB in Figure 12.14. Notice that the Life Satisfaction Index scores are stored in column C1, whereas column C2 contains the sample number of each observation (1, 2, 3, or 4). The value of the Kruskal–Wallis statistic is called H, and agrees with the previous result.

Figure 12.13

p value for KW statistic, chi-square curve with 3 *df* (example 12.9)

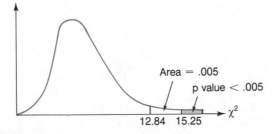

Figure 12.14

MINITAB procedure
for Kruskal–Wallis
test (example 12.9)

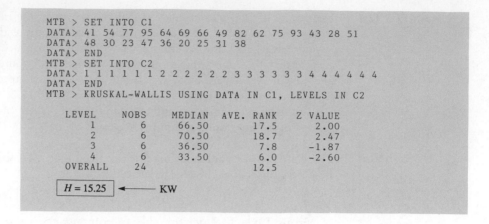

```
MTB > SET INTO C1
DATA> 41 54 77 95 64 69 66 49 82 62 75 93 43 28 51
DATA> 48 30 23 47 36 20 25 31 38
DATA> END
MTB > SET INTO C2
DATA> 1 1 1 1 1 1 2 2 2 2 2 2 3 3 3 3 3 3 4 4 4 4 4 4
DATA> END
MTB > KRUSKAL-WALLIS USING DATA IN C1, LEVELS IN C2

    LEVEL      NOBS     MEDIAN   AVE. RANK    Z VALUE
      1          6       66.50      17.5        2.00
      2          6       70.50      18.7        2.47
      3          6       36.50       7.8       -1.87
      4          6       33.50       6.0       -2.60
    OVERALL     24                  12.5
```

$H = 15.25$ ⟵ —— KW

EXERCISES

12.43   Samples of nine observations were taken from each of three populations, making a total of 27 observations. When the observations were pooled and then ranked from smallest to largest, the sum of the ranks for the first, second, and third samples were 120, 148, and 110, respectively. At a significance level of .05, do the data indicate a difference in mean value of the three populations?

12.44   The following table lists observations from a sample of four populations. Is there reason to believe that the four populations have different means using the Kruskal–Wallis procedure? Use a 1% significance level.

Population 1	Population 2	Population 3	Population 4
101	104	104	105
110	99	102	110
120	86	100	120
105	105	111	121
100	110	103	127
107	120	102	118
106	114		112
	110		

12.45   A survey was taken of the starting salaries of students who had completed a degree in business administration at either Oceanspray College, Stanton University, or Hillside College. Do the following data indicate, at the 0.05 level, a difference in the mean starting salaries of students with a degree in business administration from one of the three schools?

Oceanspray College	Stanton University	Hillside College
27,100	24,500	22,500
25,300	27,250	23,000
22,450	26,700	26,000
26,800	27,000	25,750
25,100	26,500	21,630
21,350	22,300	22,500
22,500	27,600	23,650
25,000	28,150	24,180
		22,750

12.46 Thirty new employees were selected to test two training programs. Ten of them (group A) were randomly selected for a self-paced training program. Another ten (group B) were randomly selected for a classroom training program. The remaining employees (group C) were not given any training. After the completion of the experiment, the manager evaluated the 30 employees on their productivity over a 2-week span. The following ranks were given by the manager, with the highest rankings being given to those who were not productive.

Program A	Program B	Program C
6	5	1
22	21	10
25	15	11
26	4	13
20	8	2
30	12	3
16	18	7
23	24	14
28	27	17
9	29	19

From the data, is there a difference in the productivity of the three groups? Use a .05 significance level.

12.47 The number of hours it takes three workers to complete a task is given in the following table. The task is assigned to each worker four times. Do the data indicate a significant difference in the time it takes each worker to complete the task? Use a significance level of .05.

Worker 1	Worker 2	Worker 3
3.1	3.4	3.5
3.4	3.3	3.2
3.0	3.4	3.3
3.1	3.2	3.1

12.48 The number of cars passing each of three different intersections in Crossroads City between 5:00 P.M. and 5:30 P.M. is given in the following table for randomly selected days. Fifteen days were randomly selected and then the amount of traffic was recorded for 5 of the 15 days at each intersection. Test the null hypothesis that there is no difference in the amount of traffic at each intersection between 5:00 P.M. and 5:30 P.M. Use a 10% significance level.

Intersection 1	Intersection 2	Intersection 3
440	480	433
420	392	406
530	386	427
401	456	338
454	427	397

12.49 A manager believes that the higher-salaried employees in a certain company are more satisfied with their job than are the lower-salaried employees. A sample of 10 employees from each of the salary levels indicated by the following table was taken. Is there a significant difference in the satisfaction level, measured on a scale of 1 to 10 (10 being a perfectly satisfied employee), for the three groups? Use a significance level of 5%.

$25,000 to $40,000	$40,001 to $60,000	Over $60,000	$25,000 to 40,000	$40,001 to $60,000	Over $60,000
4	7	8	9	3	9
3	8	7	1	4	10
7	6	6	8	9	3
6	7	7	7	6	8
5	9	5	6	7	7

12.50 A chemist was interested in knowing whether three different drugs used for insomnia were equally effective. Three groups of mice, with ten mice to a group, were used. Each group of mice was given the adult-equivalent dosage for one of the three drugs. The time in minutes it took for the mice to fall asleep was recorded. Using the following data, test the null hypothesis that each drug is equally effective in reducing sleep latency. Use a significance level of .05.

Drug 1	Drug 2	Drug 3	Drug 1	Drug 2	Drug 3
32	38	31	41	35	30
35	37	33	28	39	28
40	42	29	34	40	33
30	44	34	39	41	37
33	37	31	28	42	35

SUMMARY

The **analysis of variance (ANOVA)** procedure is a method of detecting differences between the means of two or more normal populations. The various populations represent the levels of a factor under observation. The factor might consist of, for example, different locations (does the crime rate differ among five cities?), groups of people (is there a difference between males and females?), or time periods (is average attendance the same during each day of the week?).

Samples for this analysis must be obtained independently of each other. The ANOVA technique measures sources of variation among the sample data by computing various **sums of squares**. The variation from one level (population) to the next is measured by the factor sum of squares (SS(factor)), which is large when there is great variation among the sample means. The variation within the samples is measured by the error sum of squares (SS(error)). Each of these SS has a corresponding degrees of freedom (df), which is divided into the SS to produce a mean square, MS.

The ratio of MS(factor) to MS(error) produces an F statistic that is used to test for equal means within the various populations. If the F value is large (significant), then we conclude that the population means are not all the same, which implies that the factor of interest does have a significant effect on the variable under observation, called the **dependent variable**.

When we analyze the effect of a single factor, we perform a one-factor ANOVA and use a **completely randomized design**. The results of this analysis, including the various sums of squares, mean squares, and degrees of freedom, are summarized in an ANOVA table. If the ANOVA procedure concludes that the population means are not the same, a follow-up analysis can be conducted to determine which of the population means are unequal. This analysis is a **multiple comparisons** procedure, and should only be performed in the event that the ANOVA null hypothesis of equal means is rejected.

When samples are not obtained in an independent manner, a **randomized block design** often can be used to test for differences in the population means. Again, there is a single factor of interest, but, to determine the effect of this factor, the sample data are organized into blocks. For this situation, the samples are not independently obtained, but data within the same block may be gathered from the same city or person or at the same point in time. By including a block effect in the ANOVA procedure, we can analyze the factor of interest (the population means) using an F test. In addition, another F statistic can be used for determining whether there is a significant block effect within the sample data.

The other experimental design that was discussed was the **two-way factorial design**, where the researcher investigates the effect of two factors that are of interest. Observations are obtained for each combination of factor levels, called **treatments**. For such a design, it is necessary to obtain two or more independent replicates for each treatment. The two-way factorial design allows the researcher to investigate the effect of each factor individually, as well as the combined effect of the two factors, referred to as **interaction** effect.

A nonparametric procedure of analyzing the completely randomized design is the **Kruskal–Wallis test**. This technique does not require the assumption of normally distributed populations, necessary when using the one-way ANOVA procedure. It is extremely useful when using small samples from populations where there is reason to suspect that the assumption of normality may not hold.

REVIEW EXERCISES

12.51 A random sample is taken from each of four independent and normally distributed populations with a common variance. Do the data provide sufficient evidence to reject the null hypothesis that the means of the four populations are the same with a significance level of .05?

Population 1	Population 2	Population 3	Population 4
10.0	11.3	9.6	10.4
12.1	11.4	9.7	10.5
11.2	11.1	9.6	10.0
9.5	10.4	10.5	10.1
10.1	10.9	11.1	9.3
13.8	10.1	9.5	11.2
10.2	9.8	9.9	11.3
9.8	9.1	10.8	9.9
	11.8		10.1
	12.1		10.3

12.52 To compare the effectiveness of three motivational lectures, 21 employees hired in the past seven months were randomly divided into three groups. Each group heard one lecture. The increase in productivity of the 21 employees was measured over the two weeks following the lectures. The coded values for the increase in productivity were:

Lecture 1:	5	6	7	4	6	5	6
Lecture 2:	3	4	8	3	5	4	4
Lecture 3:	1	2	6	7	2	4	3

(a)   State the null and alternative hypotheses for this experiment.
(b)   What is the between-sample variation?
(c)   What is the within-sample variation?
(d)   Test the null hypothesis at the .05 significance level and find the p value.

12.53 For the data in exercise 12.52, find the 95% CI for each of the three pairwise differences in the mean increase of productivity after the three lectures.

12.54 To test the merit of relaxation therapy, eight men over age 40 were selected. Their heart rate was recorded before and after the therapy. The results were:

Subject:	1	2	3	4	5	6	7	8
Before	74	75	79	81	73	85	88	70
After	70	68	69	72	69	70	82	62

(a)   Use a paired t test to test for the difference in the mean heart rate before and after the therapy at the .05 level.
(b)   Use the F test in the randomized block design to test for the difference in the mean heart rate before and after the therapy at the .05 level. What is the relationship between the paired t test and the F test?

12.55 Assume four samples of size 10 are taken from four normally distributed populations with a common variance. The total sum of squares is 221.6 and the error mean square is 3.7. Test the null hypothesis that there is no difference in the mean of each population at the 0.05 level. Find the p value.

12.56 Assume that, in exercise 12.55, you are given the additional information that the sample means of population 1 and 2 are 33.5 and 37.9, respectively. Find the 95% CI for the difference in the means of populations 1 and 2.

12.57 Complete the following table for a randomized block design.

Source	df	SS	MS	F
Factor	9			
Blocks		250		
Error	18	149		
Total		589		

12.58 Fifteen university campuses of similar size were selected to determine which of the three methods of advertising a blood-donation drive was most effective. Five randomly selected campuses advertised in the university newspaper (method 1).

Another five advertised only by posters and signs around campus (method 2). The remaining five had each professor credit five points to the student's last test if the student contributed (method 3). The table gives the percentage of the student body that contributed.

Method 1	.10	.15	.19	.21	.25
Method 2	.20	.18	.20	.23	.19
Method 3	.29	.20	.25	.30	.25

Do these data provide sufficient evidence at the .01 level of significance to reject the null hypothesis that there is no difference in the effectiveness of the three methods of advertising? Assume the populations are normally distributed, and the samples are independent.

12.59 Sample data from three normally distributed populations were generated by a computer program for a simulation study. Do the data provide evidence to indicate that at least two of the population variances are not equal? Use a .05 significance level.

Sample 1	Sample 2	Sample 3
38	45	28
37	47	27
35	43	31
40	44	30
39	42	31
35	44	32
34	45	29
33	47	30
32	45	31
39	43	32
37	42	31
36	44	29
35	45	30

12.60 A microbiologist wants to know whether there is any difference in the time in hours it takes to make yogurt from three different starters. Seven batches of yogurt were made with each of the starters. The following data give the times it took to make each batch. At the .05 significance level, do the data indicate a difference in the time it takes to make yogurt from the three starters? Do *not* assume normally distributed populations; apply the Kruskal–Wallis procedure.

Lactobaccillus Acidophilus	Bulgarius	Mixture of Acidophilus and Bulgarius
6.8	6.1	7.3
6.3	6.4	6.1
7.4	5.7	6.4
6.1	6.5	7.2
8.2	6.9	7.4
7.3	6.3	6.5
6.9	6.7	6.8

12.61 Repeat exercise 12.58 without assuming normally distributed independent populations. Use the Kruskal–Wallis test by ranking the data.

12.62 The following table contains data on visual acuity scores in a randomized block experiment. Lower scores represent better visual acuity. Type of method represents the block factor.

	Visual Acuity Measurement	
Type of Vision	**Method 1**	**Method 2**
Binocular	0.09	1.05
Monocular	0.23	1.45

(a) Compute the ANOVA table for a randomized block design.

(b) Using $\alpha = .05$, determine if there is a significant difference in visual acuity between binocular and monocular vision.

(c) State the p value for the test.

12.63 Four universities cooperated in an experiment to determine the reactions of college students to three different leadership styles. A professor, who had acting ability and training in drama, taught three different classes of the same size at each university as a temporary substitute instructor. One class was taught in an authoritarian military style, a second class in a democratic fashion, and the third class in a libertarian style. The proportion of students having a positive reaction to the instructor is given below.

		Leadership Style	
University	**Military**	**Democratic**	**Libertarian**
1	0.6	0.7	0.5
2	0.7	0.6	0.6
3	0.5	0.7	0.4
4	0.3	0.6	0.3

(a) Are the samples independent?

(b) At $\alpha = .05$, decide if there is a significant difference in students' approval of the three leadership styles.

(c) State the p value for the test.

(d) Is there a significant difference between the four universities in the approval of the students? Use $\alpha = .05$.

12.64 The following data give scores obtained by respondents on a social conformity scale, for three groups. It is assumed the populations are normal and independent, and the observations are randomly obtained. High scores represent more conformity.

High School	**Undergraduate**	**Graduate**
18	36	55
21	60	42
16	21	68
45	31	33
20	40	48

(a)   Compute the ANOVA table.
(b)   Is there a significant difference, on the average, between the conformity scores of the three groups? Use $\alpha = .05$. Find the p value.
(c)   What is the 95% confidence interval for the mean conformity score of undergraduates?
(d)   What is the 99% confidence interval for the difference between mean scores for the graduates and the undergraduates?
(e)   Using a significance level of .05, perform a multiple comparisons procedure, if appropriate.

12.65   The study in exercise 12.64 has been modified to cover the same students' conformity scores from high school through graduate school. Fifteen students were initially chosen, but only 10 actually went all the way to graduate school. Since the same students were used, a "repeated measures" design was obtained, with three scores for each student. Assume that the populations are normally distributed. The table below lists the conformity scores for the 10 students who completed the study.

Student	High School	Undergraduate	Graduate
1	25	30	30
2	16	35	48
3	17	18	20
4	30	25	20
5	35	30	32
6	28	29	28
7	29	30	35
8	30	40	48
9	27	29	35
10	40	30	32

(a)   Compute the ANOVA table.
(b)   Test the hypothesis that the means for the three classes are equal, at $\alpha = .05$. Can you conclude that, on the average, conformity scores remain stable, or do they tend to change as education progresses?
(c)   Is there a significant difference among the students' means? Use $\alpha = .05$.

12.66   Repeat exercise 12.64 without the assumption of normality. Rank the data and apply the Kruskal–Wallis procedure. Is there a significant difference between the three groups' mean conformity scores at $\alpha = .05$?

12.67   Exercise 12.64 was further modified to take into account the influence of sex. Thus, a 3×2 two-way factorial design was implemented. It was decided to have 3 replicates for each cell. The conformity scores are given in the table below.

Sex	High School	Undergraduate	Graduate
Male	16, 17, 25	18, 25, 30	20, 30, 42
Female	18, 20, 28	20, 28, 30	30, 41, 55

(a)   Compute the ANOVA table.
(b)   At $\alpha = .10$, test for a significant difference among the three classes.

(c)   At $\alpha = .10$, is there a difference between the conformity scores of males and females?

(d)   Test for interaction between class and sex, at $\alpha = .10$.

(e)   Find the p values for the three hypothesis tests above. Do the conclusions change at $\alpha = .05$ or at $\alpha = .01$?

12.68   An industrial psychologist was hired as a consultant by a corporation to help reduce stress levels in factory workers. One of the experiments that the psychologist conducted dealt with the effect of different combinations of illumination and noise on stress levels. Three noise levels (20 decibels, 40 db, and 60 db) and three illumination intensities (60 watt bulb, 120 watts, and 200 watts) were selected. Two workers were randomly assigned to each treatment of noise–illumination. The workers were a fairly homogeneous group. After the workers had spent one week in the treatment environment, the change in stress levels was measured. The table below lists the changes in stress levels in coded form, with higher numbers indicating higher stress levels.

	Illumination Intensity		
Noise Intensity	60 W	120 W	200 W
20 db	4.0, 5.1	1.3, 2.2	5.2, 6.1
40 db	4.8, 3.0	3.9, 4.6	5.5, 6.3
60 db	9.3, 7.5	7.2, 6.0	8.1, 6.7

(a)   Compute the ANOVA table for the above data.

(b)   Test for a noise effect at $\alpha = .05$.

(c)   Test for an illumination effect at $\alpha = .05$.

(d)   Test for an interaction between noise and illumination at $\alpha = .05$.

(e)   Find the p values for the above tests.

12.69   Ten children in an elementary school were given four different puzzles to solve. The order in which each child solved the puzzles was randomly determined. A rest period was provided between each puzzle solving session. The time taken to solve the puzzle (in seconds) was noted.

Child	Puzzle 1	Puzzle 2	Puzzle 3	Puzzle 4
1	20	21	19	23
2	25	22	26	23
3	37	40	31	35
4	105	100	95	101
5	90	95	88	100
6	10	12	13	11
7	38	35	31	36
8	45	40	42	42
9	41	42	36	38
10	10	12	11	9

A MINITAB computer printout provided the following analysis. Can it be concluded at a .05 significance level that the time it takes to solve the puzzles is different, that is, the puzzles are not equally difficult?   N.O.

```
MTB > read into c1-c3
DATA> 20 1 1
DATA> 25 1 2
DATA> 37 1 3
DATA> 105 1 4
DATA> 90 1 5
DATA> 10 1 6
DATA> 38 1 7
DATA> 45 1 8
DATA> 41 1 9
DATA> 10 1 10
DATA> 21 2 1
DATA> 22 2 2
DATA> 40 2 3
DATA> 100 2 4
DATA> 95 2 5
DATA> 12 2 6
DATA> 35 2 7
DATA> 40 2 8
DATA> 42 2 9
DATA> 12 2 10
DATA> 19 3 1
DATA> 26 3 2
DATA> 31 3 3
DATA> 95 3 4
DATA> 88 3 5
DATA> 13 3 6
DATA> 31 3 7
DATA> 42 3 8
DATA> 36 3 9
DATA> 11 3 10
DATA> 23 4 1
DATA> 23 4 2
DATA> 35 4 3
DATA> 101 4 4
DATA> 100 4 5
DATA> 11 4 6
DATA> 36 4 7
DATA> 42 4 8
DATA> 38 4 9
DATA> 9 4 10
DATA> end
     40 ROWS READ
MTB > twoway anova, data in c1, puzzle in c2, child in c3;
SUBC> additive.

ANALYSIS OF VARIANCE   C1

SOURCE       DF          SS         MS
C2            3       56.50      18.83
C3            9    35231.00    3914.56
ERROR        27      214.00        7.93
TOTAL        39    35501.50
```

(handwritten annotations): F; C2 row: 2.371; C3 row: 493.6; ERROR DF 27 circled; $F_{.05,3,927} = 2.96 \Rightarrow F^{**} < 2.96$; $F < 2.96$; FTR H_0.

12.70 A MINITAB program using the ANOVA procedure to test the equality of savings rates for three ethnic groups provided the following computer printout. The data represent the percentage of gross income saved by the individuals. At the 10% significance level, state your conclusion regarding the equality of savings rate for the three ethnic groups.

```
MTB > read into c1 c2
DATA> 12 1
DATA> 10 1
DATA> 18 1
DATA> 7 1
DATA> 20 1
DATA> 14 1
DATA> 20 1
DATA> 15 1
DATA> 14 2
DATA> 7 2
DATA> 11 2
DATA> 15 2
DATA> 16 2
DATA> 18 2
DATA> 15 2
DATA> 13 2
DATA> 10 3
DATA> 11 3
DATA> 12 3
DATA> 6 3
DATA> 5 3
DATA> 13 3
DATA> 15 3
DATA> 14 3
DATA> end
     24 ROWS READ
MTB > oneway anova, data in c1, group in c2

ANALYSIS OF VARIANCE ON C1
SOURCE      DF          SS          MS          F
C2           2        61.6        30.8       1.98
ERROR       21       327.4        15.6
TOTAL       23       389.0
                                     INDIVIDUAL 95 PCT CI'S FOR MEAN
                                     BASED ON POOLED STDEV
LEVEL        N        MEAN       STDEV    ----+---------+---------+---------+--
    1        8      14.500       4.721                (---------*---------)
    2        8      13.625       3.378             (---------*---------)
    3        8      10.750       3.615    (---------*---------)
                                          ----+---------+---------+---------+--
POOLED STDEV =       3.948               9.0      12.0      15.0      18.0
```

CASE STUDY

Can Auditors Serve The Role of Corporate Police Officers?

The late 1980s have seen a host of business scandals, such as check floating scams, blatant insider trading, nepotism in bank loans, and so on. However, corporate irregularity is not a recent phenomenon, and public concern about corporate accountability has existed for some time. The 1978 report of the Commission on Auditors' Responsibilities, Congressional passage of the Foreign Corrupt Practices Act of 1977, Securities and Exchange Commission guidelines, and various such forces have pushed onto managers and auditors greater responsibility for preventing and detecting corporate hanky-panky. In this context, one can easily visualize auditors as playing the role of corporate policemen and acting as a deterrent to such behavior. One might also assume that the more aggressive the auditor is perceived to be, the greater will be the deterrent effect. But is this actually the case?

A study by Uecker, Brief and Kinney (1981) addressed precisely this issue. The study was funded by a grant from the Peat, Marwick Mitchell Foundation. Middle- and upper-level managers located throughout the United States were selected, through the researchers' own contacts with business and industry and the assistance of Peat, Marwick Mitchell & Co. A packet of "in-basket exercises" and "manipulation check questionnaires" was mailed to each of 143 managers; 104 were received back. Of these 104, eighteen were invalidated, leaving 86 usable responses. Although the researchers reported their analysis based on these 86 responses, for purposes of this case study we shall work with a slightly reduced set of 72 observations. (The reason for this is to make certain calculations simpler and more straightforward, without detracting from the essence of the experiment.)

The experimental design chosen by the researchers was a 2 × 2 factorial analysis of variance, which permits an assessment of main effects and interaction effects of the independent variables. The independent variables (i.e., factors or treatments) were perceived "aggressiveness" of the internal auditor (IA) and the external auditor (EA). The perception of high or low aggressiveness was created by means of certain communications and memos included in the in-basket exercises, and was further validated by a seven-point Likert scale in the manipulations check questionnaire. The purpose of the questionnaire was to check the internal validity of the experiment.

The researchers postulated the following hypotheses:

H_1: Increasing the perceived "aggressiveness" of internal auditing activities in the organization decreases the occurrence of corporate irregularities.

H_2: Increasing the perceived "aggressiveness" of external auditing activities in the client organization decreases the occurrence of corporate irregularities.

With the manipulation of the independent variables set up (as high vs. low aggressiveness of IA and EA), it was necessary now to measure the response of the dependent variable, which had to capture the occurrence of corporate irregularity. This was a major purpose of the in-basket exercises, so called because they created a very realistic simulation of an executive's in-basket. For this experiment, the in-basket put the manager in a sort of dilemma.

Acting in the capacity of president, he or she had to decide on the size of allowance for writing off a bad debt relating to a loan of $84,000 made by the company to a senior corporate officer, the marketing vice-president. The most "honest" estimate was to write off at least $75,600, since the officer was not likely to be able to repay more than one-tenth of the original loan. The problem was that a large allowance for the write-off would reduce the net income before taxes (NIBT). Without any write-off, the projected NIBT was $755,000 versus a target of $742,500, so there was only $12,500 room to "play with." If more than $12,500 was written off, the NIBT would not be met, the president would lose a promised bonus of $20,000, and face problems with stockholders. If less than $75,600 was written off, the president would be guilty of fraudulently overstating the NIBT.

Faced with this dilemma, would the decision-maker be influenced by the chances of getting caught, that is, by the perception of high or low aggressiveness on the part of the internal and external auditors? The table below summarizes the results. Note that if more than one person chose a specific amount as a write-off allowance, this is indicated in the frequency column.

SUMMARY STATISTICS FOR AMOUNT OF ALLOWANCE FOR DOUBTFUL COLLECTIONS FOR EACH EXPERIMENTAL CONDITION

| | Internal Auditor (IA) | | | |
	Low Aggressiveness		High Aggressiveness	
External Auditor (EA) — **Low Aggressiveness**	$\bar{x} = \$56,450$		$\bar{x} = \$54,228$	
	Frequency	Allowance	Frequency	Allowance
	5	$ 0	5	$ 0
	1	31,500	1	12,500
	1	60,000	1	40,000
	1	75,600	1	75,600
	1	76,000	1	76,000
	7	84,000	8	84,000
	1	85,000	1	100,000
	1	100,000		
High Aggressiveness	$\bar{x} = \$47,311$		$\bar{x} = \$63,033$	
	Frequency	Allowance	Frequency	Allowance
	6	$ 0	4	$ 0
	1	6,000	1	60,000
	1	40,000	1	75,000
	1	74,000	1	75,600
	1	75,600	2	80,000
	1	76,000	7	84,000
	2	80,000	2	88,000
	5	84,000		

Source: Wilfred C. Uecker, Arthur P. Brief, and William R. Kinney, Jr., Perception of the internal and external auditor as a deterrent to corporate irregularities. Accounting Rev. LVI (3): 465–478 (July 1981) Table 5. Slightly modified to achieve the same number of replicates in each treatment cell. Reprinted with permission.

Case Study Questions

1. What is the dependent variable used to measure corporate irregularity?

2. Would aggressiveness of EA and IA be assumed to be independent for the factorial ANOVA design?

3. How many replicates are there in each cell?

4. If the data had been collected only on the basis of EA alone (or IA alone), what kind of experimental design is that?

5. Compute the ANOVA table for the above factorial experiment.

6. Consider hypothesis H_1 formulated by the researchers. Do the results suggest that perceived aggressiveness of the internal auditor affects the amount of the write-off allowance, at 5% significance? Report the p value for the test.

7. Consider hypothesis H_2 formulated by the researchers. What do the results indicate about the influence of the external auditor? Use $\alpha = .05$ for the test. Report the p value.

8. Show a diagrammatic presentation of any possible interaction between the EA and IA variables.

9. Is there a significant EA $\times$ IA interaction, at $\alpha = 0.05$? Report the p value.

10. Viewing the internal and external auditors as "police officers in corporate society," do they act as a deterrent to managers contemplating irregular corporate activities, on the basis of the above study?

COMPUTER EXERCISES USING THE DATABASE

Using a sample of 100 observations, complete the following exercises.

1. Using a significance level of .05, determine if there is a difference in life satisfaction (using variable LSINDEX) for the four marital categories of women attending Brookhaven.

2. Using a significance level of .05, determine if there is a difference in the verbal GRE scores for the four marital categories of men attending Brookhaven.

3. Determine the effect of sex and marital category on the graduate grade point average of students at Brookhaven. Use a significance level of .05. Use only the first four replicates at each of the eight treatments. Be sure to discuss any interaction effect, and include a multiple comparisons analysis, if appropriate.

4. Using the ages of the male students attending Brookhaven, define 3 groups:

Group	Age
1	Under 29
2	29 and under 33
3	33 or over

Is there a difference in the graduate grade point average among these three groups, using a significance level of .05?

**SPSSX
APPENDIX
(MAINFRAME
VERSION)**

**Solution to
Example 12.4**

Example 12.4 was concerned with using the one-factor ANOVA procedure when testing for a difference in two or more population means. The purpose was to test for a difference in job knowledge among three groups. The groups were (1) employees with a high school degree only, (2) employees with a bachelor's degree only, and (3) employees with a master's degree. The SPSSX program listing in Figure 12.15 was used to request the ANOVA table and in particular the F value and the corresponding p value.

Figure 12.15

SPSSX program
listing used to
request ANOVA
table, F value, and
p value for F

```
TITLE   JOB KNOWLEDGE
DATA LIST FREE / EDUC SCORE
BEGIN DATA
1 81
1 84
1 69
1 85
1 84
1 95
2 94
2 83
2 86
2 81
2 78
3 88
3 89
3 78
3 85
END DATA
PRINT / EDUC SCORE
ONEWAY SCORE BY EDUC(1,3)
```

The TITLE command names the SPSSX run.

The DATA LIST command gives each variable a name and describes the data as being in free form.

The BEGIN DATA command indicates to SPSSX that the input data immediately follow.

The next 15 lines contain the data values, which represent the level of education and the respective score on the test. The first line indicates that the first person has a level 1 (highschool) education and scored 81 on the exam measuring job knowledge.

The END DATA statement indicates the end of the input data.

The PRINT command requests a printout of the input data.

The ONEWAY command specifies a one-way analysis of variance model. SCORE is the dependent variable and EDUC is the factor of interest. EDUC has minimum and maximum values of 1 and 3, respectively.

Figure 12.16 contains the output obtained by executing the program listing in Figure 12.15.

SPSSX APPENDIX (MAINFRAME VERSION)

Figure 12.16

Output obtained by executing the SPSSX program listing in Figure 12.15

```
- - - - - - - - - - - - - - - - - - - - - - - - - - - - - - O N E W A Y - - - - - - -

    Variable   SCORE
 By Variable   EDUC
                              ANALYSIS OF VARIANCE

                            SUM OF          MEAN            F       F
        SOURCE       D.F.   SQUARES        SQUARES        RATIO   PROB.
BETWEEN GROUPS        2     10.8000         5.4000         .1130   .8940
WITHIN GROUPS        12    573.2000        47.7667
TOTAL                14    584.0000
                                                          F *     p value
```

SPSS/PC+ APPENDIX (MICRO VERSION)

Solution to Example 12.4

Example 12.4 was concerned with using the one-factor ANOVA procedure when testing for a difference in two or more population means. The purpose was to test for a difference in job knowledge among three groups. The groups were (1) employees with a high school degree only, (2) employees with a bachelor's degree only, and (3) employees with a master's degree. The SPSS/PC+ program listing in Figure 12.17 was used to request the ANOVA table and in particular the F value and the corresponding p value.

Figure 12.17

SPSS/PC+ program listing used to request ANOVA table, F value, and p value for F

```
TITLE   JOB KNOWLEDGE.
DATA LIST FREE/EDUC SCORE.
BEGIN DATA.
1 81
1 84
1 69
1 85
1 84
1 95
2 94
2 83
2 86
2 81
2 78
3 88
3 89
3 78
3 85
END DATA.
LIST VAR=EDUC SCORE.
ONEWAY SCORE BY EDUC(1,3).
```

The TITLE command names the SPSS/PC+ run.

The DATA LIST command gives each variable a name and describes the data as being in free form.

The BEGIN DATA command indicates to SPSS/PC+ that the input data immediately follow.

The next 15 lines contain the data values, which represent the level of education and the respective score on the test. The first line indicates that the first person had a level 1 (high school) education and scored 81 on the exam measuring job knowledge.

The END DATA statement indicates the end of the input data.

The LIST command requests a printout of the input data.

The ONEWAY command specifies a one-way analysis of variance model. SCORE is the dependent variable and EDUC is the factor of interest. EDUC has minimum and maximum values of 1 and 3, respectively.

Figure 12.18 contains the output obtained by executing the program listing in Figure 12.17.

Figure 12.18

Output obtained by executing the SPSS/PC+ program listing in Figure 12.17

```
                                            JOB KNOWLEDGE

              - - - - - - - - - - O N E W A Y - - - - - - - - - -

        Variable   SCORE
      By Variable  EDUC

                             Analysis of Variance
                              Sum of        Mean          F        F
          Source        D.F.  Squares       Squares     Ratio    Prob.
Between Groups           2      10.8000       5.4000     .1130    .8940

Within Groups           12     573.2000      47.7667

Total                   14     584.0000                  F *     p value
```

SPSSX APPENDIX (MAINFRAME VERSION)

Solution in Section 12.4

The quality-of-life example was based on a randomized block design. The purpose was to analyze 10 different assessments of life quality in three cities. The three cities represented the factor of interest and the 10 individuals represented 10 blocks. Each observation consisted of an individual's assessment of a particular city (the dependent variable). The SPSSX program listing in Figure 12.19 was used to request the ANOVA table and in particular the sums of squares for error, block, factor, and total. We were also interested in the F values for the two hypothesis tests along with the corresponding p values.

Figure 12.19

SPSSX program listing used to request the randomized block ANOVA table

```
TITLE QUALITY OF LIFE
DATA LIST FREE/ PERSON CITY SCORES
BEGIN DATA
1  1  68
1  2  72
1  3  65
2  1  40
2  2  43
2  3  42
3  1  82
3  2  89
3  3  84
4  1  56
4  2  60
4  3  50
5  1  70
5  2  75
5  3  68
6  1  80
6  2  91
6  3  86
7  1  47
7  2  58
7  3  50
8  1  55
8  2  68
8  3  52
9  1  78
9  2  77
9  3  75
10 1  53
10 2  65
10 3  60
END DATA
PRINT / PERSON CITY SCORES
ANOVA SCORES BY CITY(1,3) PERSON(1,10)
OPTIONS 3
```

The TITLE command names the SPSSX run.

The DATA LIST command gives each variable a name and describes the data as being in free form.

SPSSX APPENDIX (MAINFRAME VERSION)

The BEGIN DATA command indicates to SPSSX that the input data immediately follow.

The next 30 lines contain the data values. In this example, each line of data represents one rater's scoring of city quality. For example, the first line represents the quality of city 1 as measured by rater 1 (this value was 68).

The END DATA statement indicates the end of the input data.

The PRINT command requests a printout of the input data.

The ANOVA command requests the ANOVA model. SCORES is the dependent variable, CITY is the factor of interest (ranging from 1 to 3), and PERSON represents the blocking variable (ranging from 1 to 10).

The OPTIONS 3 statement specifies that there is no interaction between the factor and blocking variables.

Figure 12.20 contains the output obtained by executing the program listing in Figure 12.19.

Figure 12.20

Output obtained by executing the SPSSX program listing in Figure 12.19

```
          * * *  A N A L Y S I S   O F   V A R I A N C E  * * *

               SCORES
         BY    CITY
               PERSON
                                              SS ( factor )                    F1*

                                                 SUM OF                  MEAN                 SIGNIF
SOURCE OF VARIATION                              SQUARES       DF       SQUARE          F      OF F

MAIN EFFECTS                                    6009.167       11      546.288     56.147    0.000
   CITY                                          304.200        2      152.100     15.633    0.000
   PERSON               SS ( blocks ) ---------5704.967        9      633.885     65.150    0.000

EXPLAINED                                       6009.167       11      546.288     56.147    0.000

RESIDUAL                SS ( error ) ----------- 175.133       18        9.730

TOTAL                                           6184.300       29      213.252

     30 CASES WERE PROCESSED.                          SS ( total )            F2*
      0 CASES (   0.0 PCT) WERE MISSING.
```

SPSS/PC+ APPENDIX (MICRO VERSION)

Solution in Section 12.4

The quality-of-life example was based on a randomized block design. The purpose was to analyze 10 different assessments of life quality in three cities. The three cities represented the factor of interest and the 10 individuals represented 10 blocks. Each observation consisted of an individual's assessment of a particular city (the dependent variable). The SPSS/PC+ program listing in Figure 12.21 was used to request the ANOVA table and in particular the sums of squares for error, block, factor, and total. We were also interested in the F values for the two hypothesis tests along with the corresponding p values.

Figure 12.21

SPSS/PC+ program listing used to request the randomized block ANOVA table

```
TITLE QUALITY OF LIFE.
DATA LIST FREE/ PERSON CITY SCORES.
BEGIN DATA.
1  1  68
1  2  72
1  3  65
2  1  40
2  2  43
2  3  42
3  1  82
3  2  89
3  3  84
4  1  56
4  2  60
4  3  50
5  1  70
5  2  75
5  3  68
6  1  80
6  2  91
6  3  86
7  1  47
7  2  58
7  3  50
8  1  55
8  2  68
8  3  52
9  1  78
9  2  77
9  3  75
10 1  53
10 2  65
10 3  60
END DATA.
LIST VAR=PERSON CITY SCORES.
ANOVA SCORES BY CITY(1,3) PERSON(1,10)
  /OPTIONS 3.
```

SPSS/PC+ APPENDIX (MICRO VERSION)

The TITLE command names the SPSS/PC+ run.

The DATA LIST command gives each variable a name and describes the data as being in free form.

The BEGIN DATA command indicates to SPSS/PC+ that the input data immediately follow.

The next 30 lines contain the data values. In this example, each line of data represents one rater's scoring of city quality. For example, the first line represents the quality of city 1 as measured by rater 1 (this value was 68).

The END DATA statement indicates the end of the input data.

The LIST command requests a printout of the input data.

The ANOVA command requests the ANOVA model. SCORES is the dependent variable, CITY is the factor of interest (ranging from 1 to 3), and PERSON represents the blocking variable (ranging from 1 to 10).

Figure 12.22

Output obtained by executing the SPSS/PC+ program listing in Figure 12.21

The OPTIONS 3 statement specifies that there is no interaction between the factor and blocking variables.

Figure 12.22 contains the output obtained by executing the program listing in Figure 12.21.

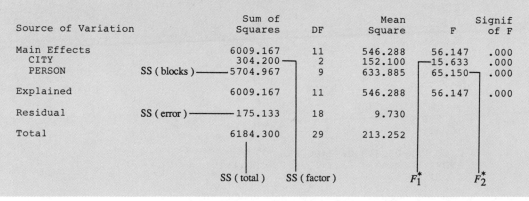

QUALITY OF LIFE

$*\ *\ *$ A N A L Y S I S   O F   V A R I A N C E $*\ *\ *$

BY   SCORES
     CITY
     PERSON

Source of Variation		Sum of Squares	DF	Mean Square	F	Signif of F
Main Effects		6009.167	11	546.288	56.147	.000
CITY		304.200	2	152.100	15.633	.000
PERSON	SS (blocks)	5704.967	9	633.885	65.150	.000
Explained		6009.167	11	546.288	56.147	.000
Residual	SS (error)	175.133	18	9.730		
Total		6184.300	29	213.252		

SS (total)   SS (factor) F_1^* F_2^*

SPSSX APPENDIX (MAINFRAME VERSION)

Solution to Example 12.8

Example 12.8 used a two-way factorial design to determine the effects of sex and marital status on an individual's life satisfaction. The purpose was to determine the effects of these two factors as well as test for a possible interaction effect between these two variables. Each observation consisted of an individual's score on the Life Satisfaction Index questionnaire. The SPSSX program listing in Figure 12.23 was used to request the ANOVA table, F values, and corresponding p values.

Figure 12.23

SPSSX program listing used to request the two-way ANOVA table

```
TITLE LIFE SATISFACTION
DATA LIST FREE/SEX MARITAL INDEX
BEGIN DATA
1  1  58
1  1  65
1  1  60
1  2  47
1  2  56
1  2  40
1  3  72
1  3  86
1  3  90
1  4  81
1  4  75
1  4  68
2  1  67
2  1  58
2  1  74
2  2  75
2  2  81
2  2  77
2  3  52
2  3  61
2  3  48
2  4  51
2  4  42
2  4  35
END DATA
PRINT / SEX MARITAL INDEX
ANOVA INDEX BY SEX(1,2) MARITAL(1,4)
```

The TITLE command names the SPSSX run.

The DATA LIST command gives each variable a name and describes the data as being in free form.

The BEGIN DATA command indicates to SPSSX that the input data immediately follow.

The next 24 lines contain the data values. In this example, each line represents one individual's life satisfaction score. For example, the first line represents the score (58) for a never-married (MARITAL = 1) male (SEX = 1).

The END DATA statement indicates the end of the input data.

The PRINT command requests a printout of the input data.

SPSSX APPENDIX (MAINFRAME VERSION)

The ANOVA command requests the ANOVA analysis procedure. INDEX is the dependent variable, SEX (ranging from 1 to 2) and MARITAL (ranging from 1 to 4) are the two factors of interest.

Figure 12.24

Output obtained by executing the SPSSX program listing in Figure 12.23

Figure 12.24 contains the output obtained by executing the program listing in Figure 12.23.

```
        * * *   A N A L Y S I S   O F   V A R I A N C E   * * *

               INDEX
          BY   SEX
               MARITAL
                                      SSA
                                SUM OF              MEAN            SIGNIF
SOURCE OF VARIATION             SQUARES    DF      SQUARE      F     OF F

MAIN EFFECTS                    521.167     4     130.292    2.664   0.071
     SEX                        247.042─    1     247.042    5.050   0.039
     MARITAL                    274.125─    3      91.375    1.868   0.176

2-WAY INTERACTIONS             3943.125     3    1314.375   26.870   0.000
     SEX      MARITAL          3943.125─    3    1314.375   26.870   0.000

EXPLAINED                      4464.292     7     637.756   13.038   0.000

RESIDUAL         SS ( error )── 782.667    16      48.917

TOTAL            SS ( total )──5246.958    23     228.129


      24 CASES WERE PROCESSED.
       0 CASES (  0.0 PCT) WERE MISSING.        SSB

                                         SSAB
```

SPSS/PC+ APPENDIX (MICRO VERSION)

Solution to Example 12.8

Example 12.8 used a two-way factorial design to determine the effects of sex and marital status on an individual's life satisfaction. The purpose was to determine the effects of these two factors as well as test for a possible interaction effect between these two variables. Each observation consisted of an individual's score on the Life Satisfaction Index questionnaire. The SPSS/PC+ program listing in Figure 12.25 was used to request the ANOVA table, F values, and corresponding p values.

SPSS/PC+ APPENDIX (MICRO VERSION)

Figure 12.25

SPSS/PC+ program listing used to request the two-way ANOVA table

```
TITLE LIFE SATISFACTION.
DATA LIST FREE/SEX MARITAL INDEX.
BEGIN DATA.
1  1  58
1  1  65
1  1  60
1  2  47
1  2  56
1  2  40
1  3  72
1  3  86
1  3  90
1  4  81
1  4  75
1  4  68
2  1  67
2  1  58
2  1  74
2  2  75
2  2  81
2  2  77
2  3  52
2  3  61
2  3  48
2  4  51
2  4  42
2  4  35
END DATA.
LIST VAR=SEX MARITAL INDEX.
ANOVA INDEX BY SEX(1,2) MARITAL(1,4).
```

The TITLE command names the SPSS/PC+ run.

The DATA LIST command gives each variable a name and describes the data as being in free form.

The BEGIN DATA command indicates to SPSS/PC+ that the input data immediately follow.

The next 24 lines contain the data values. In this example, each line represents one individual's life satisfaction score. For example, the first line represents the score (58) for a never-married (MARITAL = 1) male (SEX = 1).

The END DATA statement indicates the end of the input data.

The LIST command requests a printout of the input data.

The ANOVA command requests the ANOVA analysis procedure. INDEX is the dependent variable, SEX (ranging from 1 to 2) and MARITAL (ranging from 1 to 4) are the two factors of interest.

Figure 12.26

Output obtained
by executing the
SPSS/PC+ program
listing in Figure
12.25

Figure 12.26 contains the output obtained by executing the program listing in Figure 12.25.

LIFE SATISFACTION

* * * A N A L Y S I S   O F   V A R I A N C E * * *

```
              INDEX
        BY    SEX
              MARITAL
```

Source of Variation	Sum of Squares	DF	Mean Square	F	Signif of F
Main Effects	521.167	4	130.292	2.664	.071
SEX	247.042	1	247.042	5.050	.039
MARITAL	274.125	3	91.375	1.868	.176
2-way Interactions	3943.125	3	1314.375	26.870	.000
SEX MARITAL	3943.125	3	1314.375	26.870	.000
Explained	4464.292	7	637.756	13.038	.000
Residual	782.667	16	48.917		
Total	5246.958	23	228.129		

SSA (points to Sum of Squares column)

SS (error) —— 782.667

SS (total) —— 5246.958

SSB

SSAB

SPSSX APPENDIX (MAINFRAME VERSION)

Solution to Example 12.9

Example 12.9 used the nonparametric Kruskal–Wallis test to determine whether there was a difference in the life satisfaction for four marital populations. Data were collected by observing the score on the Life Satisfaction Index questionnaire for six randomly selected women in each of the four groups. The SPSSX program listing in Figure 12.27 was used to compute the Kruskal–Wallis statistic and the corresponding *p* value.

Figure 12.27

SPSSX program listing used to determine the Kruskal–Wallis statistic and the *p* value for example 12.9

```
TITLE LIFE SATISFACTION USING KW
DATA LIST FREE / INDEX MARITAL
BEGIN DATA
41  1
54  1
77  1
95  1
64  1
69  1
66  2
49  2
82  2
62  2
75  2
93  2
43  3
28  3
51  3
48  3
30  3
23  3
47  4
36  4
20  4
25  4
31  4
38  4
END DATA
PRINT / INDEX MARITAL
NPAR TESTS K-W = INDEX BY MARITAL(1,4)
```

The TITLE command names the SPSSX run.

The DATA LIST command gives each variable a name and describes the data as being in free form.

The BEGIN DATA command indicates to SPSSX that input data immediately follow.

The next 24 lines contain the data values; each line represents the life satisfaction score for each person and the marital category (1, 2, 3, 4).

The END DATA statement indicates the end of the data.

The PRINT command prints the values of INDEX and MARITAL.

The NPAR TESTS K–W command indicates that we wish to perform a Kruskal–Wallis test, analyzing INDEX by MARITAL.

SPSSX APPENDIX (MAINFRAME VERSION)

Figure 12.28

Output obtained by executing the SPSSX program listing in Figure 12.27

Figure 12.28 shows the output obtained by executing the program listing in Figure 12.27.

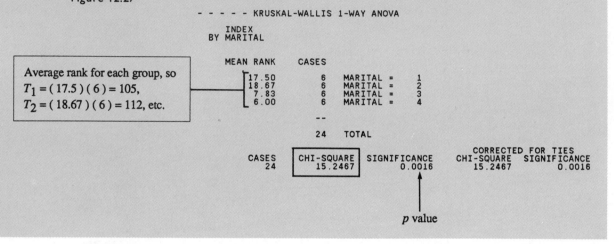

Average rank for each group, so $T_1 = (17.5)(6) = 105$, $T_2 = (18.67)(6) = 112$, etc.

```
- - - - - KRUSKAL-WALLIS 1-WAY ANOVA

     INDEX
BY MARITAL

     MEAN RANK        CASES
       17.50            6      MARITAL  =    1
       18.67            6      MARITAL  =    2
        7.83            6      MARITAL  =    3
        6.00            6      MARITAL  =    4

                       --
                       24      TOTAL

                                                    CORRECTED FOR TIES
        CASES     CHI-SQUARE    SIGNIFICANCE    CHI-SQUARE    SIGNIFICANCE
         24         15.2467         0.0016         15.2467        0.0016
```

p value

SPSS/PC+ APPENDIX (MICRO VERSION)

Solution to Example 12.9

Example 12.9 used the nonparametric Kruskal–Wallis test to determine whether there was a difference in the life satisfaction for four marital populations. Data were collected by observing the score on the Life Satisfaction Index questionnaire for six randomly selected women in each of the four groups. The SPSS/PC+ program listing in Figure 12.29 was used to compute the Kruskal–Wallis statistic and the corresponding *p* value.

Figure 12.29

SPSS/PC+ program
listing used to
determine the
Kruskal–Wallis
statistic and the
p value for example
12.9

```
TITLE LIFE SATISFACTION USING KW.
DATA LIST FREE / INDEX MARITAL.
BEGIN DATA.
41  1
54  1
77  1
95  1
64  1
69  1
66  2
49  2
82  2
62  2
75  2
93  2
43  3
28  3
51  3
48  3
30  3
23  3
47  4
36  4
20  4
25  4
31  4
38  4
END DATA.
LIST VAR=INDEX MARITAL.
NPAR TESTS K-W = INDEX BY MARITAL(1,4).
```

The TITLE command names the SPSS/PC+ run.

The DATA LIST command gives each variable a name and describes the data as being in free form.

The BEGIN DATA command indicates to SPSS/PC+ that input data immediately follow.

The next 24 lines contain the data values; each line represents the life satisfaction score for each person and the marital category (1, 2, 3, 4).

The END DATA statement indicates the end of the data.

The LIST command prints the values of INDEX and MARITAL.

The NPAR TESTS K–W command indicates that we wish to perform a Kruskal–Wallis test, analyzing INDEX by MARITAL.

Figure 12.30 (p. 664) shows the output obtained by executing the program listing in Figure 12.29.

SPSS/PC+ APPENDIX (MICRO VERSION)

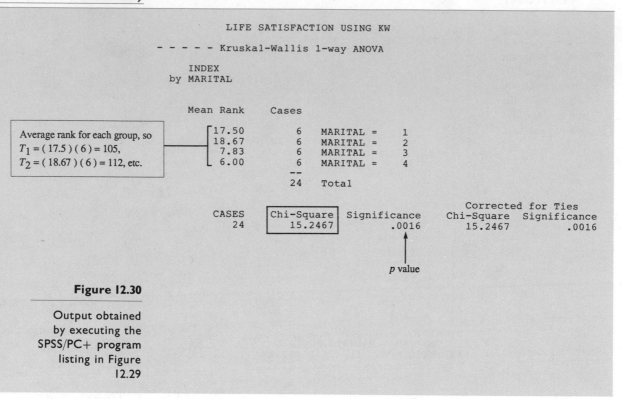

LIFE SATISFACTION USING KW

- - - - - Kruskal-Wallis 1-way ANOVA

INDEX
by MARITAL

	Mean Rank	Cases		
	17.50	6	MARITAL =	1
	18.67	6	MARITAL =	2
	7.83	6	MARITAL =	3
	6.00	6	MARITAL =	4
		--		
		24	Total	

Average rank for each group, so
$T_1 = (17.5)(6) = 105$,
$T_2 = (18.67)(6) = 112$, etc.

				Corrected for Ties	
CASES	Chi-Square	Significance		Chi-Square	Significance
24	15.2467	.0016		15.2467	.0016

p value

Figure 12.30

Output obtained by executing the SPSS/PC+ program listing in Figure 12.29

SAS APPENDIX (MAINFRAME VERSION)

Solution to Example 12.4

Example 12.4 was concerned with using the one-factor ANOVA procedure when testing for a difference in two or more population means. The purpose was to test for a difference in job knowledge among three groups. The groups were (1) employees with a high school degree only, (2) employees with a bachelor's degree only, and (3) employees with a master's degree. The SAS program listing in Figure 12.31 was used to request the ANOVA table and in particular the F value and the corresponding p value.

SAS APPENDIX (MAINFRAME VERSION)

Figure 12.31

SAS program listing used to request ANOVA table, F value, and p value for F table

```
TITLE 'JOB KNOWLEDGE';
DATA EDUCEXP;
INPUT EDUC SCORE;
CARDS;
1 81
1 84
1 69
1 85
1 84
1 95
2 94
2 83
2 86
2 81
2 78
3 88
3 89
3 78
3 85
PROC PRINT;
PROC ANOVA;
  CLASS EDUC;
  MODEL SCORE=EDUC;
```

The TITLE command names the SAS run.

The DATA command gives the data a name.

The INPUT command names the variables and specifies the correct order during input.

The CARDS statement indicates to SAS that the input data immediately follow.

The next 15 lines contain the data values, which represent the level of education and the corresponding score on the test. The first line indicates that the first person had a level 1 (high school) education and scored 81 on the exam measuring job knowledge.

The PROC PRINT command directs SAS to list the input data.

The PROC ANOVA command requests the ANOVA analysis procedure.

The CLASS subcommand identifies EDUC as the factor of interest. The MODEL subcommand indicates that the exam score is the dependent variable and the education level is the independent variable (that is, the factor of interest).

Figure 12.32 (p. 666) contains the output obtained by executing the program listing in Figure 12.31.

```
                                          JOB KNOWLEDGE
                                  ANALYSIS OF VARIANCE PROCEDURE
DEPENDENT VARIABLE: SCORE
SOURCE                DF      SUM OF SQUARES        MEAN SQUARE        F VALUE        PR > F       R-SQUARE          C.V.
MODEL                  2         10.80000000         5.40000000           0.11        0.8940       0.018493        8.2278
ERROR                 12        573.20000000        47.76666667                     ROOT MSE                   SCORE MEAN
CORRECTED TOTAL       14        584.00000000                                      6.91134333                  84.00000000

SOURCE                DF            ANOVA SS    F VALUE      PR > F
EDUC                   2         10.80000000       0.11      0.8940
```

F^* — p value

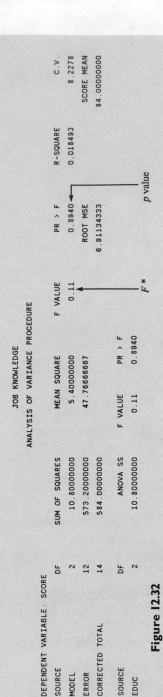

Figure 12.32

Output obtained by executing the SAS program listing in Figure 12.31

SAS APPENDIX (MICRO VERSION)

Solution to Example 12.4

Example 12.4 was concerned with using the one-factor ANOVA procedure when testing for a difference in two or more population means. The purpose was to test for a difference in job knowledge among three groups. The groups were (1) employees with a high school degree only, (2) employees with a bachelor's degree only, and (3) employees with a master's degree. The SAS program listing in Figure 12.33 was used to request the ANOVA table and in particular the F value and the corresponding p value.

Figure 12.33

SAS program listing used to request ANOVA table, F value, and p value for F table

```
TITLE 'JOB KNOWLEDGE';
DATA EDUCEXP;
INPUT EDUC SCORE;
CARDS;
1 81
1 84
1 69
1 85
1 84
1 95
2 94
2 83
2 86
2 81
2 78
3 88
3 89
3 78
3 85
PROC PRINT;
PROC ANOVA;
  CLASS EDUC;
  MODEL SCORE=EDUC;
RUN;
```

The TITLE command names the SAS run.

The DATA command gives the data a name.

The INPUT command names the variables and specifies the correct order during input.

The CARDS statement indicates to SAS that the input data immediately follow.

The next 15 lines contain the data values, which represent the level of education and the corresponding score on the test. The first line indicates that the first person had a level 1 (high school) education and scored 81 on the exam measuring job knowledge.

The PROC PRINT command directs SAS to list the input data.

SAS APPENDIX (MICRO VERSION)

The PROC ANOVA command requests the ANOVA analysis procedure.

The CLASS subcommand identifies EDUC as the factor of interest. The MODEL subcommand indicates that the exam score is the dependent variable and the education level is the independent variable (that is, the factor of interest).

The RUN statement tells the SAS system to execute the previous SAS statements.

Figure 12.34

Output obtained by executing the SAS program listing in Figure 12.33

Figure 12.34 contains the output obtained by executing the program listing in Figure 12.33.

```
                    JOB KNOWLEDGE


              Analysis of Variance Procedure
                 Class Level Information

            Class     Levels     Values

            EDUC          3       1 2 3


    Number of observations in data set = 15

            Analysis of Variance Procedure
```

Dependent Variable: SCORE

Source	DF	Sum of Squares	Mean Square	F Value	Pr > F
Model	2	10.80000000	5.40000000	0.11	0.8940
Error	12	573.20000000	47.76666667		
Corrected Total	14	584.00000000		$F*$	p value

R-Square	C.V.	Root MSE	SCORE Mean
0.018493	8.2277897	6.9113433	84.00000000

```
            Analysis of Variance Procedure
```

Dependent Variable: SCORE

Source	DF	Anova SS	Mean Square	F Value	Pr > F
EDUC	2	10.80000	5.40000	0.11	0.8940

SAS APPENDIX (MAINFRAME VERSION)

Solution in Section 12.4

The quality-of-life example was based on a randomized block design. The purpose was to analyze 10 different assessments of life quality in three cities. The three cities represented the factor of interest and the 10 individuals represented 10 blocks. Each observation consisted of an individual's assessment of a particular city (the dependent variable). The SAS program listing in Figure 12.35 was used to request the ANOVA table and in particular the sums of squares for error, block, factor, and total. We were also interested in the F values for the two hypothesis tests along with the corresponding p values.

Figure 12.35

SAS program listing used to request the randomized block ANOVA table

```
TITLE 'QUALITY OF LIFE';
DATA QUALITY;
INPUT PERSON CITY SCORES;
CARDS;
1  1  68
1  2  72
1  3  65
2  1  40
2  2  43
2  3  42
3  1  82
3  2  89
3  3  84
4  1  56
4  2  60
4  3  50
5  1  70
5  2  75
5  3  68
6  1  80
6  2  91
6  3  86
7  1  47
7  2  58
7  3  50
8  1  55
8  2  68
8  3  52
9  1  78
9  2  77
9  3  75
10 1  53
10 2  65
10 3  60
PROC PRINT;
PROC ANOVA;
  CLASS CITY PERSON;
  MODEL SCORES = CITY PERSON;
```

SAS APPENDIX (MAINFRAME VERSION)

The TITLE command names the SAS run.

The DATA command gives the data a name.

The INPUT command names the variables and specifies the correct order during input.

The CARDS command indicates to SAS that the input data immediately follow.

The next 30 lines contain the data values. In this example, each line of data represents one rater's scoring of city quality. For example, the first line represents the quality of city 1 as measured by rater 1 (this value was 68).

The PROC PRINT command directs SAS to list the input data.

The PROC ANOVA command requests the ANOVA analysis procedure.

The CLASS subcommand identifies two variables in the model; CITY is the factor of interest and PERSON represents the blocking variable. The MODEL subcommand indicates that SCORES is the dependent variable and that the analysis will consider the effects of CITY and PERSON.

Figure 12.36

Output obtained by executing the SAS program listing in Figure 12.35

Figure 12.36 contains the output obtained by executing the program listing in Figure 12.35.

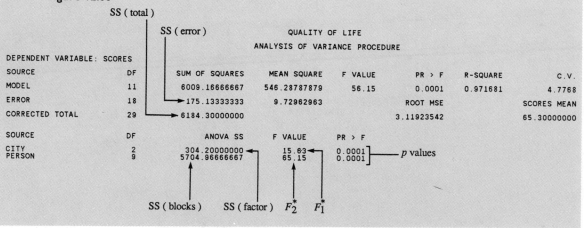

SAS APPENDIX (MICRO VERSION)

Solution in Section 12.4

The quality-of-life example was based on a randomized block design. The purpose was to analyze 10 different assessments of life quality in three cities. The three cities represented the factor of interest and the 10 individuals represented 10 blocks. Each observation consisted of an individual's assessment of a particular city (the dependent variable). The SAS program listing in Figure 12.37 was used to request the ANOVA table and in particular the sums of squares for error, block, factor, and total. We were also interested in the F values for the two hypothesis tests along with the corresponding p values.

Figure 12.37

SAS program listing used to request the randomized block ANOVA table

```
TITLE 'QUALITY OF LIFE';
DATA QUALITY;
INPUT PERSON CITY SCORES;
CARDS;
1  1  68
1  2  72
1  3  65
2  1  40
2  2  43
2  3  42
3  1  82
3  2  89
3  3  84
4  1  56
4  2  60
4  3  50
5  1  70
5  2  75
5  3  68
6  1  80
6  2  91
6  3  86
7  1  47
7  2  58
7  3  50
8  1  55
8  2  68
8  3  52
9  1  78
9  2  77
9  3  75
10 1  53
10 2  65
10 3  60
PROC PRINT;
PROC ANOVA;
  CLASS CITY PERSON;
  MODEL SCORES = CITY PERSON;
RUN;
```

**SAS
APPENDIX
(MICRO
VERSION)**

```
                        QUALITY OF LIFE

                Analysis of Variance Procedure
                   Class Level Information

          Class     Levels    Values

          CITY         3      1 2 3

          PERSON      10      1 2 3 4 5 6 7 8 9 10


        Number of observations in data set = 30



                Analysis of Variance Procedure

Dependent Variable: SCORES
                            Sum of         Mean
Source              DF      Squares       Square      F Value    Pr > F

Model               11   6009.1666667   546.2878788    56.15     0.0001

Error               18    175.1333333     9.7296296

Corrected Total     29   6184.3000000

          R-Square          C.V.       Root MSE          SCORES Mean

          0.971681       4.7767771     3.1192354         65.30000000



                Analysis of Variance Procedure

Dependent Variable: SCORES

Source              DF     Anova SS     Mean Square    F Value    Pr > F

CITY                 2    304.20000     152.10000       15.63     0.0001
PERSON               9   5704.96667     633.88519       65.15     0.0001
```

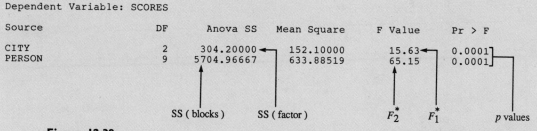

SS (error)
SS (total)

SS (blocks)   SS (factor) F_2^* F_1^* p values

Figure 12.38

Output obtained by
executing the SAS
program listing in
Figure 12.37

**SAS
APPENDIX
(MICRO
VERSION)**

The TITLE command names the SAS run.

The DATA command gives the data a name.

The INPUT command names the variables and specifies the correct order during input.

The CARDS command indicates to SAS that the input data immediately follow.

The next 30 lines contain the data values. In this example, each line of data represents one rater's scoring of city quality. For example, the first line represents the quality of city 1 as measured by rater 1 (this value was 68).

The PROC PRINT command directs SAS to list the input data.

The PROC ANOVA command requests the ANOVA analysis procedure.

The CLASS subcommand identifies two variables in the model; CITY is the factor of interest and PERSON represents the blocking variable. The MODEL subcommand indicates that SCORES is the dependent variable and that the analysis will consider the effects of CITY and PERSON.

The RUN statement tells the SAS system to execute the previous SAS statements.

Figure 12.38 contains the output obtained by executing the program listing in Figure 12.37.

**SAS
APPENDIX
(MAINFRAME
VERSION)**

**Solution to
Example 12.8**

Example 12.8 used a two-way factorial design to determine the effects of sex and marital status on an individual's life satisfaction. The purpose was to determine the effects of these two factors as well as test for a possible interaction effect between these two variables. Each observation consisted of an individual's score on the Life Satisfaction Index questionnaire. The SAS program listing in Figure 12.39 (p. 674) was used to request the ANOVA table, F, values and corresponding p values.

**SAS
APPENDIX
(MAINFRAME
VERSION)**

Figure 12.39

SAS program listing
used to request the
two-way ANOVA
table

```
TITLE 'LIFE SATISFACTION';
DATA SATIS;
INPUT SEX MARITAL INDEX;
CARDS;
1 1 58
1 1 65
1 1 60
1 2 47
1 2 56
1 2 40
1 3 72
1 3 86
1 3 90
1 4 81
1 4 75
1 4 68
2 1 67
2 1 58
2 1 74
2 2 75
2 2 81
2 2 77
2 3 52
2 3 61
2 3 48
2 4 51
2 4 42
2 4 35
PROC PRINT;
PROC ANOVA;
 CLASS SEX MARITAL;
 MODEL INDEX = SEX MARITAL SEX*MARITAL;
```

The TITLE command names the SAS run.

The DATA command gives the data a name.

The INPUT command names the variables and specifies the correct order during input.

The CARDS command indicates to SAS that the input data immediately follow.

The next 24 lines contain the data values. In this example, each line represents one individual's life satisfaction score. For example, the first line represents the score (58) for a never-married (MARITAL = 1) male (SEX = 1).

The PROC PRINT command directs SAS to list the input data.

The PROC ANOVA command requests the ANOVA analysis procedure.

The CLASS subcommand identifies two factors in the model, SEX and MARITAL. The MODEL subcommand indicates that INDEX is the dependent variable and that the analysis will consider the effects of SEX, MARITAL and the interaction between them (SEX*MARITAL).

Figure 12.40 contains the output obtained by executing the program listing in Figure 12.39.

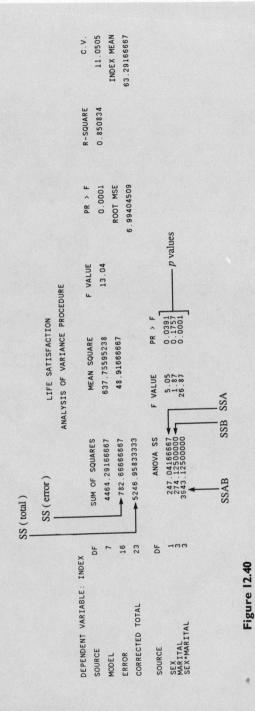

Figure 12.40

Output obtained by executing the SAS program listing in Figure 12.39

**SAS
APPENDIX
(MICRO
VERSION)**

**Solution to
Example 12.8**

Example 12.8 used a two-way factorial design to determine the effects of sex and marital status on an individual's life satisfaction. The purpose was to determine the effects of these two factors as well as test for a possible interaction effect between these two variables. Each observation consisted of an individual's score on the Life Satisfaction Index questionnaire. The SAS program listing in Figure 12.41 was used to request the ANOVA table, F values, and corresponding p values.

Figure 12.41

SAS program listing
used to request the
two-way ANOVA
table

```
TITLE 'LIFE SATISFACTION';
DATA SATIS;
INPUT SEX MARITAL INDEX;
CARDS;
1 1 58
1 1 65
1 1 60
1 2 47
1 2 56
1 2 40
1 3 72
1 3 86
1 3 90
1 4 81
1 4 75
1 4 68
2 1 67
2 1 58
2 1 74
2 2 75
2 2 81
2 2 77
2 3 52
2 3 61
2 3 48
2 4 51
2 4 42
2 4 35
PROC PRINT;
PROC ANOVA;
 CLASS SEX MARITAL;
 MODEL INDEX = SEX MARITAL SEX*MARITAL;
RUN;
```

The TITLE command names the SAS run.

The DATA command gives the data a name.

The INPUT command names the variables and specifies the correct order during input.

The CARDS command indicates to SAS that the input data immediately follow.

The next 24 lines contain the data values. In this example, each line represents one individual's life satisfaction score. For example, the first line represents the score (58) for a never-married (MARITAL = 1) male (SEX = 1).

The PROC PRINT command directs SAS to list the input data.

The PROC ANOVA command requests the ANOVA analysis procedure.

The CLASS subcommand identifies two factors in the model, SEX and MARITAL. The MODEL subcommand indicates that INDEX is the dependent variable and that the analysis will consider the effects of SEX, MARITAL and the interaction between them (SEX*MARITAL).

The RUN statement tells the SAS system to execute the previous SAS statements.

Figure 12.42 (p. 678) contains the output obtained by executing the program listing in Figure 12.41.

SAS APPENDIX (MICRO VERSION)

Figure 12.42

Output obtained by executing the SAS program listing in Figure 12.41

LIFE SATISFACTION

Analysis of Variance Procedure
Class Level Information

Class	Levels	Values
SEX	2	1 2
MARITAL	4	1 2 3 4

Number of observations in data set = 24

Analysis of Variance Procedure

Dependent Variable: INDEX

Source	DF	Sum of Squares	Mean Square	F Value	Pr > F
Model	7	4464.2916667	637.7559524	13.04	0.0001
Error	16	782.6666667	48.9166667		
Corrected Total	23	5246.9583333			

SS (error)
SS (total)

R-Square	C.V.	Root MSE	INDEX Mean
0.850834	11.050499	6.9940451	63.29166667

Analysis of Variance Procedure

Dependent Variable: INDEX

Source	DF	Anova SS	Mean Square	F Value	Pr > F
SEX	1	247.041667	247.041667	5.05	0.0391
MARITAL	3	274.125000	91.375000	1.87	0.1757
SEX*MARITAL	3	3943.125000	1314.375000	26.87	0.0001

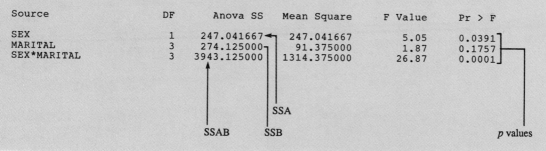

SSA

SSAB   SSB   *p* values

SAS APPENDIX (MAINFRAME VERSION)

Solution to Example 12.9

Example 12.9 used the nonparametric Kruskal–Wallis test to determine whether there was a difference in the life satisfaction for four marital populations. Data were collected by observing the score on the Life Satisfaction Index questionnaire for six randomly selected women in each of the four groups. The SAS program listing in Figure 12.43 was used to compute the Kruskal–Wallis statistic and the corresponding p value.

Figure 12.43

SAS program listing used to determine the Kruskal–Wallis statistic and the p value for example 12.9

```
TITLE  'LIFE SATISFACTION USING KW';
DATA SATIS;
 INPUT INDEX MARITAL;
CARDS;
41 1
54 1
77 1
95 1
64 1
69 1
66 2
49 2
82 2
62 2
75 2
93 2
43 3
28 3
51 3
48 3
30 3
23 3
47 4
36 4
20 4
25 4
31 4
38 4
PROC PRINT;
PROC NPAR1WAY WILCOXON;
 VAR INDEX;
 CLASS MARITAL;
```

The TITLE command names the SAS run.

The DATA command gives the data a name.

The INPUT command names and gives the correct order for the variables in the data to follow.

The CARDS command indicates to SAS that the input data immediately follow.

The next 24 lines contain the data values; each line represents the life satisfaction score for each person and the marital category (1, 2, 3, 4).

The PROC PRINT command directs SAS to print the input data.

The following command set was used to obtain the Kruskal–Wallis statistic:

PROC NPAR1WAY WILCOXON;

VAR INDEX;

CLASS MARITAL;

Using the above command set, the differences in average life satisfaction can be analyzed for the four marital categories.

Figure 12.44 shows the output obtained by executing the program listing in Figure 12.43.

Figure 12.44

Output obtained by executing the SAS program listing in Figure 12.43

```
                          LIFE SATISFACTION USING KW
              ANALYSIS FOR VARIABLE INDEX CLASSIFIED BY VARIABLE  MARITAL

                          WILCOXON SCORES (RANK SUMS)

                                    SUM OF     EXPECTED     STD DEV        MEAN
             LEVEL          N       SCORES     UNDER H0     UNDER H0       SCORE
                     1      6       105.00       75.00        15.00        17.50
                     2      6       112.00       75.00        15.00        18.67
                     3      6        47.00       75.00        15.00         7.83
                     4      6        36.00       75.00        15.00         6.00

              KRUSKAL-WALLIS TEST (CHI-SQUARE APPROXIMATION)
              CHISQ=  15.25      DF= 3      PROB > CHISQ=0.0016
```

Rank Totals

p value

SAS APPENDIX (MICRO VERSION)

Solution to Example 12.9

Example 12.9 used the nonparametric Kruskal–Wallis test to determine whether there was a difference in the life satisfaction for four marital populations. Data were collected by observing the score on the Life Satisfaction Index questionnaire for six randomly selected women in each of the four groups. The SAS program listing in Figure 12.45 was used to compute the Kruskal–Wallis statistic and the corresponding p value.

Figure 12.45

SAS program listing used to determine the Kruskal–Wallis statistic and the p value for example 12.9

```
TITLE  'LIFE SATISFACTION USING KW';
DATA SATIS;
INPUT INDEX MARITAL;
CARDS;
41 1
54 1
77 1
95 1
64 1
69 1
66 2
49 2
82 2
62 2
75 2
93 2
43 3
28 3
51 3
48 3
30 3
23 3
47 4
36 4
20 4
25 4
31 4
38 4
PROC PRINT;
PROC NPAR1WAY WILCOXON;
 VAR INDEX;
 CLASS MARITAL;
RUN;
```

The TITLE command names the SAS run.

The DATA command gives the data a name.

The INPUT command names and gives the correct order for the variables in the data to follow.

The CARDS command indicates to SAS that the input data immediately follow.

The next 24 lines contain the data values; each line represents the life satisfaction score for each person and the marital category (1, 2, 3, 4).

The PROC PRINT command directs SAS to print the input data.

The following command set was used to obtained the Kruskal–Wallis statistic:

PROC NPAR1WAY WILCOXON;

VAR INDEX;

CLASS MARITAL;

Using the above command set, the differences in average life satisfaction can be analyzed for the four marital categories.

The RUN statement tells the SAS system to execute the previous SAS statements.

Figure 12.46 shows the output obtained by executing the program listing in Figure 12.45.

Figure 12.46

Output obtained by executing the SAS program listing in Figure 12.45

LIFE SATISFACTION USING KW

NPAR1WAY PROCEDURE

Wilcoxon Scores (Rank Sums) for Variable INDEX
Classified by Variable MARITAL

MARITAL	N	Sum of Scores	Expected Under H0	Std Dev Under H0	Mean Score
1	6	105.0	75.0	15.0000000	17.5000000
2	6	112.0	75.0	15.0000000	18.6666667
3	6	47.0	75.0	15.0000000	7.8333333
4	6	36.0	75.0	15.0000000	6.0000000

Rank Totals

Kruskal-Wallis Test (Chi-Square Approximation)
CHISQ= 15.247   DF= 3   Prob > CHISQ=   0.0016

↑
p value

13

APPLICATIONS OF THE CHI-SQUARE STATISTIC

A LOOK BACK/INTRODUCTION

We have examined several topics in both descriptive and inferential statistics. The descriptive area introduced you to the numeric (for example, mean, median, and variance) and the graphic (for example, histogram and scatter diagram) methods of describing data. In inferential statistics, we discussed point estimation, confidence intervals, and tests of hypothesis. In the remaining chapters, we will turn our attention to other applications of the material from these earlier chapters.

Chapter 10 introduced you to nonparametric statistics, which make few assumptions about the population from which your sample is obtained. The chi-square techniques discussed in this chapter are another example of nonparametric statistics.

In Chapter 9, we introduced the chi-square (χ^2) distribution. We used this distribution to test that the variance of a normal distribution was equal to a specified value. The shape of the chi-square distribution is skewed right and is nonnegative. The shape of a chi-square curve and areas (probabilities) under such a curve are shown in Table A-6. The test statistic in Chapter 9 had a chi-square distribution. This chapter will introduce you to additional applications of statistics by using the chi-square distribution to answer such questions as

> Do reported percentages of California homeowners belonging to one of four possible marital and family categories accurately describe the percentages for Los Angeles?

> Are a person's age and sex associated with the occurrence of a particular mental disorder?

13.1   CHI-SQUARE GOODNESS-OF-FIT TESTS

The Binomial Situation

The binomial situation was introduced in Chapter 6, where the four following conditions had to be satisfied:

1 The experiment consists of n repetitions, called trials.
2 The trials are independent.
3 Each trial has two (and only two) outcomes, referred to as success and failure.
4 The probability of a success for each trial is p, where p remains the same for each trial. For a large finite population, p is the proportion of successes in this population.

Consequently, the binomial distribution applies to applications where there are only two possible outcomes, such as

The person selected is a male or a female.

A student does or does not have a reading deficiency.

A new-car buyer buys either an American-made car or a foreign-made car.

Inferences for the Binomial Situation

Estimating the binomial parameter p was covered in Chapter 11. We obtained a random sample of size n and observed the number of successes x. The estimator of p, the proportion of successes in the population, was $\hat{p} = x/n =$ the proportion of successes in the sample. We also discussed hypothesis testing for p. For example, we discussed a binomial situation in which a remarried couple obtained a divorce within 3 years (with probability p) or they did not. The hypothetical value of p was .3, and we determined whether the results of the sample, 61 divorces in 150 marriages, indicated a departure from this percentage. Here $\hat{p} = 61/150 = .4067$. So 40.67% of the remarried couples divorced within 3 years of the remarriage. Is this a large enough percentage for us to conclude that p is different from .3, or is this large value of p just due to the fact that we tested a sample and not the entire population—that is, is this a sampling error?

The resulting value of the test statistic was

$$Z^* = \frac{\hat{p} - .30}{\sqrt{\dfrac{(.3)(.7)}{150}}}$$

$$= \frac{.4067 - .3}{\sqrt{.0014}} = 2.852$$

By comparing $Z^* = 2.852$ with the value 1.96 in Table A-4, we rejected H_0 using $\alpha = .05$; that is, this particular divorce rate was *not* 30% as reported. The corresponding p value was .0044.

Another Test for H_0: $p = p_0$ versus H_a: $p \neq p_0$

There is another test for a two-tailed test on p. This new test will extend easily to a situation in which there are more than two possible outcomes for each trial: the **multinomial situation**.

To demonstrate this new testing procedure, a **chi-square goodness-of-fit** test, look at the divorce rate example. Note that the population consists of two **categories**—remarrieds divorcing within 3 years (category 1) and remarrieds *not* divorcing in the first 3 years of remarriage (category 2). Let p_1 = the proportion of remarrieds divorcing within 3 years in the population and p_2 = the proportion of remarried couples *not* obtaining a divorce in the first 3 years of the remarriage in the population. In the previous solution, $p_1 = p$ and $p_2 = 1 - p$.

We observed 61 sample values in category 1 (divorced) and 89 in category 2. So define

$$O_1 = 61$$
$$O_2 = 89$$

How many people do we expect to see in each category if H_0 is true? The hypotheses here can be written

$$H_0: \quad p_1 = .3, p_2 = .7$$
$$H_a: \quad p_1 \neq .3, p_2 \neq .7$$

This means that if H_0 is true, then, on the average, 30% of the sample values should be divorced (category 1) and 70% should not be divorced (category 2). Define

E_1 = expected number of sample values in category 1 if H_0 is true
$\quad = (150)(.3) = 45$

E_2 = expected number of sample values in category 2 if H_0 is true
$\quad = (150)(.7) = 105$

We next define a test statistic that has an approximate chi-square distribution.

$$\chi^2 = \sum \frac{(O - E)^2}{E} \qquad \text{(13-1)}$$

where the summation is over all categories ($= 2$ here). In previous uses of this distribution, its shape depended on the sample size, specified by the degrees of freedom (df). Now the shape depends on the number of categories, where

$$df = \text{number of categories} - 1$$

For the binomial situation this is

$$df = 2 - 1 = 1$$

Therefore, for any binomial application, the test statistic in equation 13-1 has a chi-square distribution with 1 df.

EXAMPLE 13.1   Analyze the divorce rate data using the chi-square test statistic and a significance level of $\alpha = .05$.

Solution   **Step 1** The hypotheses are

$$H_0: \quad p_1 = .3, p_2 = .7$$

$$H_a: \quad p_1 \neq .3, p_2 \neq .7$$

Step 2 The test statistic is

$$\chi^2 = \sum \frac{(O - E)^2}{E}$$

$$= \frac{(O_1 - E_1)^2}{E_1} + \frac{(O_2 - E_2)^2}{E_2}$$

Step 3 If H_0 is not true (H_a is true), we expect the observed values to be different from the expected values. This would result in a large value for χ^2, so the procedure is to reject H_0 if the chi-square test statistic lies in the right tail. Consequently, the test procedure is to

$$\text{Reject } H_0 \text{ if } \chi^2 > \chi^2_{.05,1}$$

where $\chi^2_{.05,1}$ is the χ^2 value having a right-tail area of .05 with 1 df. Using Table A-6, this is 3.84. Therefore,

$$\text{Reject } H_0 \text{ if } \chi^2 > 3.84$$

Step 4 We have

$$O_1 = 61, E_1 = 45$$

$$O_2 = 89, E_2 = 105$$

(Note that $O_1 + O_2 = E_1 + E_2 = n = 150$.) The calculated value of the test statistic is

$$\chi^{2*} = \frac{(61 - 45)^2}{45} + \frac{(89 - 105)^2}{105}$$

$$= 5.69 + 2.44$$

$$= 8.13$$

This value is larger than 3.84, so we reject H_0.

Step 5   We conclude, as before, that the proportion of remarrieds obtaining a divorce within 3 years (p_1) is not 30%.   □

The p value for example 13.1 using the chi-square analysis is shown in Figure 13.1; it is the shaded area to the right of 8.13. Using Table A-6, all we can say is that this value is less than .005. The actual value is .0044 (calculated using a statistical software package). This is the same p value as obtained when Z^* was used to perform this test of hypothesis.

In the previous discussion, we observe some quite fascinating (would you believe, mildly interesting?) parallels with Chapter 12. In Chapter 12, we noted that, when using the F test from the ANOVA procedure to test H_0: $\mu_1 = \mu_2$, we obtained an F value that was the square of the t value obtained using the corresponding t test. Also, the value from the F table used to define the rejection region was the square of the corresponding t value. This relationship held only when testing the equality of two means. Finally, the p values from the two tests were identical; it made no difference which test we used for a two-tailed test on two means because the results were the same for both procedures. However, the ANOVA technique also could be used for comparing the means for more than two populations.

Using the results from the divorce rate example in this chapter, we again find that:

1    $\chi^{2*} = 8.13 = (2.852)^2 = (Z^*)^2$.

2   The table values for the rejection region for Z test $= 1.96$ and for χ^2 test $= 3.84 = (1.96)^2$.

3   The p value for each test was the same.

So again we have two testing procedures that produce identical conclusions. The chi-square test, however, extends easily to the multinomial situation. The chi-square goodness-of-fit test is an extension of the Z test used to test a binomial parameter. Furthermore, there is a definite relationship between the standard normal distribution (Z) and the chi-square distribution: the square of Z always is a chi-square random variable with 1 df.

Figure 13.1

p value for example 13.1 using the chi-square analysis

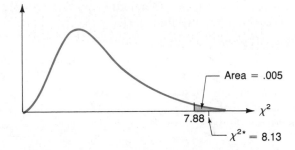

Testing H_0: $p = p_0$ versus H_a: $p \neq p_0$

Using Z Test

Test statistic

$$Z = \frac{\hat{p} - p_0}{\sqrt{\dfrac{p_0(1 - p_0)}{n}}}$$

Rejection region

Reject H_0 if $|Z| > Z_{\alpha/2}$
(use Table A-4)

Using χ^2 Test

Test statistic

$$\chi^2 = \frac{(O_1 - E_1)^2}{E_1} + \frac{(O_2 - E_2)^2}{E_2}$$

$$= \sum \frac{(O - E)^2}{E}$$

Rejection region

Reject H_0 if $\chi^2 > \chi^2_{\alpha,1}$
(use Table A-6)

The Multinomial Situation

The multinomial situation is identical to the binomial situation, except that there are k possible outcomes on each trial rather than two. Here k is any integer that is at least 2.

Suppose that a recent survey indicated the following profile of California home owners:

Category	Percent
Married with children	52
Married without children	38
Unmarried male	7
Unmarried female	3
	100

A city planner in Los Angeles questions whether these percentages apply to the homeowners in Los Angeles. Obtaining a random sample of 500 homeowners, she observes the following frequencies:

Category	Frequency
Married with children	233
Married without children	207
Unmarried male	42
Unmarried female	18
	500

The assumptions necessary for a multinomial experiment are:

1 The experiment consists of n independent repetitions (trials).
2 Each trial outcome falls in exactly one of k categories.
3 The probabilities of the k outcomes are denoted $p_1, p_2, \ldots, p_k$, where these probabilities (proportions) remain the same on each trial. Also, $p_1 + p_2 + \cdots + p_k = 1$.

For this situation, we can define k random variables as the k observed values, where

O_1 = the observed number of sample values in category 1
O_2 = the observed number of sample values in category 2
$\vdots$
O_k = the observed number of sample values in category k

For this example, $n = 500$ trials, where each trial consists of obtaining a homeowner and observing in which of the $k = 4$ categories this person lies. Assuming these homeowners are obtained in a random manner (not 500 homeowners in a particular neighborhood), then these trials are independent. Also,

p_1 = the proportion of California homeowners that are married with children
p_2 = the proportion of California homeowners that are married without children
p_3 = the proportion of California homeowners that are unmarried males
p_4 = the proportion of California homeowners that are unmarried females

The four random variables here are

O_1 = the observed number of married homeowners with children
O_2 = the observed number of married homeowners without children
O_3 = the observed number of single male homeowners
O_4 = the observed number of single female homeowners

Thus, this example fits the assumptions for the multinomial situation.

Hypothesis Testing for a Multinomial Situation

The hypotheses for the Los Angeles homeowner example would be

H_0: $p_1 = .52$, $p_2 = .38$, $p_3 = .07$, $p_4 = .03$,
H_a: at least one of the p_i's is incorrect.

Notice that H_a is not $p_1 \neq .52$, $p_2 \neq .38$, $p_3 \neq .07$, $p_4 \neq .03$. This is too strong and is not the opposite of H_0.

Let $p_{1,0}$ be any specified value of p_1, $p_{2,0}$ any specified value of p_2, and so on. The multinomial goodness-of-fit hypotheses are

$$H_0: \quad p_1 = p_{1,0}, \, p_2 = p_{2,0}, \, \ldots , \, p_k = p_{k,0}$$
$$H_a: \quad \text{at least one of the } p_i\text{'s is incorrect}$$

Using the observed value $(O_1, O_2, \ldots)$, the point estimators here are

$$\hat{p}_1 = \text{estimate of } p_1 = O_1/n$$
$$\hat{p}_2 = \text{estimate of } p_2 = O_2/n$$

and so on.

To test H_0 versus H_a, we use the previously stated chi-square statistic. To define the rejection region, notice that, when H_a is true, we would expect the O's and E's to be "far apart" because the E's are determined by assuming that H_0 is true. In other words, if H_a is true, the chi-square test statistic should be large. Consequently, we always reject H_0 when χ^{2*} lies in the right tail when using this particular statistic.

To test H_0 versus H_a, compute

$$\chi^2 = \sum \frac{(O - E)^2}{E} \qquad \qquad \textbf{(13-2)}$$

where

1 The summation is across all categories (outcomes).
2 The O's are the observed frequencies in each category using the sample.
3 The E's are the expected frequencies in each category if H_0 is true, so

$$E_1 = np_{1,0}$$
$$E_2 = np_{2,0}$$
$$E_3 = np_{3,0}$$
$$\vdots$$

4 The df for the chi-square statistic are $k - 1$, where k is the number of categories.

To carry out the test,

$$\text{Reject } H_0 \text{ if } \chi^2 > \chi^2_{\alpha, df}$$

Notice that the hypothetical proportions (probabilities) for each of the categories are specified in H_0. Consequently, we will complete the analysis by concluding that at least one of the proportions is incorrect (we reject H_0) or that there is not enough evidence to conclude that these proportions are incorrect (we fail to reject H_0). We do not accept H_0; we never conclude that these specified proportions are correct. We act like the juror who acquits a defendant, not because he or she is convinced that this person is innocent but rather because there was not sufficient evidence for conviction.

When we introduced the ANOVA technique, we mentioned that this procedure allowed us to determine whether many population means were equal using a single test. This was preferable to using many t tests to test the equality of two means, one pair at a time; these tests are not independent, and the overall significance level is difficult to determine. We encounter the same situation here. It is much better to use a chi-square goodness-of-fit test to test all of the proportions at once rather than using many Z tests to test the individual proportions.

EXAMPLE 13.2

What do the observed number of homeowners in the Los Angeles sample tell us about the profile of these homeowners? Do they conform to the percentages reported for the entire state? Use a significance level of $\alpha = .05$.

Solution

Step 1 The hypotheses under investigation are

$$H_0: \quad p_1 = .52, \; p_2 = .38, \; p_3 = .07, \; p_4 = .03$$
$$H_a: \quad \text{at least one of these } p_i\text{'s is incorrect.}$$

Step 2 The test statistic will be

$$\chi^2 = \sum \frac{(O - E)^2}{E}$$

where the summation is over the four categories.

Step 3 The test procedure here is to

$$\text{Reject } H_0 \text{ if } \chi^2 > \chi^2_{\alpha, df}$$

The df are (number of categories) $- 1$, so $df = 4 - 1 = 3$. The chi-square value from Table A-6 is $\chi^2_{.05,3} = 7.81$, and we

$$\text{Reject } H_0 \text{ if } \chi^2 > 7.81$$

Step 4 The observed values are

$$O_1 = 233, \quad O_2 = 207, \quad O_3 = 42, \quad O_4 = 18$$

The expected values when H_0 is true are obtained by multiplying $n = 500$ by each of the proportions in H_0. So,

$$E_1 = (500)(.52) = 260$$

$$E_2 = (500)(.38) = 190$$

$$E_3 = (500)(.07) = 35$$

$$E_4 = (500)(.03) = 15$$

The computed value of the chi-square test statistic is

$$\chi^{2*} = \frac{(233 - 260)^2}{260} + \frac{(207 - 190)^2}{190} + \frac{(42 - 35)^2}{35} + \frac{(18 - 15)^2}{15}$$

$$= 6.32$$

Because 6.32 does not exceed 7.81, we fail to reject H_0.

Step 5 There is insufficient evidence to suggest that the Los Angeles home-owner profile differs from the state profile. In other words, the observed values were "close enough" to the expected values under H_0 to let this hypothesis stand.

□

In example 13.2, the proportions under investigation were directly specified. We can also use the chi-square statistic when the proportions are implied.

EXAMPLE 13.3   Suppose we are interested in determining if children exhibit a color preference when given the opportunity to select from rooms painted in three different colors: yellow, orange, and blue. Each child is given a small project to do alone, shown each of the three rooms (identical, except for color) and told to select that room that he or she likes the best. A sample of 300 children produced the following results: yellow, 85; orange, 110; blue, 105. What can we conclude using $\alpha = .01$?

Solution   **Step 1** Let $p_1 =$ proportion of all children preferring the yellow room, $p_2 =$ proportion preferring orange, and $p_3 =$ proportion preferring blue. If there is no color preference, then each of these proportions will be $1/3$. This provides the values for H_0:

H_0: $p_1 = 1/3, p_2 = 1/3, p_3 = 1/3$

H_a: at least one of these proportions is incorrect

Steps 2, 3 The test procedure will be to

$$\text{Reject } H_0 \text{ if } \chi^2 > \chi^2_{.01,2} = 9.21$$

Figure 13.2

Shaded area is p value for example 13.3

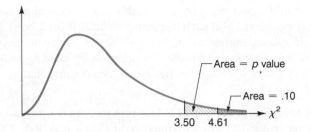

because $df = k - 1 = 3 - 1 = 2$. The value of χ^2 is determined from equation 13-2.

Step 4 The observed and expected values are

O	E (if H_0 is true)	$\hat{p}$
$O_1 = 85$	$E_1 = (300)(1/3) = 100$	$85/300 = .283$
$O_2 = 110$	$E_2 = (300)(1/3) = 100$	$110/300 = .367$
$O_3 = 105$	$E_3 = (300)(1/3) = 100$	$105/300 = .350$
$\overline{300}$	$\overline{300}$	$\overline{1.0}$

$$\chi^{2*} = \frac{(85 - 100)^2}{100} + \frac{(110 - 100)^2}{100} + \frac{(105 - 100)^2}{100}$$
$$= 2.25 + 1.0 + .25$$
$$= 3.50$$

So we fail to reject H_0 because $3.50 < 9.21$.

Step 5 There is not a significant color preference among the children for these three colors. □

The p value for the results in example 13.3 is shown in Figure 13.2. Using Table A-6 and 2 df, the smallest value in the right-hand table is 4.61, with a corresponding right-tail area of .10. All you can say using this table is that the p value is $>.10$. At any rate, it is large and would lead you to fail to reject H_0 for the most common values of α.

Pooling Categories

When using the chi-square procedure of comparing observed and expected values, we determine the difference between these two values for each category, square it, and divide by the expected value E. If the value of E is very small (say, less than 5), then this produces an extremely large contribution to the final χ^2 value from this category. In other words, this small expected value produces an inflated chi-square value, with the result that we reject H_0 when perhaps we should not have. To prevent this from occurring, we use the rule: When using equation 13-2, each expected value E should be at least 5.*

If you encounter an application where one or more of the expected values is less than 5, you can handle this situation by pooling your categories such that each of the new categories has an expected value that is at least 5.

EXAMPLE 13.4

The analysis in example 13.2 was repeated for a much smaller California community using 100 homeowners, rather than $n = 500$. The following mixture was observed (number of homeowners):

Category	Frequency
Married with children	54
Married without children	28
Unmarried male	12
Unmarried female	6

Do the data from this community appear to fit the state profile of 52% in the first category, and so on? Use $\alpha = .05$.

Solution

Category	Observed (O)	Expected (E), if H_0 is true
Married with children	$O_1 = 54$	$E_1 = (100)(.52) = 52$
Married without children	$O_2 = 28$	$E_2 = (100)(.38) = 38$
Unmarried male	$O_3 = 12$	$E_3 = (100)(.07) = 7$
Unmarried female	$O_4 = 6$	$E_4 = (100)(.03) = 3$
	100	100

Notice that the last expected value is less than 5. The category needs to be pooled (combined) with another category. If we pool the two unmarried categories together, we obtain a new category, which we will label Unmarried. The new summary is

Category	Observed (O)	Expected (E), if H_0 is true
Married with children	54	52
Married without children	28	38
Unmarried	18 (= 12 + 6)	10 (= 7 + 3)

Now each of the expected values is at least 5, and we can continue the analysis. The hypotheses using the three categories will be

H_0: $p_1 = .52, p_2 = .38, p_3 = .10$

H_a: at least one of the proportions is incorrect

---

* This rule is somewhat arbitrary, but commonly used. Another procedure used for pooling requires that all the expected values be 3 or more, while yet another requires that no more than 20% of the expected values be less than 5 with none less than 1.

The value of p_3 represents the proportion of all unmarried homeowners in this community. The hypothetical value of p_3 becomes $.07 + .03 = .10$.

The computed chi-square value is now

$$\chi^{2*} = \frac{(54 - 52)^2}{52} + \frac{(28 - 38)^2}{38} + \frac{(18 - 10)^2}{10}$$

$$= .077 + 2.63 + 6.4$$

$$= 9.11$$

This exceeds the Table A-6 value of $\chi^2_{.05,2} = 5.99$, so we reject H_0 and conclude that we do have a significant departure from the state mixture of homeowners in this community. The largest contributor to this chi-square value is from the unmarried category, which exceeds expectations by 8 homeowners (80% more than expected if H_0 is true). □

Testing a Hypothesis about a Distributional Form (Optional)

In this discussion of the goodness-of-fit test, we will examine such questions as:

Is it true that these data came from a binomial distribution with probability of success p equal to .2?

Is there any reason to doubt the assumption that the scores on a national achievement test are normally distributed?

The first question concerns a **discrete** distribution (the binomial). The second question is concerned with whether the data came from a particular **continuous** (the normal) distribution. We will demonstrate how to apply the chi-square goodness-of-fit test to these situations in the illustrations that follow.

Suppose that the adult population of a particular county is made up of 20% minorities. There is some concern that jury selection has not conformed to this percentage; that is, less than 20% of jury members were enlisted from minority members. A recent study contained a random sample of 100 juries, consisting of 12 members each. The results were:

Number of Minorities Per Jury	Number of Juries
0	9
1	27
2	29
3	18
>3	17
	100

We wish to know whether these data appear to come from a binomial distribution with $p = .2$, using $\alpha = .05$.

Your immediate reaction may well be that this problem does not fit a multinomial situation. However, there are 100 independent trials, each trial consisting of randomly selecting twelve jurors. Also, we can set up 5 categories here, namely:

Category 1   Observe 0 minority members in the 12 jurors (probability p_1)

Category 2   Observe 1 minority member in the 12 jurors (probability p_2)

Category 3   Observe 2 minority members in the 12 jurors (probability p_3)

Category 4   Observe 3 minority members in the 12 jurors (probability p_4)

Category 5   Observe >3 minority members in the 12 jurors (probability p_5)

So we do have a multinomial situation here. The hypotheses can be stated as

H_0:   the data follow a binomial distribution with $p = .2$

H_a: the data do not follow a binomial distribution with $p = .2$

If H_0 is true, then how often should we observe zero minority members in 12 jurors? Each multinomial trial fits a binomial situation, where a success consists of a juror being a minority member. We repeat the trial 12 times and count the number of successes (minorities). According to Table A-1, with $n = 12$ and probability of success $p = .2$, you should observe zero minority members out of 12 jurors 6.9% of the time. So, if H_0 is true, $p_1 = .069$. Similarly, if H_0 is true, we should see one minority member out of 12 jurors 20.6% of the time. In other words, $p_2 = .206$. Therefore, another way to state your hypotheses is

H_0: $p_1 = .069$, $p_2 = .206$, $p_3 = .283$, $p_4 = .236$, $p_5 = .206$

H_a: at least one of these p_i's is incorrect

We obtain p_5 by finding the probability of more than three successes in 12 trials. This is

$$1 - (\text{probability of 3 or less}) = 1 - (p_1 + p_2 + p_3 + p_4) = 1 - .794 = .206$$

So this is a multinomial test of hypothesis in disguise. Next, we compute the expected values E:

Category	Observed (O)	Expected (E), if H_0 is true
0 minorities	9	$(100)(.069) =$ 6.9
1 minority	27	$(100)(.206) =$ 20.6
2 minorities	29	$(100)(.283) =$ 28.3
3 minorities	18	$(100)(.236) =$ 23.6
>3 minorities	17	$(100)(.206) =$ 20.6
	100	100

Make sure that all the E's (do not worry about the O's) are at least 5. In this case, all E's are greater than 5, so no pooling of categories is necessary. Also, note that we do not round off the expected values, because they are **averages**.

To define the rejection region, we notice that there are $k = 5$ categories. So the degrees of freedom in the chi-square statistic is $df = k - 1 = 4$. Also, $\chi^2_{.05,4} = 9.49$, and so the test procedure will be

$$\text{Reject } H_0 \text{ if } \chi^2 > 9.49$$

The value of our test statistic is

$$\chi^{2*} = \frac{(9 - 6.9)^2}{6.9} + \frac{(27 - 20.6)^2}{20.6} + \cdots + \frac{(17 - 20.6)^2}{20.6}$$
$$= 4.60$$

Because 4.60 is less than 9.49, we fail to reject H_0 and conclude that there is no evidence to suggest that these data have violated the binomial assumption. These 100 observations suggest that we have no reason to suspect racial discrimination in the jury selection procedure.

In summary, suppose you are trying to perform a test of hypothesis on a binomial parameter, p—say, $H_0: p = .2$.

1 If we are using a single sample of size n, then the results of Chapter 11 apply; we are dealing with a binomial experiment.

2 If we are using many samples of size n, then the results of this section apply; this problem can be expressed as a multinomial experiment. The chi-square goodness-of-fit procedure allows us to perform a test on the population proportion p, as well as determine if the population follows a binomial distribution. The previous example provides an illustration of this type of situation.

EXAMPLE 13.5

The scores on a national achievement test were of considerable interest in Chapter 8, when it was assumed that these scores followed a normal distribution with a mean of $\mu = 500$ points and a standard deviation of $\sigma = 75$ points. A particular private school questions whether the scores of their students follow this distribution, and so decided to test the hypotheses

H_0: The scores on the national achievement test from this school follow a normal distribution with $\mu = 500$ and $\sigma = 75$

H_a: The scores do not follow a normal distribution with $\mu = 500$ and $\sigma = 75$

The (ordered) scores from a random sample of 100 students from this school are contained in Table 13.1 (p. 698). What can you conclude, using a significance level of .05?

Table 13.1

Scores on national achievement test at private school							
288	347	397	405	406	410	417	418
432	434	443	448	448	459	463	467
471	477	478	480	485	486	497	498
500	501	504	506	508	509	512	520
526	527	529	529	530	531	533	535
545	546	550	551	553	554	554	556
558	558	562	562	563	568	569	572
573	579	581	584	584	589	591	592
596	597	601	602	606	608	611	612
612	614	617	618	620	621	627	628
630	631	635	637	638	647	657	661
662	666	674	676	679	686	695	704
714	735	747	888				

Solution

Notice that the values for the mean and standard deviation are specified in H_0. Another procedure you can use is to *not* specify these values and simply test to determine if the data come from some normal population. This will be illustrated in the next example.

A frequency distribution of the data in Table 13.1 using six classes is shown below. Notice that we used two open ended classes, "under 450" and "650 or more."

Class	Frequency
Under 450	13
450 and under 500	11
500 and under 550	18
550 and under 600	24
600 and under 650	20
650 or more	14
	100

Figure 13.3

Areas under the Z curve for example 13.5

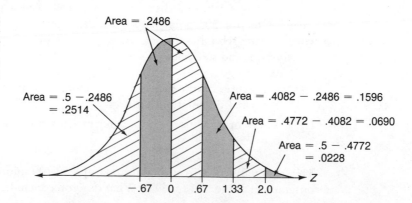

The next step is to convert each of the class limits to a Z score by subtracting the mean and dividing by the standard deviation. Here we assume that H_0 is true, that is $\mu = 500$ and $\sigma = 75$. For example, the Z score for 450 would be

$$Z = \frac{450 - 500}{75} = -.67$$

and the value for 500 is

$$Z = \frac{500 - 500}{75} = 0$$

Continuing in this way, the remaining Z scores can be obtained.

Test result	450	500	550	600	650
Z score	$-.67$	0	.67	1.33	2.00

In the first class (under 450), we observed 13 values. How many would you *expect* in this class if H_0 is true? The chances of obtaining a score under 450 is the same as observing a Z score under $-.67$. According to Figure 13.3 and Table A-4, this is .2514. So, 25.14% of the values should be under 450, which in this case is $(100)(.2514) = 25.14$ values. For this class, we observed $O_1 = 13$ values and expected $E_1 = 25.14$ values.

For the second class (450 and under 500), we observed 11 values. The chances of obtaining a test result in this range (if H_0 is true) are the same as the chances that Z is between $-.67$ and 0. Using Figure 13.3, this value is .2486. Thus we would expect $(100)(.2486) = 24.86$ values between 450 and 500. For this class, we observed $O_2 = 11$ values and expected $E_2 = 24.86$ values. Continuing this procedure, the remaining expected values can be obtained.

Range of Test Values	Range of Z Scores	Corresponding Area (see Figure 13.3)	Observed (O)	Expected (E)
Under 450	Less than $-.67$	.2514	13	$(100)(.2514) = 25.14$
450 and under 500	$-.67$ to 0	.2486	11	$(100)(.2486) = 24.86$
500 and under 550	0 to .67	.2486	18	$(100)(.2486) = 24.86$
550 and under 600	.67 to 1.33	.1596	24	$(100)(.1596) = 15.96$
600 and under 650	1.33 to 2	.0690	$34\begin{cases} 20 \\ 14 \end{cases}$	$\left.\begin{array}{l}(100)(.0690) = 6.90 \\ (100)(.0228) = 2.28\end{array}\right\}9.18$
650 or more	2 or more	.0228		
			100	100

Due to the small expected value in the last class, it needs to be pooled with the preceding class, leaving $k = 5$ classes. Using the same chi-square statistic in equation 13-2, the computed value is

$$\chi^{2*} = \frac{(13 - 25.14)^2}{25.14} + \frac{(11 - 24.86)^2}{24.86} + \frac{(18 - 24.86)^2}{24.86}$$

$$+ \frac{(24 - 15.96)^2}{15.96} + \frac{(34 - 9.18)^2}{9.18}$$

$$= 5.862 + 7.727 + 1.893 + 4.050 + 67.106$$

$$= 86.64$$

We will reject H_0 if the computed chi-square value exceeds $\chi^2_{.05,4} = 9.49$, and so we strongly reject H_0 in this situation. We conclude that the test scores do not follow a normal distribution with $\mu = 500$ and $\sigma = 75$. Be careful at this point, however; we are *not* concluding that the data came from a nonnormal population. The data may in fact be normally distributed (as the next example will demonstrate), but the values of μ and σ specified in H_0 may be incorrect.

□

Distribution Form with Unknown Parameters In the previous two examples, H_0 not only stated a particular distribution (binomial or normal) but also specified a value of the necessary parameters ($p = .2$ in the jury illustration and $\mu = 500$, $\sigma = 75$ in example 13.5). Your only concern often is whether the data follow a particular distribution and the value of the corresponding parameters is not preset.

EXAMPLE 13.6   Do the data in example 13.5 come from a normal population? That is, are the test scores normally distributed?

Solution H_0:   The national achievement test scores at this school follow a normal distribution

H_a: The scores do not follow a normal distribution.

The procedure for testing H_0 versus H_a is very similar to that used in example 13.5, except now we must estimate μ and σ and use these estimates in the calculation of the Z scores. The other change here is in the degrees of freedom for the chi-square statistic, where now

$$df = (\text{number of classes}) - 1 - (\text{number of estimated parameters})$$

For the binomial situation, you would estimate one parameter, p, and so $df = k - 1 - 1 = k - 2$, where k is the number of classes. For example 13.6, we are forced to estimate two parameters, μ and σ, and so $df = k - 1 - 2 = k - 3$. Using the data in Table 13.1, the estimate of μ is $\bar{X} = 559.0$ and the estimate of σ is $s = 92.75$.

Figure 13.4

Areas under the Z curve for example 13.6

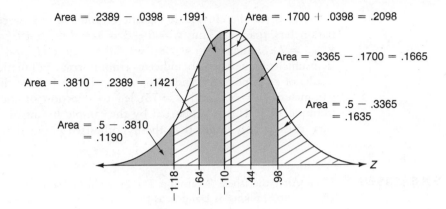

Using the same frequency distribution in example 13.5,* the Z score for a test result of 450 is $Z = (450 - 559.0)/92.75 = -1.18$. The Z score for 500 is $(500 - 559.0)/92.75 = -.64$. The remaining Z scores can be determined in a similar manner.

Test score	450	500	550	600	650
Z score	-1.18	$-.64$	$-.10$	$.44$	$.98$

Next, we partition the Z curve into sections (similar to Figure 13.3) using the preceding Z scores. The resulting areas are shown in Figure 13.4. Using these areas, we can obtain the corresponding expected values.

Class	Observed Frequency	Expected Frequency
Under 450	13	$(100)(.1190) = 11.90$
450 and under 500	11	$(100)(.1421) = 14.21$
500 and under 550	18	$(100)(.1991) = 19.91$
550 and under 600	24	$(100)(.2098) = 20.98$
600 and under 650	20	$(100)(.1665) = 16.65$
650 or more	14	$(100)(.1635) = 16.35$
	100	100

Notice that there is no need to pool classes here, since all the expected values exceed 5. Using equation 13-2, the computed value of the chi-square statistic is

$$\chi^{2*} = \frac{(13 - 11.90)^2}{11.90} + \frac{(11 - 14.21)^2}{14.21} + \cdots + \frac{(14 - 16.35)^2}{16.35}$$

$$= 2.46.$$

* It is not necessary to use the same classes here. Any frequency distribution providing a good mix of expected frequencies can be used. However, the value of χ^2 can be expected to change if the categories are different from those given.

The df for the chi-square statistic are now $k - 3$, since we estimated two parameters, μ and σ. Thus $df = 6 - 3 = 3$, and the test is to reject H_0 if $\chi^{2*} > \chi^2_{.05,3} = 7.81$. Since $2.46 < 7.81$, we fail to reject H_0 and conclude that there is insufficient evidence to indicate a nonnormal population. Comparing the results of examples 13.5 and 13.6, we are led to believe that the values of the parameters ($\mu = 500$ and $\sigma = 75$) led to rejection of the null hypothesis in example 13.5. This is supported by the sample means of $\bar{X} = 559.0$ for these data, considerably larger than the hypothesized population mean of 500. $\square$

EXERCISES

13.1 After the label, "Warning: The Surgeon General Has Determined That Cigarette Smoking Is Dangerous to Your Health" was added to cigarette packages, a chapter of the American Heart Association wished to know whether the proportion of the population that were current smokers in a particular metropolitan area was equal to 35%. A random sample of 100 observations was taken and the estimate of the proportion of smokers was .31. Let the significance level equal .05.

(a) Test the belief that the population proportion of smokers is equal to 35%. Use the Z test from Section 11.2.

(b) Use the chi-square goodness-of-fit test to test the hypothesis in question (a).

(c) What is the relationship between the test statistics in questions (a) and (b)?

13.2 A car insurance company owner believes that 20% of drivers under 25 years of age have been in exactly one automobile accident in the past two years. She also believes that 15% of the drivers under 25 have been in exactly two automobile accidents in the past two years. Finally, she believes that 10% of the drivers under 25 have been in more than two accidents in the past two years. A survey of 300 randomly selected drivers under 25 years of age was taken. Test the beliefs using the following data and letting the significance level be .10.

No Accident	1 Accident	2 Accidents	>2 Accidents
153	68	51	28

13.3 In exercise 13.2, the equality of proportions for any two categories can be tested using the Z test in Section 11.2. Is there any difference in testing the equality of two proportions one pair at a time and testing all of the proportions at once using the chi-square goodness-of-fit test? Is the overall significance level the same in both cases?

13.4 According to the U.S. Bureau of Census figures, the following national profile of poverty emerges for 1985:

Share of Poor by Age

Age Group	Share
0–4	12.6%
5–17	26.7%
18–21	7.7%
22–44	29.7%
45–64	12.8%
65 and over	10.5%
	100.0%

Calgon City officials in the department of Human Resources conducted a random survey of their files. From 50 records selected, the following distribution of the poor was noted for the city:

Age Group	Number of Persons
0–4	3
5–17	15
18–21	5
22–44	20
45–64	4
65 and over	3
	50

(a) At a 10% significance level, is there sufficient evidence to indicate that Calgon City's distribution of poor by age differs significantly from the national profile?

(b) Find the p value for the test.

(c) Would the conclusion change at $\alpha = .05$ or $\alpha = .01$?

13.5 Depression is fairly common among residents of nursing homes in Australia and the United States. Snowdon and Donnelly (1986) at the Prince of Wales Hospital in Randwick, NSW, Australia, conducted a study of six nursing homes in Sydney. They used a relatively new instrument called the Geriatric Depression Scale (GDS), which is believed to be a reliable and valid depression screening scale for elderly persons. They also assessed cognitive impairment using the Mental Status Questionnaire (MSQ). The table below provides the distribution of MSQ errors, wherein 0 to 2 MSQ errors is classified as "little or no" cognitive impairment, and 3 to 7 and 8 to 10 MSQ errors signify "moderate" and "severe" cognitive impairment, respectively. The GDS scores are also given (although you will not need them for this exercise) as a matter of interest.

MSQ Errors	Number in Group	GDS Mean	GDS S.D.
0 to 2	73	9.21	5.82
3 to 7	77	10.48	6.39
8 to 10	56	10.61	5.94
Total	206		

Source: John Snowdon and Neil Donnelly, A study of depression in nursing homes. J. Psychiatr. Res. (Great Britain) **20** (4): 331 (1986), Table 4.

(a) Are the patients evenly distributed in the three categories of low, moderate, and severe cognitive impairment, as measured by the MSQ error scores? Conduct a test at a 5% significance level.

(b) State the p value for the test.

(c) Would the conclusion be different if the test was conducted at a 10% or 1% level of significance?

13.6 In a discussion on the proper way to obtain a random sample, some students asked why they should bother with a random number generator or table. Why not just call out numbers in a haphazard manner? Would that not generate random numbers? The instructor decided to put this to the test at once. The class of 50 students were instructed to each choose a single digit number at random and write it down on a piece of paper, without looking at their neighbors. The instructor pointed out that each digit from 0 to 9 should occur with roughly equal frequency to qualify for randomness. The fifty slips of paper were collected and a tally was made. The following results were noted.

Digit	0	1	2	3	4	5	6	7	8	9
No. of times selected	2	5	4	5	6	7	6	4	5	6

At $\alpha = .01$, test the claim that calling out numbers in a haphazard manner does lead to randomness in the selection of single digits. What do you conclude?

13.7 A soft-drink company believes that people are particular about the type of sweetener used in the soft drinks. The manager of the marketing department believes that 50% of the people prefer sugar, 35% prefer aspartame, 10% prefer saccharin, and 5% have no preference. Thirty people who regularly drank sweetened soft drinks were randomly selected. Using the following data and a significance level of 5%, test the manager's claim. (Do any of the categories need to be pooled?)

Sugar	Aspartame	Saccharin	No preference
12	11	5	2

13.8 A nutritionist surveyed 200 randomly selected athletes to find out how concerned the athletes were about their food intake. These athletes responded that they were very, somewhat, little, or not at all concerned about the quality of the food they ate; the results were 65, 50, 45, and 40, respectively. Test the hypothesis that athletes are equally divided among the four categories. Use a 10% significance level.

13.9 A private foundation interested in the welfare of war veterans estimates that of those Vietnam veterans diagnosed as suffering from some physical or mental problem, one out of five has lymphoma, a form of cancer. It is believed that lymphoma is related to exposure to a group of chemicals called phenoxy acids, more popularly known as the herbicide and defoliant "Agent Orange." A random survey is taken of 20 veterans' hospitals. At each hospital, 10 cases are randomly selected, and the number of lymphoma cases is recorded, as shown below.

Number of Lymphoma Cases (in each sample of 10 cases)	Number of Hospitals
0	2
1	5
2	7
3	4
4	1
5	1
6	0
7 or more	0
	20

Thus, a total of 200 observations were made (20 hospitals times a sample of 10 cases observed at each hospital).

(a) Conduct a test to determine if the above data follow a binomial distribution with $p = 0.20$ (i.e., 1 out of 5 are lymphoma cases). Use the significance level of 0.05.

(b) Find the p value for the test.

13.10 Repeat exercise 13.9, but this time assume that the probability of lymphoma cases (p) is unknown.

(a) Calculate $\hat{p}$, your estimate for p.

(b) Do the degrees of freedom for the chi-square statistic change?

(c) What is the result of the hypothesis test, at $\alpha = .05$?

(d) Does the p value as compared to exercise 13.9(b) change?

13.11 On each flight of Astral Airways, 12 randomly selected passengers are asked if they would be willing to pay a 5% air fare increase to fly on an airline that had an open bar in the airplane. Results of the survey from 50 different flights were as follows:

Yes Answers	Flights
0	3
1	10
2	13
3	12
4	7
5	3
6	2
More than 6	0

Use a chi-square goodness-of-fit test to determine whether these data came from a binomial distribution. Let the significance level be .05. (*Hint:* p must be estimated using (total number of people who said yes)/(total number asked).)

13.12 Sports traumatology is an area of sports medicine dealing with injuries. Beskin and others (1987) evaluated the surgical results of 42 patients with spontaneous rupture of the Achilles tendon. The patients were treated between 1973 and 1984. Their age distribution was as shown below:

PATIENTS WITH ACHILLES TENDON RUPTURE

Age (at injury)	Number of Patients
0–9	0
10–19	1
20–29	7
30–39	18
40–49	8
50–59	6
60–69	0
70–79	2
Total	42

Source: J. L. Beskin, R. A. Sanders, S. C. Hunter, and J. C. Hughston, Surgical repair of Achilles tendon ruptures. *J. Sports Med.* **15** (1): 1–8 (1987), Fig. 1.

(a)　Conduct the chi-square test at a 5% significance level to determine if the age (at injury) of patients with Achilles tendon rupture follows a normal distribution. What is your conclusion?

(b)　State the p value for the test.

13.13　Refer back to the hypothetical distribution of scores on the TOEFL (Test of English as a Foreign Language) in exercise 3.69. The scores are reproduced below for ease of reference.

TOEFL Score	Number of Candidates
340 and under 380	3
380 and under 420	7
420 and under 460	14
460 and under 500	22
500 and under 540	24
540 and under 580	16
580 and under 620	10
620 and under 660	4
	100

(a)　Conduct a chi-square test to determine if the above data follow a normal distribution. Use $\alpha = .05$.

(b)　Find the p value for the test.

13.14　A psychologist studying sexual value systems started with a preliminary hypothesis that sexual permissiveness would gradually increase with age, reach a peak, and then slack off. The underlying assumption was that adolescents would tend to be more romantic initially ("there's only one person for me in the whole world") but would begin to widen the net as they reach college age. After marriage, individuals would tend to revert to notions of loyalty in order to make the marriage stable. A survey was conducted in which a questionnaire was used to determine a cutoff score above which the individual was classed as sexually permissive. The same questionnaire was used for all age groups and the same number of people in each age group were tested. The results are given below. Note that the figures do not describe actual behavior, but subjective values.

Age Group	Number of Persons Classified as Sexually Permissive
11–13	16
14–16	20
17–19	40
20–22	36
23–25	40
26–28	39
29–31	32
32–34	40
35–37	35
38–40	22
	320

The psychologist is interested in whether these scores follow a normal distribution. Do the above data contradict this hypothesis? Use a 10% significance level.

13.15 The Parapsychology Institute conducts experiments to establish whether an individual has extrasensory perception (ESP). In a typical experiment one person (the transmitter) concentrates on cards with one of two special symbols printed on them, and the ESP subject (acting as the receiver) tries to write down which symbol the sender is looking at. Of course, the persons are not allowed to communicate. The logic behind the experiment is that someone guessing at random would, in the long run, have only a 50% chance of being correct. Thus, someone who could consistently get a rating of say, 70% correct answers is obviously not achieving accuracy purely by chance. Since no physical communication between transmitter and receiver is allowed, the receiver could be considered to possess some special ability or "power" called ESP. One individual who claimed to possess ESP was put through a series of tests of the sort described above. Each test consisted of 6 trials (cards read) and the number of correct responses was noted. For example, one test might result in TTTFTF (where T = true and F = false) giving four correct responses out of 6 trials. The subject was put through 128 tests like these. The results are summarized below.

X = number of correct responses out of 6 trials	Frequency with which X was observed in the 128 tests
0	2
1	13
2	29
3	42
4	28
5	13
6	1
	128

(a) The institute uses a rule of thumb that anyone who achieves a 70% accuracy rate probably has ESP abilities. Do the above data for 128 tests fit a binomial distribution with a probability of success of .70 ($p = .70$)? Use $\alpha = .01$. State the p value.

(b) Repeat the chi-square test, but this time check if the data will fit a binomial distribution with $p = .50$. You will recall that a success rate of 50% would be achieved just by random guessing, without any ESP.

(c) Based on p values and the results of the above tests, how "strongly" could you say whether the subject has ESP?

13.2   CHI-SQUARE TESTS OF INDEPENDENCE

In the previous section, we classified each member of the population into one of many categories. This was a one-dimensional situation because each member was classified using only one criterion such as marital status. In this section, we extend this idea into a two-dimensional situation in which each element in the population is classified according to two criteria, such as sex and income level (high, medium, or low). The question of interest will be, are these two variables (classifications) independent? For example, if sex and income level are not independent, then perhaps there is sex discrimination present in the salary structure of a company. If a person's salary was not related to sex, then these two classifications would be independent.

In example 13.3, we considered some color preferences among a sample of school children and found no significant difference among the proportions of children preferring each of the three colored rooms. Suppose that the researcher decided to take a closer look at this sample by considering the sex of each child. These results can be summarized in a **contingency** (or **cross-tab**) table. The table consists of **cells**, where each cell contains the **frequency** of students in the sample that satisfy each of the various cross-classifications.

COLOR PREFERENCE

		Yellow	Orange	Blue	Total
SEX	**Male**	35	52	73	160
	Female	50	58	32	140
	Total	85	110	105	300

This is a 2×3 contingency table. It shows that there were 35 students who were both male and preferred the yellow room. In Chapter 5, we selected a person at random from this group of 300 and determined various probabilities, such as the probability that this person is both a male and prefers blue. Here we will not select a person at random. Instead, we will view these data as the results of a particular experiment and attempt to determine whether the variables—sex and color preference—are independent for this application. Put another way, is the color preference of the boys the same as that for the girls? The hypotheses are

H_0:   The classifications (sex and color preference) are independent

H_a: The classifications are dependent

This problem can also be viewed as a multinomial experiment containing 300 trials and $(2)(3) = 6$ possible categories for each trial outcome.

Deriving a Test of Hypothesis for Independent Classifications

Calculating the Expected Values   We want to decide whether the data about the color preference exhibit random variation or a pattern of some type due to a dependency between sex and color preference. If these classifications are independent (H_0 is true), how many students would you expect in each cell? Consider the upper right cell, which shows males who prefer blue. The expected number of sample observations in this cell is $300 \cdot P$(sampled student is a male and prefers blue). Assuming independence, this is $300 \cdot P$(sampled student is a male) $\cdot P$(sampled student prefers blue) using the multiplicative rule for independent events discussed in Chapter 5.

What is P(sampled student is a male)? We do not know, because we do not have enough information to determine what percentage of all students are male. However, from these data, we can estimate this probability using the percentage of males in the sample. This is $160/300 = .533$.

Similarly, P(sampled student prefers blue) can be estimated by the fraction of students selecting the blue room in the sample—namely, $105/300$. So, our estimate of the expected number of observations for this cell is

$$\hat{E} = 300 \cdot \frac{160}{300} \cdot \frac{105}{300} = \frac{(160)(105)}{300}$$

$$= 56$$

So, for this cell, the observed frequency is $O = 73$, and our estimate of the expected frequency (if H_0 is true) is $\hat{E} = 56$. In general,

$$\hat{E} = \frac{(\text{row total for this cell}) \cdot (\text{column total for this cell})}{n}$$

where $n =$ total sample size.

A summary of the calculations can be tabulated as

Sex	Color Preference	Observed (O)	Expected ($\hat{E}$), if H_0 is true	
Male	Yellow	35	$160 \cdot 85/300$ =	45.33
	Orange	52	$160 \cdot 110/300$ =	58.67
	Blue	73	$160 \cdot 105/300$ =	56.00
Female	Yellow	50	$140 \cdot 85/300$ =	39.67
	Orange	58	$140 \cdot 110/300$ =	51.33
	Blue	32	$140 \cdot 105/300$ =	49.00
		100		300

The easiest way to represent these 12 values is to place the expected value in parentheses alongside the observed value in each cell:

		Yellow	Orange	Blue
Sex	Male	35 (45.33)	52 (58.67)	73 (56.00)
	Female	50 (39.67)	58 (51.33)	32 (49.00)
	Total	85	110	105

Pooling   At this point, you need to check your expected values. If any of them is less than 5, you need to combine the column (or row) in which this small value occurs with another column (or row). This is to comply with the earlier assumption that all expected values for this new column (row) are obtained by summing the values for the two columns (rows).

The Test Statistic   The test statistic for testing H_0: the classifications are independent versus H_a: the classifications are dependent is the usual chi-square statistic, which in this case compares each observed frequency in the table with the corresponding expected frequency estimate.

$$\chi^2 = \sum \frac{(O - \hat{E})^2}{\hat{E}} \qquad\qquad (13\text{-}3)$$

where the summation is over all cells of the contingency table.

Degrees of Freedom   For the multinomial situation, the degrees of freedom for the chi-square statistic were $k - 1$, where $k =$ the number of categories (outcomes). For this situation, there were k values of $(O - \hat{E})$. However, because the sum of the observed frequencies is the same as the sum of the expected frequencies, the sum of the k values of $(O - \hat{E})$ is always zero. This means that, of these k values, only $k - 1$ are free to vary. This resulted in $k - 1$ *df* for the chi-square statistic.

Take a close look at the observed and expected frequencies in the contingency table for sex and color preference. Notice that (1) for each row, sum of O's = sum of $\hat{E}$'s, and (2) for each column, sum of O's = sum of $\hat{E}$'s. In general, if classification 1 has c categories and classification 2 has r categories, you construct an **$r \times c$ contingency table** (Figure 13.5). Of the c values of $(O - \hat{E})$ in each

Figure 13.5

Expected value estimates for an $r \times c$ contingency table

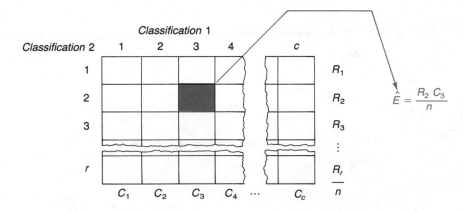

row, only $c - 1$ are free to vary. Similarly, only $r - 1$ of the values in each column are free to assume any value. So, for this contingency table, only $(r - 1)(c - 1)$ values are free to vary. Therefore, for the chi-square test of independence,

$$df = (r - 1)(c - 1) \qquad \textbf{(13-4)}$$

Testing Procedure When H_0 is not true, the expected frequencies and observed frequencies will be very different, producing a large χ^2 value. We again reject H_0 if the value of the test statistic falls in the right-tail rejection region, so we

$$\text{Reject } H_0 \text{ if } \chi^2 > \chi^2_{\alpha,df}$$

where $df = (r - 1)(c - 1)$.

In summary, the chi-square test for independence hypotheses are

H_0: the (row and column) classifications are independent (not related)

H_a: the classifications are dependent (related or associated in some way)

The test statistic is

$$\chi^2 = \sum \frac{(O - \hat{E})^2}{\hat{E}}$$

where

1 The summation is over all cells of the contingency table consisting of r rows and c columns.

2 O is the observed frequency in this cell.

3 $\hat{E}$ is the estimated expected frequency for this cell.

$$\hat{E} = \frac{\left(\begin{array}{c}\text{total of row in} \\ \text{which the cell lies}\end{array}\right) \cdot \left(\begin{array}{c}\text{total of column in} \\ \text{which the cell lies}\end{array}\right)}{(\text{total of all cells})}$$

4 The degrees of freedom for the chi-square statistic are $df = (r - 1)(c - 1)$.

The test procedure is (using Table A-6):

$$\text{Reject } H_0 \text{ if } \chi^2 > \chi^2_{\alpha,df}$$

We can now return to our question of whether sex and color preference are independent. Step 1 (statement of hypotheses) and step 2 (definition of test statistic) of our five-step procedure have been discussed already. Assume that a significance level of $\alpha = .1$ was specified. For step 3, the df are $(2 - 1)(3 - 1) = 2$. Using Table A-6, $\chi^2_{.1,2} = 4.61$. So we will reject H_0 if $\chi^2 > 4.61$. For step 4, referring to the contingency table,

$$\chi^{2*} = \frac{(35 - 45.33)^2}{45.33} + \frac{(52 - 58.67)^2}{58.67} + \frac{(73 - 56.00)^2}{56.00} + \frac{(50 - 39.67)^2}{39.67}$$

$$+ \frac{(58 - 51.33)^2}{51.33} + \frac{(32 - 49.00)^2}{49.00}$$

$$= 2.354 + .758 + 5.161 + 2.690 + .867 + 5.898$$

$$= 17.73$$

This exceeds the table value of 4.61, so we reject H_0. We thus conclude that the sex and color preference are not independent (step 5).

If the results of the chi-square test lead to a conclusion that the classifications are not independent, a closer look at the individual terms in the chi-square statistic can often reveal what the relationship is between these two variables. Examining the six terms here, we obtain four large values; namely 2.354 (male/yellow), 2.690 (female/yellow), 5.161 (male/blue), and 5.898 (female/blue). We obtained more boys (and fewer girls) preferring the blue room than we would expect if there was no dependency. Similarly, there were fewer boys (and more girls) preferring the yellow room.

This example serves as a good illustration of how examining one category (such as color preference) can fail to show any difference among subcategories (such as yellow, orange, and blue) but when examined along with another category (such as sex), patterns emerge. Such a technique is often useful in detecting job discrimination within companies. For example, no sex discrimination may be evident in a sample, but when examined along with race or age categories, certain discriminatory practices can be identified.

We can find the p value for this also, given $\chi^{2*} = 17.73$. Using a χ^2 curve with 2 df, the area to the right of 17.73, using Table A-6, is $<.005$. The p value indicates the **strength** of the dependency between two classifications. The smaller the p value is, the more you tend to support the alternative hypothesis, which indicates a stronger dependency between the two variables. For the illustration, $p < .005$ so we conclude that the sex and color preference are strongly related.

EXAMPLE 13.7   In example 12.4, a researcher attempted to determine whether an employee's educational level had an effect on his or her job performance. An exam was given to a sample of the employees, and we used the ANOVA procedure to test for a difference among the three groups: (1) those with a high-school diploma only, (2) those with a bachelor's degree only, and (3) those with a master's degree.

The researcher decided to expand this procedure by testing 120 employees; rather than recording the exam scores, she rated each person's exam performance as high, average, or low. The results of this study are:

	High	Average	Low	Total
Master's	4	20	11	35
Bachelor's	12	18	15	45
High School Diploma	9	22	9	40
Total	25	60	35	120

Does job performance as measured by the exam appear to be related to the level of an employee's education, at this particular firm? Use $\alpha = .05$.

Solution

Step 1 This calls for a chi-square test of independence, with hypotheses

H_0: exam performance is independent of educational level

H_a: these classifications are dependent

Steps 2, 3 Your test statistic is the chi-square statistic in equation 13-3. The table of frequencies here is a 3×3 contingency table, which means that the degrees of freedom are $df = (3 - 1)(3 - 1) = 4$. From Table A-6, we determine that $\chi^2_{.05,4} = 9.49$, so the testing procedure is to

Reject H_0 if $\chi^2 > 9.49$

Step 4 Computing the expected frequency estimates in the usual way, we arrive at the following table:

	High	Average	Low	Total
Master's	4 (7.29)	20 (17.5)	11 (10.21)	35
Bachelor's	12 (9.38)	18 (22.5)	15 (13.12)	45
High School Diploma	9 (8.33)	22 (20.0)	9 (11.67)	40
Total	25	60	35	120

To illustrate the calculations, the 11.67 in the lower right cell is $(40 \cdot 35)/120$.

The computed chi-square value is

$$\chi^{2*} = \frac{(4 - 7.29)^2}{7.29} + \frac{(20 - 17.5)^2}{17.5} + \cdots + \frac{(9 - 11.67)^2}{11.67}$$

$$= 4.67$$

This value is <9.49, and so we fail to reject H_0.

Step 5 We see no evidence of a relationship between job performance and level of education.

Figure 13.6

MINITAB solution to example 13.7 (test for independence)

```
MTB > READ INTO C1-C3
DATA> 4 20 11
DATA> 12 18 15
DATA> 9 22 9
DATA> END
      3 ROWS READ
MTB > CHISQUARE USING C1-C3

   Expected counts are printed below observed counts

              C1        C2        C3      Total
      1        4        20        11        35
              7.3      17.5      10.2  ←──────── Observed ( O )
                                          ←────── Expected ( E )
      2       12        18        15        45
              9.4      22.5      13.1

      3        9        22         9        40
              8.3      20.0      11.7

   Total      25        60        35       120

   ChiSq =  1.49  +   0.36  +   0.06  +
            0.74  +   0.90  +   0.27  +
            0.05  +   0.20  +   0.61  = 4.67
   df = 4
```

We do not conclude that these data demonstrate that the two classifications are clearly independent because this amounts to accepting H_0. We are simply unable to demonstrate that a relationship exists. □

A MINITAB solution to example 13.7 is contained in Figure 13.6. Notice that the format of this table is similar to that of the one we constructed, with the expected value (assuming H_0) and the observed value shown for each cell.

COMMENTS

In example 13.7, the researcher recorded the exam performance as high, average, or low rather than listing the actual exam score. Why would anyone take interval/ratio data (the exam scores) and convert them to seemingly weaker ordinal data (the exam performance classifications)? Do you lose useful information by doing this? When using the ANOVA procedure, we were forced to assume that these data came from normal populations with equal variances. In this chapter, aside from the randomness of the sample, no assumptions regarding the populations were necessary. So, by converting the exam scores to a form suitable for a contingency table and using the chi-square test of independence, we can avoid the assumptions of normality and equal variances.

This is another application of **nonparametric statistics**, often called distribution-free statistics. The beauty of these procedures is that they require only very weak assumptions regarding the populations. However, if the data do satisfy the requirements of the ANOVA procedure (or nearly so), the ANOVA F test is a better procedure to use in this situation. This is because the chi-square test is less sensitive to real differences (is less powerful) than the ANOVA test, especially for small sample sizes.

Marginal Totals (Test of Homogeneity)

A slightly different interpretation of the previous chi-square procedure occurs when we determine in advance the number of observations to be sampled within each column (or row). In the previous discussion, the row and column totals were random variables because we had no way of knowing what they would be before the sample was obtained. In this discussion, the contingency table is the same, except that the column (or row) totals are predetermined.

A researcher in Riverdale is interested in determining the usage of marijuana by high school students in the four area schools: Johnson High School, Truman High School, King High School, and Jefferson High School. He restricts the sample to senior high students (those classified as sophomore, junior or senior) and obtains confidential replies to a question regarding the number of times each student uses marijuana. The possible replies were (1) never tried, it, (2) use it somewhat (no more than twice a month), or (3) use it often (more than twice per month). Of particular interest is whether marijuana usage differs across the four schools.

The Johnson and Jefferson High Schools are considerably larger than the other two, so the researcher obtains a random sample of 200 students from each of these two schools and 100 from each of the two smaller schools. The results of the survey are:

	Johnson H.S.	Truman H.S.	King H.S.	Jefferson H.S.	Total
Never	40	38	41	45	164
Some	92	35	38	98	263
Often	68	27	21	57	173
Total	200	100	100	200	600

In the previous tests of independence, we had a single population, where each member was classified according to two criteria, such as age and sex. Now we have four distinct populations, namely, the senior high students in each of the four schools. Consequently, we obtained a random sample from each one. The column totals (sample sizes) were determined in advance. This differs from our previous examples, where we had no idea what the row or column totals would be before the sample was obtained.

The question of interest here becomes, is marijuana usage the same in each of the four schools? In other words, we are trying to determine whether these four populations can be viewed as belonging to the same population (in terms of this criterion). Identical populations are said to be **homogeneous**. Consequently, the test of hypothesis here is a **test of homogeneity** as well as a test for independence. The null hypothesis can be written as

H_0:   the four populations are homogeneous in their usage
of marijuana

or as

H_0:   high school and marijuana usage are independent classifications

The procedure for analyzing a contingency table is the same whether or not the column (or row) totals are fixed in advance.

A MINITAB solution using $\alpha = .05$ is provided in Figure 13.7. The expected cell frequencies are computed by finding

$$\hat{E} = \frac{(\text{row total})(\text{column total})}{600}$$

The computed chi-square value is

$$\chi^{2*} = \frac{(40 - 54.7)^2}{54.7} + \frac{(38 - 27.3)^2}{27.3} + \cdots + \frac{(57 - 57.7)^2}{57.7} = 24.73$$

The degrees of freedom here are $(3 - 1)(4 - 1) = 6$. This means that we reject H_0 if $\chi^2 > 12.59$, where 12.59 is $\chi^2_{.05,6}$. The computed value (24.73) exceeds the tabled value, so we reject H_0. We conclude that these four populations are not homogeneous. The level of marijuana usage is not identical at each of the four schools. We can also say that the location and level of marijuana usage

Figure 13.7

MINITAB solution
to test if school
populations are
homogeneous

```
MTB > READ INTO C1-C4
DATA> 40 38 41 45
DATA> 92 35 38 98
DATA> 68 27 21 57
DATA> END
     3 ROWS READ
MTB > CHISQUARE USING C1-C4

Expected counts are printed below observed counts
              C1       C2       C3       C4     Total
     1        40       38       41       45      164
            54.7     27.3     27.3     54.7

     2        92       35       38       98      263
            87.7     43.8     43.8     87.7

     3        68       27       21       57      173
            57.7     28.8     28.8     57.7

Total       200      100      100      200      600

ChiSq =    3.93 +   4.16 +   6.83 +   1.71 +
           0.21 +   1.78 +   0.78 +   1.22 +
           1.85 +   0.12 +   2.13 +   0.01 = 24.73
df = 6
```

classifications are not independent. Examining the individual terms of the chi-square value in Figure 13.7, we note that the larger schools (Johnson and Jefferson) had a lower proportion of students that never use marijuana. At Johnson High School, for example, if these classifications were independent, we would expect 54.7 students to never have tried marijuana; instead we observed 40. The same argument applies to Jefferson High School.

From Table A-6, the p value here is $<.005$. Because of this extremely small value, we can conclude that, at these four schools, students vary considerably in their usage of marijuana. The small p value also implies that there is an extremely strong dependence between the two classifications.

EXERCISES

13.16 Suppose you are interested in determining whether there is a relationship between one's educational preference (major) and one's sex. A random sample of 172 students at Hamilton College yields the following data:

Major	Female	Male	Total
Liberal Arts	35	25	60
Home Economics	6	9	15
Physics	18	21	39
Business	26	32	58
Total	85	87	172

(a) Formulate the necessary hypothesis.

(b) Using the chi-square test, test the hypothesis to determine whether educational preference is independent of sex, using a significance level of .10.

(c) Find and interpret the p value for the chi-square test.

13.17 The table below gives the number of police officers for various cities, classified by sex.

City	Male Officers	Female Officers	Total
Detroit	4180	876	5056
Dallas	2024	272	2296
Fort Worth	764	87	851
Houston	4197	421	4618
Los Angeles	6392	568	6960
Denver	1258	112	1370
Boston	1682	118	1800
San Antonio	1336	74	1410

Do the above data indicate that the cities are not homogeneous with respect to proportions of male and female police officers at a 5% significance level? State the p value.

13.18 A real-estate firm wanted to know whether the type of house purchased is associated with the amount of education of the head of the household. Fox and Jones Construction builds four styles of homes. A random sample of 175 homeowners who own Fox and Jones houses was taken and the educational level of the household was noted.

Type of House	No College Degree	Bachelor's Degree	Master's Degree	Doctoral Degree	Total
1	12	5	1	0	18
2	13	10	8	2	33
3	10	20	25	10	65
4	2	18	30	9	59

Do the data provide sufficient evidence to indicate that the type of house owned is related to the educational level of the head of the household? Use a .05 level of significance.

13.19 Anthropologists describe "child circulation" as the transfer of nurturance responsibility for a child from one adult to another. Fonseca (1986) conducted an analysis of decisions by women in urban squatter settlements in Brazil to place their children with the state juvenile authorities, *Fundacão Estadual para o Bem-Estar do Menor* (FEBEM). The author's field work took place between April 1981 and March 1983 in the Vila do Cachorro Sentado, a shantytown in a middle-class area of Porto Alegre in southern Brazil. Consider the following data:

INCIDENCE OF PARTICIPATION IN CHILD CIRCULATION ACTIVITIES

Age of Female	Participant	Nonparticipant	Total
14 to 19 years old	40	28	68
20 to 34 years old	38	16	54
Over 35 years old	20	2	22
Total	98	46	144

Source: Claudio Fonseca, Orphanages, foundlings, and foster mothers: The system of child circulation in a Brazilian squatter settlement. *Anthropological Quarterly* **59** (1): 18, (January 1986), Table 1.

(a) Are age of the women and incidence of participation in child circulation activities independent or related? Perform a chi-square test at 10% significance.

(b) State your conclusion, and report the *p* value.

(c) Would your conclusion change, at 5% significance or at 1% significance?

13.20 A national skiing magazine surveys 500 skiers. It prepares the following tabulation by sex and skill level. Do these data provide sufficient evidence to indicate that skiing ability is related to gender? Use a 5% significance level.

Skiing Ability	Male	Female	Total
Advanced	98	22	120
Intermediate	189	147	336
Beginner	30	14	44
			500

13.21 An aspiring politician decided to sample 300 citizens from each of two major cities to find out whether the two populations were homogeneous with regard to their opinion on gun control. Do the following data indicate a lack of homogeneity? Use a 10% significance level.

City	Favor Gun Control	Against Gun Control
A	126	174
B	148	152

13.22 Rosenthal and others (1986) reported on 9 preschool age children with a history of suicidal behaviors, who were admitted to the University of Massachusetts Medical Center between April 1981 and July 1983. As a part of the study, these 9 suicidal inpatients were assessed for type of death-seeking behavior, and were compared in this respect with another group of 16 outpatient preschoolers. The results are summarized below:

Death-Seeking Behavior	Inpatients	Outpatients
Setting self on fire	1	1
Drug ingestion	2	4
Jumping from high places	2	5
Cutting/stabbing	3	1
Running into fast traffic	1	5
Putting rope around neck	1	0
Total	10	16

Source: P. A. Rosenthal, S. Rosenthal, M. B. Doherty, and D. Santora, Suicidal thoughts and behaviors in depressed hospitalized preschoolers. *Am. J. Psychotherapy* **40** (2): 204 (April 1986), Table 1.

(a) Conduct a chi-square test of homogeneity to decide whether the inpatient and outpatient preschoolers differ significantly in their death-seeking behaviors. Use $\alpha = .05$.

(b) State your conclusion, and the p value for the test.

13.23 A research team conducts a study to see whether voting for the candidates in the recent local election is homogeneous within age group (given in years). One hundred voters were randomly selected from each of five age classifications. Do the data indicate that voting is homogeneous with respect to age group? Use a 5% significance level.

Age	Candidate A	Candidate B	Candidate C	Total
Less than 25	48	22	30	100
25 and <35	55	20	25	100
35 and <45	50	28	22	100
45 and <55	45	21	34	100
55 or older	49	21	30	100

13.24 In the field of organizational psychology, extensive study has been made of different leadership styles. One researcher refers to two extremes as authoritarian vs. democratic; another refers to task-oriented vs. people-oriented; yet others have their own labels for these qualities. Whatever the label, do these different styles affect the morale of the subordinates? To address this issue, a researcher established a ranking scale for worker morale, based on interviews, and grouped the workers into low, acceptable, and high morale categories. These were cross-classified against the leadership style of the supervisor. The contingency table below summarizes the results.

Worker Morale	Leadership Style Authoritarian	Democratic	Total
Low	10	5	15
Acceptable	8	12	20
High	6	9	15
Total	24	26	50

(a) Apply the chi-square test of independence on the above data, at a 5% significance level.

(b) State the p value for your test.

(c) Is worker morale related to the supervisor's leadership style, or are these qualities independent?

13.25 An immigration attorney was investigating which industries to target for obtaining new clients who might have problems with changes in the immigration laws. Five industries were selected. Twenty workers were chosen in each industry, and their visa status was verified. The data are summarized below:

Visa status	Industry A	B	C	D	E	Total
Illegal alien	8	10	5	10	1	34
Legal resident	4	2	6	4	9	25
U.S. citizen	8	8	9	6	10	41
Total	20	20	20	20	20	100

(a) Are the five industries homogeneous with respect to the visa status of its workers? Use $\alpha = .05$.

(b) State the p value.

SUMMARY

When performing a two-tailed test of hypothesis on a binomial parameter (for example, $p = .75$) we can use a chi-square test statistic. The advantage of this approach is that it extends easily to the **multinomial situation**, where each trial can result in any specified number of outcomes. An example is the roll of a single die, which has six possible outcomes on each roll.

For the multinomial situation, the probability of observing each possible outcome may be specified (such as 1/6 for each outcome in the single die illustration). To test the hypothesis, a random sample of observations is obtained, and a chi-square test statistic is evaluated either to reject or to fail to reject this set of probabilities (percentages). Such a test is referred to as a **chi-square goodness-of-fit** test. The form of this chi-square test statistic is

$$\chi^2 = \sum \frac{(O - E)^2}{E}$$

where

1 O represents the observed frequency of observations in a particular category (such as the observed number of 3's in 60 rolls of a single die).

2 E is the expected frequency for this category. For example, we would expect to see $60 \cdot 1/6 = 10$ values of 3 in the die illustration.

3 The chi-square value is obtained by summing over all categories of the multinomial random variable.

4 Categories must be combined (pooled) together whenever an expected value (E) for a particular category is less than 5.

The chi-square goodness-of-fit procedure can be used to determine whether a certain set of sample data came from a specified probability distribution. For example, you might attempt to determine whether a given set of data came from a normal distribution. By collecting a random sample, you can compare the observed values in each class with what you would expect if the null hypothesis—H_0: the data are from a normal distribution—is true. If the calculated chi-square value is significantly large (in the right tail), this hypothesis will be rejected. Whenever any of the parameters for this distribution are unknown (such as μ and σ for the normal illustration), they can be estimated using the sample data. The degrees of freedom of the chi-square test statistic are reduced by one for each estimated parameter.

Finally, this chi-square statistic can be used to test whether two classifications (such as sex and color preference) used to define a contingency table are independent. This is the **chi-square test of independence**. The expected value within each cell of the contingency table is determined under the assumption that H_0 is true, where H_0 claims that the row and column classifications are independent. This also leads to a right-tailed rejection region using the chi-square statistic.

This procedure can be used as a **test for homogeneity** when fixed sample sizes are used for each row or column of the table. If the column totals are fixed in advance, this test will determine whether the populations defined by the column categories are homogeneous (identical) with respect to the variable defining the rows. A similar argument applies when the row totals are predetermined. The test statistic used for a test of homogeneity is the same chi-square statistic used in the test of independence.

REVIEW EXERCISES

13.26 The manager of the Grandiose Hotel guarantees that a customer's room will be ready at 6:00 P.M. if a reservation is made. Otherwise, the customer stays at the hotel for free. The manager believes that this policy should be continued; he believes that a room is not available on time only 5% of the time. A random sample of 200 past reservations was selected and the estimate of the proportion of times when a room was not available on time was .065.

(a)   Letting the significance level be .05, test the belief that the proportion of occurrences when a room is not available on time is .05. Use the Z test from Section 11.2.

(b)   Use the chi-square goodness of fit test instead of the Z test in question (a).

(c)   What is the relationship between the test statistic in questions (a) and (b)?

13.27   A student wanted to know whether the answers a, b, c, d, and e on the standardized departmental multiple choice test occurred equally as often. Several old departmental tests were randomly selected and the occurrences of each answer were tabulated.

Answer	Times Used As Answer
a	39
b	26
c	43
d	42
e	25

Test the belief that all answer choices occur equally often. Use a significance level of .10.

13.28   A large department store in New York City has five entrances and exits. It is believed that the proportion of shoppers entering or leaving the store is approximately the same for each of the five doorways on any single day. The number of customers entering or leaving the store is tallied at each doorway for 3 randomly selected days.

Doorways	Customers
1	150
2	123
3	126
4	163
5	152

Do the data justify the statement that all five entrances and exits are used equally often? Use a 5% significance level.

13.29   The police department in a suburb of Los Angeles believes that 50% of the rapes occur between the hours of 10 P.M. and midnight, 20% occur between 6 P.M. and 10 P.M. and 20% occur between midnight and 3 A.M. One hundred randomly selected rape cases were selected and the times were noted.

Time	Number of Occurrences
6 P.M. to 10 P.M.	24
10 P.M. to midnight	44
Midnight to 3 A.M.	22
Other times	10

Do the data justify the percentages given by the police department? Use a 1% significance level.

13.30 The Consumer Nutrition Awareness Group's data on degree of compliance with food-labeling laws in four major supermarket chains is shown below.

	Supermarket Chain				
Food Labeling Compliance	A	B	C	D	Total
Full	60	50	72	53	235
Substantial	30	45	18	44	137
Inadequate	35	30	35	28	128
Total	125	125	125	125	500

(a) Are the four supermarket chains similar with regard to food labeling compliance, at a 1% significance level?

(b) What is the p value for the test?

13.31 Consider the following cross-classification of child prodigies and precocious geniuses by talent and emotional stability:

	Emotional Stability		
Talent	Stable	Unstable	Total
Musical genius	10	7	17
Math wizard	15	5	20
Artist	3	5	8
Poet	2	3	5
Total	30	20	50

(a) Are talent and emotional stability independent qualities? Use the chi-square test at $\alpha = .05$.

(b) Find the p value for the test.

13.32 Intermarriage is a long-standing topic of interest in the study of race and ethnic relations. One consequence of ethnic intermarriage is new cohorts who are themselves of mixed ethnic origins. Lieberson and Waters (1985) investigated marriage patterns among persons of mixed ethnic ancestry. Their results were reported in a multiplicity of tables, which cannot all be reproduced here. However, the following table might provide some of the flavor of their study:

	Ethnic Ancestry of Wife		
Ethnic Ancestry of Husband	Italian only	Italian/ Polish	Polish only
Italian	50	28	8
Polish	6	17	40
Neither	45	57	52

Source: Stanley Lieberson and Mary Waters, Ethnic mixtures in the United States. Sociol. Social Res. **70** (1): 43–52 (October 1985), Table 4.

(a) Are ethnic ancestry of husband and wife in the above table related or independent? Conduct a test at a 5% level of significance.

(b) State your conclusion, and report the p value.

13.33 Attacks of dizziness and falls occur frequently in the elderly. One cause of such attacks is vertebrobasilar insufficiency secondary to cervical spondylosis. Spondylosis is an abnormal immobility of the cervical vertebrae. Neck movements lead to pressure on, and reduced blood flow through, the vertebral arteries, causing brain-stem ischemia (insufficient flow of oxygen-rich blood to an organ). Adams *et al.* (1986) examined radiographs (x rays) of the cervical spine of 32 elderly patients diagnosed as having vertebrobasilar insufficiency secondary to cervical spondylosis. These 32 symptomatic patients with a history of dizziness or falls related to neck movements were compared to a control group of 32 asymptomatic subjects matched for age and sex. The x rays were rated on a scale of 0 (normal) to 3 (severe pathology) based on the degree of spondylosis and bone growth. The authors reported the following distribution:

Relationship of Radiological Score of Severity of Spondylosis Between the Asymptomatic and Symptomatic Subjects

Symptoms	Normal or Mild	Moderate	Severe	Total
Asymptomatic	10	10	12	32
Symptomatic	5	10	17	32

Source: K. R. H. Adams, M. W. Yung, Michael Lye, and G. H. Whitehouse. Are cervical spine radiographs of value in elderly patients with vertebrobasilar insufficiency? *Age and Aging: J. Br. Geriatrics Soc.* **15** (1): 57–59 (January 1986).

(a) The purpose of the study was to determine if routine cervical spine x rays were useful for the kind of patients already described above (which would be indicated by a significant difference between the symptomatic and asymptomatic groups). Alternatively, since radiological evidence of cervical spondylosis was fairly widespread among old people, even nonsymptomatic persons, the x rays would not be of much value. Conduct the proper test at 5% significance. Report the p value.

(b) Does the study suggest routine cervical spine radiographs as being useful?

(c) Why were the control subjects (asymptomatics) matched for age and sex?

13.34 A computer generates 100 observations from a normally distributed population with mean 35 and standard deviation of 2. The results of the 100 observations generated are:

Interval	Observation Frequency
Less than 32	6
32 but less than 33	9
33 but less than 34	12
34 but less than 35	23
35 but less than 36	19
36 but less than 37	15
37 but less than 38	11
38 or more	5

Use the chi-square goodness-of-fit procedure to test that there is enough evidence to support the conclusion that the generated numbers did not come from a normally distributed population with mean = 35 and standard deviation = 2. Use a 1% significance level.

13.35 Researchers at Yale and Harvard have predicted that white women born in the mid 1950's (so called "baby-boom" women) who remain unmarried by age 30 have only a 20% chance of ever getting married. Suppose that 10 baby-boom women are selected in each of 20 American towns, cities, and rural communities. Assume that these 200 (20 times 10) women are representative of the "baby-boom" population. Followup studies are done for many years. Finally, at the end of the experiment, the number of married women in each group of 10 is noted for all the 20 cities.

City	No. of married women (out of 10)	City	No. of married women (out of 10)
1	2	11	4
2	1	12	0
3	1	13	2
4	0	14	1
5	2	15	1
6	3	16	2
7	1	17	3
8	2	18	2
9	3	19	5
10	4	20	6

(a) Make up a discrete frequency distribution, where X = the number of women who got married in each group of 10, and frequency is the number of times that X was seen in the 20 cities.

(b) Test whether the data indicate a binomially distributed variable, with p = .20. Use a 5% significance level.

(c) State the p value.

13.36 The Meyers-Briggs Type Indicator (MBTI) is a personality scale that can be used to classify qualities like Extrovert (E), Introvert (I), Intuitive (N), Feeling (F), Sensing (S), Thinking (T), Judging (J), Perceptive (P). Thus, EN means Extrovert–Intuitive, SP means Sensing–Perceptive. Consider the following hypothetical frequencies for a cross-tabulation of four types of personality using the MBTI against profession.

Profession	Personality (MBTI)				Total
	EN	IF	SP	JT	
Computer Programmer	4	6	5	6	21
Accountant	3	7	5	5	20
Marketeer	9	3	7	4	23
Educator	5	5	5	5	20
Total	21	21	22	20	84

(a) From the above table, can you conclude at a 1% significance level, whether personality type and profession are related?

(b) State the p value for your test.

13.37 A volunteer working with church groups and other organizations to assist refugees from Central America estimates that only 1 in 5 applications for asylum has a chance of getting approval from the U.S. Immigration and Naturalization Service. Samples of 5 Central American refugees at a time are taken from church groups and various other assistance organizations. In each sample, the number of applications for asylum ultimately approved are noted. Twenty such samples are taken, to obtain a total of 100 observations. The results are summarized below, where X = the number of applications approved in each lot of 5:

X	Frequency
0	6
1	9
2	2
3	1
4	2
5	0

(a) Test whether the above data follow a binomial distribution with $p = .20$ (approval rate of 20%). Use $\alpha = .05$.

(b) Does your test support the volunteer's belief?

(c) Suppose we do not have any preconceived idea of the success rate for asylum applications. Do the above data follow a binomial distribution? Use a 5% significance level for the test. (*Hint:* You have to estimate p yourself.)

(d) What is your estimate for the success rate of asylum applications by Central American refugees?

13.38 A pollster asked 120 Democrats and Democrat-leaning Independents whom they would like to see nominated as the Democratic party's candidate for president in 1988. The poll was repeated in January, April, July, and October 1986. The following data are adapted from newspaper reports.

Candidate	Jan.	Apr.	July	Oct.
Gary Hart	46	40	34	32
Mario Cuomo	23	26	22	27
Lee Iacocca	17	16	26	26
Jesse Jackson	15	16	17	14
Others	19	22	21	21
Total	120	120	120	120

From the above data, does it appear that the relative support for the various candidates has remained stable from January to October? (*Hint:* Test for homogeneity.) Use $\alpha = .05$.

13.39 A sample of households classified as having incomes below the poverty level revealed the following distribution of persons by age:

Age	Frequency
0 to <5	27
5 to <18	53
18 to <22	16
22 to <45	60
45 to <65	26
65 and above	18
	200

(a) Compute the mean and standard deviation for the above distribution.

(b) Using a 10% significance level, determine with a chi-square test whether the data fits a normal population.

(c)   Find the p value for the test.

(d)   Does your conclusion change if $\alpha = .05$ or $\alpha = .01$?

13.40   The Cranberry Mountain Visitor Center in the Monongahela National Forest in Richwood, West Virginia, includes 53,000 acres of back country, 36,000 of which are part of the National Wilderness Preservation System, and the 750-acre Cranberry Glades Botanical Area, which contains the largest bogs in the state. The Visitor Center is looked upon as being the pulse of the Monongahela National Forest. In a study of population density and crowding, Burrus-Bammel and Bammel (1986) noted the following pattern of arrivals of visitors at the Cranberry Mountain Visitor Center between July 24 and August 5, 1983, where "party size" is the number of persons per vehicle and "number" is the number of such vehicle arrivals:

Party Size	Number	Percentage
1	25	6.5%
2	165	43.0
3	62	16.1
4	83	21.6
5	21	5.5
6	14	3.6
7	3	0.8
8	4	1.0
9	4	1.0
10	1	0.3
11 or more	2	0.6
	384	100.0%

Source: L. L. Burrus-Bammel and G. Bammel, Visiting patterns and effects of density at a Visitors' Center. J. Environ. Educ. 18 (1): 8 (Fall 1986), Table 2.

(a)   Test whether the above data on arrival patterns follows a Poisson distribution, at a 5% significance level.

(b)   State your conclusion, and report the p value for the test.

13.41   Is domestic violence more likely to occur on certain days of the week (such as weekends), or is it uniformly spread over all days of the week? Use the following data to conduct a hypothesis test at a 5% significance level, and state your conclusion.

Day of the Week	No. of Reports of Domestic Violence (from Police Records)
Monday	8
Tuesday	7
Wednesday	9
Thursday	8
Friday	12
Saturday	13
Sunday	13
Total	70

13.42 To illustrate the effect of stereotyped images on sexual preferences, an instructor gave the male students in his class three photographs of the same girl. The first photograph showed the girl with her "natural" look, without makeup. The second photograph showed the girl with her hair dyed black and wearing glasses, but generally well groomed and with her makeup. The third photograph showed the girl with her hair dyed blonde, with makeup and a glamorous hair style. The male students were asked to rate the girls for a date as first, second, and third preference. They were not told it was the same girl in all three photographs. The results were recorded as shown below:

	Photo 1	Photo 2	Photo 3
Number of male students showing first preference for a date	8	14	16

Since this was a small sample, the instructor told the class that there would be some random deviations from expectations. The expectation was that no one photograph should have higher preference over the others, since they were all of the same girl. If there was a significant deviation from a uniform distribution, the class would assume this might be due to unconscious stereotypes influencing the choices.

(a) Are the above choices uniformly distributed, or is there a significant deviation from that expectation? Use $\alpha = .05$.

(b) Interpret your result in light of the above discussion.

13.43 Laboratory and radiographic tests are an important factor in the growth of health care expenses. Since physical examinations and investigation of patients' histories tend to be time-consuming, physicians might tend to prefer common office tests, such as the EKG, CBC, and UA tests. Epstein, Krock, and McNeil (1984) investigated how doctors perceived the profitability of such tests. One aspect of their study, which covered practicing physicians in Massachusetts, was whether the perception of profitability was different for patients with different types of insurance coverage. The results for EKG (electrocardiogram) and UA (urinalysis) tests are given in two separate 2×3 contingency tables below:

	Type of Insurance	PERCEPTION OF PROFITABILITY ("Appropriateness of Reimbursement")	
		Appropriate or More Than Appropriate	Less Than Appropriate
Table 1 (EKG)	Medicare	63	18
	Blue Cross/Blue Shield	61	23
	Medicaid	7	70
Table 2 (UA)	Medicare	38	15
	Blue Cross/Blue Shield	41	15
	Medicaid	8	43

Source: A. M. Epstein, S. J. Krock, and B. J. McNeil, Office laboratory tests: Perceptions of profitability. Medical Care 22 (2): 163 (February 1984), Table 3.

(a) The researchers analyzed the above data using the chi-square statistic and concluded that perceptions of profitability were significantly different for the different insurance coverages. Perform the necessary tests on Tables 1 and 2 (two separate tests) to decide if the researchers' conclusions are supported. Use a 1% significance level. (*Hint:* To find differences, you can test for similarities.)

(b) "Collapse" (i.e., combine) the two tables into one 2 × 3 table, and repeat the test. (For example, the category "less than appropriate" would be aggregated as 70 + 43 = 113 for Medicaid, and so on.)

(c) Comment on (a) and (b) above. For example, would it be possible (theoretically and not necessarily in this particular instance) to find the chi-square statistic significant in (b), yet not significant in Table 1 or 2? Explain why.

CASE STUDY

Recreational Home Owners and the Environment

Areas with natural resources and recreational amenities attract tourism and often become prime targets for new home developments, especially where land is relatively inexpensive. Frequently, these areas are in environmentally sensitive locations. New recreational home developments, such as a second home used for vacation purposes on a seasonal basis, would have an impact on the environment. Some studies have been carried out on the environmental impact of such developments, but these have concentrated on seasonal homes.

Gartner (1987) broadened the scope of his investigation to cover three segments of recreational home development, rather than just one: permanent home owners, seasonal home owners, and recreational lot owners. These three segments were surveyed in Michigan for information on location preferences, reasons for property purchase, frequency of visitation, attitudes and knowledge concerning future developments, and land use controls.

The idea was to examine possible differences in these home-owner segments, and to consider if these differences would help explain the pattern of environmental impact. For example, the author pointed out that a larger proportion of permanent home owners favor lessening of present land use controls, and more seasonal home owners favor stricter land use controls, than the other two groups, respectively. Thus, it might be that the relative flexibility of permanent home owners, who benefit economically from tourism, could lead to extensive environmental damage occurring before the permanent home-owner segment is affected to a degree that causes them to take corrective political action. Gartner's article provided almost a dozen tables of data, but we will be able to look at only a few of them, reproduced below.

Table I

Number and percentage of respondents within each property-owner segment

Type of Development	Number	Percent
Permanent Home Owner	600	29.9
Seasonal Home Owner	738	36.8
Recreational Lot Owner	668	33.3
Total	2006	100%

Table 2

Location of property relative to water by type of development

Type of Development	Lake or Pond	River	No Water	Total
Recreational Lot	178	29	453	660
Percentage	27.0	4.4	68.6	33.5
Seasonal Home	380	46	293	719
	52.8	6.4	40.7	36.6
Permanent Home	159	31	401	591
	26.9	5.2	67.8	30.0
Total	717	106	1147	1970
	36.4	5.4	58.2	100.0

Table 3

Median family income by type of home development

| Family Income (1978 Dollars) | TYPE OF DEVELOPMENT | | |
	Recreational Lot %	Seasonal Home %	Permanent Home %
$0–$5,999	6.4	5.2	17.2
$6,000–$9,999	10.1	9.4	24.4
$10,000–$14,999	12.4	12.9	22.5
$15,000–$25,000	31.1	29.7	21.6
Over $25,000	40.0	42.7	14.3

Source: William C. Gartner, Environmental impacts of recreational home developments. *Ann. Tourism Res.* **14**: 38–57, 1987, Tables 1, 2, 3, and 5.

Case Study Questions

1. Refer to Table 1. Are the three property-owner segments uniformly distributed, or does any one group predominate? Conduct a chi-square test at 5% significance, and report the p value.

2. The author gave a contingency table of the three property-owner segments cross-classified against five reasons for the property acquisition. The five reasons were (a) as an investment or retirement home, (b) to get out of the city, (c) to hunt and fish, (d) obtained as an inheritance, and (e) other reasons. Although the table itself is not reproduced below, you are hereby informed that the chi-square computed test statistic was 462.36731. At a 1% significance level, do we have sufficient evidence to reject the hypothesis that the three segments were similar in their reasons for the property acquisition? What is the p value for your test?

3. Refer to Table 2. Gartner quoted another writer, Kevin Walter, in the context of "weed-choked lakes and algae-fouled beaches [that] have become commonplace in the state," to describe cultural eutrophication, the "acceleration of the natural aging process of a lake due to man's influence." Conduct a chi-square test of homogeneity on the variables in Table 2, at a 1% level of significance. Are the types of development homogeneous regarding location of property relative to water?

4. On the basis of your analysis in item (3), do you observe significant differences in the preference for lakefront living among the three property owner groups?

5. The author investigated whether income levels among the segments could help explain any differences in preference for lakefront living. Refer to Table 3. It is an intuitively appealing argument that a second home or recreational lot purchase requires greater discretionary income than purchase of a permanent home. If this argument were correct, would type of development and family income be related in Table 3, or would they be independent?

6. Conduct a chi-square test of independence on the data in Table 3, using a 1% significance level, and state your conclusions.

COMPUTER EXERCISES USING THE DATABASE

Using a sample of 100 observations, complete the following exercises.

1. Determine whether the quantitative GRE scores (variable QGRE) follow a normal distribution. Let the significance level be .05.

2. Based upon a significance level of .05, test whether the verbal GRE scores (variable VGRE) follow a normal distribution with a mean of 475 and a standard deviation of 100.

3. Are the marital category and sex classifications independent for students attending Brookhaven? Use a significance level of .05.

4. Define the following age categories for males attending Brookhaven:

   Category 1: age is less than 29
   Category 2: age is at least 29 but under 33
   Category 3: age is at least 33

   Using a significance level of .05, does the age classification appear to be related to the marital classification for the male population attending Brookhaven?

SPSSX APPENDIX (MAINFRAME VERSION)

Solution to Example 13.4

Example 13.4 was concerned with a multinomial goodness-of-fit test. One of the expected values was less than 5, and it was necessary to pool the categories and create a new category with an expected value of at least 5. The problem was to determine whether the percentage of homeowners in each category fit the stated proportions. The percentages in the null hypothesis were .52, .38, and .10. The SPSSX program listing in Figure 13.8 was used to request a computed chi-square value for these pooled categories.

**SPSSX
APPENDIX
(MAINFRAME
VERSION)**

Figure 13.8

SPSSX program
listing used to
request a chi-square
value for goodness-
of-fit test

```
TITLE HOMEOWNER PROFILE
DATA LIST FREE/TYPE FREQ
BEGIN DATA
1 54
2 28
3 18
END DATA
VALUE LABELS TYPE 1 'MARRIED W. CHILDREN' 2 'MARRIED W/O CHILDREN'
    3 'UNMARRIED'
PRINT / TYPE FREQ
WEIGHT BY FREQ
NPAR TESTS CHISQUARE = TYPE/ EXPECTED = 0.52,0.38,0.10
```

The TITLE command names the SPSSX run.

The DATA LIST command gives each variable a name and describes the data as being in free form.

The BEGIN DATA command indicates to SPSSX that the input data immediately follow.

The next four lines are the data values, with each line representing a type (1 = MARRIED W. CHILDREN, and so on) and the observed number of people in this category.

The END DATA statement indicates the end of the input data.

The VALUE LABELS statement assigns the labels MARRIED W. CHILDREN to type 1 records, MARRIED W/O CHILDREN to type 2 records, and UNMARRIED to type 3 records.

The PRINT command requests a printout of the input data.

The WEIGHT command is used to weight the cases by the number of observed people.

Figure 13.9

Output obtained by
executing the SPSSX
program listing in
Figure 13.8

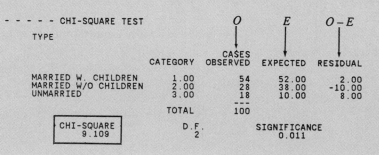

- - - - - CHI-SQUARE TEST		O	E	$O-E$
TYPE				
	CATEGORY	CASES OBSERVED	EXPECTED	RESIDUAL
MARRIED W. CHILDREN	1.00	54	52.00	2.00
MARRIED W/O CHILDREN	2.00	28	38.00	-10.00
UNMARRIED	3.00	18	10.00	8.00
	TOTAL	100		
CHI-SQUARE 9.109	D.F. 2		SIGNIFICANCE 0.011	

The NPAR TESTS CHISQUARE = statement requests a chi-square test between the observed and expected values. For instance, 0.52 is the expected percentage for the variable labeled MARRIED W. CHILDREN.

Figure 13.9 contains the output obtained by executing the program listing in Figure 13.8.

**SPSS/PC+
APPENDIX
(MICRO
VERSION)**

**Solution to
Example 13.4**

Example 13.4 was concerned with a multinomial goodness-of-fit test. One of the expected values was less than 5, and it was necessary to pool the categories and create a new category with an expected value of at least 5. The problem was to determine whether the percentage of homeowners in each category fit the stated proportions. The percentages in the null hypothesis were .52, .38, and .10. The SPSS/PC+ program listing in Figure 13.10 was used to request a computed chi-square value for these pooled categories.

The TITLE command names the SPSS/PC+ run.

The DATA LIST command gives each variable a name and describes the data as being in free form.

The BEGIN DATA command indicates to SPSS/PC+ that the input data immediately follow.

The next four lines are the data values, with each line representing a type (1 = MARRIED W. CHILDREN, and so on) and the observed number of people in this category.

Figure 13.10

SPSS/PC+ program listing used to request a chi-square value for goodness-of-fit test

```
TITLE HOMEOWNER PROFILE.
DATA LIST FREE / TYPE FREQ.
BEGIN DATA.
1 54
2 28
3 18
END DATA.
VALUE LABELS TYPE 1 'MARRIED W. CHILDREN'
     2 'MARRIED W/O CHILDREN' 3 'UNMARRIED'.
LIST / TYPE FREQ.
WEIGHT BY FREQ.
NPAR TESTS CHISQUARE = TYPE/ EXPECTED = 0.52,0.38,0.10.
```

SPSS/PC+
APPENDIX
(MICRO
VERSION)

The END DATA statement indicates the end of the input data.

The VALUE LABELS statement assigns the labels MARRIED W. CHILDREN to type 1 records, MARRIED W/O CHILDREN to type 2 records, and UNMARRIED to type 3 records.

The LIST command requests a printout of the input data.

The WEIGHT command is used to weight the cases by the number of observed people.

The NPAR TESTS CHISQUARE = statement requests a chi-square test between the observed and expected values. For instance, 0.52 is the expected percentage for the variable labeled MARRIED W. CHILDREN.

Figure 13.11 contains the output obtained by executing the program listing in Figure 13.10.

Figure 13.11

Output obtained
by executing the
SPSS/PC+ program
listing in Figure
13.10

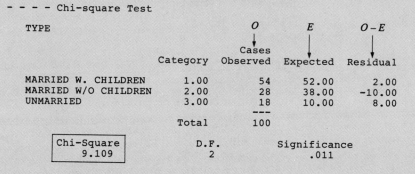

```
                  HOMEOWNER PROFILE

- - - - - Chi-square Test

    TYPE                                    O          E         O-E
                                            ↓          ↓          ↓
                                          Cases
                              Category   Observed   Expected   Residual

    MARRIED W. CHILDREN         1.00        54       52.00       2.00
    MARRIED W/O CHILDREN        2.00        28       38.00     -10.00
    UNMARRIED                   3.00        18       10.00       8.00
                                          ---
                              Total        100

        ┌──────────────┐
        │  Chi-Square  │           D.F.            Significance
        │    9.109     │            2                  .011
        └──────────────┘
```

SPSSX
APPENDIX
(MAINFRAME
VERSION)

Solution to
Example 13.7

Example 13.7 was concerned with a chi-square test of independence. The problem was to determine whether an employee's educational level had an effect on job performance, as measured by an exam. The null hypothesis was that there is no relationship between educational level and job performance. The SPSSX program listing in Figure 13.12 was used to request the chi-square and p value statistics from the data obtained by testing 120 employees.

Figure 13.12

SPSSX program
listing used to
request the
chi-square value and
p value for a test of
independence

```
TITLE EDUCATIONAL LEVEL VERSUS JOB KNOWLEDGE
DATA LIST FREE/ LEVEL PERFORM COUNT
BEGIN DATA
1 1 4
1 2 20
1 3 11
2 1 12
2 2 18
2 3 15
3 1 9
3 2 22
3 3 9
END DATA
VALUE LABELS LEVEL 1 'MASTERS' 2 'BACHELOR' 3 'HIGH SCHOOL'/
             PERFORM 1 'HIGH' 2 'AVERAGE' 3 'LOW'
PRINT / LEVEL PERFORM COUNT
WEIGHT BY COUNT
CROSSTABS TABLES=LEVEL BY PERFORM
STATISTICS 1
OPTIONS 14
```

The TITLE command names the SPSSX run.

The DATA LIST command gives each variable a name and describes the data as being in free form.

The BEGIN DATA command indicates to SPSSX that the input data immediately follow.

The next nine lines are the data values, representing an educational code, a job performance code, and the number of observations that compose the two adjacent categories.

The END DATA statement indicates the end of the input data.

The VALUE LABELS statement assigns codes to different categories (LEVEL and PERFORM) of data. These are positional categories. The first position of the input data stream is the LEVEL category where 1 indicates a master's degree, 2 indicates a bachelor's degree and 3 indicates a high-school degree. For the second position of the data stream, 1 indicates a high score, 2 is an average score, and 3 is a low score.

The PRINT command requests a printout of the input data.

The WEIGHT command requests that the data be weighted by the variable COUNT. This variable indicates the total number of employees who satisfy the criteria in the adjacent LEVEL and PERFORM categories. In the first row of data, 4 employees had master's degrees and scored high on the exam.

The CROSSTABS command produces a cross-tabulation of the variables LEVEL and PERFORM.

The STATISTICS 1 command requests the chi-square test.

The OPTIONS 14 command requests the expected frequencies to be printed.

Figure 13.13 contains the output obtained by executing the program listing in Figure 13.12.

Figure 13.13

Output obtained by executing the SPSSX program listing in Figure 13.12

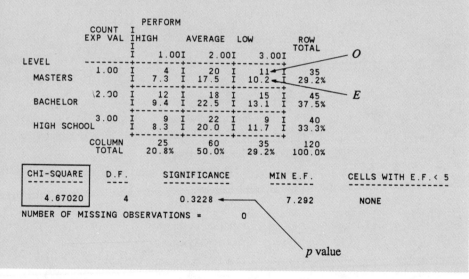

CHI-SQUARE = 4.67020   D.F. = 4   SIGNIFICANCE = 0.3228 ← *p* value   MIN E.F. = 7.292   CELLS WITH E.F. < 5 = NONE

Solution to Example 13.7

Example 13.7 was concerned with a chi-square test of independence. The problem was to determine whether an employee's educational level had an effect on job performance, as measured by an exam. The null hypothesis

SPSS/PC+ APPENDIX (MICRO VERSION)

was that there is no relationship between educational level and job performance. The SPSS/PC+ program listing in Figure 13.14 was used to request the chi-square and p value statistics from the data obtained by testing 120 employees.

Figure 13.14

SPSS/PC+ program listing used to request the chi-square value and p value for a test of independence

```
TITLE EDUCATIONAL LEVEL VERSUS JOB KNOWLEDGE.
DATA LIST FREE/ LEVEL PERFORM COUNT.
BEGIN DATA.
1 1 4
1 2 20
1 3 11
2 1 12
2 2 18
2 3 15
3 1 9
3 2 22
3 3 9
END DATA.
VALUE LABELS LEVEL 1 'MASTERS' 2 'BACHELOR' 3 'HIGH SCHOOL'/
             PERFORM 1 'HIGH' 2 'AVERAGE' 3 'LOW'.
LIST / LEVEL PERFORM COUNT.
WEIGHT BY COUNT.
CROSSTABS TABLES=LEVEL BY PERFORM
 /STATISTICS 1 /OPTIONS 14.
```

The TITLE command names the SPSS/PC+ run.

The DATA LIST command gives each variable a name and describes the data as being in free form.

The BEGIN DATA command indicates to SPSS/PC+ that the input data immediately follow.

The next nine lines are the data values, representing an educational code, a job performance code, and the number of observations that compose the two adjacent categories.

The END DATA statement indicates the end of the input data.

The VALUE LABELS statement assigns codes to different categories (LEVEL and PERFORM) of data. These are positional categories. The first position of the input data stream is the LEVEL category where 1 indicates a master's degree, 2 indicates a bachelor's degree and 3 indicates a high-school degree. For the second position of the data stream, 1 indicates a high score, 2 is an average score, and 3 is a low score.

The LIST command requests a printout of the input data.

**SPSS/PC+
APPENDIX
(MICRO
VERSION)**

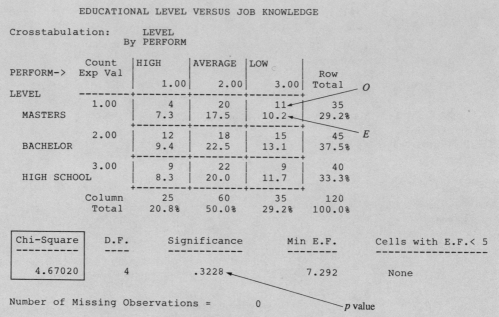

```
                    EDUCATIONAL LEVEL VERSUS JOB KNOWLEDGE

        Crosstabulation:      LEVEL
                          By PERFORM

                    Count  |HIGH    |AVERAGE |LOW    |
          PERFORM->  Exp Val|        |        |       |    Row
                           |   1.00 |   2.00 |   3.00|   Total     O
          LEVEL      --------+--------+--------+--------+
                      1.00  |    4   |   20   |   11   |    35
             MASTERS        |   7.3  |  17.5  |  10.2  |   29.2%
                           +--------+--------+--------+
                      2.00  |   12   |   18   |   15   |    45      E
             BACHELOR       |   9.4  |  22.5  |  13.1  |   37.5%
                           +--------+--------+--------+
                      3.00  |    9   |   22   |    9   |    40
          HIGH SCHOOL       |   8.3  |  20.0  |  11.7  |   33.3%
                           +--------+--------+--------+
                    Column      25       60       35       120
                    Total     20.8%    50.0%    29.2%   100.0%
```

Chi-Square	D.F.	Significance	Min E.F.	Cells with E.F.< 5
4.67020	4	.3228	7.292	None

Number of Missing Observations =   0   *p* value

Figure 13.15

Output obtained
by executing the
SPSS/PC+ program
listing in Figure
13.14

The WEIGHT command requests that the data be weighted by the variable COUNT. This variable indicates the total number of employees who satisfy the criteria in the adjacent LEVEL and PERFORM categories. In the first row of data, 4 employees had master's degrees and scored high on the exam.

The CROSSTABS command produces a cross-tabulation of the variables LEVEL and PERFORM.

The STATISTICS 1 command requests the chi-square test.

The OPTIONS 14 command requests the expected frequencies to be printed.

Figure 13.15 contains the output obtained by executing the program listing in Figure 13.14.

SAS APPENDIX (MAINFRAME VERSION)

Solution to Example 13.7

Example 13.7 was concerned with a chi-square test of independence. The problem was to determine whether an employee's educational level had an effect on job performance, as measured by an exam. The null hypothesis was that there is no relationship between educational level and job performance. The SAS program listing in Figure 13.16 was used to request the chi-square and p value statistics from the data obtained by testing 120 employees.

Figure 13.16

SAS program listing used to request the chi-square value and p value for a test of independence

```
TITLE  'EDUCATIONAL LEVEL VERSUS JOB KNOWLEDGE';
DATA EXAM PERFORM;
 INPUT LEVEL $ PERFORM $ COUNT@@;
CARDS;
MASTERS HIGH 4 MASTERS AVERAGE 20 MASTERS LOW 11
BACHELORS HIGH 12 BACHELORS AVERAGE 18 BACHELORS LOW 15
HIGHSCHOOL HIGH 9 HIGHSCHOOL AVERAGE 22 HIGHSCHOOL LOW 9
PROC PRINT;
PROC FREQ;
 WEIGHT COUNT;
 TABLES LEVEL*PERFORM/CHISQ;
```

The TITLE command names the SAS run.

The DATA command gives the data a name.

The INPUT command names and gives the correct order for the variables during input. The $ indicates that both LEVEL and PERFORM are character data. The @@ indicates that each card image contains two additional sets of data.

The CARDS command indicates to SAS that the input data immediately follow.

The next three lines are the data values. The first line, for example, indicates that 4 employees with a master's degree scored high on the exam, 20 scored average, and 11 scored low.

SAS APPENDIX (MAINFRAME VERSION)

The PROC PRINT command directs SAS to list the input data.

The PROC FREQ command and WEIGHT COUNT subcommand specify that the values of the variable COUNT are relative weights for the observations.

The TABLES subcommand produces a cross-tabulation of the variables LEVEL and PERFORM.

The CHISQ command generates the chi-square statistic.

Figure 13.17 contains the output obtained by executing the program listing in Figure 13.16.

Figure 13.17

Output obtained by executing the SAS program listing in Figure 13.16

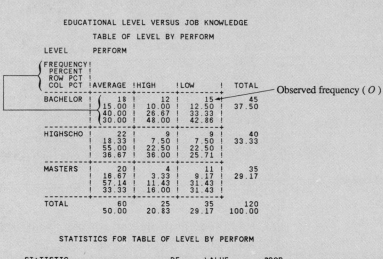

**SAS
APPENDIX
(MICRO
VERSION)**

**Solution to
Example 13.7**

Example 13.7 was concerned with a chi-square test of independence. The problem was to determine whether an employee's educational level had an effect on job performance, as measured by an exam. The null hypothesis was that there is no relationship between educational level and job performance. The SAS program listing in Figure 13.18 was used to request the chi-square and p value statistics from the data obtained by testing 120 employees.

Figure 13.18

SAS program listing used to request the chi-square value and p value for a test of independence

```
TITLE  'EDUCATIONAL LEVEL VERSUS JOB KNOWLEDGE';
DATA EXAM PERFORM;
 INPUT LEVEL $ PERFORM $ COUNT@@;
CARDS;
MASTERS HIGH 4 MASTERS AVERAGE 20 MASTERS LOW 11
BACHELORS HIGH 12 BACHELORS AVERAGE 18 BACHELORS LOW 15
HIGHSCHOOL HIGH 9 HIGHSCHOOL AVERAGE 22 HIGHSCHOOL LOW 9
PROC PRINT;
PROC FREQ;
 WEIGHT COUNT;
 TABLES LEVEL*PERFORM/CHISQ;
RUN;
```

The TITLE command names the SAS run.

The DATA command gives the data a name.

The INPUT command names and gives the correct order for the variables during input. The $ indicates that both LEVEL and PERFORM are character data. The @@ indicates that each card image contains two additional sets of data.

The CARDS command indicates to SAS that the input data immediately follow.

The next three lines are the data values. The first line, for example, indicates that 4 employees with a master's degree scored high on the exam, 20 scored average, and 11 scored low.

The PROC PRINT command directs SAS to list the input data.

The PROC FREQ command and WEIGHT COUNT subcommand specify that the values of the variable COUNT are relative weights for the observations.

The TABLES subcommand produces a cross-tabulation of the variables LEVEL and PERFORM.

SAS APPENDIX (MICRO VERSION)

The CHISQ command generates the chi-square statistic.

The RUN command tells the SAS system to execute the previous SAS statements.

Figure 13.19 contains the output obtained by executing the program listing in Figure 13.18.

Figure 13.19

Output obtained by executing the SAS program listing in Figure 13.18

```
                    EDUCATIONAL LEVEL VERSUS JOB KNOWLEDGE

                         TABLE OF LEVEL BY PERFORM

            LEVEL       PERFORM

           Frequency|
            Percent |
            Row Pct |
            Col Pct |AVERAGE |HIGH    |LOW     |  Total
           ---------+--------+--------+--------+
            BACHELOR|     18 |     12 |     15 |     45
                    |  15.00 |  10.00 |  12.50 |  37.50
                    |  40.00 |  26.67 |  33.33 |
                    |  30.00 |  48.00 |  42.86 |
           ---------+--------+--------+--------+
            Total         60       25       35      120
                       50.00    20.83    29.17   100.00
           (Continued)
```
Observed frequency (O)

```
                         TABLE OF LEVEL BY PERFORM

            LEVEL       PERFORM

           Frequency|
            Percent |
            Row Pct |
            Col Pct |AVERAGE |HIGH    |LOW     |  Total
           ---------+--------+--------+--------+
            HIGHSCHO|     22 |      9 |      9 |     40
                    |  18.33 |   7.50 |   7.50 |  33.33
                    |  55.00 |  22.50 |  22.50 |
                    |  36.67 |  36.00 |  25.71 |
           ---------+--------+--------+--------+
            Total         60       25       35      120
                       50.00    20.83    29.17   100.00
           (Continued)
```

SAS APPENDIX (MICRO VERSION)

```
                        TABLE OF LEVEL BY PERFORM

            LEVEL       PERFORM

            Frequency|
             Percent |
             Row Pct |
             Col Pct |AVERAGE |HIGH    |LOW     | Total
            ---------+--------+--------+--------+
            MASTERS  |    20  |     4  |    11  |    35
                     |  16.67 |  3.33  |  9.17  | 29.17
                     |  57.14 | 11.43  | 31.43  |
                     |  33.33 | 16.00  | 31.43  |
            ---------+--------+--------+--------+
            Total         60       25       35      120
                       50.00    20.83    29.17   100.00
```

STATISTICS FOR TABLE OF LEVEL BY PERFORM

Statistic	DF	Value	Prob
Chi-Square	4	4.670	0.323 ←— *p* value
Likelihood Ratio Chi-Square	4	4.986	0.289
Mantel-Haenszel Chi-Square	1	1.094	0.296
Phi		0.197	
Contingency Coefficient		0.194	
Cramer's V		0.139	

Sample Size = 120   computed value of
                                   chi-square statistic

14

SIMPLE LINEAR REGRESSION

A LOOK BACK/INTRODUCTION

Chapter 4 introduced bivariate data. These are sample data on two variables, such as age and income, which can be graphically represented using a scatter diagram. We also measured the strength of the linear relationship between these two variables using the coefficient of correlation. Our discussion was descriptive; we measured and illustrated how these two variables reacted together in a sample from a bivariate distribution. If you believe that a "significant" linear relationship exists, your next step is to construct the best-fitting line through the sample points.

We now turn our attention to the question of what we are estimating when using a bivariate sample. How can we determine whether a significant linear relationship exists? We will answer this by introducing the concept of a **statistical model** and the assumptions behind it. Various tests of hypothesis will examine the adequacy of this model (Is it a good one?), and an assortment of confidence intervals will measure the reliability of the corresponding estimates using this model.

14.1 BIVARIATE DATA AND CORRELATION

In bivariate data, each observation consists of data on two variables. For example, you obtain a sample of people and record their ages (X) and cholesterol levels (Y). Or, for each month, you record the average interest rate (X) and the number of new housing starts (Y). These data are paired.

Figure 14.1

Scatter diagram of
university data. (A)
Scatter diagram of
sample data. (B) Line
through sample data.

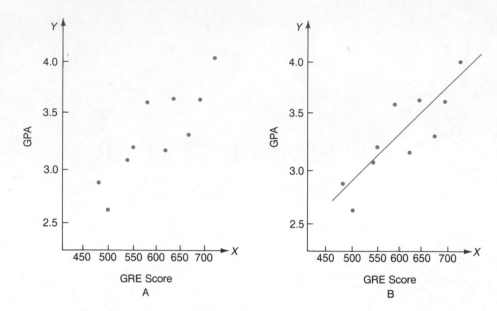

Suppose that the admissions officer at Brookhaven University is interested in the relationship between the verbal score on the Graduate Record Exam (GRE) and the first year grade point average (GPA) of graduate students majoring in the behavioral and social sciences. A random sample of 10 such students is obtained with the following results:

GRE score	580	500	670	480	710	550	640	540	620	690
First Year GPA	3.54	2.62	3.30	2.90	4.00	3.21	3.57	3.05	3.15	3.61

Figure 14.1a provides a scatter diagram of these data. In this graph, each observation is represented by a point. The underlying pattern here appears to be that larger verbal GRE scores (X) are associated with larger GPAs (Y). This means that X and Y have a **positive relationship**. A **negative relationship** occurs when Y decreases as X increases; for example, when Y is the demand for a particular consumer product and X is the selling price.

We next try to determine whether we can estimate this relationship by means of a straight line. One possible line is sketched in Figure 14.1b, which passes among these points and has a positive slope. To measure the strength of the linear relationship between these two variables, we determine the coefficient of correlation.

Coefficient of Correlation

It is often difficult to determine whether a *significant* linear relationship exists between X and Y by inspecting a scatter diagram of the data. A second procedure is to include a measure of this linearity—the sample coefficient of correlation.

It is computed from the sample data by combining these pairs of values into a single number. Thus, the sample **coefficient of correlation, r,** measures the strength of the linear relationship that exists within a sample of n bivariate data. Its value is given by

$$r = \frac{\Sigma(x - \bar{x})(y - \bar{y})}{\sqrt{\Sigma(x - \bar{x})^2}\sqrt{\Sigma(y - \bar{y})^2}} \qquad \text{(14-1)}$$

$$= \frac{\Sigma xy - (\Sigma x)(\Sigma y)/n}{\sqrt{\Sigma x^2 - (\Sigma x)^2/n}\sqrt{\Sigma y^2 - (\Sigma y)^2/n}} \qquad \text{(14-2)}$$

where Σx = sum of X values, Σx^2 = sum of X^2 values, Σy = sum of Y values, Σy^2 = sum of Y^2 values, Σxy = sum of XY values, $\bar{x} = \Sigma x/n$, and $\bar{y} = \Sigma y/n$.

Possible values for r range from -1 to 1. A value of -1 indicates a perfect negative linear relationship, 0 indicates no linear relationship, and 1 indicates a perfect positive linear relationship. Scatter diagrams for various values of r are shown in Chapter 4, Figure 4.4, page 111.

Sums of Squares

We will introduce a shorthand notation at this point, related to the notation in Chapter 12 for ANOVA. Let

$$\begin{aligned} \text{SS}_X &= \text{sum of squares for } X \\ &= \Sigma(x - \bar{x})^2 \qquad \text{(14-3)} \\ &= \Sigma x^2 - (\Sigma x)^2/n \end{aligned}$$

$$\begin{aligned} \text{SS}_Y &= \text{sum of squares for } Y \\ &= \Sigma(y - \bar{y})^2 \qquad \text{(14-4)} \\ &= \Sigma y^2 - (\Sigma y)^2/n \end{aligned}$$

$$\begin{aligned} \text{SS}_{XY} &= \text{sum of squares for } XY \\ &= \Sigma(x - \bar{x})(y - \bar{y}) \qquad \text{(14-5)} \\ &= \Sigma xy - (\Sigma x)(\Sigma y)/n \end{aligned}$$

Using this notation, we can write r as

$$r = \frac{\text{SS}_{XY}}{\sqrt{\text{SS}_X}\sqrt{\text{SS}_Y}} \qquad \text{(14-6)}$$

EXAMPLE 14.1

Determine the sample correlation coefficient for the university data in Figure 14.1.

Solution

Calculations can be organized as follows:

Student	X (GRE)	Y (GPA)	XY	X^2	Y^2
1	580	3.54	2053.2	336,400	12.5316
2	500	2.62	1310.0	250,000	6.8644
3	670	3.30	2211.0	448,900	10.8900
4	480	2.90	1392.0	230,400	8.4100
5	710	4.00	2840.0	504,100	16.0000
6	550	3.21	1765.5	302,500	10.3041
7	640	3.57	2284.8	409,600	12.7449
8	540	3.05	1647.0	291,600	9.3025
9	620	3.15	1953.0	384,400	9.9225
10	690	3.61	2490.9	476,100	13.0321
	5980	32.95	19,947.4	3,634,000	110.002

Using the totals from this table,

$$SS_X = 3,634,000 - (5980)^2/10 = 57,960$$

$$SS_Y = 110.002 - (32.95)^2/10 = 1.4318$$

$$SS_{XY} = 19,947.4 - (5980)(32.95)/10 = 243.3$$

The value of the sample correlation coefficient is

$$r = \frac{SS_{XY}}{\sqrt{SS_X}\sqrt{SS_Y}}$$

$$= \frac{243.3}{\sqrt{57,960}\sqrt{1.4318}} = \frac{243.3}{288.07}$$

$$= .8446$$

Interpreting the Correlation Coefficient

Having calculated a value of r, you next need to interpret the result. In example 14.1, is $r = .8446$ large enough to conclude that a significant linear relationship exists between the verbal GRE score and GPA? In Chapter 4, we used a table to answer this question. We did not discuss the assumptions behind this table, or the probability of concluding that a linear relationship exists when in fact it does not. Both topics will be discussed later in this chapter, when we introduce the idea of satatistical modeling. We will outline another test of hypothesis that enables you to determine if the value of r leads to a conclusion that a significant (positive or negative) linear relationship exists between the two variables.

Covariance

Another commonly used measure of the association between two variables X and Y is the sample covariance, written $\text{cov}(X, Y)$. It is similar to the correlation between these two variables. For one thing, the covariance and correlation always have the same sign. Consequently, if large values of X are associated with large values of Y, then both the covariance and correlation are positive. Similarly, both values are negative whenever large values of X are associated with small values of Y. For any two variables X and Y the sample **covariance** between these variables is

$$\text{cov}(X, Y) = \frac{1}{n-1} \sum (x - \bar{x})(y - \bar{y}) \qquad \textbf{(14-7)}$$

$$\text{cov}(X, Y) = \frac{1}{n-1} \text{SS}_{XY} \qquad \textbf{(14-8)}$$

In example 14.1, the covariance between GRE score (X) and GPA (Y) is

$$\text{cov}(X, Y) = \frac{1}{n-1} \text{SS}_{XY}$$
$$= \tfrac{1}{9}(243.3) = 27.03$$

To see how the sample covariance and sample correlation (r) are related, let

$$s_X = \text{standard deviation of the } X \text{ values}$$
$$= \sqrt{\frac{\text{SS}_X}{n-1}}$$

and

$$s_Y = \text{standard deviation of the } Y \text{ values}$$
$$= \sqrt{\frac{\text{SS}_Y}{n-1}}$$

Then

$$r = \text{correlation between } X \text{ and } Y$$
$$= \frac{\text{cov}(X, Y)}{s_X s_Y} \qquad \textbf{(14-9)}$$

In example 14.1,

$$s_X = \sqrt{\frac{57{,}960}{9}} = 80.250$$

$$s_Y = \sqrt{\frac{1.4318}{9}} = .3988$$

and so

$$r = \frac{27.03}{(80.250)(.3988)} = .8446 \quad \text{(as before)}$$

The correlation between two variables is used more often than is the co-variance because r always ranges from -1 to 1. The covariance, on the other hand, has no limits and can assume any value. Furthermore, the units of measurement for a covariance are difficult to interpret. For example, the previously calculated covariance is 27.03 GPA · GRE—a somewhat meaningless unit of measurement. So, in a sense, the correlation is a scaled version of the covariance and has no units of measurement (a nice feature). To illustrate, the sample correlation between body weight and height will be the same whether you use the English or the metric systems to obtain the sample data. However, the sample covariance will *not* be the same for these two situations.

As a final look at these two measures, you can consider the correlation between two variables to be the covariance between the **standardized** variables. By defining

$$X' = \frac{X - \bar{X}}{s_X}$$

and

$$Y' = \frac{Y - \bar{Y}}{s_Y}$$

then it can be shown that

$$\text{cov}(X', Y') = \text{correlation between } X \text{ and } Y$$
$$= r$$

Least Squares Line

If we believe that two variables do exhibit an underlying linear pattern, how can we determine a straight line that best passes through these points? So far, we have demonstrated only the calculations necessary to compute a correlation coefficient. We will next illustrate how to construct a line through a set of points exhibiting a linear pattern; we will look at the assumptions behind this procedure in the next section.

Look at the scatter diagram in Figure 14.1b, which shows one possible line through these points. This diagram, as well as the vertical distances from each point to the line ($d_1, d_2, \ldots$) are contained in Figure 14.2.

Figure 14.2

Vertical distances from line L to university data (example 14.1) represented by $d_1, d_2, \ldots, d_{10}$

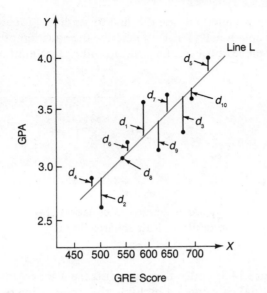

Is line L the best line through these points? Because we would like the distances $d_1, d_2, \ldots, d_{10}$ to be small, we define the best line to be the one that minimizes

$$\Sigma d^2 = d_1^2 + d_2^2 + d_3^2 + \cdots + d_{10}^2 \qquad \textbf{(14-10)}$$

We square each distance because some of these distances are positive (the point lies above line L) and some are negative (the point lies below line L). If we did not square each distance d, the positive d's might cancel out the negative ones. This means that using $(d_1 + d_2 + \cdots + d_{10})$ as a measure of fit is not a good idea. A better method is to determine which line makes equation 14-10 as small as possible; this is called the **least squares line**. Deriving this line in general requires the use of calculus (derivatives, in particular).*

* For the mathematically curious, we provide a condensed derivation of these coefficients. To minimize Σd^2, first write this expression as

$$\begin{aligned} f(b_0, b_1) &= \Sigma d^2 = \Sigma(y - \hat{y})^2 \\ &= \Sigma(y - b_0 - b_1 x)^2 \end{aligned}$$

because $\hat{y} = b_0 + b_1 x$.

To minimize this function, determine the partial derivatives with respect to b_0 (written f_{b_0}) and with respect to b_1 (written as f_{b_1}). These are

$$f_{b_0} = 2\Sigma(y - b_0 - b_1 x)(-1) = -2[\Sigma y - nb_0 - b_1\Sigma x]$$

$$f_{b_1} = 2\Sigma(y - b_0 - b_1 x)(-x) = -2[\Sigma xy - b_0\Sigma x - b_1\Sigma x^2]$$

Setting $f_{b_0} = f_{b_1} = 0$ and solving for b_0 and b_1 results in equations 14-11 and 14-12.

Because we intend to use this line to predict Y for a particular value of X, we use the notation $\hat{Y}$ (Y hat) to describe the equation of the line. We can now define, for the least squares line, the b_0 and b_1 that minimize $(d_1^2 + d_2^2 + \cdots + d_n^2)$, given by

$$b_1 = \frac{SS_{XY}}{SS_X} \qquad (14\text{-}11)$$

$$b_0 = \bar{y} - b_1\bar{x} \qquad (14\text{-}12)$$

where SS_X and SS_{XY} are as defined in equations 14-3 and 14-5. Also $\bar{x} = \Sigma x/n$ and $\bar{y} = \Sigma y/n$. The resulting least squares line is

$$\hat{Y} = b_0 + b_1 X$$

In Figure 14.3, notice that each distance d is actually $Y - \hat{Y}$, and consists of the **residual** encountered by using the straight line to estimate the value of Y at this point. So

$$\Sigma d^2 = \Sigma(y - \hat{y})^2$$

This term is the **sum of squares of error** (or residual sum of squares) and is written **SSE**. Consequently, the least squares line is the one that makes SSE as small as possible.

$$SSE = \Sigma d^2 = \Sigma(y - \hat{y})^2 \qquad (14\text{-}13)$$

Figure 14.3

The least squares line for example 14.1. Each $d = Y - \hat{Y}$ represents the residual encountered by using the straight line to estimate the value of Y at the corresponding point.

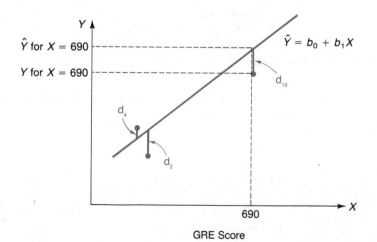

GRE Score

There is another method of determining SSE when using the least squares line, which avoids having to determine the value of $\hat{Y}$ at each point.

$$SSE = SS_Y - \frac{(SS_{XY})^2}{SS_X} \qquad (14\text{-}14)$$

EXAMPLE 14.2 Determine the least squares line for the university data we used in example 14.1. What is the SSE?

Solution Using the calculations from example 14.1, $SS_{XY} = 243.3$, $SS_X = 57,960$, and $SS_Y = 1.4318$. This leads to

$$b_1 = \frac{SS_{XY}}{SS_X}$$

$$= \frac{243.3}{57,960} = .004198$$

and

$$b_0 = \bar{y} - b_1 \bar{x}$$
$$= 3.295 - (.004198)(598) = .785$$

because

$$\bar{y} = \Sigma y/n$$
$$= 32.95/10 = 3.295$$

and

$$\bar{x} = \Sigma x/n$$
$$= 5980/10 = 598$$

So the equation of the best (least squares) line through these points is

$$\hat{Y} = .785 + .0042X$$

This equation tells us that in the sample data, an increase of 10 points in the GRE score (X increases by 10) is accompanied by an increase of .042 in the student's GPA (Y increases by .042), on the average. For this illustration (and many others in practice), the intercept b_0 has no real meaning because it corresponds to a GRE score of zero. Furthermore, a GRE score of zero is considerably outside the range of the GRE scores in the sample. It is unsafe to assume that the linear relationship between X and Y present over the range of sample GRE values (480 to 710) exists outside this range—in particular, all the way to a GRE score of zero. The slope b_1 generally is the more informative value.

Figure 14.4

Least squares line
for university data
(example 14.2)

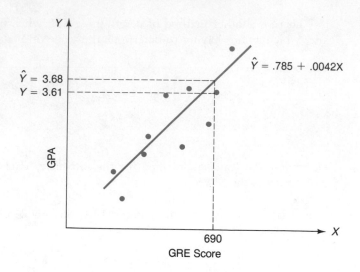

In Figure 14.4, the actual value of Y (in the sample data) for $X = 690$ is $Y = 3.61$ (the last pair of X, Y values). The predicted value of Y using the least squares line would be

$$\hat{Y} = .785 + (.0042)(690) = 3.68$$

The residual at this point is

$$\text{Residual} = Y - \hat{Y} = 3.61 - 3.68 = -.07$$

Repeating this for the other nine points leads to the following results. Notice that the sum of the residuals when using the least squares line is zero. This is always true.

X	Y	$\hat{Y}$	$Y - \hat{Y}$	$(Y - \hat{Y})^2$
580	3.54	3.22	.32	.1024
500	2.62	2.88	−.26	.0676
670	3.30	3.60	−.30	.0900
480	2.90	2.80	.10	.0100
710	4.00	3.77	.23	.0529
550	3.21	3.09	.12	.0144
640	3.57	3.47	.10	.0100
540	3.05	3.05	0	0
620	3.15	3.39	−.24	.0576
690	3.61	3.68	−.07	.0049
			0	.410

SSE

As you can see, calculating the SSE ($=.410$) using the table and equation 14-13 is tedious. Using equation 14-14 instead leads to

$$SSE = 1.4318 - \frac{(243.3)^2}{57,960}$$

$$= 1.4318 - 1.0213$$

$$= .4105 \text{ (the same as before)}$$

Remember, however, that equation 14-13 applies to any line that you choose to construct through these points, whereas equation 14-14 applies only to the SSE for the least squares line.   □

In example 14.2, we attempted to predict a student's first year GPA (Y) using the corresponding verbal GRE score (X). The variable Y is the **dependent variable** and X is the **independent variable**. By passing a straight line through the sample points with Y as the dependent variable, we are **regressing** Y on X. In linear regression, you regress the dependent variable Y, which you are trying to predict, on the independent (or predictor) variable X.

EXERCISES   14.1 The quantity of blood expelled from the heart with each ventricular contraction is termed stroke volume or stroke output. For an individual resting comfortably and motionlessly in bed the normal, average stroke volume is 75 milliliters of blood. Medical researchers were studying the relationship between age and stroke volume. Consider the following data from one experiment.

Age in Years (X)	Stroke Volume in ml of Blood (Y)
20	74
25	76
30	77
35	74
40	71
45	72
50	70
55	68
60	67
65	64
70	62

(a) Calculate the sample correlation coefficient and interpret its meaning.

(b) Compute the sample covariance between X and Y.

(c) Obtain the least squares line for the above data.

(d) What is the residual sum of squares?

14.2 The "discomfort index" is a composite of the inflation rate and the unemployment rate. A social worker had some suspicions about whether the discomfort index might be related to the crime rate in a large city. The following data are for 10 randomly selected months.

X Discomfort Index	Y Crime Rate (Coded)
11.2	16
10.9	15
18.0	24
14.0	22
21.3	25
18.8	23
17.5	19
12.6	18
15.0	21
16.8	21

(a) Compute the least squares regression line for the above data.
(b) Compute the error sum of squares.

14.3 Use the data from exercise 4.7 in Chapter 4 to compute the following:
(a) Covariance between X and Y.
(b) Least squares line with affectivity score as the independent variable and effectiveness index as the dependent variable.

14.4 The use of birth control in developing countries has been studied by demographers. One factor that has been investigated was socioeconomic status (SES) as measured by income level. In the following data, the population has been grouped into seven income levels coded from 1 to 7, with 7 being the highest income group. The use of birth control is measured by the percentage of married couples using some form of contraceptive technique.

Socioeconomic Status	Use of Birth Control
1	25
2	36
3	44
4	51
5	60
6	67
7	62

(a) Compute the sample correlation coefficient.
(b) Compute the covariance.
(c) Compute the least squares line, using SES as the independent variable.

14.5 Because $b_0 = \bar{y} - b_1\bar{x}$ we can replace b_0 in $\hat{Y} = b_0 + b_1 X$ by $\bar{y} - b_1\bar{x}$. Hence, we have $\hat{Y} = \bar{y} + b_1(X - \bar{x})$. From this equation, show that the point $(\bar{x}, \bar{y})$ falls on the least squares line.

14.6 Compare the formulas for the sample correlation r and the slope of the least squares line b_1, and verify that $b_1 = r\sqrt{SS_Y/SS_X}$. What can we say about the sign of r and b_1?

14.7 Lucky Jack's retail store advertises sale items each month to increase its sales. The manager believes that there is a linear relationship between the amount spent on advertising (X, in thousands of dollars) and the amount of merchandise sold (Y, in thousands). Data were collected for 10 months.

Y	X
21	5.3
16	3.8
13	3.1
12	2.9
18	4.4
20	4.9
23	5.1
24	5.4
14	3.2
19	5.1

(a) Calculate the line of best fit.

(b) Use the prediction equation to predict the amount of merchandise sold if $4700 is spent on advertising.

14.8 An industrial engineer collected data to study the relationship between the intensity of illumination on the shop floor and the output of the workers. Ten levels of illumination, coded from 1 to 10, were studied and the output of workers was noted at each level. Output was measured as the number of items produced per hour.

Illumination Intensity	Output of Workers
1	70
2	70
3	75
4	88
5	91
6	94
7	100
8	92
9	90
10	85

(a) Obtain the least squares line for the above data by regressing output of workers on illumination intensity.

(b) Plot the data in a scatter diagram and comment on any pattern you observe.

14.9 The "dryness index" (a composite measure of temperature, rainfall and humidity) developed in exercise 4.37 was applied to a much larger metropolitan area, with the following results:

X (Dryness Index)	Y (Daily Fire Alarms)
85	2
88	1
91	1
110	10
101	6
86	3
95	5
120	14

(a) Compute the correlation coefficient.

(b) Compute the least squares equation.

14.10 In most developing countries, suicide is an important problem. One of the major causes is the economic pressure. The following are the unemployment rate (X) and the suicide rate (Y) for a certain third world country over the past 12 years. X is the number of unemployed per 100 people and Y represents the number of suicides per 10,000 people.

Y	X	Y	X
15.4	12.4	15.4	13.8
19.6	17.8	18.6	16.9
14.3	12.7	12.8	10.5
10.1	7.2	16.4	13.8
13.3	12.1	13.3	11.5
16.7	14.3	9.9	7.0

(a) Calculate the least squares line. Interpret the coefficients of the least squares line in the context of the problem.

(b) Calculate the error sum of squares.

(c) Graph the data and draw the least squares line through them.

14.2   THE SIMPLE LINEAR REGRESSION MODEL

When we construct a straight line through a set of data points, we are attempting to predict the behavior of a dependent variable Y using a straight line equation with one predictor (independent) variable X. Examples 14.1 and 14.2 examined the relationship at Brookhaven University between the first year grade point average (Y) of a graduate student and the corresponding score on the verbal portion of the Graduate Record Exam (X).

Another application would be attempting to predict the utility expenditures (Y) of a home using the square footage of the home (X) as the independent variable. We expect that, for larger homes, the utility expenditures will be higher. In other words, we expect a positive relationship for this situation.

Regression analysis is a method of studying the relationship between two (or more) variables, one purpose being to arrive at a method for predicting a value of the dependent variable. In **simple linear regression**, you use only one predictor variable, X, to describe the behavior of the dependent variable Y. Also, the relationship between X and Y is assumed to be basically linear.

You have learned the mechanics of constructing a line through a set of bivariate sample values. We are now ready to introduce the concept of a statistical model.

Defining the Model

Return to example 14.2 and Figure 14.4. This set of sample data contained a value of $X = 690$ and $Y = 3.61$. Consider the population of all students receiving a verbal score of 690 on the GRE. Will they all have the same GPA? Unless cloning is very popular at Brookhaven University, certainly not. Does this mean that the straight line predictor is of no use? The answer, again, is no, because very few things in this world are that perfectly predictable. When you use the straight line to predict the GPA, you should be aware that there will be a certain amount of error present in this estimate. This is similar to the situation dealing with estimating the mean μ of a population where the sample mean $\bar{X}$ always estimates this parameter with a certain amount of inherent error.

When we elect to use a straight line predictor, we employ a **statistical model** of the form

$$Y = \beta_0 + \beta_1 X + e \qquad (14\text{-}15)$$

where $\beta_0 + \beta_1 X$ is the assumed line about which all values of X and Y will fall, called the **deterministic** portion of the model, and e is the error component, referred to as the **random** part of the model.

In other words, there exists some (unknown) line about which all X, Y values can be expected to fall. Notice that we said "about which," not "on which"—hence the necessity of the error term e, which is the unexplained error that is part of the simple linear model. Because this model considers only one independent variable, the effect of the other predictor variables (perhaps unknown to the analyst) is contained in this error term.

We emphasize that the deterministic portion, $\beta_0 + \beta_1 X$, refers to the straight line for the population and will remain unknown. However, by obtaining a random sample of bivariate data from this population, we are able to estimate the unknown parameters, β_0 and β_1. Thus b_0 is the **intercept** of the sample regression line and is the estimate of the population intercept β_0. The value of b_0 can be calculated using equation 14-12. Similarly, b_1 is the **slope** of the sample regression line and is the estimate of the population slope β_1. The value of b_1 can be calculated using equation 14-11.

Assumptions for the Simple Linear Regression Model

We can construct a least squares line through any set of sample points, whether or not the pattern is linear. We could construct a least squares line through a set of sample data exhibiting no linear pattern at all. However, to have an effective predictor and a model that will enable us to make statistical decisions, certain assumptions are necessary.

We treat the values of X as fixed (nonrandom) quantities when using the simple linear regression model. For any given value of X, the only source of variation comes from the error component e, which is a random variable. In fact, there are many random variables here, one for each possible value of X. The assumptions used with this model are concerned with the nature of these random variables.

Figure 14.5

Illustration
of assumption 1
(see text)

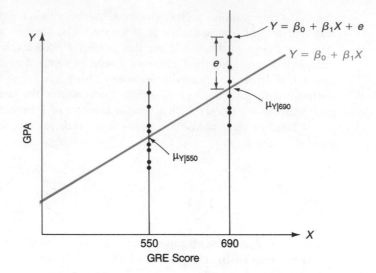

The first three assumptions are concerned with the behavior of the error component for a fixed value of X. The fourth assumption deals with the manner in which the error components (random variables) affect each other.

Assumption 1 *The mean of each error component is zero.* This is the key assumption behind simple linear regression. Look at Figure 14.5, where we once again examine a value of $X = 690$. If we consider all students at Brookhaven who have a GRE of 690, we have already decided they will not all have the same first year GPA. In fact, the GPA values will be scattered about the (unknown) line $Y = \beta_0 + \beta_1 X$, with some values lying above the line (e is positive) and some falling below it (e is negative). Consider the average of all Y values with $X = 690$. This is written as

$$\mu_{Y|690}$$

which is the mean of Y given $X = 690$. Our assumption here is that $\mu_{Y|690}$ lies on this line; that is, for any value of X, $\mu_{Y|X}$ lies on the line $Y = \beta_0 + \beta_1 X$ (such as $\mu_{Y|550}$ in Figure 14.5). This is an alternative way to say that the error is zero, on the average.

Assumption 2 *Each error component (random variable) follows an approximate normal distribution.* In our sample of 10 students, we had one student with $X = 690$ and $Y = 3.61$. Figure 14.5 illustrates what we might expect if we were to examine other students with a GRE score of 690. We assume here that if we were to obtain, say, 100 students with this score, a histogram of the resulting errors (e) would be bell-shaped in appearance. So we would expect a concentration of errors near zero (from assumption 1), with one-half of them positive and one-half of them negative.

Figure 14.6

A violation
of assumption 3
(see text)

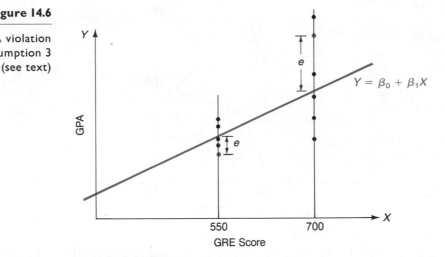

Assumption 3 *The variance of the error component,* σ_e^2, *is the same for each value of X*. For each value of X, the errors illustrated in Figure 14.5 have so far been assumed to follow a normal distribution, with a mean of zero. So each error e is from such a normal population. The variance of this population is σ_e^2. The assumption here is that σ_e^2 does not change as the value of X changes. This is the assumption of **homoscedasticity**. A situation where this assumption is violated is illustrated in Figure 14.6, where we once again consider what might occur if we were to obtain (we will not, actually) many values of Y for $X = 550$ and also for $X = 700$. If Figure 14.6 were the result, assumption 3 would be violated because the errors would be much larger (in absolute value) for the students scoring 700 than they would for the students scoring 550. This is **heteroscedasticity**, which does pose a problem when we try to infer results from a linear regression equation.

You might argue that, proportionally, the errors for $X = 700$ seem about the same as those for $X = 550$, which means that you would expect larger errors for larger values of X here. If this is the case, then the confidence intervals and tests of hypothesis that we are about to develop for the simple linear regression model are not appropriate. There are methods of "repairing" this situation, by applying a transformation to the dependent variable Y, such as $\sqrt{Y}$ or $\log(Y)$. By using this "new" dependent variable rather than the original Y, the resulting errors often will exhibit a nearly constant variance. Such transformations, however, are beyond the scope of this text.

A summary of the first three assumptions is shown in Figure 14.7. Note that the distribution of errors is identical for each illustrated value of X; namely, it is a normal distribution with a mean of zero and variance σ_e^2.

Assumption 4 *The errors are independent of each other*. This implies that the error encountered for one value of Y is unaffected by the error for any other value

Figure 14.7

Illustration of
assumptions 1, 2, 3
(see text)

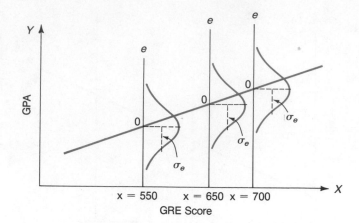

of Y. To illustrate, consider a situation where data for the two variables, X and Y, are recorded annually. The dependent variable Y might be the total number of new home constructions for the year and the independent variable X could be the average interest rate. You may want to restrict your population to a particular city, county, or state. Your sample then consists of data collected from n different years. Often for such data, the value of the error term in, say, 1982 is very much affected by what it was in 1981. If the error was a large positive value in 1981, there may be a good chance it will have a similar value in 1982. In other words, the residuals are not independent, which is a violation of assumption 4.

We can draw two conclusions from these assumptions. First, each value of the dependent variable Y is a normal random variable with mean $= \beta_0 + \beta_1 X$ and variance σ_e^2. Second, the error components come from the same normal population, regardless of the value of X. In other words, it makes sense to examine the residuals resulting from each value of X in the sample, to construct a histogram of these residuals, and to determine whether its appearance is bell-shaped (normal), centered at zero. A key assumption when using simple linear regression is that the errors follow a normal distribution with a mean of zero. Constructing a histogram of the sample residuals provides a convenient method of determining whether this assumption is reasonable for a particular application.

We will further discuss methods of analyzing the validity of each of these assumptions in Chapter 15, where we will learn how to use more than one independent variable in a linear regression equation.

Estimating the Error Variance σ_e^2

The variance of the error components σ_e^2 measures the variation of the error terms resulting from the simple linear regression model. The value of σ_e^2 severely affects our ability to use this model as an effective predictor for a given situation.

Suppose, for example, that σ_e^2 is very large in Figure 14.7. This means that, if we were to obtain many observations (GPA values, Y) for a *fixed* value of X (say, GRE score = 650), these Y values would vary a great deal. This decreases the accuracy of our model; we would prefer that these values were grouped closely about the mean, $\mu_{Y|650}$.

In practice, σ_e^2 typically is unknown and must be estimated from the sample. To estimate this variance, we first determine the sum of squares of error, SSE, using SSE $= \Sigma(y - \hat{y})^2$ or equation 14-14. Estimating β_0 and β_1 for the simple regression model results in a loss of 2 degrees of freedom, leaving $n - 2$ *df* for estimating the error variance. Consequently,

$$s^2 = \hat{\sigma}_e^2 = \text{estimate of } \sigma_e^2$$
$$= \frac{\text{SSE}}{n - 2} \tag{14-16}$$

where

$$\text{SSE} = \Sigma(y - \hat{y})^2$$
$$= \text{SS}_Y - \frac{(\text{SS}_{XY})^2}{\text{SS}_X}$$

We can determine the estimate of σ_e^2 and σ_e for the university data in example 14.2, where we calculated the value of SSE to be .4105. Our estimate of σ_e^2 is

$$s^2 = \frac{\text{SSE}}{n - 2} = \frac{.4105}{8} = .0513$$

and so $s = \sqrt{.0513} = .2265$ provides an estimate of σ_e. The values of s^2 and s are a measure of the variation of the Y values about the least squares line.

COMMENTS

We know from the empirical rule that approximately 95% of the data from a normal population should lie within two standard deviations of the mean. For this example, this implies that approximately 95% of the residuals should lie within 2(.2265) = .453 of the mean. In the table in example 14.2, the sample residuals are in the fourth column. Their sum is always zero, when using the least squares line; therefore, their mean is zero. So, approximately 95% of the residuals should be no larger (in absolute value) than .453. In fact, all of them are less than .453—not a surprising result, given that we had only 10 values to work with.

EXERCISES

14.11 The following data show the number of total annual bankruptcy petitions filed in the northern district of a southern state (in thousands) and the size of permanent staff at the U.S. Bankruptcy Court for that district.

X (Bankruptcies, in thousands)	Y (Permanent staff at U.S. Bankruptcy Court)
2.1	15
3.8	18
4.1	18
10.0	59
3.2	14
3.9	18
6.1	24

(a)   Compute the least squares line.

(b)   Identify the values of the slope and the intercept for the simple linear regression model.

(c)   Estimate the variance of the error for the model.

(d)   Find the residuals $(Y - \hat{Y})$ for all the Y values.

14.12   Is an increase in crime accompanied by increased cocaine arrests? A newspaper investigated this issue. As part of its research, it obtained the following statistics:

X (Cocaine Arrests)	Y (Burglaries & Robberies, in thousands)
190	103.820
195	104.325
322	108.506
539	101.244
902	106.177
1059	121.183
1124	140.256

(a)   Obtain the least squares equation.

(b)   Estimate the variance of the error in the model.

(c)   Assuming there is a significant correlation between X and Y, would it be logical to conclude that cocaine arrests *cause* burglaries?

14.13   Jeffrey Boyd and Eve Moscicki, researchers in the epidemiology and psychopathology branch of the clinical research division at the National Institute of Mental Health, reported in a study that the rising rate of teenage suicides in recent decades has been accompanied by a rapid rise in the number of civilian firearms in the United States. The following data are adapted from graphs published in their article:

Year	(Y) Suicides per 100,000 population by Use of Firearms, Males Aged 15–19	(X) Civilian Firearms Production, in Millions of Guns
1960	3.0	1.4
1965	3.8	2.4
1970	4.6	5.1
1975	7.0	5.8
1978	8.0	4.9

Source: J. H. Boyd and E. K. Moscicki, Firearms and youth suicide. *Am. J. Public Health.*
76 (10): 1240–1242 (October 1986).

(a) Obtain the least-squares regression model, using civilian firearms production as the predictor variable.

(b) Estimate the variance of the error.

(c) Find the residuals for all the Y values.

(d) Do the residuals follow the empirical rule for 95% of the error values?

(e) Do you think a regression analysis of this kind can be used to prove a causal relationship, i.e., that increased firearms production causes increased teenage suicides by the use of firearms?

14.14 A sociologist adapted data from the Internal Revenue Service to paint a portrait of charity in America. The following table provides the distribution of charitable contributions by individuals in the United States:

(X) Median Adjusted Gross Income (in thousands of dollars)	(Y) Percentage in Group Making Charitable Contributions (claiming itemized deductions)
5	17.0
7.5	36.0
12.5	40.5
17.5	38.5
25.0	29.2
40.0	14.0
75.0	4.2
100.0	1.5

(a) Obtain the least squares line for the above data.

(b) Identify the values of the intercept, the slope, and the variance of the error for the simple linear regression model.

(c) Find the residuals for all the Y values.

(d) If the correlation between X and Y above was very strong, would it then be correct to conclude that an increase in income causes people to become less charitable?

14.15 An experiment was performed in which different amounts (in milligrams) of an antihypertension drug were given to a rat. The decrease in systolic blood pressure (Y) was recorded for X number of milligrams. The following sample statistics were collected from 25 observations.

$$SS_{XY} = -10.3 \qquad SS_Y = 15.3 \qquad SS_X = 13.2$$

Determine an estimate for the variance of the error component for the linear model $Y = \beta_0 + \beta_1 X + e$.

14.16 What assumptions need to be made about the error component of a linear model in order that statistical inference can be used?

14.17 The following is a list of sample errors $(Y - \hat{Y})$ from a linear regression application:

2.1, $-.3$, 1.4, -2.8, -3.9, 4.2, 3.6, 4.3, 1.8, -2.7, $-.8$, 1.2, -1.1, .9, -4.5, -5.2, -1.3, .5, .9, $-.6$, 1.5, 2.1, -2.2, .9

Do the data appear to conform to the empirical rule that approximately 95% of the errors should lie within two standard deviations of the mean? Construct a histogram for the results.

14.18   Let X be the distance an employee lives from his or her job. Let Y be the average time that it takes the employee to drive to work. Data from 30 employees gave the following sample statistics.

$$SS_{XY} = 8.4 \qquad SS_X = 9.4 \qquad SS_Y = 12.2$$

(a)   Find the estimate of the error variance.

(b)   Find the interval in which approximately 68.26% of the error values should fall.

14.19   The following are residuals resulting from a regression analysis:

$Y - \hat{Y}$	X
.2	1
−.2	1.5
−.5	1.75
.6	2.00
−.5	2.50
−.8	3.00
1.0	4.00
−1.5	5.00
−1.7	6.00
2.1	7.00
−2.5	8.00
3.8	9.00

From these data, where Y is the dependent variable and X is the independent variable, does it appear that any of the standard assumptions of regression analysis are violated?

14.20   Why is $\Sigma(y - \hat{y})^2$ used in estimating the variance of the error term instead of $\Sigma(y - \hat{y})$?

14.3   INFERENCE ON THE SLOPE β_1

Performing a Test of Hypothesis on the Slope of the Regression Line

Under the assumptions of the simple linear regression model outlined in the previous section, we are now in a position to determine whether a linear relationship exists between the variables X and Y. Examining the estimate of the slope b_1 will provide information as to the nature of this relationship.

Consider the population slope β_1. Three possible situations are demonstrated in Figure 14.8. What can you say about using X as a predictor of Y in Figure 14.8a? When $\beta_1 = 0$, the population line is perfectly horizontal. As a result, the value Y is the same for each value of X, and so X is not a good predictor of Y; the value of X provides no information regarding the value of Y. In the event $\beta_1 = 0$, the best predictor of Y is given by $\hat{Y} = \bar{y}$ and so $\beta_1 \neq 0$ is equivalent to saying that $\hat{Y}$ (using X as a predictor) is superior to using the sample mean ($\hat{Y} = \bar{y}$) as a predictor.

Figure 14.8

Three possible population slopes (β_1)

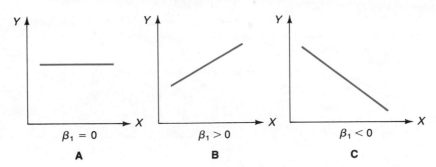

$\beta_1 = 0$ $\beta_1 > 0$ $\beta_1 < 0$

A   **B**   **C**

To determine whether X provides information in predicting Y, the hypotheses would be

H_0: $\beta_1 = 0$ (X provides no information)

H_a: $\beta_1 \neq 0$ (X does provide information)

Other Alternative Hypotheses

If we are attempting to demonstrate that a significant positive linear relationship exists between X and Y, the appropriate alternative hypothesis would be H_a: $\beta_1 > 0$. For example, do the data in example 14.1 support the hypothesis that students with large GRE scores are associated with larger GPAs?

When the purpose of the analysis is to determine whether a negative linear relationship exists between X and Y, the alternative hypothesis should be H_a: $\beta_1 < 0$. For example, you would expect such a relationship between the number of new housing starts (Y) and the interest rate (X). As the interest rate increases, you would expect the number of new houses under construction to decrease.

The Test Statistic

We use the point estimate of β_1 (that is, b_1) in the test statistic to determine the nature of β_1. What is b_1? A constant? A variable? Suppose that we obtained a different set of data and recalculated b_1. The new value would not be exactly the same as the previous value, which implies that b_1 is actually a variable. To be more precise, under the assumptions of the previous section, b_1 is a normal random variable with mean $= \beta_1$ and variance $= \sigma_{b_1}^2 = (\sigma_e^2)/(SS_X)$. Notice that b_1 is, on the average, equal to β_1; that is b_1 is an *unbiased* estimator of β_1. The variance $\sigma_{b_1}^2$ is a parameter describing the variation in the b_1 values if we were to obtain random samples of n observations indefinitely.

If we replace the unknown σ_e^2 by its estimate s^2, then the estimated variance of b_1 is $s_{b_1}^2 = s^2/SS_X$. As a result,

$$t = \frac{b_1 - \beta_1}{s/\sqrt{SS_X}} = \frac{b_1 - \beta_1}{s_{b_1}} \qquad \text{(14-17)}$$

has a t distribution with $n - 2$ df. If the null hypothesis is $H_0: \beta_1 = 0$, the test statistic becomes

$$t = \frac{b_1}{s/\sqrt{SS_X}} \qquad (14\text{-}18)$$

A summary of the testing procedure is shown in the accompanying box. As usual, for a one-tailed test, the null hypothesis can be written as an inequality ($\leqslant 0$ or $\geqslant 0$), or as an equality using the boundary condition ($= 0$).

Test of Hypothesis on the Slope of the Regression Line

Two-Tailed Test

$H_0: \quad \beta_1 = 0$

$H_a: \quad \beta_1 \neq 0$

Test statistic

$$t = \frac{b_1}{s_{b_1}}$$

where $s_{b_1} = s/\sqrt{SS_X}$ and the $df = n - 2$

Test

$$\text{Reject } H_0 \text{ if } |t| > t_{\alpha/2,\, n-2}$$

One-Tailed Test

$H_0: \quad \beta_1 = 0 \ (\leqslant 0)$ $H_0: \quad \beta_1 = 0 \ (\geqslant 0)$

$H_a: \quad \beta_1 > 0$ $H_a: \quad \beta_1 < 0$

Test statistic   **Test statistic**

$$t = \frac{b_1}{s_{b_1}} \qquad\qquad\qquad t = \frac{b_1}{s_{b_1}}$$

where $s_{b_1} = s/\sqrt{SS_X}$ and the   where $s_{b_1} = s/\sqrt{SS_X}$ and the

$df = n - 2$ $df = n - 2$

Test   **Test**

$$\text{Reject } H_0 \text{ if } t > t_{\alpha,\, n-2} \qquad \text{Reject } H_0 \text{ if } t < -t_{\alpha,\, n-2}$$

EXAMPLE 14.3

Is there sufficient evidence, using the university data in example 14.1, to conclude that a significant positive relationship exists between verbal GRE score (X) and first year GPA (Y)? Use $\alpha = .05$.

Solution

Step 1 The hypotheses indicated here are

$$H_0: \quad \beta_1 = 0$$
$$H_a: \quad \beta_1 > 0$$

Step 2 The test statistic is

$$t = \frac{b_1}{s_{b_1}}$$

which has a t distribution with $n - 2 = 8 \; df$.

Step 3 The testing procedure will be to

$$\text{Reject } H_0 \text{ if } t > t_{.05,8} = 1.860$$

This rejection region is sketched in Figure 14.9.

Step 4 We previously determined that $SS_X = 57{,}960$, $b_1 = .0042$, and $s = .2265$. The calculated test statistic is then

$$t^* = \frac{.0042}{.2265/\sqrt{57{,}960}} = \frac{.0042}{.000941} = 4.46$$

where $s_{b_1} = .000941$. Because $4.46 > 1.86$, we reject H_0.

Step 5 Based on these ten observations, we conclude that a positive linear relationship does exist between the verbal GRE score and GPA.

Figure 14.9

t curve with 8 df showing rejection region (shaded) for example 14.3

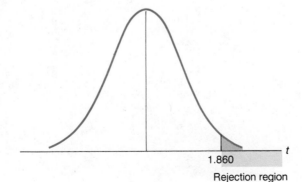

1.860

Rejection region

A MINITAB solution using the university data is shown in Figure 14.10. This output contains nearly all the calculations performed so far. In particular, note that

1 The least squares equation is $\hat{Y} = .785 + .0042X$.

2 The standard deviation of b_1 is $s_{b_1} = .000941$.

3 The value of the test statistic is $t^* = b_1/s_{b_1} = 4.46$.

4 The standard deviation of the error component is $s = .2265$.

5 The value of SSE is .4105, contained in the ANOVA table (construction of this table will be discussed in Chapter 15).

6 The column of estimated Y's ($\hat{Y}$'s) and the corresponding residuals are in the column labeled $Y - \hat{Y}$.
□

Figure 14.10

MINITAB solution
to example 14.3

```
MTB > SET INTO C1
DATA> 580 500 670 480 710 550 640 540 620 690
DATA> END
MTB > SET INTO C2
DATA> 3.54 2.62 3.30 2.90 4.00 3.21 3.57 3.05 3.15 3.61
DATA> END
MTB > BRIEF 3
MTB > REGRESS Y IN C2 USING 1 PREDICTOR IN C1

The regression equation is                    ── least squares line
C2 = 0.785 + 0.00420 C1

Predictor        Coef        Stdev        t-ratio
Constant        0.7848       0.5672        1.38
C1              0.0041977    0.0009410 ←sb₁   4.46 ←

s = 0.2265       R-sq = 71.3%     R-sq(adj) = 67.7%
```

$$t = \frac{b_1}{s_{b_1}}$$
(Example 14.3)

```
Analysis of Variance

SOURCE          DF          SS            MS
Regression       1        1.0213        1.0213
Error            8        0.4105        0.0513  ← s²
Total            9        1.4318
                                SSE

Obs.     C1       C2      Fit  Stdev.Fit  Residual   St.Resid
  1     580     3.5400   3.2194   0.0736    0.3206      1.50
  2     500     2.6200   2.8836   0.1168   -0.2636     -1.36
  3     670     3.3000   3.5972   0.0986   -0.2972     -1.46
  4     480     2.9000   2.7997   0.1321    0.1003      0.55
  5     710     4.0000   3.7651   0.1274    0.2349      1.25
  6     550     3.2100   3.0935   0.0847    0.1165      0.55
  7     640     3.5700   3.4713   0.0818    0.0987      0.47
  8     540     3.0500   3.0515   0.0901   -0.0015     -0.01
  9     620     3.1500   3.3873   0.0746   -0.2373     -1.11
 10     690     3.6100   3.6812   0.1124   -0.0712     -0.36
                          ↑         ↑         ↑
                          Ŷ        sŷ        Y-Ŷ
```

Figure 14.11

Scatter diagram of
age and absenteeism
data from example
14.4 with a summary
of the calculations

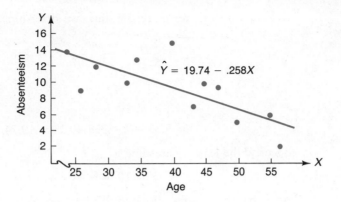

$$SS_X = 1342.67 \qquad SS_Y = 164.67 \qquad SS_{XY} = -346.33$$

$$\bar{x} = 40.33 \qquad \bar{y} = 9.33$$

$$r = \frac{SS_{XY}}{\sqrt{SS_X}\ \sqrt{SS_Y}} = -.737$$

EXAMPLE 14.4

The principal at Wilson Elementary School is interested in knowing whether
a relationship exists between the age of the teachers and the number of days
they are absent. She does not know whether younger or older teachers are more
likely to have high absenteeism. A random sample of 12 teachers was selected,
and the following data were obtained. Absenteeism is given as the number of
days absent during the recently completed school year.

Age (X)	Absenteeism (Y)
24	14
57	2
33	10
45	10
55	6
26	9
50	5
43	7
29	12
35	13
47	9
40	15

A scatter diagram of these 12 observations is provided in Figure 14.11, with
a summary of the calculations. Using $\alpha = .10$, do you think that a teacher's age
provides useful information for predicting absenteeism?

Solution   To derive the least squares regression line, we determine

$$b_1 = \frac{SS_{XY}}{SS_X} = \frac{-346.33}{1342.67} = -.258$$

and

$$b_0 = \bar{y} - b_1\bar{x}$$
$$= 9.33 - (-.258)40.33 = 19.74$$

Consequently, the least squares line is

$$\hat{Y} = 19.74 - .258X$$

Notice that the apparent slope of this line is negative. As the following test of hypothesis will conclude, this slope is significant. Consequently, higher absenteeism is associated with the younger teachers. According to these data, each additional year of age is accompanied by a decrease of .258 day in annual absenteeism, on the average.

To carry out a test of hypothesis, we follow the usual five-step procedure.

Step 1   Because the suspected direction of the relationship between these two variables (positive or negative) is unknown before the data are obtained, a two-tailed test is appropriate. The hypotheses are

$$H_0: \quad \beta_1 = 0$$
$$H_a: \quad \beta_1 \neq 0$$

Step 2   The test statistic is $t = b_1/s_{b_1}$, which has $n - 2 = 10 \; df$.

Step 3   The test procedure will be to

$$\text{Reject } H_0 \text{ if } |t| > t_{.10/2,10} = t_{.05,10} = 1.812$$

Step 4   Based on the data summary in Figure 14.11, and using equation 14-14,

$$SSE = SS_Y - \frac{(SS_{XY})^2}{SS_X}$$
$$= 164.67 - \frac{(-346.33)^2}{1342.67}$$
$$= 164.67 - 89.33$$
$$= 75.34$$

Consequently,

$$s^2 = \frac{SSE}{n-2}$$
$$= \frac{75.34}{10} = 7.534$$

and so

$$s_{b_1} = \frac{s}{\sqrt{SS_X}}$$

$$= \frac{\sqrt{7.534}}{\sqrt{1342.67}} = .0749$$

This means that the computed value of the test statistic is

$$t^* = \frac{b_1}{s_{b_1}}$$

$$= \frac{-.258}{.0749} = -3.44$$

Because $|t^*| = 3.44$ exceeds the table value of 1.812, we reject H_0 in support of H_a.

Step 5   Our conclusion is that age is a useful (although imperfect) predictor of absenteeism.   □

One thing to keep in mind is that statistical significance does not always imply practical significance. In other words, rejection of H_0: $\beta_1 = 0$ (statistical significance) does not mean that precise prediction (practical significance) will follow. It does demonstrate to the researcher that, within the sample data at least, this particular independent variable has an association with the dependent variable.

Confidence Interval for β_1

Following our usual procedure of providing a confidence interval with a point estimate, we use the t distribution of the previous test statistic and equation 14-17 to define a confidence interval for β_1. The narrower this confidence interval is, the more faith we have in our estimate of β_1 and in our model as an accurate, reliable predictor of the dependent variable. A $(1 - \alpha) \cdot 100\%$ confidence interval for β_1 is

$$b_1 - t_{\alpha/2, n-2}s_{b_1} \quad \text{to} \quad b_1 + t_{\alpha/2, n-2}s_{b_1}$$

EXAMPLE 14.5   Construct a 90% confidence interval for the population slope β_1 using the university data in example 14.1.

Solution   All the necessary calculations have been completed; $b_1 = .0042$ and $s_{b_1} = .000941$ (from example 14.3). Using $t_{.05,8} = 1.860$, the resulting confidence interval is

$$.0042 - (1.860)(.000941) \quad \text{to} \quad .0042 + (1.860)(.000941)$$

$$= .0042 - .0018 \quad \text{to} \quad .0042 + .0018$$

$$= .0024 \quad \text{to} \quad .0060$$

Figure 14.12

Curvilinear
relationship. The
horizontal line is
the least squares
line.

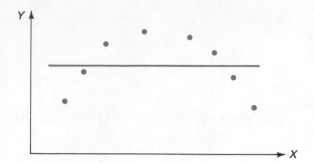

So we are 90% confident that the value of the actual slope (β_1) is within .0018 of the estimated slope ($b_1 = .0042$). The large width of the interval is due in part to the lack of information (small sample size) used to derive the estimates; a larger sample would decrease the width of this confidence interval.   □

COMMENTS   A failure to reject H_0 when performing a hypothesis test on β_1 does not always indicate that no relationship exists between the two variables. Some form of nonlinear relationship may exist between these variables. For example, in Figure 14.12, there is clearly a strong curved (**curvilinear**) relationship between X and Y. However, the least squares line through these points is flat, leading to a small t value and a failure to reject H_0. Furthermore, the sample correlation coefficient r for these data is zero.

Of course, you may fail to reject H_0 as the result of a type II error. In other words, you failed to reject H_0 when in fact a significant linear relationship does exist. This situation is more apt to occur when using a small sample to test the null hypothesis.

More often, a failure to reject H_0 will occur when there is no visible relationship between the two variables within the sample data. To determine whether there is no relationship, or that there is a nonlinear one, you should inspect either a scatter diagram of the data, a scatter diagram of the residuals, or, better yet, both. The latter diagram is a picture of the residuals ($Y - \hat{Y}$) plotted against the independent variable X. Residual plots will be discussed further in Chapter 15.

In many situations, the analyst has the opportunity to select the values of the independent variable X before the sample is obtained. At first glance, it might appear that the accuracy of our model is unaffected by the X values. This is partially true but not completely true. Because a narrow confidence interval for β_1 lends credibility to our model, we may choose to decrease the width of this confidence interval by decreasing s_{b_1}. Now, $s_{b_1} = s/\sqrt{SS_X}$, so, if we make SS_X large, the resulting s_{b_1} will be small. Therefore, given the opportunity, select a set of X values having a large variance. You can accomplish this by choosing a great many X values on the lower end of your range of interest, a large number of values at the upper end, and some values in between to detect any curvature that exists (as in Figure 14.12).

EXERCISES

14.21 Is the height of fathers related to the height of sons? A geneticist obtained the following sample data to address this issue:

Father's Height (in inches)	Son's Height (in inches)
62	65.2
63	66.1
64	67.0
65	66.2
66	68.0
67	68.1
68	69.2
69	70.0
70	69.5
71	70.8
72	71.0
73	72.5

(a) Compute the simple linear regression model for the above data, using father's height as the independent variable.

(b) Conduct a hypothesis test to determine whether the model has a significant positive slope, at a 5% significance level.

(c) Can you conclude whether there is a significant positive relationship between the heights of fathers and sons?

(d) Between what values would the slope fall, 90% of the time? Determine this confidence interval.

(e) What is the standard deviation of the distribution of the slope?

(f) What is the standard deviation of the error in the model?

(g) State the p value for the above hypothesis test.

14.22 Use the data in exercise 14.2 to answer the following:

(a) Does the least squares line for the data express a significant linear relationship, i.e., is there a significant relationship (positive or negative) between the discomfort index and the crime rate? Use the p value technique.

(b) Construct a 95% confidence interval for the slope of the least squares equation.

14.23 Use the data in exercise 14.9 to answer the following:

(a) Does the least squares equation for the data constitute a significant linear regression model? Use $\alpha = .05$. State the p value.

(b) Is there a significant *positive* relationship between the dryness index and number of daily fire alarms? Use $\alpha = .05$. State the p value.

(c) What is the standard deviation of the slope?

(d) Construct a 99% confidence interval for the slope.

14.24 The regression equation $\hat{Y} = 2.3 + 1.5X$ was arrived at by fitting a least squares line to 25 data points. The standard deviation (error) of the estimate of the slope was found to be .812. Test the null hypothesis at the .01 significance level that the slope of the line is equal to zero.

14.25 In exercise 14.10, find the 95% confidence interval for the slope of the regression equation used to predict the suicide rate from the unemployment rate.

14.26 The life of a lawn mower engine can be extended by frequent oil changes. An experiment was conducted in which 20 lawn mowers were used over many years with different time intervals between oil changes. Let X be the number of hours of operation between oil changes. Let Y be the number of years that the engine was able to perform adequately.

X	Y	X	Y
11.25	12.1	23.5	8.8
15.5	11.8	24.0	7.1
17.5	11.5	24.5	7.2
20.5	10.1	25.0	5.8
19.5	9.9	25.5	6.1
18.5	9.7	26.0	5.4
21.5	10.1	26.5	4.8
22.0	9.5	27.0	4.6
22.5	9.2	28.0	4.8
23.0	8.4	30.0	4.1

(a) Graph the data and the least squares line.

(b) Is there sufficient evidence to conclude, at the 10% significance level, that a negative relationship exists between Y and X? What is the critical region?

14.27 Industrial activity and motor traffic have contributed to the contamination of the environment by heavy metals such as lead and cadmium. One way to monitor atmospheric pollution is to do it indirectly, by measuring the influence of air pollution on plants. The use of biological materials as metal indicators has been reported to be a relatively cheap and reliable procedure. Hernandez and others (1987) have reported the following data on lead concentration, in parts per million, found in rosebay leaves (*Nerium oleander*) taken from along the roadside at 15 different locations in urban areas of Madrid, Spain. They also measured the daily motor-car traffic intensity at each location.

Place of Collection	Lead (ppm)	Traffic Intensities/day
C.S.	33.03	79,000
N.M.	58.25	109,000
N.S.	42.95	90,000
A.S.	60.23	104,000
M.A.J.S.	23.23	57,000
R.G.S.	52.67	111,000
C.R.S.	48.35	123,000
E.S.	74.25	128,000
C.C.P.	14.17	1,000
A.M.S.	38.31	103,000
R.S.	34.83	120,000
C.C.S.	54.73	119,000
R.P.	13.63	1,000
R.E.S.	66.73	130,000
S.S./M.M.S	65.63	136,000

Source: L. M. Hernandez, Ma. C. Rico, Ma. J. Gonzalez, and Ma. A. Hernan, Environmental contamination by lead and cadmium in plants from urban area of Madrid, Spain. *Bull. Environ. Contamination Toxicol.* **38** (2): 203–208 (February 1987), Table 3.

(a)   Plot a scatter diagram of the data.

(b)   Compute the linear regression equation, with traffic intensity as the predictor and lead concentration as the dependent variable.

(c)   Conduct a hypothesis test at 10% significance to decide whether a significant positive linear relationship exists between the two variables.

(d)   State the p value for the test. Does your conclusion change at the 5% or 1% significance level?

(e)   Comment on the two locations (CCP and RP) with very low traffic intensities. Are these observations outliers that should be ignored and eliminated, or are they useful observations?

(f)   Does the above analysis support the idea that biological material such as plant leaves are useful metal indicators for measuring atmospheric pollution?

14.28 A medical researcher was interested in the amount of weight loss caused by a particular diuretic. In a controlled experiment with 18 rats, the amount of weight loss was recorded after 1 month of daily dose of the diuretic. Let X be the amount, in milligrams, of diuretic given. Let Y be the weight loss in pounds.

X	Y	X	Y
0.10	0.05	0.30	0.42
0.10	0.08	0.35	0.43
0.15	0.11	0.35	0.42
0.15	0.13	0.40	0.44
0.20	0.19	0.40	0.47
0.20	0.21	0.45	0.51
0.25	0.35	0.45	0.52
0.25	0.31	0.50	0.54
0.30	0.41	0.50	0.53

Is there sufficient evidence to conclude that a significant positive relationship exists between the amount of diuretic given and the amount of weight loss? Use a significance level of 10%. Find a 90% confidence interval for the slope of the regression equation used to predict Y.

14.29 Among adults, the relationship between age (X) and psychomotor abilities (Y) is likely to be negative, because proficiency declines with age. A researcher studying this relationship obtained the following statistics from a sample of 25 observations.

$$SS_{XY} = -10.8$$

$$SS_X = 11.2$$

$$SS_Y = 20.5$$

(a)   Compute the slope of the simple linear regression equation.

(b)   Test whether the regression model suggests a significant *negative* relationship between age and psychomotor abilities. Use $\alpha = .05$.

(c)   State the p value for the test.

(d)   Construct the 90% confidence interval for the slope of the regression equation.

14.30 Drehman, Pelton, and Noack (1986), researchers in materials science and engineering, used differential scanning calorimetry and transmission electron microscopy to study the nucleation and growth kinetics of Pd–U–Si transformation from a metallic glass state to a single-phase quasicrystalline material. The transformation was achieved by a process called isothermic annealing at temperatures ranging from 400 to 455°C, and the corresponding nucleation rate was obtained. The latter is related to the number of nuclei per unit volume; its computational details need not concern us here. Consider the following data:

(X) Annealing Temperature (°C)	(Y) Nucleation Rate (log I in nuclei/cm³)
400	10.92
410	11.12
410	11.13
420	11.33
430	11.65
440	11.83
440	11.76
445	11.77
450	11.89
455	11.91

Source: A. J. Drehman, A. R. Pelton, and M. A. Noack, Nucleation and growth of quasicrystalline Pd–U–Si from the glassy state. *J. Mat. Res.* **I** (6): 743 (Nov./Dec. 1986), Table II.

(a) Compute the least-squares line.

(b) Test for a significant positive relationship between X and Y, at $\alpha = .05$.

(c) State your conclusion about the relationship between annealing temperature and nucleation rate, and report the p value.

(d) Would the relationship hold at temperatures of, say, 500 or 600°C?

(e) Construct a 95% confidence interval for the slope of the linear regression equation.

14.4   MEASURING THE STRENGTH OF THE MODEL

We have already used the sample coefficient of correlation r as a measure of the amount of linear association within a sample of bivariate data. The value of r is given by

$$r = \frac{SS_{XY}}{\sqrt{SS_X}\sqrt{SS_Y}} \qquad (14\text{-}19)$$

The possible range for r is -1 to 1.

Comparing the equations for r and b_1, we see that

$$r = b_1 \sqrt{\frac{SS_X}{SS_Y}}$$

Because SS_X and SS_Y are always greater than zero, r and b_1 have the same sign. Thus, if a positive relationship exists between X and Y, then both r and b_1 will be greater than zero. Similarly, they are both less than zero if the relationship is negative.

When you determine r, you use a sample of observations; r is a statistic. What does r estimate? It is actually an estimate of ρ (rho, pronounced "roe"), the **population correlation coefficient**. To grasp what ρ is, imagine obtaining all possible X, Y values and using equation 14-19 to determine a correlation. The resulting value is ρ.

The population slope β_1 and ρ are closely related. In particular, $\beta_1 = 0$ if and only if $\rho = 0$. This leads to another method of determining whether the simple linear regression model (using X to predict Y) is satisfactory. The hypotheses are

H_0: $\rho = 0$ (no linear relationship exists between X and Y)

H_a: $\rho \neq 0$ (linear relationship does exist)

The test statistic uses the point estimate of ρ (that is, r) and is defined by

$$t = \frac{r}{\sqrt{\dfrac{1 - r^2}{n - 2}}} \qquad\qquad (14\text{-}20)$$

where n = the number of observations in the sample. This is also a t statistic with $n - 2$ df. Although equations 14-18 and 14-20 appear to be unrelated, the two are algebraically equivalent and their values are always the same.

Thus, the t test for H_0: $\beta_1 = 0$ and H_0: $\rho = 0$ produce identical results, provided both tests use the same level of significance. These tests are therefore redundant; they both produce the same conclusion. Remember, if you have already computed the sample correlation coefficient r, equation 14-20 offers a much easier method of determining whether the simple linear model is statistically significant. Notice also in equation 14-20 that the significance of the t value depends upon the sample size n. As a result, if the sample size is large enough, then virtually any value of r can produce a "significantly large" value of t.

EXAMPLE 14.6

Use equation 14-20 to determine whether a positive linear relationship exists between X = verbal GRE score and Y = first year GPA, based on the data from example 14.1. Use $\alpha = .05$.

Solution   The hypotheses to be used here are $H_0: \rho \leqslant 0$ versus $H_a: \rho > 0$. In example 14.1, we found that $r = .8446$. This leads to a computed test statistic value of

$$t^* = \frac{r}{\sqrt{\dfrac{1 - r^2}{n - 2}}}$$

$$= \frac{.8446}{\sqrt{\dfrac{1 - (.8446)^2}{8}}}$$

$$= \frac{.8446}{.1893}$$

$$= 4.46$$

Because this value is the same as the one obtained in example 14.3 (testing $H_0: \beta_1 \leqslant 0$ versus $H_a: \beta_1 > 0$), we draw the same conclusion. A positive linear relationship does exist between these two variables. In other words, r is large enough to justify this conclusion.   □

Remember, there is no harm in using equation 14-20 as a substitute for equation 14-18 with $H_0: \beta_1 = 0$ (or $\leqslant 0$, or $\geqslant 0$), particularly if you have already determined the value of r.

Danger of Assuming Causality

In Chapter 4, when we introduced the correlation coefficient, we cautioned you that high statistical correlation does not imply **causality**. Even if the correlation between X and Y is extremely high (say, $r = .95$), a unit increase in X does not necessarily cause an increase in Y. All we know is that, based upon this set of sample data, a unit increase in X is associated with a change in Y of approximately b_1 units.

To illustrate this point, consider the variables X = percentage of gray hairs a person has and Y = blood pressure. For these two variables, we would expect a large positive correlation. So even though the percentage of gray hairs may be a good predictor of blood pressure, this does not imply that an additional gray hair *causes* your blood pressure to rise. It is much more likely that a third factor—age—caused both the percentage of gray hairs and the blood pressure to increase.

Coefficient of Determination

In our earlier discussion of ANOVA techniques, we used the expression $SS(\text{total}) = \Sigma(y - \bar{y})^2$ to measure the tendency of a set of observations to group about the mean. If this value was large, then the observations (data) contained much variation and were not all clustered about the mean $\bar{y}$.

In the simple linear regression model, $SS_Y = \Sigma(y - \bar{y})^2$ is computed in the same way, and (as before) measures the total variation in the values of the dependent variable.

SS_Y = total variation of the dependent variable observations

When comparing the sum of squares of error, SSE, to the total variation, SS_Y, we use the ratio SSE/SS_Y. If all Y values are equal to their respective $\hat{Y}$ values, there is a perfect fit, with SSE = 0 and $r = 1$ or -1. Our model explains 100% of this total variation, and the unexplained variation is zero.

In general, SSE/SS_Y (expressed as a percentage) is the percentage of unexplained variation. Recall from equations 14-14 and 14-19 that

$$SSE = SS_Y - \frac{(SS_{XY})^2}{SS_X}$$

and

$$r^2 = \frac{SS_{XY}^2}{SS_X SS_Y}$$

and thus

$$r^2 = 1 - \frac{SSE}{SS_Y} = \frac{SS_Y - SSE}{SS_Y}$$

As a result, r^2 may be interpreted as a measure of the explained variation in the dependent variable using the simple linear model; r^2 is the **coefficient of determination**.

$$r^2 = \text{coefficient of determination}$$
$$= 1 - \frac{SSE}{SS_Y} = \frac{SS_Y - SSE}{SS_Y} \qquad \textbf{(14-21)}$$

Thus,

r^2 = the percentage of explained variation in the dependent variable using the simple linear regression model

For this model, we can determine r^2 simply by squaring the coefficient of correlation. In Chapter 15, we will predict the dependent variable Y using more than one predictor (independent) variable. To derive the coefficient of determination for this case, we must first calculate SSE and then use equation 14-21. So, although this definition may appear to be unnecessary, it will enable us to compute this value when we use a multiple linear regression model.

EXAMPLE 14.7

What percentage of the total variation of the first year GPA values is explained by means of the single predictor, verbal score on the GRE, using the university data from example 14.1?

Solution

We previously calculated r to be .8446, so the coefficient of determination is

$$r^2 = (.8446)^2 = .713$$

Therefore, we have accounted for 71.3% of the total variation in the GPA values by using the GRE score as a predictor of GPA.

Notice that we could have determined this value by using the calculations from examples 14.1 and 14.2, where

$$r^2 = 1 - \frac{SSE}{SS_Y}$$

$$= 1 - \frac{.4105}{1.4318} = .713$$

□

Total Variation, SS_Y

In Chapter 12, when discussing the ANOVA procedure, the total variation in the observations, measured by SS(total), was partitioned into two sums of squares—namely, SS(factor) and SS(error). The resulting equation was

$$SS(total) = SS(factor) + SS(error)$$

In a similar fashion, we can partition the total variation of the Y values in linear regression, measured by SS_Y, into two other sums of squares. In Figure 14.13, notice that the value of $y - \bar{y}$ can be written as the sum of two deviations, namely

$$y - \bar{y} = (\hat{y} - \bar{y}) + (y - \hat{y})$$

By squaring and summing over all the data points in the sample, we can show that*

$$\Sigma(y - \bar{y})^2 = \Sigma(\hat{y} - \bar{y})^2 + \Sigma(y - \hat{y})^2$$

The summation on the left of the equal sign is SS_Y. The second summation on the right is the sum of squares of error, SSE. The first summation on the right is defined to be the **sum of squares of regression, SSR**.

$$\Sigma(\hat{y} - \bar{y})^2 = SSR$$

As a result, we have

$$SS_Y = SSR + SSE \qquad \textbf{(14–22)}$$

* This result follows since it can be shown that $\Sigma(\hat{y} - \bar{y})(y - \hat{y}) = 0$ when using the least squares line.

Figure 14.13

Splitting $(y - \bar{y})$ into two deviations, $(\hat{y} - \bar{y}) + (y - \hat{y})$

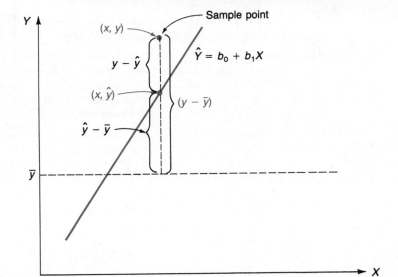

The regression sum of squares, SSR, measures the variation in the Y values that would exist if differences in X were the only cause of differences among the Y's. If this were the case, then all the (X, Y) points would lie exactly on the regression line. In practice, this does not happen when using a simple linear regression model. Otherwise, we would have a deterministic phenomenon, not an object of statistical investigation. Consequently, the sample points can be assumed to lie about the regression line, rather than on this line. This variation about the regression line is measured by the error sum of squares, SSE.

EXERCISES

14.31 The following data collected by an agriculture researcher show the annual yield of wheat (in bushels per acre) and the annual rainfall (in inches).

Yield of Wheat (Y)	Rainfall (X)
28.0	9.7
35.1	19.0
23.8	8.2
25.6	11.1
20.1	6.9
30.2	13.6
28.5	13.0
33.7	15.0

(a) Compute the sample correlation between rainfall and yield of wheat.

(b) Is the correlation significant? Conduct a hypothesis test on H_0: $\rho = 0$ at a 5% significance level.

(c) State the p value for the test.

(d)   Compute the coefficient of determination.

(e)   What percentage of variation in yield is explained by rainfall?

14.32   In exercise 14.22, you tested for a significant relationship between the discomfort index and the crime rate by testing the slope of the regression line. Repeat the test, but this time state the hypothesis in terms of the population correlation coefficient (H_0: $\rho = 0$) and use formula 14-20 to compute the test statistic. Compare the results with those obtained for exercise 14.22.

14.33   The sales manager of a real estate firm believes that experience is the best predictor for determining the yearly sales of the various salespeople in the real estate industry. Data were collected from 15 salespeople. Let X be the number of years of prior experience. Let Y be the annual sales (in thousands).

Y	X
50	1.3
161	5.1
195	6.2
172	5.4
132	3.9
133	4.1
181	6.1
69	1.9
78	2.2
124	3.4
131	7.1
64	2.1
80	4.5
110	3.8
127	4.4

Using a 10% significance level, test whether the population correlation coefficient between the variables X and Y is zero.

14.34   Lord and others (1987), under a research grant from the National Institute of Handicapped Research, Washington, D.C., have analyzed upper extremity strength and functional capability in 28 boys with Duchenne muscular dystrophy (DMD), a neuromuscular disease. They utilized the Brooke upper extremity functional grading scale (FG), which ranges from FG 1 (subject can abduct the arms in a full circle until they touch above the head) to FG 6 (subject cannot raise hands to mouth and has no useful function of the hands), with varying degrees of functional capability falling in between. In one part of the analysis, functional grade (FG) was regressed on age (in years) for boys in the 10 to 13 year old age group. The results were:

Regression equation: FG $= -4.342 + 0.605$ (AGE)
Sample size: $n = 10$
Coefficient of determination: $r^2 = 0.49$

Source: J. P. Lord, M. M. Portwood, J. S. Lieberman, W. M. Fowler, and P. Berck. Upper extremity functional rating for patients with Duchenne muscular dystrophy. *Arch. Phys. Med. Rehabilitation* **68** (3): 151–154. (March 1987).

(a)   What is the correlation coefficient between FG (functional grade) and Age?

(b) Does the above equation represent a significant linear relationship between FG and Age? Conduct a hypothesis test at a 10% significance level.

(c) State the p value for the test.

(d) If the variance of the error in the regression is equal to 1.1025, what is the value of

  (i) The error sum of squares, SSE?

  (ii) The regression sum of squares, SSR?

  (iii) The total sum of squares, SS_YT?

14.35 Chemical physicists Lambrakos and Peterson (1987) reported using a technique called a Firsov inversion procedure to analyze differential cross sections for scattering of high-velocity neutral helium atoms from helium gas. Their article provided, among other things, the repulsive potential energy of a pair of ground state helium atoms for internuclear separations of 0.5 to 1.0 Å. The following table gives the potential energy function obtained from experimentally determined differential scattering cross sections, with beam energy = 1.0 keV.

(X) Internuclear Separation (Å)	(Y) Potential Energy (eV)
.8071	7.804
.7927	8.299
.7781	8.836
.7631	9.426
.7479	10.08
.7323	10.80
.7165	11.61
.7002	12.53
.6837	13.57
.6667	14.77
.5494	16.16
.6444	16.59
.6394	17.05
.6344	17.53
.6294	18.05
.6242	18.60
.6192	19.20
.6140	19.84
.6089	20.54
.6037	21.32
.5985	22.18

Source: S. G. Lambrakos and N. C. Peterson. Determination of an experimental He_2 potential energy function by inversion of differential cross sections. *J. Chem. Phys.* **86** (5): 2730 (March 1987), Table II.

(a) Compute the correlation coefficient.

(b) Test for a significant negative relationship between X and Y, at $\alpha = .05$.

(c) State your conclusion about the relationship between internuclear separation and potential energy.

(d) State the p value for the test.

(e) What is the coefficient of determination?

14.36 For the data in exercise 14.26, test that there is no linear relationship between the number of hours of operation between oil changes and the number of years that the engine was able to perform adequately. Use equation 14–20 and test with a 10% significance level. Is the result the same as in exercise 14.26?

14.37 Ten cards numbered 1 through 10 are shuffled and a person is asked to pick one card. The card is replaced and the deck is reshuffled. Then the person is asked to draw a second card. If the second card is higher than the first, the dealer gives 85 cents to the player. If the second card is not higher, then the player gives $1.15 to the dealer. A sample of 15 pairs of draws is taken to see whether there is any correlation between the first and the second cards.

(a) Would you expect to observe significant correlation here? Why or why not?

(b) Find the coefficient of determination for the following data and test using a 5% significance level that there is no correlation between the first (X) and second (Y) cards. Interpret the value of the coefficient of determination.

X	Y
7	3
3	10
8	2
5	8
2	7
7	9
9	4
1	1
10	5
3	6
4	3
6	1
7	8
8	4
2	6

14.38 A sample of 35 observations is taken and a sample correlation coefficient is computed to be $r = .48$. Do the data provide sufficient evidence to reject the null hypothesis of no correlation? Use a 1% significance level.

14.39 A group of researchers were studying how teenagers' occupational aspirations are influenced by various factors. Obviously, a great many variables were investigated. However, suppose that we simplify the research somewhat and consider $Y =$ respondent's occupational aspirations and $X =$ best friend's occupational aspirations. In other words, the explanatory variable X hopefully provides some measure of peer influence on aspirations. The following statistics are obtained for a sample of 45 pairs of observations:

$$SS_{XY} = 91.125 \qquad SS_X = 196.69 \qquad SSS_Y = 208.61$$

(a) Compute the sample correlation coefficient.

(b) Test for a significant positive correlation between X and Y. State your hypothesis in terms of ρ (rho), and use $\alpha = .05$.

(c) Compute the coefficient of determination.

(d) What percentage of the variation in respondents' aspirations is explained by peer influence?

14.40 In this chapter, the focus has been on the correlation between two variables. However, in "real life," researchers frequently investigate many variables. In such cases, it is useful to know the interrelationship among all the possible pairs of these variables. This information is summarized in a correlation matrix. A study by Milby, Mizes, and Giles (1986) will illustrate this point. The study investigated the validity of five measures of phobic behavior during a process of desensitization therapy. The following table is the correlation matrix for these five measures, and provides the Pearson r values (correlation coefficients) along with the exact p values in parentheses.

	Latency to Visualization	SUDS at Visualization	Latency to Relaxation	SUDS at Relaxation	Clarity of Visualization
Latency to Visualization	1.000				
SUDs at Visualization	−0.185 ($p = .346$)	1.000			
Latency to Relaxation	0.356 ($p = .063$)	−0.057 ($p = .773$)	1.000		
SUDs at Relaxation	−0.331 ($p = .085$)	0.386 ($p = .043$)	−0.099 ($p = .618$)	1.000	
Clarity of Visualization	−0.868 ($p = .000001$)	−0.392 ($p = .039$)	−0.885 ($p = .000001$)	−0.186 ($p = .343$)	1.000

Source: J. B. Milby, J. S. Mizes, and T. R Giles. Assessing the process of desensitization therapy: Five practical measures. *J. Psychopathol. Behavioral Assessment* **8** (3): 241–251 (1986), Table III.

(a) What is the coefficient of correlation between SUDs at relaxation and SUDs at visualization, and is it significant at a 5% significance level?

(b) Which correlations are significant at
   (i) $\alpha = .10$?
   (ii) $\alpha = .05$?
   (iii) $\alpha = .01$?

(c) Which two variables have the strongest correlation between them?

ESTIMATION AND PREDICTION
14.5   USING THE SIMPLE LINEAR MODEL

We have concentrated on predicting a value of the dependent variable Y for a given value of X. In the previous examples, we used a student's GRE score X to predict his or her GPA (Y). Notice in Figure 14.5 that we can also use the least squares line to estimate the *average* value of Y for a specified value of X. So we can use this line to handle two different situations.

Situation 1 The regression equation $\hat{Y} = b_0 + b_1 X$ estimates the *average* value of Y for a specified value of the independent variable X. For $X = x_0$, this would be written $\mu_{Y|x_0}$ (the mean of Y given $X = x_0$).

For example, the least squares line passing through the university data in example 14.1 is $\hat{Y} = .785 + .0042X$. The average GPA for all students in the population with a GRE score of 690 is $\mu_{Y|690}$. Its estimate is provided by the corresponding value on the least squares line, namely

$$\hat{Y} = .785 + .0042(690)$$
$$= 3.68$$

So the estimate of the average GPA of all such students is 3.68 (Figure 14.4).

Situation 2 An *individual* predicted value of Y also uses the regression equation $\hat{Y} = b_0 + b_1 X$ for a specified value of X. This is denoted Y_{x_0} for $X = x_0$.

For example, assume that Fred Jenkins is a first year graduate student at Brookhaven with a verbal GRE score of 690. A prediction of his GPA (Y_{690}) is also

$$\hat{Y} = .785 + .0042(690)$$
$$= 3.68$$

We see that the least squares line can be used to estimate average values (situation 1) or predict individual values (situation 2). Since $\mu_{Y|690}$ is a parameter, we use $\hat{Y}$ to *estimate* this value. On the other hand, Y_{690} represents a particular value of a dependent (random) variable and so $\hat{Y}$ is used to *predict* this value. In the first situation, we can determine a *confidence* interval for $\mu_{Y|690}$; in the second, a prediction interval for Y_{690}.

Confidence Interval for $\mu_{Y|x_0}$ (Situation 1)

We have already established that the point estimate of $\mu_{Y|x_0}$ is the corresponding value of $\hat{Y}$. The reliability of this estimate depends on (1) the number of observations in the sample, (2) the amount of variation in the sample, and (3) the value of $X = x_0$. A confidence interval for $\mu_{Y|x_0}$ takes all three factors into consideration.

A $(1 - \alpha) \cdot 100\%$ confidence interval for $\mu_{Y|x_0}$ is

$$\hat{Y} - t_{\alpha/2,\, n-2} s \sqrt{\frac{1}{n} + \frac{(x_0 - \bar{x})^2}{SS_X}} \quad \text{to} \quad \hat{Y} + t_{\alpha/2,\, n-2} s \sqrt{\frac{1}{n} + \frac{(x_0 - \bar{x})^2}{SS_X}}$$

$$(14\text{-}23)$$

EXAMPLE 14.8   Determine a 95% confidence interval for the average GPA of all students having a verbal GRE score of 610, using the data from example 14.1.

Solution   We previously determined that $n = 10$, $\bar{x} = 598$, $SS_X = 57,960$, and $s = .2265$. The point estimate for the average GPA, $\mu_{Y|610}$, is

$$\hat{Y} = .785 + .0042(610)$$
$$= 3.35$$

Obtaining $t_{.025,8} = 2.306$ from Table A-5, the 95% confidence interval for $\mu_{Y|610}$ is

$$3.35 - (2.306)(.2265)\sqrt{\frac{1}{10} + \frac{(610 - 598)^2}{57,960}}$$

$$\text{to}\quad 3.35 + (2.306)(.2265)\sqrt{\frac{1}{10} + \frac{(610 - 598)^2}{57,960}}$$

$$= 3.35 - (2.306)(.2265)(.320)\quad \text{to}\quad 3.35 + (2.306)(.2265)(.320)$$

$$= 3.35 - .17\quad \text{to}\quad 3.35 + .17$$

$$= 3.18\quad \text{to}\quad 3.52$$

We are thus 95% confident that the average GPA for students with a GRE score of 610 is between 3.18 and 3.52.   □

Using MINITAB to Construct Confidence Intervals   The MINITAB solution for the university problem is contained in Figure 14.10. To construct confidence intervals the column of interest is labeled as Stdev Fit, which, when translated, means the standard deviation of the predicted Y. Writing this as $s_{\hat{Y}}$,

$$s_{\hat{Y}} = s\sqrt{\frac{1}{n} + \frac{(x_0 - \bar{x})^2}{SS_X}}$$

For each value of X in the sample (say, x_0), the corresponding confidence interval for $\mu_{Y|x_0}$ is

$$\hat{Y} - t \cdot s_{\hat{Y}}\quad \text{to}\quad \hat{Y} + t \cdot s_{\hat{Y}}$$

where $t = t_{\alpha/2, n-2}$, as before, and $\hat{Y}$ is contained in the column to the left of the standard deviations. For values of X not in the sample, you can (1) approximate this confidence interval by using the value of $s_{\hat{Y}}$ corresponding to an X value near this particular value or (2) use the computer procedure that will be discussed in Chapter 15, which will provide an exact value for $s_{\hat{Y}}$ belonging to this particular X value.

Using the MINITAB output in Figure 14.10, we can find the confidence intervals corresponding to X values of 480, 580, and 670. The remaining seven confidence intervals are constructed in a similar manner.

For $X = 480$, the 95% confidence interval is

$$2.80 - (2.306)(.1321)\quad \text{to}\quad 2.80 + (2.306)(.1321)\quad \text{or}\quad 2.50\quad \text{to}\quad 3.10$$

Figure 14.14

95% confidence
intervals for the
university data
derived from
MINITAB output
shown in Figure
14.10

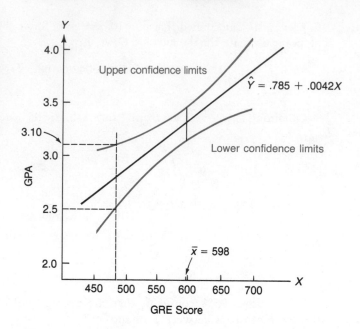

For $X = 580$, the confidence interval is

$$3.22 - (2.306)(.0736) \quad \text{to} \quad 3.22 + (2.306)(.0736) \qquad \text{or} \qquad 3.05 \quad \text{to} \quad 3.39$$

For $X = 670$, the confidence interval is

$$3.60 - (2.306)(.0986) \quad \text{to} \quad 3.60 + (2.306)(.0986) \qquad \text{or} \qquad 3.37 \quad \text{to} \quad 3.83$$

Notice that the confidence intervals are much wider for $X = 480$ and $X = 670$ than for $X = 580$.

By connecting the upper end of the confidence intervals for all 10 data points and connecting the lower limits, we obtain Figure 14.14. Equation 14-23 indicates that the confidence interval is narrowest when $(x_0 - \bar{x})^2 = 0$; that is, at $X = x_0 = \bar{x}$. For values of X to the left or right of $\bar{x}$, the confidence interval is wider. In other words, the farther x_0 is from $\bar{x}$, the less reliable is the estimate.

The Danger of Extrapolation Extrapolation is calculating an estimate corresponding to a value of X outside the range of the data used to derive the prediction equation (the least squares line). For example, in Figure 14.14, the least squares line could be used to estimate the average GPA for students with a GRE score of 250. Although we can estimate $\mu_{Y|250}$, the corresponding confidence interval for this parameter will be extremely wide, which means that the point estimate $\hat{Y}$ has little practical value.

To use the simple regression model effectively for estimation, you need to stay within the range of the sampled values for the independent variable X.

This is **interpolation**. If you use values outside this range, you need to be aware that, given another set of data, you would quite likely obtain a considerably different estimate. Furthermore, you have no assurance that the linear relationship still holds outside the range of your sample data.

Prediction Interval for Y_{x_0} (Situation 2)

The procedure of predicting individual values is used in a variety of applications. The regression equation can be used to **forecast** (predict) a future value of the dependent variable for a particular value of the independent variable. When attempting to predict a single value of the dependent variable Y using the simple linear regression model, we begin, as before, with $\hat{Y}$. Substituting $X = x_0$ into the regression equation provides the best prediction of Y_{x_0}. For example, if Ms. Johnson has a verbal GRE score of 610, our best guess as to her first year GPA (using this particular model) is $\hat{Y}$ for $X = 610$. From the results of example 14.8, this is 3.35.

We do not use the term confidence interval for this procedure because what we are estimating (Y_{x_0}) is not a parameter. It is a value of a random variable, so we use the term **prediction interval**.

The variability of the error in predicting a single value of Y is more than that for estimating the average value of Y (situation 1). It can be shown that an estimate of the variance of the error, when using $\hat{Y}$ to predict an individual Y for $X = x_0$, is

$$s^2\left(1 + \frac{1}{n} + \frac{(x_0 - \bar{x})^2}{SS_X}\right) \qquad \textbf{(14-24)}$$

This result can be used to construct a $(1 - \alpha) \cdot 100\%$ prediction interval for Y_{x_0}, as follows:

$$\hat{Y} - t_{\alpha/2,\,n-2}s\sqrt{1 + \frac{1}{n} + \frac{(x_0 - \bar{x})^2}{SS_X}}$$
$$\text{to} \quad \hat{Y} + t_{\alpha/2,\,n-2}s\sqrt{1 + \frac{1}{n} + \frac{(x_0 - \bar{x})^2}{SS_X}} \qquad \textbf{(14-25)}$$

Notice that the only difference between this prediction interval and the confidence interval in equation 14-23 is the inclusion of "1+" under the square root sign. The other two terms under the square root are usually quite small, so this "1+" has a large effect on the width of the resulting interval.

Be aware that our warning about extrapolating outside the range of the data applies here as well. In equations 14-24 and 14-25, the distance from the mean $(x_0 - \bar{x})$ is squared, which increases the risk of predicting beyond the range of the sampled data.

EXAMPLE 14.9

We previously determined that Ms. Johnson has a verbal GRE score of 610, and so the best prediction of her first year GPA is $\hat{Y} = 3.35$. Determine a 95% prediction interval for this situation.

Solution

We can use the calculations from example 14.8 to derive the prediction interval for Y_{610}. The result is

$$3.35 - (2.306)(.2265)\sqrt{1 + \frac{1}{10} + \frac{(610 - 598)^2}{57,960}}$$

$$\text{to} \quad 3.35 + (2.306)(.2265)\sqrt{1 + \frac{1}{10} + \frac{(610 - 598)^2}{57,960}}$$

$$= 3.35 - (2.306)(.2265)(1.050) \quad \text{to} \quad 3.35 + (2.306)(.2265)(1.050)$$

$$= 3.35 - .55 \quad \text{to} \quad 3.35 + .55$$

$$= 2.80 \quad \text{to} \quad 3.90$$

Comparing this interval to the confidence interval for $\mu_{Y|610}$ in example 14.8, we see that individual GPA predictions are considerably less accurate than estimations for the average GPA. Of course, we could reduce the width of this interval by obtaining additional data. Expecting accurate results from a sample of 10 observations is being a bit optimistic. □

Using MINITAB for Constructing Prediction Intervals We can use the MINITAB output in Figure 14.10 for example 14.9. The values in the column labeled Stdev.Fit assume that $\hat{Y}$ is estimating $\mu_{Y|x}$; we previously used $s_{\hat{Y}}$ as a symbol for this standard deviation.

A prediction interval for a value of $X = x_0$ in the sample is provided by

$$\hat{Y} - t\sqrt{s_{\hat{Y}}^2 + s^2} \quad \text{to} \quad \hat{Y} + t\sqrt{s_{\hat{Y}}^2 + s^2}$$

where $t = t_{\alpha/2, n-2}$ is obtained from Table A-5 and $\hat{Y}$ is obtained from the column labeled FIT. For values of X not in the sample, you can derive a prediction interval by using the computer procedure to be discussed in Chapter 15. The following calculations determine the prediction intervals for three of the sample X values, namely, $X = 480, 580,$ and 670.

For $X = 480$, the 95% prediction interval is

$$2.80 - 2.306\sqrt{(.1321)^2 + (.2265)^2}$$

$$\text{to} \quad 2.80 + 2.306\sqrt{(.1321)^2 + (.2265)^2}$$

$$= 2.80 - .60 \quad \text{to} \quad 2.80 + .60$$

$$= 2.20 \quad \text{to} \quad 3.40$$

For $X = 580$, the 95% prediction interval is

$$3.22 - 2.306\sqrt{(.0736)^2 + (.2265)^2}$$
$$\text{to} \quad 3.22 + 2.306\sqrt{(.0736)^2 + (.2265)^2}$$
$$= 3.22 - .55 \quad \text{to} \quad 3.22 + .55$$
$$= 2.67 \quad \text{to} \quad 3.77$$

For $X = 670$, the 95% prediction interval is

$$3.60 - 2.306\sqrt{(.0986)^2 + (.2265)^2}$$
$$\text{to} \quad 3.60 + 2.306\sqrt{(.0986)^2 + (.2265)^2}$$
$$= 3.60 - .57 \quad \text{to} \quad 3.60 + .57$$
$$= 3.03 \quad \text{to} \quad 4.17$$

(*Note:* This interval can be changed to (3.03, 4.00) since a GPA over 4.00 is impossible.)

Figure 14.15 shows the prediction intervals for all ten data points; the upper and lower limits have been connected. The increased width of a prediction interval versus a confidence interval is quite apparent from this graph. Also, as with that of a confidence interval, the width of a prediction interval increases as the value of X strays from $\bar{x}$. Also illustrated are the confidence and prediction intervals corresponding to $X = 480$.

Figure 14.15

95% prediction and confidence intervals for the university data

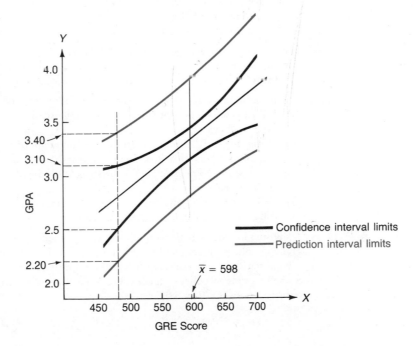

EXERCISES   **14.41** According to a study by the U.S. Census Bureau ("Marital Status and Living Arrangements: March 1985"), the following are median ages at first marriage for men and women since 1890:

Year	Men	Women
1890	26.1	22.0
1900	25.9	21.9
1910	25.1	21.6
1920	24.6	21.2
1930	24.3	21.3
1940	24.3	21.5
1950	22.8	20.3
1960	22.8	20.3
1970	23.2	20.8
1980	24.7	22.0
1981	24.8	22.3
1982	25.2	22.5
1983	25.4	22.8
1984	25.4	23.0
1985	25.5	23.3

(a) Compute the simple linear regression equation for the above data. Let X = the age of men.

(b) Compute the correlation coefficient.

(c) Compute the coefficient of determination.

(d) Compute the t statistic for testing for a significant slope.

(e) Compute the 95% confidence interval for the average age of women at first marriage when men's age is 24.6.

(f) Compute the 95% prediction interval for the age of women at first marriage when men's age is 24.6.

14.42 Use a computer package (such as MINITAB, SAS, SPSS, etc.) to process the data in exercise 14.41, and answer questions (a) to (f) in that exercise. Verify that your answers match (except for negligible differences due to rounding).

14.43 Use the data in exercise 14.1 to answer the following questions:

(a) What is the point estimate for the average stroke volume of persons whose age is 50 years?

(b) What is the 95% interval estimate of the range of values of average stroke volume of persons whose age is 50 years?

(c) What is the point estimate for the stroke volume of an individual whose age is 50 years?

(d) What is the 95% interval estimate of the range of values of stroke volume of an individual whose age is 50 years?

14.44 A statistics instructor believes that there is a strong correlation between a student's grade in college algebra and a student's grade in introductory statistics. The following statistics were collected from 20 randomly selected students. Let X be the student's grade in college algebra. Let Y be the student's grade in introductory statistics (A = 4.0, B = 3.0, and so on).

$$SS_{XY} = 31.8 \qquad SS_X = 32.3 \qquad SS_Y = 45.4$$

$$\bar{x} = 2.3 \qquad \bar{y} = 2.5 \qquad n = 20$$

(a) Determine a 95% confidence interval for the average grade in introductory statistics for students who make a B (equal to 3.0) in college algebra.

(b) Find the 95% prediction interval for the grade in introductory statistics for a student who obtained a B in college algebra.

14.45 For the data in exercise 14.28, find a 99% prediction interval for the monthly weight loss of an individual rat that has a daily dose of 0.3 milligrams of the diuretic.

14.46 For the data in exercise 14.26, find the 99% confidence interval for the average number of years that a lawn mower will be able to function properly if the number of hours of operation between oil changes is 23 hours. What is the standard deviation for a predicted value of the number of years that a single lawn mower will be able to perform adequately, for which the number of hours of operation between oil changes is 23?

14.47 For a fixed value of X, which interval is larger, the confidence interval for the mean value of Y at X (equation 14-23) or the prediction interval for a predicted value of Y at X (equation 14-25)? What value can you assign to X to achieve the smallest confidence interval for the mean value of Y at X or for the predicted value of Y at X?

14.48 Use the data in exercise 14.21 to answer the following:

(a) What is the point estimate for the average height of sons whose fathers are 5 feet 6 inches tall?

(b) What is the 99% interval estimate of the range of values for the average height of sons whose fathers are 5 feet 6 inches tall.

(c) What is the point estimate for the height of an individual whose father is 5 feet 6 inches tall?

(d) What is the 99% interval estimate of the range of values for the height of an individual whose father is 5 feet 6 inches tall?

14.49 A clinical psychologist retained by a firm of personnel consultants has been given the following data pertaining to an employee's score on Randall and Cantrell's job aptitude test and that employee's evaluation (on a scale of 1 to 10) by the supervisor.

X (Job Aptitude Score)	Y (Supervisor's Rating)
92	9.3
70	6.0
35	2.2
49	6.6
80	7.2
60	5.0
73	7.0
65	5.5
94	9.0
55	7.0

(a)   Determine if a significant positive relationship exists between the job aptitude test score and the supervisor's evaluation rating of the employee. Use a 5% significance level.

(b)   What percentage of the variation in the supervisor's ratings is explained by the job aptitude test score?

(c)   Determine the 95% confidence interval for the average rating for employees with an aptitude score of 80.

(d)   Determine the 95% prediction interval for the rating of an individual employee with an aptitude score of 80.

(e)   As an optional, extra exercise, verify your results using MINITAB (or some other computer package).

14.50   Using the following sample statistics, find a 90% confidence interval for the mean value of Y at $X = .5$.

$$SS_{XY} = 31.4$$

$$SS_X = 40$$

$$SS_Y = 45$$

$$\bar{x} = 7.5$$

$$\bar{y} = 10.0$$

$$n = 31$$

AN ALTERNATIVE TO THE LEAST SQUARES
14.6   LINE: A NONPARAMETRIC APPROACH (OPTIONAL)

The least squares line discussed in previous sections has an interesting property: it passes through the point $(\bar{x}, \bar{y})$, where $\bar{x}$ is the mean of the X values and $\bar{y}$ is the mean of the Y values. However, when the sample data contain one or two outliers (points at a considerable distance from the remaining points), this can have a drastic effect on the least squares line. Such an outlier, in an effort to minimize the squared distances, will "pull" the line toward this point.

An alternative to this procedure, is a line of "best fit" that passes through the **medians** of the X and Y values, rather than through the means. Consequently, we will force the point (median of X values, median of Y values) to lie on the new line called the **regression line through the medians**. This regression line will be much less affected by the presence of an outlier or two, and will do a better job of "passing through" the remaining points. We will write the regression line through the median as

$$\hat{Y} = B_0 + B_1 X$$

Determination of B_0 and B_1

The sample data consist of n points, say point 1, point 2, ..., point n. For points i and j, the slope of the line connecting them is

$$S_{ij} = \frac{Y_i - Y_j}{X_i - X_j} \qquad (14\text{-}26)$$

To determine the slope of the sample regression line B_1, we begin by determining all possible slopes S_{ij} for which the denominator is nonzero, that is, $X_i \neq X_j$. If no two X values are the same, there will be $n(n-1)/2$ such S_{ij} values. Next, we arrange the S_{ij} values in order, from smallest to largest. The *median* of these slopes is B_1, that is

$$B_1 = \text{median of slopes, } S_{ij} \qquad (14\text{-}27)$$

Finally,

$$B_0 = (\text{median of } Y \text{ values}) - B_1(\text{median of } X \text{ values})$$

To illustrate, consider five points, one of which (namely $X = 25$, $Y = 55$) appears to be an outlier.

X	5	8	12	20	25
Y	10	15	18	23	55

The corresponding least squares line using these five values is (see Figure 14.16):

$$\hat{Y} = -2.34 + 1.896X$$

To determine the regression line through the medians, we determine the $(5)(4)/2 = 10$ slopes, S_{ij}. These are

$$\frac{15-10}{8-5}, \frac{18-10}{12-5}, \frac{23-10}{20-5}, \frac{55-10}{25-5}, \frac{18-15}{12-8}, \frac{23-15}{20-8}, \frac{55-15}{25-8},$$

$$\frac{23-18}{20-12}, \frac{55-18}{25-12}, \frac{55-23}{25-20}$$

That is,

$$1.67, 1.14, .87, 2.25, .75, .67, 2.35, .625, 2.85, 6.4$$

Figure 14.16

Least squares
regression line
through the means
and regression line
through the medians

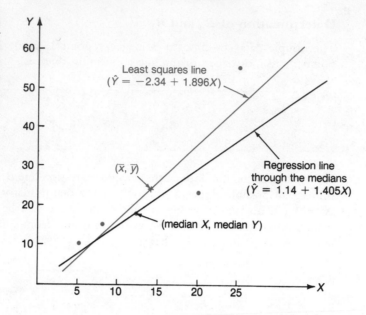

Next, we arrange these values in order:

$$.625, .67, .75, .87, \mathbf{1.14}, \mathbf{1.67}, 2.25, 2.35, 2.85, 6.4$$

The median of these 10 values is

$$B_1 = \frac{1.14 + 1.67}{2} = 1.405$$

Finally, the median of the X values is 12 and the median of the Y values is 18. Consequently,

$$B_0 = 18 - (1.405)(12) = 1.14$$

The regression line through the medians is

$$\hat{Y} = 1.14 + 1.405X$$

The least squares regression line and the regression line through the medians are illustrated in Figure 14.16. Notice how the median regression line is less affected by the presence of the outlier in the upper right corner.

EXAMPLE 14.10

Use the university data in example 14.1 to determine the regression line through the medians.

Table 14.1

The 45 slopes, S_{ij}, for the university data (example 14.10)

−.0140	−.0098	−.0090	−.0027	−.0009
.0005	.0006	.0008	.0008	.0012
.0018	.0019	.0021	.0025	.0029
.0030	.0034	.0035	.0037	.0040
.0040	.0042	.0044	.0044	.0048
.0049	.0052	.0052	.0056	.0061
.0064	.0066	.0066	.0068	.0094
.0108	.0110	.0115	.0118	.0122
.0155	.0160	.0175	.0195	.0210

Solution

The first step is to determine the $(10)(9)/2 = 45$ slopes, S_{ij}. The ordered slopes are contained in Table 14.1. The median value (boxed in color) is $B_1 = .0044$. Also, the median X value is 600 and the median Y value is 3.26, and so $B_0 = 3.26 - (.0044)(600) = .62$. The resulting regression line through the medians is

$$\hat{Y} = .62 + .0044X$$

Notice that this line is very similar to the least squares line derived in example 14.2 ($\hat{Y} = .785 + .0042X$), due to the lack of any significant outliers in the sample data. □

Confidence Interval for the Slope β_1

In deriving a confidence interval for the slope of the (population) regression line β_1 in Section 14.3 a critical assumption was that the residuals were normally distributed. An alternative to this procedure is to remove this assumption by deriving a *nonparametric* confidence interval for β_1. This confidence interval can be used as a follow-up to either the least squares regression line procedure or the regression line through the medians procedure.

To derive this confidence interval, we begin by examining the ordered slopes S_{ij}. If the regression line through the medians is being used, these values have already been determined in the calculation of B_1. Suppose that n_1 slopes have been derived. If no two X values are equal, then we previously stated that

$$n_1 = n(n-1)/2$$

where n is the number of data points in the sample.

To construct, say, a 90% confidence interval for β_1, we consult Table A-14, using the column labeled $V_{1-(.10/2)} = V_{.95}$. In general, for a $(1-\alpha) \cdot 100\%$ confidence interval, use the column labeled $V_{1-\alpha/2}$ in Table A-14. This confidence interval procedure is based on a nonparametric technique developed by M. G. Kendall about 1940, hence the title of Table A-14.

Next, using the sample size, n in the far left of the table, find the corresponding value (called n_2) from the table. Define

$$k = \tfrac{1}{2}(n_1 - n_2)$$

and round this value downward if it is not an integer.

Finally, let (L = lower, U = upper)

S_L = the kth slope, S_{ij}, starting from the *smallest* value

S_U = the kth slope, S_{ij}, starting from the *largest* value

The $(1 - \alpha) \cdot 100\%$ confidence interval for β_1 is then (S_L, S_U).

EXAMPLE 14.11

Using the university data from example 14.1 and the calculations in example 14.10, construct a 90% confidence interval for the slope of the regression line.

Solution

The $n_1 = 45$ ordered slopes, S_{ij}, are contained in Table 14.1. Using Table A-14 with $n = 10$ under the $V_{.95}$ column, we find that $n_2 = 19$. Thus

$$k = \tfrac{1}{2}(45 - 19) = 13$$

Consequently, S_L is the 13th value in Table 14.1 beginning with the smallest value. This value is boxed in Table 14.1, where $S_L = .0021$. The upper confidence limit S_U is also boxed and is the 13th value beginning with the largest value. This is $S_U = .0066$. The 90% confidence interval for β_1 is then .0021 to .0066. This is nearly the same interval obtained in example 14.5, which relied on the assumption of a normal distribution for the residuals. Don't be misled, however—in many applications you will observe a considerable difference in these two intervals.

EXERCISES

14.51 A work study expert was trying to optimize the speed of work on an assembly line. In one experiment, "gaps" were intentionally created in the tempo of the assembly line to permit the workers to catch up or re-establish the rhythm. The following data show the productivity (number of items produced per day) for different tempos.

X (Time between Gaps, in minutes)	Y (Productivity, No. of items per day)
10	100
15	105
20	108
25	95
30	90
35	87
40	85
45	84

(a) Compute the regression line through the medians.
(b) Use a computer package like MINITAB (or use hand calculations) to compute the least squares regression line. Compare the result with part (a) above.
(c) Compute the 95% confidence interval for the slope without assuming a normal distribution for the residuals (i.e., use the nonparametric approach).
(d) For the sake of comparison, compute the 95% confidence interval for the slope, assuming the residuals are normally distributed.

14.52 Use the data in exercise 14.2 to do the following:
(a) Compute the regression line through the medians. Compare your result with the least squares line obtained in exercise 14.2(a).
(b) Compute the 95% confidence interval for the slope, using the nonparametric approach. Compare your results with the confidence interval obtained in exercise 14.22(b).

14.53 Use the data in exercise 14.9 to do the following:
(a) Compute the regression line through the medians. Compare your result with the least squares equation of exercise 14.9(b).
(b) Compute the 95% confidence interval for the slope, without assuming normality of the residuals (i.e., use the nonparametric approach). Compare your result with the confidence interval obtained in exercise 14.23(d).

14.54 Consider the data in exercise 14.4. Note that the use of birth control for the lowest income category (group 1) could be looked at as an outlier.
(a) If the regression line is obtained by the least squares method, in which direction is it pulled by the outlier, i.e., does this outlier tend to increase or decrease the value of the slope (β_1)?
(b) Similarly, would the regression line through the medians tend to increase or decrease the slope, in comparison to the least squares line?
(c) Verify your answers to (a) and (b) by computing the regression line by both methods, and plotting the two lines in a graph.
(d) Verify algebraically and graphically that the least squares line passes through $(\bar{x}, \bar{y})$ and the regression line through the medians passes through (median of X, median of Y).

14.55 The following data show the number of hours that a random sample of college students studied for an examination in statistics, and their grades on the examination:

X (Number of Hours)	Y (Exam Grade)
3	33
11	81
13	72
10	70
8	56
5	52
19	95
15	85

(a) Compute the regression line through the medians.
(b) Compute the 99% confidence interval for β_1, without assuming a normal distribution of the residuals. (Use the nonparametric method.)

14.56 Use the data in exercise 14.1 to do the following:

   (a)   Compute the regression line through the medians.

   (b)   Compute the 95% confidence interval for the slope, using the nonparametric approach.

SUMMARY

When dealing with a pair of variables (say, X and Y), we generally are interested in determining whether the variables are related in some manner. If a relationship does exist, perhaps the **independent** variable (X) can be used to predict values of the **dependent** variable (Y). If a significant linear relationship exists within the sample data, both the direction (positive or negative) and the strength of this linear relationship can be measured using the sample **coefficient of correlation**, r. The sample correlation is an estimate of the population coefficient of correlation, ρ. Another commonly used measure of association between two variables is the sample **covariance**.

Whenever a sample of bivariate data contains a significant linear pattern, we can determine the **least squares** line through the data points. This generates an equation that can be used to predict values of the dependent variable. To describe accurately the assumptions behind this procedure, we introduced the concept of a **statistical model** consisting of a **deterministic** portion (the straight line) and a **random error** component. This model can be written as $Y = \beta_0 + \beta_1 X + e$, where $\beta_0 + \beta_1 X$ is the deterministic component and e represents the random error component. When we perform any test of hypothesis regarding the underlying bivariate population, we must be careful to satisfy the necessary assumptions behind this procedure. These assumptions will be examined more closely in Chapter 15.

By regressing Y on X, we are able to determine the least squares line, $\hat{Y} = b_0 + b_1 X$. The value of b_0 is the **intercept** of the least squares line and estimates the population intercept β_0. The **slope** of the least squares line b_1 estimates the population slope β_1. One question of interest is: If we regress X on Y (that is, switch the independent and dependent variables), can we rearrange the previous equation and say that $\hat{X} = -(b_0/b_1) + (1/b_1)Y$? The answer is no, although the coefficient of correlation r is the same in either case. Consequently, constructing a least squares line is not a good idea in the event it is not obvious which variable is the independent variable.

Various methods for determining the utility of the model as a predictor of the dependent variable include: (1) a t test for detecting a significant slope b_1—a value of $\beta_1 = 0$ indicates that X has no predictive ability; (2) a t test for determining whether the sample correlation r is significantly large—a value of $\rho = 0$ indicates that there is no linear relationship between the two variables; and (3) a confidence interval for the slope β_1. The two t tests appear to be quite different, but their computed values (and df) are identical; there is no point in performing both tests.

Another measure of how well the model provides estimates that fit the sample data is given by the **coefficient of determination**, r^2. For simple linear regression (one independent variable), this is the square of the correlation coefficient. Another definition of the coefficient of determination also can be used to examine more than one independent variable (called multiple linear regression), namely, $r^2 = 1 - (SSE/SS_Y)$. Here, SSE is the sum of squared errors and SS_Y represents the total variation in the sample Y values. For example, if $r^2 = .85$, then 85% of the variation in the sample Y values has been explained using this model.

The value of $\hat{Y}$ from the least squares regression line at a specific value of X (say, $X = x_0$), can be used to estimate the *average* value of Y, given this value of X (written $\mu_{Y|x_0}$). The value of $\hat{Y}$ centers a confidence interval for $\mu_{Y|x_0}$. Similarly, we can use the $\hat{Y}$ value to center a prediction interval for an *individual* value of the dependent variable, given this specific value of X (written Y_{x_0}). The value of $\hat{Y}$ can be used to *estimate* the value of $\mu_{Y|x_0}$ or *predict* the value of Y_{x_0}.

An alternative to the least squares line was discussed that passed through the point determined by the median of the X values and the median of the Y values. This line is less affected by the presence of outliers than is the least squares line which passes through the point determined by the mean of the X and Y values. A nonparametric procedure for determining a confidence interval for the population slope β_1 was presented which does not rely on the assumptions of normally distributed residuals.

Summary of Linear Regression and Correlation Formulas

Description	Formula
Correlation between two variables	$r = \dfrac{SS_{XY}}{\sqrt{SS_X}\sqrt{SS_Y}}$

where

$$SS_{XY} = \Sigma xy - (\Sigma x)(\Sigma y)/n$$
$$SS_X = \Sigma x^2 - (\Sigma x)^2/n$$
$$SS_Y = \Sigma y^2 - (\Sigma y)^2/n$$

| Least squares line | $\hat{Y} = b_0 + b_1 X$ |

where

$$b_1 = SS_{XY}/SS_X$$

and

$$b_0 = \bar{y} - b_1 \bar{x}$$

Description	Formula
Estimate of the residual variance	$\hat{\sigma}_e^2 = s^2 = \dfrac{\text{SSE}}{n-2}$

where

$$\text{SSE} = \Sigma(y - \hat{y})^2 = \text{SS}_Y - \frac{(\text{SS}_{XY})^2}{\text{SS}_X}$$

Description	Formula
t statistic for detecting a significant slope	$t = \dfrac{b_1}{s_{b_1}} \qquad (df = n - 2)$

where

$$s_{b_1} = s/\sqrt{\text{SS}_X}$$

Description	Formula
Confidence interval for the slope β_1	$b_1 - t_{\alpha/2,\,n-2}s_{b_1}$ to $\quad b_1 + t_{\alpha/2,\,n-2}s_{b_1}$
t statistic for detecting a significant correlation	$t = \dfrac{r}{\sqrt{\dfrac{1-r^2}{n-2}}} \qquad (df = n-2)$
Coefficient of determination	$r^2 =$ square of correlation coefficient $= 1 - \dfrac{\text{SSE}}{\text{SS}_Y}$

Description	Formula
Confidence interval for the average value of Y at a specific value of X (say, x_0)	$\hat{Y} \pm t_{\alpha/2,\,n-2}s\sqrt{\dfrac{1}{n} + \dfrac{(x_0 - \bar{x})^2}{\text{SS}_X}}$

or, for MINITAB,

$$\hat{Y} \pm t_{\alpha/2,\,n-2}s_{\hat{Y}}$$

Description	Formula
Prediction interval for a particular value of Y at a specific value of X (say, x_0)	$\hat{Y} \pm t_{\alpha/2,\,n-2}s\sqrt{1 + \dfrac{1}{n} + \dfrac{(x_0 - \bar{x})^2}{\text{SS}_X}}$

or, for MINITAB,

$$\hat{Y} \pm t_{\alpha/2,\,n-2}\sqrt{s_{\hat{Y}}^2 + s^2}$$

REVIEW EXERCISES

14.57 Cheng and Hom (1986) utilized Hofstede's four-dimensional classification of culture in a cross-cultural study of how cultural dimensions can moderate the impact of various determinants of organizational commitment. Hofstede's four cultural dimensions are *Power Distance Index* (PDI), which measures the extent to which unequal distribution of power is accepted in an organization; *Uncertainty Avoidance Index* (UAI), which indicates the extent to which uncertainty and ambiguous situations are avoided by establishing formal rules and beliefs in absolute truth; *Individualism-Collectivism* (IDV), which relates to the degree of people's self-orientation versus group orientation; and *Masculinity and Femininity* (MAS), which measures the extent to which the dominant values in society are "masculine" (i.e., the acquisition of money and things, etc.) versus "feminine" (i.e., caring for others, quality of life, or people). A 1985 survey of managers employed by overseas subsidiaries of an American manufacturing corporation resulted in the following data:

COUNTRY SCORES FOR HOFSTEDE'S CULTURAL VARIABLES

Country	PDI	UAI	IDV	MAS
Austria	11	70	55	79
Belgium	65	94	75	54
Canada	39	48	80	52
Denmark	18	23	74	16
England	35	35	89	66
Finland	33	59	63	26
Germany	35	65	67	66
Greece	60	112	35	57
Italy	50	75	76	70
Netherlands	38	53	80	14
Norway	31	50	69	8
Portugal	63	104	27	31
Spain	57	86	51	42
Sweden	31	29	71	5
Switzerland	34	58	68	70

Source: C. Cheng and P. Hom. In need of blind men: Case illustration of testing organizational commitment cross-culturally. *Proceedings of the 1986 Annual Meeting of the Decision Sciences Institute in Hawaii*, Nov. 23–25, 1986, Vol. 1, pp. 120–122.

(a) Although the focus of Cheng and Hom's study was how these cultural variables influenced organizational commitment, we can examine them for their own sake. Obtain the correlation coefficient between each of the four dimensions. (A computer package would be useful here.)

(b) Which relationships are statistically significant, at $\alpha = .05$?

(c) Try to describe the significant relationships in words; for example, a strong positive correlation between PDI and UAI might suggest that organizations with unequal distribution of power have a greater tendency to avoid uncertainty with established formal procedures and doing things "by the book" and so on.

14.58 A journalist collected data on SAT (Scholastic Aptitude Test) scores for various school districts. The following list contains scores on the verbal and math portion of the SAT in 1986:

School	SAT Verbal Score	SAT Math Score
Arlington	444	475
Cedar Hill	449	481
DeSoto	418	461
Duncanville	444	478
Garland	421	454
Grand Prairie	400	444
Grapevine	435	471
Highland Park	468	504
Irving	430	466
Lancaster	397	429
Lewisville	420	462
Mesquite	442	471
Plano	461	507
Richardson	436	484
Rockwall	432	467

(a) Use a computer package like MINITAB, or a hand calculator, to compute the simple linear regression equation (least squares line). Let the SAT verbal score be the predictor variable.

(b) What percentage of variation in the Math scores is explained by variation in the verbal scores?

(c) Is the relationship between the two variables significantly linear, at $\alpha = .05$?

(d) State the p value for the test.

(e) What is the variance of the error in the model?

14.59 The following data were collected for a certain regression analysis.

$$SS_{XY} = -138.6$$

$$SS_X = 112.3$$

$$SS_Y = 325.2$$

$$\bar{x} = 86.2$$

$$\bar{y} = 112.9$$

$$n = 41$$

(a) Find the least squares line.

(b) Test, using a 1% significance level, that the population correlation coefficient between the variables X and Y is negative.

(c) Find the 90% confidence interval for the slope of the regression equation.

14.60 Use the data in exercise 14.58 to answer the following:

(a) What is the predicted value (point estimate) of the math score on the SAT for a school which obtained a verbal score of 430?

(b) What is the 95% confidence interval for the average SAT Math score of schools that obtained a verbal score of 430?

(c) What is the 95% confidence interval for the slope of the regression line?

14.61 A professor of medical sociology, who specialized in the study of longevity, was telling his class in a lecture that while gender, race, heredity, diet, exercise, and nonuse of tobacco were obvious factors influencing longevity, they might be surprised to learn that satisfaction in marriage was also a significant factor. He put the following data on the board:

Marital Satisfaction (on a scale of 1 to 10)	Longevity (gain in life span over the norm)
1	−0.5
2	−0.2
3	0.1
4	1.0
5	1.2
6	2.1
7	1.5
8	3.8
9	2.7
10	3.5

You are required to do the following:
(a) Compute the least squares line for the data.
(b) What is the residual sum of squares?
(c) What is the error sum of squares?
(d) What is the coefficient of determination?
(e) What percentage in variation in longevity is accounted for in the regression model?
(f) Test the hypothesis $H_0: \beta_1 \leqslant 0$ at a 5% significance level. State the p value for the test.
(g) Is there a significant positive relationship between marital satisfaction and longevity?

14.62 Use the data in exercise 14.61 to do the following:
(a) Compute the regression line through the medians.
(b) Compute the 90% confidence interval for the slope, using the nonparametric approach.

14.63 A regression line is fitted to a set of data and the values of $Y - \hat{Y}$ are calculated. Does the following set of sample errors appear to conform to the empirical rule that 95% of the data should lie within two standard deviations of the mean? Construct a histogram for the residuals. Comment on the shape of the histogram.

1.1, −0.8, 2.6, 1.5, 0.2, −0.4, 0.8, −1.8, −2.3, 0.9, −2.7, 3.1, −1.0, 0.9, 4.5, −3.4, −0.1, −0.2, 2.4, −1.7, −0.7

14.64 A regression analysis, where X = number of persons attending a conference and Y = number of extra meals served at the conference dining facility, yielded the following computations from a sample of 24 pairs of observations:

$$\Sigma x = 5462 \qquad \Sigma x^2 = 1,388,856 \qquad \Sigma xy = 1,110,550$$

$$\Sigma y = 4478 \qquad \Sigma y^2 = 914,032$$

(a) Obtain the least squares line.

(b) Test for a significant positive relationship, at $\alpha = .05$, between X and Y. State the hypothesis in terms of ρ (rho), the population correlation coefficient.

14.65 Consider the following selected information from a computer printout of a regression analysis:

Predictor	Coefficient	Std. Dev.	t Ratio
Constant	2.82033	0.32156	8.7708
X	0.87090	0.12496	6.9694

$$DF = 18$$

Index of determination (R-SQ) = 0.7296

Analysis of Variance

Source	SS	DF	MS
Regression	111.201	1	111.201
Error	41.206	18	2.289
Total	152.407	19	

Answer the following questions:
(a) Write the regression equation.
(b) What percentage of variation in Y is explained in the model?
(c) What is the sample size n used for the above regression analysis?
(d) Find the value of
     (i)   The total sum of squares, SS_Y
     (ii) The error sum of squares, SSE
     (iii) The regression sum of squares, SSR
(e) What is the range of values between which β_1 will fall, 95% of the time?
(f) Does the model represent a significant linear relationship between X and Y? Conduct a hypothesis test on $H_0: \beta_1 = 0$ at a 5% significance level.
(g) State your conclusion and the p value.

CASE STUDY

Nitrogen Fertilizer, Soft Red Winter Wheat, and Foliar Disease

The use of statistical techniques in agricultural experiments has a long history. A classical example is the study of grain yields obtained from the application of different fertilizer rates. Many such experiments lend themselves readily to an ANOVA design, where fertilizer rate is the treatment, and yield is the dependent variable. Since there are some underlying similarities in linear regression and analysis of variance, the use of regression techniques is also not uncommon in agronomic studies. We will look at one such set of experiments conducted from 1981 to 1984 at the Northeast Research Station, St. Joseph, Louisiana, under the auspices of the Louisiana State University Agricultural Center.

Boquet and Johnson (1987) conducted experiments to determine the effects of fertilizer N (nitrogen) with and without P (phosphorus) and K (potassium) applications on grain yield and quality of soft red winter wheat (SRWW). The experiments were conducted on a Sharkey clay soil, technically described as "very fine, montmorillonitic, thermic Vertic Haplaquent." Each year, wheat was planted in November as a doublecrop following soybean at a seeding rate of 67 kg per hectare. The preceding soybean crop received no fertilizers. As can be seen in the following tables, the fertilizer applications consisted of 14 different treatments, namely, single applications in February at different N rates, split applications in the fall prior to planting with the remainder in February, and five N–P–K combinations. An unfertilized treatment was included as a check in each experiment. The nitrogen was in the form of ammonium nitrate, the phosphorus was applied as triple superphosphate, and the potassium as potassium chloride. In addition to grain yield in kilograms per hectare, data were also obtained for 1000-kernel weight, kernel number per spike, crude grain protein, and incidence (in percentages) of the foliar diseases known as powdery mildew (*Erysiphe graminis* D.C.) and leaf rust (*Puccinia recondita* Rab. ex Desm). We will not use all the data, but they are provided anyway for the sake of completeness.

Table I

Response of SRWW grown on Sharkey clay soil to N–P–K fertilization, 1981–1982

Fertilizer Timing and Rate (kg/ha)	Grain Yield (kg/ha)	1000-Kernel Wt. (g)	Kernel No. (tiller^{-1})	FOLIAR DISEASE INFECTION	
				Leaf Rust (%)	Powdery Mildew (%)
February N					
0	4070	37.3	32	20	11
34	4220	37.7	34	30	31
56	4165	34.0	34	27	30
78	4350	34.9	35	30	37
101	4005	32.7	34	29	45
Fall preplant + February N					
17 + 17	4165	36.7	33	26	14
17 + 39	4215	36.3	34	32	28
17 + 62	4380	36.4	35	32	39
17 + 84	4080	34.0	35	34	46
Fall preplant N–P–K + February N					
0–20–37 + 56	4490	36.4	35	26	25
0–20–0 + 56	4320	34.8	35	21	14
0–0–37 + 56	4295	33.8	34	31	37
17–20–37 + 39	4330	34.3	34	27	37
17–20–37 + 62	4415	35.2	35	29	29
LSD (0.05)	335	2.5	NS	NS	20
CV (%)	6.0	5.0	7.1	23.7	46
SE	128	0.8	1.2	3.3	6.8

Table 2

Response of SRWW grown on Sharkey clay soil to N–P–K fertilization, 1982–1983

Fertilizer Timing and Rate (kg/ha)	Grain Yield (kg/ha)	1000-Kernel Wt. (g)	Kernel No. (tiller^{-1})	FOLIAR DISEASE INFECTION Leaf Rust (%)	Powdery Mildew (%)	Grain Protein (g/kg)
February N						
0	1975	29.5	31	35	0	107
34	2410	30.9	35	47	0	110
56	2775	33.0	35	62	4	106
78	2985	33.1	33	62	5	103
101	3400	32.7	34	85	5	107
Fall preplant + February N						
17 + 17	2075	31.3	33	32	0	100
17 + 39	2580	33.4	32	45	0	95
17 + 62	2710	32.5	34	62	4	100
17 + 84	3350	32.8	34	86	3	105
Fall preplant N–P–K + February N						
0–20–37 + 56	2600	32.9	35	67	0	107
0–20–0 + 56	2525	31.9	34	32	0	107
0–0–37 + 56	2690	33.4	33	50	1	107
17–20–37 + 39	2215	33.1	33	57	2	102
17–20–37 + 62	2730	34.6	33	50	2	105
LSD (0.05)	350	1.5	2.8	19	2.0	7.0
CV (%)	9.4	3.3	6.0	24.0	5.9	4.8
SE	124	0.5	1.0	6.1	0.6	2.4

Table 3

Response of SRWW grown on Sharkey clay soil to N–P–K fertilization, 1983–1984

Fertilizer Timing and Rate (kg/ha)	Grain Yield (kg/ha)	1000-Kernel Wt. (g)	Kernel No. (tiller^{-1})	FOLIAR DISEASE INFECTION Leaf Rust (%)	Powdery Mildew (%)	Grain Protein (g/kg)
February N						
0	2900	34.3	31	5	2	119
34	3355	34.4	35	22	3	118
56	3795	34.8	33	20	12	116
78	4180	33.3	36	15	16	119
101	4405	33.0	35	14	19	120
Fall preplant + February N						
17 + 17	3125	34.4	33	6	2	115
17 + 39	3800	33.5	35	15	5	117
17 + 62	4000	33.2	36	16	10	116
17 + 84	4445	34.8	37	21	14	120
Fall preplant N–P–K + February N						
0–20–37 + 56	4015	34.4	35	9	9	118
0–20–0 + 56	3690	34.3	33	21	4	115
0–0–37 + 56	3580	34.7	35	11	14	115
17–20–37 + 39	3740	33.4	34	21	8	118
17–20–37 + 62	3940	34.0	35	8	13	118
LSD (0.05)	295	1.2	3.0	NS	NS	NS
CV (%)	5.5	2.4	6.7	104	98	3.8
SE	105	0.4	1.2	7.6	4.4	2.2

Source: Reproduced from Donald J. Boquet and Catherine C. Johnson, Fertilizer effects on yield, grain composition, and foliar disease of doublecrop soft red winter wheat. *Agronomy J.* **79** (1): 35–141 (1987) by permission of the American Society of Agronomy, Inc.

Case Study Questions

1. Using only the single-application February N data for the 3 years 1981–1984, perform a regression analysis to obtain a linear equation, where Y = grain yield in kg/ha and X = nitrogen rate in kg/ha. You will have 15 observations, for N rates of 0, 34, 56, 78 and 101 kg/ha.

2. Repeat item (1), but this time eliminate the 1981–1982 data, and restrict yourself to the set of 10 observations for the 2-year period 1982–1984.

3. Conduct the appropriate hypothesis tests to determine whether the equations in (1) and (2) above represent a significant positive relationship between N rate and yield, at a 5% significance level. Report p values.

4. Notwithstanding that the 3-year relationship based on the 1981–1984 data might be statistically significant, examine the data for each year carefully and explain how the 1981–1982 results could be misleading and should be excluded in favor of the results of the 2-year period 1982–1984.

5. Assuming the equation for the 1982–1984 period is an accurate predictor of yield,
   (a) What is the predicted yield of SRWW on any given plot receiving a February N fertilizer rate of 56 kg per hectare?
   (b) What is the 95% confidence interval for average SRWW yield on plots receiving a February N fertilizer rate of 56 kg/ha?

6. Within the range of the observed data, a higher N rate seems to be associated with higher yields. Would you expect this relationship to continue at even higher N rates?

7. The authors found a strong association between incidence of foliar diseases and N rate. The association was strongest in the periods when the disease occurrence overall was highest, which was 1982–1983 for leaf rust and 1981–1982 for powdery mildew. In these years, the regression equations for percentage of disease infection (Y) plotted against N rate in kg/ha (X) were: $\hat{Y} = 33.5 + 0.46X$ with $r^2 = 0.77$ (leaf rust), and $\hat{Y} = 14.5 + 0.30X$ with $r^2 = 0.42$ (powdery mildew).
   (a) What are the correlation coefficients in each case?
   (b) Is the relationship between incidence of the two diseases and N rate positive or negative for the periods given above?
   (c) Is the relationship in each case significant at $\alpha = .05$? (*Note:* $n = 5$.)
   (d) What implication does this relationship have for the issue raised in question (6) above?

COMPUTER EXERCISES USING THE DATABASE

Using a sample of 100 observations, complete the following exercises.

1. Determine the sample correlation between age (variable AGE) and graduate GPA (variable GR-GPA). Test H_0: $\rho = 0$ at a significance level of .05.

2. Consider the model $Y = \beta_0 + \beta_1 X + e$, where X is a student's undergraduate GPA and Y is the student's graduate GPA at Brookhaven University. Using a significance level of .10, test H_0: $\beta_1 = 0$ versus H_a: $\beta_1 > 0$. Would you suggest using this model? Discuss why or why not.

3. In exercise 2, what is the coefficient of determination? Interpret the meaning of this value.

4. Construct a confidence interval for the average graduate GPA for all students attending Brookhaven with an undergraduate GPA of 3.15. What is the corresponding prediction interval for an individual student having an undergraduate GPA of 3.15?

5. Someone made the observation that older males at Brookhaven appeared to be more satisfied in life than younger males. Using variables LSINDEX, AGE, and SEX, explore this statement. What is your conclusion at a significance level of .10?

SPSSX APPENDIX (MAINFRAME VERSION)

Solution to Example 14.2

Example 14.2 was concerned with computing the simple linear regression equation. The problem was to determine the relationship between first year grade point average (GPA) in graduate school and the performance on the verbal portion of the GRE exam. The SPSSX program listing in Figure 14.17 was used to request the regression coefficients, SSE as well as other statistics. The regression analysis should always include a scatter diagram of the sample data.

Figure 14.17

SPSSX program listing used to request the SSE, b_0, and b_1 values

```
TITLE FIRST YEAR GPA USING VERBAL GRE SCORE
DATA LIST FREE / GRE GPA
BEGIN DATA
580 3.54
500 2.62
670 3.30
480 2.90
710 4.00
550 3.21
640 3.57
540 3.05
620 3.15
690 3.81
END DATA
PRINT / GRE GPA
REGRESSION VARIABLES=GRE,GPA/DEPENDENT=GPA/ENTER/
   SCATTERPLOT (GPA,GRE)/
```

The TITLE command names the SPSSX run.

The DATA LIST command gives each variable a name and describes the data as being in free form.

The BEGIN DATA command indicates to SPSSX that the input data immediately follow.

The next 10 lines contain the data values representing verbal GRE scores and corresponding GPAs. For example, the first line of data (580, 3.54) represents a student with a GRE score of 580 whose first year graduate GPA was 3.54.

**SPSSX
APPENDIX
(MAINFRAME
VERSION)**

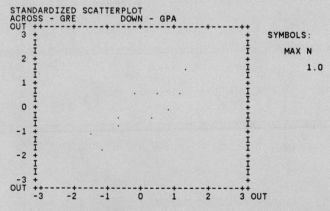

```
                          * * * *   M U L T I P L E    R E G R E S S I O N   * * * *

LISTWISE DELETION OF MISSING DATA
EQUATION NUMBER 1    DEPENDENT VARIABLE..    GPA
BEGINNING BLOCK NUMBER  1.  METHOD:  ENTER

VARIABLE(S) ENTERED ON STEP NUMBER  1..    GRE

MULTIPLE R          .84456        ANALYSIS OF VARIANCE
R SQUARE            .71328                            DF     SUM OF SQUARES     MEAN SQUARE
ADJUSTED R SQUARE   .67744        REGRESSION           1           1.02131        1 02131
STANDARD ERROR      .22653        RESIDUAL             8            .41054          .05132

                                  F =    19.90151      SIGNIF F =  .0021        SSE

------------------ VARIABLES IN THE EQUATION ------------------

VARIABLE      b₁         B         SE B        BETA          T   SIG T

GRE                    .004198   9.4096E-04    .844557     4.461  .0021
(CONSTANT)             .784762    .567235                  1.383  .2039

END BLOCK NUMBER   1    ALL REQUESTED VARIABLES ENTERED.
```

Below the table, annotations: b_0 pointing to .784762; s_{b_1} pointing to SE B; $t = \dfrac{b_1}{s_{b_1}}$ pointing to T column.

```
                          * * * * * * * * * * * * * * * * * * * * * * * * * * * *

STANDARDIZED SCATTERPLOT
ACROSS - GRE          DOWN - GPA
OUT ++-----+-----+-----+-----+-----+-----++
  3 +                                      +      SYMBOLS:
    I                                      I
    I                                      I         MAX N
  2 +                                      +
    I                     .                I          1.0
    I                                      I
  1 +                                      +
    I                                      I
    I           .        .                 I
  0 +                     .                +
    I                                      I
    I           .                          I
 -1 +      .                               I
    I                                      I
    I                                      I
 -2 +                                      +
    I                                      I
    I                                      I
 -3 +                                      +
OUT ++-----+-----+-----+-----+-----+-----++
    -3    -2    -1     0     1     2     3 OUT
```

Figure 14.18

Output obtained by
executing the SPSSX
program in Figure
14.17

**SPSSX
APPENDIX
(MAINFRAME
VERSION)**

The END DATA statement indicates the end of the input data.

The PRINT command requests a printout of the input data.

The REGRESSION statement defines the variables, GRE and GPA, as the regression variables, and specifies that GPA is to be the dependent variable. The ENTER subcommand indicates that all independent variables (only one here) are to be entered into the regression analysis. The SCATTERPLOT subcommand requests a scatter diagram of the sample data with GPA on the vertical axis and GRE on the horizontal axis. The data for each variable are standardized by subtracting the mean and dividing by the standard deviation.

Figure 14.18 contains the output obtained by executing the program listing in Figure 14.17.

**SPSS/PC+
APPENDIX
(MICRO
VERSION)**

**Solution to
Example 14.2**

Example 14.2 was concerned with computing the simple linear regression equation. The problem was to determine the relationship between first year grade point average (GPA) in graduate school and the performance on the verbal portion of the GRE exam. The SPSS/PC+ program listing in Figure 14.19 was used to request the regression coefficients, SSE as well as other

Figure 14.19

SPSS/PC+ program listing used to request the SSE, b_0, and b_1 values

```
TITLE FIRST YEAR GPA USING VERBAL GRE SCORE.
DATA LIST FREE / GRE GPA.
BEGIN DATA.
580 3.54
500 2.62
670 3.30
480 2.90
710 4.00
550 3.21
640 3.57
540 3.05
620 3.15
690 3.61
END DATA.
LIST / GRE GPA.
REGRESSION VARIABLES=GRE,GPA/DEPENDENT=GPA/ENTER/
   SCATTERPLOT (GPA,GRE).
```

statistics. The regression analysis should always include a scatter diagram of the sample data.

The TITLE command names the SPSS/PC+ run.

The DATA LIST command gives each variable a name and describes the data as being in free form.

Figure 14.20

Output obtained
by executing the
SPSS/PC+ program
in Figure 14.19

```
                    FIRST YEAR GPA USING VERBAL GRE SCORE
              * * * *   M U L T I P L E   R E G R E S S I O N   * * * *
Equation Number 1      Dependent Variable..    GPA


Variable(s) Entered on Step Number
   1..    GRE


Multiple R              .84456
R Square                .71328  ⟵ r²
Adjusted R Square       .67744
Standard Error          .22653                        SSE

Analysis of Variance
                        DF        Sum of Squares      Mean Square
Regression              1               1.02131          1.02131
Residual                8                .41054           .05132

F =      19.90151       Signif F =  .0021
------------------------------------------------------------------
                    FIRST YEAR GPA USING VERBAL GRE SCORE
              * * * *   M U L T I P L E   R E G R E S S I O N   * * * *

Equation Number 1      Dependent Variable..    GPA


------------------ Variables in the Equation ------------------

Variable               B          SE B         Beta          T   Sig T

GRE          4.197723E-03  9.40959E-04       .84456      4.461   .0021
(Constant)        .78476       .56724                    1.383   .2039


End Block Number    1   All requested variables entered.
```

$b_1 = .0042$ $b_0 = .78476$ s_{b_1}

$$t = \frac{b_1}{s_{b_1}}$$

**SPSS/PC+
APPENDIX
(MICRO
VERSION)**

The BEGIN DATA command indicates to SPSS/PC+ that the input data immediately follow.

The next 10 lines contain the data values representing verbal GRE scores and corresponding GPAs. For example, the first line of data (580, 3.54) represents a student with a GRE score of 580 whose first year graduate GPA was 3.54.

The END DATA statement indicates the end of the input data.

The LIST command requests a printout of the input data.

The REGRESSION statement defines the variables, GRE and GPA, as the regression variables, and specifies that GPA is to be the dependent variable. The ENTER subcommand indicates that all independent variables (only one here) are to be entered into the regression analysis. The SCATTERPLOT subcommand requests a scatter diagram of the sample data with GPA on the vertical axis and GRE on the horizontal axis. The data for each variable are standardized by subtracting the mean and dividing by the standard deviation.

Figure 14.20 contains the output obtained by executing the program listing in Figure 14.19.

Figure 14.20

continued

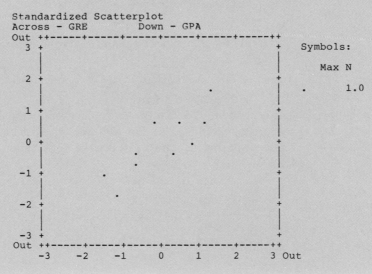

FIRST YEAR GPA USING VERBAL GRE SCORE

SAS APPENDIX (MAINFRAME VERSION)

Solution to Example 14.2

Example 14.2 was concerned with computing the simple linear regression equation. The problem was to determine the relationship between first year grade point average (GPA) in graduate school and the performance on the verbal portion of the GRE exam. The SAS program listing in Figure 14.21 was used to request the regression coefficients, SSE as well as other statistics. The regression analysis should always include a scatter diagram of the sample data.

Figure 14.21

SAS program listing used to request the SSE, b_0, and b_1 values

```
TITLE 'FIRST YEAR GPA USING VERBAL GRE SCORE';
DATA UNIVER;
 INPUT GRE GPA;
CARDS;
580 3.54
500 2.62
670 3.30
480 2.90
710 4.00
550 3.21
640 3.57
540 3.05
620 3.15
690 3.61
PROC PRINT;
PROC PLOT DATA = UNIVER;
  PLOT GPA*GRE;
PROC REG;
 MODEL GPA=GRE;
```

The TITLE command names the SAS run.

The DATA command gives the data a name.

The INPUT command names and gives the correct order for the data during the data input.

The CARDS command indicates to SAS that the input data immediately follow.

The next 10 lines contain the data values representing verbal GRE scores and corresponding GPAs. For example, the first line of data (580, 3.54) represents a student with a GRE score of 580 and a first year graduate GPA of 3.54.

The PROC PRINT command requests a printout of the input data.

The PROC PLOT requests a scatter diagram of the sample data, named in the previous DATA statement. The PLOT GPA*GRE statement will plot the GPA values on the vertical axis and the GRE values on the horizontal axis.

The PROC REG command and MODEL subcommand indicate that GPA and GRE are the regression variables, with GPA being the dependent variable and GRE the independent variable.

Figure 14.22 contains the output obtained by executing the program listing in Figure 14.21.

Figure 14.22

Output obtained by executing the SAS program in Figure 14.21

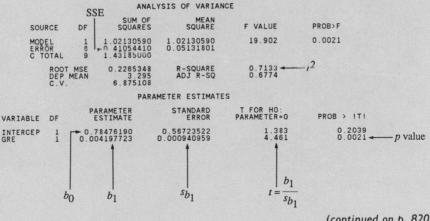

(continued on p. 820)

**SAS
APPENDIX
(MAINFRAME
VERSION)**

Figure 14.22

continued

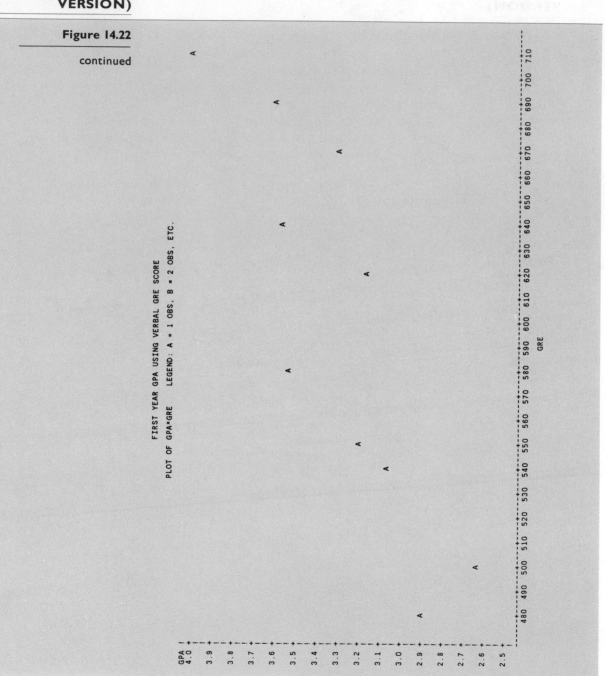

SAS APPENDIX (MICRO VERSION)

Solution to Example 14.2

Example 14.2 was concerned with computing the simple linear regression equation. The problem was to determine the relationship between first year grade point average (GPA) in graduate school and the performance on the verbal portion of the GRE exam. The SAS program listing in Figure 14.23 was used to request the regression coefficients, SSE as well as other statistics. The regression analysis should always include a scatter diagram of the sample data.

Figure 14.23

SAS program listing used to request the SSE, b_0, and b_1 values

```
TITLE 'FIRST YEAR GPA USING VERBAL GRE SCORE';
DATA UNIVER;
  INPUT GRE GPA;
CARDS;
580 3.54
500 2.62
670 3.30
480 2.90
710 4.00
550 3.21
640 3.57
540 3.05
620 3.15
690 3.61
PROC PRINT;
PROC PLOT DATA=UNIVER;
  PLOT GPA*GRE;
PROC REG;
  MODEL GPA=GRE;
RUN;
```

The TITLE command names the SAS run.

The DATA command gives the data a name.

The INPUT command names and gives the correct order for the data during the data input.

The CARDS command indicates to SAS that the input data immediately follow.

The next 10 lines contain the data values representing verbal GRE scores and corresponding GPAs. For example, the first line of data (580, 3.54) represents a student with a GRE score of 580 and a first year graduate GPA of 3.54.

**SAS
APPENDIX
(MICRO
VERSION)**

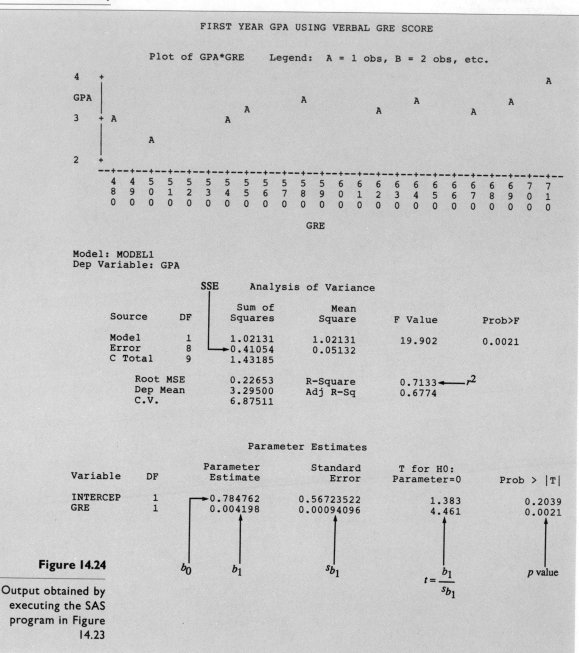

FIRST YEAR GPA USING VERBAL GRE SCORE

Plot of GPA*GRE   Legend: A = 1 obs, B = 2 obs, etc.

Model: MODEL1
Dep Variable: GPA

Analysis of Variance

Source	DF	Sum of Squares	Mean Square	F Value	Prob>F
Model	1	1.02131	1.02131	19.902	0.0021
Error	8	0.41054	0.05132		
C Total	9	1.43185			

Root MSE	0.22653	R-Square	0.7133	
Dep Mean	3.29500	Adj R-Sq	0.6774	
C.V.	6.87511			

Parameter Estimates

Variable	DF	Parameter Estimate	Standard Error	T for H0: Parameter=0	Prob > |T|
INTERCEP	1	0.784762	0.56723522	1.383	0.2039
GRE	1	0.004198	0.00094096	4.461	0.0021

b_0 b_1 s_{b_1} $t = \dfrac{b_1}{s_{b_1}}$ p value

r^2

Figure 14.24

Output obtained by executing the SAS program in Figure 14.23

The PROC PRINT command requests a printout of the input data.

The PROC PLOT requests a scatter diagram of the sample data, named in the previous DATA statement. The PLOT GPA*GRE statement will plot the GPA values on the vertical axis and the GRE values on the horizontal axis.

The PROC REG command and MODEL subcommand indicate that GPA and GRE are the regression variables, with GPA being the dependent variable and GRE the independent variable.

The RUN command tells the SAS system to execute the previous SAS statements.

Figure 14.24 contains the output obtained by executing the program listing in Figure 14.23.

15

MULTIPLE LINEAR REGRESSION

A LOOK BACK/INTRODUCTION

We used the technique of simple linear regression in Chapter 14 to explain the behavior of a dependent variable using a single predictor (independent) variable. For example, one can attempt to explain a first year graduate student's grade point average (GPA) using the student's verbal score on the Graduate Record Exam (GRE) as a predictor variable.

To define this procedure in statistical terms, we introduced the concept of a statistical model. This model consists of two parts. The first is the deterministic component. This was assumed to be $Y = \beta_0 + \beta_1 X$ (a straight line), implying that the underlying pattern for the X and Y variables is linear. If a simple linear regression model is appropriate for the university illustration, a scatter diagram of the GPA values (Y) and the corresponding GRE scores (X) should reveal a basic linear pattern. We never expect all of the sample data to lie exactly on a straight line; we realize that with any statistical model there is error involved. This makes up the random component. The actual model used for simple linear regression is $Y = \beta_0 + \beta_1 X + e$, where e represents the distance from the actual Y value to the line passing through all X, Y values. The value of e is the error and represents the random component of the model. The assumptions behind the use of this model were concerned about the behavior of these error terms—are they normally distributed, centered at zero, with the same variance? Are they independent?

In the university example, it seems reasonable to assume that there are other variables that may serve as useful predictors of graduate student performance, such as undergraduate GPA, age, and type of undergraduate major. We will next look at statistical models used to predict the dependent variable (such as

Y = student GPA) as a function of more than one independent variable. The concept and assumptions are the same as before—now we are merely concerned with more than one predictor variable. When we include these additional variables, the predictive ability of the model should be significantly improved. This procedure is called multiple linear regression and is a very useful statistical technique.

15.1   THE MULTIPLE LINEAR REGRESSION MODEL

Prediction Using More Than One Variable

To explain or predict the behavior of a certain dependent variable using more than one predictor variable, we use a multiple linear regression model. The form of this model is

$$Y = \beta_0 + \beta_1 X_1 + \beta_2 X_2 + \cdots + \beta_k X_k + e \qquad \text{(15-1)}$$

where $X_1, X_2, \ldots, X_k$ are the k independent (predictor) variables, and e is the error associated with this model.

Notice that equation 15-1 is similar to that used in the simple linear regression model, except that the deterministic component is now

$$\beta_0 + \beta_1 X_1 + \cdots + \beta_k X_k \qquad \text{(15-2)}$$

rather than $\beta_0 + \beta_1 X$. Once again the error term e is included to provide for deviations about this component.

What is the appearance of the deterministic portion in equation 15-2? In Chapter 14, where we discussed simple linear regression, this was a straight line. For the multiple case, this is more difficult (usually impossible) to represent graphically. If your model contains two predictor variables, X_1 and X_2, the deterministic component becomes a plane, as shown in Figure 15.1 (p. 828). Consequently, the key assumption behind the use of this particular model is that the Y values will lie in this plane, on the average, for any particular values of X_1 and X_2.

In Chapter 14, we examined the relationship between the GPA of a first-year graduate student in the social or behavioral sciences (Y) and the student's verbal score on the Graduate Record Exam (GRE) (X). The results were:

Least squares line: $\hat{Y} = .785 + .0042X$

Correlation between X and Y: $r = .8446$

Coefficient of determination: $r^2 = .713$

There is a significant linear relationship

We now want to include two additional variables in the model. The admissions officer at Brookhaven University suspects that two other variables, quantitative score on the GRE and undergraduate GPA, may improve the prediction of graduate student performance. This results in three independent variables

$X_1 = $ verbal score on GRE

$X_2 = $ quantitative score on GRE

$X_3 = $ undergraduate GPA

The same 10 students were used in the study, but data were collected on the two additional variables, X_2 and X_3.*

Student	Y (First Year GPA)	X_1 (Verbal Score on GRE)	X_2 (Quant. Score on GRE)	X_3 (Undergraduate GPA)
1	3.54	580	720	3.82
2	2.62	500	660	2.67
3	3.30	670	580	3.16
4	2.90	480	520	3.31
5	4.00	710	630	3.60
6	3.21	550	690	3.42
7	3.57	640	700	3.51
8	3.05	540	530	2.75
9	3.15	620	490	3.21
10	3.61	690	530	3.70

The data configuration now has 4 columns (including Y) and 10 rows (called "observations"). Our task is to use the data on all three variables (X_1, X_2, and X_3) to provide a better estimate of graduate student GPA (Y).

The Least Squares Estimates Using Figure 15.1, we will proceed as we did for simple regression and determine an estimate of the β's that makes the sum of squares of the residuals as small as possible. A **residual** is defined as the difference between an actual Y value and its estimate, that is, $Y - \hat{Y}$. In other words, we attempt to find the $b_0, b_1, \ldots, b_k$ that minimize the sum of squares

* A sample of size 10 is unrealistically small in practice.

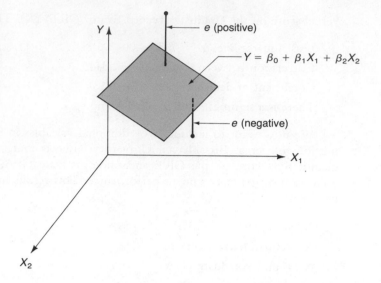

of error,

$$SSE = \Sigma(Y - \hat{Y})^2 \qquad\qquad \textbf{(15-3)}$$

where now $\hat{Y} = b_0 + b_1X_1 + b_2X_2 + \cdots + b_kX_k$ and $b_0, b_1, \ldots, b_k$ are called the **least squares estimates** of $\beta_0, \beta_1, \ldots, \beta_k$.

By determining the estimated regression coefficients ($b_0, b_1, \ldots, b_k$) that minimize SSE, rather than $\Sigma(Y - \hat{Y})$, we once again avoid the problem of positive errors canceling out negative ones. Another advantage of this procedure is that, by means of a little calculus, one can show that a fairly simple expression exists for these sample regression coefficients. Because this expression involves the use of matrix notation, we will omit this result.*

There is only one way to solve a multiple regression problem in practice, and that is with the help of a computer. All computer packages determine the values of $b_0, b_1, \ldots, b_k$ in the same way—namely, by minimizing SSE. As a result, these values will be identical (except for numerical rounding errors) regardless of which computer program you use.

* Information on this expression is presented in W. Mendenhall and T. Sincich, *A Second Course in Business Statistics: Regression Analysis*, 2d ed. (San Francisco: Dellen, 1986); J. Neter, W. Wasserman, and M. Kutner, *Applied Linear Regression Models* (Homewood, IL: Richard D. Irwin, 1983); M. S. Younger, *Handbook for Linear Regression*, 2d ed. (North Scituate, MA: Duxbury, 1986).

In the example attempting to predict the graduate GPA using the three predictor variables, the prediction equation is

$$\hat{Y} = b_0 + b_1 X_1 + b_2 X_2 + b_3 X_3$$

where

$\hat{Y}$ = predicted GPA

X_1 = verbal score on GRE

X_2 = quantitative score on GRE

X_3 = undergraduate GPA

and b_0, b_1, b_2, and b_3 are the least squares estimates of β_0, β_1, β_2, and β_3.

Figure 15.2 contains the MINITAB solution using the data we presented. According to this output, the best prediction equation (in the least squares sense) of a graduate GPA value is

$$\hat{Y} = -.395 + .0031 X_1 + .00059 X_2 + .4457 X_3$$

So this solution minimizes SSE. But what is the SSE here? We need to determine how well this equation "fits" the 10 observations in the data set.

Figure 15.2

MINITAB multiple regression solution for predicting first-year graduate GPA using three predictor variables (see text)

```
MTB > READ INTO C1-C4
DATA> 3.54 580 720 3.82
DATA> 2.62 500 660 2.67
DATA> 3.30 670 580 3.16
DATA> 2.90 480 520 3.31
DATA> 4.00 710 630 3.60
DATA> 3.21 550 690 3.42
DATA> 3.57 640 700 3.51
DATA> 3.05 540 530 2.75
DATA> 3.15 620 490 3.21
DATA> 3.61 690 530 3.70
DATA> END
      10 ROWS READ
MTB > REGRESS Y IN C1 USING 3 PREDICTORS IN C2-C4

The regression equation is
C1 = - 0.395 + 0.00310 C2 +0.000590 C3 + 0.446 C4

Predictor        Coef            Stdev       t-ratio
Constant       -0.3950           0.5813        -0.68
C2              0.0031028 ←b1    0.0007987      3.88
C3              0.0005900 ←b2    0.0006699      0.88
C4              0.4457 ←b3       0.1764         2.53

s = 0.1594       R-sq = 89.4%      R-sq(adj) = 84.0%

Analysis of Variance

SOURCE       DF        SS           MS
Regression    3     1.27940      0.42647
Error         6     0.15245      0.02541
Total         9     1.43185   SSE
```

Consider the first student, where $X_1 = 580$, $X_2 = 720$, and $X_3 = 3.82$. The predicted GPA here is

$$\hat{Y} = -.395 + (.0031)(580) + (.00059)(720) + (.4457)(3.82)$$
$$= 3.53$$

Consequently, the predicted GPA is 3.53. The actual GPA for this observation is 3.54, so the sample residual here is $Y - \hat{Y} = 3.54 - 3.53 = .01$.

Using this procedure on the remaining nine observations, we find the following results:

Y	$\hat{Y}$	$Y - \hat{Y}$	$(Y - \hat{Y})^2$
3.54	3.53	.01	.0001
2.62	2.74	-.12	.0144
3.30	3.43	-.13	.0169
2.90	2.88	.02	.0004
4.00	3.78	.22	.0484
3.21	3.24	-.03	.0009
3.57	3.57	0	0
3.05	2.82	.23	.0529
3.15	3.25	-.10	.0100
3.61	3.71	-.10	.0100
		0	$.154 \approx$ SSE

The computed value for the error sum of squares is SSE $= .154$. This value also is contained in the MINITAB output in Figure 15.2. The MINITAB value for SSE is .15245, which differs slightly from the previous result because the computer uses much more accurate calculations than those in the table. SSE $= .15245$ is more accurate, so we will use this value in the remaining discussion.

This implies that for any other values of b_0, b_1, b_2, and b_3, if we were to find the corresponding $\hat{Y}$'s and the resulting SSE $= \Sigma(Y - \hat{Y})^2$ using these values, this new SSE would be larger than .15245. Thus, $b_0 = -.395$, $b_1 = .0031$, $b_2 = .00059$, and $b_3 = .4457$ minimize the error sum of squares, SSE. Put still another way, these values of b_0, b_1, b_2, and b_3 provide the **best fit** to our data.

Using only the verbal GRE score (X_1) as a predictor, in Chapter 14 we found the SSE to be .4105 in our table. By including the additional two variables, the SSE has been reduced from .4105 to .15245 (a 63% reduction). It appears that either the quantitative GRE score (X_2), undergraduate GPA (X_3), or both contribute, perhaps significantly, to the prediction of Y.

Interpreting the Regression Coefficients   When using a multiple linear regression equation, such as $Y = \beta_0 + \beta_1 X_1 + \beta_2 X_2 + \beta_3 X_3 + e$, what does β_3 represent? Very simply, it reflects the change in Y that can be expected to accompany a change of one unit in X_3 provided all other variables (namely, X_1 and X_2) are held constant.

In the previous example, the sample estimate of β_3 was $b_3 = .4457$. Can we expect an increase of .4457 every time X_3 (the undergraduate GPA) increases by one if X_1 and X_2 are held constant? This type of argument is filled with problems, as we will demonstrate later. The primary problem is that a change in one of the predictor variables (such as X_3) very often is accompanied by a change in one of the other predictors (say, X_1) in the sample observations. Consequently, variables X_1 and X_3 are related in some manner, such as $X_1 = 1 + 5X_3$. In other words, a situation in which, say, X_3 changed and the others remained constant would not be observed within the sample data.

In the other case, the predictor variables are totally unrelated. In this situation, a unit change in, say, X_3 can be expected to be accompanied by a change of β_3 in the dependent variable.

In general, it is not safe to assume that the predictor variables are unrelated. As a result, the b's usually do not reflect the true "partial effects" of the predictor variables and you should avoid such conclusions. Section 15.4 will discuss methods of dealing with this type of situation.

The Assumptions behind the Multiple Linear Regression Model

The form of the multiple linear regression model is given by equation 15-1, which contains a linear combination of the k predictor (independent) variables as well as the error component e. The assumptions for the case of $k > 1$ predictors are exactly the same as for $k = 1$ independent variable (simple linear regression). These assumptions, discussed in Chapter 14, are:

1 The errors follow a normal distribution, centered at zero, with common variance σ_e^2.

2 The errors are (statistically) independent.

For the case of $k = 2$ predictor variables, this can be represented graphically, as shown in Figures 15.1 and 15.3. Using Figure 15.3, consider the situation in which $X_1 = 500$ and $X_2 = 650$. If you were to obtain repeated values of Y having these values of X_1 and X_2, you would obtain some Y's above the plane and some below. The assumption is that the average value of Y with $X_1 = 500$ and $X_2 = 650$ lies on the plane. Moreover, these errors are normally distributed.

The final part of assumption 1 is that the variation about this plane does not depend on the values of X_1 and X_2. You should see roughly the same amount of variation if you obtain repeated values of Y corresponding to $X_1 = 550$ and $X_2 = 600$ as you observed for $X_1 = 500$ and $X_2 = 650$. The variance of these errors, if you could observe Y indefinitely, is σ_e^2.

Finally, assumption 2 means that the error encountered at, say, $X_1 = 550$ and $X_2 = 600$ is not affected by the error at any other point, such as $X_1 = 500$ and $X_2 = 650$. The error associated with one pair of X_1, X_2 values has no effect on any other error.

Figure 15.3

The errors in
multiple linear
regression ($k = 2$)

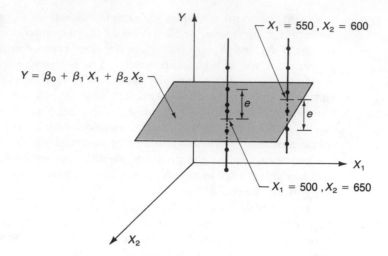

An Estimate of σ_e^2   When using a straight line to model a relationship between Y and a single predictor, the estimate of σ_e^2 was given by equation 14-16, where

$$s^2 = \hat{\sigma}_e^2 = \frac{\text{SSE}}{n - 2}$$

In general, for k predictors and n observations, the estimate of this variance is

$$s^2 = \hat{\sigma}_e^2 = \frac{\text{SSE}}{n - (k + 1)}$$
$$= \frac{\text{SSE}}{n - k - 1}$$

(15-4)

The value of s^2 is critical in determining the reliability and usefulness of the model as a predictor. If $s^2 = 0$, then SSE $= 0$, which implies that $Y = \hat{Y}$ for each of the observations in the sample data. This rarely happens in practice, but it does point out that a small s^2 is desirable. As s^2 increases, you can expect more error when predicting a value of Y for specified values of $X_1, X_2, \ldots, X_k$. In the next section, we will use s^2 as a key to determining whether the model is satisfactory and which of the independent variables are useful in the prediction of the dependent variable.

The square root of this estimated variance is the **residual standard deviation**.

$$s = \sqrt{\frac{\text{SSE}}{n - k - 1}} \qquad \textbf{(15-5)}$$

In the MINITAB solution in Figure 15.2, the value of s is shown in the box containing $s = .1594$.

EXAMPLE 15.1

Determine the estimate of σ_e^2 and the residual standard deviation of the university data in the previous illustration.

Solution

The illustration contained $n = 10$ observations and $k = 3$ predictor variables. The resulting error sum of squares was SSE $= .15245$ (from Figure 15.2). Therefore,

$$s^2 = \hat{\sigma}_e^2$$

$$= \frac{.15245}{10 - 3 - 1} = \frac{.15245}{6} = .0254$$

Also,

$$s = \sqrt{.0254} = .1594$$

That is, the residual standard deviation is .1594.

If a particular regression model meets all the required assumptions, then the next question of interest is whether this set of independent variables provides an accurate method of predicting the dependent variable Y. The next section shows how to calculate the predictive ability of your model and determine which variables contribute significantly to an accurate prediction of Y. □

EXERCISES

15.1 Researchers studying the relationship between mortality rates and pollution levels developed a regression model, expressed as:

$$\hat{Y} = 201.35 + 0.316X_1 + 0.542X_2 + 0.09X_3$$

where

$$Y = \text{mortality rate}$$

$$X_1 = \text{emission of nitrous and sulfur oxides}$$

$$X_2 = \text{water pollution index}$$

$$X_3 = \text{population density}$$

(a) If the emission of nitrous and sulfur oxides is 0.86 g/km, the water pollution index is 15.8, and population density is 254 persons per square mile, what is the predicted value of the mortality rate?

(b) If the water pollution index increases to 32.5, while the other predictor variables remain the same as in part (a), what would be the net change in the mortality rate?

15.2 For the model described in exercise 15.1, the following were the residuals obtained from 20 observations:

-1.8, 1.5, -3.7, 7.0, -3.1, 5.2, 4.3, 0.9, -0.3, -4.2, 3.2, 2.7, -1.7, -6.5, 1.0, -2.5, -1.0, 0.0, -1.2, 0.2

Do the residuals $Y - \hat{Y}$ appear to conform to the empirical rule that about 95% of the numbers should lie within two standard deviations of the mean?

15.3 The following demographic data are from an abbreviated set of figures provided by Gunst and Mason (1980), who compiled these from a much larger data base from Loether, McTavish, and Voxland (1974). Infant mortality is infant deaths per 1000 live births; literacy rate represents the percentage of population aged 15 and over who are literate; population density is measured as population per square kilometer; and concentration of physicians is number of inhabitants per physician. Use the data to predict infant mortality (Y) with $X_1 =$ literacy rate, $X_2 =$ population density, and $X_3 =$ physician concentration.

Country	Infant Mortality	Literacy Rate	Population Density	Physician Concentration
Austria	37.5	98.5	84	695
Barbados	60.4	91.1	548	3000
British Guiana	67.1	74.0	3	3900
Bulgaria	45.1	85.0	72	740
Canada	27.3	97.5	2	900
Guatemala	91.9	29.4	36	6400
Malaysia	68.9	38.4	54	6400
Mexico	77.7	50.0	18	1700
New Zealand	22.8	98.5	9	700
Norway	20.2	98.5	11	946
West Germany	33.8	98.5	217	798
USSR	35.0	95.0	10	578
Trinidad	45.4	73.8	168	2300
Sweden	16.6	98.5	17	1089
Romania	75.7	89.0	78	788

Source: H. J. Loether, D. G. McTavish, and P. M. Voxland, *Statistical Analysis for Sociologists: A Student Manual* (Boston: Allyn and Bacon, 1974). Data selected from an abbreviated set provided in R. F. Gunst and R. L. Mason, *Regression Analysis and Its Applications; A Data-Oriented Approach* (New York: Dekker, 1980), p. 358.

(a) Use a computerized statistical package to obtain the least-squares prediction equation for the above data.
(b) Find the value of the SSE.
(c) Estimate the residual standard deviation.
(d) What is the predicted infant mortality for a country with a literacy rate of 85, a population density of 150 per square kilometer, and concentration of physicians equal to 875 inhabitants per physician?

15.4 What assumptions need to be made about the error component of a multiple linear regression model in order that the results of statistical inference can be used?

15.5 The linear regression $\hat{Y} = 5.63 + 0.220X_1 - 0.166X_2$ was obtained by a computer analysis of a set of 10 observations. The data used to derive this equation are:

X_1:	15	8	18	6	20	25	10	7	7	30
X_2:	6	10	5	12	9	10	13	15	4	4
Y:	7.2	5.4	7.7	5.6	9.1	8.8	6.2	4.3	7.0	12.5

(a) Compute the error sum of squares, SSE.
(b) Estimate the variance of the error.
(c) Interpret the coefficients of the regression equation, i.e., What do the numbers 0.220 and −0.166 tell us?

15.6 Using the multiple regression model $\hat{Y} = 7.5 + 2.5X_1 - 1.6X_2 + 5.7X_3$ where do you expect the average value of Y to fall for $X_1 = 3$, $X_2 = 4.1$, and $X_3 = 5.6$?

15.7 A regression analysis was performed for data with three independent variables. The following residuals were found for the 20 observations of the dependent variable:

5.4, 8.1, −7.4, 2.5, −3.5, −4.1, −8.1, 6.5, 4.3, −7.8, 2.8, 7.1, −6.2, 5.6, −5.1, 2.9, 2.8, −7.2, 8.3, −6.9

Find the residual standard deviation.

15.8 A veterinarian uses three independent variables to predict the weight of a Doberman Pinscher from birth to 2 years of age. The variables are $X_1 = $ age, $X_2 = $ height, $X_3 = X_1 X_2$. Thirty-five observations are used to find the least squares prediction equation. The veterinarian's residual standard deviation for the collected data was 4.25. What is the value of SSE?

HYPOTHESIS TESTING AND CONFIDENCE
15.2 INTERVALS FOR THE β PARAMETERS

Multiple linear regression is a popular tool in the application of statistical techniques to business decisions. However, this modeling procedure does not always result in an accurate and reliable predictor. When the independent variables that you have selected account for very little of the variation in the values of the dependent variable, the model (as is) serves no useful purpose.

The first thing we will demonstrate is how to determine whether your overall model is satisfactory. We will begin by summarizing a regression analysis in an ANOVA table, much as we did in Chapter 12.

The ANOVA Table

The summary ANOVA table contains the usual headings:

Source	df	SS	MS	F
Regression	k	SSR	MSR	MSR/MSE
Residual	$n - k - 1$	SSE	MSE	
Total	$n - 1$	SST		

where n = number of observations and k = number of independent variables

$$
\begin{aligned}
\text{SST} &= \text{total sum of squares} \\
&= \text{SS}_Y \qquad\qquad\qquad\qquad\qquad\qquad \textbf{(15-6)} \\
&= \Sigma(Y - \bar{Y})^2 = \Sigma Y^2 - (\Sigma Y)^2/n
\end{aligned}
$$

$$
\begin{aligned}
\text{SSE} &= \text{sum of squares for error} \\
&= \Sigma(Y - \hat{Y})^2 \qquad\qquad\qquad\qquad\quad \textbf{(15-7)}
\end{aligned}
$$

$$
\begin{aligned}
\text{SSR} &= \text{sum of squares for regression} \\
&= \Sigma(\hat{Y} - \bar{Y})^2 \qquad\qquad\qquad\qquad\quad \textbf{(15-8)} \\
&= \text{SST} - \text{SSE}
\end{aligned}
$$

$$
\begin{aligned}
\text{MSR} &= \text{mean square for regression} \\
&= \text{SSR}/k \qquad\qquad\qquad\qquad\qquad\quad \textbf{(15-9)}
\end{aligned}
$$

$$
\begin{aligned}
\text{MSE} &= \text{mean square for error} \\
&= \text{SSE}/(n - k - 1) \qquad\qquad\qquad\quad \textbf{(15-10)}
\end{aligned}
$$

Practically all computer packages will provide you with this ANOVA summary as part of the standard output. The ANOVA section of the MINITAB solution for the university model is highlighted in Figure 15.4(A).

Notice that

$$
\begin{aligned}
\text{SST} &= \text{SS}_Y \\
&= (3.54^2 + 2.62^2 + \cdots + 3.61^2) - (3.54 + 2.62 + \cdots + 3.61)^2/10 \\
&= 1.43185
\end{aligned}
$$

This is the same value of SS_Y we obtained for the same example in Chapter 14, when we used only verbal GRE score (X_1) as the predictor variable. This is hardly surprising because this value is strictly a function of the Y values and is unaffected by the model that you are using to predict Y. The total sum of squares (SST) measures the total variation in the values of the dependent variable.

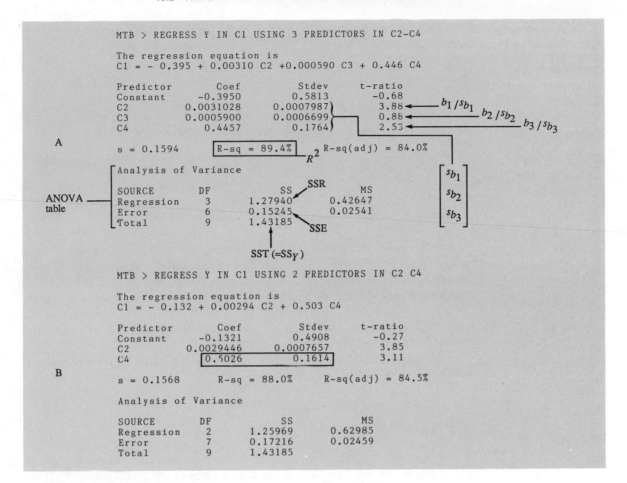

Figure 15.4

MINITAB output (see Figure 15.2). (A) Prediction equation and ANOVA table using X_1, X_2, and X_3. (B) Prediction equation and ANOVA table using X_1 and X_3.

Its value is the same, regardless of which predictor variables are included in the model.

The df for the regression source of variation are $k =$ the number of predictor variables in the analysis. The df for the error sum of squares are $n - k - 1$ where $n =$ the number of observations in the sample data.

As in the case of simple linear regression, the sum of squares of regression (SSR) measures the variation explained by the model—the variation in the Y values that would exist if differences in the values of the predictor variables were the only cause of differences among the Y's. On the other hand, the sum of squares of error (SSE) represents the variation unexplained by the model. The easiest way to determine the sum of squares of regression is to subtract SSR = SST − SSE.

The error mean square is $MSE = SSE/(n - k - 1) = .15245/(10 - 3 - 1) = .0254$. This is the same as the estimate of σ_e^2 determined in example 15.1. So

$$s^2 = \hat{\sigma}_e^2 = MSE$$

A Test for H_0: all β's $= 0$

We have yet to make use of the F value calculated in the ANOVA table, where

$$F = \frac{MSR}{MSE} \qquad (15\text{-}11)$$

When using the simple regression model, we previously argued that one way to determine whether X is a significant predictor of Y is to test $H_0: \beta_1 = 0$, where β_1 is the coefficient of X in the model $Y = \beta_0 + \beta_1 X + e$. If you reject H_0, the conclusion is that the independent variable X *does* contribute significantly to the prediction of Y. For example, in example 14.3, we concluded that the verbal GRE score (X_1) was a useful predictor of graduate GPA (Y) using the simple linear model, by rejecting $H_0: \beta_1 = 0$.

We use a similar test as the first step in the multiple regression analysis, where we examine the hypotheses

$H_0: \quad \beta_1 = \beta_2 = \cdots = \beta_k = 0$
$H_a: \quad$ at least one of the β's $\neq 0$.

If we reject H_0, we can conclude that at least one (but maybe not all) of the independent variables contributes significantly to the prediction of Y. If we fail to reject H_0, we are unable to demonstrate that any of the independent variables (or combination of them) help explain the behavior of the dependent variable Y. For example, in our university example, if we were to fail to reject H_0, this would imply that we are unable to demonstrate that the variation in the graduate GPA values (Y) can be explained by the effect of verbal GRE score, quantitative GRE score, and undergraduate GPA.

Test Statistic for H_0 versus H_a

The test statistic used to determine whether our multiple regression model contains at least one significant explanatory variable is the F statistic from the preceding ANOVA table.

When testing H_0: all β's $= 0$ (this is a poor set of predictor variables) versus H_a: at least one $\beta \neq 0$ (at least one of these variables is a good predictor), the

Figure 15.5

F curve with *k* and $n - k - 1$ *df*. The shaded area is the rejection region.

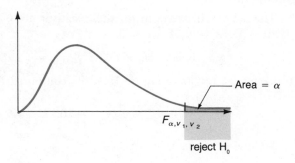

test statistic is

$$F = \frac{\text{MSR}}{\text{MSE}}$$

which has an F distribution with k and $n - k - 1$ *df*. The expression $n - k - 1$ can be written as $n - (k + 1)$, where $k + 1$ is the number of coefficients (β's) estimated.

Notice that the *df* for the F statistic come directly from the ANOVA table. The testing procedure will be to

$$\text{Reject } H_0 \text{ if } F > F_{\alpha, v_1, v_2}$$

where (1) $v_1 = k$, $v_2 = n - k - 1$, and (2) F_{α, v_1, v_2} is the corresponding F value in Table A-7, having a right-tail area $= \alpha$ (Figure 15.5).

EXAMPLE 15.2

Using the university data and the model we developed, what can you say about the predictive ability of the independent variables, verbal GRE score (X_1), quantitative GRE score (X_2), and undergraduate GPA (X_3)? Use $\alpha = .10$.

Solution   **Step 1**   The hypotheses are

$$H_0: \quad \beta_1 = \beta_2 = \beta_3 = 0$$
$$H_a: \quad \text{at least one } \beta \neq 0$$

Remember that our hope here is to reject H_0. If you are unable to demonstrate that any of your independent variables have any predictive ability, then you will fail to reject H_0.

Step 2   The test statistic is

$$F = \frac{\text{MSR}}{\text{MSE}}$$

The mean squares are obtained from the ANOVA summary of the regression analysis (see Figure 15.4(A)).

Step 3   The degrees of freedom for the F statistic are $k = 3$ and $n - k - 1 = 10 - 3 - 1 = 6$. So we will

$$\text{Reject } H_0 \text{ if } F > F_{.10,3,6} = 3.29$$

Step 4   Using the results in Figure 15.4, the computed F value is

$$F^* = \frac{.42647}{.02541} = 16.78$$

Because $F^* > 3.29$, we reject H_0.

Step 5   The three independent variables, as a group, constitute a significant predictor of graduate GPA. This does not imply that all three variables have significant predictive ability; however, at least one of them does. The next section will show how you can tell which of these predictor variables significantly contributes to the prediction of graduate GPA.

$\square$

A Test for H_0: $\beta_i = 0$

Assuming that you rejected the null hypotheses that all of the β's are zero, the next logical question would be, which of the independent variables contributes to the prediction of Y?

In example 15.2, we rejected the null hypothesis, so at least one of these three independent variables affects the variation of the 10 graduate GPAs in the sample. To determine the contribution of each variable, we will perform three separate t tests.

H_0: $\beta_1 = 0$ (X_1 does not contribute)
H_a: $\beta_1 \neq 0$ (X_1 does contribute)

H_0: $\beta_2 = 0$ (X_2 does not contribute)
H_a: $\beta_2 \neq 0$ (X_2 does contribute)

H_0: $\beta_3 = 0$ (X_3 does not contribute)
H_a: $\beta_3 \neq 0$ (X_3 does contribute)

One-tailed tests also can be used here, but we will demonstrate this procedure using two-tailed tests. This means that we are testing to see whether this particular X contributes to the prediction of Y, but we are not concerned about the direction (positive or negative) of this relationship.

When the verbal GRE score (X_1) was the only predictor of graduate GPA (Y), we used a t test to determine whether the simple linear regression model was adequate. In example 14.3, the value of the test statistic was derived, where

$$t = \frac{b_1}{s_{b_1}}$$

Also, b_1 is the estimate of β_1 in the simple regression model, and s_{b_1} is the (estimated) standard deviation of b_1.

All computer packages provide both the estimated coefficient (b_1) and its standard deviation (s_{b_1}). In example 14.3, the computed value of this t statistic was $t^* = 4.46$. This led us to conclude that the verbal GRE score was a good predictor of graduate GPA because a significant positive relationship existed between these two variables.

We use the same t statistic procedure to test the effect of the individual variables in a multiple regression model. When examining the effect of an individual independent variable X_i on the prediction of a dependent variable, the hypotheses are

$$H_0: \quad \beta_i = 0$$
$$H_a: \quad \beta_i \neq 0$$

The test statistic is

$$t = \frac{b_i}{s_{b_i}} \qquad \text{(15-12)}$$

where (1) b_i is the estimate of β_i, (2) s_{b_i} is the (estimated) standard deviation of b_i, and (3) the df for the t statistic are $n - k - 1$.

The test of H_0 versus H_a is to

$$\text{Reject } H_0 \text{ if } |t| > t_{\alpha/2, n-k-1}$$

where $t_{\alpha/2, n-k-1}$ is obtained from Table A-5.

We can now re-examine the university data in example 15.2.

X_1 = Verbal GRE Score   Consider the hypotheses

$$H_0: \quad \beta_1 = 0$$
$$H_a: \quad \beta_1 \neq 0$$

As in example 15.2, we will use $\alpha = .10$.

According to Figure 15.4(A), $b_1 = .0031028$ and $s_{b_1} = .0007987$. Also contained in the output is the computed value of

$$t^* = \frac{b_1}{s_{b_1}} = \frac{.0031028}{.0007987} = 3.88$$

Why is this value of t^* not the same as the value of t calculated previously in Chapter 14 for this variable, when the verbal GRE score was the only predictor of Y? When there are three predictors in the model, t^* for this predictor is 3.88. When this variable is the only predictor in the model, $t^* = 4.46$. The difference in the two values is simply that $t^* = 3.88$ provides a measure of the contribution of $X_1 =$ verbal GRE score, given that X_2 and X_3 already have been included in the model. A large value of t^* indicates that X_1 contributes significantly to the prediction of Y, even if X_2 and X_3 have been included previously as predictors.

The hypotheses can better be stated as

H_0: verbal GRE score does not contribute to the prediction of graduate GPA, given that quantitative GRE score and undergraduate GPA already have been included in the model

H_a: verbal GRE score does contribute to this prediction, given that quantitative GRE score and undergraduate GPA already have been included in the model

or as

H_0: $\beta_1 = 0$ (if X_2 and X_3 are included)

H_a: $\beta_1 \neq 0$

Because $t^* = 3.88$ exceeds the table value of $t_{\alpha/2, n-k-1} = t_{.05, 10-3-1} = t_{.05,6} = 1.943$, we conclude that the verbal GRE score contributes significantly to the prediction of graduate GPA and should be kept in the model.

$X_2 =$ Quantitative GRE Score   Using a similar argument, the following test of hypothesis will determine the contribution of quantitative GRE score X_2 as a predictor of the graduate GPA, given that X_1 and X_3 already have been included. The hypotheses here are

H_0: $\beta_2 = 0$ (if X_1 and X_3 are included)

H_a: $\beta_2 \neq 0$

According to Figure 15.4(A), the computed t statistic here is

$$t^* = \frac{b_2}{s_{b_2}}$$

$$= \frac{.00059}{.0006699} = .88$$

Because $t^* = .88$ does *not* exceed $t_{.05,6} = 1.943$, we fail to reject H_0. We conclude that, given the values of $X_1 =$ verbal GRE score and $X_3 =$ undergraduate GPA, a student's quantitative score on the GRE appears not to contribute to the prediction of his or her first year graduate GPA. This means that X_2 can be removed from the final prediction equation, leaving only X_1 and X_3. As a word of warning, you should *not* simply remove this term from the equation containing all three variables. Since the predictor variables are typically related in some manner, the sample regression coefficients (b_0, b_1, . . .) change as variables are added to or deleted from the model. Referring to Figure 15.4(A), the prediction equation using X_1 and X_3 is not $\hat{Y} = -.395 + .0031X_1 + .4457X_3$. Instead, the coefficients of X_1 and X_3 should be derived by repeating the analysis using only these two variables. According to Figure 15.4(B), the prediction equation is $\hat{Y} = -.132 + .00294X_1 + .503X_3$.

X_3 = Undergraduate GPA   To test

H_0: $\beta_3 = 0$ (if X_1 and X_2 are included)
H_a: $\beta_3 \neq 0$

we once again use the *t* statistic.

$$t = \frac{b_3}{s_{b_3}}$$

Using Figure 15.4(A), the computed value of this statistic is

$$t^* = \frac{.4457}{.1764} = 2.53$$

This exceeds $t_{.05,6} = 1.943$, and so the undergraduate GPA provides useful information in predicting a graduate student GPA. We conclude that we should keep X_3 in the model.

A Confidence Interval for β_i

Using what you believe to be the "best" model you can easily construct a $(1 - \alpha) \cdot 100\%$ confidence interval for β_i based upon the previous *t* statistic:

$$b_i - t_{\alpha/2, n-k-1} s_{b_i} \quad \text{to} \quad b_i + t_{\alpha/2, n-k-1} s_{b_i} \qquad \textbf{(15-13)}$$

Once again, k represents the number of predictor variables used to estimate β_i.

EXAMPLE 15.3   Suppose we decide to retain only X_1 = verbal GRE score and X_3 = undergraduate GPA in the prediction equation. Referring to Figure 15.4(B), construct a 90% confidence interval for β_3, the coefficient for X_3.

Solution   Since this model contains $k = 2$ predictor variables, we first find $t_{\alpha/2, n-k-1} = t_{.05,7} = 1.895$. Based upon the MINITAB results in Figure 15.4(B), the

confidence interval for β_3 is

$$.5026 - (1.895)(.1614) \quad \text{to} \quad .5026 + (1.895)(.1614)$$
$$= .5026 - .3058 \quad \text{to} \quad .5026 + .3058$$
$$= .1968 \quad \text{to} \quad .8084$$

Therefore, we are 90% confident that the estimate of β_3 (that is, $b_3 = .5026$) is within .3058 of the actual value of β_3. Notice that this is an extremely wide confidence interval. As usual, increasing the sample size would help to reduce the width of this confidence interval.   □

EXAMPLE 15.4   Fieldmore Institute is interested in predicting the score on a recently introduced instrument designed to measure the level of intimacy for married individuals, called the Spousal Intimacy Scale. Fifty spouses were randomly selected for the study. As a first attempt, they used X_1 = number of years in the marriage as a predictor of Y = score on the Spousal Intimacy Scale. Possible values on this scale range from a minimum of 10 to a maximum of 100. However, the correlation between X_1 and Y turned out to be only .370, so an effort was made to improve the model by including an additional predictor variable. They then

 Figure 15.6

MINITAB solution
for example 15.4

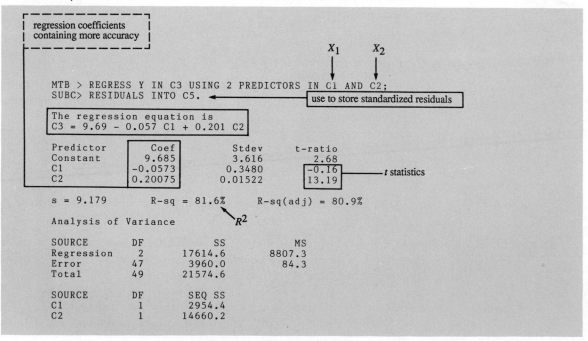

included the individual's score on an instrument that measured life satisfaction (which we will call the Life Satisfaction Index) as another independent variable.
Let

Y = score on Spousal Intimacy Scale

X_1 = number of years in the marriage

X_2 = score on Life Satisfaction Index

The model used for this application is

$$Y = \beta_0 + \beta_1 X_1 + \beta_2 X_2 + e$$

The data collected from the 50 individuals are contained in the following table. The least squares equation and the complete MINITAB output are contained in Figure 15.6.

Spouse	Y (Spousal Intimacy Scale)	X_1 (Number of Years in Marriage)	X_2 (Life Satisfaction Index)
1	49	5	274
2	77	3	287
3	83	12	373
4	29	13	114
5	12	9	93
6	29	16	115
7	49	9	131
8	70	16	333
9	49	11	214
10	59	6	245
11	38	14	159
12	51	5	228
13	29	3	100
14	38	5	102
15	24	4	74
16	29	8	273
17	21	3	59
18	47	10	191
19	12	2	69
20	66	16	298
21	48	12	157
22	97	13	394
23	46	7	95
24	36	13	117
25	57	6	138
26	76	6	280
27	56	13	267

(Continued)

Spouse	Y (Spousal Intimacy Scale)	X_1 (Number of Years in Marriage)	X_2 (Life Satisfaction Index)
28	40	7	230
29	79	8	320
30	92	15	358
31	69	6	322
32	67	9	304
33	39	4	208
34	80	10	338
35	29	4	70
36	80	12	343
37	37	12	220
38	35	9	166
39	45	12	243
40	81	8	313
41	78	15	283
42	24	1	67
43	61	14	258
44	42	5	189
45	45	6	182
46	66	13	288
47	52	4	176
48	61	9	181
49	33	11	131
50	29	8	101

Determine the estimated score on the Spousal Intimacy Scale for an individual married 10 years and scoring 310 (out of a maximum possible of 400) on the Life Satisfaction Index. What can you say about the predictive ability of these two independent variables? Use a significance level of $\alpha = .10$.

Solution　The least squares equation from Figure 15.6 is

$$\hat{Y} = 9.685 - .0573X_1 + .20075X_2$$

The predicted score on the Spousal Intimacy Scale is

$$\hat{Y} = 9.685 - .0573(10) + .20075(310)$$
$$= 9.685 - .573 + 62.232$$
$$= 71.3$$

that is, approximately a score of 71. Note that, similar to the argument in Chapter 14, 71 also serves as an estimate of the average score on the Spousal Intimacy Scale for all individuals having $X_1 = 10$ and $X_2 = 310$. This will be explored further in Section 15.5 where we discuss the construction of a confidence interval for an *average* score or a prediction interval for an *individual* score.

The first test of hypothesis will determine whether these two variables as a group provide a useful model for predicting the Spousal Intimacy Scale scores.

$$H_0: \quad \beta_1 = \beta_2 = 0$$
$$H_a: \quad \beta_1 \neq 0, \beta_2 \neq 0, \text{ or both} \neq 0$$

Using the ANOVA table in Figure 15.6, the value of the F statistic is

$$F^* = \frac{\text{MSR}}{\text{MSE}} = \frac{8807.3}{84.3} = 104.5$$

The *df* here are $v_1 = k = 2$ and $v_2 = n - k - 1 = 50 - 2 - 1 = 47$. Because $F_{.10,2,47}$ is not in Table A-7, we use the nearest value, $F_{.10,2,40} = 2.44$. The computed F^* exceeds this value, so we reject H_0 and conclude that at least one of these two independent variables is a significant predictor of Spousal Intimacy Scale scores. □

The *t* Tests

Because we rejected H_0 in example 15.4, the next step is to examine the *t* tests to determine which of the two independent variables are useful predictors. The *t* value from Table A-5 is $t_{\alpha/2, n-k-1} = t_{.05,47} \cong 1.684$. The computed *t* values in Figure 15.6 lead to the following conclusions. First, the *t* value for X_1 = number of years in the marriage is $t^* = -.16$. The absolute value of t^* is less than 1.684, which means that, given the presence of X_2 in the model, X_1 does not contribute useful information to the prediction of scores on the Spousal Intimacy Scale. It can be removed from the model without seriously affecting the accuracy of the resulting prediction. Second, for X_2 = score on the Life Satisfaction Index, the computed *t* value is $t^* = 13.19$. Now, $|t^*| = 13.19 > 1.684$, which means that an individual's general satisfaction in life is an excellent predictor of his or her level of marital intimacy. It was the contribution of this variable, and not of X_1, that produced the extremely large F value we obtained.

As we have seen, a quick glance at the computer output allows you to determine whether your model is useful as a whole and, furthermore, which variables are useful predictors. But beware—the analysis is not over! Before you form your conclusions from this analysis and make critical decisions based on several tests of hypotheses, you need to be sure that none of the assumptions of the multiple linear regression model (discussed earlier) has been violated. We will discuss this problem in the final section of this chapter, where we will conclude the analysis by examining the sample residuals, $Y - \hat{Y}$.

The use of *t* tests allows you to determine the predictive contribution of each independent variable, provided you want to examine the contribution of one such variable while assuming that the remaining variables are included in the equation. The next section will show you how to extend this procedure to a situation in which you wish to determine the contribution of any set of predictor variables by using a single test.

EXERCISES

15.9 The job placement center at Ozark Technological University would like to predict the starting salaries (given in thousands of dollars) for the college graduates in the engineering department. Two variables are used: The variable X_1 represents the student's overall grade point average (GPA). The variable X_2 represents the number of years of prior job-related experience. Data for randomly selected graduating students are:

Y	X_1	X_2
27.1	3.7	0
23.3	2.9	1.1
21.4	2.4	1.5
24.2	3.2	0.5
26.1	3.6	0.8
19.8	2.7	0
22.8	3.1	0
20.5	2.2	2.1
32.3	3.8	2.5
18.1	2.1	1.4
22.5	3.0	0.3
23.8	3.4	0.5
20.9	2.8	0
20.0	2.5	1.0
27.8	3.3	2.1

The least squares equation for the data is:

$$\hat{Y} = 2.189 + 6.5144X_1 + 1.9259X_2$$

(a) Find the F value using an ANOVA table to test the hypothesis that $\beta_1 = 0$ and $\beta_2 = 0$ at the 5% level of significance.

(b) Interpret the coefficients in the context of the problem.

15.10 Complete the following ANOVA table to test the usefulness of a model with five independent variables that attempted to explain the variation in the dependent variable:

Source	df	SS	MS	F
Regression				
Error		180		
Total	50	215		

15.11 The Department of Public Safety of Tennessee wanted to convince people to drive more slowly on wet highways. A highway patrol officer slammed the brakes at 12 different speeds on a wet highway and on a dry highway to measure the distance in feet that it took to stop; 24 observations were collected. Let X_1 be the speed in miles per hour. Let X_2 be 1 if the highway surface is wet and 0 if the surface is dry. A computer package gave the following sample statistics:

$$b_0 = 6.3 \qquad b_1 = 3.1 \qquad b_2 = 16.5 \qquad s_{b_1} = 1.12 \qquad s_{b_2} = 2.54$$

(a) Given that X_2 is in the model, does the variable X_1 contribute to predicting the dependent variable at the .05 significance level?

(b) Given that X_1 is in the model, does the variable X_2 contribute to predicting the dependent variable at the .05 significance level?

15.12 The following sample statistics were computed for a regression analysis:

$$b_0 = 10.2 \qquad b_1 = 5.6 \qquad b_2 = 100.4 \qquad s_{b_1} = 1.04 \qquad s_{b_2} = 17.95$$

Assume that twenty observations were taken.

(a) Test that X_2 significantly contributes to the prediction of Y given that X_1 is in the model. Use a 5% significance level.

(b) Find a 95% CI for β_1.

15.13 Return to the infant mortality model of exercise 15.3, and continue the analysis as follows:

(a) Do the three variables X_1 through X_3 taken as a whole contribute significantly to prediction of infant mortality? Use a 5% significance level, and state the p value for the test.

(b) Does the literacy rate (X_1) contribute significantly to the model, given that X_2 and X_3 are already included? Use the 5% significance level, and state the p value for the test.

(c) Test the individual contribution of the other two variables, X_2 and X_3, just as you did for X_1 in part (b) above.

(d) Construct the following confidence intervals:
  (i)   90% CI for β_1.
  (ii) 95% CI for β_2.
  (iii) 99% CI for β_3.

15.14 The model $Y = 3.2 + 6.1X_1 + 5.2X_2$ was calculated to fit 20 data points pertaining to the growth rate of a hog. The variable X_1 represents the daily food consumption of the hog and X_2 represents the age of the hog. If the standard deviation of the estimate of β_1 is 2.5, what is a 95% CI for the parameter β_1?

15.15 Complete the following ANOVA table to test the null hypothesis that the independent variables are not useful predictors of the dependent variable.

Source	df	SS	MS	F
Regression				
Error		55	2.75	
Total	27	255		

15.16 A criminologist set out to investigate whether firearms availability could help to account for a rising rate of murders in an American city. Fisher (1976) obtained data on homicides per 100,000 population in Detroit, along with 13 other variables that might influence or be associated with the homicide rate. The data covered the period 1961–1973. To simplify our analysis, we will only examine three of the 13 explanatory variables studied by Fisher. Thus, the data are:

Y	X_1	X_2	X_3
8.60	260.35	11.0	178.15
8.90	269.80	7.0	156.41
8.52	272.04	5.2	198.02
8.89	272.96	4.3	222.10
13.07	272.51	3.5	301.92
14.57	261.34	3.2	391.22
21.36	268.89	4.1	665.56
28.03	295.99	3.9	1131.21
31.49	319.87	3.6	837.60
37.39	341.43	7.1	794.90
46.26	356.59	8.4	817.74
47.24	376.69	7.7	583.17
52.33	390.19	6.3	709.59

where

Y = number of homicides per 100,000 population

X_1 = number of fulltime police officers per 100,000 population

X_2 = unemployment rate for the population

X_3 = number of handgun licenses issued per 100,000 population

Source: J. C. Fisher, Homicide in Detroit: The role of firearms. *Criminology* **14**: 387–400 (1976).

(a)   Use a computerized statistical package to perform a regression analysis on the above data. Obtain:
   (i)   The least-squares equation
   (ii)   The error sum of squares
   (iii)   The residual standard deviation
   (iv)   The F statistic to test the validity of the entire model
   (v)   The t statistic to test the contribution of the unemployment rate to the prediction of homicides

(b)   Conduct a hypothesis test to evaluate the validity of the entire model. Use $\alpha = .05$.

(c)   Conduct a hypothesis test to evaluate whether the number of firearms licenses is a significantly useful predictor of homicides (given that the other two variables are already included in the model). Use $\alpha = .05$.

(d)   Construct a 95% confidence interval for β_3.

15.17   The least squares line of $\hat{Y} = 3.4 + 1.2X_1 + 4.3X_2$ was obtained. The sample residuals of the 20 observations used in fitting the regression line are:

4.1, −3.2, 1.5, 6.7, 6.4, 3.8, −4.2, −2.4, 1.6, −8.7, −3.1, 1.2, −5.1, 2.1, 0.6, 5.4, 3.4, −7.1, −6.2, 3.2

Given that the value of SST is 510 test the null hypothesis that the variables X_1 and X_2 contribute to predicting the variation in the dependent variable. Use a 5% significance level.

15.18 The F test discussed in this section is used to test the significance of the multiple linear equation in a global fashion, i.e., the significance of all the predictor variables taken together, as a whole. The t test is used in the chapter to test the contribution of an individual predictor within a multiple regression model, whereas in Chapter 14 the t test was used to test the entire model (which had only one predictor in any case, since it was a simple linear regression model).

(a) Show how the F test could also be used to test the significance of a simple linear regression model (H_0: $\beta_1 = 0$).

(b) Show what the ANOVA table would look like in such cases (give a generalized form).

(c) What would be the degrees of freedom for the critical value of F in a simple regression test situation?

15.19 Given below is selected information from a computer printout of a multiple regression analysis:

Predictor	Coefficient	S. D.	t Ratio
Constant	− .50	.198	−2.525
X1	2.40	.256	9.375
X2	2.95	.210	14.048

R Square = 0.9381

Analysis of Variance

Source	SS	df	MS	F
Model	295.30	2	147.65	128.39
Residual	19.50	17	1.15	
Total	314.80	19		

Answer the following questions:

(a) Write the multiple regression equation.

(b) What percentage of variation in Y is explained in the model?

(c) What is the sample size n used for the above regression analysis?

(d) Find the vlaue of:

    (i) The total sum of squares

    (ii) The error sum of squares

    (iii) The regression sum of squares

    (iv) The F statistic to test H_0: $\beta_1 = \beta_2 = 0$

    (v) The critical value of F for (iv) above

    (vi) The t statistic to test H_0: $\beta_2 = 0$

    (vii) The critical value of t for (vi) above

    (viii) The estimated variance of the error in the model

15.20 If the residual standard deviation for a set of data is 3.82 and the total sum of squares is 269, what is the F value for testing that a model with five independent variables does not contribute to predicting the variation in the dependent variable? Assume that 15 observations of the dependent variable were taken.

15.3   DETERMINING THE PREDICTIVE ABILITY OF CERTAIN INDEPENDENT VARIABLES

We can extend the procedure we used to examine the contribution of each independent variable, one at a time, using a t test.

Assume that a researcher has developed a regression model to predict an individual's score on a personality assessment instrument that measures his or her self motivation. The multiple linear regression model contains eight independent variables, three of which (say, X_6, X_7, X_8) describe the physical attributes of each individual (say, height, weight, and age). Can all three of these variables be removed from the analysis without seriously affecting the predictive ability of the model?

To answer this question, we return to a statistic, described in Chapter 14, that measures how well the model captures the variation in the values of your dependent variable.

Coefficient of Determination

The total variation of the sampled dependent variable is determined by

$$SST = \text{total sum of squares}$$
$$= SS_Y$$
$$= \Sigma(Y - \bar{Y})^2$$
$$= \Sigma Y^2 - (\Sigma Y)^2/n$$

where $n = $ number of observations. To determine what percentage of this variation has been explained by the predictor variables in the regression equation, we determine the **coefficient of determination, R^2**.

$$R^2 = 1 - \frac{SSE}{SST} \qquad (15\text{-}14)$$

The range for R^2 is 0 to 1. If $R^2 = 1$, then 100% of the total variation has been explained because, in this case, $SSE = \Sigma(Y - \hat{Y})^2 = 0$, and so $Y = \hat{Y}$ for each observation in the sample; that is, the model provides a perfect predictor. This will not occur in practice, but the main point is that a large value of R^2 is desirable for a regression application. It should be mentioned that $R^2 = 1$ whenever the number of observations (n) is equal to the number of estimated coefficients ($k + 1$). This does not mean that you have a "wonderful" model; rather, you have inadequate data. As a result, you need to guard against using too small a sample in your regression analysis; a general rule of thumb would be to use a sample containing at least three times as many (unique) observations as the number of predictor variables (k) in the model.

H_0: all β's = 0

A test statistic for testing H_0: all β's = 0 was introduced in equation 15-11, which used the ratio of two mean squares from the ANOVA table. Another way to calculate this F value is to use

$$F = \frac{R^2/k}{(1 - R^2)/(n - k - 1)} \qquad (15\text{-}15)$$

This version of the F statistic is used to answer the question, is the value of R^2 significantly large? If H_0 is rejected, then the answer is yes, and so this group of predictor variables has at least some predictive ability for predicting Y.

The F value computed in this way will be exactly the same as the one computed using $F = \text{MSR}/\text{MSE}$, except for possible roundoff error. This will be illustrated in example 15.5.

Once again, remember that *statistical* significance does not always imply *practical* significance. A large value of R^2 (rejecting H_0) does not imply that precise prediction (practical significance) will follow. However, it does inform the researcher that these predictor variables, as a group, are associated with the dependent variable.

EXAMPLE 15.5   In example 14.7, we determined that X = verbal score on the GRE explained 71.3% of the total variation of the graduate GPAs in the sample. What percentage is explained using all three predictors (verbal GRE score, quantitative GRE score, and undergraduate GPA)?

Solution   The coefficient of determination using X_1 only is 71.3. Using the MINITAB solution in Figure 15.4(A), the coefficient of determination using X_1, X_2 and X_3 is

$$R^2 = 1 - \frac{\text{SSE}}{\text{SST}}$$

$$= 1 - \frac{.15245}{1.43185} = .8935$$

Consequently, 89.35% of this variation has been explained using the three independent variables.

The F value determined in example 15.2 for testing H_0: $\beta_1 = \beta_2 = \beta_3 = 0$ can be duplicated using equation 15-15 because

$$F = \frac{.8935/3}{(1 - .8935)/(10 - 3 - 1)} = \frac{.8935/3}{.1065/6}$$

$$= 16.78 \quad \text{(as before)} \qquad \square$$

COMMENTS   In example 15.5, notice that the value of R^2 increased when we went from using one independent variable to using three. As you add variables to your regression model, R^2 never decreases. However, the increase may not be a *significant* one. If adding 10 more predictor variables to your model causes R^2 to increase from .91 to .92, this is not a significant increase. Therefore, do not include these ten variables; they clutter up your model and add, most likely, spurious predictive ability to it.

How can we tell if adding (or removing) a certain set of X variables causes a significant increase (or decrease) in R^2?

The Partial F Test

Consider the situation in which the researcher is trying to determine whether to retain three variables (X_6 = height, X_7 = weight, X_8 = age) as predictors of a person's score on an instrument to measure the level of self-motivation. We know one thing—R^2 will be higher with these three variables included in the model. If we do not observe a significant increase, however, our advice would be to remove these variables from the analysis. To determine the extent of this increase, we use another F test.

We define two models—one contains X_6, X_7, X_8, and one does not.

Complete model: Uses all predictor variables, including X_6, X_7, and X_8

Reduced model: Uses the same predictor variables as the complete model except X_6, X_7, and X_8

Also, let

R_c^2 = the value of R^2 for the complete model

R_r^2 = the value of R^2 for the reduced model

Do X_6, X_7, and X_8 contribute to the prediction of Y? We will test

H_0: $\beta_6 = \beta_7 = \beta_8 = 0$ (they do not contribute)

H_a: at least one of the β's $\neq 0$ (at least one of them does contribute)

The test statistic here is

$$F = \frac{(R_c^2 - R_r^2)/v_1}{(1 - R_c^2)/v_2} \qquad \textbf{(15-16)}$$

where v_1 = number of β's in H_0, and $v_2 = n - 1 -$ (number of X's in the complete model).

For this illustration, $v_1 = 3$ because there are three β's contained in H_0. Assuming that there are eight variables in the complete model, then $v_2 = n - 1 - 8 = n - 9$. Here, n is the total number of observations (rows) in your data. This F statistic measures the partial effect of these three variables; it is a **partial F statistic**.

Equation 15-16 resembles the F statistic given in equation 15-15, which we used to test H_0: all β's = 0. If all the β's are zero, then the reduced model consists of only a constant term and the resulting R^2 will be zero; that is, $R_r^2 = 0$. Setting $R_r^2 = 0$ in equation 15-16 produces equation 15-15, where $v_1 = k$ and $v_2 = n - k - 1$.

These variables (as a group) contribute significantly if the computed partial F value in equation 15-16 exceeds F_{α, v_1, v_2} from Table A-7.

EXAMPLE 15.6

The researcher gathered data from 30 individuals using all eight of the independent variables. These data were entered into a computer, and a multiple linear regression analysis was performed. The resulting R^2 was .857.

Next, variables X_6, X_7, and X_8 were omitted, and a second regression analysis was performed. The resulting R^2 was .824. Do the variables X_6, X_7, and X_8 (height, weight, and age) appear to have any predictive ability? Use $\alpha = .10$.

Solution

Here, $n = 30$ and

$$R_c^2 = .857 \text{ (complete model)}$$
$$R_r^2 = .824 \text{ (reduced model)}$$

Based on the previous discussion, the value of the partial F statistic is

$$F^* = \frac{(.857 - .824)/3}{(1 - .857)/(30 - 1 - 8)}$$
$$= \frac{.033/3}{.143/21}$$
$$= 1.62$$

The procedure is to reject H_0: $\beta_6 = \beta_7 = \beta_8 = 0$ if $F^* > F_{.10,3,21} = 2.36$. The computed F value does not exceed the table value, so we fail to reject H_0. We conclude that these variables should be removed from the analysis because including them in the model fails to produce a significantly larger R^2.   □

The partial F test also can be used to determine the effect of adding a *single* variable to the model.

EXAMPLE 15.7

Using the university data analyzed in example 15.2, determine whether $X_3 =$ undergraduate GPA contributes to the prediction of graduate GPA, given that $X_1 =$ verbal GRE score and $X_2 =$ quantitative GRE score are included in the model. Use a significance level of $\alpha = .10$.

Solution

We will test the hypotheses

$$H_0: \quad \beta_3 = 0 \text{ (if } X_1 \text{ and } X_2 \text{ are included)}$$
$$H_a: \quad \beta_3 \neq 0.$$

The complete model uses X_1, X_2, and X_3. Using example 15.5,

$$R_c^2 = .8935$$

 Figure 15.7

MINITAB solution using verbal GRE score (X_1) and quantitative GRE score (X_2) as predictors

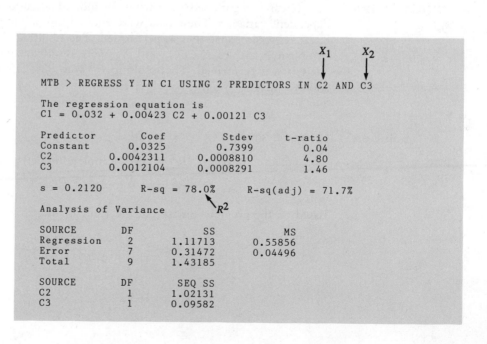

The reduced model uses X_1 and X_2 only. Figure 15.7 shows the MINITAB output for this, and

$$R_r^2 = .780$$

The value of the partial F statistic is

$$F^* = \frac{(.8935 - .780)/1}{(1 - .8935)/(10 - 1 - 3)}$$

$$= \frac{.1135/1}{.1065/6} = 6.4$$

The one in the numerator indicates that there is one β in H_0; subtracting the 3 in the denominator indicates that there are three X's in the complete model.

This value does exceed $F_{.10,1,6} = 3.78$, and so X_3 = undergraduate GPA does (as we concluded from the earlier t test) significantly improve the model's predictive ability when included with X_1 and X_2. In other words, there is a significant increase in R^2 (from .780 to .894) when X_3 is added to the model, and, as a result, our conclusion would be to retain this variable in the model.

□

COMMENTS   Both example 15.7 and the t test for X_3 discussed earlier on page 843, dealt with testing $H_0: \beta_3 = 0$ versus $H_a: \beta_3 \neq 0$. Both tests attempted to determine whether X_3 should be included as a predictor given that X_1 and X_2 were already included as predictor variables. Note two things: (1) the partial F value $= 6.4 = (2.53)^2 = (t \text{ value})^2$, and (2) the p value using the t test (not shown) = the p value using the F test (not shown).

We can see that these tests are identical: They result in exactly the same p value and the same conclusion. This means that, to determine the predictive ability of an individual independent variable, we can compute the partial F statistic or the somewhat simpler t statistic. Some computer packages use the F statistics to summarize the individual predictors, whereas others (such as MINITAB) use the t values to measure the influence of each predictor. You should use whatever is provided (the F statistic or t statistic) to measure the partial effect of each variable; both sets of statistics accomplish the same thing and reach the same conclusion.

Using Curvilinear Models: Polynomial Regression

Is there a linear relationship between the annual income of families in Riverport and their annual food expenditure? A city planner in Riverport is investigating this relationship and suspects that as income increases, so does food expenditure but after a certain point will slow down; that is, food expenditure will increase as income increases but at a slower rate.

Data were gathered from 15 randomly selected Riverport families. All units are in thousands of dollars.

Family	Y (Annual Food Exp.)	X (Annual Income)
1	7	34
2	8	44
3	9	60
4	9	53
5	4	26
6	6	30
7	8	39
8	7	36
9	9	47
10	10	72
11	10	56
12	2	24
13	9	63
14	10	66
15	9	49

The scatter diagram is shown in Figure 15.8. The city planner seems to have a point—the amount spent on food does appear to level off after a certain level of income.

Does the simple linear model $Y = \beta_0 + \beta_1 X + e$ capture the relationship between income (X) and food expenditure (Y)? Although Y does increase as X

Figure 15.8

MINITAB scatter diagram of data for Riverport example

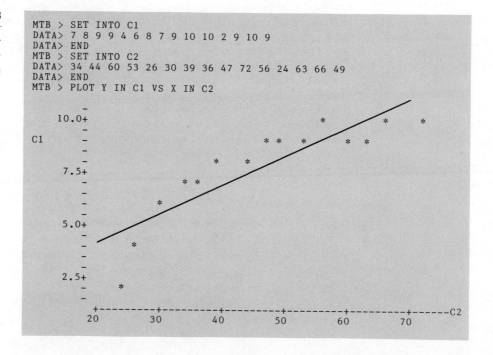

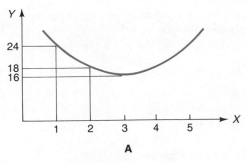

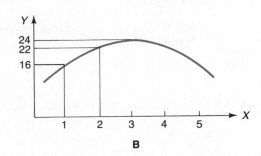

A

B

Figure 15.9

Quadratic curves.
(A) Graph of
$Y = 34 - 12X + 2X^2$.
(B) Graph of
$Y = 6 + 12X - 2X^2$.

increases here, the linear model does not capture the "slowing down" of Y for larger values of X. The least squares line (sketched in Figure 15.8) will underpredict a family's food expenditure for the middle range of X but overpredict this value for small or large incomes.

Figure 15.9 shows **quadratic curves** rather than straight lines. If we include X^2 in the model, we can describe the curved relationship that seems to exist between income and food expenditure. More specifically, the left half of Figure 15.9(B) closely resembles the shape of the scatter diagram in Figure 15.8. Consider the model

$$Y = \beta_0 + \beta_1 X + \beta_2 X^2 + e \qquad (15\text{-}17)$$

Is this a linear regression model? At first glance, it would appear not to be. However, by the word linear we really mean that the model is linear in the unknown β's, not X. In equation 15-17, there are no terms such as β_1^2, $\beta_1\beta_2$, $\sqrt{\beta_0}$, and so on. So the model is linear in the β's, and this is a (multiple) linear regression application.

The model in equation 15-17 is a **curvilinear model** and is an example of **polynomial regression**. Such models are very useful when a particular independent variable and dependent variable exhibit a definite increasing and/or decreasing relationship that is nonlinear.

Solving for β_0, β_1, and β_2 Equation 15-17 represents a multiple regression model containing two predictors, namely, $X_1 = X$ and $X_2 = X^2$. The data for the model then are

Y	X_1	X_2
7	34	1156 ($= 34^2$)
8	44	1936 ($= 44^2$)
9	60	3600
$\vdots$	$\vdots$	$\vdots$
10	66	4356
9	49	2401

These data for Y, X_1, and X_2 will be your input to the multiple linear regression computer program. You can simplify this task by letting the computer build the $X_2 = X^2$ column of data by squaring the entries in the $X_1 = X$ column.

EXAMPLE 15.8

Look at the MINITAB solution using the model $Y = \beta_0 + \beta_1 X + \beta_2 X^2 + e$ for the Riverport example, shown in Figure 15.10.

1 Predict the food expenditure for a family earning $40,000.
2 What do the F and t tests tell you about this model? Use $\alpha = .10$.

Solution 1

The predicted food expenditure for $X_1 = 40$ (thousand) is

$$\hat{Y} = -7.87 + .576(40) - .0047(40)^2$$
$$= 7.65$$

That is, it is approximately $7650.

Solution 2

We will first examine the F test. Our first test of hypothesis determines whether the overall model has predictive ability.

$$H_0: \quad \beta_1 = \beta_2 = 0$$

$$H_a: \quad \text{at least one of the } \beta\text{'s} \neq 0$$

 Figure 15.10

MINITAB solution using $\hat{Y} = b_0 + b_1 X + b_2 X^2$

```
                contains X              contains X²
                    ↓                       ↓
MTB > MULT C2 BY C2, PUT INTO C3
MTB > REGRESS Y IN C1 USING 2 PREDICTORS IN C2 AND C3

The regression equation is
C1 = - 7.87 + 0.576 C2 - 0.00470 C3

Predictor      Coef          Stdev        t-ratio
Constant      -7.870         1.897         -4.15
C2             0.57643       0.08550        6.74      ← t values
C3            -0.0046991     0.0008987     -5.23

s = 0.6754      R-sq = 92.6%     R-sq(adj) = 91.4%
                              ↖R²
Analysis of Variance

SOURCE        DF          SS            MS
Regression     2        68.925        34.463
Error         12         5.475         0.456
Total         14        74.400
```

Using the R^2 value from Figure 15.10 and equation 15-15,

$$F = \frac{.926/2}{.074/(15 - 2 - 1)} = \frac{.926/2}{.074/12}$$

$$= 75.08$$

As we might have suspected, the model does have significant predictive ability; $F^* = 75.08$ exceeds $F_{\alpha,k,n-k-1} = F_{.10,2,12} = 2.81$ from Table A-7.

Now we want to look at the t tests (same as partial F tests). Here, we examine each variable in the model, namely, X and X^2. The t value from Table A-5 is $t_{.10,12} = 1.356$ for a one-tailed test. We want to determine first whether X_1 = income should be included in the model. Increased income should be associated with increased food expenditures, so β_1 should be greater than zero. As a result, we will use a one-tailed procedure to test $H_0: \beta_1 \leqslant 0$ versus $H_a: \beta_1 > 0$.

According to Figure 15.10, the computed t statistic is $t^* = b_1/(\text{standard deviation of } b_1) = 6.74$. $t^* = 6.74 > 1.356$, which means that the income variable should be retained as a predictor of food expenditure.

Next, we want to determine whether X_2 = (income)^2 contributes significantly to the prediction of food expenditure. We are asking whether including the quadratic term was necessary. If this model is the correct one, then, according to Figure 15.9(B), β_2 not only should be unequal to zero, but also, more specifically, should be less than zero. This follows since if the food expenditures do, in fact, level off after a certain level of income, the curve should resemble the left half of the quadratic curve in Figure 15.9(B).

The appropriate hypotheses are

$$H_0: \quad \beta_2 = 0 \text{ (or } \geqslant 0)$$
$$H_a: \quad \beta_2 < 0.$$

We will reject H_0 if $t < -t_{.10,12} = -1.356$.

From Figure 15.10, we see that $t^* = b_2/(\text{standard deviation of } b_2) = -5.23$. This lies in the rejection region, so we conclude that β_2 is < 0, which means that the quadratic term, X^2, contributes significantly, and in the correct direction.

□

COMMENTS   There are three things you should note about the curvilinear model.

1   Curvilinear models often are used for situations in which the rate of increase or decrease in the dependent variable is not constant when plotted against a particular independent variable. The use of X^2 (and in some cases X^3) in your model allows you to capture this nonlinear relationship between your variables.

2   There are other methods available for modeling a nonlinear relationship, including

$$Y = \beta_0 + \beta_1(1/X) + \text{error}$$

Figure 15.11

Error resulting from
extrapolation. See
text and Figure
15.9B.

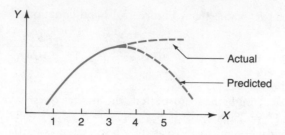

and

$$Y = \beta_0 + \beta_1 e^{-X} + \text{error}$$

These models are also (simple, here) linear regressions; they are linear in the
unknown parameters. These models, unlike the quadratic one discussed previously, involve a **transformation** of the independent variable X. When replacing X by the transformed X (such as $1/X$ or e^{-X}) in the model, one has many
other curvilinear models that may better fit a set of sample data displaying a
nonlinear pattern.

3   Avoid using the model $Y = \beta_0 + \beta_1 X + \beta_2 X^2 + e$ for values of X outside the
range of data used in the analysis. Extrapolation is extremely dangerous when
using this modeling technique. Consider Figure 15.9(B) and suppose values of X
between 1 and 3 were used to derive the estimates of β_0, β_1, and β_2. Figure
15.11 shows the results. For values of X larger than 3, the predicting equation
will turn down, whereas the actual relationship will probably continue to level
off. So this model works for interpolation (for values of X between 1 and 3, here)
but is extremely unreliable for extrapolation.

EXERCISES

15.21   If all the values of the dependent variable Y fell on the plane $\hat{Y} = 2 + 5.2X_1 +
10X_2$ and the variables X_1 and X_2 were used in a least squares fit to the data,
what would be the value of the coefficient of determination?

15.22   If the residual standard deviation is 1.5 for a regression model with five independent variables and 26 observations, what is the value of R^2, assuming that
SST is 231.95? Interpret the value of R^2.

15.23   A botanist would like to study the growth rate of a variety of grass that will with-
stand the dry heat of southern Arizona. Three variables are used:

$$X_1 = \text{gallons of water per week}$$

$$X_2 = \text{temperature in Fahrenheit}$$

$$X_3 = \text{kilograms of fertilizer per month}$$

After observing 50 growth rates under different values of X_1, X_2, and X_3, the botanist wishes to test that X_2 and X_3 do not contribute to predicting the variation in the growth rates. The coefficient of determination for the model involving Y and X_1 is .71. The coefficient of determination for the complete model with X_1, X_2, and X_3 is .82. Do the additional independent variables X_2 and X_3 contribute significantly to the model? Use a 5% significance level.

15.24 The dean of the college of business at Fargo University would like to see whether several variables affect a student's grade point average. Thirty first-year students were randomly selected and data were collected on the following variables:

Y = grade point average for the first year

X_1 = average time spent per month at fraternity or sorority functions

X_2 = average time spent per month working part time

X_3 = total number of hours of coursework attempted

The SSE for the least squares line involving only Y and X_1 was found to be 5.21. The SSE for the complete model was found to be 4.31. The SST is 24.1. At the 5% significance level, test the null hypothesis that the independent variables X_2 and X_3 do not contribute to predicting the variation in Y given that X_1 is already in the model.

15.25 A multiple regression analysis run on a computer resulted in the following equation:

$$\hat{Y} = 1.7542 - 2.3454X_1 + 3.7200X_2$$

with an R^2 value of 0.8672. A total of 20 observations were used in the analysis.
(a) Is the model as a whole significant at $\alpha = .05$?
(b) State the p value for the test.

15.26 In exercise 14.8, you might have noticed that the output of the workers increases with illumination levels up to a point, but later on, it begins to decline. The data are reproduced here for quick reference.

Y (output of worker)	X (illumination intensity)
70	1
70	2
75	3
88	4
91	5
94	6
100	7
92	8
90	9
85	10

A curvilinear (quadratic) effect is suspected.

(a) It is desired to obtain a model of the form $Y = \beta_0 + \beta_1 X + \beta_2 X^2 + e$. Use a computerized statistical package to obtain this model. What is the regression equation?

(b) Is the model significant, at $\alpha = .10$?

(c) Does the X^2 term contribute significantly to the model, at $\alpha = .10$?

(d) Obtain another model of the form $Y = \beta_0 + \beta_1 X + \beta_2 X^2 + \beta_3 X^3$. Does the X^3 term contribute significantly to this model, at $\alpha = .10$?

(e) Compare the simple linear model $Y = \beta_0 + \beta_1 X + e$ and the curvilinear model in (d). The R^2 of the model will increase due to the extra variables, but is the increase in R^2 due to the inclusion of X^2 and X^3 significant? Use the partial F test, at $\alpha = .10$.

15.27 An independent research firm wants to investigate what motivates people to commit suicide. The variables used in this experiment are as follows:

Y = number of suicides

X_1 = number of psychiatric admissions per 1000 population

X_2 = number of persons below the poverty level (in millions)

X_3 = the number of unemployed persons (in millions)

X_4 = the number of divorces (in thousands)

X_5 = the number of crimes in the country (in millions)

Yearly data were collected for 15 years. The coefficient of determination for the model involving only the independent variables X_1 and X_3 is .53. The coefficient of determination for the model involving all five independent variables is .75. Do the variables X_2, X_4, and X_5 contribute to predicting the variation in the number of yearly suicides given that X_1 and X_3 are in the model? Use a 5% level of significance.

15.28 The reliability (Y) of a silicon chip is believed to be related to the temperature at which the silicon wafer is "cooked" (X_1) and the percentage of impurity allowed into the pure silicon (X_2). A regression analysis run with 50 observations on a model including X_1 and X_2 as predictors yielded an SSE of 1215.80 and SST of 4863.21. A second run also included the variables $\sqrt{X_1}$ and $\sqrt{X_2}$ along with X_1 and X_2. The SSE for this larger model was 877.11. Do the variables $\sqrt{X_1}$ and $\sqrt{X_2}$ contribute to predicting the reliability of the chip? Use a 5% significance level.

15.29 (a) If a regression model has $r = .80$, what is the percentage of variation in Y explained in the model?

(b) In terms of "the variation in Y explained by the model," it can be said that a correlation of $r = .80$ is four times stronger than a correlation of $r = .40$. Explain why.

(c) Again in terms of "explained variation in Y," a correlation of $r = .60$ is how many times as strong as $r = .20$?

15.30 Hoaglin and Welsch (1978) conducted a study involving the strength of wood beams in relation to moisture content and specific gravity. While their analysis was somewhat more complex than the material in this chapter, we can still use their data, in the manner of a construction engineer trying to predict the strength of wood beams. The following data were collected for 10 beams:

Beam	Strength	Moisture	Specific Gravity
1	11.14	11.1	0.499
2	12.74	8.9	0.558
3	13.13	8.8	0.604
4	11.51	8.9	0.441
5	12.38	8.8	0.550
6	12.60	9.9	0.528
7	11.13	10.7	0.418
8	11.70	10.5	0.480
9	11.02	10.5	0.406
10	11.41	10.7	0.467

Source: D. C. Hoaglin and R. F. Welsch, Hat matrix in regression and ANOVA. *The American Statistician* **32** (1): 19 (Feb. 1978).

(a) Use a computerized statistical package to find the least-squares equation.

(b) Obtain the F statistic to test the usefulness of the whole model, by using the ANOVA table.

(c) Compute the same F statistic as in (b) by using the R^2 of the model. Are your results for (b) and (c) identical, allowing for negligible rounding differences?

(d) Is the model as a whole significant at $\alpha = .05$?

(e) Does the moisture content contribute significantly in this model? Use $\alpha = .05$.

(f) Does the specific gravity contribute significantly in this model? Use $\alpha = .05$.

(g) As an extra exercise, investigate the model $Y = \beta_0 + \beta_1 X_1 + \beta_2 X_2 + \beta_3 X_1^2 + \beta_4 X_2^2 + e$. Do the extra two (quadratic) variables contribute significantly, i.e., is the curvilinear model a better one than the model obtained in part (a)? Use $\alpha = .05$ for the test.

15.4 THE PROBLEM OF MULTICOLLINEARITY

Another possible title for this section is "What do the individual b_i's tell you?" We will discuss one of the common problems in the use (or misuse) of multiple linear regression—namely, trying to extract more information from the results than they actually contain.

We will examine the validity of such statements as "Because $b_1 = 10$, increasing X_1 by one while holding X_2 constant will result in an increase of 10 in Y."

Assume that a sample of 10 teachers at Johnson High School was examined in an effort to determine the ability of age (X_1) and years of experience (X_2)

to predict a teacher's salary (Y). The following data were obtained:

Employee	Y	X_1	X_2
1	37	52	33
2	25	47	21
3	32	38	14
4	20	25	3
5	30	44	18
6	42	55	30
7	22	36	8
8	27	40	15
9	23	32	7
10	34	50	27

First, we can ask, how well does X_1 (age) predict Y (salary)?

 Figure 15.12

MINITAB solution to
$\hat{Y}$ (salary) =
$b_0 + b_1$ (age)

```
MTB > READ INTO C1-C3
DATA> 37 52 33
DATA> 25 47 21
DATA> 32 38 14
DATA> 20 25 3
DATA> 30 44 18
DATA> 42 55 30
DATA> 22 36 8
DATA> 27 40 15
DATA> 23 32 7
DATA> 34 50 27
DATA> END
     10 ROWS READ
MTB > REGRESS Y IN C1 USING 1 PREDICTOR IN C2

The regression equation is
C1 = 2.29 + 0.642 C2

Predictor      Coef          Stdev        t-ratio
Constant       2.291         5.862         0.39
C2             0.6422        0.1368        4.69  ◄──── t value

s = 3.886      R-sq = 73.4%      R-sq(adj) = 70.0%

Analysis of Variance

SOURCE      DF          SS            MS
Regression   1        332.79        332.79
Error        8        120.81         15.10
Total        9        453.60


MTB > CORRELATION BETWEEN C1 AND C2

Correlation of C1 and C2 = 0.857
```

instructions for obtaining the sample
correlation between any two variables

Y X_1

Figure 15.13

MINITAB solution to $\hat{Y}$ (salary) $= b_0 + b_1 \cdot$ (years of experience)

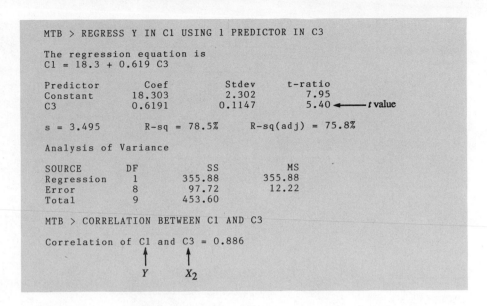

```
MTB > REGRESS Y IN C1 USING 1 PREDICTOR IN C3

The regression equation is
C1 = 18.3 + 0.619 C3

Predictor        Coef        Stdev       t-ratio
Constant       18.303        2.302          7.95
C3             0.6191       0.1147          5.40  ◄──── t value

s = 3.495        R-sq = 78.5%      R-sq(adj) = 75.8%

Analysis of Variance

SOURCE         DF          SS          MS
Regression      1       355.88      355.88
Error           8        97.72       12.22
Total           9       453.60

MTB > CORRELATION BETWEEN C1 AND C3

Correlation of C1 and C3 = 0.886
                          ↑        ↑
                          Y        X₂
```

A MINITAB solution using the model $Y = \beta_0 + \beta_1$ (age) $+ e$ is shown in Figure 15.12. Notice the computed t value. Now, $k = 1$ because this model considers only one independent variable, so the tabulated value for comparison (using $\alpha = .10$) is $t_{\alpha/2, n-k-1} = t_{.05, 10-1-1} = t_{.05, 8} = 1.860$. The value of $t^* = 4.69$ is considerably larger than 1.86, so X_1 (age) is an excellent predictor of Y (salary).

What is the correlation between X_1 and Y? It seems reasonable that this would be quite large because age has been shown to be a good predictor. In fact, according to Figure 15.12, this value is .857. So there is a positive relationship between age and salary, as one would expect.

Next, we determine how well X_2 (years of experience) predicts Y (salary). The solution using $Y = \beta_0 + \beta_1$(years of experience) $+ e$ is shown in Figure 15.13. Once again, the computed t value $= t^* = 5.40$ is much larger than $t_{.05, 8} = 1.860$. Also, the correlation between these two variables is .886. This is not surprising; we might expect teachers with more years of experience to have higher salaries. Consequently, a significant positive relationship appears to exist between these two variables. Finally, we turn to the question, how well do both X_1 (age) and X_2 (years of experience) predict salary? The model here is $Y = \beta_0 + \beta_1 X_1 + \beta_2 X_2 + e$. The least squares solution is shown in Figure 15.14.

$$\hat{Y} = 19.2 - .034X_1 + .650X_2$$

A few seemingly bizarre things show up here.

The coefficient of X_1 is $b_1 = -.034$. This would appear to indicate that larger values of X_1 (older teachers) produce smaller salaries. But we know from

Figure 15.14

MINITAB solution to $\hat{Y}$ (salary) $= b_0 + b_1 \cdot$ (age) $+ b_2 \cdot$ (years of experience)

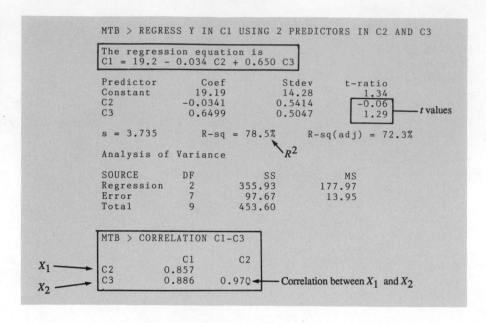

our first analysis that the opposite is true. We would have expected a positive value of b_1 here, and so the coefficient of X_1 appears to have the wrong sign.

The small t values also are puzzling. The value of the F statistic (using Figure 15.14) is

$$F^* = \frac{R^2/2}{(1 - R^2)/(10 - 1 - 2)} = \frac{.785/2}{.215/7}$$
$$= 12.8$$

As before, you can compute this value using the ANOVA table, where

$$F^* = \frac{\text{MSR}}{\text{MSE}} = \frac{177.97}{13.95}$$
$$= 12.8$$

Using $\alpha = .10$, this is much larger than $F_{.10,2,7} = 3.26$, and so the model does provide a very good predictor of Y. The coefficient of determination is $R^2 = .785$; it explains 78.5% of the total variation in the 10 salary values.

The t values are very small; both are smaller in absolute value than $t_{\alpha/2, n-k-1} = t_{.05, 10-2-1} = t_{.05,7} = 1.895$. Does this imply that both predictors are weak and should be removed from the model? Certainly not, as our previous analyses made clear.

This is the problem of **multicollinearity**. In multiple regression models, it is desirable for each independent variable X to be highly correlated with Y, but it is not desirable for the X's to be highly correlated with each other. In many applications of multiple linear regression, the independent variables typically have a certain amount of pairwise correlation. Extremely high correlation between any pair of variables can cause a variety of problems, as we will show.

The (sample) correlation between X_1 and X_2 is

$$r = \frac{\Sigma X_1 X_2 - (\Sigma X_1)(\Sigma X_2)/n}{\sqrt{\Sigma X_1^2 - (\Sigma X_1)^2/n}\sqrt{\Sigma X_2^2 - (\Sigma X_2)^2/n}}$$

This value, using Figure 15.14, is $r = .970$. Notice in the data set that nearly every time X_1 increases, so does X_2; X_1 and X_2 are highly correlated. As a result, these data contain a great deal of multicollinearity.

Implications

First of all, the correlation of X_1 and X_2 explains the small t values. Remember that each t value describes the contribution of that particular independent variable after all other independent variables have been included in the model. X_1 is very nearly a linear function of X_2 (as evidenced by $r = .970$), so it contributes very little to the prediction of Y, given that X_2 is in the model. The same argument applies to X_2. This means that neither X_1 nor X_2 is a strong predictor given that the other variable is included—not that each one is a weak predictor by itself.

The second implication of the multicollinearity is that the situation "increasing X_1 by one while holding X_2 constant" never occurred in the sample data. In the past, as X_1 increased by one, X_2 always changed also, because X_1 and X_2 are so highly correlated.

Finally, the sample coefficients (b_1, b_2) of our independent variables have very large variances. If we took another sample from this population, the values of b_1 and b_2 probably would change dramatically—which is not a good situation. In fact, as this example has demonstrated, these coefficients can even have the "wrong" sign; that is, a sign different from that obtained when regressing X_1 or X_2 alone on Y.

Eliminating the Effects of Multicollinearity

The easiest way out of this dilemma is to remove some of the correlated predictors from the model. For this illustration, we should remove either X_1 or X_2 (but not both). Our best bet would be to retain X_2 = years of experience because it has the highest correlation with Y.

One method of eliminating correlated predictor variables is to use a **stepwise** selection procedure. This technique of selecting the variables to be used in a multiple linear regression equation will be discussed in the next section. Essentially, it selects variables one at a time and generally (although not always) does not insert into the regression equation a variable that is highly correlated with a variable already in the equation. In the previous example, a stepwise procedure would have selected variable X_2 (the single best predictor of Y) and then informed the user that X_1 did not significantly improve the prediction of Y, given that X_2 is already included in the prediction equation.

Whenever you perform a multiple regression analysis, it is always a good idea to examine the pairwise correlations between all of your variables, including the dependent variable. In this way, you often can detect easily the two independent variables that are contributing to the multicollinearity problem. Since multicollinearity can involve more than two predictor variables, an examination of pairwise correlations does not always reveal potential multicollinearity problems. These pairwise correlations can be obtained using a single command with most computer packages. The MINITAB command to generate a table (often called a matrix) of pairwise correlations is shown in the bottom box in Figure 15.14. This output indicates that the correlation between Y and X_1 is .857, between Y and X_2 is .886, and between X_1 and X_2 is .970.

Other, more advanced methods of detecting and treating this situation are beyond the scope of this text. One of the more popular procedures is ridge regression.*

We have seen that the problem of multicollinearity enters into our regression analysis when an independent variable is highly correlated with one or more other independent variables. Multicollinearity produces inflated regression coefficients that can even have the wrong sign. Also, the resulting t statistics can be small, making it difficult to determine the predictive ability of an individual variable. Therefore, $b_1, b_2, \ldots$ tell us nothing about the partial effect of each variable, unless we can demonstrate that there is no correlation among our predictor variables. In regression applications, correlation among the independent variables is far from unusual. As a final note, perfect multicollinearity will be present whenever one of your predictor variables is a sum of two or more other predictor variables. For example, you cannot use SAT math score, SAT verbal score, and SAT total score (math + verbal) in the same prediction equation, since they are linearly related.

EXERCISES

15.31 What might cause the following situation to occur for a regression model with two independent variables? The t values for both β_1 and β_2 are nonsignificant. However, the F test for both $\beta_1 = \beta_2 = 0$ is highly significant.

15.32 If it is known that multicollinearity exists between three independent variables, how would you choose the independent variables that should remain in the model?

15.33 Re-examine the data in exercise 15.3 and investigate whether multicollinearity might be a problem due to strong correlation between one or more of the predictor variables. What does your analysis indicate? (*Hint:* Obtain the correlation matrix.)

* For an excellent discussion of this topic, see J. Neter, W. Wasserman, and M. Kutner. *Applied Linear Regression Models* (Homewood, IL: Richard D. Irwin, 1983).

15.34 A least squares equation was fit to a set of data for an experiment and was found to be $\hat{Y} = 30 - 501X_1 + 300X_2$. The experiment was repeated and a new set of data from the same population was fit with the least squares line $\hat{Y} = -20 + 309X_1 - 151X_2$. Is there any explanation for these two different prediction equations?

15.35 Consider the following set of data of 12 emerging growth-oriented companies. Y represents the growth rate of a company for the current year, X_1 represents the growth rate of the company for the previous year, and X_2 represents the percent of the market that does not use the company's product or a similar product. All values are percentages.

Y	X_1	X_2
20	10	30
24	12	35
18	15	25
33	30	40
27	19	32
20	24	20
30	15	60
36	42	38
47	45	40
35	32	32
28	24	31
32	20	50

(a) Find the coefficient of determination for the model with only X_1 included in the model.

(b) Find the coefficient of determination for the model with only X_2 in the model.

(c) The coefficient of determination for the complete model is .896. Does it appear from observing the values of the coefficient of determination in questions (a) and (b) that both variables X_1 and X_2 should stay in the model?

15.36 The longevity model discussed in exercise 14.61 was extended to also include job satisfaction as a predictor of gain in longevity, in addition to marital satisfaction. The following data resulted for 10 persons observed:

Y = Longevity (life span gain)	X_1 = Marital Satisfaction (scale of 1 to 10)	X_2 = Job Satisfaction (scale of 1 to 10)
−0.5	1.5	1.2
−0.2	2.1	2.0
0.1	3.0	3.1
1.0	4.5	4.4
1.2	5.3	5.2
2.1	6.0	6.6
1.5	6.9	7.0
3.8	8.1	8.2
2.7	9.0	9.4
3.5	9.9	10.0

Use a computerized statistical package to assist you with the following:

(a) Find the R^2 for a simple regression model using only X_1 as a predictor.

(b) Find the R^2 for a simple regression model using only X_2 as a predictor.

(c) Find the R^2 for a complete multiple regression model using both X_1 and X_2 as predictors.

(d) Does it appear from observing the R^2 value above that both X_1 and X_2 should stay in the model? (Use a 5% significance level.) If only one predictor should be used rather than two, which one should be retained?

(e) What is the correlation between X_1 and X_2? In light of the above analysis, does it indicate that multicollinearity is a serious problem in the multiple regression model?

15.37   Looking at the regression equation:

$$\hat{Y} = 25 + 3.5X_1 - 1.8X_2$$

a mathematics student interpreted the values as follows: $b_1 = 3.5$ means that Y increases by 3.5 for each unit increase in X_1 while X_2 is held constant. Similarly, $b_2 = -1.8$ means that Y decreases by 1.8 for each unit increase in X_2, while X_1 is held constant. While this sounds very good in theory, what is the practical problem in trying to interpret the b values in this fashion?

15.38   Refer to exercise 15.9. Find the correlation between students' overall grade point average and years of job-related experience. Does multicollinearity appear to be a concern?

15.5   ADDITIONAL TOPICS IN MULTIPLE LINEAR REGRESSION

The Use of Dummy Variables

The use of **dummy** or **indicator variables** in regression analysis allows you to include qualitative variables in the model. For example, if you wanted to include a teacher's sex as a predictor variable in a regression model, define

$$X_1 = \begin{cases} 1 \text{ if female} \\ 0 \text{ if male} \end{cases}$$

Note that the choice of which sex is assigned the value of 1, male or female, is arbitrary. The estimated value of Y will be the same, regardless of which coding procedure is used.

Returning to the data we used in example 15.2, the admissions office at Brookhaven University decided to use the graduate student's undergraduate major as a predictor in the model. For this model she drops the quantitative GRE score (variable, X_2) as a predictor variable since it was earlier demonstrated to not contribute significantly towards the prediction of a graduate student's GPA. Taking note of the student's undergraduate major led to the following data:

Student	Y First Year Graduate GPA	X_1 Verbal GRE Score	X_3 Undergraduate GPA	Undergraduate Major
1	3.54	580	3.82	Psychology
2	2.62	500	2.67	Sociology
3	3.30	670	3.16	Sociology
4	2.90	480	3.31	Sociology
5	4.00	710	3.60	Sociology
6	3.21	550	3.42	Education
7	3.57	640	3.51	Psychology
8	3.05	540	2.75	Sociology
9	3.15	620	3.21	Psychology
10	3.61	690	3.70	Education

Using these data, we can construct the necessary dummy variables and determine whether they contribute significantly to the prediction of GPA (Y). One way to code the student's major field would be to define

$$X_4 = \begin{cases} 0 \text{ if student was a Psychology major} \\ 1 \text{ if student was a Sociology major} \\ 2 \text{ if student was an Education major} \end{cases}$$

However, this type of coding has many problems. First, because $0 < 1 < 2$, the codes imply that a Psychology major is "smaller" than a Sociology major, which is "smaller" than an Education major. Furthermore, any difference between a Psychology major and an Education major receives twice the weight (because $2 - 0 = 2$) of any difference between a Psychology major and a Sociology major, or a Sociology major and an Education major. So this coding transforms data that are actually nominal to data that are interval, a much stronger type. A better procedure is to use the necessary number of dummy variables (coded 0 or 1) to represent the student's major.

We needed one dummy variable with two categories (male and female) to specify a person's sex. To represent the three undergraduate majors, we use two dummy variables, by letting

$$X_4 = \begin{cases} 1 \text{ if student was a Psychology major} \\ 0 \text{ otherwise} \end{cases}$$

and

$$X_5 = \begin{cases} 1 \text{ if student was a Sociology major} \\ 0 \text{ otherwise} \end{cases}$$

Note that, as with the male/female dummy variable, this coding is arbitrary as far as the prediction $\hat{Y}$ is concerned. We could have assigned $X_4 = 0$ and $X_5 = 0$ for a Psychology major, with $X_4 = 1$ for a Sociology major and $X_5 = 1$ for an Education major.

Why wasn't a dummy variable defined for the Education major? It is not necessary to develop a third dummy variable here because we have the following scheme:

Student Major In	X_4	X_5
Psychology	1	0
Sociology	0	1
Education	0	0

In fact, it can be shown that a third dummy variable is not only unnecessary, it is very important that you not include it. If you attempted to use three such dummy variables in your model, you would receive a message in your computer output informing you that "no solution exists" for this model. Suppose we had introduced a third dummy variable (say, X_6) that was equal to 1 if the student majored in Education as an undergraduate. For each observation in the sample we would have

$$X_6 = 1 - X_4 - X_5$$

Whenever any one predictor variable is a linear function (including a constant term) of one or more other predictors, then mathematically no solution exists for the least squares coefficients since you have multicollinearity at its worst. To arrive at a usable equation, any such predictor variable must not be included.

The resulting model here is $Y = \beta_0 + \beta_1 X_1 + \beta_3 X_3 + \beta_4 X_4 + \beta_5 X_5 + e.$* The final array of data (ready for input into a computer program) is

Row	Y	X_1	X_3	X_4	X_5
1	3.54	580	3.82	1	0
2	2.62	500	2.67	0	1
3	3.30	670	3.16	0	1
4	2.90	480	3.31	0	1
5	4.00	710	3.60	0	1
6	3.21	550	3.42	0	0
7	3.57	640	3.51	1	0
8	3.05	540	2.75	0	1
9	3.15	620	3.21	1	0
10	3.61	690	3.70	0	0

where Y = first-year graduate GPA, X_1 = verbal GRE score, X_3 = undergraduate GPA, X_4 = 1 if undergraduate major was Psychology, and X_5 = 1 if undergraduate major was Sociology.

* Models that include dummy variables typically contain terms that reflect any **interaction** between the dummy variables and the other quantitative variables. For this model, this would amount to adding four additional terms to the model; namely $X_1 X_4$, $X_1 X_5$, $X_3 X_4$, and $X_3 X_5$. Such a model would require a slightly larger sample size (n) than that used in this illustration, since the model now contains $k = 8$ predictor variables.

A MINITAB solution is shown in Figure 15.15. To determine whether the particular undergraduate major has any effect on the prediction of graduate GPA, we test

$$H_0: \quad \beta_4 = \beta_5 = 0 \text{ (if } X_1 \text{ and } X_3 \text{ are included)}$$
$$H_a: \quad \text{at least one } \beta \neq 0$$

In the complete model, the variables are X_1, X_3, X_4, and X_5, and, from Figure 15.15(A),

$$R_c^2 = .901$$

Figure 15.15

MINITAB solution for university data with dummy variables. (A) Solution using all four variables. (B) Solution using verbal GRE (X_1) and undergraduate GPA (X_3).

```
        MTB > READ INTO C1-C5
        DATA> 3.54 580 3.82 1 0
        DATA> 2.62 500 2.67 0 1
        DATA> 3.30 670 3.16 0 1
        DATA> 2.90 480 3.31 0 1
        DATA> 4.00 710 3.60 0 1
        DATA> 3.21 550 3.42 0 0
        DATA> 3.57 640 3.51 1 0
        DATA> 3.05 540 2.75 0 1
        DATA> 3.15 620 3.21 1 0
        DATA> 3.61 690 3.70 0 0
        DATA> END
   A         10 ROWS READ
        MTB > REGRESS Y IN C1 USING 4 PREDICTORS IN C2-C5

        The regression equation is
        C1 = - 0.584 + 0.00285 C2 + 0.625 C3 + 0.058 C4 + 0.167 C5

        Predictor       Coef        Stdev      t-ratio
        Constant      -0.5842      0.6821       -0.86
        C2          0.0028519    0.0008266       3.45
        C3             0.6253      0.2121        2.95
        C4             0.0582      0.1537        0.38
        C5             0.1670      0.1652        1.01

        s = 0.1681      R-sq = 90.1%    R-sq(adj) = 82.2%
```

$$R_c^2$$

```
        MTB > REGRESS Y IN C1 USING 2 PREDICTORS IN C2 AND C3

        The regression equation is
        C1 = - 0.132 + 0.00294 C2 + 0.503 C3

   B    Predictor       Coef        Stdev      t-ratio
        Constant      -0.1321      0.4908       -0.27
        C2          0.0029446    0.0007657       3.85
        C3             0.5026      0.1614        3.11

        s = 0.1568      R-sq = 88.0%    R-sq(adj) = 84.5%
```

$$R_r^2$$

In the reduced model, the variables are X_1 and X_3 only, and, from Figure 15.15(B),

$$R_r^2 = .880$$

At first glance, it does not appear that X_4 and X_5 produced a significant increase in R^2. The partial F test will determine whether this is true.

$$F = \frac{(R_c^2 - R_r^2)/(\text{number of } \beta\text{'s in } H_0)}{(1 - R_c^2)/[n - 1 - (\text{number of } X\text{'s in the complete model})]}$$

$$= \frac{(.901 - .880)/2}{(1 - .901)/(10 - 1 - 4)}$$

$$= \frac{.021/2}{.099/5} = .53$$

Using $\alpha = .10$, this is considerably less than $F_{.10,2,5} = 3.78$, so there is no evidence that the undergraduate major dummy variables significantly improve the prediction of graduate GPA.

In this example, the dummy variables were not significant predictors in the model. However, do not let this mislead you. In many applications, dummy variables representing location, sex, yes/no situations, marital status, and many other variables can have a tremendous effect on improving the results of a multiple regression model.

Stepwise Procedures

Assume you wish to predict the time required for a newly introduced anesthetic to take effect using, among other techniques, a multiple linear regression model. Your strategy is to consider any variable that you think could have an effect on this length of time. You have identified 12 such variables, including weight and sex.

One possibility is to include all of these in your model and to use the t tests to decide which variables are significant predictors. However, this procedure invites multicollinearity because your model is more apt to include correlated predictors, which will severely hinder the interpretation of your model. In particular, two independent variables that are very highly correlated may both have small t values (as we saw in the teacher salary example), causing you possibly to discard both of them from the model. This is not the right thing to do because possibly you should have retained one of them.

A better way to proceed here is to use one of the several stepwise selection procedures. These techniques either choose or eliminate variables, one at a time, in an effort not to include those variables that either have no predictive ability or are highly correlated with other predictor variables. A word of caution— these procedures do not provide a guarantee against multicollinearity; however, they greatly reduce the chances of including a large set of correlated independent variables.

These procedures consist of three different selection techniques: (1) forward regression, (2) backward regression, and (3) stepwise regression.

Forward Regression The forward regression method of model selection puts variables into the equation, one at a time, beginning with that variable having the highest correlation (or R^2) with Y. For the sake of argument, call this variable X_1.

Next, it examines the remaining variables for the variable that, when included with X_1, has the highest R^2. That predictor (with X_1) is inserted into the model. This procedure continues until adding the "best" remaining variable at that stage results in an insignificant increase in R^2, according to the partial F test.

Backward Regression Backward regression is the opposite of forward regression: It begins with all variables in the model and, one by one, removes them. It begins by finding the "worst" variable—the one that, when removed from the complete model, causes the smallest decrease in R^2. If the decrease is insignificant, this variable is removed, and the process continues.

The variable, among those remaining in the model, that causes the smallest decrease in the new R^2 is considered next. You continue this procedure of removing variables until a significant drop in R^2 is obtained, at which point you replace this significant predictor and terminate the selection.

Will the model resulting from a backward regression be the same as that obtained using forward regression? Not necessarily; usually, however, the resulting models are very similar. Of course, if two variables are highly correlated, the forward procedure could choose one of the correlated predictors, whereas the backward procedure could choose the other.

Stepwise Regression Stepwise regression is a modification of forward regression. It is the most popular and flexible of the three selection techniques. It proceeds exactly as does forward regression, except that, at each stage it can remove any variable whose partial F value indicates that this variable does not contribute, given the present set of independent variables in the model. As with forward regression, it stops when the "best" variable among those remaining produces an insignificant increase in R^2.

Figure 15.16 illustrates this procedure using a sample of data (not shown) for the illustration concerned with predicting the time required for an anesthetic to take effect. Data from all 12 independent variables, as well as from Y, are used as input to a stepwise regression program. One possible outcome from this analysis is shown by Figure 15.16.

The stepwise solution for the data we used to predict graduate student GPA is contained in the end-of-chapter MINITAB appendix. As we previously determined, $X_2 =$ quantitative score on the GRE exam does not contribute significantly, and so the resulting prediction equation includes only $X_1 =$ verbal GRE score and $X_3 =$ undergraduate GPA. This equation is

$$\hat{Y} = -.132 + .00294X_1 + .50X_3$$

Using Dummy Variables in Forward or Stepwise Regression We emphasized that $C - 1$ dummy variables should be used to represent C categories

Figure 15.16

Possible solution
using anesthetic data

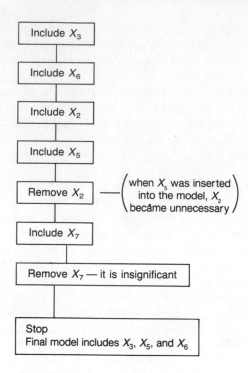

if all of the dummy variables were to be inserted into the regression equation. When using a forward or stepwise regression procedure, this may not be the best way to proceed, as the following illustration will point out.

Suppose you are using nine dummy (indicator) variables to represent ten hospitals. The dependent variable is patient length of stay in the hospital, and the purpose is to determine which hospital (or hospitals) exhibits very long or very short patient length of time hospitalized. If a forward or stepwise selection procedure is used, then including one of these dummy variables would indicate that specifying this particular hospital significantly improves the prediction of length of patient hospitalization. In other words, this is an indication that the length of patient hospitalization for this hospital is not just average but is much higher (its coefficient will be positive) or lower (its coefficient will be negative).

When you use the forward or stepwise techniques, you probably will not include all nine dummy variables in the model. Your ability to predict length of hospitalization (Y) is unaffected by not defining a tenth dummy variable, and in fact, as pointed out earlier, the regression analysis will not accept all 10 dummy variables.

For this situation, however, there is the danger of not detecting an extremely high or low hospitalization length in the tenth hospital that did not receive a dummy variable. When including these variables one at a time in the regression equation using a forward or stepwise procedure, we can allow the regression model to examine the effect of all ten hospitals. We do this by defining 10 such dummy variables, one for each hospital.*

Because a forward regression procedure generally will not attempt to include all 10 dummy variables, you are able to investigate the existence of high or low length of hospitalization in each of the 10 hospitals. When using dummy variables in a forward or stepwise regression procedure, it is perfectly acceptable to use C such variables to represent C categories.

Checking the Assumptions: Examination of the Residuals

When you use a multiple linear regression model, you should keep two things in mind. First, no assumptions are necessary to derive the least squares estimates of $\beta_1, \beta_2, \beta_3, \ldots$. The regression coefficients $b_1, b_2, b_3, \ldots$ determined by a computer solution are the "best" estimates, in the least squares sense.

Second, several key assumptions are required to construct CIs and perform any test of hypothesis. If these assumptions are violated, you may still have an accurate prediction $\hat{Y}$, but the validity of these inference procedures will be very questionable.

Your final step in any regression analysis should be to verify your assumptions.

Assumption 1 *The errors are normally distributed, with a mean of zero.*

An easy method to determine whether the errors follow a normal distribution, centered at zero, is to let the computer construct a histogram of the sample residuals $(Y - \hat{Y})$. Since the residuals *always* sum to zero, the residual histogram is typically centered at zero. This plot should reveal whether the distribution of residuals is severely skewed.

Consider the 50 residuals resulting from the analysis in example 15.4. The computer solution for this problem is shown in Figure 15.6. Notice that the RESIDUALS subcommand following the REGRESS command can be used to store the residuals in any column (such as column C5 in Figure 15.6).

A MINITAB histogram of these values is shown in Figure 15.17. The distribution of these residuals appears to be centered at zero and, except for a slight lack of observations in the 2nd and 5th classes, is bell-shaped (normally distributed) in appearance. Remember that an exact normal distribution is not necessary here; problems arise only when the distribution is severely skewed and does not resemble a normal distribution.

More sophisticated methods of checking this assumption involve the use of a *probability plot* or a *chi-square goodness-of-fit test*. We will not discuss the probability plot technique here, except to say that you plot the residuals in a specialized type of graph. If the resulting graph is basically linear in appearance, the normality assumption has been verified. The goodness-of-fit test was discussed in Chapter 13, where we used a chi-square statistic to test the hypothesis

* This problem is discussed in D. Dorsett and J. T. Webster, Guidelines for variable selection problems when dummy variables are used, *The American Statistician* 37 (4): 337 (1983).

```
MTB > HISTOGRAM OF C5

    Histogram of C5    N = 50

    Midpoint    Count
        -15       6    ******
        -10       2    **
         -5      12    ************
          0      13    *************
          5       6    ******
         10       7    *******
         15       2    **
         20       2    **
```

that a particular set of data (in this case, the regression residuals) came from a specific distribution. Several examples in Chapter 13 discussed how to use the chi-square test for a suspected normal population.

If you have reason to believe that this assumption of your model has been violated, then you need to search for another model. This model may include additional predictor variables that have been overlooked. Another possibility is to transform the dependent variable (such as using $\sqrt{Y}$ rather than Y) or transforming one or more of the predictor variables. As your model tends to "improve," you should observe the residuals tending toward a normal distribution.

Assumption 2   *The variance of the errors remains constant.*

For example, you should not observe larger errors associated with larger values of $\hat{Y}$.

When the residuals $(Y - \hat{Y})$ are plotted against the predicted values $(\hat{Y})$, one hopes to observe no pattern (a "shotgun blast" appearance) in this graph, as in Figure 15.18(A). Remember—the assumptions are essentially that the errors consist of what engineers call noise, with no observable pattern.

A common violation of the assumption of equal variances occurs when the value of the residual increases as $\hat{Y}$, or an individual predictor, increases. This is illustrated in Figure 15.18(B). In this figure, the variance of the residual is increasing with $\hat{Y}$. This has a serious effect on the validity of the hypothesis tests developed in this chapter, which determine the strength of the regression model and the individual predictors.

When you encounter a violation of this type, you need to resort to more advanced modeling techniques, including weighted least squares or transformations of your dependent variable.*

Assumption 3   *The errors are independent.*

To examine these assumptions after the regression equation has been determined involves using the residual from each of the sample observations. For

* See J. Neter, W. Wasserman, and M. Kutner. *Applied Linear Regression Models* (Homewood, IL: Richard D. Irwin, 1983).

Figure 15.18

Examination of the residuals. (A) The shotgun effect (no violation of assumptions I and 2). (B) A violation of the equal variance assumption (assumption 2).

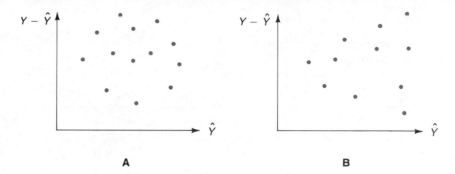

given values of $X_1, X_2, \ldots, X_k$, the actual error is

$$e = Y - (\beta_0 + \beta_1 X_1 + \beta_2 X_2 + \cdots + \beta_k X_k)$$

The β's are unknown, so we estimate the error by using the residual for this particular observation,

$$Y - \hat{Y} = Y - (b_0 + b_1 X_1 + b_2 X_2 + \cdots + b_k X_k)$$

The residuals for the university data are shown in the third column of the table, labeled $Y - \hat{Y}$, on page 830. In general, a close examination of these values will reveal any departures from the regression assumptions.

When your regression data consist of data collected over time (called time series data), your errors often are not independent. This type of data has the appearance

Time	Y	X_1	X_2 $\cdots$	X_k
1978	×	×	×	×
1979	×	×	×	×
1980	×	×	×	×
⋮	⋮	⋮	⋮	⋮

Also remember that the error component includes the effect of missing variables in your model. In many applications, there is a positive relationship between time-related predictor variables, such as population and wages, because they both increase over time. This can produce a set of residuals in your regression analysis that are not independent of one another but, instead, display a pattern similar to that illustrated in Figure 15.19. This plot contains the sample residuals on the vertical axis and time on the horizontal axis. If this assumption were not violated here, we should observe the shotgun appearance. Instead we notice that adjacent residuals have roughly the same value, and so are correlated with each other. This is **autocorrelation**. To be more specific, the pattern in Figure 15.19 is one of positive autocorrelation. Negative autocorrelation exists when most of the neighboring residuals are very unequal in size.

The amount of autocorrelation that exists in your residuals is measured by the **Durbin–Watson statistic**. It ranges from zero to four, with a value near

Figure 15.19

Autocorrelated
errors

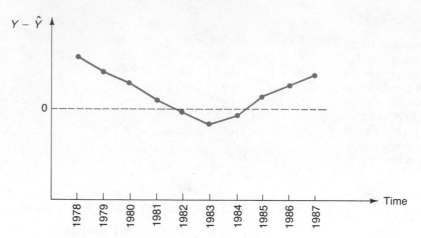

zero indicating strong positive autocorrelation and a value close to four meaning that there is a significant negative autocorrelation. A value near two indicates that there is no (or very little) autocorrelation—the ideal situation. The textbook by Neter, Wasserman and Kutner footnoted in this section discusses the calculation of this statistic and its use in detecting autocorrelated residuals.

Prediction Using Multiple Regression

Once a regression equation has been derived, its primary application generally is to derive predicted values of the dependent variable. Computer packages provide an easy method of deriving such an estimate. To illustrate, consider the regression equation we developed for the university data. For this illustration, we will include X_2 = quantitative score on the GRE although, as we demonstrated in example 15.2, this variable could be dropped from the model without any significant loss in the prediction of graduate GPA. The resulting prediction equation was

$$\hat{Y} = -.395 + .0031X_1 + .00059X_2 + .4457X_3$$

Consider a situation in which

$$X_1 = \text{verbal GRE score} = 520$$

$$X_2 = \text{quantitative GRE score} = 600$$

$$X_3 = \text{undergraduate GPA} = 3.55$$

The predicted graduate GPA (Y) here is

$$\hat{Y} = -.395 + .0031(520) + .00059(600) + .4457(3.55)$$
$$= 3.15$$

We can derive this predicted value using MINITAB, SAS, or SPSS. When using the latter two packages, we can derive a predicted value by adding one additional row to our input data containing these specific values of the predictor variables and a missing value for the dependent variable value. The computer routine will ignore this row when deriving the regression equation (and all subsequent tests of hypotheses) but will attach this row when listing the predicted values. The procedure when using MINITAB is slightly different, as described below.

Using MINITAB The MINITAB solution for the preceding illustration is shown in Figure 15.20. By using the subcommand PREDICT following the

Figure 15.20

Prediction for new data using MINITAB. For the input data, see Figure 15.2.

```
MTB > BRIEF 3
MTB > REGRESS Y IN C1 USING 3 PREDICTORS IN C2-C4;
SUBC> PREDICT FOR 520 600 3.55.

The regression equation is
C1 = - 0.395 + 0.00310 C2 +0.000590 C3 + 0.446 C4

Predictor        Coef        Stdev      t-ratio
Constant       -0.3950       0.5813       -0.68
C2            0.0031028    0.0007987       3.88
C3            0.0005900    0.0006699       0.88
C4              0.4457       0.1764        2.53

s = 0.1594     R-sq - 89.4%    R-sq(adj) = 84.0%

Analysis of Variance

SOURCE        DF         SS          MS
Regression     3      1.27940     0.42647
Error          6      0.15245     0.02541
Total          9      1.43185

SOURCE        DF       SEQ SS
C2             1      1.02131
C3             1      0.09582
C4             1      0.16228

Continue? YES
Obs.      C2         C1      Fit  Stdev.Fit   Residual  St.Resid
  1      580      3.5400   3.5321   0.1119     0.0079      0.07
  2      500      2.6200   2.7359   0.1216    -0.1159     -1.12
  3      670      3.3000   3.4346   0.0887    -0.1346     -1.02
  4      480      2.9000   2.8765   0.1302     0.0235      0.26
  5      710      4.0000   3.7843   0.0915     0.2157      1.65
  6      550      3.2100   3.2430   0.0815    -0.0330     -0.24
  7      640      3.5700   3.5683   0.0835     0.0017      0.01
  8      540      3.0500   2.8190   0.0967     0.2310      1.82
  9      620      3.1500   3.2486   0.0885    -0.0986     -0.74
 10      690      3.6100   3.7078   0.1017    -0.0978     -0.80

     Fit  Stdev.Fit        95% C.I.            95% P.I.
  3.1548    0.1060    ( 2.8954, 3.4142)   ( 2.6863, 3.6233)
```

$\hat{Y}$ using $X_1 = 520, X_2 = 600, X_3 = 3.55$

REGRESS command, you can easily derive a predicted value for the input values of the independent variables. The output will contain the predicted (fitted) Y value, the standard deviation of the predicted Y, a 95% confidence interval, and a 95% prediction interval. The resulting predicted value for $X_1 = 520$, $X_2 = 600$, and $X_3 = 3.55$ is once again $\hat{Y} = 3.15$.

Using SPSS A solution for this illustration using SPSS is presented at the end of the chapter. A numeric value (such as -9) is used for the missing Y value, and then SPSS is informed that such a value represents a missing value using the MISSING VALUES command. This command instructs SPSS that a value of -9 appearing as a value for GPA (the Y variable) should be interpreted as a missing value. The resulting predicted value of $\hat{Y} = 3.15$ appears at the end of the list of predicted values.

Using SAS Similar to SPSS, SAS predicts values by including an additional row containing a missing value for the dependent variable. Any single letter (A to Z) can be used to represent the missing value; SAS is informed that such a character represents a missing value by using the MISSING statement after the DATA statement. This row is then automatically ignored during subsequent calculations, but, once again, the predicted value of $\hat{Y} = 3.15$ is generated for this set of X values at the end of the list of predicted values.

Confidence and Prediction Intervals

In the preceding illustration, what does $\hat{Y} = 3.15$ estimate? For ease of notation, let X_0 represent the set of X values used for this estimate; that is, $X_0 = (520, 600, 3.55)$, where $X_1 = 520$, $X_2 = 600$, and $X_3 = 3.55$. This value of $\hat{Y}$ estimates (1) the *average* GPA of all graduate students at Brookhaven with this specific set of X values, written $\mu_{Y|X_0}$ and (2) the GPA for an individual graduate student having this specific set of X values, written Y_{X_0}.

Using the notation from Chapter 14, let

$$s_{\hat{Y}} = \text{the standard deviation of the predicted } Y \text{ mean}$$

These values can be computed and included in the output by each of the computer packages. To determine the reliability of this particular point estimate $\hat{Y}$, you can (1) derive a confidence interval for $\mu_{Y|X_0}$ if your intent is to estimate the average value of Y given X_0 (not the usual situation) or (2) derive a prediction interval for Y_{X_0} if the purpose is to forecast an individual value of Y given this specific set of values for the predictor variables.

These intervals are summarized as follows. A $(1 - \alpha) \cdot 100\%$ confidence interval for $\mu_{Y|X_0}$ is

$$\hat{Y} - t_{\alpha/2, n-k-1} s_{\hat{Y}} \quad \text{to} \quad \hat{Y} + t_{\alpha/2, n-k-1} s_{\hat{Y}} \qquad \textbf{(15-18)}$$

A $(1 - \alpha) \cdot 100\%$ prediction interval for Y_{X_0} is

$$\hat{Y} - t_{\alpha/2, n-k-1}\sqrt{s^2 + s_{\hat{Y}}^2} \quad \text{to} \quad \hat{Y} + t_{\alpha/2, n-k-1}\sqrt{s^2 + s_{\hat{Y}}^2} \quad \textbf{(15-19)}$$

where $s^2 = MSE$.

Using MINITAB [$X_0 = (520, 600, 3.55)$] Prediction and confidence intervals are easy to derive using MINITAB by using the PREDICT subcommand as illustrated in Figure 15.20. The resulting confidence interval (rounded to two decimal places) for the average graduate student GPA, given X_0, is (2.90, 3.41). As usual, the 95% prediction interval for an individual student with this set of X values, namely (2.69, 3.62), is wider than the corresponding CI.

Using SPSS [$X_0 = (520, 600, 3.55)$] These intervals are not directly available on this package but can be obtained easily using equations 15-18 and 15-19 and the regression output. SPSS provides the estimated value $\hat{Y}$ and the standard deviation of the predicted value $s_{\hat{Y}}$. Referring to the discussion at the end of the chapter, these values are $\hat{Y} = 3.1548$ and $s_{\hat{Y}} = .1060$.

Using equation 15-18, a 95% confidence interval for $\mu_{Y|X_0}$ is derived by first using Table A-5 to obtain $t_{\alpha/2, n-k-1} = t_{.025, 10-3-1} = t_{.025, 6} = 2.447$. The resulting confidence interval is

$$3.1548 - (2.447)(.1060) \quad \text{to} \quad 3.1548 + (2.447)(.1060)$$
$$= 3.1548 - .259 \quad \text{to} \quad 3.1548 + .259$$
$$= 2.90 \quad \text{to} \quad 3.41$$

Consequently, we have estimated the average graduate GPA for students with $X_1 = 520$, $X_2 = 600$, and $X_3 = 3.55$ to within .259 of the actual mean with 95% confidence.

The prediction interval from equation 15-19 is derived by using $MSE = .02541$ from the computer solution to obtain

$$3.1548 - 2.447\sqrt{.02541 + (.1060)^2} \quad \text{to} \quad 3.1548 + 2.447\sqrt{.02541 + (.1060)^2}$$
$$= 3.1548 - .468 \quad \text{to} \quad 3.1548 + .468$$
$$= 2.69 \quad \text{to} \quad 3.62$$

This means that we have predicted the first year graduate GPA of an individual student with $X_1 = 520$, $X_2 = 600$, and $X_3 = 3.55$ to within .468 of the actual value with 95% confidence. Once again, this result agrees to two decimal places with that obtained using MINITAB.

Using SAS [$X_0 = (520, 600, 3.55)$] SAS provides an easy method of determining these intervals by including CLI (for an individual Y_{X_0}) or CLM (for a mean $\mu_{Y|X_0}$) or both in the final SAS statement. According to the output at the end of the chapter, the 95% confidence interval for the average GPA, given X_0, is (2.90, 3.41), with a corresponding prediction interval of (2.69, 3.62).

EXERCISES

15.39 The mortality–pollution model described in exercise 15.1 is to be extended by the inclusion of two qualitative factors, residential location and sex. Residential location can be one of three categories: Urban, Suburban, and Rural. Sex, of course, is either Male or Female.

(a) How many extra variables are needed to code the qualitative data, using dummy variables?

(b) Express the complete new model in general form, defining what X_1, X_2, X_3, and so on, mean.

15.40 You may recall having previously encountered the work of Georgotas *et al.* (1987) in relation to platelet monoamine oxidase (MAO) activity, which is believed to have possible usefulness as a biological marker of psychiatric disorders. Patients were treated with an antidepressant, either nortriptyline or phenelzine, the latter being an MAO inhibitor. Consider the following data for 14 patients who did not respond well to the antidepressant treatment, where baseline MAO activity is in nmol product/hr/mg protein and Hamilton scores are a measure of depression.

Patient	Age (Years)	Sex	Treatment	Baseline MAO Activity	Hamilton Score Baseline	Hamilton Score Final
1	72	F	Nortriptyline	45.3	28	25
2	67	M	Phenelzine	39.9	22	19
3	61	F	Nortriptyline	3.5	34	24
4	50	F	Phenelzine	41.2	15	22
5	76	F	Nortriptyline	36.9	26	19
6	63	M	Nortriptyline	43.8	26	15
7	68	F	Phenelzine	45.7	19	18
8	65	F	Phenelzine	37.3	22	11
9	63	F	Phenelzine	28.3	32	11
10	66	F	Nortriptyline	17.5	24	19
11	56	F	Phenelzine	18.7	19	19
12	62	F	Nortriptyline	21.3	19	13
13	67	M	Nortriptyline	4.4	31	25
14	59	F	Phenelzine	48.9	20	20

Source: A. Georgotas, R. E. McCue, E. Friedman, and T. Cooper, Prediction of response to nortriptyline and phenelzine by platelet MAO activity. *Am J. Psychiatry* **144** (3): 338–340 (March 1987).

(a) Identify the qualitative and the quantitative variables.

(b) If the final Hamilton score is the dependent variable, and age, sex, treatment, baseline MAO activity, and baseline Hamilton score are the predictors, state the general form of the multiple regression model.

(c) Use a computer package to obtain the multiple regression equation.

(d) Test the significance of the model at $\alpha = .01$, and report the p value for the test.

(e) Is sex a significant predictor in this model, at $\alpha = .01$?

(f) Obtain a simple regression model, using only baseline MAO activity to predict the final Hamilton score. Call this the reduced model.

(g) Use the partial F test at 1% significance to determine which of the two models is "better," the complete multiple regression model or the reduced simple regression model.

15.41 Limnological models are mathematical models used for processing quantitative and qualitative data derived from physical, chemical, and biological aspects of lakes and ponds. An example is a study conducted at the Lake St. George Field Station in Ontario, Canada, by Knoechel and Holtby (1986). The aim of the study was to develop a model for estimating filtering rates of crustacean zooplankton to better understand energy flow and plankton dynamics in aquatic ecosystems. Filtering rates may be influenced by factors like species identity, animal size, food particle size, food concentration, and temperature. The traditional assumption had been that species identity was the primary determinant of filtering capability, but the above researchers proposed a model in which body size was the fundamental predictor variable. Body size was measured as the animal's body length in millimeters, and filtering rate was in ml per animal per day. The researchers did a logarithmic transformation on the data, i.e., $Y = $ log filtering rate and $X = $ log length. Stepwise multiple regression gave the following results, where species identity is coded as dummy variables to represent the lake's dominant cladoceran species (*Eubosmina*, *Chydorus*, *Bosmina*, *Ceriodaphnia* and *Daphnia*), and day and night are coded as $0 = $ day, $1 = $ night.

Variable	B	r^2	Δr^2	t
		Full model		
Log length	1.828	.930	.930	28.6
Eubosmina	−.796	.941	.011	−12.8
Chydorus	.756	.951	.010	−12.1
Bosmina	−.620	.961	.010	−10.6
Ceriodaphnia	−.327	.968	.007	−6.9
Constant	1.2011			77.7
		Without species		
Log length	2.5722	.930	.930	48.2
Temp.	.0186	.936	.006	4.0
Day/night	−.0770	.938	.002	−2.3
Constant	.7926			7.8

Note: "Stepwise multiple regression models of log filtering rate as a function of log length, temperature, and dummy variables for each species and for day/night effects. Regression coefficients (B), explained variance (r^2), stepwise increase in explained variance (Δr^2), t statistics, listed in order of entry for all statistically significant ($p < .05$) variables. Results are presented for both a full model and for a submodel with species dummy variables removed."

Source: R. Knoechel and L. B. Holtby, Construction and validation of a body-length-based model for the prediction of cladoceran community filtering rates. *Limnology and Oceanography* **31** (1): 1–16 (January 1986).

(a)   State the regression equation for the full model, with species identity variables. Define your variables.

(b)   State the regression equation for the submodel, without species identity but including the environmental variables temperature and day/night.

(c)   What percentage of the variation in (log) filtering rates is accounted for by (log) body length alone?

(d)   What is the increase in explained variation obtained by the inclusion of the species identity variables?

(e)   What is the increase in explained variation obtained by the inclusion of temperature and day/night variables?

(f)   Although the increase in explained variation in the above two cases is statistically significant, it is quantitatively rather a minor improvement. Hence, the authors finally settled for the simple model, $\hat{Y} = 1.068 + 2.480X$, where Y is the log filtering rate and X is the log body length, with an R^2 of 93% for $n = 213$. Is this model significant at a 1% significance level?

(g)   Why does the regression coefficient of log length turn out different each time, first 1.828, then 2.5722, and finally 2.480? Should it remain the same?

15.42   A number of researchers have helped to develop Fishbein's "theory of behavioral intentions," various versions of which have been quite useful in increasing an understanding of behaviors relating to social health, such as blood donations, birth control pills usage, and participation in alcoholic treatment programs. Pomazal and Brown (1977) applied this theory to the issue of drug use motivation, specifically, to marijuana smoking. According to the theory of behavioral intention, a person's intention to engage in a behavior is determined by three major factors: (1) personal attitude toward the act (Aact), (2) perceived social norms regarding the behavior (SN), and (3) personal moral norms regarding the behavior (MN). From a sample of 101 university students, the following correlations were obtained:

Variable	Aact (X_1)	SN (X_2)	MN (X_3)	BI (Y)
Aact (X_1)	—			
SN (X_2)	0.46	—		
MN (X_3)	0.39	0.26	—	
BI (Y)	0.75	0.38	0.49	—

Source: R. J. Pomazal and J. D. Brown, Understanding drug use motivation: A new look at a current problem. Health Social Behavior **18**: 212–222 (June 1977), Table 2.

(a)   Which independent variable, Aact, SN, or MN, would be selected first in a forward regression procedure? Why?

(b)   If only one independent variable could be chosen as a predictor of BI, which one would it be? Why?

(c)   The authors express the theory in algebraic form as

$$BI = [Aact]w_1 + [SN]w_2 + [MN]w_3$$

where w_1, w_2, and w_3 are the weights reflecting the importance of each component in determining the behavioral intention. In the context of regression, what are these weights known as, and what do they represent?

(d) For the model in (c) above, the authors reported $R^2 = .6084$. Is the model significant at $\alpha = .01$? Report the p value.

15.43 Agoraphobia is an abnormal fear of open spaces and might be considered to be the "opposite" of claustrophobia, which is an abnormal fear of closed spaces. Anxiety sensitivity is quite common in agoraphobia. In a study designed to investigate the validity of the "Reiss-Epstein-Gursky Anxiety Sensitivity Index" (ASI), McNally and Lorenz (1987) obtained scores on the ASI, as well as a number of other instruments, from 48 agoraphobic patients. The correlations among these measures of psychopathology are given in the correlation matrix table below, where correlations in parentheses are nonsignificant ($p > .05$), the rest being significant at $p < .05$ through $p < .001$. ASI = Anxiety Sensitivity Index, BSQ = Body Sensations Questionnaire, ACQ = Agoraphobic Cognitions Questionnaire, FQ-T = Fear Questionnaire Total score, FQ-Ag = Fear Questionnaire—Agoraphobia Subscale, a measure of agoraphobic avoidance, BDI = Beck Depression Inventory, FSS-II = Fear Survey Schedule-II, and TMAS = "Taylor Manifest Anxiety Scale, a general measure of anxiety.

CORRELATIONS AMONG MEASURES OF PSYCHOPATHOLOGY

	BSQ	ACQ	FQ-T	FQ-Ag	BDI	FSS-II	TMAS
ASI	0.64	0.66	0.47	(0.20)	0.63	0.65	0.46
BSQ		0.63	0.30	(0.16)	0.44	0.38	0.30
ACQ			0.33	(0.17)	0.48	0.42	0.33
FQ-T				0.79	0.46	0.64	0.38
FQ-Ag					0.22	0.48	(0.10)
BDI						0.68	0.81
FSS-II							0.53

Source: R. J. McNally and M. Lorenz, Anxiety sensitivity in agoraphobics. *J. Behavior Therapy Exp. Psychiatry* **18** (1): 3–11 (March 1987).

(a) In a stepwise regression procedure in which FSS-II was the dependent variable and ASI, ACQ, BSQ, and TMAS were the independent variables, which predictor would be selected first?

(b) To validate the ASI, the authors had to show that it was not just another measure of anxiety itself. To quote: "The scale must measure the *fear* of anxiety symptoms, rather than the *frequency* or *intensity* of these symptoms." The authors argued that the ASI should explain *additional* variation on a dependent measure of fearfulness (namely, FSS-II) over and above the variation explained by some measure of general anxiety (namely, TMAS). Regression of FSS-II as the dependent variable using TMAS resulted in $R^2 = 25.8\%$, with a computed F value of 10.76 and $df = (1, 31)$ for this model. Introduction of ASI into the model boosted the explained variation by 19.0%, giving $R^2 = 44.8\%$, with computed $F = 12.14$ and $df = (2, 30)$ for this model. Use the partial F test to decide if these results support the validity of the ASI, i.e., is the increase in R^2 significant? Use a 1% significance level.

(c) When FSS-II was regressed using only ASI, an R^2 of 41.3% was obtained, with computed $F = 21.80$, $df = (1, 31)$. When the stepwise procedure was continued, only the ASI remained; TMAS, ACQ, and BSQ did not contribute additional explanatory power beyond that provided by ASI, and were therefore not included. Since ASI, ACQ and BSQ were presumed to be measuring the same construct, thereby duplicating information, does this suggest the presence of multicollinearity?

15.44 The least squares regression equation

$$\hat{Y} = 1.5 + 3.5X_1 + 7.5X_2 - 150X_3$$

has the following t values for the independent variables:

Null hypothesis	t statistic
$\beta_1 = 0$	4.5
$\beta_2 = 0$	1.89
$\beta_3 = 0$	1.52

Twenty observations were used in calculating the least squares equation. In the first stage of a backward-selection procedure, which independent variable would be eliminated first? Use a 5% level of significance.

15.45 Describe the main difference between the forward-selection procedure and the stepwise-selection procedure in regression analysis.

15.46 If a statistician would like to include dummy variables to indicate one of three cities and also one of four salespeople in a regression model, how many dummy variables would be needed in the model?

15.47 Which of the standard assumptions of regression appear to have been violated from the following table, which lists the dependent variable and the residual values?

Y	$Y - \hat{Y}$
1.5	0.12
2.1	−.70
3.5	−.91
4.0	1.02
4.5	−1.18
5.0	−1.45
5.5	1.61
6.0	1.79
7.0	2.40
7.0	2.10

15.48 How should a plot of the residuals $(Y - \hat{Y})$ plotted against the predicted values $\hat{Y}$ look if the standard assumptions of regression are satisfied?

15.49 In exercise 15.3, you obtained a point estimate for infant mortality when the literacy rate was 85.0, the population density was 150 per sq km and there were 875 inhabitants per physician. Continue the analysis using those same values for the predictor variables:

(a) Obtain the 95% confidence interval for the average infant mortality at locations with $X_0 = (85.0, 150, 875)$.

(b) Obtain the 95% prediction interval for infant mortality at an individual location where $X_0 = (85.0, 150, 875)$.

(c) Compare the answers for (a) and (b), and comment on what you notice about them.

15.50 A criminologist and a psychologist teamed up to study the behavior of juries in criminal cases. One aspect that they investigated was the possibility of latent racial discrimination in jury verdicts. They considered a model of the form

$$Y = \beta_0 + \beta_1 X_1 + \beta_2 X_2 + \beta_3 X_3$$

where

Y = percentage of cases resulting in guilty verdicts

X_1 = (1 if the jury is all-white)
       (0 otherwise)

X_2 = (1 if the defendant is black)
       (0 otherwise)

X_3 = (1 if the victim of the crime is white)
       (0 otherwise)

Suppose that the model obtained from 124 cases analyzed is:

$$\hat{Y} = 34.3 + 8.6X_1 + 14.4X_2 + 3.6X_3$$

(a) Interpret the meaning of the coefficients b_1, b_2, and b_3. (*Hint:* Plug in 0 or 1 into the equation and see what happens to $\hat{Y}$.)

(b) What is the average conviction rate for cases where a mixed jury tries a nonblack defendant and the victim is white?

(c) What is the average conviction rate for cases where an all-white jury tries a black defendant and the victim is nonwhite?

15.51 Explain the difference between a CI for the mean value of Y at particular values of the independent variables and a prediction interval for a future value of Y at particular values of the independent variables. Will the prediction interval for Y always be larger than the corresponding CI for particular values of the independent variables?

15.52 A researcher at the Institute for Human Fitness is interested in how a runner's present body fat, X_1 (in percent), and a runner's resting pulse, X_2 (in beats per minute), determine the time Y (in minutes) that it takes the runner to finish 10,000 meters. A random sample of 18 runners was selected and the following data was collected.

Y	X_1	X_2	Y	X_1	X_2
38.0	10.4	56	35.4	9.6	53
40.2	12.4	51	30.7	8.1	43
34.3	9.6	47	31.2	7.6	39
32.7	8.3	50	33.7	6.5	53
37.8	13.6	47	36.3	10.8	50
39.1	11.5	58	37.4	12.6	49
32.5	6.8	42	34.1	10.4	52
32.3	5.7	50	33.8	9.6	54
30.2	8.1	42	32.4	8.2	47

Use a computerized statistical package to answer the following questions.

(a) Find a 90% confidence interval for the mean value of Y at $X_1 = 9.0$ and $X_2 = 50$.

(b) Interpret the confidence interval in question (a) in the context of this exercise.

(c) Find a 90% prediction interval for the value of Y at $X_1 = 9.0$ and $X_2 = 50$.

(d) Interpret the prediction interval in question (c) in the context of this exercise.

15.53 The operations manager in charge of a production process is interested in the amount of time in minutes, Y, that it takes an assembly-line worker to perform a certain task relative to his or her score, X, on an aptitude test. The proposed model is

$$Y = \beta_0 + \beta_1 X + \beta_2 X^2 + e$$

Twelve assembly line workers were randomly selected with the following results:

Y	X
49	58
37	67
12	95
60	43
33	72
22	83
19	85
17	89
50	52
67	41
39	67
35	70

(a) Using a computerized statistical package and the proposed model, construct a 99% confidence interval for the mean time that it takes assembly-line workers to complete the task if the aptitude score is 80.

(b) Using a computerized statistical package and the proposed model, construct a 99% prediction interval for the time it takes a worker to complete the task if the aptitude score of that worker is 80.

(c) Compare the answers in questions (a) and (b).

15.54 Refer to the computer printout given in exercise 15.19. Suppose that at $X_1 = 20$ and $X_2 = 10$, the standard deviation of $\hat{Y}$ is 1.85.

(a) Construct the 95% confidence interval for the average value of Y at $X_1 = 20$ and $X_2 = 10$.

(b) Construct the 95% prediction interval for Y at $X_1 = 20$ and $X_2 = 10$.

15.55 Refer to exercise 15.1, part (b). It is a reasonable assumption, and one that could be supported by empirical observation, that the high population density areas are the ones which will have higher levels of air and water pollution. Thus, there is likely to be a fair amount of correlation among X_1, X_2, and X_3.

(a) If so, what criticism could you make of question 15.1 (b)?

(b) If the correlation between population density and pollution were particularly strong, what problem might arise in the model?

(c) How could such a problem be avoided? Would stepwise regression guarantee that the problem mentioned in (b) will be avoided, or would the stepwise procedure help to minimize the "risk" of this problem?

SUMMARY

Multiple linear regression offers you a method of predicting (or modeling) the behavior of a particular **dependent** variable (Y) using two or more **independent (predictor)** variables. As in the case of simple linear regression, which uses one predictor variable, the regression coefficients are those that minimize

$$\text{SSE} = \text{sum of squares of error}$$
$$= \Sigma(Y - \hat{Y})^2$$

To use this technique properly, you must pay special attention to the assumptions behind it. These are that (1) the regression errors follow a normal distribution, centered at zero, with a common variance, and (2) the errors are statistically independent. An estimate of this common variance is

$$\hat{\sigma}_e^2 = s^2 = \text{MSE} = \frac{\text{SSE}}{n - k - 1}$$

To determine the adequacy of the regression model, you can test the entire set of predictor variables using an F test with k and $n - k - 1$ degrees of freedom,

$$F = \frac{\text{MSR}}{\text{MSE}} = \frac{R^2/k}{(1 - R^2)/(n - k - 1)}$$

The contribution of an individual predictor variable (say, X_i) can be tested using a t statistic with $n - k - 1$ degrees of freedom:

$$t = \frac{b_i}{s_{b_i}}$$

where s_{b_i} represents the estimated standard deviation of b_i. Here, b_i is the least squares estimate of the population parameter β_i and centers the confidence interval for this parameter.

The **coefficient of determination, R^2**, describes the percentage of the total variation in the sample Y values explained by this set of predictor variables. To determine the contribution of a particular subset of the predictor variables— say, X_2 and X_4—R^2 is computed with X_2 and X_4 included and then with X_2 and X_4 excluded from the regression equation. A **partial F test** is then used to determine whether the resulting decrease in R^2 is significant.

When a **curvilinear** pattern exists between two variables, X and Y, this nonlinear relationship often can be modeled by including an X^2 term in the regression equation. The resulting equation is

$$\hat{Y} = b_0 + b_1 X + b_2 X^2$$

This type of model often works well in situations where Y (for example, performance on an exam) appears to increase more slowly as the independent variable X (for example, the amount of time spent preparing for the exam) continues to increase.

The problem of **multicollinearity** arises in the application of multiple linear regression when two or more independent variables are highly correlated.

The resulting regression equation contains coefficients that are highly inflated (have a large variance) with t statistics that are extremely small, despite the fact that one or more of these seemingly insignificant variables are very useful predictors. An easy means of correcting this problem is to remove certain variables from the regression equation or to use a stepwise procedure.

Stepwise techniques allow you to insert variables one at a time into the equation (**forward regression**), remove them one at a time after initially including all variables in the equation (**backward regression**), or perform a combination of the two by inserting variables one at a time but removing a variable that has become redundant at any stage (**stepwise regression**). Once the variables for the model have been selected, **residual plots** should be obtained to examine the underlying assumptions that are necessary in a regression analysis.

Dummy variables can be used in a regression application to represent the categories of a qualitative variable (such as city). If all dummy variables are to be inserted into the equation, then $C - 1$ such variables should be defined to represent C categories. If a forward or stepwise selection procedure is used to define the final regression equation, then a better procedure is to define C dummy variables to represent this situation.

Use of a computer package is essential in the derivation of a multiple regression equation. In this chapter, we used MINITAB, SPSS, and SAS. They provide the sampling coefficients (b_0, b_1, b_2, . . .), the statistics necessary to perform any test of hypothesis, and those needed for the prediction and confidence intervals for any specific set of predictor variable values. A **confidence interval** is derived whenever the predicted Y value is used to estimate the average value of the dependent variable for a specific set of X values. When the predicted Y value is used to estimate an individual value of Y for a specific set of X values, a **prediction interval** can be used to place bounds on the actual Y value.

Formulas Used in Multiple Linear Regression

H_0: all β's $= 0$

$$F = \frac{\text{MSR}}{\text{MSE}}$$

H_a: at least one $\beta \neq 0$

$$= \frac{R^2/k}{(1 - R^2)/(n - k - 1)}$$

$(df = k$ and $n - k - 1)$

H_0: $\beta_i = 0$

H_a: $\beta_i \neq 0$

$$t = \frac{b_i}{s_{b_i}}$$

(or H_a: $\beta_i > 0$)

(or H_a: $\beta_i < 0$)

$(df = n - k - 1)$

Coefficient of determination

$$R^2 = 1 - (\text{SSE/SST})$$

where

$$SST = \Sigma(Y - \bar{Y})^2$$
$$= \Sigma Y^2 - (\Sigma Y)^2/n$$

and

$$SSE = \Sigma(Y - \hat{Y})^2$$

H_0: $X_i, X_{i+1}, \ldots, X_j$ do not contribute

$$F = \frac{(R_c^2 - R_r^2)/v_1}{(1 - R_c^2)/v_2}$$

H_a: at least one of them contributes

where (1) R_c^2 is the R^2 including the variables in H_0 (the complete model), (2) R_r^2 is the R^2 excluding the variables in H_0 (the reduced model), (3) v_1 = the number of β's in H_0, (4) $v_2 = n - 1 -$ (the number of X's in the complete model), and (5) the degrees of freedom for the F statistic are v_1 and v_2.

REVIEW EXERCISES

15.56 A company has opened several ice-skating rinks and would like to know what factors affect the attendance at the rinks. The manager believes that the following variables affect attendance:

$$X_1 = \text{temperature (forecasted high)}$$

$$X_2 = \text{wind speed (forecasted high)}$$

$$X_3 = 1 \text{ if weekend and 0 otherwise}$$

$$X_4 = X_1 X_2$$

The following least squares model was found from 30 days of data:

$$\hat{Y} = 250 + 4.8X_1 - 30X_2 + 1.3X_3 + 35X_4$$

(a) What is the predicted attendance on a weekend if the forecasted high temperature is 28°F and the forecasted high wind speed is 12 mph?

(b) If the coefficient of determination for the model is 0.67, test that the overall model contributes to predicting the attendance at the ice-skating rinks. Use a 5% significance level.

(c) If the standard deviation of the estimate of the coefficient of X_2 is 2.01, does the variable wind speed contribute to predicting the variation in attendance, assuming that the variables X_1, X_3, and X_4 are in the model?

15.57 The manager of Stay Trim Health Studios would like to determine the average number of times per month a member attends the health studio (Y). The following independent variables were used in the analysis:

$$X_1 = \text{weight at initial visit (in pounds)}$$

$$X_2 = X_1^2$$

$$X_3 = \text{age at initial visit}$$

$$X_4 = \text{length of membership (in years)}$$

$$X_5 = \begin{matrix} (1 \text{ if employed}) \\ (0 \text{ if not}) \end{matrix}$$

The manager collected the following data:

Y	X_1	X_3	X_4	X_5
11	202	30	1	1
9	180	22	2	1
7	130	19	1	0
14	175	32	4	1
12	225	41	2	1
19	191	52	5	1
7	142	40	1	1
11	208	33	2	1
13	245	35	1	0
15	215	24	3	0
11	185	43	2	1
12	165	27	3	1
12	195	38	1	0
11	217	42	1	1
10	205	40	1	1

The least squares equation was found to be:

$$\hat{Y} = -11.218 + 0.15178X_1 - 0.0003X_2 + 0.08286X_3 + 1.9138X_4 - 2.299X_5$$

(a) Does the overall model contribute significantly to predicting the monthly attendance at Stay Trim Health Studios? Use a 10% significance level.

(b) Does weight squared contribute significantly to predicting the monthly attendance, assuming that the variables X_1, X_3, X_4, and X_5 are in the model? Use a 10% significance level.

(c) Find a 95% confidence interval for the coefficient of age.

(d) Find a 95% confidence interval for the coefficient of length of membership.

(e) Use the model to predict the monthly attendance of a 35-year-old member who weighs 200 lb, has a 2-year membership, and is currently employed.

(f) Construct a histogram for the residuals of the complete model.

15.58 Complete the following ANOVA table for testing whether a model with five independent variables contributes significantly to the prediction of the dependent variable:

Source	df	SS	MS	F
Regression		95.6		
Error	20	159.0		
Total				

15.59 Given below is selected information from a computer printout of a multiple regression analysis:

Analysis of Variance

	df	Sum of Squares	Mean Squares	F Ratio
Regression	2	86.091	43.0455	8.649
Error	17	84.609	4.9770	

Variable	Coefficient	St. Dev. Coeff.	T Ratio
Intercept	.10255		
X_1	.68808	.195	3.529
X_2	.39844	.295	1.351

Answer the following questions.
(a) Write down the multiple regression equation.
(b) What percentage of variation in Y is explained by the model?
(c) What is the size of the sample (n) used in the above regression analysis?
(d) Find the value of
    (i) The total sum of squares
    (ii) The error sum of squares
    (iii) The estimated variance of the error in the model
(e) Is the model as a whole significant at $\alpha = .05$?
(f) Does X_1 contribute to this model? Use $\alpha = .05$.
(g) Does X_2 contribute to this model? Use $\alpha = .05$.
(h) Suppose that at $X_1 = 15$ and $X_2 = 20$, the standard deviation of $\hat{Y}$ is 2.65. Construct the 95% confidence interval for the average value of Y at $X_0 = (15, 20)$.
(i) Also, construct the 95% prediction interval for Y at $X_0 = (15, 20)$.

15.60 A computerized statistical routine printed the following correlation matrix for a data set of 65 observations collected by an educational researcher.

	Y EA	X_1 IQ	X_2 n-Ach	X_3 Age	X_4 SES
Y	1.00	.82	.59	.70	.33
X_1	.82	1.00	.44	.80	.41
X_2	.59	.44	1.00	.36	.35
X_3	.70	.80	.36	1.00	.06
X_4	.33	.41	.35	.06	1.00

EA = Education Attainment, a composite weighted measure of GPA and standardized test scores.

IQ = Intelligence quotient, a standard measure of mental ability.

n-Ach = need for achievement, a measure of motivation.

Age = chronological age in years.

SES = Socio-Economic Status.

(a)   Interpret the correlation coefficients.

(b)   Which variables should be carefully screened for possible multicollinearity?

(c)   In a forward regression procedure, which variable will be included in the first stage of the procedure?

(d)   If the R^2 for the complete model is .75, is the model as a whole significant? Use $\alpha = .05$.

Source: Modified and adapted from E. J. Pedhazur, *Multiple Regression in Behavioral Research: Explanation and Prediction*, 2d ed. Chapter 5 (New York: CBS College Publishing/Holt Rinehart and Winston, 1982.)

15.61   A psychotherapist found a very high correlation between the number of hours of psychotherapy received by patients, and the suicide rate. Thus, the more psychotherapy one has received, the greater is the likelihood of suicide. The same psychotherapist also claims that psychotherapy probably reduces a depressed person's likelihood of suicide. How could you reconcile these apparently contradictory positions? (*Hint:* Some researchers make a distinction between *predictive* models used for forecasting, and *structural* models used for establishing explanatory cause–effect relationships.)

15.62   A public opinion pollster was studying the attitude of women ages 18 and above toward abortion. The following data are illustrative only, and are based on two factors: religious affiliation and political affiliation. The data represent scores on an attitude scale, with a higher score indicating a more favorable attitude.

Political Affiliation	Religious Affiliation		
	Catholic	Protestant	Other
Democratic	4	5	7
	5	7	8
	5	7	8
	6	9	10
	7	12	12
Republican	5	5	6
	5	5	6
	6	7	7
	6	8	7
	7	8	8
Independent	3	4	2
	4	8	7
	5	9	9
	8	9	9
	9	10	11

(a)   State the general form of a multiple regression model for the above data. Let Y = score on attitude scale for abortion. Use dummy variables to code religious affiliation and political affiliation, and define your variables.

(b) Use a computerized statistical package to run a multiple regression analysis on the above data. Obtain the following for the model:
  (i) The least squares equation
  (ii) The variance of the error in the model
  (iii) The test statistic to determine the validity of the entire model
  (iv) The coefficient of determination for the model

(c) At a 5% significance level, test whether the model as a whole is useful. Are religious and political affiliation significant predictors of attitude toward abortion?

(d) Run another multiple regression on the above data, using only religious affiliation as a predictor, and answer questions (b(i–iv)) and (c) for this model.

(e) Run a third multiple regression using only political affiliation as a predictor and answer the same questions as in (d).

(f) Perform a hypothesis test to determine if political affiliation contributes significantly to the model, given that religious affiliation is already included in the model. Use $\alpha = .05$. (*Hint:* the partial F test is involved.)

(g) Perform a hypothesis test to determine if religious affiliation contributes significantly to the model, given that political affiliation is already included in the model. Use $\alpha = .05$.

(h) Compare the models in (b), (d), and (e) to the kind of models presented in the chapter on ANOVA. Can you see how the model in (b) resembles a two-way (3×3) factorial design with 5 replicates in each cell? Which type of ANOVA design do (d) and (e) resemble?

15.63 Suppose that in an experiment on visual acuity, psychologists used the following variables as predictors of Y, the number of correct responses on a test of visual discrimination:

$$X_1 = \text{amount of practice time allowed before test (in minutes)}$$

$$X_2 = X_1^2$$

$$X_3 = \text{height in inches}$$

$$X_4 = \text{measure of obesity (body fat)}$$

Consider the following data:

Y	X_1	X_3	X_4
31	2	66	11.8
29	2	68	16.8
38	2	64	19.9
40	5	62	7.5
32	5	72	13.7
42	5	70	30.4
31	8	61	12.8
26	8	65	11.8
20	8	67	9.2
18	11	62	6.4
15	11	60	22.1
21	11	64	10.5

(a) Use a computerized statistical package to obtain the regression model

$$Y = \beta_0 + \beta_1 X_1 + \beta_2 X_2 + \beta_3 X_3 + \beta_4 X_4 + e$$

Obtain (i)   The regression equation
      (ii)   The test statistic to determine the usefulness of the entire model
      (iii)   The test statistic to determine if the curvilinear effect is significant, i.e., $H_0: \beta_2 = 0$

(b) Is the model as a whole useful at $\alpha = .05$? State the p value for the test.
(c) Does the variable X_1^2 (i.e., X_2) contribute to this model, given all the others are already included? Use $\alpha = .05$ and state the p value.
(d) What is the predicted value for Y (number of correct responses) for an individual 5.5 feet tall with an obesity measure of 15.0 given 6 minutes of practice time?
(e) What is the 95% prediction interval for the individual mentioned in (d)?
(f) What is the 95% confidence interval for the average number of correct responses for all individuals fitting the description in (d)?

15.64 A realtor wanted to explore the feasibility of using multiple regression analysis in appraising the value of single-family homes within a certain community. The following variables were used:

$$Y = \text{selling price of a house (in dollars)}$$

$$X_1 = \text{total living area (in square feet)}$$

$$X_2 = \begin{cases} 1 \text{ if in neighborhood 1} \\ 0 \text{ if not} \end{cases}$$

$$X_3 = \begin{cases} 1 \text{ if in neighborhood 2} \\ 0 \text{ if not} \end{cases}$$

$$X_4 = \begin{cases} 1 \text{ if lot size is larger than the typical house lot} \\ 0 \text{ if not} \end{cases}$$

The data are as follows:

Y	X_1	X_2	X_3	X_4
63,000	2020	1	0	1
36,000	980	1	0	0
44,000	1230	0	0	1
37,000	980	0	1	0
28,000	640	0	1	0
28,000	720	0	1	0
56,000	2400	1	0	1
28,600	670	0	1	0
31,350	640	0	1	0
49,400	1910	0	0	1
31,000	900	1	0	0
56,000	1890	1	0	0
63,500	1900	0	0	1
49,000	2080	1	0	1
63,000	1900	0	0	1

Using a computerized statistical package, find the following:
(a) The least squares equation.
(b) The 95% confidence interval for the coefficient of total living area.
(c) The 95% prediction interval for selling price given that $X_1 = 1800$, $X_2 = 1$, $X_3 = 0$, and $X_4 = 0$.
(d) The overall F test for the model and the resulting conclusion using a 5% significance level.

15.65 Suppose the following data were obtained by an industrial psychologist investigating job performance and its relationship to some variables.

Y	X_1	X_2	X_3	X_4
54	49	20	15	2
37	18	1	10	1
48	22	1	8	4
61	30	5	15	3
72	25	2	21	3
68	27	2	26	4
59	19	0	10	3
80	38	13	35	5
47	39	14	25	2
69	44	20	18	4
74	36	16	36	3
87	35	11	24	4
35	52	28	30	1
74	31	10	12	2
81	36	12	18	3

In the above table,

$$Y = \text{job performance rating of employee}$$

$$X_1 = \text{age, in years}$$

$$X_2 = \text{number of years of experience}$$

$$X_3 = \text{skill level}$$

$$X_4 = \text{employee's status (from 1 to 5)}$$

(a) Use a computer package to perform a multiple regression analysis on the above data. Obtain the following:
(i)   The regression equation
(ii) The estimated variance of the error in the model
(iii) The coefficient of determination
(iv) The F statistic to test the entire model
(v)   The four t statistics to test the contribution of each predictor, given the other three are included in the model
(b) Is the overall regression significant, at $\alpha = .10$? State the p value for the test.
(c) Is skill level (X_3) a useful predictor in this model?
(d) Perform similar tests as in (c) for age (X_1), experience (X_2), and status (X_4) to determine whether each predictor contributes significantly, given the

others are included in the model. Use $\alpha = .10$, and state the p value in each case.

(e)   Obtain the correlation matrix, and investigate the possibility of multicollinearity in the model.

(f)   Construct the 90% confidence interval for the average performance rating for employees who are 35 years old, with 10 years of experience, a skill level of 22 and a status of 3.

(g)   Construct the 99% prediction interval for an individual employee who fits the description in (f).

CASE STUDY

**Predicting
Exhaust
Emissions of
Diesel Trucks**

In the United States, the Federal government is charged with the responsibility of laying down vehicle exhaust emissions standards. As technology improves and newer engine configurations and vehicle models are introduced, emissions standards have to be updated to reflect these changes. Vehicle emissions standards are, of course, just a small part of the much larger considerations of environmental pollution control, with which agencies such as the Environmental Protection Agency (EPA) are concerned.

For this case study, we will examine data from a study by Hare and Bradow (1977), which was conducted to obtain prediction equations for exhaust emissions of light-duty, diesel-powered trucks. The data set consists of 44 observations made on a single truck as three environmental factors were varied: humidity (HUMID), temperature (TEMP), and the atmospheric or barometric pressure (BPRES). Humidity was measured in grains H_2O/lb_m dry air, temperature in degrees Fahrenheit, and barometric pressure in inches Hg. The specific exhaust emissions measured were those of nitrous oxides, NO and NO_2 in g/km. This is abbreviated in the data set as NOX (which seems quite appropriate considering the noxious nature of the gases.)

The nitrous oxide emissions served as the dependent variable, with humidity, temperature, and barometric pressure as predictors. The idea behind this attempt to construct a model of nitrous oxide emissions was that the prediction equations obtained could be used to take any given observed emissions reading and convert it to a set of humidity, temperature, and pressure levels representing standard environmental conditions.

Case Study Questions

1. Use a computer package to obtain the correlation matrix for all the variables. Which predictor has the strongest relationship with NOX?

2. Although computer packages can automatically carry out stepwise regression procedures, it is instructive to perform a similar process manually. Since we have only three independent variables, the task is quite manageable.

   (a)   Perform a simple regression of NOX using the predictor that is most strongly correlated with NOX. Note the R^2.

   (b)   Next, include the second most strongly correlated predictor and perform another regression. Note the R^2 for this two-predictor model.

Nitrous oxide emissions	No.	NOX	HUMID	TEMP	BPRES
	1	.81	74.9	78.4	29.08
	2	.96	44.6	72.5	29.37
	3	.96	34.3	75.2	29.28
	4	.94	42.4	67.3	29.29
	5	.99	10.1	76.4	29.78
	6	1.11	13.2	67.1	29.40
	7	1.09	17.1	77.8	29.51
	8	.77	73.7	77.4	29.14
	9	1.01	21.5	67.6	29.50
	10	1.03	33.9	77.2	29.36
	11	.96	47.8	86.6	29.35
	12	1.12	21.9	78.1	29.63
	13	1.01	13.1	86.5	29.65
	14	1.10	11.1	88.1	29.65
	15	.86	78.4	78.1	29.43
	16	.85	69.1	76.7	29.28
	17	.70	96.5	78.7	29.08
	18	.79	108.7	87.9	28.98
	19	.95	61.4	68.3	29.35
	20	.85	91.3	70.6	29.03
	21	.79	96.8	71.0	29.05
	22	.77	95.9	76.1	29.04
	23	.76	83.6	78.3	28.87
	24	.79	75.9	69.4	29.07
	25	.77	108.7	75.4	28.99
	26	.82	78.6	85.7	29.02
	27	1.01	33.9	77.3	29.43
	28	.94	49.2	77.3	29.43
	29	.86	75.7	86.4	29.06
	30	.79	128.8	86.8	28.96
	31	.81	82.4	87.1	29.12
	32	.87	122.6	86.2	29.15
	33	.86	124.7	87.2	29.09
	34	.82	120.0	87.5	29.09
	35	.91	139.5	87.7	28.99
	36	.89	105.4	86.1	29.21
	37	.87	90.7	87.0	29.16
	38	.85	142.2	87.1	28.99
	39	.85	136.5	85.0	29.09
	40	.70	138.9	85.9	29.16
	41	.82	89.7	86.7	29.15
	42	.84	92.6	85.3	28.18
	43	.84	147.6	87.3	29.10
	44	.85	141.4	86.3	29.06

Source: C. T. Hare and R. L. Bradow, Light duty diesel emission correction factors for ambient conditions. Paper No. 770717, Society of Automotive Engineers Off-Highway Vehicle Meeting, MECCA Milwaukee, Sept. 12–15, 1977. Submitted by Hare as final report to the Environmental Protection Agency under Contract 68-02-1777, April 1977. Reproduced in R. F. Gunst and R. L. Mason, *Regression Analysis and Its Application: A Data-Oriented Approach* (New York: Dekker, 1980), p. 362. Reprinted from *Regression Analysis and Its Application: A Data-Oriented Approach*, p. 362, by courtesy of Marcel Dekker, Inc.

(c)   Perform a partial F test (or, if you like, a t test) to determine if the inclusion of this variable significantly improved the R^2 for the model. (Use a 5% significance level for all the tests.)

(d)   Include the third predictor, and repeat (c).

(e)   Is this three-predictor model significant?

3.   You will now have a model with all three predictor variables. Assuming that, at each stage, the additional predictor did improve the R^2 significantly, can you be sure that you have the "best" model? Or is it at least theoretically possible to have a two-predictor model that is better than the three-predictor model?

4.   Since there are two more combinations of the predictors (taken in pairs) that have been ignored, go ahead and use them in two more regressions. Most likely, the R^2 for any of these will not be greater than that for the model you obtained in 2(c), but in case it is, you can compare this two-predictor model to 2(d) to decide which model is better. Hopefully, this exercise, tedious though it might seem at first, will give you a better appreciation of the way multiple regression works.

5.   Examine the correlation matrix to detect any potential for multicollinearity. What do you look for?

6.   Obtain a plot of the residuals $(Y - \hat{Y})$ against the predicted Y values $(\hat{Y})$. What do you look for? How would you detect heteroscedasticity (a violation of the assumption of constant variance)?

7.   Print a histogram of the residuals. What shape would you like to see? If it is not this shape, what is the problem?

8.   Suppose you have a situation with humidity $= 90.5$, temperature $= 87.5°F$, and pressure $= 29.10$.

(a)   What is the predicted value (point estimate) of NOX?

(b)   What is the 95% confidence interval for the average nitrous oxide emissions under those environmental conditions?

(c)   If a diesel truck under these conditions gave NOX readings of 1.10, would that be considered to be excessively high emissions?

(d)   Would a prediction interval be more tolerant of NOX emissions than a confidence interval, for a given level of confidence?

COMPUTER EXERCISES USING THE DATABASE

Using a sample of 100 observations, complete the following exercises.

1.   Consider a multiple regression model, where the dependent variable is graduate GPA, and the predictor variables are VGRE = verbal GRE score, QGRE = quantitative GRE score, AGE, and SEX. Using these four predictors, what percentage of the variation in the graduate GPA values has been explained?

2.   Referring to exercise 1, determine the contribution of variables AGE and SEX by testing $H_0: \beta_6 = \beta_7 = 0$ at a significance level of .10.

3.   Set up three dummy variables to represent marital category (say, X_9, X_{10}, and X_{11}). Using the same dependent variable as in exercise 1, determine the value of R^2 with

VGRE, QGRE, X_5, X_6, and X_7 as predictor variables. What is the computed F value used to test H_0: all β's $= 0$? What is your conclusion at a significance level of .10?

4. Examine the residual plot in exercise 3. Do the regression assumptions appear to be satisfied?

5. Perform a stepwise regression using variables X_5, X_6, X_7 (defined in exercise 3) along with variables VGRE, QGRE, AGE, and SEX. The dependent variable is $Y = $ graduate GPA. What model would you suggest here?

6. Define the dependent variable to be $Y = $ score on the Life Satisfaction Index questionnaire. Using variables X_5, X_6, X_7 (defined in exercise 3), along with variables AGE and SEX, perform a stepwise analysis. What is the predicted value of Y for a 35-year-old divorced male?

SPSSX APPENDIX (MAINFRAME VERSION)

Solution to Example 15.2

At the beginning of the chapter, we computed the regression equation for predicting the performance of a first-year graduate student, based on the three predictor variables: verbal score on the GRE, quantitative score on the GRE, and the undergraduate GPA. The SPSSX program listing in Figure 15.21 requests a multiple regression solution.

Figure 15.21

SPSSX program listing used to request a multiple regression analysis of the university data

```
TITLE    UNIVERSITY EXAMPLE USING TWO PREDICTORS
DATA LIST FREE/ GRADGPA VGRE QGRE UNDGPA
BEGIN DATA
0.54 500 720 0.02
2.62 500 660 2.67
3.30 670 580 3.16
2.90 480 520 3.31
4.00 710 630 3.60
3.21 550 690 3.42
3.57 640 700 3.51
3.05 540 530 2.75
3.15 620 490 3.21
3.61 690 530 3.70
END DATA
PRINT / GRADGPA VGRE QGRE UNDGPA
REGRESSION VARIABLES=GRADGPA,VGRE,QGRE,UNDGPA/
  DEPENDENT=GRADGPA/ENTER/
```

The TITLE command names the SPSSX run.

The DATA LIST command gives each variable a name and describes the data as being in free form.

The BEGIN DATA command indicates to SPSSX that the input data immediately follow.

The next 10 lines contain the data values and represent the four variables to be considered in the regression analysis. The first line of data represents a student with a graduate GPA of 3.54, a verbal GRE score of 580, a quantitative GRE score of 720, and an undergraduate GPA of 3.82.

The END DATA statement indicates the end of the input data.

The PRINT command requests a printout of the input data.

The REGRESSION statement requests that the independent variables VGRE, QGRE, and UNDGPA be entered in the regression equation to predict the dependent variable, GRADGPA.

Figure 15.22 contains the output obtained by executing the program listing in Figure 15.21.

Figure 15.22

SPSSX output
obtained by
executing the
program listing in
Figure 15.21

```
* * * *   M U L T I P L E   R E G R E S S I O N   * * * *

LISTWISE DELETION OF MISSING DATA

EQUATION NUMBER 1    DEPENDENT VARIABLE..    GRADGPA

BEGINNING BLOCK NUMBER  1.  METHOD: ENTER

VARIABLE(S) ENTERED ON STEP NUMBER   1..    UNDGPA          p value for F test
                                     2..    QGRE
                                     3..    VGRE                            SSE

MULTIPLE R              .94527       ANALYSIS OF VARIANCE
R SQUARE               .89353                        DF   SUM OF SQUARES   MEAN SQUARE
ADJUSTED R SQUARE      .84030       REGRESSION        3          1.27940       .42647
STANDARD ERROR         .15940       RESIDUAL          6           .15245       .02541

                                    F =    16.78488    SIGNIF F =  .0025

----------------- VARIABLES IN THE EQUATION -----------------

VARIABLE          B         SE B        BETA        T    SIG T

UNDGPA         .445704     .176362     .425316    2.527   .0448      MSE = s²
QGRE       5.90007E-04  6.6987E-04     .126143     .881   .4123
VGRE          .003103   7.9875E-04     .624270    3.885   .0081
(CONSTANT)   -.394954     .581347                 -.679   .5222

END BLOCK NUMBER   1    ALL REQUESTED VARIABLES ENTERED.
```

Regression coefficients

SPSS/PC+ APPENDIX (MICRO VERSION)

Solution to Example 15.2

At the beginning of the chapter, we computed the regression equation for predicting the performance of a first-year graduate student, based on the three predictor variables: verbal score on the GRE, quantitative score on the GRE, and the undergraduate GPA. The SPSS/PC+ program listing in Figure 15.23 requests a multiple regression solution.

Figure 15.23

SPSS/PC+ program listing used to request a multiple regression analysis of the university data

```
TITLE   UNIVERSITY EXAMPLE USING TWO PREDICTORS.
DATA LIST FREE/ GRADGPA VGRE QGRE UNDGPA.
BEGIN DATA.
3.54 580 720 3.82
2.62 500 660 2.67
3.30 670 580 3.16
2.90 480 520 3.31
4.00 710 630 3.60
3.21 550 690 3.42
3.57 640 700 3.51
3.05 540 530 2.75
3.15 620 490 3.21
3.61 690 530 3.70
END DATA.
LIST / GRADGPA VGRE QGRE UNDGPA.
REGRESSION VARIABLES=GRADGPA,VGRE,QGRE,UNDGPA/
   DEPENDENT=GRADGPA/ENTER.
```

The TITLE command names the SPSS/PC+ run.

The DATA LIST command gives each variable a name and describes the data as being in free form.

The BEGIN DATA command indicates to SPSS/PC+ that the input data immediately follow.

The next 10 lines contain the data values and represent the four variables to be considered in the regression analysis. The first line of data represents a student with a graduate GPA of 3.54, a verbal GRE score of 580, a quantitative GRE score of 720, and an undergraduate GPA of 3.82.

The END DATA statement indicates the end of the input data.

The LIST command requests a printout of the input data.

The REGRESSION statement requests that the independent variables VGRE, QGRE, and UNDGPA be entered in the regression equation to predict the dependent variable, GRADGPA.

Figure 15.24 contains the output obtained by executing the program listing in Figure 15.23.

**SPSS/PC+
APPENDIX
(MICRO
VERSION)**

```
--------------------------------------------------------------------------------
              UNIVERSITY EXAMPLE USING TWO PREDICTORS

          * * * *   M U L T I P L E   R E G R E S S I O N   * * * *

Equation Number 1    Dependent Variable..    GRADGPA


Variable(s) Entered on Step Number
      1..    UNDGPA
      2..    QGRE
      3..    VGRE


Multiple R              .94527
R Square                .89353
Adjusted R Square       .84030
Standard Error          .15940

Analysis of Variance
                    DF      Sum of Squares      Mean Square
Regression           3           1.27940           .42647
Residual             6            .15245           .02541

F =       16.78488       Signif F =  .0025
--------------------------------------------------------------------------------
              UNIVERSITY EXAMPLE USING TWO PREDICTORS

          * * * *   M U L T I P L E   R E G R E S S I O N   * * * *

Equation Number 1    Dependent Variable..    GRADGPA


------------------ Variables in the Equation ------------------

Variable              B          SE B         Beta        T    Sig T

UNDGPA           .44570        .17636       .42532     2.527   .0448
QGRE       5.900068E-04  6.69869E-04       .12614      .881   .4123
VGRE       3.102825E-03  7.98746E-04       .62427     3.885   .0081
(Constant)      -.39495        .58135                 -.679   .5222


End Block Number   1   All requested variables entered.
```

Annotations on the figure:

- R^2 (pointing to .89353 R Square)
- SSE (pointing to .15245)
- MSE $= s^2$ (pointing to .02541)
- p value for F test (pointing to Signif F = .0025)
- p values for t tests (pointing to .0448, .4123, .0081)
- Regression coefficients (pointing to the B column values .44570, 5.900068E-04, 3.102825E-03, -.39495)

Figure 15.24

SPSS/PC+ output obtained by executing the program listing in Figure 15.23

SPSSX APPENDIX (MAINFRAME VERSION)

Stepwise Solution to University Example

SPSSX can be used for determining the stepwise solution used in predicting the performance of a first-year graduate student, based on the three predictor variables: verbal score on the GRE, quantitative score on the GRE, and the undergraduate GPA. The SPSSX program listing in Figure 15.25 requests a stepwise regression solution. Notice that the word STEPWISE is substituted for ENTER and that a new line has been added. Instead of forcing all variables into the equation with the ENTER command, STEPWISE selects the variables that meet the entry criteria. The CRITERIA = PIN(0.1) statement specifies that each independent variable must produce a significant increase in R^2 at a significance level of .1 before it is allowed to enter into the regression equation. The POUT (.15) statement dictates that any variable in the model *not* producing a significant decrease in R^2, at a significance level of .15, when dropped from the model, should be removed. The value contained in POUT must exceed the value contained in PIN.

Figure 15.25

SPSSX program listing used to request a stepwise regression analysis of the university data

```
TITLE    UNIVERSITY EXAMPLE USING STEPWISE
DATA LIST FREE/ GRADGPA VGRE QGRE UNDGPA
BEGIN DATA
3.54 580 720 3.82
2.62 500 660 2.67
3.30 670 580 3.16
2.90 480 520 3.31
4.00 710 630 3.60
3.21 550 690 3.42
3.57 640 700 3.51
3.05 540 530 2.75
3.15 620 490 3.21
3.61 690 530 3.70
END DATA
PRINT / GRADGPA VGRE QGRE UNDGPA
REGRESSION VARIABLES=GRADGPA,VGRE,QGRE,UNDGPA/
   CRITERIA=PIN(0.1),POUT(0.15)/
   DEPENDENT=GRADGPA/STEPWISE/
```

The TITLE command names the SPSSX run.

The DATA LIST command gives each variable a name and describes the data as being in free form.

The BEGIN DATA command indicates to SPSSX that the input data immediately follow.

The next 10 lines contain the data values and represent the four variables to be considered in the regression analysis. The first line of data represents a student with a graduate GPA of 3.54, a verbal GRE score of 580, a quantitative GRE score of 720, and an undergraduate GPA of 3.82.

The END DATA statement indicates the end of the input data.

The PRINT command requests a printout of the input data.

Figure 15.26

SPSSX output obtained by executing the program listing in Figure 15.25

* * * * M U L T I P L E   R E G R E S S I O N * * * *

LISTWISE DELETION OF MISSING DATA

EQUATION NUMBER 1   DEPENDENT VARIABLE.. GRADGPA

BEGINNING BLOCK NUMBER 1.   METHOD: STEPWISE

VARIABLE(S) ENTERED ON STEP NUMBER 1.. VGRE ———1st variable entered

MULTIPLE R	.84456
R SQUARE	.71328
ADJUSTED R SQUARE	.67744
STANDARD ERROR	.22653

ANALYSIS OF VARIANCE

	DF	SUM OF SQUARES	MEAN SQUARE
REGRESSION	1	1.02131	1.02131
RESIDUAL	8	.41054	.05132

F =   19.90151   SIGNIF F = .0021

------------- VARIABLES IN THE EQUATION -------------

VARIABLE	B	SE B	BETA	T	SIG T
VGRE	.004198	9.4096E-04	.844557	4.461	.0021
(CONSTANT)	.784762	.567235		1.383	.2039

--------- VARIABLES NOT IN THE EQUATION ---------

VARIABLE	BETA IN	PARTIAL	MIN TOLER	T	SIG T
QGRE	.258780	.483117	.999325	1.460	.1877
UNDGPA	.479637	.762010	.723701	3.113	.0170

Enter UNDGPA into the equation because the value is < PIN, where PIN = .1

* * * * * * * * * * * * * * * *

VARIABLE(S) ENTERED ON STEP NUMBER 2.. UNDGPA ———2nd variable entered

MULTIPLE R	.93796
R SQUARE	.87977
ADJUSTED R SQUARE	.84541
STANDARD ERROR	.15682

ANALYSIS OF VARIANCE

	DF	SUM OF SQUARES	MEAN SQUARE
REGRESSION	2	1.25969	.62985
RESIDUAL	7	.17216	.02459

F =   25.60978   SIGNIF F = .0006

------------- VARIABLES IN THE EQUATION -------------

VARIABLE	B	SE B	BETA	T	SIG T
VGRE	.002945	7.6572E-04	.592440	3.846	.0063
UNDGPA	.502629	.161444	.479637	3.113	.0170
(CONSTANT)	-.132097	.490843		-.269	.7956

--------- VARIABLES NOT IN THE EQUATION ---------

VARIABLE	BETA IN	PARTIAL	MIN TOLER	T	SIG T
QGRE	.126143	.338367	.626513	.881	.4123

Do not enter QGRE because this value is > PIN, where PIN = .1

Final equation : $\hat{Y} = -.132097 + .002945$ (verbal GRE) + .5026 (undergraduate GPA)

The REGRESSION statement requests that the independent variables VGRE, QGRE, and UNDGPA be considered in the regression equation to predict the dependent variable, GRADGPA.

Figure 15.26 contains the output obtained by executing the program listing in Figure 15.25.

**SPSS/PC+
APPENDIX
(MICRO
VERSION)**

**Stepwise
Solution to
University
Example**

SPSS/PC+ can be used for determining the stepwise solution used in predicting the performance of a first-year graduate student, based on the three predictor variables: verbal score on the GRE, quantitative score on the GRE, and the undergraduate GPA. The SPSS/PC+ program listing in Figure 15.27 requests a stepwise regression solution. Notice that the word STEPWISE is substituted for ENTER and that a new line has been added. Instead of forcing all variables into the equation with the ENTER command, STEPWISE selects the variables that meet the entry criteria. The CRITERIA = PIN(0.1) statement specifies that each independent variable must produce a significant

Figure 15.27

SPSS/PC+ program
listing used to
request a stepwise
regression analysis of
the university data

```
TITLE    UNIVERSITY EXAMPLE USING STEPWISE.
DATA LIST FREE/ GRADGPA VGRE QGRE UNDGPA.
BEGIN DATA.
3.54 580 720 3.82
2.62 500 660 2.67
3.30 670 580 3.16
2.90 480 520 3.31
4.00 710 630 3.60
3.21 550 690 3.42
3.57 640 700 3.51
3.05 540 530 2.75
3.15 620 490 3.21
3.61 690 530 3.70
END DATA.
LIST / GRADGPA VGRE QGRE UNDGPA.
REGRESSION VARIABLES=GRADGPA,VGRE,QGRE,UNDGPA/
  CRITERIA=PIN(0.1),POUT(0.15)/
  DEPENDENT=GRADGPA/STEPWISE.
```

SPSS/PC+
APPENDIX
(MICRO
VERSION)

Figure 15.28

SPSS/PC+ output
obtained by
executing the
program listing in
Figure 15.27   UNIVERSITY EXAMPLE USING STEPWISE

```
* * * *   M U L T I P L E    R E G R E S S I O N   * * * *

Equation Number 1     Dependent Variable..    GRADGPA


Variable(s) Entered on Step Number
   1..    VGRE  ◄─────────────────── 1st variable entered


Multiple R              .84456
R Square                .71328
Adjusted R Square       .67744
Standard Error          .22653

Analysis of Variance
                   DF        Sum of Squares      Mean Square
Regression          1              1.02131          1.02131
Residual            8               .41054           .05132

F =      19.90151       Signif F =  .0021
------------------------------------------------------------------------------
                UNIVERSITY EXAMPLE USING STEPWISE

     * * * *   M U L T I P L E    R E G R E S S I O N   * * * *

Equation Number 1     Dependent Variable..    GRADGPA


------------------ Variables in the Equation ------------------

Variable                B          SE B         Beta          T   Sig T

VGRE        4.197723E-03  9.40959E-04       .84456       4.461   .0021
(Constant)       .78476       .56724                     1.383   .2039


------------- Variables not in the Equation -------------

Variable      Beta In   Partial   Min Toler        T   Sig T

QGRE          .25878    .48312     .99933       1.460   .1877
UNDGPA        .47964    .76201     .72370       3.113   .0170
                                                          ▲
```

Enter UNDGPA into the equation because
the value is < PIN, where PIN = .1

SPSS/PC+ APPENDIX (MICRO VERSION)

```
                    UNIVERSITY EXAMPLE USING STEPWISE

           * * * *   M U L T I P L E   R E G R E S S I O N   * * * *

Equation Number 1     Dependent Variable..    GRADGPA


Variable(s) Entered on Step Number
   2..     UNDGPA ◄───────────────── 2nd variable entered


Multiple R              .93796
R Square                .87977
Adjusted R Square       .84541
Standard Error          .15682

Analysis of Variance
                     DF        Sum of Squares      Mean Square
Regression            2             1.25969            .62985
Residual              7              .17216            .02459

F =      25.60978       Signif F =   .0006
-----------------------------------------------------------------------------
                    UNIVERSITY EXAMPLE USING STEPWISE

           * * * *   M U L T I P L E   R E G R E S S I O N   * * * *

Equation Number 1     Dependent Variable..    GRADGPA


------------------ Variables in the Equation ------------------

Variable              B          SE B         Beta         T    Sig T

VGRE          2.944620E-03  7.65721E-04     .59244      3.846   .0063
UNDGPA              .50263       .16144      .47964      3.113   .0170
(Constant)         -.13210       .49084                  -.269   .7956


------------- Variables not in the Equation -------------

Variable    Beta In   Partial   Min Toler        T   Sig T

QGRE         .12614    .33837      .62651       .881   .4123


End Block Number    1    PIN =      .100 Limits reached.
```

Final equation : $\hat{Y} = -0.1321 + .002945\,(\text{verbal GRE})$
$+ .5026\,(\text{undergraduate GPA})$

Do not enter QGRE because this value is > PIN, where PIN = .1

Figure 15.28

continued

increase in R^2 at a significance level of .1 before it is allowed to enter into the regression equation. The POUT (.15) statement dictates that any variable in the model *not* producing a significant decrease in R^2, at a significance level of .15, when dropped from the model, should be removed. The value contained in POUT must exceed the value contained in PIN.

The TITLE command names the SPSS/PC+ run.

The DATA LIST command gives each variable a name and describes the data as being in free form.

The BEGIN DATA command indicates to SPSS/PC+ that the input data immediately follow.

The next 10 lines contain the data values and represent the four variables to be considered in the regression analysis. The first line of data represents a student with a graduate GPA of 3.54, a verbal GRE score of 580, a quantitative GRE score of 720, and an undergraduate GPA of 3.82.

The END DATA statement indicates the end of the input data.

The LIST command requests a printout of the input data.

The REGRESSION statement requests that the independent variables VGRE, QGRE, and UNDGPA be considered in the regression equation to predict the dependent variable, GRADGPA.

Figure 15.28 contains the output obtained by executing the program listing in Figure 15.27.

**Prediction Using
University Data**

In the university example in Section 15.5, we wished to determine the graduate GPA (Y) for values of $X_1 = 520$, $X_2 = 600$, and $X_3 = 3.55$. Of course, one way to do this is to insert them manually into the computer generated regression equation contained in the previous SPSSX example. An easier way is to attach these values at the end of the input data with a numeric value for Y (we used -9 here) which will be identified as a missing value using the MISSING VALUES GRADGPA(-9) statement. SPSSX ignores this row of data when it computes the regression equation but then includes this row when it summarizes the predicted values. The predicted values as well as their standard errors are calculated in Figure 15.29 and included in the output using the CASEWISE = ALL PRED SEPRED statement. This informs SPSSX that you would like to see the predicted values (and their standard errors) for all of the cases in the input data, including the row(s) with the missing Y value.

Figure 15.29

SPSSX program listing used to request predicted values for the input data and an additional row of predictor variable values

```
TITLE    PREDICTION FOR UNIVERSITY EXAMPLE
DATA LIST FREE/ GRADGPA VGRE QGRE UNDGPA
MISSING VALUES GRADGPA(-9)
BEGIN DATA
3.54 580 720 3.82
2.62 500 660 2.67
3.30 670 580 3.16
2.90 480 520 3.31
4.00 710 630 3.60
3.21 550 690 3.42
3.57 640 700 3.51
3.05 540 530 2.75
3.15 620 490 3.21
3.61 690 530 3.70
-9   520 600 3.55
END DATA
PRINT / GRADGPA VGRE QGRE UNDGPA
REGRESSION VARIABLES=GRADGPA,VGRE,QGRE,UNDGPA/
   DEPENDENT=GRADGPA/ENTER/
   CASEWISE=ALL PRED SEPRED/
```

The TITLE command names the SPSSX run.

The DATA LIST command gives each variable a name and describes the data as being in free form.

The BEGIN DATA command indicates to SPSSX that the input data immediately follow.

The next 10 lines contain the data values and represent the four variables to be considered in the regression analysis. The first line of data represents a student with a graduate GPA of 3.54, a verbal GRE score of 580, a quantitative GRE score of 720, and an undergraduate GPA of 3.82.

The END DATA statement indicates the end of the input data.

The PRINT command requests a printout of the input data.

The REGRESSION statement requests that the independent variables VGRE, QGRE, and UNDGPA be entered in the regression equation to predict the dependent variable, GRADGPA.

SPSSX APPENDIX (MAINFRAME VERSION)

Figure 15.30

SPSSX output obtained by executing the program listing in Figure 15.29

Figure 15.30 contains the output obtained by executing the program listing in Figure 15.29.

```
                                * * * *   M U L T I P L E   R E G R E S S I O N   * * * *

LISTWISE DELETION OF MISSING DATA

EQUATION NUMBER 1    DEPENDENT VARIABLE..    GRADGPA

BEGINNING BLOCK NUMBER   1.  METHOD:  ENTER

VARIABLE(S) ENTERED ON STEP NUMBER    1..      UNDGPA
                                      2..      QGRE
                                      3..      VGRE


MULTIPLE R           .94527        ANALYSIS OF VARIANCE
R SQUARE             .89353                          DF      SUM OF SQUARES      MEAN SQUARE
ADJUSTED R SQUARE    .84030        REGRESSION         3             1.27940          .42647
STANDARD ERROR       .15940        RESIDUAL           6              .15245          .02541

                                   F =      16.78488       SIGNIF F =  .0025


----------------- VARIABLES IN THE EQUATION ------------------

VARIABLE            B           SE B         BETA         T    SIG T

UNDGPA          .445704       .176362      .425316     2.527   .0448
QGRE         5.90007E-04   6.6987E-04      .126143      .881   .4123
VGRE            .003103    7.9875E-04      .624270     3.885   .0081
(CONSTANT)     -.394954       .581347                  -.679   .5222

END BLOCK NUMBER   1    ALL REQUESTED VARIABLES ENTERED.



EQUATION NUMBER 1    DEPENDENT VARIABLE..    GRADGPA

CASEWISE PLOT OF STANDARDIZED RESIDUAL

*: SELECTED    M: MISSING

              -3.0            0.0            3.0
    CASE #     O:..............:..............:O        *PRED      *SEPRED
       1       .              .     *         .         3.5321       .1119
       2       .          *   .               .         2.7359       .1216
       3       .          *   .               .         3.4346       .0887
       4       .            * .               .         2.8765       .1302
       5       .              .    *          .         3.7843       .0915
       6       .           *  .               .         3.2430       .0815
       7       .            * .               .         3.5683       .0835
       8       .              .   *           .         2.8190       .0967
       9       .         *    .               .         3.2486       .0885
      10       .            * .               .         3.7078       .1017
      11       .            * .               .         3.1548       .1060
    CASE #     O:..............:..............:O        *PRED      *SEPRED
              -3.0            0.0            3.0
```

If $X_1 = 520, X_2 = 600, X_3 = 3.55$

then $\hat{Y} = 3.1548$ and $s\hat{Y} = .1060$

**SPSS/PC+
APPENDIX
(MICRO
VERSION)**

**Prediction Using
University Data**

In the university example in Section 15.5, we wished to determine the graduate GPA (Y) for values of $X_1 = 520$, $X_2 = 600$, and $X_3 = 3.55$. Of course, one way to do this is to insert them manually into the computer generated regression equation contained in the previous SPSS/PC+ example. An easier way is to attach these values at the end of the input data with a numeric value for Y (we used -9 here) which will be identified as a missing value using the MISSING VALUES GRADGPA(-9) statement. SPSS/PC+ ignores this row of data when it computes the regression equation but then includes this row when it summarizes the predicted values. The predicted values as well as their standard errors are calculated in Figure 15.31 and included in the output using the CASEWISE = ALL PRED SEPRED statement. This informs SPSS/PC+ that you would like to see the predicted values (and their standard errors) for all of the cases in the input data, including the row(s) with the missing Y value.

Figure 15.31

SPSS/PC+ program
listing used to
request predicted
values for the input
data and an
additional row of
predictor variable
values

```
TITLE    PREDICTION FOR UNIVERSITY EXAMPLE.
DATA LIST FREE/ GRADGPA VGRE QGRE UNDGPA.
MISSING VALUES GRADGPA(-9).
BEGIN DATA.
3.54 580 720 3.82
2.62 500 660 2.67
3.30 670 580 3.16
2.90 480 520 3.31
4.00 710 630 3.60
3.21 550 690 3.42
3.57 640 700 3.51
3.05 540 530 2.75
3.15 620 490 3.21
3.61 690 530 3.70
-9   520 600 3.55
END DATA.
LIST / GRADGPA VGRE QGRE UNDGPA.
REGRESSION VARIABLES=GRADGPA,VGRE,QGRE,UNDGPA/
  DEPENDENT=GRADGPA/ENTER/
  CASEWISE=ALL PRED SEPRED.
```

The TITLE command names the SPSS/PC+ run.

The DATA LIST command gives each variable a name and describes the data as being in free form.

The BEGIN DATA command indicates to SPSSX that the input data immediately follow.

SPSS/PC+
APPENDIX
(MICRO
VERSION)

```
------------------------------------------------------------------------
              PREDICTION FOR UNIVERSITY EXAMPLE

        * * * *    M U L T I P L E    R E G R E S S I O N    * * * *

Equation Number 1     Dependent Variable..    GRADGPA


Variable(s) Entered on Step Number
    1..     UNDGPA
    2..     QGRE
    3..     VGRE


Multiple R              .94527
R Square                .89353
Adjusted R Square       .84030
Standard Error          .15940

Analysis of Variance
                     DF      Sum of Squares      Mean Square
Regression            3            1.27940          .42647
Residual              6             .15245          .02541-

F =      16.78488      Signif F =  .0025
------------------------------------------------------------------------
              PREDICTION FOR UNIVERSITY EXAMPLE

        * * * *    M U L T I P L E    R E G R E S S I O N    * * * *

Equation Number 1     Dependent Variable..    GRADGPA


------------------ Variables in the Equation ------------------

Variable            B          SE B         Beta        T    Sig T

UNDGPA           .44570       .17636       .42532      2.527  .0448
QGRE        5.900068E-04  6.69869E-04     .12614       .881  .4123
VGRE        3.102825E-03  7.98746E-04     .62427      3.885  .0081
(Constant)      -.39495      .58135                   -.679  .5222


End Block Number   1   All requested variables entered.
------------------------------------------------------------------------
```

Figure 15.32

SPSS/PC+ output
obtained by
executing the
program listing in
Figure 15.31

The next 10 lines contain the data values and represent the four variables to be considered in the regression analysis. The first line of data represents a student with a graduate GPA of 3.54, a verbal GRE score of 580, a quantitative GRE score of 720, and an undergraduate GPA of 3.82.

The END DATA statement indicates the end of the input data.

**SPSS/PC+
APPENDIX
(MICRO
VERSION)**

```
              PREDICTION FOR UNIVERSITY EXAMPLE

        * * * *   M U L T I P L E   R E G R E S S I O N   * * * *

Equation Number 1     Dependent Variable..   GRADGPA


Casewise Plot of Standardized Residual

*: Selected    M: Missing

              -3.0        0.0        3.0
   Case #    O:.......:........:O          *PRED      *SEPRED
       1     .           *          .       3.5321      .1119
       2     .     *                .       2.7359      .1216
       3     .        *   .         .       3.4346      .0887
       4     .            *         .       2.8765      .1302
       5     .              *       .       3.7843      .0915
       6     .          *.          .       3.2430      .0815
       7     .           *          .       3.5683      .0835
       8     .            .   *     .       2.8190      .0967
       9     .        *   .         .       3.2486      .0885
      10     .        *   .         .       3.7078      .1017
   Case #    O:.......:........:O          *PRED      *SEPRED
              -3.0        0.0        3.0
```

```
              PREDICTION FOR UNIVERSITY EXAMPLE


Casewise Plot of Standardized Residual

*: Selected    M: Missing

              -3.0        0.0        3.0
   Case #    O:.......:........:O          *PRED      *SEPRED
      11     .           .          .      3.1548      .1060
   Case #    O:.......:........:O          *PRED      *SEPRED
              -3.0        0.0        3.0
```

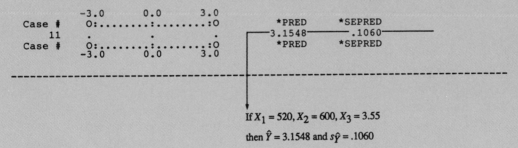

If $X_1 = 520, X_2 = 600, X_3 = 3.55$

then $\hat{Y} = 3.1548$ and $s\hat{y} = .1060$

Figure 15.32

continued

The LIST command requests a printout of the input data.

The REGRESSION statement requests that the independent variables VGRE, QGRE, and UNDGPA be entered in the regression equation to predict the dependent variable, GRADGPA.

Figure 15.32 contains the output obtained by executing the program listing in Figure 15.31.

**SAS
APPENDIX
(MAINFRAME
VERSION)**

**Solution to
Example 15.2**

At the beginning of the chapter, we computed the regression equation for predicting the performance of a first-year graduate student, based on the three predictor variables: verbal score on the GRE, quantitative score on the GRE, and the undergraduate GPA. The SAS program listing in Figure 15.33 requests a multiple regression solution.

Figure 15.33

SAS program listing
used to request a
multiple regression
analysis of the
university data

```
TITLE   'UNIVERSITY EXAMPLE USING TWO PREDICTORS';
DATA UNIV;
  INPUT GRADGPA VGRE QGRE UNDGPA;
CARDS;
3.54 580 720 3.82
2.62 500 660 2.67
3.30 670 580 3.16
2.90 480 520 3.31
4.00 710 630 3.60
3.21 550 690 3.42
3.57 640 700 3.51
3.05 540 530 2.75
3.15 620 490 3.21
3.61 690 530 3.70
PROC PRINT;
PROC REG;
  MODEL GRADGPA=VGRE QGRE UNDGPA;
```

The TITLE command names the SAS run.

The DATA command gives the data a name.

The INPUT command names and gives the correct order for the data during the data input.

The CARDS command indicates to SAS that the input data immediately follow.

The next 10 lines contain the data values and represent the four variables to be considered in the regression analysis. The first line of data represents a student with a graduate GPA of 3.54, a verbal GRE score of 580, a quantitative GRE score of 720, and an undergraduate GPA of 3.82.

The PROC PRINT command requests a printout of the input data.

The PROC REG command and MODEL subcommand request that the independent variables VGRE, QGRE, and UNDGPA be entered in the regression equation to predict the dependent variable, GRADGPA.

Figure 15.34 contains the output obtained by executing the program listing in Figure 15.33.

**SAS
APPENDIX
(MAINFRAME
VERSION)**

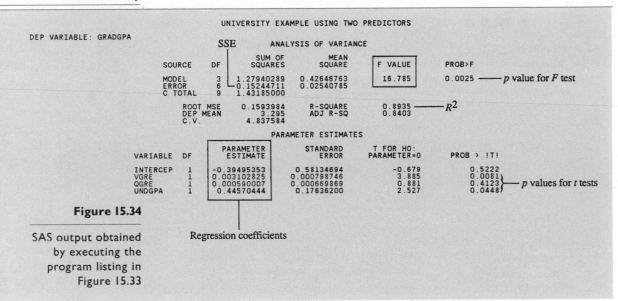

Figure 15.34

SAS output obtained
by executing the
program listing in
Figure 15.33

**SAS
APPENDIX
(MICRO
VERSION)**

**Solution to
Example 15.2**

At the beginning of the chapter, we computed the regression equation for predicting the performance of a first-year graduate student, based on the three predictor variables: verbal score on the GRE, quantitative score on the GRE, and the undergraduate GPA. The SAS program listing in Figure 15.35 requests a multiple regression solution.

The TITLE command names the SAS run.

The DATA command gives the data a name.

The INPUT command names and gives the correct order for the data during the data input.

The CARDS command indicates to SAS that the input data immediately follow.

SAS APPENDIX (MICRO VERSION)

Figure 15.35

SAS program listing used to request a multiple regression analysis of the university data

```
TITLE    'UNIVERSITY EXAMPLE USING TWO PREDICTORS';
DATA UNIV;
  INPUT GRADGPA VGRE QGRE UNDGPA;
CARDS;
3.54 580 720 3.82
2.62 500 660 2.67
3.30 670 580 3.16
2.90 480 520 3.31
4.00 710 630 3.60
3.21 550 690 3.42
3.57 640 700 3.51
3.05 540 530 2.75
3.15 620 490 3.21
3.61 690 530 3.70
PROC PRINT;
PROC REG;
  MODEL GRADGPA=VGRE QGRE UNDGPA;
RUN;
```

Figure 15.36

SAS output obtained by executing the program listing in Figure 15.35

UNIVERSITY EXAMPLE USING TWO PREDICTORS

Model: MODEL1
Dep Variable: GRADGPA

Analysis of Variance

Source	DF	Sum of Squares	Mean Square	F Value	Prob>F
Model	3	1.27940	0.42647	16.785	0.0025
Error	6	0.15245	0.02541		
C Total	9	1.43185			

SSE → (points to 0.15245)

F Value box; *p* value for *F* test (points to 0.0025)

Root MSE	0.15940	R-Square	0.8935	← R^2
Dep Mean	3.29500	Adj R-Sq	0.8403	
C.V.	4.83758			

UNIVERSITY EXAMPLE USING TWO PREDICTORS

Parameter Estimates

Variable	DF	Parameter Estimate	Standard Error	T for H0: Parameter=0	Prob > |T|
INTERCEP	1	-0.394954	0.58134694	-0.679	0.5222
VGRE	1	0.003103	0.00079875	3.885	0.0081
QGRE	1	0.000590	0.00066987	0.881	0.4123
UNDGPA	1	0.445704	0.17636200	2.527	0.0448

Regression coefficients (points to Parameter Estimate box)

p values for *t* tests (points to Prob > |T| column)

SAS APPENDIX (MICRO VERSION)

The next 10 lines contain the data values and represent the four variables to be considered in the regression analysis. The first line of data represents a student with a graduate GPA of 3.54, a verbal GRE score of 580, a quantitative GRE score of 720, and an undergraduate GPA of 3.82.

The PROC PRINT command requests a printout of the input data.

The PROC REG command and MODEL subcommand request that the independent variables VGRE, QGRE, and UNDGPA be entered in the regression equation to predict the dependent variable, GRADGPA.

The RUN command tells the SAS system to execute the previous SAS statements.

Figure 15.36 contains the output obtained by executing the program listing in Figure 15.35.

SAS APPENDIX (MAINFRAME VERSION)

Stepwise Solution to University Example

SAS can be used for determining the stepwise solution used in predicting the performance of a first-year graduate student, based on the three predictor variables: verbal score on the GRE, quantitative score on the GRE, and the undergraduate GPA. The SAS program listing in Figure 15.37 requests a

Figure 15.37

SAS program listing used to request a stepwise regression analysis of the university data

```
TITLE   'UNIVERSITY EXAMPLE USING STEPWISE';
DATA UNIV;
 INPUT GRADGPA VGRE QGRE UNDGPA;
CARDS;
3.54 580 720 3.82
2.62 500 660 2.67
3.30 670 580 3.16
2.90 480 520 3.31
4.00 710 630 3.60
3.21 550 690 3.42
3.57 640 700 3.51
3.05 540 530 2.75
3.15 620 490 3.21
3.61 690 530 3.70
PROC PRINT;
PROC STEPWISE;
 MODEL GRADGPA=VGRE QGRE UNDGPA/
       SLENTRY=0.1;
```

Figure 15.38

SAS output obtained by executing the program listing in Figure 15.37

UNIVERSITY EXAMPLE USING STEPWISE

STEPWISE REGRESSION PROCEDURE FOR DEPENDENT VARIABLE GRADGPA

NOTE: SLSTAY HAS BEEN SET TO .15 FOR THE STEPWISE TECHNIQUE.

STEP 1 VARIABLE VGRE ENTERED   R SQUARE = 0.7132716   C(P) = 10.15815860

	DF	SUM OF SQUARES	MEAN SQUARE	F	PROB>F
REGRESSION	1	1.02130590	1.02130590	19.90	0.0021
ERROR	8	0.41054410	0.05131801		
TOTAL	9	1.43185000			

	B VALUE	STD ERROR	TYPE II SS	F	PROB>F
INTERCEPT	0.78476190				
VGRE	0.00419772	0.00094096	1.02130590	19.90	0.0021

BOUNDS ON CONDITION NUMBER: 1,   1

STEP 2 VARIABLE UNDGPA ENTERED   R SQUARE = 0.8976548   C(P) = 2.77577156

	DF	SUM OF SQUARES	MEAN SQUARE	F	PROB>F
REGRESSION	2	1.25969220	0.62984610	25.61	0.0006
ERROR	7	0.17215780	0.02459397		
TOTAL	9	1.43185000			

	B VALUE	STD ERROR	TYPE II SS	F	PROB>F
INTERCEPT	-0.13209674				
VGRE	0.00294462	0.00076572	0.36370208	14.79	0.0063
UNDGPA	0.50262874	0.16144366	0.23838630	9.69	0.0170

BOUNDS ON CONDITION NUMBER: 1.381787,   5.527146

NO OTHER VARIABLES MET THE 0.1000 SIGNIFICANCE LEVEL FOR ENTRY INTO THE MODEL.

SUMMARY OF STEPWISE REGRESSION PROCEDURE FOR DEPENDENT VARIABLE GRADGPA

STEP	VARIABLE ENTERED	VARIABLE REMOVED	NUMBER IN	PARTIAL R**2	MODEL R**2	C(P)	F	PROB>F
1	VGRE		1	0.7133	0.7133	10.1582	19.9015	0.0021
2	UNDGPA		2	0.1665	0.8798	2.7758	9.6929	0.0170

Final equation : $\hat{Y} = -.132097 + .002945\,(\text{verbal GRE}) + .5026\,(\text{undergraduate GPA})$

SAS APPENDIX (MAINFRAME VERSION)

stepwise regression solution. Notice that the word STEPWISE is substituted for REG and that a new line has been added. Instead of forcing all variables into the equation with the REG command, STEPWISE selects the variables that meet the entry criteria. The SLENTRY = 0.1 statement specifies the significance level for entering a variable into the regression equation.

The TITLE command names the SAS run.

The DATA command gives the data a name.

The INPUT command names and gives the correct order for the data during the data input.

The CARDS command indicates to SAS that the input data immediately follow.

The next 10 lines contain the data values and represent the four variables to be considered in the regression analysis. The first line of data represents a student with a graduate GPA of 3.54, a verbal GRE score of 580, a quantitative GRE score of 720, and an undergraduate GPA of 3.82.

The PROC PRINT command requests a printout of the input data.

The PROC STEPWISE command and MODEL subcommand request that the independent variables VGRE, QGRE, and UNDGPA be considered in the regression equation to predict the dependent variable, GRADGPA.

Figure 15.38 contains the output obtained by executing the program listing in Figure 15.37.

SAS APPENDIX (MICRO VERSION)

Stepwise Solution to University Example

SAS can be used for determining the stepwise solution used in predicting the performance of a first-year graduate student, based on the three predictor variables: verbal score on the GRE, quantitative score on the GRE, and the undergraduate GPA. The SAS program listing in Figure 15.39 requests a stepwise regression solution. Notice that the word STEPWISE is substituted for

SAS APPENDIX (MICRO VERSION)

Figure 15.39

SAS program listing used to request a stepwise regression analysis of the university data

```
   TITLE     'UNIVERSITY EXAMPLE USING STEPWISE';
DATA UNIV;
 INPUT GRADGPA VGRE QGRE UNDGPA;
CARDS;
3.54 580 720 3.82
2.62 500 660 2.67
3.30 670 580 3.16
2.90 480 520 3.31
4.00 710 630 3.60
3.21 550 690 3.42
3.57 640 700 3.51
3.05 540 530 2.75
3.15 620 490 3.21
3.61 690 530 3.70
PROC PRINT;
PROC STEPWISE;
 MODEL GRADGPA=VGRE QGRE UNDGPA/
       SLENTRY=0.1;
RUN;
```

REG and that a new line has been added. Instead of forcing all variables into the equation with the REG command, STEPWISE selects the variables that meet the entry criteria. The SLENTRY = 0.1 statement specifies the significance level for entering a variable into the regression equation.

The TITLE command names the SAS run.

The DATA command gives the data a name.

The INPUT command names and gives the correct order for the data during the data input.

The CARDS command indicates to SAS that the input data immediately follow.

The next 10 lines contain the data values and represent the four variables to be considered in the regression analysis. The first line of data represents a student with a graduate GPA of 3.54, a verbal GRE score of 580, a quantitative GRE score of 720, and an undergraduate GPA of 3.82.

The PROC PRINT command requests a printout of the input data.

The PROC STEPWISE command and MODEL subcommand request that the independent variables VGRE, QGRE, and UNDGPA be considered in the regression equation to predict the dependent variable, GRADGPA.

The RUN command tells the SAS system to execute the previous SAS statements.

SAS APPENDIX (MICRO VERSION)

Figure 15.40

SAS output obtained by executing the program listing in Figure 15.39

Figure 15.40 contains the output obtained by executing the program listing in Figure 15.39.

UNIVERSITY EXAMPLE USING STEPWISE

Stepwise Procedure for Dependent Variable GRADGPA

| Step 1   Variable VGRE Entered | | R-square = 0.71327716 | C(p) = 10.15815860 |

	DF	Sum of Squares	Mean Square	F	Prob>F
Regression	1	1.02130590	1.02130590	19.90	0.0021
Error	8	0.41054410	0.05131801		
Total	9	1.43185000			

Variable	Parameter Estimate	Standard Error	Type II Sum of Squares	F	Prob>F
INTERCEP	0.78476190	0.56723522	0.09822438	1.91	0.2039
VGRE	0.00419772	0.00094096	1.02130590	19.90	0.0021

Bounds on condition number: 1.0000,   1.0000

| Step 2   Variable UNDGPA Entered | | R-square = 0.87976548 | C(p) = 2.77577156 |

	DF	Sum of Squares	Mean Square	F	Prob>F
Regression	2	1.25969220	0.62984610	25.61	0.0006
Error	7	0.17215780	0.02459397		
Total	9	1.43185000			

Variable	Parameter Estimate	Standard Error	Type II Sum of Squares	F	Prob>F
INTERCEP	-0.13209674	0.49084301	0.00178126	0.07	0.7956
VGRE	0.00294462	0.00076572	0.36370208	14.79	0.0063
UNDGPA	0.50262874	0.16144366	0.23838630	9.69	0.0170

Bounds on condition number: 1.3818,   5.5271

Final equation : $\hat{Y} = -.132097 + .002945$ (verbal GRE) $+ .5026$ (undergraduate GPA)

All variables in the model are significant at the 0.1500 level. No other variables met the 0.1000 significance level for entry into the model.

Summary of Stepwise Procedure for Dependent Variable GRADGPA

Step	Variable Entered Removed	Number In	Partial R**2	Model R**2	C(p)	F	Prob>F
1	VGRE	1	0.7133	0.7133	10.1582	19.9015	0.0021
2	UNDGPA	2	0.1665	0.8798	2.7758	9.6929	0.0170

SAS APPENDIX (MAINFRAME VERSION)

Prediction Using University Data

In the university example in Section 15.5, we wished to determine the graduate GPA (Y) for values of $X_1 = 520$, $X_2 = 600$, and $X_3 = 3.55$. Of course, one way to do this is to insert them manually into the computer generated regression equation contained in the previous SAS example. An easier way is to attach these values at the end of the input data along with any single character from A to Z for the value of the dependent variable, indicating to SAS that this value is missing. Which character you use is arbitrary, but it should be specified in the MISSING statement immediately following the DATA statement. SAS ignores this row when it computes the regression equation but includes this row when it summarizes the predicted values.

The predicted values as well as the corresponding confidence intervals and prediction intervals are calculated and included in the output by inserting /CLI CLM in the final MODEL statement. The CLI command generates the prediction intervals for an individual, whereas CLM will produce confidence intervals for the mean. The row(s) containing the missing value(s) will be included in the summary.

The listing in Figure 15.41 is used to generate the predicted value, confidence intervals, and prediction intervals.

Figure 15.41

SAS program listing used to request predicted values for the input data and an additional row of predictor variable values

```
TITLE   'PREDICTION FOR UNIVERSITY EXAMPLE';
DATA UNIV;
  MISSING A;
  INPUT GRADGPA VGRE QGRE UNDGPA;
CARDS;
3.54 580 720 3.82
2.62 500 660 2.67
3.30 670 580 3.16
2.90 480 520 3.31
4.00 710 630 3.60
3.21 550 690 3.42
3.57 640 700 3.51
3.05 540 530 2.75
3.15 620 490 3.21
3.61 690 530 3.70
A    520 600 3.55
PROC PRINT;
PROC REG;
  MODEL GRADGPA=VGRE QGRE UNDGPA/ CLI CLM;
```

The TITLE command names the SAS run.

The DATA command gives the data a name.

The INPUT command names and gives the correct order for the data during the data input.

The CARDS command indicates to SAS that the input data immediately follow.

SAS APPENDIX (MAINFRAME VERSION)

The next 10 lines contain the data values and represent the four variables to be considered in the regression analysis. The first line of data represents a student with a graduate GPA of 3.54, a verbal GRE score of 580, a quantitative GRE score of 720, and an undergraduate GPA of 3.82.

The PROC PRINT command requests a printout of the input data.

The PROC REG command and MODEL subcommand request that the independent variables VGRE, QGRE, and UNDGPA be entered in the regression equation to predict the dependent variable, GRADGPA.

Figure 15.42

SAS output obtained by executing the program listing in Figure 15.41

Figure 15.42 contains the output obtained by executing the program listing in Figure 15.41.

PREDICTION FOR UNIVERSITY EXAMPLE

DEP VARIABLE: GRADGPA

ANALYSIS OF VARIANCE

SOURCE	DF	SUM OF SQUARES	MEAN SQUARE	F VALUE	PROB>F
MODEL	3	1.27940289	0.42646763	16.785	0.0025
ERROR	6	0.15244711	0.02540785		
C TOTAL	9	1.43185000			

ROOT MSE	0.1593984	R-SQUARE	0.8935
DEP MEAN	3.295	ADJ R-SQ	0.8403
C.V.	4.837584		

PARAMETER ESTIMATES

	VARIABLE	DF	PARAMETER ESTIMATE	STANDARD ERROR	T FOR HO: PARAMETER=0	PROB > !T!
$s_{\hat{Y}}$	INTERCEP	1	-0.39495353	0.58134694	-0.679	0.5222
	VGRE	1	0.003102825	0.000798746	3.885	0.0081
	QGRE	1	0.000590007	0.000669869	0.881	0.4123
	UNDGPA	1	0.44570444	0.17636200	2.527	0.0448

$\hat{Y}$ $Y-\hat{Y}$

OBS	ACTUAL	PREDICT VALUE	STD ERR PREDICT	LOWER95% MEAN	UPPER95% MEAN	LOWER95% PREDICT	UPPER95% PREDICT	RESIDUAL
1	3.5400	3.5321	0.1119	3.2583	3.8059	3.0555	4.0086	.0079193
2	2.6200	2.7359	0.1216	2.4384	3.0334	2.2453	3.2265	-0.1159
3	3.3000	3.4346	0.0887	3.2176	3.6516	2.9882	3.8809	-0.1346
4	2.9000	2.8765	0.1302	2.5579	3.1951	2.3729	3.3801	0.0235
5	4.0000	3.7843	0.0915	3.5603	4.0083	3.3345	4.2341	0.2157
6	3.2100	3.2430	0.0815	3.0437	3.4423	2.8050	3.6810	-0.0330
7	3.5700	3.5683	0.0835	3.3641	3.7725	3.1280	4.0085	.0017183
8	3.0500	2.8190	0.0967	2.5823	3.0557	2.3627	3.2752	0.2310
9	3.1500	3.2486	0.0885	3.0320	3.4652	2.8025	3.6948	-0.0986
10	3.6100	3.7078	0.1017	3.4590	3.9566	3.2452	4.1704	-0.0978
11	A	3.1548	0.1060	2.8954	3.4141	2.6864	3.6232	

SUM OF RESIDUALS	1.28786E-14
SUM OF SQUARED RESIDUALS	0.1524471
PREDICTED RESID SS (PRESS)	0.4069532

CI for $\mu_{Y \mid X}$

Prediction interval for Y_X

$\hat{Y}$ and $s_{\hat{Y}}$ for $X_1 = 520, X_2 = 600, X_3 = 3.55$

SAS APPENDIX (MICRO VERSION)

Prediction Using University Data

In the university example in Section 15.5, we wished to determine the graduate GPA (Y) for values of $X_1 = 520$, $X_2 = 600$, and $X_3 = 3.55$. Of course, one way to do this is to insert them manually into the computer generated regression equation contained in the previous SAS example. An easier way is to attach these values at the end of the input data along with any single character from A to Z for the value of the dependent variable, indicating to SAS that this value is missing. Which character you use is arbitrary, but it should be specified in the MISSING statement immediately following the DATA statement. SAS ignores this row when it computes the regression equation but includes this row when it summarizes the predicted values.

The predicted values as well as the corresponding confidence intervals and prediction intervals are calculated and included in the output by inserting /CLI CLM in the final MODEL statement. The CLI command generates the prediction intervals for an individual, whereas CLM will produce confidence intervals for the mean. The row(s) containing the missing value(s) will be included in the summary.

The listing in Figure 15.43 is used to generate the predicted value, confidence intervals, and prediction intervals.

The TITLE command names the SAS run.

The DATA command gives the data a name.

Figure 15.43

SAS program listing used to request predicted values for the input data and an additional row of predictor variable values

```
TITLE    'PREDICTION FOR UNIVERSITY EXAMPLE';
DATA UNIV;
 MISSING A;
 INPUT GRADGPA VGRE QGRE UNDGPA;
CARDS;
3.54 580 720 3.82
2.62 500 660 2.67
3.30 670 580 3.16
2.90 480 520 3.31
4.00 710 630 3.60
3.21 550 690 3.42
3.57 640 700 3.51
3.05 540 530 2.75
3.15 620 490 3.21
3.61 690 530 3.70
A    520 600 3.55
PROC PRINT;
PROC REG;
 MODEL GRADGPA=VGRE QGRE UNDGPA/ CLI CLM;
RUN;
```

SAS APPENDIX (MICRO VERSION)

Model: MODEL1
Dep Variable: GRADGPA

Analysis of Variance

Source	DF	Sum of Squares	Mean Square	F Value	Prob>F
Model	3	1.27940	0.42647	16.785	0.0025
Error	6	0.15245	0.02541		
C Total	9	1.43185			

Root MSE	0.15940	R-Square	0.8935
Dep Mean	3.29500	Adj R-Sq	0.8403
C.V.	4.83758		

PREDICTION FOR UNIVERSITY EXAMPLE

Parameter Estimates

Variable	DF	Parameter Estimate	Standard Error	T for H0: Parameter=0	Prob > |T|
INTERCEP	1	-0.394954	0.58134694	-0.679	0.5222
VGRE	1	0.003103	0.00079875	3.885	0.0081
QGRE	1	0.000590	0.00066987	0.881	0.4123
UNDGPA	1	0.445704	0.17636200	2.527	0.0448

$\hat{Y}$ ↓   CI for $\mu_{Y|X}$ Prediction interval for Y_X

Obs	GRADGPA	Predict Value	Lower95% Mean	Upper95% Mean	Lower95% Predict	Upper95% Predict
1	3.5400	3.5321	3.2583	3.8059	3.0555	4.0086
2	2.6200	2.7359	2.4384	3.0334	2.2453	3.2265
3	3.3000	3.4346	3.2176	3.6516	2.9882	3.8809
4	2.9000	2.8765	2.5579	3.1951	2.3729	3.3801
5	4.0000	3.7843	3.5603	4.0083	3.3345	4.2341
6	3.2100	3.2430	3.0437	3.4423	2.8050	3.6810
7	3.5700	3.5683	3.3641	3.7725	3.1280	4.0085
8	3.0500	2.8190	2.5823	3.0557	2.3627	3.2752
9	3.1500	3.2486	3.0320	3.4652	2.8025	3.6948
10	3.6100	3.7078	3.4590	3.9566	3.2452	4.1704
11	A	3.1548	2.8954	3.4141	2.6864	3.6232

Sum of Residuals	-3.9968E-15
Sum of Squared Residuals	0.1524
Predicted Resid SS (Press)	0.4070

$\hat{Y}$ for $X_1 = 520, X_2 = 600, X_3 = 3.55$

Figure 15.44

SAS output obtained by executing the program listing in Figure 15.43

The INPUT command names and gives the correct order for the data during the data input.

The CARDS command indicates to SAS that the input data immediately follow.

SAS APPENDIX (MICRO VERSION)

The next 10 lines contain the data values and represent the four variables to be considered in the regression analysis. The first line of data represents a student with a graduate GPA of 3.54, a verbal GRE score of 580, a quantitative GRE score of 720, and an undergraduate GPA of 3.82.

The PROC PRINT command requests a printout of the input data.

The PROC REG command and MODEL subcommand request that the independent variables VGRE, QGRE, and UNDGPA be entered in the regression equation to predict the dependent variable, GRADGPA.

The RUN command tells the SAS system to execute the previous SAS statements.

Figure 15.44 (p. 931) contains the output obtained by executing the program listing in Figure 15.43.

 MINITAB APPENDIX

MINITAB Instructions for Multiple and Stepwise Regression

The MINITAB REGRESS command is illustrated in many of the examples contained in this chapter. This command has features that allow you to store (1) the residuals and (2) the predicted Y values ($\hat{Y}$).

> informs MINITAB that subcommands are needed

EXAMPLE

REGRESS Y IN C1 USING 2 PREDICTORS IN C2 C3;

RESIDUALS INTO C4;

> informs MINITAB that there are no more subcommands

FITS INTO C5.

abbreviated statement: REGR C1 2 C2 C3;
                          RESID C4;
                          FITS C5.

MINITAB APPENDIX

These commands will store the residuals in column C4 and the predicted Y values in column C5. These two columns can be manipulated, printed, and so forth, just like any other column in your analysis.

If you wish to have the predicted values and the residuals printed out, type the command BRIEF 3 at some point prior to the REGRESS command. This is illustrated in Figure 15.45 using the university data.

Figure 15.45

Multiple regression analysis of the university data

```
MTB > BRIEF 3
MTB > REGRESS Y IN C1 USING 3 PREDICTORS IN C2-C4

The regression equation is
C1 = - 0.395 + 0.00310 C2 +0.000590 C3 + 0.446 C4

Predictor        Coef        Stdev      t-ratio
Constant       -0.3950       0.5813       -0.68
C2            0.0031028     0.0007987      3.88
C3            0.0005900     0.0006699      0.88
C4              0.4457        0.1764       2.53

s = 0.1594      R-sq = 89.4%      R-sq(adj) = 84.0%

Analysis of Variance

SOURCE         DF          SS           MS
Regression      3       1.27940      0.42647
Error           6       0.15245      0.02541  ◄──── MSE
Total           9       1.43185

SOURCE         DF        SEQ SS
C2              1       1.02131
C3              1       0.09582
C4              1       0.16228

Continue? YES
Obs.       C2          C1        Fit   Stdev.Fit   Residual   St.Resid
  1       580       3.5400     3.5321    0.1119      0.0079       0.07
  2       500       2.6200     2.7359    0.1216     -0.1159      -1.12
  3       670       3.3000     3.4346    0.0887     -0.1346      -1.02
  4       480       2.9000     2.8765    0.1302      0.0235       0.26
  5       710       4.0000     3.7843    0.0915      0.2157       1.65
  6       550       3.2100     3.2430    0.0815     -0.0330      -0.24
  7       640       3.5700     3.5683    0.0835      0.0017       0.01
  8       540       3.0500     2.8190    0.0967      0.2310       1.82
  9       620       3.1500     3.2486    0.0885     -0.0986      -0.74
 10       690       3.6100     3.7078    0.1017     -0.0978      -0.80
```

MINITAB APPENDIX

Stepwise Regression

MINITAB also performs stepwise regression using the command

STEPWISE REGRESSION OF Y IN C1

USING PREDICTORS C2 C3 C4 . . .

your predictor
variables for
consideration

abbreviated statement: STEP C1 C2 C3 C4 . . .

This procedure will enter a variable if its corresponding (partial) F value exceeds FENTER = 4 and remove any variable whose (partial) F value falls below FREMOVE = 4. An illustration of this procedure using the university data is shown in Figure 15.46. Only two of the three variables being considered were selected (using the default values of FREMOVE = FENTER = 4). The resulting equation is

$$\hat{Y} = -0.1321 + 0.00294X_1 + 0.50X_3$$

Remarks

1 Changing the values of FENTER and/or FREMOVE

EXAMPLE

MTB > STEPWISE REGRESSION OF Y IN C1

USING C2, C3, C4, C5, C6;

SUBC > FENTER = 3.5;

SUBC > FREMOVE = 3.5.

(the output from the
STEPWISE command)

2 To perform a forward selection, you simply don't allow any variable to be removed once it is included in the model. Setting FREMOVE = 0 will accomplish this. The procedure ends when the (partial) F statistic for an entering variable is below FENTER.

3 Similarly, you can perform backward regression by
   (a) First using the subcommand ENTER

   SUBC > ENTER C2-C6;

   all of your predictor
   variables

(this enters all of your predictor variables into the model).

MINITAB APPENDIX

(b) Next, use

$$SUBC > FENTER = 10,000. \text{ (or any such large value)}$$

This procedure stops when no variable in the model has an F value less than FREMOVE.

Figure 15.46

MINITAB stepwise regression procedure

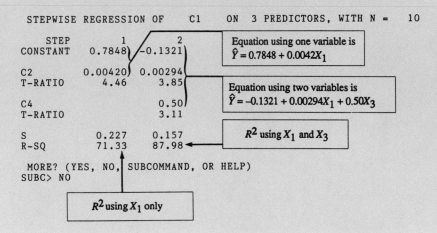

APPENDIXES

Appendix A
Tables

Table A-1 Binomial Probabilities $[_nC_x p^x (1-p)^{n-x}]$

n	x	0.01	0.05	0.10	0.20	0.30	0.40	*P* 0.50	0.60	0.70	0.80	0.90	0.95	0.99	x
2	0	980	902	810	640	490	360	250	160	090	040	010	002	0+	0
	1	020	095	180	320	420	480	500	480	420	320	180	095	020	1
	2	0+	002	010	040	090	160	250	360	490	640	810	902	980	2
3	0	970	857	729	512	343	216	125	064	027	008	001	0+	0+	0
	1	029	135	243	384	441	432	375	288	189	096	027	007	0+	1
	2	0+	007	027	096	189	288	375	432	441	384	243	135	029	2
	3	0+	0+	001	008	027	064	125	216	343	512	729	857	970	3
4	0	961	815	656	410	240	130	062	026	008	002	0+	0+	0+	0
	1	039	171	292	410	412	346	250	154	076	026	004	0+	0+	1
	2	001	014	049	154	265	346	375	346	265	154	049	014	001	2
	3	0+	0+	004	026	076	154	250	346	412	410	292	171	039	3
	4	0+	0+	0+	002	008	026	062	130	240	410	656	815	961	4
5	0	951	774	590	328	168	078	031	010	002	0+	0+	0+	0+	0
	1	048	204	328	410	360	259	156	077	028	006	0+	0+	0+	1
	2	001	021	073	205	309	346	312	230	132	051	008	001	0+	2
	3	0+	001	008	051	132	230	312	346	309	205	073	021	001	3
	4	0+	0+	0+	006	028	077	156	259	360	410	328	204	048	4
	5	0+	0+	0+	0+	002	010	031	078	168	328	590	774	951	5
6	0	941	735	531	262	118	047	016	004	001	0+	0+	0+	0+	0
	1	057	232	354	393	303	187	094	037	010	002	0+	0+	0+	1
	2	001	031	098	246	324	311	234	138	060	015	001	0+	0+	2
	3	0+	002	015	082	185	276	312	276	185	082	015	002	0+	3
	4	0+	0+	001	015	060	138	234	311	324	246	098	031	001	4
	5	0+	0+	0+	002	010	037	094	187	303	393	354	232	057	5
	6	0+	0+	0+	0+	001	004	016	047	118	262	531	735	941	6
7	0	932	698	478	210	082	028	008	002	0+	0+	0+	0+	0+	0
	1	066	257	372	367	247	131	055	017	004	0+	0+	0+	0+	1
	2	002	041	124	275	318	261	164	077	025	004	0+	0+	0+	2
	3	0+	004	023	115	227	290	273	194	097	029	003	0+	0+	3
	4	0+	0+	003	029	097	194	273	290	227	115	023	004	0+	4

From Mosteller, Rourke, & Thomas, *Probability with Statistical Applications*, © 1970, Addison-Wesley, Reading, Mass. Reprinted with permission.

Table A-1 (continued)

n	x	0.01	0.05	0.10	0.20	0.30	0.40	P 0.50	0.60	0.70	0.80	0.90	0.95	0.99	x
	5	0+	0+	0+	004	025	077	164	261	318	275	124	041	002	5
	6	0+	0+	0+	0+	004	017	055	131	247	367	372	257	066	6
	7	0+	0+	0+	0+	0+	002	008	028	082	210	478	698	932	7
8	0	923	663	430	168	058	017	004	001	0+	0+	0+	0+	0+	0
	1	075	279	383	336	198	090	031	008	001	0+	0+	0+	0+	1
	2	003	051	149	294	296	209	109	041	010	001	0+	0+	0+	2
	3	0+	005	033	147	254	279	219	124	047	009	0+	0+	0+	3
	4	0+	0+	005	046	136	232	273	232	136	046	005	0+	0+	4
	5	0+	0+	0+	009	047	124	219	279	254	147	033	005	0+	5
	6	0+	0+	0+	001	010	041	109	209	296	294	149	051	003	6
	7	0+	0+	0+	0+	001	008	031	090	198	336	383	279	075	7
	8	0+	0+	0+	0+	0+	001	004	017	058	168	430	663	923	8
9	0	914	630	387	134	040	010	002	0+	0+	0+	0+	0+	0+	0
	1	083	299	387	302	156	060	018	004	0+	0+	0+	0+	0+	1
	2	003	063	172	302	267	161	070	021	004	0+	0+	0+	0+	2
	3	0+	008	045	176	267	251	164	074	021	003	0+	0+	0+	3
	4	0+	001	007	066	172	251	246	167	074	017	001	0+	0+	4
	5	0+	0+	001	017	074	167	246	251	172	066	007	001	0+	5
	6	0+	0+	0+	003	021	074	164	251	267	176	045	008	0+	6
	7	0+	0+	0+	0+	004	021	070	161	267	302	172	063	003	7
	8	0+	0+	0+	0+	0+	004	018	060	156	302	387	299	083	8
	9	0+	0+	0+	0+	0+	0+	002	010	040	134	387	630	914	9
10	0	904	599	349	107	028	006	001	0+	0+	0+	0+	0+	0+	0
	1	091	315	387	268	121	040	010	002	0+	0+	0+	0+	0+	1
	2	004	075	194	302	233	121	044	011	001	0+	0+	0+	0+	2
	3	0+	010	057	201	267	215	117	042	009	001	0+	0+	0+	3
	4	0+	001	011	088	200	251	205	111	037	006	0+	0+	0+	4
	5	0+	0+	001	026	103	201	246	201	103	026	001	0+	0+	5
	6	0+	0+	0+	006	037	111	205	251	200	088	011	001	0+	6
	7	0+	0+	0+	001	009	042	117	215	267	201	057	010	0+	7
	8	0+	0+	0+	0+	001	011	044	121	233	302	194	075	004	8
	9	0+	0+	0+	0+	0+	002	010	040	121	268	387	315	091	9
	10	0+	0+	0+	0+	0+	0+	001	006	028	107	349	599	904	10
11	0	895	569	314	086	020	004	0+	0+	0+	0+	0+	0+	0+	0
	1	099	329	384	236	093	027	005	001	0+	0+	0+	0+	0+	1
	2	005	087	213	295	200	089	027	005	001	0+	0+	0+	0+	2
	3	0+	014	071	221	257	177	081	023	004	0+	0+	0+	0+	3
	4	0+	001	016	111	220	236	161	070	017	002	0+	0+	0+	4
	5	0+	0+	002	039	132	221	226	147	057	010	0+	0+	0+	5
	6	0+	0+	0+	010	057	147	226	221	132	039	002	0+	0+	6
	7	0+	0+	0+	002	017	070	161	236	220	111	016	001	0+	7
	8	0+	0+	0+	0+	004	023	081	177	257	221	071	014	0+	8
	9	0+	0+	0+	0+	001	005	027	089	200	295	213	087	005	9
	10	0+	0+	0+	0+	0+	001	005	027	093	236	384	329	099	10
	11	0+	0+	0+	0+	0+	0+	0+	004	020	086	314	569	895	11
12	0	886	540	282	069	014	002	0+	0+	0+	0+	0+	0+	0+	0
	1	107	341	377	206	071	017	003	0+	0+	0+	0+	0+	0+	1
	2	006	099	230	283	168	064	016	002	0+	0+	0+	0+	0+	2
	3	0+	017	085	236	240	142	054	012	001	0+	0+	0+	0+	3
	4	0+	002	021	133	231	213	121	042	008	001	0+	0+	0+	4

Table A-1 (continued)

n x	0.01	0.05	0.10	0.20	0.30	0.40	P 0.50	0.60	0.70	0.80	0.90	0.95	0.99	x
12 5	0+	0+	004	053	158	227	193	101	029	003	0+	0+	0+	5
6	0+	0+	0+	016	079	177	226	177	079	016	0+	0+	0+	6
7	0+	0+	0+	003	029	101	193	227	158	053	004	0+	0+	7
8	0+	0+	0+	001	008	042	121	213	231	133	021	002	0+	8
9	0+	0+	0+	0+	001	012	054	142	240	236	085	017	0+	9
10	0+	0+	0+	0+	0+	002	016	064	168	283	230	099	006	10
11	0+	0+	0+	0+	0+	0+	003	017	071	206	377	341	107	11
12	0+	0+	0+	0+	0+	0+	0+	002	014	069	282	540	886	12
13 0	878	513	254	055	010	001	0+	0+	0+	0+	0+	0+	0+	0
1	115	351	367	179	054	011	002	0+	0+	0+	0+	0+	0+	1
2	007	111	245	268	139	045	010	001	0+	0+	0+	0+	0+	2
3	0+	021	100	246	218	111	035	006	001	0+	0+	0+	0+	3
4	0+	003	028	154	234	184	087	024	003	0+	0+	0+	0+	4
5	0+	0+	006	069	180	221	157	066	014	001	0+	0+	0+	5
6	0+	0+	001	023	103	197	209	131	044	006	0+	0+	0+	6
7	0+	0+	0+	006	044	131	209	197	103	023	001	0+	0+	7
8	0+	0+	0+	001	014	066	157	221	180	069	006	0+	0+	8
9	0+	0+	0+	0+	003	024	087	184	234	154	028	003	0+	9
10	0+	0+	0+	0+	001	006	035	111	218	246	100	021	0+	10
11	0+	0+	0+	0+	0+	001	010	045	139	268	245	111	007	11
12	0+	0+	0+	0+	0+	0+	002	011	054	179	367	351	115	12
13	0+	0+	0+	0+	0+	0+	0+	001	010	055	254	513	878	13
14 0	869	488	229	044	007	001	0+	0+	0+	0+	0+	0+	0+	0
1	123	359	356	154	041	007	001	0+	0+	0+	0+	0+	0+	1
2	008	123	257	250	113	032	006	001	0+	0+	0+	0+	0+	2
3	0+	026	114	250	194	085	022	003	0+	0+	0+	0+	0+	3
4	0+	004	035	172	229	155	061	014	001	0+	0+	0+	0+	4
5	0+	0+	008	086	196	207	122	041	007	0+	0+	0+	0+	5
6	0+	0+	001	032	126	207	183	092	023	002	0+	0+	0+	6
7	0+	0+	0+	009	062	157	209	157	062	009	0+	0+	0+	7
8	0+	0+	0+	002	023	092	183	207	126	032	001	0+	0+	8
9	0+	0+	0+	0+	007	041	122	207	196	086	008	0+	0+	9
10	0+	0+	0+	0+	001	014	061	155	229	172	035	004	0+	10
11	0+	0+	0+	0+	0+	003	022	085	194	250	114	026	0+	11
12	0+	0+	0+	0+	0+	001	006	032	113	250	257	123	008	12
13	0+	0+	0+	0+	0+	0+	001	007	041	154	356	359	123	13
14	0+	0+	0+	0+	0+	0+	0+	001	007	044	229	488	869	14
15 0	860	463	206	035	005	0+	0+	0+	0+	0+	0+	0+	0+	0
1	130	366	343	132	031	005	0+	0+	0+	0+	0+	0+	0+	1
2	009	135	267	231	092	022	003	0+	0+	0+	0+	0+	0+	2
3	0+	031	129	250	170	063	014	002	0+	0+	0+	0+	0+	3
4	0+	005	043	188	219	127	042	007	001	0+	0+	0+	0+	4
5	0+	001	010	103	206	186	092	024	003	0+	0+	0+	0+	5
6	0+	0+	002	043	147	207	153	061	012	001	0+	0+	0+	6
7	0+	0+	0+	014	081	177	196	118	035	003	0+	0+	0+	7
8	0+	0+	0+	003	035	118	196	177	081	014	0+	0+	0+	8
9	0+	0+	0+	001	012	061	153	207	147	043	002	0+	0+	9
10	0+	0+	0+	0+	003	024	092	186	206	103	010	001	0+	10
11	0+	0+	0+	0+	001	007	042	127	219	188	043	005	0+	11
12	0+	0+	0+	0+	0+	002	014	063	170	250	129	031	0+	12

Table A-1 (continued)

n	x	0.01	0.05	0.10	0.20	0.30	0.40	P 0.50	0.60	0.70	0.80	0.90	0.95	0.99	x
13		0+	0+	0+	0+	0+	0+	003	022	092	231	267	135	009	13
14		0+	0+	0+	0+	0+	0+	001	005	031	132	343	366	130	14
15		0+	0+	0+	0+	0+	0+	0+	0+	005	035	206	463	860	15
16	0	852	440	185	028	003	0+	0+	0+	0+	0+	0+	0+	0+	0
	1	138	371	329	113	023	003	0+	0+	0+	0+	0+	0+	0+	1
	2	010	146	274	211	073	015	002	0+	0+	0+	0+	0+	0+	2
	3	0+	036	142	246	146	047	008	001	0+	0+	0+	0+	0+	3
	4	0+	006	051	200	204	101	028	004	0+	0+	0+	0+	0+	4
	5	0+	001	014	120	210	162	067	014	001	0+	0+	0+	0+	5
	6	0+	0+	003	055	165	198	122	039	006	0+	0+	0+	0+	6
	7	0+	0+	0+	020	101	189	175	084	018	001	0+	0+	0+	7
	8	0+	0+	0+	006	049	142	196	142	049	006	0+	0+	0+	8
	9	0+	0+	0+	001	018	084	175	189	101	020	0+	0+	0+	9
	10	0+	0+	0+	0+	006	039	122	198	165	055	003	0+	0+	10
	11	0+	0+	0+	0+	001	014	067	162	210	120	014	001	0+	11
	12	0+	0+	0+	0+	0+	004	028	101	204	200	051	006	0+	12
	13	0+	0+	0+	0+	0+	001	008	047	146	246	142	036	0+	13
	14	0+	0+	0+	0+	0+	0+	002	015	073	211	274	146	010	14
	15	0+	0+	0+	0+	0+	0+	0+	003	023	113	329	371	138	15
	16	0+	0+	0+	0+	0+	0+	0+	0+	003	028	185	440	852	16
17	0	843	418	167	022	002	0+	0+	0+	0+	0+	0+	0+	0+	0
	1	145	374	315	096	017	002	0+	0+	0+	0+	0+	0+	0+	1
	2	012	158	280	191	058	010	001	0+	0+	0+	0+	0+	0+	2
	3	001	042	156	239	124	034	005	0+	0+	0+	0+	0+	0+	3
	4	0+	008	060	209	187	080	018	002	0+	0+	0+	0+	0+	4
	5	0+	001	018	136	208	138	047	008	001	0+	0+	0+	0+	5
	6	0+	0+	004	068	178	184	094	024	003	0+	0+	0+	0+	6
	7	0+	0+	001	027	120	193	148	057	010	0+	0+	0+	0+	7
	8	0+	0+	0+	008	064	161	186	107	028	002	0+	0+	0+	8
	9	0+	0+	0+	002	028	107	186	161	064	008	0+	0+	0+	9
	10	0+	0+	0+	0+	010	057	148	193	120	027	001	0+	0+	10
	11	0+	0+	0+	0+	003	024	094	184	178	068	004	0+	0+	11
	12	0+	0+	0+	0+	001	008	047	138	208	136	018	001	0+	12
	13	0+	0+	0+	0+	0+	002	018	080	187	209	060	008	0+	13
	14	0+	0+	0+	0+	0+	0+	005	034	124	239	156	042	001	14
	15	0+	0+	0+	0+	0+	0+	001	010	058	191	280	158	012	15
	16	0+	0+	0+	0+	0+	0+	0+	002	017	096	315	374	145	16
	17	0+	0+	0+	0+	0+	0+	0+	0+	002	022	167	418	843	17
18	0	834	397	150	018	002	0+	0+	0+	0+	0+	0+	0+	0+	0
	1	152	376	300	081	013	001	0+	0+	0+	0+	0+	0+	0+	1
	2	013	168	284	172	046	007	001	0+	0+	0+	0+	0+	0+	2
	3	001	047	168	230	105	025	003	0+	0+	0+	0+	0+	0+	3
	4	0+	009	070	215	168	061	012	001	0+	0+	0+	0+	0+	4
	5	0+	001	022	151	202	115	033	004	0+	0+	0+	0+	0+	5
	6	0+	0+	005	082	187	166	071	014	001	0+	0+	0+	0+	6
	7	0+	0+	001	035	138	189	121	037	005	0+	0+	0+	0+	7
	8	0+	0+	0+	012	081	173	167	077	015	001	0+	0+	0+	8
	9	0+	0+	0+	003	039	128	186	128	039	003	0+	0+	0+	9

Table A-1 (continued)

n	x	0.01	0.05	0.10	0.20	0.30	0.40	P 0.50	0.60	0.70	0.80	0.90	0.95	0.99	x
	10	0+	0+	0+	001	015	077	167	173	081	012	0+	0+	0+	10
	11	0+	0+	0+	0+	005	037	121	189	138	035	001	0+	0+	11
	12	0+	0+	0+	0+	001	014	071	166	187	082	005	0+	0+	12
	13	0+	0+	0+	0+	0+	004	033	115	202	151	022	001	0+	13
	14	0+	0+	0+	0+	0+	001	012	061	168	215	070	009	0+	14
	15	0+	0+	0+	0+	0+	0+	003	025	105	230	168	047	001	15
	16	0+	0+	0+	0+	0+	0+	001	007	046	172	284	168	013	16
	17	0+	0+	0+	0+	0+	0+	0+	001	013	081	300	376	152	17
	18	0+	0+	0+	0+	0+	0+	0+	0+	002	018	150	397	834	18
19	0	826	377	135	014	001	0+	0+	0+	0+	0+	0+	0+	0+	0
	1	159	377	285	068	009	001	0+	0+	0+	0+	0+	0+	0+	1
	2	014	179	285	154	036	005	0+	0+	0+	0+	0+	0+	0+	2
	3	001	053	180	218	087	018	002	0+	0+	0+	0+	0+	0+	3
	4	0+	011	080	218	149	047	007	0+	0+	0+	0+	0+	0+	4
	5	0+	002	027	164	192	093	022	002	0+	0+	0+	0+	0+	5
	6	0+	0+	007	096	192	145	052	008	0+	0+	0+	0+	0+	6
	7	0+	0+	001	044	152	180	096	024	002	0+	0+	0+	0+	7
	8	0+	0+	0+	017	098	180	144	053	008	0+	0+	0+	0+	8
	9	0+	0+	0+	005	051	146	176	098	022	001	0+	0+	0+	9
	10	0+	0+	0+	001	022	098	176	146	051	005	0+	0+	0+	10
	11	0+	0+	0+	0+	008	053	144	180	098	017	0+	0+	0+	11
	12	0+	0+	0+	0+	002	024	096	180	152	044	001	0+	0+	12
	13	0+	0+	0+	0+	0+	008	052	145	192	096	007	0+	0+	13
	14	0+	0+	0+	0+	0+	002	022	093	192	164	027	002	0+	14
	15	0+	0+	0+	0+	0+	0+	007	047	149	218	080	011	0+	15
	16	0+	0+	0+	0+	0+	0+	002	018	087	218	180	053	001	16
	17	0+	0+	0+	0+	0+	0+	0+	005	036	154	285	179	014	17
	18	0+	0+	0+	0+	0+	0+	0+	001	009	068	285	377	159	18
	19	0+	0+	0+	0+	0+	0+	0+	0+	001	014	135	377	826	19
20	0	818	358	122	012	001	0+	0+	0+	0+	0+	0+	0+	0+	0
	1	165	377	270	058	007	0+	0+	0+	0+	0+	0+	0+	0+	1
	2	016	189	285	137	028	003	0+	0+	0+	0+	0+	0+	0+	2
	3	001	060	190	205	072	012	001	0+	0+	0+	0+	0+	0+	3
	4	0+	013	090	218	130	035	005	0+	0+	0+	0+	0+	0+	4
	5	0+	002	032	175	179	075	015	001	0+	0+	0+	0+	0+	5
	6	0+	0+	009	109	192	124	037	005	0+	0+	0+	0+	0+	6
	7	0+	0+	002	054	164	166	074	015	001	0+	0+	0+	0+	7
	8	0+	0+	0+	022	114	180	120	036	004	0+	0+	0+	0+	8
	9	0+	0+	0+	007	065	160	160	071	012	0+	0+	0+	0+	9
	10	0+	0+	0+	002	031	117	176	117	031	002	0+	0+	0+	10
	11	0+	0+	0+	0+	012	071	160	160	065	007	0+	0+	0+	11
	12	0+	0+	0+	0+	004	036	120	180	114	022	0+	0+	0+	12
	13	0+	0+	0+	0+	001	015	074	166	164	054	002	0+	0+	13
	14	0+	0+	0+	0+	0+	005	037	124	192	109	009	0+	0+	14
	15	0+	0+	0+	0+	0+	001	015	075	179	175	032	002	0+	15
	16	0+	0+	0+	0+	0+	0+	005	035	130	218	090	013	0+	16
	17	0+	0+	0+	0+	0+	0+	001	012	072	205	190	060	001	17
	18	0+	0+	0+	0+	0+	0+	0+	003	028	137	285	189	016	18
	19	0+	0+	0+	0+	0+	0+	0+	0+	007	058	270	377	165	19
	20	0+	0+	0+	0+	0+	0+	0+	0+	001	012	122	358	818	20

Table A-2 Values of e^{-a}

a	e^{-a}	a	e^{-a}	a	e^{-a}	a	e^{-a}
0.00	1.000000	2.60	.074274	5.10	.006097	7.60	.000501
0.10	.904837	2.70	.067206	5.20	.005517	7.70	.000453
0.20	.818731	2.80	.060810	5.30	.004992	7.80	.000410
0.30	.740818	2.90	.055023	5.40	.004517	7.90	.000371
0.40	.670320	3.00	.049787	5.50	.004087	8.00	.000336
0.50	.606531	3.10	.045049	5.60	.003698	8.10	.000304
0.60	.548812	3.20	.040762	5.70	.003346	8.20	.000275
0.70	.496585	3.30	.036883	5.80	.003028	8.30	.000249
0.80	.449329	3.40	.033373	5.90	.002739	8.40	.000225
0.90	.406570	3.50	.030197	6.00	.002479	8.50	.000204
1.00	.367879	3.60	.027324	6.10	.002243	8.60	.000184
1.10	.332871	3.70	.024724	6.20	.002029	8.70	.000167
1.20	.301194	3.80	.022371	6.30	.001836	8.80	.000151
1.30	.272532	3.90	.020242	6.40	.001661	8.90	.000136
1.40	.246597	4.00	.018316	6.50	.001503	9.00	.000123
1.50	.223130	4.10	.016573	6.60	.001360	9.10	.000112
1.60	.201897	4.20	.014996	6.70	.001231	9.20	.000101
1.70	.182684	4.30	.013569	6.80	.001114	9.30	.000091
1.80	.165299	4.40	.012277	6.90	.001008	9.40	.000083
1.90	.149569	4.50	.011109	7.00	.000912	9.50	.000075
2.00	.135335	4.60	.010052	7.10	.000825	9.60	.000068
2.10	.122456	4.70	.009095	7.20	.000747	9.70	.000061
2.20	.110803	4.80	.008230	7.30	.000676	9.80	.000056
2.30	.100259	4.90	.007447	7.40	.000611	9.90	.000050
2.40	.090718	5.00	.006738	7.50	.000553	10.00	.000045
2.50	.082085						

Table A-3 Poisson Probabilities $\left[\dfrac{e^{-\mu}\mu^x}{x!}\right]$

x	\multicolumn{10}{c}{μ}

x	0.005	0.01	0.02	0.03	0.04	0.05	0.06	0.07	0.08	0.09
0	0.9950	0.9900	0.9802	0.9704	0.9608	0.9512	0.9418	0.9324	0.9231	0.9139
1	0.0050	0.0099	0.0192	0.0291	0.0384	0.0476	0.0565	0.0653	0.0738	0.0823
2	0.0000	0.0000	0.0002	0.0004	0.0008	0.0012	0.0017	0.0023	0.0030	0.0037
3	0.0000	0.0000	0.0000	0.0000	0.0000	0.0000	0.0000	0.0001	0.0001	0.0001

x	0.1	0.2	0.3	0.4	0.5	0.6	0.7	0.8	0.9	1.0
0	0.9048	0.8187	0.7408	0.6703	0.6065	0.5488	0.4966	0.4493	0.4066	0.3679
1	0.0905	0.1637	0.2222	0.2681	0.3033	0.3293	0.3476	0.3595	0.3659	0.3679
2	0.0045	0.0164	0.0333	0.0536	0.0758	0.0988	0.1217	0.1438	0.1647	0.1839
3	0.0002	0.0011	0.0033	0.0072	0.0126	0.0198	0.0284	0.0383	0.0494	0.0613
4	0.0000	0.0001	0.0002	0.0007	0.0016	0.0030	0.0050	0.0077	0.0111	0.0153

Source: Robert Parsons, *Statistical Analysis: A Decision Making Approach*, 2d ed. (New York: Harper & Row, 1978). Reproduced with permission.

Table A-3 (continued)

x	0.1	0.2	0.3	0.4	0.5	0.6	0.7	0.8	0.9	1.0
5	0.0000	0.0000	0.0000	0.0001	0.0002	0.0004	0.0007	0.0012	0.0020	0.0031
6	0.0000	0.0000	0.0000	0.0000	0.0000	0.0000	0.0001	0.0002	0.0003	0.0005
7	0.0000	0.0000	0.0000	0.0000	0.0000	0.0000	0.0000	0.0000	0.0000	0.0001

x	1.1	1.2	1.3	1.4	1.5	1.6	1.7	1.8	1.9	2.0
0	0.3329	0.3012	0.2725	0.2466	0.2231	0.2019	0.1827	0.1653	0.1496	0.1353
1	0.3662	0.3614	0.3543	0.3452	0.3347	0.3230	0.3106	0.2975	0.2842	0.2707
2	0.2014	0.2169	0.2303	0.2417	0.2510	0.2584	0.2640	0.2678	0.2700	0.2707
3	0.0738	0.0867	0.0998	0.1128	0.1255	0.1378	0.1496	0.1607	0.1710	0.1804
4	0.0203	0.0260	0.0324	0.0395	0.0471	0.0551	0.0636	0.0723	0.0812	0.0902
5	0.0045	0.0062	0.0084	0.0111	0.0141	0.0176	0.0216	0.0260	0.0309	0.0361
6	0.0008	0.0012	0.0018	0.0026	0.0035	0.0047	0.0061	0.0078	0.0098	0.0120
7	0.0001	0.0002	0.0003	0.0005	0.0008	0.0011	0.0015	0.0020	0.0027	0.0034
8	0.0000	0.0000	0.0001	0.0001	0.0001	0.0002	0.0003	0.0005	0.0006	0.0009
9	0.0000	0.0000	0.0000	0.0000	0.0000	0.0000	0.0001	0.0001	0.0001	0.0002

x	2.1	2.2	2.3	2.4	2.5	2.6	2.7	2.8	2.9	3.0
0	0.1225	0.1108	0.1003	0.0907	0.0821	0.0743	0.0672	0.0608	0.0050	0.0498
1	0.2572	0.2438	0.2306	0.2177	0.2052	0.1931	0.1815	0.1703	0.1596	0.1494
2	0.2700	0.2681	0.2652	0.2613	0.2565	0.2510	0.2450	0.2384	0.2314	0.2240
3	0.1890	0.1966	0.2033	0.2090	0.2138	0.2176	0.2205	0.2225	0.2237	0.2240
4	0.0992	0.1082	0.1169	0.1254	0.1336	0.1414	0.1488	0.1557	0.1622	0.1680
5	0.0417	0.0476	0.0538	0.0602	0.0668	0.0735	0.0804	0.0872	0.0940	0.1008
6	0.0146	0.0174	0.0206	0.0241	0.0278	0.0319	0.0362	0.0407	0.0455	0.0504
7	0.0044	0.0055	0.0068	0.0083	0.0099	0.0118	0.0139	0.0163	0.0188	0.0216
8	0.0011	0.0015	0.0019	0.0025	0.0031	0.0038	0.0047	0.0057	0.0068	0.0081
9	0.0003	0.0004	0.0005	0.0007	0.0009	0.0011	0.0014	0.0018	0.0022	0.0027
10	0.0001	0.0001	0.0001	0.0002	0.0002	0.0003	0.0004	0.0005	0.0006	0.0008
11	0.0000	0.0000	0.0000	0.0000	0.0000	0.0001	0.0001	0.0001	0.0002	0.0002
12	0.0000	0.0000	0.0000	0.0000	0.0000	0.0000	0.0000	0.0000	0.0000	0.0001

x	3.1	3.2	3.3	3.4	3.5	3.6	3.7	3.8	3.9	4.0
0	0.0450	0.0408	0.0369	0.0334	0.0302	0.0273	0.0247	0.0224	0.0202	0.0183
1	0.1397	0.1304	0.1217	0.1135	0.1057	0.0984	0.0915	0.0850	0.0789	0.0733
2	0.2165	0.2087	0.2008	0.1929	0.1850	0.1771	0.1692	0.1615	0.1539	0.1465
3	0.2237	0.2226	0.2209	0.2186	0.2158	0.2125	0.2087	0.2046	0.2001	0.1954
4	0.1734	0.1781	0.1823	0.1858	0.1888	0.1912	0.1931	0.1944	0.1951	0.1954
5	0.1075	0.1140	0.1203	0.1264	0.1322	0.1377	0.1429	0.1477	0.1522	0.1563
6	0.0555	0.0608	0.0662	0.0716	0.0771	0.0826	0.0881	0.0936	0.0989	0.1042
7	0.0246	0.0278	0.0312	0.0348	0.0385	0.0425	0.0466	0.0508	0.0551	0.0595
8	0.0095	0.0111	0.0129	0.0148	0.0169	0.0191	0.0215	0.0241	0.0269	0.0298
9	0.0033	0.0040	0.0047	0.0056	0.0066	0.0076	0.0089	0.0102	0.0116	0.0132
10	0.0010	0.0013	0.0016	0.0019	0.0023	0.0028	0.0033	0.0039	0.0045	0.0053
11	0.0003	0.0004	0.0005	0.0006	0.0007	0.0009	0.0011	0.0013	0.0016	0.0019
12	0.0001	0.0001	0.0001	0.0002	0.0002	0.0003	0.0003	0.0004	0.0005	0.0006
13	0.0000	0.0000	0.0000	0.0000	0.0001	0.0001	0.0001	0.0001	0.0002	0.0002
14	0.0000	0.0000	0.0000	0.0000	0.0000	0.0000	0.0000	0.0000	0.0000	0.0001

Table A-3 (continued)

x	4.1	4.2	4.3	4.4	4.5	4.6	4.7	4.8	4.9	5.0
0	0.0166	0.0150	0.0136	0.0123	0.0111	0.0101	0.0091	0.0082	0.0074	0.0067
1	0.0679	0.0630	0.0583	0.0540	0.0500	0.0462	0.0427	0.0395	0.0365	0.0337
2	0.1393	0.1323	0.1254	0.1188	0.1125	0.1063	0.1005	0.0948	0.0894	0.0842
3	0.1904	0.1852	0.1798	0.1743	0.1687	0.1631	0.1574	0.1517	0.1460	0.1404
4	0.1951	0.1944	0.1933	0.1917	0.1898	0.1875	0.1849	0.1820	0.1789	0.1755
5	0.1600	0.1633	0.1662	0.1687	0.1708	0.1725	0.1738	0.1747	0.1753	0.1755
6	0.1093	0.1143	0.1191	0.1237	0.1281	0.1323	0.1362	0.1398	0.1432	0.1462
7	0.0640	0.0686	0.0732	0.0778	0.0824	0.0869	0.0914	0.0959	0.1002	0.1044
8	0.0328	0.0360	0.0393	0.0428	0.0463	0.0500	0.0537	0.0575	0.0614	0.0653
9	0.0150	0.0168	0.0188	0.0209	0.0232	0.0255	0.0280	0.0307	0.0334	0.0363
10	0.0061	0.0071	0.0081	0.0092	0.0104	0.0118	0.0132	0.0147	0.0164	0.0181
11	0.0023	0.0027	0.0032	0.0037	0.0043	0.0049	0.0056	0.0064	0.0073	0.0082
12	0.0008	0.0009	0.0011	0.0014	0.0016	0.0019	0.0022	0.0026	0.0030	0.0034
13	0.0002	0.0003	0.0004	0.0005	0.0006	0.0007	0.0008	0.0009	0.0011	0.0013
14	0.0001	0.0001	0.0001	0.0001	0.0002	0.0002	0.0003	0.0003	0.0004	0.0005
15	0.0000	0.0000	0.0000	0.0000	0.0001	0.0001	0.0001	0.0001	0.0001	0.0002

x	5.1	5.2	5.3	5.4	5.5	5.6	5.7	5.8	5.9	6.0
0	0.0061	0.0055	0.0050	0.0045	0.0041	0.0037	0.0033	0.0030	0.0027	0.0025
1	0.0311	0.0287	0.0265	0.0244	0.0225	0.0207	0.0191	0.0176	0.0162	0.0149
2	0.0793	0.0746	0.0701	0.0659	0.0618	0.0580	0.0544	0.0509	0.0477	0.0446
3	0.1348	0.1293	0.1239	0.1185	0.1133	0.1082	0.1033	0.0985	0.0938	0.0892
4	0.1719	0.1681	0.1641	0.1600	0.1558	0.1515	0.1472	0.1428	0.1383	0.1339
5	0.1753	0.1748	0.1740	0.1728	0.1714	0.1697	0.1678	0.1656	0.1632	0.1606
6	0.1490	0.1515	0.1537	0.1555	0.1571	0.1584	0.1594	0.1601	0.1605	0.1606
7	0.1086	0.1125	0.1163	0.1200	0.1234	0.1267	0.1298	0.1326	0.1353	0.1377
8	0.0692	0.0731	0.0771	0.0810	0.0849	0.0887	0.0925	0.0962	0.0998	0.1033
9	0.0392	0.0423	0.0454	0.0486	0.0519	0.0552	0.0586	0.0620	0.0654	0.0688
10	0.0200	0.0220	0.0241	0.0262	0.0285	0.0309	0.0334	0.0359	0.0386	0.0413
11	0.0093	0.0104	0.0116	0.0129	0.0143	0.0157	0.0173	0.0190	0.0207	0.0225
12	0.0039	0.0045	0.0051	0.0058	0.0065	0.0073	0.0082	0.0092	0.0102	0.0113
13	0.0015	0.0018	0.0021	0.0024	0.0028	0.0032	0.0036	0.0041	0.0046	0.0052
14	0.0006	0.0007	0.0008	0.0009	0.0011	0.0013	0.0015	0.0017	0.0019	0.0022
15	0.0002	0.0002	0.0003	0.0003	0.0004	0.0005	0.0006	0.0007	0.0008	0.0009
16	0.0001	0.0001	0.0001	0.0001	0.0001	0.0002	0.0002	0.0002	0.0003	0.0003
17	0.0000	0.0000	0.0000	0.0000	0.0000	0.0001	0.0001	0.0001	0.0001	0.0001

x	6.1	6.2	6.3	6.4	6.5	6.6	6.7	6.8	6.9	7.0
0	0.0022	0.0020	0.0018	0.0017	0.0015	0.0014	0.0012	0.0011	0.0010	0.0009
1	0.0137	0.0126	0.0116	0.0106	0.0098	0.0090	0.0082	0.0076	0.0070	0.0064
2	0.0417	0.0390	0.0364	0.0340	0.0318	0.0296	0.0276	0.0258	0.0240	0.0223
3	0.0848	0.0806	0.0765	0.0726	0.0688	0.0652	0.0617	0.0584	0.0552	0.0521
4	0.1294	0.1249	0.1205	0.1162	0.1118	0.1076	0.1034	0.0992	0.0952	0.0912
5	0.1579	0.1549	0.1519	0.1487	0.1454	0.1420	0.1385	0.1349	0.1314	0.1277
6	0.1605	0.1601	0.1595	0.1586	0.1575	0.1562	0.1546	0.1529	0.1511	0.1490

Table A-3 (continued)

					μ					
x	6.1	6.2	6.3	6.4	6.5	6.6	6.7	6.8	6.9	7.0
7	0.1399	0.1418	0.1435	0.1450	0.1462	0.1472	0.1480	0.1486	0.1489	0.1490
8	0.1066	0.1099	0.1130	0.1160	0.1188	0.1215	0.1240	0.1263	0.1284	0.1304
9	0.0723	0.0757	0.0791	0.0825	0.0858	0.0891	0.0923	0.0954	0.0985	0.1014
10	0.0441	0.0469	0.0498	0.0528	0.0558	0.0588	0.0618	0.0649	0.0679	0.0710
11	0.0245	0.0265	0.0285	0.0307	0.0330	0.0353	0.0377	0.0401	0.0426	0.0452
12	0.0124	0.0137	0.0150	0.0164	0.0179	0.0194	0.0210	0.0227	0.0245	0.0264
13	0.0058	0.0065	0.0073	0.0081	0.0089	0.0098	0.0108	0.0119	0.0130	0.0142
14	0.0025	0.0029	0.0033	0.0037	0.0041	0.0046	0.0052	0.0058	0.0064	0.0071
15	0.0010	0.0012	0.0014	0.0016	0.0018	0.0020	0.0023	0.0026	0.0029	0.0033
16	0.0004	0.0005	0.0005	0.0006	0.0007	0.0008	0.0010	0.0011	0.0013	0.0014
17	0.0001	0.0002	0.0002	0.0002	0.0003	0.0003	0.0004	0.0004	0.0005	0.0006
18	0.0000	0.0001	0.0001	0.0001	0.0001	0.0001	0.0001	0.0002	0.0002	0.0002
19	0.0000	0.0000	0.0000	0.0000	0.0000	0.0000	0.0000	0.0001	0.0001	0.0001

x	7.1	7.2	7.3	7.4	7.5	7.6	7.7	7.8	7.9	8.0
0	0.0008	0.0007	0.0007	0.0006	0.0006	0.0005	0.0005	0.0004	0.0004	0.0003
1	0.0059	0.0054	0.0049	0.0045	0.0041	0.0038	0.0035	0.0032	0.0029	0.0027
2	0.0208	0.0194	0.0180	0.0167	0.0156	0.0145	0.0134	0.0125	0.0116	0.0107
3	0.0492	0.0464	0.0438	0.0413	0.0389	0.0366	0.0345	0.0324	0.0305	0.0286
4	0.0874	0.0836	0.0799	0.0764	0.0729	0.0696	0.0663	0.0632	0.0602	0.0573
5	0.1241	0.1204	0.1167	0.1130	0.1094	0.1057	0.1021	0.0986	0.0951	0.0916
6	0.1468	0.1445	0.1420	0.1394	0.1367	0.1339	0.1311	0.1282	0.1252	0.1221
7	0.1489	0.1486	0.1481	0.1474	0.1465	0.1454	0.1442	0.1428	0.1413	0.1396
8	0.1321	0.1337	0.1351	0.1363	0.1373	0.1382	0.1388	0.1392	0.1395	0.1396
9	0.1042	0.1070	0.1096	0.1121	0.1144	0.1167	0.1187	0.1207	0.1224	0.1241
10	0.0740	0.0770	0.0800	0.0829	0.0858	0.0887	0.0914	0.0941	0.0967	0.0993
11	0.0478	0.0504	0.0531	0.0558	0.0585	0.0613	0.0640	0.0667	0.0695	0.0722
12	0.0283	0.0303	0.0323	0.0344	0.0366	0.0388	0.0411	0.0434	0.0457	0.0481
13	0.0154	0.0168	0.0181	0.0196	0.0211	0.0227	0.0243	0.0260	0.0278	0.0296
14	0.0078	0.0086	0.0095	0.0104	0.0113	0.0123	0.0134	0.0145	0.0157	0.0169
15	0.0037	0.0041	0.0046	0.0051	0.0057	0.0062	0.0069	0.0075	0.0083	0.0090
16	0.0016	0.0019	0.0021	0.0024	0.0026	0.0030	0.0033	0.0037	0.0041	0.0045
17	0.0007	0.0008	0.0009	0.0010	0.0012	0.0013	0.0015	0.0017	0.0019	0.0021
18	0.0003	0.0003	0.0004	0.0004	0.0005	0.0006	0.0006	0.0007	0.0008	0.0009
19	0.0001	0.0001	0.0001	0.0002	0.0002	0.0002	0.0003	0.0003	0.0003	0.0004
20	0.0000	0.0000	0.0001	0.0001	0.0001	0.0001	0.0001	0.0001	0.0001	0.0002
21	0.0000	0.0000	0.0000	0.0000	0.0000	0.0000	0.0000	0.0000	0.0001	0.0001

x	8.1	8.2	8.3	8.4	8.5	8.6	8.7	8.8	8.9	9.0
0	0.0003	0.0003	0.0002	0.0002	0.0002	0.0002	0.0002	0.0002	0.0001	0.0001
1	0.0025	0.0023	0.0021	0.0019	0.0017	0.0016	0.0014	0.0013	0.0012	0.0011
2	0.0100	0.0092	0.0086	0.0079	0.0074	0.0068	0.0063	0.0058	0.0054	0.0050
3	0.0269	0.0252	0.0237	0.0222	0.0208	0.0195	0.0183	0.0171	0.0160	0.0150
4	0.0544	0.0517	0.0491	0.0466	0.0443	0.0420	0.0398	0.0377	0.0357	0.0337
5	0.0882	0.0849	0.0816	0.0784	0.0752	0.0722	0.0692	0.0663	0.0635	0.0607
6	0.1191	0.1160	0.1128	0.1097	0.1066	0.1034	0.1003	0.0972	0.0941	0.0911
7	0.1378	0.1358	0.1338	0.1317	0.1294	0.1271	0.1247	0.1222	0.1197	0.1171
8	0.1395	0.1392	0.1388	0.1382	0.1375	0.1366	0.1356	0.1344	0.1332	0.1318
9	0.1256	0.1269	0.1280	0.1290	0.1299	0.1306	0.1311	0.1315	0.1317	0.1318

Table A-3 (continued)

					μ					
x	8.1	8.2	8.3	8.4	8.5	8.6	8.7	8.8	8.9	9.0
10	0.1017	0.1040	0.1063	0.1084	0.1104	0.1123	0.1140	0.1157	0.1172	0.1186
11	0.0749	0.0776	0.0802	0.0828	0.0853	0.0878	0.0902	0.0925	0.0948	0.0970
12	0.0505	0.0530	0.0555	0.0579	0.0604	0.0629	0.0654	0.0679	0.0703	0.0728
13	0.0315	0.0334	0.0354	0.0374	0.0395	0.0416	0.0438	0.0459	0.0481	0.0504
14	0.0182	0.0196	0.0210	0.0225	0.0240	0.0256	0.0272	0.0289	0.0306	0.0324
15	0.0098	0.0107	0.0116	0.0126	0.0136	0.0147	0.0158	0.0169	0.0182	0.0194
16	0.0050	0.0055	0.0060	0.0066	0.0072	0.0079	0.0086	0.0093	0.0101	0.0109
17	0.0024	0.0026	0.0029	0.0033	0.0036	0.0040	0.0044	0.0048	0.0053	0.0058
18	0.0011	0.0012	0.0014	0.0015	0.0017	0.0019	0.0021	0.0024	0.0026	0.0029
19	0.0005	0.0005	0.0006	0.0007	0.0008	0.0009	0.0010	0.0011	0.0012	0.0014
20	0.0002	0.0002	0.0002	0.0003	0.0003	0.0004	0.0004	0.0005	0.0005	0.0006
21	0.0001	0.0001	0.0001	0.0001	0.0001	0.0002	0.0002	0.0002	0.0002	0.0003
22	0.0000	0.0000	0.0000	0.0000	0.0001	0.0001	0.0001	0.0001	0.0001	0.0001

x	9.1	9.2	9.3	9.4	9.5	9.6	9.7	9.8	9.9	10.0
0	0.0001	0.0001	0.0001	0.0001	0.0001	0.0001	0.0001	0.0001	0.0001	0.0000
1	0.0010	0.0009	0.0009	0.0008	0.0007	0.0007	0.0006	0.0005	0.0005	0.0005
2	0.0046	0.0043	0.0040	0.0037	0.0034	0.0031	0.0029	0.0027	0.0025	0.0023
3	0.0140	0.0131	0.0123	0.0115	0.0107	0.0100	0.0093	0.0087	0.0081	0.0076
4	0.0319	0.0302	0.0285	0.0269	0.0254	0.0240	0.0226	0.0213	0.0201	0.0189
5	0.0581	0.0555	0.0530	0.0506	0.0483	0.0460	0.0439	0.0418	0.0398	0.0378
6	0.0881	0.0851	0.0822	0.0793	0.0764	0.0736	0.0709	0.0682	0.0656	0.0631
7	0.1145	0.1118	0.1091	0.1064	0.1037	0.1010	0.0982	0.0955	0.0928	0.0901
8	0.1302	0.1286	0.1269	0.1251	0.1232	0.1212	0.1191	0.1170	0.1148	0.1126
9	0.1317	0.1315	0.1311	0.1306	0.1300	0.1293	0.1284	0.1274	0.1263	0.1251
10	0.1198	0.1210	0.1219	0.1228	0.1235	0.1241	0.1245	0.1249	0.1250	0.1251
11	0.0991	0.1012	0.1031	0.1049	0.1067	0.1083	0.1098	0.1112	0.1125	0.1137
12	0.0752	0.0776	0.0799	0.0822	0.0844	0.0866	0.0888	0.0908	0.0928	0.0948
13	0.0526	0.0549	0.0572	0.0594	0.0617	0.0640	0.0662	0.0685	0.0707	0.0729
14	0.0342	0.0361	0.0380	0.0399	0.0419	0.0439	0.0459	0.0479	0.0500	0.0521
15	0.0208	0.0221	0.0235	0.0250	0.0265	0.0281	0.0297	0.0313	0.0330	0.0347
16	0.0118	0.0127	0.0137	0.0147	0.0157	0.0168	0.0180	0.0192	0.0204	0.0217
17	0.0063	0.0069	0.0075	0.0081	0.0088	0.0095	0.0103	0.0111	0.0119	0.0128
18	0.0032	0.0035	0.0039	0.0042	0.0046	0.0051	0.0055	0.0060	0.0065	0.0071
19	0.0015	0.0017	0.0019	0.0021	0.0023	0.0026	0.0028	0.0031	0.0034	0.0037
20	0.0007	0.0008	0.0009	0.0010	0.0011	0.0012	0.0014	0.0015	0.0017	0.0019
21	0.0003	0.0003	0.0004	0.0004	0.0005	0.0006	0.0006	0.0007	0.0008	0.0009
22	0.0001	0.0001	0.0002	0.0002	0.0002	0.0002	0.0003	0.0003	0.0004	0.0004
23	0.0000	0.0001	0.0001	0.0001	0.0001	0.0001	0.0001	0.0001	0.0002	0.0002
24	0.0000	0.0000	0.0000	0.0000	0.0000	0.0000	0.0000	0.0001	0.0001	0.0001

Table A-4 Areas of the Standard Normal Distribution

The entries in this table are the probabilities that a standard normal random variable is between 0 and z (the shaded area).

z	0.00	0.01	0.02	0.03	0.04	0.05	0.06	0.07	0.08	0.09
						Second Decimal Place in z				
0.0	0.0000	0.0040	0.0080	0.0120	0.0160	0.0199	0.0239	0.0279	0.0319	0.0359
0.1	0.0398	0.0438	0.0478	0.0517	0.0557	0.0596	0.0636	0.0675	0.0714	0.0753
0.2	0.0793	0.0832	0.0871	0.0910	0.0948	0.0987	0.1026	0.1064	0.1103	0.1141
0.3	0.1179	0.1217	0.1255	0.1293	0.1331	0.1368	0.1406	0.1443	0.1480	0.1517
0.4	0.1554	0.1591	0.1628	0.1664	0.1700	0.1736	0.1772	0.1808	0.1844	0.1879
0.5	0.1915	0.1950	0.1985	0.2019	0.2054	0.2088	0.2123	0.2157	0.2190	0.2224
0.6	0.2257	0.2291	0.2324	0.2357	0.2389	0.2422	0.2454	0.2486	0.2517	0.2549
0.7	0.2580	0.2611	0.2642	0.2673	0.2704	0.2734	0.2764	0.2794	0.2823	0.2852
0.8	0.2881	0.2910	0.2939	0.2967	0.2995	0.3023	0.3051	0.3078	0.3106	0.3133
0.9	0.3159	0.3186	0.3212	0.3238	0.3264	0.3289	0.3315	0.3340	0.3365	0.3389
1.0	0.3413	0.3438	0.3461	0.3485	0.3508	0.3531	0.3554	0.3577	0.3599	0.3621
1.1	0.3643	0.3665	0.3686	0.3708	0.3729	0.3749	0.3770	0.3790	0.3810	0.3830
1.2	0.3849	0.3869	0.3888	0.3907	0.3925	0.3944	0.3962	0.3980	0.3997	0.4015
1.3	0.4032	0.4049	0.4066	0.4082	0.4099	0.4115	0.4131	0.4147	0.4162	0.4177
1.4	0.4192	0.4207	0.4222	0.4236	0.4251	0.4265	0.4279	0.4292	0.4306	0.4319
1.5	0.4332	0.4345	0.4357	0.4370	0.4382	0.4394	0.4406	0.4418	0.4429	0.4441
1.6	0.4452	0.4463	0.4474	0.4484	0.4495	0.4505	0.4515	0.4525	0.4535	0.4545
1.7	0.4554	0.4564	0.4573	0.4582	0.4591	0.4599	0.4608	0.4616	0.4625	0.4633
1.8	0.4641	0.4649	0.4656	0.4664	0.4671	0.4678	0.4686	0.4693	0.4699	0.4706
1.9	0.4713	0.4719	0.4726	0.4732	0.4738	0.4744	0.4750	0.4756	0.4761	0.4767
2.0	0.4772	0.4778	0.4783	0.4788	0.4793	0.4798	0.4803	0.4808	0.4812	0.4817
2.1	0.4821	0.4826	0.4830	0.4834	0.4838	0.4842	0.4846	0.4850	0.4854	0.4857
2.2	0.4861	0.4864	0.4868	0.4871	0.4875	0.4878	0.4881	0.4884	0.4887	0.4890
2.3	0.4893	0.4896	0.4898	0.4901	0.4904	0.4906	0.4909	0.4911	0.4913	0.4916
2.4	0.4918	0.4920	0.4922	0.4925	0.4927	0.4929	0.4931	0.4932	0.4934	0.4936
2.5	0.4938	0.4940	0.4941	0.4943	0.4945	0.4946	0.4948	0.4949	0.4951	0.4952
2.6	0.4953	0.4955	0.4956	0.4957	0.4959	0.4960	0.4961	0.4962	0.4963	0.4964
2.7	0.4965	0.4966	0.4967	0.4968	0.4969	0.4970	0.4971	0.4972	0.4973	0.4974
2.8	0.4974	0.4975	0.4976	0.4977	0.4977	0.4978	0.4979	0.4979	0.4980	0.4981
2.9	0.4981	0.4982	0.4982	0.4983	0.4984	0.4984	0.4985	0.4985	0.4986	0.4986
3.0	0.4987	0.4987	0.4987	0.4988	0.4988	0.4989	0.4989	0.4989	0.4990	0.4990
3.1	0.4990	0.4991	0.4991	0.4991	0.4992	0.4992	0.4992	0.4992	0.4993	0.4993
3.2	0.4993	0.4993	0.4994	0.4994	0.4994	0.4994	0.4994	0.4995	0.4995	0.4995
3.3	0.4995	0.4995	0.4995	0.4996	0.4996	0.4996	0.4996	0.4996	0.4996	0.4997
3.4	0.4997	0.4997	0.4997	0.4997	0.4997	0.4997	0.4997	0.4997	0.4997	0.4998
3.5	0.4998									
4.0	0.49997									
4.5	0.499997									
5.0	0.4999997									

Reprinted with permission from *Standard Mathematical Tables*, 15th ed., © CRC Press, Inc., Boca Raton, FL.

Table A-5 Critical Values of t

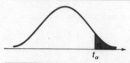

t_α

DEGREES OF FREEDOM	$t_{.100}$	$t_{.050}$	$t_{.025}$	$t_{.010}$	$t_{.005}$
1	3.078	6.314	12.706	31.821	63.657
2	1.886	2.920	4.303	6.965	9.925
3	1.638	2.353	3.182	4.541	5.841
4	1.533	2.132	2.776	3.747	4.604
5	1.476	2.015	2.571	3.365	4.032
6	1.440	1.943	2.447	3.143	3.707
7	1.415	1.895	2.365	2.998	3.499
8	1.397	1.860	2.306	2.896	3.355
9	1.383	1.833	2.262	2.821	3.250
10	1.372	1.812	2.228	2.764	3.169
11	1.363	1.796	2.201	2.718	3.106
12	1.356	1.782	2.179	2.681	3.055
13	1.350	1.771	2.160	2.650	3.012
14	1.345	1.761	2.145	2.624	2.977
15	1.341	1.753	2.131	2.602	2.947
16	1.337	1.746	2.120	2.583	2.921
17	1.333	1.740	2.110	2.567	2.898
18	1.330	1.734	2.101	2.552	2.878
19	1.328	1.729	2.093	2.539	2.861
20	1.325	1.725	2.086	2.528	2.845
21	1.323	1.721	2.080	2.518	2.831
22	1.321	1.717	2.074	2.508	2.819
23	1.319	1.714	2.069	2.500	2.807
24	1.318	1.711	2.064	2.492	2.797
25	1.316	1.708	2.060	2.485	2.787
26	1.315	1.706	2.056	2.479	2.779
27	1.314	1.703	2.052	2.473	2.771
28	1.313	1.701	2.048	2.467	2.763
29	1.311	1.699	2.045	2.462	2.756
30	1.310	1.697	2.042	2.457	2.750
40	1.303	1.684	2.021	2.423	2.704
60	1.296	1.671	2.000	2.390	2.660
120	1.289	1.658	1.980	2.358	2.617
∞	1.282	1.645	1.960	2.326	2.576

Table A-6 Critical Values of χ^2

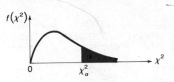

DEGREES OF FREEDOM	$\chi^2_{.995}$	$\chi^2_{.990}$	$\chi^2_{.975}$	$\chi^2_{.950}$	$\chi^2_{.900}$
1	0.0000393	0.0001571	0.0009821	0.0039321	0.0157908
2	0.0100251	0.0201007	0.0506356	0.102587	0.210720
3	0.0717212	0.114832	0.215795	0.351846	0.584375
4	0.206990	0.297110	0.484419	0.710721	1.063623
5	0.411740	0.554300	0.831211	1.145476	1.61031
6	0.675727	0.872085	1.237347	1.63539	2.20413
7	0.989265	1.239043	1.68987	2.16735	2.83311
8	1.344419	1.646482	2.17973	2.73264	3.48954
9	1.734926	2.087912	2.70039	3.32511	4.16816
10	2.15585	2.55821	3.24697	3.94030	4.86518
11	2.60321	3.05347	3.81575	4.57481	5.57779
12	3.07382	3.57056	4.40379	5.22603	6.30380
13	3.56503	4.10691	5.00874	5.89186	7.04150
14	4.07468	4.66043	5.62872	6.57063	7.78953
15	4.60094	5.22935	6.26214	7.26094	8.54675
16	5.14224	5.81221	6.90766	7.96164	9.31223
17	5.69724	6.40776	7.56418	8.67176	10.0852
18	6.26481	7.01491	8.23075	9.39046	10.8649
19	6.84398	7.63273	8.90655	10.1170	11.6509
20	7.43386	8.26040	9.59083	10.8508	12.4426
21	8.03366	8.89720	10.28293	11.5913	13.2396
22	8.64272	9.54249	10.9823	12.3380	14.0415
23	9.26042	10.19567	11.6885	13.0905	14.8479
24	9.88623	10.8564	12.4011	13.8484	15.6587
25	10.5197	11.5240	13.1197	14.6114	16.4734
26	11.1603	12.1981	13.8439	15.3791	17.2919
27	11.8076	12.8786	14.5733	16.1513	18.1138
28	12.4613	13.5648	15.3079	16.9279	18.9392
29	13.1211	14.2565	16.0471	17.7083	19.7677
30	13.7867	14.9535	16.7908	18.4926	20.5992
40	20.7065	22.1643	24.4331	26.5093	29.0505
50	27.9907	29.7067	32.3574	34.7642	37.6886
60	35.5346	37.4848	40.4817	43.1879	46.4589
70	43.2752	45.4418	48.7576	51.7393	55.3290
80	51.1720	53.5400	57.1532	60.3915	64.2778
90	59.1963	61.7541	65.6466	69.1260	73.2912
100	67.3276	70.0648	74.2219	77.9295	82.3581

Table A-6 (continued)

DEGREES OF FREEDOM	$\chi^2_{.100}$	$\chi^2_{.050}$	$\chi^2_{.025}$	$\chi^2_{.010}$	$\chi^2_{.005}$
1	2.70554	3.84146	5.02389	6.63490	7.87944
2	4.60517	5.99147	7.37776	9.21034	10.5966
3	6.25139	7.81473	9.34840	11.3449	12.8381
4	7.77944	9.48773	11.1433	13.2767	14.8602
5	9.23635	11.0705	12.8325	15.0863	16.7496
6	10.6446	12.5916	14.4494	16.8119	18.5476
7	12.0170	14.0671	16.0128	18.4753	20.2777
8	13.3616	15.5073	17.5346	20.0902	21.9550
9	14.6837	16.9190	19.0228	21.6660	23.5893
10	15.9871	18.3070	20.4831	23.2093	25.1882
11	17.2750	19.6751	21.9200	24.7250	26.7569
12	18.5494	21.0261	23.3367	26.2170	28.2995
13	19.8119	22.3621	24.7356	27.6883	29.8194
14	21.0642	23.6848	26.1190	29.1413	31.3193
15	22.3072	24.9958	27.4884	30.5779	32.8013
16	23.5418	26.2962	28.8454	31.9999	34.2672
17	24.7690	27.5871	30.1910	33.4087	35.7185
18	25.9894	28.8693	31.5264	34.8053	37.1564
19	27.2036	30.1435	32.8523	36.1908	38.5822
20	28.4120	31.4104	34.1696	37.5662	39.9968
21	29.6151	32.6705	35.4789	38.9321	41.4010
22	30.8133	33.9244	36.7807	40.2894	42.7956
23	32.0069	35.1725	38.0757	41.6384	44.1813
24	33.1963	36.4151	39.3641	42.9798	45.5585
25	34.3816	37.6525	40.6465	44.3141	46.9278
26	35.5631	38.8852	41.9232	45.6417	48.2899
27	36.7412	40.1133	43.1944	46.9630	49.6449
28	37.9159	41.3372	44.4607	48.2782	50.9933
29	39.0875	42.5569	45.7222	49.5879	52.3356
30	40.2560	43.7729	46.9792	50.8922	53.6720
40	51.8050	55.7585	59.3417	63.6907	66.7659
50	63.1671	67.5048	71.4202	76.1539	79.4900
60	74.3970	79.0819	83.2976	88.3794	91.9517
70	85.5271	90.5312	95.0231	100.425	104.215
80	96.5782	101.879	106.629	112.329	116.321
90	107.565	113.145	118.136	124.116	128.229
100	118.498	124.342	129.561	135.807	140.169

Table A-7 Percentage Points of the F Distribution
(a) $\alpha = .10$

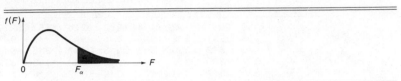

v_1 / v_2	NUMERATOR DEGREES OF FREEDOM								
	1	2	3	4	5	6	7	8	9
1	39.86	49.50	53.59	55.83	57.24	58.20	58.91	59.44	59.86
2	8.53	9.00	9.16	9.24	9.29	9.33	9.35	9.37	9.38
3	5.54	5.46	5.39	5.34	5.31	5.28	5.27	5.25	5.24
4	4.54	4.32	4.19	4.11	4.05	4.01	3.98	3.95	3.94
5	4.06	3.78	3.62	3.52	3.45	3.40	3.37	3.34	3.32
6	3.78	3.46	3.29	3.18	3.11	3.05	3.01	2.98	2.96
7	3.59	3.26	3.07	2.96	2.88	2.83	2.78	2.75	2.72
8	3.46	3.11	2.92	2.81	2.73	2.67	2.62	2.59	2.56
9	3.36	3.01	2.81	2.69	2.61	2.55	2.51	2.47	2.44
10	3.29	2.92	2.73	2.61	2.52	2.46	2.41	2.38	2.35
11	3.23	2.86	2.66	2.54	2.45	2.39	2.34	2.30	2.27
12	3.18	2.81	2.61	2.48	2.39	2.33	2.28	2.24	2.21
13	3.14	2.76	2.56	2.43	2.35	2.28	2.23	2.20	2.16
14	3.10	2.73	2.52	2.39	2.31	2.24	2.19	2.15	2.12
15	3.07	2.70	2.49	2.36	2.27	2.21	2.16	2.12	2.09
16	3.05	2.67	2.46	2.33	2.24	2.18	2.13	2.09	2.06
17	3.03	2.64	2.44	2.31	2.22	2.15	2.10	2.06	2.03
18	3.01	2.62	2.42	2.29	2.20	2.13	2.08	2.04	2.00
19	2.99	2.61	2.40	2.27	2.18	2.11	2.06	2.02	1.98
20	2.97	2.59	2.38	2.25	2.16	2.09	2.04	2.00	1.96
21	2.96	2.57	2.36	2.23	2.14	2.08	2.02	1.98	1.95
22	2.95	2.56	2.35	2.22	2.13	2.06	2.01	1.97	1.93
23	2.94	2.55	2.34	2.21	2.11	2.05	1.99	1.95	1.92
24	2.93	2.54	2.33	2.19	2.10	2.04	1.98	1.94	1.91
25	2.92	2.53	2.32	2.18	2.09	2.02	1.97	1.93	1.89
26	2.91	2.52	2.31	2.17	2.08	2.01	1.96	1.92	1.88
27	2.90	2.51	2.30	2.17	2.07	2.00	1.95	1.91	1.87
28	2.89	2.50	2.29	2.16	2.06	2.00	1.94	1.90	1.87
29	2.89	2.50	2.28	2.15	2.06	1.99	1.93	1.89	1.86
30	2.88	2.49	2.28	2.14	2.05	1.98	1.93	1.88	1.85
40	2.84	2.44	2.23	2.09	2.00	1.93	1.87	1.83	1.79
60	2.79	2.39	2.18	2.04	1.95	1.87	1.82	1.77	1.74
120	2.75	2.35	2.13	1.99	1.90	1.82	1.77	1.72	1.68
∞	2.71	2.30	2.08	1.94	1.85	1.77	1.72	1.67	1.63

DENOMINATOR DEGREES OF FREEDOM

From M. Merrington and C. M. Thompson, "Tables of Percentage Points of the Inverted Beta (F)-Distribution," *Biometrika*, 1943, 33, 73–88. Reproduced by permission of the *Biometrika* trustees.

Table A-7 (a) (continued)

ν_2 / ν_1	10	12	15	20	24	30	40	60	120	∞
				NUMERATOR DEGREES OF FREEDOM						
1	60.19	60.71	61.22	61.74	62.00	62.26	62.53	62.79	63.06	63.33
2	9.39	9.41	9.42	9.44	9.45	9.46	9.47	9.47	9.48	9.49
3	5.23	5.22	5.20	5.18	5.18	5.17	5.16	5.15	5.14	5.13
4	3.92	3.90	3.87	3.84	3.83	3.82	3.80	3.79	3.78	3.76
5	3.30	3.27	3.24	3.21	3.19	3.17	3.16	3.14	3.12	3.10
6	2.94	2.90	2.87	2.84	2.82	2.80	2.78	2.76	2.74	2.72
7	2.70	2.67	2.63	2.59	2.58	2.56	2.54	2.51	2.49	2.47
8	2.54	2.50	2.46	2.42	2.40	2.38	2.36	2.34	2.32	2.29
9	2.42	2.38	2.34	2.30	2.28	2.25	2.23	2.21	2.18	2.16
10	2.32	2.28	2.24	2.20	2.18	2.16	2.13	2.11	2.08	2.06
11	2.25	2.21	2.17	2.12	2.10	2.08	2.05	2.03	2.00	1.97
12	2.19	2.15	2.10	2.06	2.04	2.01	1.99	1.96	1.93	1.90
13	2.14	2.10	2.05	2.01	1.98	1.96	1.93	1.90	1.88	1.85
14	2.10	2.05	2.01	1.96	1.94	1.91	1.89	1.86	1.83	1.80
15	2.06	2.02	1.97	1.92	1.90	1.87	1.85	1.82	1.79	1.76
16	2.03	1.99	1.94	1.89	1.87	1.84	1.81	1.78	1.75	1.72
17	2.00	1.96	1.91	1.86	1.84	1.81	1.78	1.75	1.72	1.69
18	1.98	1.93	1.89	1.84	1.81	1.78	1.75	1.72	1.69	1.66
19	1.96	1.91	1.86	1.81	1.79	1.76	1.73	1.70	1.67	1.63
20	1.94	1.89	1.84	1.79	1.77	1.74	1.71	1.68	1.64	1.61
21	1.92	1.87	1.83	1.78	1.75	1.72	1.69	1.66	1.62	1.59
22	1.90	1.86	1.81	1.76	1.73	1.70	1.67	1.64	1.60	1.57
23	1.89	1.84	1.80	1.74	1.72	1.69	1.66	1.62	1.59	1.55
24	1.88	1.83	1.78	1.73	1.70	1.67	1.64	1.61	1.57	1.53
25	1.87	1.82	1.77	1.72	1.69	1.66	1.63	1.59	1.56	1.52
26	1.86	1.81	1.76	1.71	1.68	1.65	1.61	1.58	1.54	1.50
27	1.85	1.80	1.75	1.70	1.67	1.64	1.60	1.57	1.53	1.49
28	1.84	1.79	1.74	1.69	1.66	1.63	1.59	1.56	1.52	1.48
29	1.83	1.78	1.73	1.68	1.65	1.62	1.58	1.55	1.51	1.47
30	1.82	1.77	1.72	1.67	1.64	1.61	1.57	1.54	1.50	1.46
40	1.76	1.71	1.66	1.61	1.57	1.54	1.51	1.47	1.42	1.38
60	1.71	1.66	1.60	1.54	1.51	1.48	1.44	1.40	1.35	1.29
120	1.65	1.60	1.55	1.48	1.45	1.41	1.37	1.32	1.26	1.19
∞	1.60	1.55	1.49	1.42	1.38	1.34	1.30	1.24	1.17	1.00

DENOMINATOR DEGREES OF FREEDOM

Table A-7 Percentage Points of the F Distribution
(b) $\alpha = .05$

ν_1 / ν_2	NUMERATOR DEGREES OF FREEDOM								
	1	2	3	4	5	6	7	8	9
1	161.4	199.5	215.7	224.6	230.2	234.0	236.8	238.9	240.5
2	18.51	19.00	19.16	19.25	19.30	19.33	19.35	19.37	19.38
3	10.13	9.55	9.28	9.12	9.01	8.94	8.89	8.85	8.81
4	7.71	6.94	6.59	6.39	6.26	6.16	6.09	6.04	6.00
5	6.61	5.79	5.41	5.19	5.05	4.95	4.88	4.82	4.77
6	5.99	5.14	4.76	4.53	4.39	4.28	4.21	4.15	4.10
7	5.59	4.74	4.35	4.12	3.97	3.87	3.79	3.73	3.68
8	5.32	4.46	4.07	3.84	3.69	3.58	3.50	3.44	3.39
9	5.12	4.26	3.86	3.63	3.48	3.37	3.29	3.23	3.18
10	4.96	4.10	3.71	3.48	3.33	3.22	3.14	3.07	3.02
11	4.84	3.98	3.59	3.36	3.20	3.09	3.01	2.95	2.90
12	4.75	3.89	3.49	3.26	3.11	3.00	2.91	2.85	2.80
13	4.67	3.81	3.41	3.18	3.03	2.92	2.83	2.77	2.71
14	4.60	3.74	3.34	3.11	2.96	2.85	2.76	2.70	2.65
15	4.54	3.68	3.29	3.06	2.90	2.79	2.71	2.64	2.59
16	4.49	3.63	3.24	3.01	2.85	2.74	2.66	2.59	2.54
17	4.45	3.59	3.20	2.96	2.81	2.70	2.61	2.55	2.49
18	4.41	3.55	3.16	2.93	2.77	2.66	2.58	2.51	2.46
19	4.38	3.52	3.13	2.90	2.74	2.63	2.54	2.48	2.42
20	4.35	3.49	3.10	2.87	2.71	2.60	2.51	2.45	2.39
21	4.32	3.47	3.07	2.84	2.68	2.57	2.49	2.42	2.37
22	4.30	3.44	3.05	2.82	2.66	2.55	2.46	2.40	2.34
23	4.28	3.42	3.03	2.80	2.64	2.53	2.44	2.37	2.32
24	4.26	3.40	3.01	2.78	2.62	2.51	2.42	2.36	2.30
25	4.24	3.39	2.99	2.76	2.60	2.49	2.40	2.34	2.28
26	4.23	3.37	2.98	2.74	2.59	2.47	2.39	2.32	2.27
27	4.21	3.35	2.96	2.73	2.57	2.46	2.37	2.31	2.25
28	4.20	3.34	2.95	2.71	2.56	2.45	2.36	2.29	2.24
29	4.18	3.33	2.93	2.70	2.55	2.43	2.35	2.28	2.22
30	4.17	3.32	2.92	2.69	2.53	2.42	2.33	2.27	2.21
40	4.08	3.23	2.84	2.61	2.45	2.34	2.25	2.18	2.12
60	4.00	3.15	2.76	2.53	2.37	2.25	2.17	2.10	2.04
120	3.92	3.07	2.68	2.45	2.29	2.17	2.09	2.02	1.96
∞	3.84	3.00	2.60	2.37	2.21	2.10	2.01	1.94	1.88

DENOMINATOR DEGREES OF FREEDOM

Table A-7 (b) (continued)

ν_2 \ ν_1	10	12	15	20	24	30	40	60	120	∞
				NUMERATOR DEGREES OF FREEDOM						
1	241.9	243.9	245.9	248.0	249.1	250.1	251.1	252.2	253.3	254.3
2	19.40	19.41	19.43	19.45	19.45	19.46	19.47	19.48	19.49	19.50
3	8.79	8.74	8.70	8.66	8.64	8.62	8.59	8.57	8.55	8.53
4	5.96	5.91	5.86	5.80	5.77	5.75	5.72	5.69	5.66	5.63
5	4.74	4.68	4.62	4.56	4.53	4.50	4.46	4.43	4.40	4.36
6	4.06	4.00	3.94	3.87	3.84	3.81	3.77	3.74	3.70	3.67
7	3.64	3.57	3.51	3.44	3.41	3.38	3.34	3.30	3.27	3.23
8	3.35	3.28	3.22	3.15	3.12	3.08	3.04	3.01	2.97	2.93
9	3.14	3.07	3.01	2.94	2.90	2.86	2.83	2.79	2.75	2.71
10	2.98	2.91	2.85	2.77	2.74	2.70	2.66	2.62	2.58	2.54
11	2.85	2.79	2.72	2.65	2.61	2.57	2.53	2.49	2.45	2.40
12	2.75	2.69	2.62	2.54	2.51	2.47	2.43	2.38	2.34	2.30
13	2.67	2.60	2.53	2.46	2.42	2.38	2.34	2.30	2.25	2.21
14	2.60	2.53	2.46	2.39	2.35	2.31	2.27	2.22	2.18	2.13
15	2.54	2.48	2.40	2.33	2.29	2.25	2.20	2.16	2.11	2.07
16	2.49	2.42	2.35	2.28	2.24	2.19	2.15	2.11	2.06	2.01
17	2.45	2.38	2.31	2.23	2.19	2.15	2.10	2.06	2.01	1.96
18	2.41	2.34	2.27	2.19	2.15	2.11	2.06	2.02	1.97	1.92
19	2.38	2.31	2.23	2.16	2.11	2.07	2.03	1.98	1.93	1.88
20	2.35	2.28	2.20	2.12	2.08	2.04	1.99	1.95	1.90	1.84
21	2.32	2.25	2.18	2.10	2.05	2.01	1.96	1.92	1.87	1.81
22	2.30	2.23	2.15	2.07	2.03	1.98	1.94	1.89	1.84	1.78
23	2.27	2.20	2.13	2.05	2.01	1.96	1.91	1.86	1.81	1.76
24	2.25	2.18	2.11	2.03	1.98	1.94	1.89	1.84	1.79	1.73
25	2.24	2.16	2.09	2.01	1.96	1.92	1.87	1.82	1.77	1.71
26	2.22	2.15	2.07	1.99	1.95	1.90	1.85	1.80	1.75	1.69
27	2.20	2.13	2.06	1.97	1.93	1.88	1.84	1.79	1.73	1.67
28	2.19	2.12	2.04	1.96	1.91	1.87	1.82	1.77	1.71	1.65
29	2.18	2.10	2.03	1.94	1.90	1.85	1.81	1.75	1.70	1.64
30	2.16	2.09	2.01	1.93	1.89	1.84	1.79	1.74	1.68	1.62
40	2.08	2.00	1.92	1.84	1.79	1.74	1.69	1.64	1.58	1.51
60	1.99	1.92	1.84	1.75	1.70	1.65	1.59	1.53	1.47	1.39
120	1.91	1.83	1.75	1.66	1.61	1.55	1.50	1.43	1.35	1.25
∞	1.83	1.75	1.67	1.57	1.52	1.46	1.39	1.32	1.22	1.00

DENOMINATOR DEGREES OF FREEDOM

Table A-7 Percentage Points of the *F* Distribution
(c) α = .025

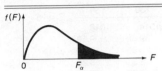

ν_2 \ ν_1	NUMERATOR DEGREES OF FREEDOM								
	1	2	3	4	5	6	7	8	9
1	647.8	799.5	864.2	899.6	921.8	937.1	948.2	956.7	963.3
2	38.51	39.00	39.17	39.25	39.30	39.33	39.36	39.37	39.39
3	17.44	16.04	15.44	15.10	14.88	14.73	14.62	14.54	14.47
4	12.22	10.65	9.98	9.60	9.36	9.20	9.07	8.98	8.90
5	10.01	8.43	7.76	7.39	7.15	6.98	6.85	6.76	6.68
6	8.81	7.26	6.60	6.23	5.99	5.82	5.70	5.60	5.52
7	8.07	6.54	5.89	5.52	5.29	5.12	4.99	4.90	4.82
8	7.57	6.06	5.42	5.05	4.82	4.65	4.53	4.43	4.36
9	7.21	5.71	5.08	4.72	4.48	4.32	4.20	4.10	4.03
10	6.94	5.46	4.83	4.47	4.24	4.07	3.95	3.85	3.78
11	6.72	5.26	4.63	4.28	4.04	3.88	3.76	3.66	3.59
12	6.55	5.10	4.47	4.12	3.89	3.73	3.61	3.51	3.44
13	6.41	4.97	4.35	4.00	3.77	3.60	3.48	3.39	3.31
14	6.30	4.86	4.24	3.89	3.66	3.50	3.38	3.29	3.21
15	6.20	4.77	4.15	3.80	3.58	3.41	3.29	3.20	3.12
16	6.12	4.69	4.08	3.73	3.50	3.34	3.22	3.12	3.05
17	6.04	4.62	4.01	3.66	3.44	3.28	3.16	3.06	2.98
18	5.98	4.56	3.95	3.61	3.38	3.22	3.10	3.01	2.93
19	5.92	4.51	3.90	3.56	3.33	3.17	3.05	2.96	2.88
20	5.87	4.46	3.86	3.51	3.29	3.13	3.01	2.91	2.84
21	5.83	4.42	3.82	3.48	3.25	3.09	2.97	2.87	2.80
22	5.79	4.38	3.78	3.44	3.22	3.05	2.93	2.84	2.76
23	5.75	4.35	3.75	3.41	3.18	3.02	2.90	2.81	2.73
24	5.72	4.32	3.72	3.38	3.15	2.99	2.87	2.78	2.70
25	5.69	4.29	3.69	3.35	3.13	2.97	2.85	2.75	2.68
26	5.66	4.27	3.67	3.33	3.10	2.94	2.82	2.73	2.65
27	5.63	4.24	3.65	3.31	3.08	2.92	2.80	2.71	2.63
28	5.61	4.22	3.63	3.29	3.06	2.90	2.78	2.69	2.61
29	5.59	4.20	3.61	3.27	3.04	2.88	2.76	2.67	2.59
30	5.57	4.18	3.59	3.25	3.03	2.87	2.75	2.65	2.57
40	5.42	4.05	3.46	3.13	2.90	2.74	2.62	2.53	2.45
60	5.29	3.93	3.34	3.01	2.79	2.63	2.51	2.41	2.33
120	5.15	3.80	3.23	2.89	2.67	2.52	2.39	2.30	2.22
∞	5.02	3.69	3.12	2.79	2.57	2.41	2.29	2.19	2.11

DENOMINATOR DEGREES OF FREEDOM

Table A-7 (c) (continued)

v_2 \ v_1	10	12	15	20	24	30	40	60	120	∞
				NUMERATOR DEGREES OF FREEDOM						
1	968.6	976.7	984.9	993.1	997.2	1001	1006	1010	1014	1018
2	39.40	39.41	39.43	39.45	39.46	39.46	39.47	39.48	39.49	39.50
3	14.42	14.34	14.25	14.17	14.12	14.08	14.04	13.99	13.95	13.90
4	8.84	8.75	8.66	8.56	8.51	8.46	8.41	8.36	8.31	8.26
5	6.62	6.52	6.43	6.33	6.28	6.23	6.18	6.12	6.07	6.02
6	5.46	5.37	5.27	5.17	5.12	5.07	5.01	4.96	4.90	4.85
7	4.76	4.67	4.57	4.47	4.42	4.36	4.31	4.25	4.20	4.14
8	4.30	4.20	4.10	4.00	3.95	3.89	3.84	3.78	3.73	3.67
9	3.96	3.87	3.77	3.67	3.61	3.56	3.51	3.45	3.39	3.33
10	3.72	3.62	3.52	3.42	3.37	3.31	3.26	3.20	3.14	3.08
11	3.53	3.43	3.33	3.23	3.17	3.12	3.06	3.00	2.94	2.88
12	3.37	3.28	3.18	3.07	3.02	2.96	2.91	2.85	2.79	2.72
13	3.25	3.15	3.05	2.95	2.89	2.84	2.78	2.72	2.66	2.60
14	3.15	3.05	2.95	2.84	2.79	2.73	2.67	2.61	2.55	2.49
15	3.06	2.96	2.86	2.76	2.70	2.64	2.59	2.52	2.46	2.40
16	2.99	2.89	2.79	2.68	2.63	2.57	2.51	2.45	2.38	2.32
17	2.92	2.82	2.72	2.62	2.56	2.50	2.44	2.38	2.32	2.25
18	2.87	2.77	2.67	2.56	2.50	2.44	2.38	2.32	2.26	2.19
19	2.82	2.72	2.62	2.51	2.45	2.39	2.33	2.27	2.20	2.13
20	2.77	2.68	2.57	2.46	2.41	2.35	2.29	2.22	2.16	2.09
21	2.73	2.64	2.53	2.42	2.37	2.31	2.25	2.18	2.11	2.04
22	2.70	2.60	2.50	2.39	2.33	2.27	2.21	2.14	2.08	2.00
23	2.67	2.57	2.47	2.36	2.30	2.24	2.18	2.11	2.04	1.97
24	2.64	2.54	2.44	2.33	2.27	2.21	2.15	2.08	2.01	1.94
25	2.61	2.51	2.41	2.30	2.24	2.18	2.12	2.05	1.98	1.91
26	2.59	2.49	2.39	2.28	2.22	2.16	2.09	2.03	1.95	1.88
27	2.57	2.47	2.36	2.25	2.19	2.13	2.07	2.00	1.93	1.85
28	2.55	2.45	2.34	2.23	2.17	2.11	2.05	1.98	1.91	1.83
29	2.53	2.43	2.32	2.21	2.15	2.09	2.03	1.96	1.89	1.81
30	2.51	2.41	2.31	2.20	2.14	2.07	2.01	1.94	1.87	1.79
40	2.39	2.29	2.18	2.07	2.01	1.94	1.88	1.80	1.72	1.64
60	2.27	2.17	2.06	1.94	1.88	1.82	1.74	1.67	1.58	1.48
120	2.16	2.05	1.94	1.82	1.76	1.69	1.61	1.53	1.43	1.31
∞	2.05	1.94	1.83	1.71	1.64	1.57	1.48	1.39	1.27	1.00

DENOMINATOR DEGREES OF FREEDOM

Table A-7 Percentage Points of the F Distribution
(d) $\alpha = .01$

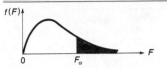

v_1 v_2	NUMERATOR DEGREES OF FREEDOM								
	1	2	3	4	5	6	7	8	9
1	4,052	4,999.5	5,403	5,625	5,764	5,859	5,928	5,982	6,022
2	98.50	99.00	99.17	99.25	99.30	99.33	99.36	99.37	99.39
3	34.12	30.82	29.46	28.71	28.24	27.91	27.67	27.49	27.35
4	21.20	18.00	16.69	15.98	15.52	15.21	14.98	14.80	14.66
5	16.26	13.27	12.06	11.39	10.97	10.67	10.46	10.29	10.16
6	13.75	10.92	9.78	9.15	8.75	8.47	8.26	8.10	7.98
7	12.25	9.55	8.45	7.85	7.46	7.19	6.99	6.84	6.72
8	11.26	8.65	7.59	7.01	6.63	6.37	6.18	6.03	5.91
9	10.56	8.02	6.99	6.42	6.06	5.80	5.61	5.47	5.35
10	10.04	7.56	6.55	5.99	5.64	5.39	5.20	5.06	4.94
11	9.65	7.21	6.22	5.67	5.32	5.07	4.89	4.74	4.63
12	9.33	6.93	5.95	5.41	5.06	4.82	4.64	4.50	4.39
13	9.07	6.70	5.74	5.21	4.86	4.62	4.44	4.30	4.19
14	8.86	6.51	5.56	5.04	4.69	4.46	4.28	4.14	4.03
15	8.68	6.36	5.42	4.89	4.56	4.32	4.14	4.00	3.89
16	8.53	6.23	5.29	4.77	4.44	4.20	4.03	3.89	3.78
17	8.40	6.11	5.18	4.67	4.34	4.10	3.93	3.79	3.68
18	8.29	6.01	5.09	4.58	4.25	4.01	3.84	3.71	3.60
19	8.18	5.93	5.01	4.50	4.17	3.94	3.77	3.63	3.52
20	8.10	5.85	4.94	4.43	4.10	3.87	3.70	3.56	3.46
21	8.02	5.78	4.87	4.37	4.04	3.81	3.64	3.51	3.40
22	7.95	5.72	4.82	4.31	3.99	3.76	3.59	3.45	3.35
23	7.88	5.66	4.76	4.26	3.94	3.71	3.54	3.41	3.30
24	7.82	5.61	4.72	4.22	3.90	3.67	3.50	3.36	3.26
25	7.77	5.57	4.68	4.18	3.85	3.63	3.46	3.32	3.22
26	7.72	5.53	4.64	4.14	3.82	3.59	3.42	3.29	3.18
27	7.68	5.49	4.60	4.11	3.78	3.56	3.39	3.26	3.15
28	7.64	5.45	4.57	4.07	3.75	3.53	3.36	3.23	3.12
29	7.60	5.42	4.54	4.04	3.73	3.50	3.33	3.20	3.09
30	7.56	5.39	4.51	4.02	3.70	3.47	3.30	3.17	3.07
40	7.31	5.18	4.31	3.83	3.51	3.29	3.12	2.99	2.89
60	7.08	4.98	4.13	3.65	3.34	3.12	2.95	2.82	2.72
120	6.85	4.79	3.95	3.48	3.17	2.96	2.79	2.66	2.56
∞	6.63	4.61	3.78	3.32	3.02	2.80	2.64	2.51	2.41

DENOMINATOR DEGREES OF FREEDOM

Table A-7 (d) (continued)

v_2 \ v_1	10	12	15	20	24	30	40	60	120	∞
				NUMERATOR DEGREES OF FREEDOM						
1	6,056	6,106	6,157	6,209	6,235	6,261	6,287	6,313	6,339	6,366
2	99.40	99.42	99.43	99.45	99.46	99.47	99.47	99.48	99.49	99.50
3	27.23	27.05	26.87	26.69	26.60	26.50	26.41	26.32	26.22	26.13
4	14.55	14.37	14.20	14.02	13.93	13.84	13.75	13.65	13.56	13.46
5	10.05	9.89	9.72	9.55	9.47	9.38	9.29	9.20	9.11	9.02
6	7.87	7.72	7.56	7.40	7.31	7.23	7.14	7.06	6.97	6.88
7	6.62	6.47	6.31	6.16	6.07	5.99	5.91	5.82	5.74	5.65
8	5.81	5.67	5.52	5.36	5.28	5.20	5.12	5.03	4.95	4.86
9	5.26	5.11	4.96	4.81	4.73	4.65	4.57	4.48	4.40	4.31
10	4.85	4.71	4.56	4.41	4.33	4.25	4.17	4.08	4.00	3.91
11	4.54	4.40	4.25	4.10	4.02	3.94	3.86	3.78	3.69	3.60
12	4.30	4.16	4.01	3.86	3.78	3.70	3.62	3.54	3.45	3.36
13	4.10	3.96	3.82	3.66	3.59	3.51	3.43	3.34	3.25	3.17
14	3.94	3.80	3.66	3.51	3.43	3.35	3.27	3.18	3.09	3.00
15	3.80	3.67	3.52	3.37	3.29	3.21	3.13	3.05	2.96	2.87
16	3.69	3.55	3.41	3.26	3.18	3.10	3.02	2.93	2.84	2.75
17	3.59	3.46	3.31	3.16	3.08	3.00	2.92	2.83	2.75	2.65
18	3.51	3.37	3.23	3.08	3.00	2.92	2.84	2.75	2.66	2.57
19	3.43	3.30	3.15	3.00	2.92	2.84	2.76	2.67	2.58	2.49
20	3.37	3.23	3.09	2.94	2.86	2.78	2.69	2.61	2.52	2.42
21	3.31	3.17	3.03	2.88	2.80	2.72	2.64	2.55	2.46	2.36
22	3.26	3.12	2.98	2.83	2.75	2.67	2.58	2.50	2.40	2.31
23	3.21	3.07	2.93	2.78	2.70	2.62	2.54	2.45	2.35	2.26
24	3.17	3.03	2.89	2.74	2.66	2.58	2.49	2.40	2.31	2.21
25	3.13	2.99	2.85	2.70	2.62	2.54	2.45	2.36	2.27	2.17
26	3.09	2.96	2.81	2.66	2.58	2.50	2.42	2.33	2.23	2.13
27	3.06	2.93	2.78	2.63	2.55	2.47	2.38	2.29	2.20	2.10
28	3.03	2.90	2.75	2.60	2.52	2.44	2.35	2.26	2.17	2.06
29	3.00	2.87	2.73	2.57	2.49	2.41	2.33	2.23	2.14	2.03
30	2.98	2.84	2.70	2.55	2.47	2.39	2.30	2.21	2.11	2.01
40	2.80	2.66	2.52	2.37	2.29	2.20	2.11	2.02	1.92	1.80
60	2.63	2.50	2.35	2.20	2.12	2.03	1.94	1.84	1.73	1.60
120	2.47	2.34	2.19	2.03	1.95	1.86	1.76	1.66	1.53	1.38
∞	2.32	2.18	2.04	1.88	1.79	1.70	1.59	1.47	1.32	1.00

DENOMINATOR DEGREES OF FREEDOM

Table A-8 Confidence Interval for a Population Proportion, Small Sample

n = 5	α = .05		α = .10	
	P_L	P_U	P_L	P_U
x = 1	0.005	0.716	0.010	0.657
2	0.053	0.853	0.076	0.811
3	0.147	0.947	0.189	0.924
4	0.284	0.995	0.343	0.990

n = 6	α = .05		α = .10	
	P_L	P_U	P_L	P_U
x = 1	0.004	0.641	0.009	0.582
2	0.043	0.777	0.063	0.729
3	0.118	0.882	0.153	0.847
4	0.223	0.957	0.271	0.937
5	0.359	0.996	0.418	0.991

n = 7	α = .05		α = .10	
	P_L	P_U	P_L	P_U
x = 1	0.004	0.579	0.007	0.521
2	0.037	0.710	0.053	0.659
3	0.099	0.816	0.129	0.775
4	0.184	0.901	0.225	0.871
5	0.290	0.963	0.341	0.947
6	0.421	0.996	0.479	0.993

n = 8	α = .05		α = .10	
	P_L	P_U	P_L	P_U
x = 1	0.003	0.527	0.006	0.471
2	0.032	0.651	0.046	0.600
3	0.085	0.755	0.111	0.711
4	0.157	0.843	0.193	0.807
5	0.245	0.915	0.289	0.889
6	0.349	0.968	0.400	0.954
7	0.473	0.997	0.529	0.994

n = 9	α = .05		α = .10	
	P_L	P_U	P_L	P_U
x = 1	0.003	0.482	0.006	0.429
2	0.028	0.600	0.041	0.550
3	0.075	0.701	0.098	0.655
4	0.137	0.788	0.169	0.749
5	0.212	0.863	0.251	0.831
6	0.299	0.925	0.345	0.902
7	0.400	0.972	0.450	0.959
8	0.518	0.997	0.571	0.994

n = 10	α = .05		α = .10	
	P_L	P_U	P_L	P_U
x = 1	0.003	0.445	0.005	0.394
2	0.025	0.556	0.037	0.507
3	0.067	0.652	0.087	0.607
4	0.122	0.738	0.150	0.696
5	0.187	0.813	0.222	0.778
6	0.262	0.878	0.304	0.850
7	0.348	0.933	0.393	0.913
8	0.444	0.975	0.493	0.963
9	0.555	0.997	0.606	0.995

n = 11	α = .05		α = .10	
	P_L	P_U	P_L	P_U
x = 1	0.002	0.413	0.005	0.364
2	0.023	0.518	0.033	0.470
3	0.060	0.610	0.079	0.564
4	0.109	0.692	0.135	0.650
5	0.167	0.766	0.200	0.729
6	0.234	0.833	0.271	0.800
7	0.308	0.891	0.350	0.865
8	0.390	0.940	0.436	0.921
9	0.482	0.977	0.530	0.967
10	0.587	0.998	0.636	0.995

n = 12	α = .05		α = .10	
	P_L	P_U	P_L	P_U
x = 1	0.002	0.385	0.004	0.339
2	0.021	0.484	0.030	0.438
3	0.055	0.572	0.072	0.527
4	0.099	0.651	0.123	0.609
5	0.152	0.723	0.181	0.685
6	0.211	0.789	0.245	0.755
7	0.277	0.848	0.315	0.819
8	0.349	0.901	0.391	0.877
9	0.428	0.945	0.473	0.928
10	0.516	0.979	0.562	0.970
11	0.615	0.998	0.661	0.996

n = 13	α = .05		α = .10	
	P_L	P_U	P_L	P_U
x = 1	0.002	0.360	0.004	0.316
2	0.019	0.454	0.028	0.410
3	0.050	0.538	0.066	0.495
4	0.091	0.614	0.113	0.573
5	0.139	0.684	0.166	0.645
6	0.192	0.749	0.224	0.713
7	0.251	0.808	0.287	0.776
8	0.316	0.861	0.355	0.834
9	0.386	0.909	0.427	0.887
10	0.462	0.950	0.505	0.934
11	0.546	0.981	0.590	0.972
12	0.640	0.998	0.684	0.996

n = 14	α = .05		α = .10	
	P_L	P_U	P_L	P_U
x = 1	0.002	0.339	0.004	0.297
2	0.018	0.428	0.026	0.385
3	0.047	0.508	0.061	0.466
4	0.084	0.581	0.104	0.540
5	0.128	0.649	0.153	0.610
6	0.177	0.711	0.206	0.675
7	0.230	0.770	0.264	0.736
8	0.289	0.823	0.325	0.794
9	0.351	0.872	0.390	0.847
10	0.419	0.916	0.460	0.896
11	0.492	0.953	0.534	0.939
12	0.572	0.982	0.615	0.974
13	0.661	0.998	0.703	0.996

Table A-8 (continued)

n = 15		α = .05		α = .10
		P_L	P_U	P_L	P_U
x =	1	0.002	0.319	0.003	0.279
	2	0.017	0.405	0.024	0.363
	3	0.043	0.481	0.057	0.440
	4	0.078	0.551	0.097	0.511
	5	0.118	0.616	0.142	0.577
	6	0.163	0.677	0.191	0.640
	7	0.213	0.734	0.244	0.700
	8	0.266	0.787	0.300	0.756
	9	0.323	0.837	0.360	0.809
	10	0.384	0.882	0.423	0.858
	11	0.449	0.922	0.489	0.903
	12	0.519	0.957	0.560	0.943
	13	0.595	0.983	0.637	0.976
	14	0.681	0.998	0.721	0.997

n = 16		α = .05		α = .10
		P_L	P_U	P_L	P_U
x =	1	0.002	0.302	0.003	0.264
	2	0.016	0.383	0.023	0.344
	3	0.040	0.456	0.053	0.417
	4	0.073	0.524	0.090	0.484
	5	0.110	0.587	0.132	0.548
	6	0.152	0.646	0.178	0.609
	7	0.198	0.701	0.227	0.667
	8	0.247	0.753	0.279	0.721
	9	0.299	0.802	0.333	0.773
	10	0.354	0.848	0.391	0.822
	11	0.413	0.890	0.452	0.868
	12	0.476	0.927	0.516	0.910
	13	0.544	0.960	0.583	0.947
	14	0.617	0.984	0.656	0.977
	15	0.698	0.998	0.736	0.997

n = 17		α = .05		α = .10
		P_L	P_U	P_L	P_U
x =	1	0.001	0.287	0.003	0.250
	2	0.015	0.364	0.021	0.326
	3	0.038	0.434	0.050	0.396
	4	0.068	0.499	0.085	0.461
	5	0.103	0.560	0.124	0.522
	6	0.142	0.617	0.166	0.580
	7	0.184	0.671	0.212	0.636
	8	0.230	0.722	0.260	0.689
	9	0.278	0.770	0.311	0.740
	10	0.329	0.816	0.364	0.788
	11	0.383	0.858	0.420	0.834
	12	0.440	0.897	0.478	0.876
	13	0.501	0.932	0.539	0.915
	14	0.566	0.962	0.604	0.950
	15	0.636	0.985	0.674	0.979
	16	0.713	0.999	0.750	0.997

n = 18		α = .05		α = .10
		P_L	P_U	P_L	P_U
x =	1	0.001	0.273	0.003	0.238
	2	0.014	0.347	0.020	0.310
	3	0.036	0.414	0.047	0.377

n = 18		α = .05		α = .10
		P_L	P_U	P_L	P_U
	4	0.064	0.476	0.080	0.439
	5	0.097	0.535	0.116	0.498
	6	0.133	0.590	0.156	0.554
	7	0.173	0.643	0.199	0.608
	8	0.215	0.692	0.244	0.659
	9	0.260	0.740	0.291	0.709
	10	0.308	0.785	0.341	0.756
	11	0.357	0.827	0.392	0.801
	12	0.410	0.867	0.446	0.844
	13	0.465	0.903	0.502	0.884
	14	0.524	0.936	0.561	0.920
	15	0.586	0.964	0.623	0.953
	16	0.653	0.986	0.690	0.980
	17	0.727	0.999	0.762	0.997

n = 19		α = .05		α = .10
		P_L	P_U	P_L	P_U
x =	1	0.001	0.260	0.003	0.226
	2	0.013	0.331	0.019	0.296
	3	0.034	0.396	0.044	0.359
	4	0.061	0.456	0.075	0.419
	5	0.091	0.512	0.110	0.476
	6	0.126	0.565	0.147	0.530
	7	0.163	0.616	0.188	0.582
	8	0.203	0.665	0.230	0.632
	9	0.244	0.711	0.274	0.680
	10	0.289	0.756	0.320	0.726
	11	0.335	0.797	0.368	0.770
	12	0.384	0.837	0.418	0.812
	13	0.435	0.874	0.470	0.853
	14	0.488	0.909	0.524	0.890
	15	0.544	0.939	0.581	0.925
	16	0.604	0.966	0.641	0.956
	17	0.669	0.987	0.704	0.981
	18	0.740	0.999	0.774	0.997

n = 20		α = .05		α = .10
		P_L	P_U	P_L	P_U
x =	1	0.001	0.249	0.003	0.216
	2	0.012	0.317	0.018	0.283
	3	0.032	0.379	0.042	0.344
	4	0.057	0.437	0.071	0.401
	5	0.087	0.491	0.104	0.456
	6	0.119	0.543	0.140	0.508
	7	0.154	0.592	0.177	0.558
	8	0.191	0.639	0.217	0.606
	9	0.231	0.685	0.259	0.653
	10	0.272	0.728	0.302	0.698
	11	0.315	0.769	0.347	0.741
	12	0.361	0.809	0.394	0.783
	13	0.408	0.846	0.442	0.823
	14	0.457	0.881	0.492	0.860
	15	0.509	0.913	0.544	0.896
	16	0.563	0.943	0.599	0.929
	17	0.621	0.968	0.656	0.958
	18	0.683	0.988	0.717	0.982
	19	0.751	0.999	0.784	0.997

Construction of Table A-8

Confidence limits for the binomial parameter can be determined using the incomplete beta distribution. A beta distribution has two parameters, say A and B. Consider a 95% confidence interval ($\alpha = .05$) for the binomial parameter, p. To find the lower confidence limit, P_L, for a given sample size (n) and observed number of successes (x), set

$$A = x \quad \text{and} \quad B = n - x + 1.$$

Then, P_L is that value of the beta random variable for which

$$\left(\begin{array}{l} \text{area under the beta distribution} \\ \text{with parameters A and B to the} \\ \text{left of } P_L \end{array} \right) = \frac{\alpha}{2}$$

$$= .025.$$

This can be determined by using a computerized incomplete beta function, as shown below.

Similarly, to find the upper limit, P_U, for a given sample size (n) and observed number of successes (x), set

$$A = x + 1 \quad \text{and} \quad B = n - x.$$

Then P_U is that value of the beta random variable for which

$$\left(\begin{array}{l} \text{area under the beta distribution} \\ \text{with parameters A and B to the} \\ \text{left of } P_U \end{array} \right) = 1 - \frac{\alpha}{2}$$

$$= .975.$$

Again, this can be determined by using a computerized incomplete beta function.

The listing of FORTRAN statements used to generate the values in Table A-8 is shown below. To change the starting value of n ($= 5$, here), the final value of n ($= 20$, here), or the two significance levels to be considered (.05 and .10, here), change one or more of the first four lines of code. The incomplete beta function subroutine used here is MDBETI, available through International Mathematical and Statistical Libraries, Inc. (IMSL), Sixth Floor, NBC Building, 7500 Bellaire Boulevard, Houston, Texas 77036.

```
0001        NST=5
0002        NEND=20
0003        ALPHA1=.05
0004        ALPHA2=.10
0005        DO 10 N=NST,NEND
0006        PRINT 500 , N,ALPHA1,ALPHA2
0007        NP1 = N + 1
0008        DO 10 J=2,N
0009        JX = J - 1
0010        T2 = ALPHA1/2.0
0011        T1 = 1.0 - T2
0012        A1 = JX + 1
```

```
0013            A2 = JX
0014            B1 = N - JX
0015            B2 = N - JX + 1
0016            CALL MDBETI(T1,A1,B1,P2,IER1)
0017            CALL MDBETI(T2,A2,B2,P1,IER2)
0018            IF(IER1.GT.0) PRINT 100 , ALPHA1,IER1
0019            IF(IER2.GT.0) PRINT 150 , ALPHA1,IER2
0020            TT2 = ALPHA2/2.0
0021            TT1 = 1.0 - TT2
0022            CALL MDBETI(TT1,A1,B1,PP2,IER1)
0023            CALL MDBETI(TT2,A2,B2,PP1,IER2)
0024            IF(IER1.GT.0) PRINT 100 , ALPHA2, IER1
0025            IF(IER2.GT.0) PRINT 150 , ALPHA2, IER2
0026            IF(J.EQ.2) PRINT 200 , JX,P1,P2,PP1,PP2
0027            IF(J.NE.2) PRINT 250 , JX,P1,P2,PP1,PP2
        C
0028      10 CONTINUE
0029     100 FORMAT('0FOR ALPHA = ',F5.2,' ERROR NUMBER',I3,
            *' OCCURRED WHEN FINDING P2')
0030     150 FORMAT('0FOR ALPHA = ',F5.2,' ERROR NUMBER',I3,
            *' OCCURRED WHEN FINDING P1')
0031     200 FORMAT('0 X = ',I2,4(4X,F5.3))
0032     250 FORMAT('0',5X,I2,4(4X,F5.3))
0033     500 FORMAT('1',///' N = ',I2,8X,'ALPHA = ',F3.2,8X,'ALPHA = ',
            *F3.2,//,13X,' P1',6X,' P2',6X,' P1',6X,' P2')
0034            STOP
0035            END
```

Table A-9 Critical Values of the Studentized Range (Q) Distribution

The values listed in the table are the critical values of Q for $\alpha = .05$ and $.01$, as a function of degrees of freedom for MS(error) and k (the number of means)

df for MS(error) (v)	α	k (Number of Means)									
		2	3	4	5	6	7	8	9	10	11
5	.05	3.64	4.60	5.22	5.67	6.03	6.33	6.58	6.80	6.99	7.17
	.01	5.70	6.98	7.80	8.42	8.91	9.32	9.67	9.97	10.24	10.48
6	.05	3.46	4.34	4.90	5.30	5.63	5.90	6.12	6.32	6.49	6.65
	.01	5.24	6.33	7.03	7.56	7.97	8.32	8.61	8.87	9.10	9.30
7	.05	3.34	4.16	4.68	5.06	5.36	5.61	5.82	6.00	6.16	6.30
	.01	4.95	5.92	6.54	7.01	7.37	7.68	7.94	8.17	8.37	8.55
8	.05	3.26	4.04	4.53	4.89	5.17	5.40	5.60	5.77	5.92	6.05
	.01	4.75	5.64	6.20	6.62	6.96	7.24	7.47	7.68	7.86	8.03
9	.05	3.20	3.95	4.41	4.76	5.02	5.24	5.43	5.59	5.74	5.87
	.01	4.60	5.43	5.96	6.35	6.66	6.91†	7.13	7.33	7.49	7.65
10	.05	3.15	3.88	4.33	4.65	4.91	5.12	5.30	5.46	5.60	5.72
	.01	4.48	5.27	5.77	6.14	6.43	6.67	6.87	7.05	7.21	7.36
11	.05	3.11	3.82	4.26	4.57	4.82	5.03	5.20	5.35	5.49	5.61
	.01	4.39	5.15	5.62	5.97	6.25	6.48	6.67	6.84	6.99	7.13
12	.05	3.08	3.77	4.20	4.51	4.75	4.95	5.12	5.27	5.39	5.51
	.01	4.32	5.05	5.50	5.84	6.10	6.32	6.51	6.67	6.81	6.94
13	.05	3.06	3.73	4.15	4.45	4.69	4.88	5.05	5.19	5.32	5.43
	.01	4.26	4.96	5.40	5.73	5.98	6.19	6.37	6.53	6.67	6.79
14	.05	3.03	3.70	4.11	4.41	4.64	4.83	4.99	5.13	5.25	5.36
	.01	4.21	4.89	5.32	5.63	5.88	6.08	6.26	6.41	6.54	6.66
15	.05	3.01	3.67	4.08	4.37	4.59	4.78	4.94	5.08	5.20	5.31
	.01	4.17	4.84	5.25	5.56	5.80	5.99	6.16	6.31	6.44	6.55
16	.05	3.00	3.65	4.05	4.33	4.56	4.74	4.90	5.03	5.15	5.26
	.01	4.13	4.79	5.19	5.49	5.72	5.92	6.08	6.22	6.35	6.46
17	.05	2.98	3.63	4.02	4.30	4.52	4.70	4.86	4.99	5.11	5.21
	.01	4.10	4.74	5.14	5.43	5.66	5.85	6.01	6.15	6.27	6.38
18	.05	2.97	3.61	4.00	4.28	4.49	4.67	4.82	4.96	5.07	5.17
	.01	4.07	4.70	5.09	5.38	5.60	5.79	5.94	6.08	6.20	6.31
19	.05	2.96	3.59	3.98	4.25	4.47	4.65	4.79	4.92	5.04	5.14
	.01	4.05	4.67	5.05	5.33	5.55	5.73	5.89	6.02	6.14	6.25
20	.05	2.95	3.58	3.96	4.23	4.45	4.62	4.77	4.90	5.01	5.11
	.01	4.02	4.64	5.02	5.29	5.51	5.69	5.84	5.97	6.09	6.19
24	.05	2.92	3.53	3.90	4.17	4.37	4.54	4.68	4.81	4.92	5.01
	.01	3.96	4.55	4.91	5.17	5.37	5.54	5.69	5.81	5.92	6.02
30	.05	2.89	3.49	3.85	4.10	4.30	4.46	4.60	4.72	4.82	4.92
	.01	3.89	4.45	4.80	5.05	5.24	5.40	5.54	5.65	5.76	5.85
40	.05	2.86	3.44	3.79	4.04	4.23	4.39	4.52	4.63	4.73	4.82
	.01	3.82	4.37	4.70	4.93	5.11	5.26	5.39	5.50	5.60	5.69
60	.05	2.83	3.40	3.74	3.98	4.16	4.31	4.44	4.55	4.65	4.73
	.01	3.76	4.28	4.59	4.82	4.99	5.13	5.25	5.36	5.45	5.53
120	.05	2.80	3.36	3.68	3.92	4.10	4.24	4.36	4.47	4.56	4.64
	.01	3.70	4.20	4.50	4.71	4.87	5.01	5.12	5.21	5.30	5.37
∞	.05	2.77	3.31	3.63	3.86	4.03	4.17	4.29	4.39	4.47	4.55
	.01	3.64	4.12	4.40	4.60	4.76	4.88	4.99	5.08	5.16	5.23

From E. S. Pearson and H. O. Hartley (eds.), *Biometrika Tables for Statisticians*, 3d ed., 1966. Reproduced by permission of *Cambridge University Press*.

Table A-10 Distribution Function for the Mann–Whitney U Statistic

This table contains the value of $P(U \leqslant U_0)$ where $n_1 \leqslant n_2$.

		n_1		
$n_2 = 3$	U_0	1	2	3
	0	.25	.10	.05
	1	.50	.20	.10
	2		.40	.20
	3		.60	.35
	4			.50

		n_1			
$n_2 = 4$	U_0	1	2	3	4
	0	.2000	.0667	.0286	.0143
	1	.4000	.1333	.0571	.0286
	2	.6000	.2667	.1143	.0571
	3		.4000	.2000	.1000
	4		.6000	.3143	.1714
	5			.4286	.2429
	6			.5714	.3429
	7				.4429
	8				.5571

Computed by M. Pagano, Dept. of Statistics, University of Florida. Reprinted by permission from *Statistics for Management and Economics*, 4th ed., by William Mendenhall and James E. Reinmuth. Copyright © 1982 by PWS-KENT Publishing Company.

Table A-10 (continued)

$n_2 = 5$

U_0	n_1				
	1	2	3	4	5
0	.1667	.0476	.0179	.0079	.0040
1	.3333	.0952	.0357	.0159	.0079
2	.5000	.1905	.0714	.0317	.0159
3		.2857	.1250	.0556	.0278
4		.4286	.1964	.0952	.0476
5		.5714	.2857	.1429	.0754
6			.3929	.2063	.1111
7			.5000	.2778	.1548
8				.3651	.2103
9				.4524	.2738
10				.5476	.3452
11					.4206
12					.5000

$n_2 = 6$

U_0	n_1					
	1	2	3	4	5	6
0	.1429	.0357	.0119	.0048	.0022	.0011
1	.2857	.0714	.0238	.0095	.0043	.0022
2	.4286	.1429	.0476	.0190	.0087	.0043
3	.5714	.2143	.0833	.0333	.0152	.0076
4		.3214	.1310	.0571	.0260	.0130
5		.4286	.1905	.0857	.0411	.0206
6		.5714	.2738	.1286	.0628	.0325
7			.3571	.1762	.0887	.0465
8			.4524	.2381	.1234	.0660
9			.5476	.3048	.1645	.0898
10				.3810	.2143	.1201
11				.4571	.2684	.1548
12				.5429	.3312	.1970
13					.3961	.2424
14					.4654	.2944
15					.5346	.3496
16						.4091
17						.4686
18						.5314

$n_2 = 7$

U_0	n_1						
	1	2	3	4	5	6	7
0	.1250	.0278	.0083	.0030	.0013	.0006	.0003
1	.2500	.0556	.0167	.0061	.0025	.0012	.0006
2	.3750	.1111	.0333	.0121	.0051	.0023	.0012
3	.5000	.1667	.0583	.0212	.0088	.0041	.0020
4		.2500	.0917	.0364	.0152	.0070	.0035
5		.3333	.1333	.0545	.0240	.0111	.0055
6		.4444	.1917	.0818	.0366	.0175	.0087
7		.5556	.2583	.1152	.0530	.0256	.0131
8			.3333	.1576	.0745	.0367	.0189
9			.4167	.2061	.1010	.0507	.0265
10			.5000	.2636	.1338	.0688	.0364

Table A-10 (continued)

	U_0	1	2	3	4	5	6	7
					n_1			
$n_2 = 7$	11				.3242	.1717	.0903	.0487
	12				.3939	.2159	.1171	.0641
	13				.4636	.2652	.1474	.0825
	14				.5364	.3194	.1830	.1043
	15					.3775	.2226	.1297
	16					.4381	.2669	.1588
	17					.5000	.3141	.1914
	18						.3654	.2279
	19						.4178	.2675
	20						.4726	.3100
	21						.5274	.3552
	22							.4024
	23							.4508
	24							.5000

	U_0	1	2	3	4	5	6	7	8
						n_1			
$n_2 = 8$	0	.1111	.0222	.0061	.0020	.0008	.0003	.0002	.0001
	1	.2222	.0444	.0121	.0040	.0016	.0007	.0003	.0002
	2	.3333	.0889	.0242	.0081	.0031	.0013	.0006	.0003
	3	.4444	.1333	.0424	.0141	.0054	.0023	.0011	.0005
	4	.5556	.2000	.0667	.0242	.0093	.0040	.0019	.0009
	5		.2667	.0970	.0364	.0148	.0063	.0030	.0015
	6		.3556	.1394	.0545	.0225	.0100	.0047	.0023
	7		.4444	.1879	.0768	.0326	.0147	.0070	.0035
	8		.5556	.2485	.1071	.0466	.0213	.0103	.0052
	9			.3152	.1414	.0637	.0296	.0145	.0074
	10			.3879	.1838	.0855	.0406	.0200	.0103
	11			.4606	.2303	.1111	.0539	.0270	.0141
	12			.5394	.2848	.1422	.0709	.0361	.0190
	13				.3414	.1772	.0906	.0469	.0249
	14				.4040	.2176	.1142	.0603	.0325
	15				.4667	.2618	.1412	.0760	.0415
	16				.5333	.3108	.1725	.0946	.0524
	17					.3621	.2068	.1159	.0652
	18					.4165	.2454	.1405	.0803
	19					.4716	.2864	.1678	.0974
	20					.5284	.3310	.1984	.1172
	21						.3773	.2317	.1393
	22						.4259	.2679	.1641
	23						.4749	.3063	.1911
	24						.5251	.3472	.2209
	25							.3894	.2527
	26							.4333	.2869
	27							.4775	.3227
	28							.5225	.3605
	29								.3992
	30								.4392
	31								.4796
	32								.5204

Table A-10 (continued)

| $n_2 = 9$ | | | | | | n_1 | | | | |
	U_0	1	2	3	4	5	6	7	8	9
	0	.1000	.0182	.0045	.0014	.0005	.0002	.0001	.0000	.0000
	1	.2000	.0364	.0091	.0028	.0010	.0004	.0002	.0001	.0000
	2	.3000	.0727	.0182	.0056	.0020	.0008	.0003	.0002	.0001
	3	.4000	.1091	.0318	.0098	.0035	.0014	.0006	.0003	.0001
	4	.5000	.1636	.0500	.0168	.0060	.0024	.0010	.0005	.0002
	5		.2182	.0727	.0252	.0095	.0038	.0017	.0008	.0004
	6		.2909	.1045	.0378	.0145	.0060	.0026	.0012	.0006
	7		.3636	.1409	.0531	.0210	.0088	.0039	.0019	.0009
	8		.4545	.1864	.0741	.0300	.0128	.0058	.0028	.0014
	9		.5455	.2409	.0993	.0415	.0180	.0082	.0039	.0020
	10			.3000	.1301	.0559	.0248	.0115	.0056	.0028
	11			.3636	.1650	.0734	.0332	.0156	.0076	.0039
	12			.4318	.2070	.0949	.0440	.0209	.0103	.0053
	13			.5000	.2517	.1199	.0567	.0274	.0137	.0071
	14				.3021	.1489	.0723	.0356	.0180	.0094
	15				.3552	.1818	.0905	.0454	.0232	.0122
	16				.4126	.2188	.1119	.0571	.0296	.0157
	17				.4699	.2592	.1361	.0708	.0372	.0200
	18				.5301	.3032	.1638	.0869	.0464	.0252
	19					.3497	.1942	.1052	.0570	.0313
	20					.3986	.2280	.1261	.0694	.0385
	21					.4491	.2643	.1496	.0836	.0470
	22					.5000	.3035	.1755	.0998	.0567
	23						.3445	.2039	.1179	.0680
	24						.3878	.2349	.1383	.0807
	25						.4320	.2680	.1606	.0951
	26						.4773	.3032	.1852	.1112
	27						.5227	.3403	.2117	.1290
	28							.3788	.2404	.1487
	29							.4185	.2707	.1701
	30							.4591	.3029	.1933
	31							.5000	.3365	.2181
	32								.3715	.2447
	33								.4074	.2729
	34								.4442	.3024
	35								.4813	.3332
	36								.5187	.3652
	37									.3981
	38									.4317
	39									.4657
	40									.5000

Table A-10 (continued)

$n_2 = 10$	U_0	1	2	3	4	5	6	7	8	9	10
							n_1				
	0	.0909	.0152	.0035	.0010	.0003	.0001	.0001	.0000	.0000	.0000
	1	.1818	.0303	.0070	.0020	.0007	.0002	.0001	.0000	.0000	.0000
	2	.2727	.0606	.0140	.0040	.0013	.0005	.0002	.0001	.0000	.0000
	3	.3636	.0909	.0245	.0070	.0023	.0009	.0004	.0002	.0001	.0000
	4	.4545	.1364	.0385	.0120	.0040	.0015	.0006	.0003	.0001	.0001
	5	.5455	.1818	.0559	.0180	.0063	.0024	.0010	.0004	.0002	.0001
	6		.2424	.0804	.0270	.0097	.0037	.0015	.0007	.0003	.0002
	7		.3030	.1084	.0380	.0140	.0055	.0023	.0010	.0005	.0002
	8		.3788	.1434	.0529	.0200	.0080	.0034	.0015	.0007	.0004
	9		.4545	.1853	.0709	.0276	.0112	.0048	.0022	.0011	.0005
	10		.5455	.2343	.0939	.0376	.0156	.0068	.0031	.0015	.0008
	11			.2867	.1199	.0496	.0210	.0093	.0043	.0021	.0010
	12			.3462	.1518	.0646	.0280	.0125	.0058	.0028	.0014
	13			.4056	.1868	.0823	.0363	.0165	.0078	.0038	.0019
	14			.4685	.2268	.1032	.0467	.0215	.0103	.0051	.0026
	15			.5315	.2697	.1272	.0589	.0277	.0133	.0066	.0034
	16				.3177	.1548	.0736	.0351	.0171	.0086	.0045
	17				.3666	.1855	.0903	.0439	.0217	.0110	.0057
	18				.4196	.2198	.1099	.0544	.0273	.0140	.0073
	19				.4725	.2567	.1317	.0665	.0338	.0175	.0093
	20				.5275	.2970	.1566	.0806	.0416	.0217	.0116
	21					.3393	.1838	.0966	.0506	.0267	.0144
	22					.3839	.2139	.1148	.0610	.0326	.0177
	23					.4296	.2461	.1349	.0729	.0394	.0216
	24					.4765	.2811	.1574	.0864	.0474	.0262
	25					.5235	.3177	.1819	.1015	.0564	.0315
	26						.3564	.2087	.1185	.0667	.0376
	27						.3962	.2374	.1371	.0782	.0446
	28						.4374	.2681	.1577	.0912	.0526
	29						.4789	.3004	.1800	.1055	.0615
	30						.5211	.3345	.2041	.1214	.0716
	31							.3698	.2299	.1388	.0827
	32							.4063	.2574	.1577	.0952
	33							.4434	.2863	.1781	.1088
	34							.4811	.3167	.2001	.1237
	35							.5189	.3482	.2235	.1399
	36								.3809	.2483	.1575
	37								.4143	.2745	.1763
	38								.4484	.3019	.1965
	39								.4827	.3304	.2179
	40								.5173	.3598	.2406
	41									.3901	.2644
	42									.4211	.2894
	43									.4524	.3153
	44									.4841	.3421
	45									.5159	.3697
	46										.3980
	47										.4267
	48										.4559
	49										.4853
	50										.5147

Table A-11 Critical Values of the Wilcoxon Signed Rank Test ($n = 5, \ldots, 50$)

1-sided	2-sided	$n = 5$	$n = 6$	$n = 7$	$n = 8$	$n = 9$	$n = 10$
$\alpha = .05$	$\alpha = .10$	1	2	4	6	8	11
$\alpha = .025$	$\alpha = .05$		1	2	4	6	8
$\alpha = .01$	$\alpha = .02$			0	2	3	5
$\alpha = .005$	$\alpha = .01$				0	2	3

1-sided	2-sided	$n = 11$	$n = 12$	$n = 13$	$n = 14$	$n = 15$	$n = 16$
$\alpha = .05$	$\alpha = .10$	14	17	21	26	30	36
$\alpha = .025$	$\alpha = .05$	11	14	17	21	25	30
$\alpha = .01$	$\alpha = .02$	7	10	13	16	20	24
$\alpha = .005$	$\alpha = .01$	5	7	10	13	16	19

1-sided	2-sided	$n = 17$	$n = 18$	$n = 19$	$n = 20$	$n = 21$	$n = 22$
$\alpha = .05$	$\alpha = .10$	41	47	54	60	68	75
$\alpha = .025$	$\alpha = .05$	35	40	46	52	59	66
$\alpha = .01$	$\alpha = .02$	28	33	38	43	49	56
$\alpha = .005$	$\alpha = .01$	23	28	32	37	43	49

1-sided	2-sided	$n = 23$	$n = 24$	$n = 25$	$n = 26$	$n = 27$	$n = 28$
$\alpha = .05$	$\alpha = .10$	83	92	101	110	120	130
$\alpha = .025$	$\alpha = .05$	73	81	90	98	107	117
$\alpha = .01$	$\alpha = .02$	62	69	77	85	93	102
$\alpha = .005$	$\alpha = .01$	55	61	68	76	84	92

1-sided	2-sided	$n = 29$	$n = 30$	$n = 31$	$n = 32$	$n = 33$	$n = 34$
$\alpha = .05$	$\alpha = .10$	141	152	163	175	188	201
$\alpha = .025$	$\alpha = .05$	127	137	148	159	171	183
$\alpha = .01$	$\alpha = .02$	111	120	130	141	151	162
$\alpha = .005$	$\alpha = .01$	100	109	118	128	138	149

1-sided	2-sided	$n = 35$	$n = 36$	$n = 37$	$n = 38$	$n = 39$	
$\alpha = .05$	$\alpha = .10$	214	228	242	256	271	
$\alpha = .025$	$\alpha = .05$	195	208	222	235	250	
$\alpha = .01$	$\alpha = .02$	174	186	198	211	224	
$\alpha = .005$	$\alpha = .01$	160	171	183	195	208	

1-sided	2-sided	$n = 40$	$n = 41$	$n = 42$	$n = 43$	$n = 44$	$n = 45$
$\alpha = .05$	$\alpha = .10$	287	303	319	336	353	371
$\alpha = .025$	$\alpha = .05$	264	279	295	311	327	344
$\alpha = .01$	$\alpha = .02$	238	252	267	281	297	313
$\alpha = .005$	$\alpha = .01$	221	234	248	262	277	292

1-sided	2-sided	$n = 46$	$n = 47$	$n = 48$	$n = 49$	$n = 50$	
$\alpha = .05$	$\alpha = .10$	389	408	427	446	466	
$\alpha = .025$	$\alpha = .05$	361	379	397	415	434	
$\alpha = .01$	$\alpha = .02$	329	345	362	380	398	
$\alpha = .005$	$\alpha = .01$	307	323	339	356	373	

From F. Wilcoxon and R. A. Wilcox, "Some Rapid Approximate Statistical Procedures," 1964. Reprinted by permission of Lederle Labs, a division of the American Cyanamid Co.

Table A-12 Random Numbers

12651	61646	11769	75109	86996	97669	25757	32535	07122	76763
81769	74436	02630	72310	45049	18029	07469	42341	98173	79260
36737	98863	77240	76251	00654	64688	09343	70278	67331	98729
82861	54371	76610	94934	72748	44124	05610	53750	95938	01485
21325	15732	24127	37431	09723	63529	73977	95218	96074	42138
74146	47887	62463	23045	41490	07954	22597	60012	98866	90959
90759	64410	54179	66075	61051	75385	51378	08360	95946	95547
55683	98078	02238	91540	21219	17720	87817	41705	95785	12563
79686	17969	76061	83748	55920	83612	41540	86492	06447	60568
70333	00201	86201	69716	78185	62154	77930	67663	29529	75116
14042	53536	07779	04157	41172	36473	42123	43929	50533	33437
59911	08256	06596	48416	69770	68797	56080	14223	59199	30162
62368	62623	62742	14891	39247	52242	98832	69533	91174	57979
57529	97751	54976	48957	74599	08759	78494	52785	68526	64618
15469	90574	78033	66885	13936	42117	71831	22961	94225	31816
18625	23674	53850	32827	81647	80820	00420	63555	74489	80141
74626	68394	88562	70745	23701	45630	65891	58220	35442	60414
11119	16519	27384	90199	79210	76965	99546	30323	31664	22845
41101	17336	48951	53674	17880	45260	08575	49321	36191	17095
32123	91576	84221	78902	82010	30847	62329	63898	23268	74283
26091	68409	69704	82267	14751	13151	93115	01437	56945	89661
67680	79790	48462	59278	44185	29616	76531	19589	83139	28454
15184	19260	14073	07026	25264	08388	27182	22557	61501	67481
58010	45039	57181	10238	36874	28546	37444	80824	63981	39942
56425	53996	86245	32623	78858	08143	60377	42925	42815	11159
82630	84066	13592	60642	17904	99718	63432	88642	37858	25431
14927	40909	23900	48761	44860	92467	31742	87142	03607	32059
23740	22505	07489	85986	74420	21744	97711	36648	35620	97949
32990	97446	03711	63824	07953	85965	87089	11687	92414	67257
05310	24058	91946	78437	34365	82469	12430	84754	19354	72745
21839	39937	27534	88913	49055	19218	47712	67677	51889	70926
08833	42549	93981	94051	28382	83725	72643	64233	97252	17133
58336	11139	47479	00931	91560	95372	97642	33856	54825	55680
62032	91144	75478	47431	52726	30289	42411	91886	51818	78292
45171	30557	53116	04118	58301	24375	65609	85810	18620	49198
91611	62656	60128	35609	63698	78356	50682	22505	01692	36291
55472	63819	86314	49174	93582	73604	78614	78849	23096	72825
18573	09729	74091	53994	10970	86557	65661	41854	26037	53296
60866	02955	90288	82136	83644	94455	06560	78029	98768	71296
45043	55608	82767	60890	74646	79485	13619	98868	40857	19415
17831	09737	79473	75945	28394	79334	70577	38048	03607	06932
40137	03981	07585	18128	11178	32601	27994	05641	22600	86064
77776	31343	14576	97706	16039	47517	43300	59080	80392	63189
69605	44104	40103	95635	05635	81673	68657	09559	23510	95875
19916	52934	26499	09821	97331	80993	61299	36979	73599	35055
02606	58552	07678	56619	65325	30705	99582	53390	46357	13244
65183	73160	87131	35530	47946	09854	18080	02321	05809	04893
10740	98914	44916	11322	89717	88189	30143	52687	19420	60061
98642	89822	71691	51573	83666	61642	46683	33761	47542	23551
60139	25601	93663	25547	02654	94829	48672	28736	84994	13071

Table A-13 Critical Values of Hartley's *H*-statistic, $\alpha = .05$

n = number of observations in each sample
k = number of samples

n	2	3	4	5	6	7	8	9	10	11	12
						k					
3	39.0	87.5	142	202	266	333	403	475	550	626	704
4	15.4	27.8	39.2	50.7	62.0	72.9	83.5	93.9	104	114	124
5	9.60	15.5	20.6	25.2	29.5	33.6	37.5	41.1	44.6	48.0	51.4
6	7.15	10.8	13.7	16.3	18.7	20.8	22.9	24.7	26.5	28.2	29.9
7	5.82	8.38	10.4	12.1	13.7	15.0	16.3	17.5	18.6	19.7	20.7
8	4.99	6.94	8.44	9.70	10.8	11.8	12.7	13.5	14.3	15.1	15.8
9	4.43	6.00	7.18	8.12	9.03	9.78	10.5	11.1	11.7	12.2	12.7
10	4.03	5.34	6.31	7.11	7.80	8.41	8.95	9.45	9.91	10.3	10.7
11	3.72	4.85	5.67	6.34	6.92	7.42	7.87	8.28	8.66	9.01	9.34
13	3.28	4.16	4.79	5.30	5.72	6.09	6.42	6.72	7.00	7.25	7.48
16	2.86	3.54	4.01	4.37	4.68	4.95	5.19	5.40	5.59	5.77	5.93
21	2.46	2.95	3.29	3.54	3.76	3.94	4.10	4.24	4.37	4.49	4.59
31	2.07	2.40	2.61	2.78	2.91	3.02	3.12	3.21	3.29	3.36	3.39
61	1.67	1.85	1.96	2.04	2.11	2.17	2.22	2.26	2.30	2.33	2.36
∞	1.00	1.00	1.00	1.00	1.00	1.00	1.00	1.00	1.00	1.00	1.00

Table A-14 Quantiles of the Kendall Test Statistic

n	$V_{.90}$	$V_{.95}$	$V_{.975}$	$V_{.990}$	$V_{.995}$
4	4	4	6	6	6
5	6	6	8	8	10
6	7	9	11	11	13
7	9	11	13	15	17
8	10	14	16	18	20
9	12	16	18	22	24
10	15	19	21	25	27
11	17	21	25	29	31
12	18	24	28	34	36
13	22	26	32	38	42
14	23	31	35	41	45
15	27	33	39	47	51
16	28	36	44	50	56
17	32	40	48	56	62
18	35	43	51	61	67
19	37	47	55	65	73
20	40	50	60	70	78
21	42	54	64	76	84
22	45	59	69	81	89
23	49	63	73	87	97
24	52	66	78	92	102
25	56	70	84	98	108
26	59	75	89	105	115
27	61	79	93	111	123
28	66	84	98	116	128
29	68	88	104	124	136
30	73	93	109	129	143
31	75	97	115	135	149
32	80	102	120	142	158
33	84	106	126	150	164
34	87	111	131	155	173
35	91	115	137	163	179
36	94	120	144	170	188
37	98	126	150	176	196
38	103	131	155	183	203
39	107	137	161	191	211
40	110	142	168	198	220
41	114	146	174	206	228
42	119	151	181	213	235
43	123	157	187	221	245
44	128	162	194	228	252
45	132	168	200	236	262
46	135	173	207	245	271
47	141	179	213	253	279
48	144	186	220	260	288
49	150	190	228	268	296
50	153	197	233	277	305
51	159	203	241	285	315
52	162	208	248	294	324
53	168	214	256	302	334
54	173	221	263	311	343
55	177	227	269	319	353
56	182	232	276	328	362
57	186	240	284	336	372
58	191	245	291	345	381
59	197	251	299	355	391
60	202	258	306	364	402

From W. J. Conover, *Practical Nonparametric Statistics*, 2/e, Table A-12, 1980, John Wiley & Sons, New York. Reprinted with permission.

Appendix B
Derivation of Minimum Total Sample Size

Claim: When obtaining two independent samples, the maximum error for the difference of the two population means, $\mu_1 - \mu_2$, is

$$E = Z_{\alpha/2} \sqrt{\frac{\sigma_1^2}{n_1} + \frac{\sigma_2^2}{n_2}} \qquad (\sigma_1, \sigma_2 \text{ known})$$

or estimated using

$$E = Z_{\alpha/2} \sqrt{\frac{s_1^2}{n_1} + \frac{s_2^2}{n_2}} \qquad (\sigma_1, \sigma_2 \text{ unknown})$$

For a specific value of E, the sample sizes, n_1 and n_2, that minimize the total sample size, $n = n_1 + n_2$, are given by

$$n_1 = \frac{Z_{\alpha/2}^2 s_1 (s_1 + s_2)}{E^2}$$

and

$$n_2 = \frac{Z_{\alpha/2}^2 s_2 (s_1 + s_2)}{E^2}$$

Proof: For ease of notation, define

$$Z = Z_{\alpha/2}$$

$$a = s_1$$

$$b = s_2$$

$$x = n_1$$

$$y = n_2$$

(For the case where the σ's are known, then $a = \sigma_1$ and $b = \sigma_2$.) Now,

$$E = Z\sqrt{\frac{a^2}{x} + \frac{b^2}{y}}$$

is fixed. Solving for y yields

$$y = \frac{Z^2 b^2 x}{E^2 x - Z^2 a^2}$$

The total sample size is $n = x + y$, and so

$$n = f(x) = x + \frac{Z^2 b^2 x}{E^2 x - Z^2 a^2}$$

To determine the value of x that minimizes $n = f(x)$, the procedure will be to find $f'(x)$, set it to zero, and solve for x.

$$f'(x) = 1 + \frac{(E^2 x - Z^2 a^2)(Z^2 b^2) - Z^2 b^2 x (E^2)}{(E^2 x - Z^2 a^2)^2}$$

$$= 1 - \frac{Z^4 a^2 b^2}{(E^2 x - Z^2 a^2)^2}$$

Now,

$$f'(x) = 0$$

iff $x^2(E^4) + x(-2Z^2 E^2 a^2) + (Z^4 a^4 - Z^4 a^2 b^2) = 0$

iff $x = \dfrac{2Z^2 E^2 a^2 \mp \sqrt{4Z^4 E^4 a^4 - 4Z^4 E^4 a^2(a^2 - b^2)}}{2E^4}$

$$= \frac{Z^2 a^2 \mp Z^2 ab}{E^2}$$

Now,

$$f''(x) = \frac{2Z^4 E^2 a^2 b^2 (E^2 x - Z^2 a^2)}{(E^2 x - Z^2 a^2)^4}$$

Consequently,

$$f''(x) > 0 \quad \text{iff } (E^2 x - Z^2 a^2) > 0$$

Letting

$$x = \frac{Z^2 a^2 + Z^2 ab}{E^2}$$

then

$$E^2 x - Z^2 a^2 = Z^2 ab > 0$$

because a and b are > 0. Letting

$$x = \frac{Z^2a^2 - Z^2ab}{E^2}$$

then

$$E^2x - Z^2a^2 = -Z^2ab < 0$$

Conclusion:

  1 $f(x)$ has a local minimum at

$$x = \frac{Z^2a(a + b)}{E^2}$$

  2 $f(x)$ has a local maximum at

$$x = \frac{Z^2a(a - b)}{E^2}$$

Because we are restricted to values of x (that is, n_1) such that $f(x) =$ total sample size is positive, and because $f(x)$ approaches ∞ as x approaches ∞, for the admissible values of x, $f(x)$ has a global minimum at

$$x = n_1 = \frac{Z^2a(a + b)}{E^2}$$

$$= \frac{Z^2s_1(s_1 + s_2)}{E^2}$$

Solving for n_2, we previously stated that

$$y = \frac{Z^2b^2x}{E^2x - Z^2a^2}$$

Substituting

$$x = \frac{Z^2s_1(s_1 + s_2)}{E^2}$$

into this expression produces

$$y = n_2 = \frac{Z^2s_2(s_1 + s_2)}{E^2}$$

Appendix C
Introduction to MINITAB—Mainframe or Microcomputer Versions

MINITAB is an easy-to-use, flexible statistical package. It was originally designed for students; over the years, it has been constantly improved. It is one of the more powerful statistical systems currently available. MINITAB will allow you to "speak" to the computer using commands that are similar to English sentences. The sequence of steps in a typical problem resembles the same steps you would take if solving the problem by hand.

MINITAB consists of a worksheet containing rows and columns. The data for each variable are stored in a particular *column*. The following discussion will provide a brief introduction of data entry and use of MINITAB commands. No distinction is made between mainframe and microcomputer MINITAB commands since they can be used on either system.

Entering the Data

Data for each of the variables in your data set are stored in columns. For example, suppose you have four scores (80, 75, 43, and 91) that you want to enter as one variable. These can be entered as shown in Figure C.1 using the SET command. Here, the four test scores are stored in column C1. Notice that (1) the data values are separated by blanks (commas are OK) and (2) at the end of the data string, the word END is entered on the next line. If your data will not fit on a single line, type as many values as you wish, enter these, continuing typing on the next line, enter another line of data, and so forth until you have entered all of your data. Your last line always will be END.

Figure C.1

Commands to enter four test scores.

```
MTB > SET INTO C1
DATA> 80 75 43 91
DATA> END
```

Suppose that you have test scores for three people (Joe, Mary, and Al), each of whom has four scores. The previous four test scores (80, 75, 43, and 91) belong to Joe. The other method of entering data is to read in your data one *row* at a time using the READ command. Each line contains a single row of data on each of the variables (three variables, here). This is shown in Figure C.2.

Figure C.2

Commands to enter four test scores for each of three students

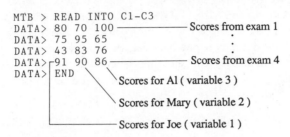

These steps will input the 12 test scores one row (exam) at a time. Note that C1-C3 means C1 through C3.

Output of the Data

To display the data, you should use the PRINT command, as shown in Figure C.3. When using the microcomputer version of MINITAB, to send your results to the printer and screen simultaneously, type

```
MTB > PAPER
```

To stop sending output to the printer at any time, type

```
MTB > NOPAPER
```

MINITAB Shorthand

One exceptionally nice feature of MINITAB is that most of the commands can be shortened to save time and effort. You are able to do this because the MINITAB system does not process the entire line that you enter—it reads only those pieces of information that it needs to know and ignores everything else! Consequently, you can misspell words or leave out unnecessary words and

Figure C.3

Commands to display (or print) data

```
MTB > PRINT C1
C1
    80     75     43     91

MTB > PRINT C1-C3
  ROW    C1     C2     C3

    1     80     70    100
    2     75     95     65
    3     43     83     76
    4     91     90     86
```

MINITAB will still execute your command. For example, in Figure C.1, you could have used the command

```
MTB > SET C1
```

That is, the word "into" was not necessary. We used it originally because it makes the statement easier to comprehend for someone who wants to know what we are doing at this step.

Many of the MINITAB commands are illustrated on the inside front cover of this book. Only those portions of each statement that are colored must be included. What you put between the colored portions is up to you—you can leave them blank (the shorthand version) or put in any words you wish to make the statement easier to understand. For example, in Figure C.1, we could have used

```
MTB > SET THE EXAM SCORES FOR JOE INTO C1
```

Here, all MINITAB needs to see is SET and C1. What you put in between is your decision.

In addition, for the commands on the inside front cover, any portion of the statement enclosed in brackets [ ] need not be included in the statement. If you want to omit this portion of the statement, you may do so. The information in the brackets generally allow you to be more specific in your input to MINITAB or informs MINITAB in which columns you would like certain information from the output to be stored.

MINITAB Subcommands

Some of the more sophisticated MINITAB commands allow you to specify further information by using one or more *subcommands*. For those commands that allow subcommands (not all of them do), you should end the main command line with a *semicolon*, ;. This informs MINITAB that subcommands will follow. Each subcommand line should end with a semicolon unless it is the last subcommand—the last one ends with a *period*. If you forget to type the period at the end of the last subcommand, simply type a period on the next line.

An example of a command utilizing subcommands (called REGRESS, discussed in Chapters 14 and 15) is shown in Figure C.4. The MINITAB solutions contained throughout the text will clarify what commands allow the use of subcommands and what these possible subcommands are.

Figure C.4

The REGRESS command, with subcommands

```
MTB > REGRESS Y IN C1 USING 2 PREDICTORS IN C2 C3;
SUBC> NOCONSTANT;
SUBC> COEF INTO C4;
SUBC> RESIDS INTO C5.
```

Informs MINITAB that this is the end of the subcommands

Informs MINITAB that one or more subcommands will follow

Appendix D
Introduction to SPSSX—
Mainframe Version

Computer packages often are used to perform various statistical analysis procedures. When used properly, these computer packages can save time and decrease
the probability of human error. The purpose of this appendix is to provide a
basic overview of one such mainframe package, the Statistical Package for the
Social Sciences (SPSSX).

To solve a statistical problem using SPSSX, you must define the data and
the format in which it is to be interpreted, and specify the statistical procedure
to be performed. For example, the data in Figure D.1 are test grades for three
students. If you wish to find the mean grade for each of the students, you
define, to SPSSX, each individual test score and specify the SPSSX procedure
to obtain mean values (CONDESCRIPTIVE). Figure D.2 shows the statements and
data required to perform this task under SPSSX.

Figure D.1

Test grades for three
students

80	70	100
75	95	65
43	83	76
91	90	86

Figure D.2

Statements and data
to find mean grades
for three students

```
TITLE  SPSSX APPENDIX
DATA LIST FREE / JOE MARY AL
BEGIN DATA
80 70 100
75 95 65
43 83 76
91 90 86
END DATA
PRINT / JOE MARY AL
CONDESCRIPTIVE  ALL
FINISH
```

Analysis of Statements

The following is an analysis of each of the statements used to obtain the mean grades for each student:

TITLE SPSSX–APPENDIX	Defines the name of the SPSSX run.
DATA LIST FREE / JOE MARY AL	Defines three variables (students) named Joe, Mary, and Al. FREE specifies the data is not in any format, other than it is to be read sequentially, starting at column one.
BEGIN DATA	Specifies that the following lines are input data lines.
80 70 100	Gives the test-score values for the first test (Joe scored 80, Mary scored 70, and Al scored 100).
75 95 65	Gives the test-score values for the second test.
43 83 76	Gives the test-score values for the third test.
91 90 86	Gives the test-score values for the fourth test.
	As shown, the values must be separated by one or more spaces, and there must be a value for all four tests for each of the three students.
END DATA	Indicates the end of the data card images.
PRINT / JOE MARY AL	Requests SPSSX to list the data corresponding to variables JOE, MARY, and AL. *In the end-of-chapter SPSSX output figures, the output resulting from the PRINT command has been omitted for clarity of presentation.*
CONDESCRIPTIVE ALL	Specifies the SPSSX procedure to be performed. In this program, the mean, as well as other statistics, will be calculated for all variables.
FINISH	Indicates the end of the SPSSX program. This statement is optional.

Figure D.3 shows the printed results of running this procedure under SPSSX. Most SPSSX procedures require additional statements, either to describe the problem or to request different options for solving the problem in different

Figure D.3

Results of running
listing in Figure D.2

```
                              80.00    70.00   100.00
                              75.00    95.00    65.00
                              43.00    83.00    76.00
                              91.00    90.00    86.00
```

```
NUMBER OF VALID OBSERVATIONS (LISTWISE) =        4.00

VARIABLE      MEAN     STD DEV    MINIMUM    MAXIMUM  VALID N   LABEL

JOE          72.250    20.614      43.00      91.00      4
MARY         84.500    10.847      70.00      95.00      4
AL           81.750    14.886      65.00     100.00      4
```

ways. All procedures, statements, and options necessary for solving the problems in this text are discussed in the end-of-chapter appendixes. For further information on SPSSX, refer to the *SPSSX Introductory Statistics Guide* (Marija Norusis, New York: McGraw-Hill, 1983).

Basic SPSSX Rules

As with all computer packages, you must observe certain rules when running SPSSX programs. The following is a list of the basic rules for SPSSX:

1 All commands must begin in column 1 and cannot exceed column 15.
2 Any additional information must appear in columns 16 through 80.
3 All command keywords must be separated by only one blank.
4 Multiple blanks are allowed beyond column 15.
5 Include the decimal point when entering decimal data (such as 38.95).

Additional rules concerning individual procedures are contained in the *SPSSX Introductory Statistics Guide*.

JCL Statements

Job Control Language (JCL) statements must be included with SPSSX statements in order for the SPSSX program to execute properly. These statements identify the user and the procedure to be performed. Typically, you need only a job statement, an execute statement (EXEC SPSSX), and a card image input statement (SYSIN DD *). The format and order of these statements may differ, so consult your computer center before attempting to run your first program.

Appendix E
Introduction to SPSS/PC+ —
Microcomputer Version

Computer packages often are used to perform various statistical analysis procedures. When used properly, these computer packages can save time and decrease the probability of human error. The purpose of this appendix is to provide a basic overview of one such microcomputer package, the Statistical Package for the Social Sciences (SPSS/PC+).

To solve a statistical problem using SPSS/PC+, you must define the data and the format in which it is to be interpreted, and specify the statistical procedure to be performed. For example, the data in Figure E.1 are test grades for three students. If you wish to find the mean grade for each of the students, you define, to SPSS/PC+, each individual test score and specify the SPSSX procedure to obtain mean values (CONDESCRIPTIVE). Figure E.2 shows the statements and data required to perform this task under SPSS/PC+.

Figure E.1

Test grades for three students

80	70	100
75	95	65
43	83	76
91	90	86

Figure E.2

Statements and data to find mean grades for three students

```
TITLE  SPSS/PC+ APPENDIX.
DATA LIST FREE / JOE MARY AL.
BEGIN DATA.
80 70 100
75 95 65
43 83 76
91 90 86
END DATA.
LIST / JOE MARY AL.
CONDESCRIPTIVE ALL.
FINISH.
```

Analysis of Statements

The following is an analysis of each of the statements used to obtain the mean grades for each student. Each statement *must* end with a period.

TITLE SPSS/PC+ APPENDIX.	Defines the name of the SPSS/PC+ run.
DATA LIST FREE / JOE MARY AL.	Defines three variables (students) named Joe, Mary, and Al. FREE specifies the data is not in any format, other than it is to be read sequentially, starting at column one.
BEGIN DATA.	Specifies that the following lines are input data lines.
80 70 100	Gives the test-score values for the first test (Joe scored 80, Mary scored 70, and Al scored 100).
75 95 65	Gives the test-score values for the second test.
43 83 76	Gives the test-score values for the third test.
91 90 86	Gives the test-score values for the fourth test.
	As shown, the values must be separated by one or more spaces, and there must be a value for all four tests for each of the three students.
END DATA.	Indicates the end of the data card images.
LIST / JOE MARY AL.	Requests SPSS/PC+ to list the data corresponding to variables JOE, MARY, and AL. *In the end-of-chapter SPSS/PC+ output figures, the output resulting from the LIST command has been omitted for clarity of presentation.*
CONDESCRIPTIVE ALL.	Specifies the SPSS/PC+ procedure to be performed. In this program, the mean, as well as other statistics, will be calculated for all variables.
FINISH.	Indicates the end of the SPSS/PC+ program. This statement is optional.

Figure E.3 shows the printed results of running this procedure under SPSS/PC+.

```
-----------------------------------------------------------------------
              SPSS/PC+ APPENDIX

     JOE      MARY       AL

     80.00    70.00    100.00
     75.00    95.00     65.00
     43.00    83.00     76.00
     91.00    90.00     86.00


Number of cases read =        4    Number of cases listed =        4


              SPSS/PC+ APPENDIX


Number of Valid Observations (Listwise) =            4.00

Variable      Mean    Std Dev   Minimum   Maximum    N  Label

JOE          72.25     20.61     43.00     91.00     4
MARY         84.50     10.85     70.00     95.00     4
AL           81.75     14.89     65.00    100.00     4
-----------------------------------------------------------------------
```

Figure E.3

Results of running
listing in Figure E.2

Most SPSS/PC+ procedures require additional statements, either to describe the problem or to request different options for solving the problem in different ways. All procedures, statements, and options necessary for solving the problems in this text are discussed in the end-of-chapter appendixes. For further information on SPSS/PC+ refer to the *SPSS/PC+ for the IBM PC/XT/AT* (Marija Norusis, Chicago: SPSS Inc., 1986).

Basic SPSS/PC+ Rules

As with all computer packages, you must observe certain rules when running SPSS/PC+ programs. The following is a list of the basic rules for SPSS/PC+.

1 All SPSS/PC+ statements must end with a period.
2 Lines containing data do not end with a period.
3 All command keywords must be separated by only one blank.
4 Multiple blanks are allowed beyond column 15.
5 Include the decimal point when entering decimal data (such as 38.95).
6 All data must be defined before requesting SPSS/PC+ to perform a statistical procedure.
7 To return to DOS, type FINISH at the end of your SPSS/PC+ session.

Additional rules concerning individual procedures are contained in the *SPSS/PC+ for the IBM PC/XT/AT*.

Obtaining Paper Output

At any point in the SPSS/PC+ session, you can obtain output on your printer by typing

SET   PRINTER = ON.

All subsequent output will be sent to the screen and the printer. To terminate the printed output, type

SET   PRINTER = OFF.

Appendix F
Introduction to SAS—
Mainframe Version

Computer packages often are used to perform various statistical analysis procedures. When used properly, these computer packages can save time and decrease the probability of human error. The purpose of this appendix is to provide a basic overview of one such mainframe package, the Statistical Analysis System (SAS).

To solve a statistical problem using SAS, you must define the data to be used and specify the statistical procedure to be performed. For example, the data in Figure F.1 lists test grades for three students. If you wish to find the mean grade for each of the students, you define, to SAS, each individual test score and specify the SAS procedure to obtain mean values (PROC MEANS). Figure F.2 shows the statements and data required to perform this task under SAS.

Figure F.1	80	70	100
Test grades for three students	75	95	65
	43	83	76
	91	90	86

Figure F.2

Statements and data to find mean grades for three students

```
TITLE   'SAS APPENDIX';
DATA   GRADES;
INPUT   JOE MARY AL;
CARDS;
80 70 100
75 95 65
43 83 76
91 90 86
PROC PRINT;
PROC MEANS;
```

Analysis of Statements

The following is an analysis of each of the statements used to obtain the mean grades for each student:

TITLE 'SAS APPENDIX'; — Defines the name of the SAS run. Enclose the title in single quotes.

DATA GRADES; — Defines a dataset named grades.

INPUT JOE MARY AL; — Defines three variables (students) named Joe, Mary, and Al.

CARDS; — Defines the input medium as card images.

80 70 100 — Gives the test-score values for the first test (Joe scored 80, Mary scored 70, and Al scored 100).

75 95 65 — Gives the test-score values for the second test.

43 83 76 — Gives the test-score values for the third test.

91 90 86 — Gives the test-score values for the fourth test.

As shown, the values must be separated by one or more spaces, and there must be a value for all four tests for each of the three students.

PROC PRINT; — Requests SAS to list the data that have been entered. *In the end-of-chapter SAS output figures, the output resulting from the PROC PRINT command has been omitted for clarity of presentation.*

PROC MEANS; — Specifies the SAS procedure to be performed.

Figure F.3 shows the printed results of running this procedure under SAS.

Most SAS procedures require additional statements, either to describe the problem or to request different options for solving the problem in different

SAS APPENDIX

OBS	JOE	MARY	AL
1	80	70	100
2	75	95	65
3	43	83	76
4	91	90	86

SAS APPENDIX

VARIABLE	N	MEAN	STANDARD DEVIATION	MINIMUM VALUE	MAXIMUM VALUE	STD ERROR OF MEAN	SUM	VARIANCE	C.V.
JOE	4	72.25000000	20.61350690	43.00000000	91.00000000	10.30675345	289.00000000	424.91666667	28.531
MARY	4	84.50000000	10.84742673	70.00000000	95.00000000	5.42371337	338.00000000	117.66666667	12.837
AL	4	81.75000000	14.88567544	65.00000000	100.00000000	7.44283772	327.00000000	221.58333333	18.209

Figure F.3

Results of running
listing in Figure F.3

ways. All procedures, statements, and options necessary for solving the problems in this text are discussed in the end-of-chapter appendices. For further information on SAS, refer to the *SAS User's Guide: Statistics Version*, 5th ed. (SAS Institute, Inc., Cary, NC, 1985).

Basic SAS Rules

As with all computer packages, you must observe certain rules when running SAS programs. The following is a list of the basic rules for SAS:

1 SAS statements must begin in column 1 and cannot exceed column 72. Any SAS statement longer than 72 columns may be continued on the next line.
2 All SAS statements must end with a semicolon (;).
3 Lines containing data do not end with a semicolon.
4 All data must be defined before a procedure can be run.
5 More than one procedure can be performed by adding more procedure statements.
6 You must include the decimal point when entering decimal data (such as 38.95).

Additional rules concerning individual procedures are contained in the *User's Guide*.

JCL Statements

Job Control Language (JCL) statements must be included with SAS statements in order for the SAS program to execute properly. These statements identify the user and the procedure to be performed. Typically, you need only a job statement, an execute statement (EXEC SAS), and a data input statement (SYSIN DD*). The format and order of these statements may differ, so consult your computer center before attempting to run your first program.

Appendix G
Introduction to SAS—
Microcomputer Version

Computer packages often are used to perform various statistical analysis procedures. When used properly, these computer packages can save time and decrease the probability of human error. The purpose of this appendix is to provide a basic overview of one such microcomputer package, the Statistical Analysis System (SAS).

To solve a statistical problem using SAS, you must define the data to be used and specify the statistical procedure to be performed. For example, the data in Figure G.1 lists test grades for three students. If you wish to find the mean grade for each of the students, you define, to SAS, each individual test score and specify the SAS procedure to obtain mean values (PROC MEANS). Figure G.2 shows the statements and data required to perform this task under SAS.

Figure G.1

Test grades for three students

80	70	100
75	95	65
43	83	76
91	90	86

Figure G.2

Statements and data to find mean grades for three students

```
TITLE 'SAS APPENDIX';
DATA GRADES;
INPUT JOE MARY AL;
CARDS;
80 70 100
75 95 65
43 83 76
91 90 86
PROC PRINT;
PROC MEANS;
RUN;
```

```
                    SAS APPENDIX


           OBS    JOE    MARY     AL

            1     80      70     100
            2     75      95      65
            3     43      83      76
            4     91      90      86




                    SAS APPENDIX
```

N Obs	Variable	N	Minimum	Maximum	Mean	Std Dev
4	JOE	4	43.00	91.00	72.25	20.61
	MARY	4	70.00	95.00	84.50	10.85
	AL	4	65.00	100.00	81.75	14.89

Figure G.3

Results of running listing in Figure G.2

Analysis of Statements

The following is an analysis of each of the statements used to obtain the mean grades for each student:

TITLE 'SAS APPENDIX';   Defines the name of the SAS run. Enclose the title in single quotes.

DATA GRADES;   Defines a dataset named grades.

INPUT JOE MARY AL;   Defines three variables (students) named Joe, Mary, and Al.

CARDS;   Defines the input medium as card images.

80 70 100   Gives the test-score values for the first test (Joe scored 80, Mary scored 70, and Al scored 100).

75 95 65   Gives the test-score values for the second test.

43 83 76   Gives the test-score values for the third test.

91 90 86   Gives the test-score values for the fourth test.

   As shown, the values must be separated by one or more spaces, and there must be a value for all four tests for each of the three students.

PROC PRINT;   Requests SAS to list the data that have been entered. *In the end-of-chapter SAS output figures, the output resulting from the PROC PRINT command has been omitted for clarity of presentation.*

PROC MEANS;   Specifies the SAS procedure to be performed.

RUN;   Tells the SAS system to execute the previous SAS statements.

Figure G.3 shows the printed results of running this procedure under SAS.

Most SAS procedures require additional statements, either to describe the problem or to request different options for solving the problem in different ways. All procedures, statements, and options necessary for solving the problems in this text are discussed in the end-of-chapter appendixes. For further information on SAS, refer to the *SAS Introductory Guide for Personal Computers Version*, 6th ed. (SAS Institute, Inc., Cary, NC, 1985).

Basic SAS Rules

As with all computer packages, you must observe certain rules when running SAS programs. The following is a list of the basic rules for SAS:

1 All SAS statements must end with a semicolon (;).
2 Lines containing data do not end with a semicolon.
3 All data must be defined before a procedure can be run.
4 More than one procedure can be performed by adding more procedure statements.
5 You must include the decimal point when entering decimal data (such as 38.95).
6 To return to DOS, at the COMMAND= prompt, enter the single letter X.

Additional rules concerning individual procedures are contained in the *SAS Introductory Guide for Personal Computers*.

Obtaining Paper Output

To obtain printed output during interactive full screen sessions, in the OUTPUT window type FILE 'PRN:'. This should be typed after the output has appeared in the window.

If submitting the SAS job from a previously created ASCII file (say, SASRUN) containing the SAS statements and data, the SAS output will be contained in a newly created file (created by SAS), called SASRUN.LST. To submit the SAS job, at the hard disk drive prompt, type

SAS   B:SASRUN

This assumes you have saved SASRUN on a floppy diskette, contained in drive B. No output is routed to the screen; instead it will be routed to the B drive into the file SASRUN.LST. This will be an ASCII file and can be printed in the usual manner (such as the DOS command PRINT B:SASRUN.LST).

Appendix H
Database Using 1250 Graduate Students Attending Brookhaven University

Variable	Label	Definition
X_1	VGRE	verbal score on graduate record exam (GRE)
X_2	QGRE	quantitative score on GRE
X_3	GR-GPA	graduate grade point average
X_4	UG-GPA	undergraduate grade point average
X_5	LSINDEX	score on Life Satisfaction Index questionnaire
X_6	AGE	age
X_7	SEX	sex, 0 = Male, 1 = Female
X_8	MARITAL	marital status 1 = never-married 2 = ever-married (married, never divorced) 3 = divorced 4 = remarried

OBS	VGRE	QGRE	GR-GPA	UG-GPA	LSINDEX	AGE	SEX	MARITAL
1	480	300	3.42	3.51	64	26	1	2
2	640	580	3.66	3.67	39	29	1	3
3	430	560	3.41	3.20	44	28	1	1
4	470	580	3.95	3.00	47	34	1	4
5	460	520	3.37	3.00	69	28	0	3
6	570	600	3.42	3.69	58	32	0	3
7	700	540	3.20	2.64	70	24	1	4
8	360	550	3.54	2.85	71	38	1	2
9	410	540	3.45	2.64	78	26	0	3
10	440	390	3.36	2.59	46	33	0	3
11	350	320	3.09	2.59	87	36	0	4
12	330	630	3.62	3.34	55	39	0	2
13	430	600	3.58	3.09	64	24	1	4
14	470	450	3.68	2.48	68	30	0	2
15	520	590	3.38	3.34	52	30	0	1
16	330	410	3.93	3.04	45	41	1	4
17	590	490	3.17	2.56	80	25	0	2
18	450	560	3.35	3.07	74	31	0	3
19	600	510	3.50	2.54	57	33	1	4
20	490	460	3.80	2.79	67	38	0	2
21	390	490	3.58	3.39	79	32	1	4
22	550	670	3.89	2.95	62	31	0	3
23	590	520	3.54	3.63	59	33	1	3
24	440	580	3.42	3.41	36	40	0	3
25	450	580	3.50	3.39	72	33	1	4
26	370	370	3.42	3.26	60	25	0	2
27	530	430	3.38	3.07	63	40	1	4
28	310	650	3.72	2.57	49	37	0	3
29	610	620	3.74	3.11	60	42	1	4
30	550	700	3.56	2.66	52	40	0	4
31	680	550	3.34	3.36	49	27	1	3
32	430	480	3.71	3.12	40	33	0	4
33	410	500	3.70	3.48	32	36	1	4
34	640	610	3.17	2.93	58	32	0	3
35	450	510	3.61	2.82	37	35	1	2
36	570	650	3.34	3.53	76	33	0	4
37	520	520	3.85	2.68	52	30	1	2
38	400	580	3.91	3.36	73	32	1	3
39	550	470	3.28	3.49	28	23	0	4
40	270	480	3.48	3.55	77	28	1	1
41	420	710	3.92	3.48	42	34	1	2
42	520	360	3.38	3.21	77	39	1	4
43	510	580	4.00	2.70	62	34	0	4
44	680	490	3.48	2.93	48	34	0	4
45	530	540	3.67	2.86	78	29	0	2
46	420	420	3.22	3.37	53	32	1	4
47	330	670	3.53	3.22	67	35	1	2
48	580	620	3.82	3.32	62	34	0	4
49	420	590	3.53	2.83	39	35	1	2
50	600	480	3.65	2.94	75	35	1	4

OBS	VGRE	QGRE	GR-GPA	UG-GPA	LSINDEX	AGE	SEX	MARITAL
51	510	520	3.37	3.67	30	30	1	3
52	370	680	3.40	3.41	70	28	1	2
53	490	510	3.59	2.86	63	38	0	1
54	610	550	3.55	2.97	88	27	0	2
55	420	580	3.18	3.70	49	22	1	1
56	300	520	3.44	3.14	85	26	1	1
57	560	600	3.74	2.89	45	30	0	4
58	370	550	3.51	2.62	37	29	0	1
59	460	610	3.60	2.94	87	29	0	3
60	390	630	3.63	3.46	31	38	1	2
61	640	420	4.00	3.23	78	28	1	2
62	770	280	3.54	2.89	33	30	0	3
63	380	670	3.21	3.66	56	26	1	1
64	390	420	3.43	3.17	82	36	0	4
65	520	300	3.59	2.61	63	41	1	4
66	630	630	3.86	3.35	37	34	0	2
67	510	710	3.71	3.66	49	33	1	4
68	540	620	3.64	2.90	45	43	0	4
69	570	420	3.88	2.71	92	32	0	2
70	450	530	3.98	2.92	66	39	0	4
71	410	520	3.40	2.99	45	38	0	4
72	520	460	3.65	2.93	50	37	0	4
73	490	450	3.93	2.40	37	33	0	4
74	540	600	4.00	2.93	57	22	0	2
75	350	440	3.77	2.86	77	37	0	4
76	680	570	3.61	3.68	30	30	1	3
77	630	630	3.62	2.73	75	28	0	1
78	420	600	3.67	2.83	53	34	0	4
79	420	650	3.32	2.86	46	29	0	4
80	430	580	3.45	2.33	52	36	0	4
81	350	540	4.00	3.42	59	34	0	3
82	490	580	3.78	2.87	63	29	0	2
83	400	460	3.53	2.63	69	34	0	3
84	630	620	3.17	2.85	36	34	0	2
85	300	460	3.59	3.67	70	32	1	4
86	460	500	3.43	3.33	67	31	1	3
87	450	620	3.35	2.97	63	28	0	2
88	330	560	3.00	3.24	61	31	1	2
89	340	600	3.89	3.06	77	37	0	2
90	500	320	3.43	2.58	38	32	0	3
91	510	370	3.57	2.68	74	31	0	3
92	340	430	3.48	3.58	77	29	1	3
93	390	630	3.55	2.73	74	30	0	3
94	440	460	3.49	2.84	87	34	0	4
95	480	620	3.88	3.62	86	31	1	2
96	520	490	3.56	2.95	28	29	0	2
97	330	470	3.35	2.43	78	30	1	3
98	470	510	3.99	2.85	74	31	0	2
99	510	540	3.73	3.76	28	29	1	3
100	570	540	3.65	3.12	78	30	1	2

OBS	VGRE	QGRE	GR-GPA	UG-GPA	LSINDEX	AGE	SEX	MARITAL
101	340	460	3.43	2.59	79	30	0	2
102	490	430	3.84	2.89	71	34	0	3
103	380	560	3.36	3.44	78	32	1	2
104	570	570	3.79	3.40	62	37	1	4
105	400	460	3.27	3.17	32	29	1	3
106	620	580	4.00	3.40	34	29	1	3
107	540	500	3.18	3.04	76	26	1	2
108	580	400	3.55	2.69	85	34	0	2
109	360	360	3.01	2.44	44	30	0	4
110	470	430	3.44	2.53	43	30	0	4
111	510	540	3.74	2.93	48	36	1	1
112	640	430	3.72	3.64	52	32	1	3
113	550	590	3.47	3.32	35	31	1	2
114	270	540	3.27	2.87	85	42	0	4
115	500	410	3.56	2.88	52	35	0	2
116	250	630	3.64	2.99	76	32	0	4
117	620	450	3.83	3.58	63	37	1	4
118	540	650	3.71	3.18	42	34	0	2
119	350	380	3.67	3.38	79	34	1	4
120	400	530	3.07	3.37	64	34	1	3
121	440	580	3.55	3.48	47	34	1	4
122	480	510	3.60	2.65	49	36	0	4
123	460	610	3.51	3.10	65	33	1	4
124	570	490	3.66	3.47	69	36	1	4
125	310	310	3.50	2.71	44	25	0	2
126	430	530	3.42	3.02	76	32	1	3
127	530	580	3.61	2.79	54	31	0	2
128	450	480	3.33	3.40	64	33	1	4
129	340	530	3.50	2.96	55	34	1	4
130	460	510	3.48	2.97	43	32	0	3
131	440	530	3.29	2.81	31	30	1	4
132	580	640	3.57	3.31	48	32	0	4
133	560	590	3.66	3.02	63	33	1	1
134	440	410	3.02	3.62	60	30	0	4
135	540	400	3.58	2.59	76	37	1	4
136	160	370	3.32	3.28	84	36	1	2
137	500	650	3.76	3.37	47	32	0	4
138	380	470	3.20	2.52	58	36	0	4
139	660	340	3.35	3.34	71	34	1	3
140	350	430	3.59	2.89	30	36	0	3
141	490	620	3.80	2.45	71	34	1	4
142	300	590	3.27	3.22	41	26	0	1
143	350	530	3.40	3.48	39	24	1	2
144	600	480	3.32	2.64	72	30	0	2
145	550	570	3.43	3.61	79	29	1	2
146	620	530	3.90	3.41	74	26	1	2
147	480	460	3.84	3.18	86	29	1	2
148	550	520	3.29	2.43	72	30	0	2
149	410	460	3.15	3.41	79	29	0	2
150	650	630	3.61		49	26	1	1

OBS	VGRE	QGRE	GR-GPA	UG-GPA	LSINDEX	AGE	SEX	MARITAL
151	410	510	3.41	2.60	61	27	0	1
152	470	410	3.56	3.20	54	31	1	4
153	290	490	3.24	3.28	32	31	1	3
154	380	390	3.19	3.37	69	24	1	2
155	410	500	3.57	2.80	53	38	0	4
156	630	400	3.57	3.34	68	35	1	4
157	440	580	3.81	3.63	44	35	0	1
158	450	540	3.52	3.37	60	32	1	4
159	430	660	3.64	2.65	75	41	1	2
160	410	500	3.61	2.66	65	37	0	3
161	170	630	3.88	3.63	28	30	1	3
162	590	630	3.87	3.42	39	30	1	3
163	500	390	3.68	3.44	60	33	1	4
164	560	470	3.09	3.15	30	30	1	1
165	380	360	3.68	2.29	32	38	0	3
166	490	730	3.64	2.79	56	27	0	1
167	310	400	3.77	2.77	78	29	0	2
168	580	470	3.44	3.62	51	29	1	1
169	360	520	3.63	3.19	35	34	0	3
170	380	640	3.52	2.68	38	31	1	4
171	600	320	3.35	3.19	62	37	1	4
172	600	440	3.55	3.50	37	31	1	3
173	560	470	3.39	3.28	64	31	1	4
174	680	670	3.96	2.92	47	37	0	4
175	430	430	3.95	3.70	60	34	0	4
176	440	540	3.70	2.52	52	31	1	2
177	510	430	3.45	3.07	90	38	0	4
178	430	440	3.52	2.86	40	40	1	4
179	630	350	3.81	2.79	56	39	1	4
180	480	670	3.39	2.83	48	25	0	3
181	490	180	3.98	3.38	42	32	1	4
182	340	550	3.58	2.75	48	32	0	4
183	420	610	3.48	3.44	64	33	1	1
184	500	460	3.40	2.44	52	31	0	4
185	440	690	3.89	3.38	47	37	1	2
186	430	510	3.66	3.86	42	22	1	4
187	570	580	3.50	2.88	35	29	0	4
188	330	550	3.46	2.85	32	31	1	4
189	540	510	3.36	3.65	49	29	0	3
190	610	420	3.42	2.68	89	36	1	3
191	360	520	3.44	3.32	60	37	0	4
192	480	600	4.00	3.24	45	34	0	1
193	600	440	3.52	3.04	36	30	1	2
194	560	370	3.30	3.57	49	27	1	2
195	380	600	3.31	2.59	32	24	0	2
196	450	550	3.93	2.64	44	30	1	2
197	440	520	3.34	2.49	50	26	1	2
198	480	670	3.65	3.17	66	34	0	2
199	410	460	3.57		72	38	0	3
200	480	520	3.01		40	27	1	3

OBS	VGRE	QGRE	GR-GPA	UG-GPA	LSINDEX	AGE	SEX	MARITAL
201	430	630	3.64	3.57	46	25	1	1
202	560	570	3.61	2.78	39	37	0	4
203	310	560	3.62	2.78	45	38	0	4
204	570	510	3.80	3.51	72	34	1	2
205	640	590	3.59	3.53	32	25	1	1
206	310	620	3.34	3.56	52	37	1	4
207	570	520	3.81	3.72	80	30	1	2
208	560	510	3.67	2.96	63	37	0	3
209	360	600	3.55	3.57	68	34	1	4
210	490	510	3.38	3.07	44	30	1	4
211	450	520	3.67	3.57	37	33	0	1
212	570	600	3.81	3.36	61	33	1	4
213	440	540	3.02	2.60	53	30	0	4
214	490	610	3.48	3.48	40	31	1	3
215	430	570	3.27	3.18	24	27	1	3
216	510	480	3.68	3.65	68	27	0	2
217	600	610	3.72	3.36	85	24	1	2
218	320	550	3.46	3.69	76	40	1	2
219	440	670	3.08	3.32	58	33	0	4
220	440	480	3.31	3.12	29	29	1	3
221	530	570	3.46	3.50	27	36	1	3
222	490	510	3.44	2.61	55	25	0	1
223	610	420	3.73	2.75	74	33	0	2
224	380	600	3.87	2.84	63	44	0	3
225	450	520	3.52	2.68	58	36	0	3
226	360	480	3.60	3.39	41	32	0	4
227	450	530	3.78	2.69	62	31	1	4
228	510	440	3.60	2.80	47	39	0	4
229	550	480	3.85	3.68	64	36	0	4
230	440	550	3.67	3.20	46	25	1	1
231	450	490	3.30	3.39	49	34	1	4
232	490	570	3.24	3.39	55	33	1	4
233	570	490	3.23	2.95	73	32	1	2
234	400	520	3.30	3.55	49	30	0	3
235	420	530	3.73	3.55	69	26	1	4
236	570	640	3.35	2.45	58	31	0	2
237	510	380	3.07	2.58	69	28	0	3
238	280	500	3.96	2.91	56	25	0	4
239	510	510	3.61	3.40	46	40	1	4
240	620	590	3.81	3.14	53	32	1	4
241	490	510	3.60	3.46	28	36	1	2
242	440	600	3.60	3.16	77	29	1	4
243	480	710	3.29	2.75	59	37	0	4
244	470	640	3.77	3.78	27	31	1	1
245	590	500	4.00	2.59	55	28	0	4
246	530	540	4.00	2.75	58	33	0	2
247	680	530	3.46	3.78	88	26	1	4
248	590	410	3.54	3.24	74	23	1	2
249	560	580	3.36	3.57	54	32	0	4
250	580	510	3.67	3.57	29	32	1	1
251	530	420	3.33	3.30	67	34	1	4
252	520	340	3.59	3.04	50	27	0	4
253	390	710	3.23	2.70	75	29	0	2
254	400	610	3.68	3.44	74	25	1	2
255	440	470	3.52	3.20	67	31	1	3
256	330	270	3.24	2.84	75	33	0	3
257	520	500	3.58	2.82	42	30	0	4
258	470	390	3.65	2.99	66	38	0	3
259	480	620	3.79	2.92	52	36	0	4
260	600	480	3.61	2.92	33	34	1	4
261	390	510	3.35	3.67	52	28	1	2
262	470	440	3.45	3.50	69	30	0	3
263	530	560	3.16	2.69	48	26	1	4
264	480	460	3.46	3.30	39	31	0	2
265	250	610	3.10	3.62	59	32	1	3
266	430	560	3.86	2.70	83	31	0	2
267	520	310	3.68	3.30	44	32	1	4
268	420	440	3.40	2.92	80	22	0	3
269	380	590	3.52	3.52	61	31	0	2
270	520	530	3.43	2.96	78	26	1	2
271	500	710	3.87	2.65	42	39	0	4
272	440	590	3.98	3.54	80	26	0	4
273	570	520	3.80	2.80	36	34	1	4
274	460	540	3.49	3.43	60	33	0	3
275	570	500	3.84	3.62	61	28	1	4
276	480	410	3.42	3.59	70	26	1	4
277	400	480	3.38	3.21	65	31	1	2
278	350	430	3.76	3.42	68	46	1	3
279	410	430	3.66	3.33	31	27	0	4
280	430	530	3.50	2.80	37	29	1	1
281	260	640	3.31	2.60	48	29	0	2
282	540	660	3.45	3.34	86	33	1	3
283	540	530	3.40	3.51	30	34	0	3
284	390	410	3.31	3.35	73	29	0	4
285	540	580	3.55	3.29	68	27	0	3
286	400	530	3.42	3.62	71	33	0	2
287	500	450	3.01	2.49	41	34	1	3
288	430	500	3.70	2.58	86	29	0	2
289	540	520	2.97	2.71	76	23	1	3
290	570	600	3.65	3.40	61	32	0	2
291	670	530	3.76	2.75	63	36	1	2
292	540	650	3.66	3.72	87	39	1	4
293	590	430	3.97	3.14	79	28	1	2
294	520	370	3.43	3.93	69	30	0	4
295	430	630	3.55	2.67	83	36	0	2
296	500	500	3.81	2.84	52	27	1	4
297	700	530	3.53	2.90	72	32	1	2
298	—	—	3.33	3.33	48	33	0	4
299	—	—	3.51	3.67	79	33	1	4
300	—	—	—	—	62	34	1	4

OBS	VGRE	QGRE	GR-GPA	UG-GPA	LSINDEX	AGE	SEX	MARITAL
351	490	620	3.75	2.96	54	34	0	3
352	540	570	3.52	3.26	32	32	1	3
353	610	570	3.59	3.17	81	33	0	2
354	240	530	3.67	2.66	57	41	0	4
355	320	460	3.45	2.65	80	27	0	2
356	470	480	3.76	2.86	57	42	0	4
357	420	490	3.55	3.58	56	37	1	4
358	490	590	3.82	3.01	67	34	0	3
359	390	390	3.32	2.79	83	31	0	2
360	470	450	3.38	2.81	84	29	0	2
361	560	570	3.59	3.42	61	30	1	2
362	560	490	3.27	3.35	74	28	1	2
363	310	330	3.54	2.87	53	37	0	3
364	500	460	4.00	3.63	39	41	1	1
365	560	410	3.69	3.40	45	32	1	3
366	530	560	3.57	3.39	34	29	1	2
367	450	700	3.78	3.59	67	34	1	4
368	750	560	3.53	2.71	47	37	0	2
369	630	470	3.63	2.86	77	26	0	3
370	450	410	3.50	2.55	89	27	0	3
371	610	470	3.36	3.47	41	27	1	3
372	440	270	3.74	2.82	76	31	0	2
373	370	570	3.52	2.89	72	38	0	4
374	570	620	3.40	2.80	76	27	0	2
375	390	520	3.70	2.60	50	36	0	4
376	410	400	3.70	2.88	81	36	0	3
377	700	560	3.46	3.00	51	36	0	4
378	340	600	3.84	3.08	34	29	1	3
379	480	570	3.43	3.84	59	34	0	4
380	360	540	3.68	2.98	58	30	1	4
381	500	540	3.64	3.56	73	29	1	3
382	350	300	3.71	3.34	54	38	0	4
383	660	510	3.80	2.71	30	30	0	4
384	370	590	3.13	2.46	45	34	0	4
385	380	430	3.62	2.71	49	33	1	1
386	570	480	3.59	3.52	33	30	0	3
387	500	320	3.51	2.62	84	30	0	3
388	400	570	3.43	3.50	33	36	1	3
389	610	680	3.97	2.55	43	34	0	4
390	680	520	3.36	2.94	78	27	0	2
391	320	420	3.51	3.31	35	35	1	3
392	500	600	3.52	3.73	65	29	1	4
393	420	480	3.26	3.26	31	36	0	4
394	650	600	4.00	2.93	46	36	0	4
395	600	620	3.71	3.00	76	32	1	1
396	550	380	3.46	3.33	36	31	1	2
397	520	480	3.65	3.47	64	29	0	4
398	750	430	3.74	2.80	46	29	1	4
399	750	730	3.75	3.75	60	34	1	4
400	360	480	3.69	2.75	78	36	0	2

OBS	VGRE	QGRE	GR-GPA	UG-GPA	LSINDEX	AGE	SEX	MARITAL
301	560	630	3.55	2.79	94	28	0	2
302	370	550	3.19	2.48	66	29	0	3
303	420	530	3.28	2.59	52	41	0	4
304	480	450	4.00	3.22	89	34	0	2
305	530	460	3.49	3.50	42	29	1	1
306	350	330	3.59	3.39	52	37	1	4
307	300	540	3.62	2.53	45	33	0	4
308	470	580	3.70	3.26	66	33	1	4
309	600	470	3.91	2.89	84	33	0	2
310	690	470	3.37	3.37	58	32	1	4
311	430	570	3.34	3.62	38	26	1	1
312	460	520	3.76	3.47	40	29	1	1
313	540	450	3.56	2.56	58	36	1	4
314	320	380	3.63	2.52	42	39	0	4
315	410	560	3.92	3.22	75	38	0	4
316	380	560	3.28	3.13	57	33	0	3
317	410	520	3.60	3.12	41	32	0	4
318	450	670	3.54	3.45	52	34	1	4
319	360	640	3.46	3.72	40	30	1	1
320	480	500	3.62	2.79	46	35	0	1
321	400	580	3.64	2.64	74	29	0	2
322	480	450	3.33	2.39	66	36	1	4
323	460	500	3.56	2.62	53	32	0	4
324	380	400	3.59	2.66	79	34	0	4
325	570	490	3.60	2.80	56	30	0	2
326	500	750	3.71	2.90	74	35	0	4
327	380	680	3.56	3.43	59	35	1	4
328	350	570	3.41	2.61	57	25	0	3
329	580	520	3.53	2.92	78	36	1	4
330	320	490	3.26	3.30	56	36	1	3
331	450	570	3.62	3.79	65	34	0	4
332	500	520	3.46	2.62	41	39	0	3
333	310	540	3.50	2.99	80	25	0	2
334	440	690	3.40	2.86	70	29	0	4
335	390	370	3.38	2.63	52	29	0	4
336	400	330	3.32	2.51	44	33	0	4
337	520	380	3.49	2.61	70	30	1	3
338	280	340	3.46	2.62	40	35	1	3
339	480	340	3.84	3.44	71	28	1	2
340	390	430	3.51	3.47	59	28	0	4
341	470	470	3.63	3.94	61	28	1	4
342	520	360	3.73	2.90	72	43	0	3
343	530	580	3.84	3.61	75	35	1	4
344	530	570	3.58	2.89	93	26	0	3
345	540	540	3.56	3.48	84	28	1	3
346	320	540	3.37	3.35	53	39	1	2
347	580	600	3.84	3.05	34	30	0	3
348	430	400	3.29	3.60	34	39	1	3
349	430	400	3.29	3.60	34	30	1	3
350	620	480	3.72	3.41	25	35	1	3

OBS	VGRE	QGRE	GR-GPA	UG-GPA	LSINDEX	AGE	SEX	MARITAL
401	530	620	3.36	2.80	42	28	0	4
402	440	570	3.44	3.01	61	26	1	4
403	560	380	3.57	2.55	68	27	0	3
404	610	400	3.58	2.67	90	28	0	2
405	490	720	3.50	3.21	62	34	1	4
406	520	560	3.81	3.80	79	33	1	2
407	380	430	3.28	2.50	79	26	1	2
408	340	640	3.25	3.26	42	28	0	1
409	580	440	4.00	3.02	49	40	1	4
410	310	640	3.54	3.54	73	32	0	2
411	510	720	4.00	3.63	60	32	0	4
412	280	490	3.34	3.02	73	28	1	2
413	520	540	3.60	3.21	67	27	1	4
414	600	520	3.72	2.75	38	34	0	2
415	380	550	3.26	2.96	44	27	1	4
416	590	480	3.67	2.76	68	30	0	2
417	390	630	3.34	3.52	46	39	1	3
418	460	520	3.61	3.09	64	31	1	4
419	420	440	3.12	2.75	36	30	1	3
420	330	530	3.17	2.64	70	34	0	3
421	620	500	3.77	2.89	77	40	0	3
422	410	560	3.94	2.82	65	43	0	2
423	590	430	3.65	2.55	49	31	0	1
424	320	570	3.38	2.81	36	31	0	4
425	660	500	3.50	3.00	87	30	0	3
426	320	590	3.28	3.22	34	29	1	3
427	440	590	3.53	2.92	50	36	1	3
428	440	550	3.22	3.44	45	28	1	1
429	580	470	3.61	2.82	70	37	1	4
430	340	430	3.53	2.72	77	34	0	4
431	620	530	3.46	2.63	57	35	1	2
432	450	440	3.41	3.54	64	28	0	3
433	530	440	3.53	2.63	53	27	1	3
434	360	530	3.26	2.88	64	32	0	3
435	600	460	3.45	2.95	79	29	0	2
436	270	570	3.56	3.46	69	29	0	1
437	610	400	3.70	3.50	47	25	1	3
438	600	420	3.18	2.47	39	30	0	1
439	450	500	3.68	3.48	65	30	1	2
440	340	560	3.53	3.60	70	32	1	4
441	410	600	3.69	3.09	68	34	0	4
442	500	640	3.79	3.42	68	39	0	4
443	430	380	3.66	2.57	71	32	1	3
444	500	360	3.38	2.73	81	28	0	2
445	460	470	3.17	2.56	82	39	0	2
446	470	480	3.44	3.07	73	29	0	1
447	440	570	3.49	3.58	38	25	1	1
448	440	680	3.88	2.70	76	29	0	4
449	300	680	3.88	2.70	58	40	0	4
450	570	550	3.76	3.26	80	30	1	2

OBS	VGRE	QGRE	GR-GPA	UG-GPA	LSINDEX	AGE	SEX	MARITAL
451	450	680	3.32	2.69	44	31	0	4
452	680	370	3.44	2.41	54	28	0	4
453	410	510	3.63	3.53	52	31	1	4
454	450	470	3.67	3.50	49	26	1	4
455	310	440	3.54	3.55	18	29	1	3
456	530	550	3.94	3.11	75	37	0	2
457	520	520	3.73	2.44	50	39	0	4
458	400	620	3.66	3.70	65	32	0	4
459	600	620	3.68	2.51	42	37	1	4
460	310	480	3.90	2.86	73	42	0	2
461	380	370	3.20	2.62	49	32	0	3
462	280	570	3.31	2.78	70	33	0	3
463	580	420	3.54	3.50	64	33	0	2
464	390	550	3.69	3.19	44	29	1	1
465	490	480	3.48	2.81	87	35	0	4
466	270	660	3.89	3.18	35	27	0	2
467	450	590	3.66	3.03	71	41	0	4
468	420	490	3.61	2.68	46	36	0	2
469	480	480	3.57	2.79	63	33	1	4
470	330	490	3.17	3.38	57	38	1	4
471	220	650	3.38	2.85	62	36	0	2
472	470	560	3.21	3.11	60	38	0	1
473	260	600	3.88	3.26	85	37	0	4
474	430	470	3.36	2.52	35	26	1	4
475	330	580	3.57	3.00	77	34	1	3
476	200	430	3.45	2.84	62	33	0	3
477	430	450	3.49	3.49	83	29	0	4
478	490	590	3.81	2.94	64	49	0	4
479	480	470	3.49	2.91	51	32	1	4
480	430	510	3.42	3.39	67	31	0	2
481	470	500	3.43	3.54	30	30	1	4
482	560	600	3.49	3.20	87	26	0	3
483	550	470	3.61	2.55	45	40	0	3
484	620	480	3.62	2.87	56	29	0	2
485	470	530	3.91	3.05	45	32	1	1
486	390	460	3.80	3.46	33	29	0	3
487	500	520	3.22	3.35	62	28	1	3
488	510	590	3.87	3.48	52	28	1	1
489	520	500	3.76	2.90	76	40	0	2
490	420	350	3.57	2.55	58	36	0	4
491	620	440	3.43	3.43	59	36	1	4
492	420	540	3.79	3.38	37	33	1	3
493	610	600	3.64	3.50	45	26	0	4
494	600	430	3.52	2.86	49	31	0	4
495	270	520	3.39	3.21	34	30	1	4
496	380	560	3.11	2.97	48	31	1	3
497	280	560	3.89	3.69	37	30	0	4
498	510	480	3.69	3.06	43	36	1	4
499	510	480	3.86	3.06	43	36	0	4
500	400	540	3.60	3.17	36	25	1	3

OBS	VGRE	QGRE	GR-GPA	UG-GPA	LSINDEX	AGE	SEX	MARITAL
501	380	490	3.12	3.06	61	36	1	4
502	270	470	3.35	2.85	70	27	0	2
503	460	380	3.66	3.66	80	29	1	2
504	400	390	3.73	2.86	48	37	0	4
505	550	580	3.76	3.60	68	31	1	4
506	430	370	3.59	2.75	76	31	0	2
507	610	290	3.46	2.98	56	38	0	4
508	450	640	3.42	3.33	66	32	1	4
509	390	520	3.47	3.19	60	31	1	4
510	360	480	3.50	3.22	75	28	1	4
511	470	360	3.51	2.53	38	37	0	2
512	490	640	3.63	3.44	38	31	1	4
513	460	330	3.59	2.52	74	28	0	1
514	580	560	3.72	3.60	40	31	0	3
515	410	570	3.42	2.65	86	30	0	2
516	620	500	3.45	2.84	35	25	1	4
517	470	540	3.58	3.49	56	29	1	2
518	430	540	3.79	3.56	67	33	0	3
519	500	440	3.59	2.99	74	25	0	4
520	490	570	3.70	2.98	29	23	0	2
521	340	500	3.40	3.23	48	29	1	3
522	510	520	3.76	3.47	32	32	1	3
523	310	350	3.26	2.63	68	37	0	3
524	350	540	3.61	3.48	61	35	0	4
525	540	490	3.78	2.81	35	31	0	4
526	370	640	3.49	3.52	48	23	1	1
527	610	490	3.93	3.34	80	36	1	4
528	650	500	3.68	2.84	83	33	0	2
529	540	520	3.50	2.88	40	26	0	2
530	390	450	3.67	3.05	76	32	0	4
531	640	540	3.71	3.22	91	30	0	2
532	440	390	3.56	3.08	57	32	1	1
533	260	480	3.44	3.60	78	28	0	4
534	400	490	3.30	3.48	59	28	0	2
535	480	510	3.35	2.80	76	32	1	4
536	350	550	3.24	3.24	53	29	1	2
537	580	570	3.51	3.15	44	33	0	3
538	570	630	3.76	2.98	56	29	1	4
539	560	470	3.85	3.46	80	32	1	4
540	590	510	3.63	3.53	64	31	0	4
541	520	420	3.79	2.71	60	40	0	2
542	460	440	3.64	2.75	54	38	0	3
543	690	470	3.82	3.22	49	40	1	4
544	510	290	3.50	3.51	61	39	1	4
545	260	500	3.55	3.56	73	38	1	3
546	420	570	3.54	3.06	77	33	0	2
547	430	630	3.75	3.50	39	34	1	4
548	310	620	3.32	2.96	23	35	0	3
549	530	510	3.51	3.47	53	29	1	1
550	580	510	3.30	2.62	53	30	0	1

OBS	VGRE	QGRE	GR-GPA	UG-GPA	LSINDEX	AGE	SEX	MARITAL
551	630	660	3.72	2.41	84	29	0	2
552	390	620	3.67	3.23	69	34	1	2
553	440	370	3.51	2.80	80	37	0	2
554	460	520	3.24	3.47	32	30	1	1
555	470	460	3.46	2.45	73	29	0	2
556	460	560	3.24	2.84	77	32	0	4
557	560	640	3.84	3.68	68	33	1	4
558	660	470	3.78	3.49	53	32	1	4
559	590	460	3.34	2.79	79	23	0	2
560	280	560	3.41	3.41	66	22	1	2
561	490	530	3.48	2.59	83	29	0	4
562	430	610	3.28	2.52	60	32	1	4
563	360	620	2.94	3.23	49	30	0	1
564	350	370	3.66	2.84	39	26	1	1
565	450	580	3.24	2.61	48	36	0	4
566	310	530	3.62	3.48	77	29	1	2
567	400	430	3.80	2.91	43	41	0	4
568	340	500	3.43	3.14	69	31	1	4
569	490	440	3.17	3.65	83	31	0	4
570	350	500	3.49	3.49	42	27	1	1
571	470	550	4.00	2.70	65	41	1	3
572	370	430	3.70	2.73	42	36	1	4
573	540	420	3.31	2.90	52	35	0	4
574	420	620	3.64	3.44	30	33	0	3
575	410	330	3.27	3.35	34	22	0	4
576	460	470	4.00	2.99	50	31	1	3
577	630	520	3.36	2.98	75	33	1	4
578	440	440	3.58	2.72	52	31	0	4
579	770	390	4.00	2.98	78	40	0	2
580	500	530	3.34	2.86	80	29	0	4
581	310	520	3.70	2.54	62	45	0	2
582	590	420	3.51	2.92	53	28	0	1
583	550	530	3.86	3.01	73	34	0	2
584	520	580	3.37	3.18	54	34	0	4
585	410	620	3.62	2.95	48	32	1	4
586	350	330	3.78	2.69	75	34	0	2
587	560	440	3.44	2.95	48	39	0	4
588	480	400	3.41	3.44	60	34	1	4
589	410	610	3.42	3.50	45	26	1	1
590	510	690	3.37	3.47	73	32	1	4
591	680	520	3.47	2.83	51	31	0	4
592	520	390	3.81	3.40	46	25	1	3
593	410	320	3.61	3.44	54	31	1	4
594	490	540	3.47	3.42	49	22	1	4
595	490	640	3.40	3.37	39	30	1	3
596	340	560	3.49	3.41	45	36	0	4
597	490	580	3.81	2.82	73	29	1	4
598	530	600	3.61	3.28	55	37	0	2
599	280	690	3.47	3.19	85	29	0	2
600	500	390	3.40	2.59	41	40	0	4

OBS	VGRE	QGRE	GR-GPA	UG-GPA	LSINDEX	AGE	SEX	MARITAL
601	450	350	3.25	3.43	63	34	1	4
602	430	570	3.41	3.29	66	36	1	4
603	570	460	3.72	3.46	62	33	1	4
604	410	580	3.78	3.52	69	28	1	2
605	430	530	3.39	2.56	91	29	0	2
606	650	470	3.03	2.47	43	30	0	4
607	540	460	3.36	3.36	32	28	1	1
608	570	370	3.66	2.83	67	44	0	2
609	390	550	3.26	2.80	74	30	0	2
610	550	450	3.56	2.99	70	36	0	3
611	500	300	3.47	2.59	89	38	0	2
612	600	530	3.28	2.75	45	32	0	4
613	470	490	3.50	3.47	62	32	1	4
614	540	420	3.17	3.28	64	35	1	4
615	480	430	3.36	3.20	81	33	0	3
616	580	440	3.37	2.91	51	28	0	1
617	310	560	3.54	2.96	71	27	0	2
618	430	580	3.88	3.54	72	25	1	2
619	270	400	3.57	2.59	44	38	0	1
620	660	370	3.63	3.45	44	29	1	4
621	520	600	3.92	3.29	55	32	0	4
622	280	460	3.21	2.42	61	25	0	4
623	240	740	3.52	3.61	73	33	1	2
624	500	560	3.37	3.08	63	26	1	4
625	560	480	3.72	2.51	60	33	0	1
626	600	410	3.53	2.80	63	32	0	2
627	340	580	3.39	3.18	50	31	1	4
628	620	610	3.80	2.60	53	39	0	4
629	310	440	3.67	3.70	36	35	1	4
630	260	500	3.31	3.21	73	27	1	3
631	220	460	3.43	2.84	60	42	0	2
632	670	420	3.27	2.65	90	29	0	1
633	320	520	3.64	2.84	43	31	1	2
634	480	630	3.65	3.79	51	27	0	3
635	470	430	3.54	3.03	50	34	1	4
636	350	700	3.42	2.66	62	33	0	2
637	580	440	3.39	3.47	70	32	0	4
638	530	600	3.64	3.29	83	28	0	2
639	450	560	3.23	2.87	44	22	0	1
640	450	460	3.11	2.52	74	27	0	3
641	500	470	3.54	2.30	52	29	0	4
642	560	450	3.70	2.56	52	38	0	4
643	450	590	4.00	2.62	43	37	1	3
644	420	590	3.69	3.39	54	34	1	3
645	440	470	3.60	3.69	70	35	0	3
646	440	500	3.52	2.87	47	34	1	4
647	280	480	3.65	3.75	36	33	0	4
648	370	530	3.77	3.40	62	36	1	2
649	440	510	3.40	2.54	80	32	0	2
650	380	530	3.40	3.18	27	29	1	3
651	650	420	3.68	3.47	35	27	1	3
652	410	490	3.34	2.77	63	32	0	3
653	410	360	3.52	2.59	84	36	0	2
654	530	550	3.72	3.53	41	28	1	1
655	230	520	3.43	3.58	62	31	1	4
656	460	500	3.42	3.19	62	31	0	4
657	390	560	3.38	3.06	77	37	1	2
658	470	250	3.17	2.68	66	37	1	4
659	700	510	3.59	2.94	84	28	0	2
660	220	660	3.55	3.14	68	34	1	4
661	510	490	3.52	2.94	88	32	0	4
662	290	400	3.49	2.56	53	34	0	2
663	660	550	3.47	2.73	70	27	0	1
664	500	610	3.78	3.46	44	30	1	4
665	300	470	3.45	2.88	41	42	0	4
666	430	480	3.33	2.90	46	33	0	4
667	380	470	3.42	2.89	75	32	0	2
668	490	400	3.45	2.81	58	28	0	4
669	330	440	3.49	3.45	53	31	1	4
670	450	570	3.61	3.36	59	33	1	4
671	480	600	3.62	2.65	59	36	0	3
672	600	350	3.76	2.66	76	31	0	2
673	410	480	3.79	3.51	35	29	1	3
674	460	460	3.24	3.27	31	29	1	3
675	430	450	3.53	3.36	36	28	0	3
676	560	500	3.64	3.85	83	36	0	3
677	470	560	3.77	2.89	66	37	0	1
678	270	300	3.59	3.48	35	26	0	3
679	280	540	3.27	3.60	36	29	1	1
680	610	670	3.31	2.86	46	27	1	4
681	400	600	4.00	3.19	44	31	1	4
682	570	490	3.71	3.39	51	35	0	1
683	540	610	3.43	3.46	47	22	1	4
684	410	540	3.74	3.10	55	40	0	1
685	330	590	3.30	3.51	40	23	0	2
686	380	560	3.48	2.30	78	30	0	4
687	480	410	3.62	2.87	75	33	1	4
688	370	500	3.30	3.40	52	28	0	4
689	520	510	3.57	3.35	48	36	0	4
690	390	580	3.76	2.78	42	37	0	3
691	560	570	3.59	3.35	63	47	0	4
692	710	370	3.85	2.82	45	36	1	4
693	600	490	3.69	2.61	63	36	0	3
694	440	530	3.56	2.91	68	30	1	3
695	500	560	3.51	3.63	38	38	0	3
696	420	480	3.93	2.83	43	32	1	4
697	440	350	3.46	3.16	57	27	0	4
698	440	350	3.39	2.87	72	32	1	2
699	300	420	3.77	2.54	80	32	0	2
700	570	700	4.00	3.29	60	38	0	3

OBS	VGRE	QGRE	GR-GPA	UG-GPA	LSINDEX	AGE	SEX	MARITAL
701	210	600	3.38	3.31	55	32	1	4
702	340	470	3.39	3.42	33	26	1	1
703	340	480	3.44	2.96	90	35	0	2
704	410	310	3.58	2.58	50	34	0	4
705	390	580	3.67	3.65	73	37	1	4
706	660	470	4.00	3.05	56	40	0	4
707	310	710	3.35	3.30	32	35	1	4
708	380	570	3.78	2.67	54	32	0	3
709	540	510	3.82	3.00	42	29	0	1
710	390	520	3.31	3.67	68	27	0	2
711	540	500	3.51	2.77	54	22	0	1
712	510	690	3.54	2.80	46	26	0	4
713	630	510	3.54	3.31	55	35	1	4
714	750	490	3.74	3.60	79	37	0	2
715	390	370	3.59	3.60	34	30	0	3
716	350	430	3.42	2.78	47	42	0	4
717	530	420	3.74	2.58	64	34	0	3
718	480	610	3.65	2.71	42	33	0	4
719	640	550	3.77	2.93	57	33	0	3
720	560	500	3.61	2.76	72	35	0	2
721	460	370	3.35	2.74	82	22	0	2
722	500	700	3.54	2.74	52	37	0	3
723	600	490	3.75	2.72	65	36	0	3
724	360	430	3.40	2.89	43	28	0	4
725	740	540	3.80	3.47	57	29	1	2
726	470	590	3.37	3.37	37	29	0	3
727	550	520	3.67	2.78	89	38	0	2
728	430	500	3.76	2.65	52	41	1	4
729	610	440	3.61	3.44	82	23	0	2
730	570	600	3.36	3.37	71	27	0	2
731	400	590	3.60	2.49	76	28	1	2
732	500	520	3.57	3.42	70	28	0	2
733	190	570	3.13	3.31	48	31	1	1
734	420	590	3.65	3.30	67	27	1	4
735	220	390	2.97	3.29	33	29	1	3
736	660	680	4.00	2.65	45	45	0	4
737	610	500	3.63	2.59	62	25	0	1
738	580	600	3.67	3.35	65	35	1	4
739	520	480	3.62	3.27	52	32	1	4
740	560	420	3.81	3.59	56	34	1	4
741	390	580	3.61	3.54	40	34	1	1
742	340	520	3.25	2.56	70	28	0	2
743	410	550	3.74	2.77	80	37	0	3
744	540	470	3.62	2.93	71	36	0	3
745	410	420	3.84	3.65	41	26	1	1
746	510	490	3.26	3.46	72	28	1	2
747	330	390	3.57	3.80	81	28	1	2
748	410	430	3.51	3.29	35	30	1	3
749	540	470	3.56	3.64	71	34	1	4
750	750	490	3.95	3.79	44	32	1	3

OBS	VGRE	QGRE	GR-GPA	UG-GPA	LSINDEX	AGE	SEX	MARITAL
751	500	480	3.42	2.29	65	27	0	1
752	240	590	3.76	3.41	56	32	1	4
753	400	450	3.53	3.54	33	30	1	3
754	400	550	3.46	2.94	57	31	1	4
755	300	400	3.15	3.33	39	29	1	3
756	460	660	3.54	3.43	73	28	1	2
757	550	470	3.50	3.35	68	26	1	4
758	600	390	3.56	2.80	72	41	0	2
759	480	560	3.71	3.09	75	33	0	2
760	270	450	3.36	2.45	81	32	0	2
761	410	440	3.34	2.68	87	31	0	2
762	510	540	3.62	3.18	80	40	0	3
763	320	510	3.54	3.60	30	29	1	4
764	410	430	3.41	2.57	37	29	0	4
765	360	640	3.50	2.85	47	41	0	3
766	610	570	3.48	2.69	66	35	0	1
767	360	680	3.55	2.58	51	27	0	3
768	550	500	3.84	3.08	45	30	1	1
769	590	470	3.87	3.46	84	28	0	3
770	430	460	3.82	2.91	61	35	0	2
771	570	330	3.65	2.94	80	41	0	2
772	370	630	3.73	2.51	49	31	0	4
773	360	470	3.60	3.52	36	29	0	3
774	600	520	3.41	3.56	50	28	0	4
775	360	400	3.37	3.17	62	32	1	4
776	530	430	3.31	2.28	82	34	1	3
777	460	420	3.47	3.59	33	26	0	2
778	420	290	3.82	3.00	54	33	0	4
779	500	480	4.00	3.75	60	32	0	2
780	430	370	3.13	3.27	48	34	0	2
781	410	380	3.50	2.49	55	33	1	2
782	460	430	3.75	2.83	56	34	0	2
783	290	510	3.51	3.36	36	29	1	1
784	550	410	3.66	3.07	68	37	1	2
785	760	430	3.60	2.73	79	38	1	1
786	430	460	3.71	3.53	36	27	0	1
787	510	380	3.72	2.55	38	31	0	4
788	530	520	3.87	3.65	41	41	1	4
789	390	490	3.65	3.10	58	36	1	4
790	460	530	3.53	2.90	34	30	1	3
791	320	410	3.53	2.59	56	36	1	4
792	560	210	3.52	2.89	51	36	1	4
793	480	540	3.56	2.85	64	35	0	4
794	720	570	3.80	2.68	62	35	0	3
795	500	380	3.41	3.49	61	33	0	1
796	500	540	3.48	3.44	71	30	0	3
797	350	450	3.59	3.19	76	32	1	3
798	410	770	3.56	2.18	65	36	1	2
799	270	470	3.34	3.41	81	30	1	4
800	600	620	3.59	3.41	41	28	1	1

OBS	VGRE	QGRE	GR-GPA	UG-GPA	LSINDEX	AGE	SEX	MARITAL
801	530	670	3.94	3.52	55	41	1	4
802	480	620	3.43	3.64	59	32	1	4
803	520	500	3.56	2.48	84	30	0	2
804	320	540	3.35	2.75	83	31	0	4
805	710	310	3.38	3.13	48	33	1	2
806	440	360	3.79	3.49	65	34	1	4
807	370	520	3.41	2.69	86	31	0	4
808	610	480	3.56	3.50	59	34	1	2
809	380	500	3.74	2.56	65	35	0	4
810	610	480	3.57	2.83	70	29	0	3
811	330	540	3.28	3.61	31	31	1	1
812	500	410	3.68	2.92	57	38	0	4
813	540	340	3.43	3.61	49	27	0	2
814	610	650	3.50	3.42	63	28	1	3
815	510	650	3.65	3.53	36	31	1	1
816	520	580	3.68	3.38	47	33	1	4
817	610	630	4.00	2.95	54	44	0	2
818	410	550	3.65	2.58	76	34	0	3
819	430	420	3.47	3.37	31	30	1	3
820	430	340	3.45	3.52	37	29	1	3
821	760	540	3.61	3.51	30	30	1	4
822	540	540	3.51	3.14	65	34	0	4
823	530	470	3.14	2.77	44	32	0	4
824	510	590	3.30	3.09	92	39	0	2
825	310	310	3.80	2.38	43	36	1	1
826	480	400	3.73	3.31	47	25	0	2
827	560	560	3.69	2.64	62	34	0	3
828	430	410	3.62	2.80	64	38	1	1
829	460	470	3.76	3.60	51	33	1	2
830	520	550	3.60	3.24	71	30	0	4
831	490	540	3.55	3.32	56	35	0	2
832	550	570	3.51	2.86	80	34	0	2
833	530	600	3.54	2.70	80	22	1	2
834	410	680	3.69	3.17	39	34	0	3
835	430	540	3.60	2.57	38	23	0	4
836	320	460	3.45	2.47	54	42	0	1
837	500	610	3.54	3.07	53	31	0	4
838	320	700	3.45	3.49	73	35	1	2
839	520	350	3.01	2.47	27	30	1	1
840	460	450	3.51	3.07	33	25	1	4
841	490	460	3.77	3.49	85	30	0	2
842	450	350	3.95	2.73	86	29	0	3
843	520	550	3.30	2.47	76	34	0	2
844	350	580	3.64	3.66	70	29	0	2
845	480	410	3.38	3.51	49	28	1	4
846	600	430	4.00	2.82	35	36	0	4
847	470	520	3.54	3.50	59	31	1	3
848	620	370	3.86	2.82	74	32	0	3
849	620	370	3.85	2.93	74	41	0	3
850	390	550	3.69	2.79	46	30	0	4
851	600	350	3.11	2.56	54	25	0	1
852	550	530	3.74	3.53	48	29	1	3
853	300	350	3.24	2.71	72	27	0	3
854	560	520	3.63	2.51	50	29	0	3
855	470	570	3.58	3.60	60	22	0	3
856	590	530	3.55	3.42	47	27	1	3
857	450	590	3.49	2.63	70	35	0	3
858	480	570	3.62	3.38	76	27	1	4
859	400	440	3.78	3.05	49	36	0	4
860	390	370	3.60	2.68	37	39	0	4
861	570	580	3.51	3.32	41	31	1	2
862	420	630	3.69	2.58	69	26	0	4
863	580	480	3.72	3.38	62	34	0	4
864	530	400	3.53	3.06	41	34	1	2
865	390	550	3.41	2.78	75	38	0	1
866	600	500	3.90	3.71	38	28	1	3
867	490	510	3.51	3.59	28	32	0	4
868	310	620	3.80	3.03	52	43	0	3
869	550	300	3.46	2.48	61	34	0	4
870	480	390	3.61	2.94	46	32	1	1
871	640	650	3.92	3.29	81	34	1	2
872	670	670	3.69	3.47	30	35	0	3
873	370	400	3.42	3.50	80	30	0	2
874	290	590	3.75	2.99	42	35	0	4
875	550	390	3.49	2.74	78	44	0	4
876	410	600	3.77	2.66	58	33	1	4
877	330	470	3.51	3.52	56	28	1	3
878	660	560	3.47	3.66	45	36	1	4
879	400	500	3.27	2.70	66	36	0	4
880	610	490	3.83	3.57	69	33	0	4
881	500	450	3.74	2.26	68	23	1	3
882	430	530	3.58	3.39	52	32	1	2
883	720	670	3.61	2.61	58	35	1	3
884	420	600	3.59	2.97	76	32	0	2
885	660	650	3.49	3.70	77	29	0	3
886	600	550	3.77	2.87	85	27	1	4
887	530	430	3.39	3.67	74	33	0	4
888	590	570	3.88	2.89	95	29	1	4
889	330	590	3.74	3.31	43	41	0	3
890	750	470	3.45	2.86	35	33	1	2
891	500	620	3.75	3.37	58	28	0	2
892	730	560	3.65	3.38	33	36	1	2
893	520	560	3.55	3.75	45	27	1	2
894	470	590	3.45	2.80	86	30	0	3
895	360	500	3.42	2.73	37	29	1	3
896	340	640	3.34	3.06	85	30	0	2
897	340	640	3.60	3.03	84	32	0	2
898	340	640	3.60	3.03	84	32	0	2
899	360	590	3.42	3.06	37	29	0	3
900	340	640	3.60	3.03	84	32	0	2

OBS	VGRE	QGRE	GR-GPA	UG-GPA	LSINDEX	AGE	SEX	MARITAL
901	270	430	3.25	3.18	59	27	1	4
902	200	590	3.58	3.38	47	27	1	1
903	350	650	3.07	2.51	57	38	0	3
904	330	460	3.52	2.77	53	40	0	4
905	380	500	3.58	3.60	52	31	1	4
906	350	420	3.78	3.34	64	34	1	4
907	630	610	4.00	2.74	54	31	0	4
908	490	580	3.64	3.42	77	28	1	4
909	700	530	3.95	2.74	83	31	0	2
910	520	460	3.70	3.91	44	30	1	3
911	480	670	3.43	2.70	75	29	0	4
912	490	490	3.70	2.59	65	35	1	4
913	390	660	3.51	2.44	41	34	0	4
914	650	490	3.73	3.38	70	37	1	4
915	440	480	3.37	3.22	54	29	1	1
916	430	500	3.41	2.64	38	30	0	4
917	650	460	3.51	3.38	34	35	1	4
918	450	550	3.33	3.20	56	33	1	4
919	510	510	3.62	2.69	65	31	0	3
920	300	460	3.34	2.75	32	35	1	4
921	470	420	3.57	3.09	59	39	1	4
922	450	310	3.85	3.30	69	32	0	3
923	530	450	3.68	3.46	83	42	0	2
924	480	400	3.54	2.35	60	32	1	4
925	620	620	3.07	3.36	66	28	1	4
926	540	530	3.44	2.62	47	34	0	1
927	370	620	3.08	2.80	57	22	1	4
928	380	360	3.21	3.50	54	33	0	4
929	430	530	3.48	3.42	75	30	1	1
930	530	420	3.59	3.41	35	29	1	2
931	430	540	3.67	2.92	35	31	0	3
932	620	530	3.71	3.25	34	41	0	3
933	600	750	3.67	3.14	45	37	1	4
934	560	500	3.44	4.00	76	33	1	2
935	640	560	4.00	3.50	38	29	0	3
936	300	490	3.66	3.29	47	35	0	1
937	530	480	3.51	2.58	55	29	1	4
938	360	510	3.30	3.50	74	37	1	2
939	350	610	3.23	3.53	52	31	1	4
940	490	480	3.64	3.15	62	34	0	4
941	620	610	3.68	3.52	39	32	1	3
942	520	600	3.63	3.60	32	34	1	3
943	510	640	3.60	2.00	84	31	1	3
944	600	580	3.78	3.56	36	32	0	4
945	540	410	3.59	2.82	28	27	1	2
946	500	530	3.60	3.28	73	32	0	4
947	580	640	3.73	2.90	56	36	1	4
948	290	440	3.24		49	35	0	1
949	510	650	3.85		46	35	1	4
950					31			4

OBS	VGRE	QGRE	GR-GPA	UG-GPA	LSINDEX	AGE	SEX	MARITAL
951	320	680	3.33	2.33	81	39	0	2
952	250	570	3.52	3.29	35	30	1	3
953	370	730	3.81	3.24	57	39	1	4
954	570	490	3.90	3.79	62	31	1	2
955	540	470	3.83	3.52	81	32	1	2
956	360	510	3.47	2.96	75	31	0	4
957	530	400	3.61	2.79	49	29	0	4
958	740	370	3.45	2.52	40	29	0	4
959	140	500	3.51	3.05	82	29	1	2
960	360	590	3.41	3.78	37	30	1	1
961	250	550	3.25	3.31	38	28	1	1
962	520	590	3.67	3.03	46	36	0	2
963	310	630	3.41	3.32	52	28	1	1
964	430	490	3.65	3.47	82	36	0	4
965	440	500	3.20	2.61	32	26	1	2
966	380	600	3.32	3.18	36	30	1	1
967	430	590	3.93	3.72	49	26	1	1
968	380	500	3.71	3.32	74	35	1	1
969	410	430	3.95	3.27	36	30	1	3
970	250	330	3.82	3.52	82	27	1	2
971	420	600	3.21	2.71	40	29	0	1
972	550	560	3.45	3.46	49	32	1	3
973	480	320	3.64	3.62	32	29	1	3
974	260	350	3.53	3.57	79	36	0	2
975	550	450	3.81	2.78	33	38	1	4
976	440	350	3.74	3.50	72	37	0	2
977	430	330	3.53	3.61	46	31	1	3
978	400	520	3.88	2.59	59	34	1	3
979	580	490	3.39	2.58	51	39	1	4
980	610	490	3.68	3.64	54	35	0	4
981	280	590	3.52	2.64	47	35	1	4
982	360	490	3.59	2.80	70	37	1	4
983	650	640	3.49	3.17	44	22	0	3
984	430	510	3.34	3.61	63	34	1	1
985	440	620	3.16	3.38	58	33	0	4
986	510	590	3.35	2.85	77	34	1	4
987	650	650	3.84	2.77	86	26	1	3
988	450	620	3.61	3.42	57	31	0	4
989	350	540	3.04	3.04	39	28	1	1
990	480	460	3.55	3.04	79	41	0	4
991	610	550	3.98	2.99	34	24	1	4
992	330	570	3.60	3.45	26	30	0	3
993	470	580	3.54	2.84	51	28	1	3
994	320	760	3.33	2.71	79	37	0	3
995	450	780	3.72	2.60	33	35	1	4
996	440	570	3.35	2.45	61	41	0	4
997	390	550	3.45	2.72	58	32	0	3
998	480	490	3.43	2.48	53	37	0	1
999	550	600	3.51	3.13	55	40	0	4
1000		440	3.90					1

OBS	VGRE	QGRE	GR-GPA	UG-GPA	LSINDEX	AGE	SEX	MARITAL
1001	250	520	3.21	3.31	69	29	1	2
1002	420	550	3.49	3.56	68	31	1	4
1003	470	500	3.47	2.51	57	35	0	4
1004	650	490	4.00	2.92	71	36	0	3
1005	300	290	3.21	2.57	50	30	0	4
1006	530	500	3.91	2.61	48	38	0	4
1007	610	550	3.71	2.34	43	40	0	4
1008	560	650	3.85	3.20	67	28	0	4
1009	440	400	3.58	2.63	52	38	1	4
1010	300	520	3.35	2.87	46	36	0	2
1011	550	440	3.85	2.96	78	34	0	1
1012	390	500	3.54	3.53	36	30	0	2
1013	520	450	3.36	2.80	81	31	1	2
1014	740	530	3.69	3.34	78	27	0	3
1015	460	650	3.56	2.39	55	36	1	4
1016	600	460	3.86	2.70	69	46	0	4
1017	460	510	3.48	3.39	70	27	0	2
1018	480	670	3.74	3.54	62	28	1	2
1019	250	640	3.64	3.54	74	33	1	2
1020	460	480	3.71	3.03	87	26	0	4
1021	380	410	3.44	2.94	77	30	0	2
1022	460	660	3.39	2.86	41	27	0	2
1023	340	560	3.31	2.99	81	33	0	2
1024	400	550	3.51	3.59	76	28	0	2
1025	670	380	3.76	2.98	75	41	0	2
1026	480	620	3.71	3.35	85	29	1	3
1027	470	280	3.56	3.44	38	35	0	4
1028	550	480	3.95	2.85	45	27	1	2
1029	640	650	3.75	3.52	73	33	1	4
1030	440	520	3.42	2.93	42	27	0	4
1031	490	610	3.73	2.64	42	36	1	4
1032	480	540	3.71	2.76	68	36	0	4
1033	500	520	3.51	3.31	41	37	0	4
1034	510	530	3.45	3.54	71	25	0	4
1035	540	460	3.66	2.77	35	28	1	1
1036	530	450	3.71	3.41	52	35	1	2
1037	410	450	3.65	2.62	46	36	0	4
1038	550	500	3.60	2.82	51	32	0	4
1039	670	510	3.72	2.82	57	36	0	4
1040	450	550	3.65	3.25	64	42	0	1
1041	420	420	3.51	2.77	85	28	1	2
1042	400	500	3.98	2.85	66	33	0	3
1043	480	560	3.51	3.06	35	28	0	3
1044	260	330	3.10	2.90	44	39	1	1
1045	390	400	3.16	3.58	41	23	1	1
1046	540	470	3.66	3.26	86	33	1	4
1047	410	530	3.47	3.44	33	27	0	3
1048	530	570	3.39	3.26	73	27	1	2
1049	760	490	3.90	3.26	—	33	0	3
1050	380	580	3.22	2.72	—	31	0	2

OBS	VGRE	QGRE	GR-GPA	UG-GPA	LSINDEX	AGE	SEX	MARITAL
1051	430	480	3.75	3.35	62	28	1	4
1052	470	360	3.80	3.52	61	41	1	4
1053	540	560	3.40	2.50	60	39	0	3
1054	410	630	3.67	3.04	60	29	0	3
1055	680	350	3.56	3.34	73	32	0	4
1056	600	500	3.92	2.94	54	37	1	4
1057	630	460	3.63	3.38	42	34	0	3
1058	400	400	3.36	3.41	71	25	1	1
1059	230	630	3.55	2.97	62	32	1	3
1060	360	420	3.19	3.21	17	27	0	4
1061	580	500	3.77	2.75	44	42	0	2
1062	300	340	3.09	2.41	74	31	1	3
1063	520	390	3.19	3.18	35	26	0	3
1064	420	630	3.47	3.25	70	33	0	2
1065	460	410	3.38	2.98	83	29	1	3
1066	470	490	3.55	3.69	32	33	0	1
1067	460	530	3.41	2.46	70	30	0	4
1068	360	550	3.08	3.37	50	31	1	2
1069	510	430	3.52	2.66	85	22	0	4
1070	410	630	3.62	2.92	54	32	0	2
1071	420	410	3.40	2.59	79	35	0	4
1072	380	650	3.73	3.32	62	37	1	4
1073	610	450	3.62	3.62	53	32	0	4
1074	500	500	4.00	3.35	60	27	1	2
1075	430	440	4.39	3.27	64	25	1	4
1076	310	480	3.16	2.17	88	33	1	4
1077	360	470	3.35	2.57	43	40	0	2
1078	560	380	3.33	2.86	49	35	0	4
1079	540	520	3.41	2.82	71	24	0	4
1080	380	530	3.39	3.16	76	33	0	3
1081	450	570	3.01	2.55	56	29	0	2
1082	530	460	3.64	3.32	44	25	0	4
1083	540	510	3.78	3.28	32	35	0	4
1084	170	520	3.16	2.49	44	38	1	1
1085	540	590	3.59	3.75	59	25	1	3
1086	330	540	3.30	3.45	56	27	0	4
1087	530	470	3.39	2.64	75	40	1	4
1088	400	650	3.63	3.29	52	27	0	2
1089	560	520	3.98	2.81	75	29	1	4
1090	660	550	3.64	3.85	32	41	0	3
1091	430	350	3.51	2.73	57	30	1	4
1092	530	490	3.59	3.56	51	35	0	3
1093	560	640	3.34	2.57	55	33	1	4
1094	570	460	3.76	3.51	58	31	0	4
1095	590	590	3.81	2.81	70	32	1	4
1096	480	470	3.38	3.44	54	29	0	2
1097	460	590	3.47	3.47	40	29	1	1
1098	310	400	3.82	2.82	73	30	1	2
1099	340	—	3.50	2.73	48	35	0	4
1100	570	400	3.57	3.90	58	33	1	4

OBS	VGRE	QGRE	GR-GPA	UG-GPA	LSINDEX	AGE	SEX	MARITAL
1101	660	530	3.50	3.46	59	31	1	4
1102	420	350	3.25	3.23	58	33	1	4
1103	520	670	3.68	2.48	58	30	0	1
1104	700	630	3.90	3.61	63	29	1	2
1105	630	630	3.49	3.50	58	27	1	2
1106	590	490	3.64	2.54	77	38	1	4
1107	650	520	3.65	3.70	76	28	0	3
1108	540	440	3.36	3.39	61	32	1	2
1109	540	530	3.88	2.86	95	32	1	4
1110	560	520	3.75	3.66	63	33	0	2
1111	580	380	3.72	2.90	83	33	1	4
1112	300	580	3.29	2.75	67	32	0	2
1113	270	510	3.27	2.79	84	38	0	2
1114	370	500	3.43	3.61	43	29	1	3
1115	330	470	3.30	3.25	40	28	0	1
1116	370	340	3.43	2.78	72	30	1	2
1117	470	360	3.40	3.51	51	34	0	4
1118	490	330	3.95	3.56	89	38	1	2
1119	470	650	3.68	3.16	42	25	1	4
1120	200	590	3.69	2.91	34	35	0	4
1121	370	430	3.49	3.28	69	27	1	1
1122	500	490	3.32	3.45	34	35	1	3
1123	540	620	3.72	3.46	59	28	0	4
1124	570	410	3.55	2.93	58	37	1	4
1125	570	520	3.53	2.66	59	30	1	4
1126	600	410	3.80	3.99	52	36	0	3
1127	400	690	3.62	2.79	37	42	1	4
1128	430	400	3.35	3.15	67	29	1	4
1129	490	550	3.72	2.93	40	33	0	3
1130	650	450	3.67	3.52	47	40	1	4
1131	460	550	3.42	2.76	83	32	0	3
1132	480	580	3.59	3.52	74	28	1	2
1133	680	600	3.88	2.79	90	28	0	2
1134	510	490	3.41	3.43	36	30	1	3
1135	430	340	3.38	3.02	48	27	0	1
1136	640	380	3.72	2.86	83	34	0	2
1137	470	500	3.40	2.78	80	32	0	2
1138	490	530	3.71	3.20	86	35	1	4
1139	520	550	3.69	3.87	70	37	0	4
1140	420	530	3.77	2.43	39	34	0	4
1141	710	570	4.00	3.08	36	37	0	3
1142	390	540	3.38	2.50	60	29	0	3
1143	530	600	3.77	3.39	60	30	1	2
1144	410	560	3.72	3.63	65	33	1	4
1145	480	530	3.69	3.06	71	29	1	3
1146	480	700	3.63	3.68	38	31	0	1
1147	500	450	3.71	2.63	46	32	0	3
1148	630	560	3.59	2.75	60	31	0	2
1149	500	560	3.59	2.75	75	34	0	3
1150	550	700	3.91	3.23	75	43	0	2

OBS	VGRE	QGRE	GR-GPA	UG-GPA	LSINDEX	AGE	SEX	MARITAL
1151	700	560	3.84	2.50	79	37	0	3
1152	520	430	3.69	2.91	61	27	0	3
1153	280	550	3.31	2.63	72	31	1	2
1154	470	460	3.68	3.57	55	36	1	4
1155	490	440	3.57	2.98	64	31	1	4
1156	600	570	3.79	3.55	15	29	1	3
1157	600	350	3.85	3.62	56	32	1	4
1158	470	460	3.68	3.37	75	29	1	2
1159	370	410	3.46	3.34	75	32	1	4
1160	540	640	3.70	3.07	82	25	0	2
1161	660	540	4.00	2.89	63	35	0	3
1162	360	520	3.38	2.95	48	36	0	3
1163	450	440	3.69	3.59	42	31	1	3
1164	540	630	3.72	2.71	63	33	0	4
1165	400	500	3.67	3.42	58	29	1	1
1166	350	470	3.45	3.36	53	32	0	2
1167	480	570	3.72	3.31	67	35	1	3
1168	490	650	3.37	2.94	36	29	1	3
1169	610	640	3.84	3.63	59	34	0	4
1170	280	540	3.61	3.44	56	32	1	4
1171	450	430	3.52	2.97	63	32	0	4
1172	460	550	3.41	2.80	56	38	1	1
1173	470	580	3.50	2.76	43	27	0	4
1174	440	490	3.80	2.59	36	44	1	4
1175	480	410	3.20	2.78	78	32	0	2
1176	500	360	3.09	2.75	43	37	1	4
1177	380	430	3.79	2.38	58	39	1	4
1178	340	620	3.23	3.59	54	28	0	3
1179	580	640	3.40	3.60	27	29	1	4
1180	410	710	3.52	3.05	55	28	1	4
1181	290	660	3.49	2.39	82	27	0	2
1182	390	500	3.49	3.25	81	26	1	3
1183	290	510	2.96	3.29	88	33	0	2
1184	510	460	3.24	2.88	32	29	0	2
1185	440	490	3.42	3.75	88	30	1	3
1186	440	560	3.85	3.44	56	39	1	4
1187	340	600	3.22	2.84	38	30	1	3
1188	570	540	3.36	2.58	60	30	0	3
1189	320	360	3.77	3.17	83	32	0	2
1190	640	490	3.07	2.74	70	28	1	4
1191	470	620	3.60	3.32	83	28	0	2
1192	780	480	3.22	3.60	58	31	0	4
1193	390	530	3.90	2.24	73	34	0	4
1194	350	680	3.48	2.73	48	32	1	2
1195	450	510	3.70	3.24	51	28	1	4
1196	510	520	3.38	3.38	32	29	1	3
1197	280	380	3.30	3.89	60	37	0	4
1198	440	610	3.69	3.53	45	28	1	1
1199	390	490	3.32	2.98	63	34	1	4
1200	410	590	3.32	2.98	63	27	0	1

OBS	VGRE	QGRE	GR-GPA	UG-GPA	LSINDEX	AGE	SEX	MARITAL
1201	440	430	3.45	3.12	70	30	1	2
1202	480	540	3.98	3.01	73	25	0	3
1203	470	700	3.64	3.04	69	32	0	3
1204	440	420	3.69	2.67	74	37	0	3
1205	470	640	3.55	2.64	85	32	0	2
1206	440	590	3.67	3.28	56	32	1	4
1207	380	630	3.57	3.50	75	32	1	3
1208	400	560	3.37	3.32	29	30	1	3
1209	340	470	3.70	3.58	35	33	1	3
1210	570	650	3.93	3.13	85	31	0	2
1211	440	580	3.66	3.60	42	30	1	3
1212	210	640	3.43	2.94	41	33	0	4
1213	500	430	3.73	3.60	39	30	1	4
1214	440	730	3.63	3.35	79	31	1	2
1215	370	390	3.56	2.71	71	47	0	4
1216	430	530	3.09	3.03	73	26	0	4
1217	440	440	3.71	2.77	45	37	1	4
1218	390	530	3.70	3.22	63	35	1	4
1219	550	610	3.79	3.29	62	40	0	2
1220	500	440	3.42	2.31	68	31	0	2
1221	480	670	3.35	2.55	79	25	1	4
1222	430	620	3.53	3.31	65	32	0	4
1223	500	410	3.60	2.99	47	37	0	4
1224	480	540	3.27	3.65	62	26	1	2
1225	530	640	3.88	3.07	80	24	0	4
1226	720	600	4.00	3.38	60	36	1	4
1227	630	520	3.55	2.58	42	35	0	4
1228	490	410	3.26	2.78	48	32	1	4
1229	630	460	3.67	2.67	85	30	0	2
1230	560	660	3.85	2.98	78	38	0	2
1231	430	370	3.42	3.31	71	34	1	4
1232	480	310	3.48	3.03	61	36	1	4
1233	440	540	3.71	2.59	51	36	0	4
1234	330	520	3.46	3.27	70	32	1	4
1235	380	500	3.18	3.53	55	35	1	4
1236	770	560	3.72	3.67	56	26	1	3
1237	630	430	3.81	3.68	27	35	1	3
1238	420	380	3.30	2.73	64	30	0	4
1239	340	530	3.24	2.55	51	38	0	4
1240	480	520	3.38	2.99	40	34	0	2
1241	450	430	3.44	3.35	67	29	1	4
1242	320	630	3.53	3.23	58	32	1	2
1243	390	660	3.43	2.46	83	27	0	2
1244	410	370	3.42	3.20	76	28	0	2
1245	440	690	3.33	2.55	79	32	1	1
1246	530	600	3.50	2.96	49	31	0	3
1247	440	520	3.61	3.09	64	33	1	3
1248	250	460	3.32	3.24	28	30	0	3
1249	410	590	3.83	2.89	78	36	0	2
1250	320	410	3.41	2.92	45	30	0	4

Appendix I
Answers to Odd-Numbered Exercises

Chapter 1

1.1 The population: the amounts of time that all children spend watching television. The sample: the set of television times for the 100 randomly selected children.

1.3 (a) represents a sample if randomly selected from all employees at GM (b) Sample (c) Sample (d) Population

1.5 Inferential statistics uses a sample to form a conclusion about a population. A census obtains the entire population and so there is no uncertainty regarding the population.

1.7 Age is ratio, sex is nominal, race is nominal, GPA is interval, classification is nominal.

1.9 How the United States finished (1st, 2nd, 3rd, . . .) in ten different Olympic track and field events.

Chapter 2

2.1 (b) and (c) using $K = 8$, $CW = 10$ (d) classes are under 40, 40–49, 50–59, . . . (f) relative frequencies are .049, .122, .098, . . .

2.3 (a) Relative frequencies are .38, .31, .21, .06, .03, .01 (b) lower class limits are 0, 2, 4, 6, 8 and the upper class limits are 2, 4, 6, 8, 10 (c) class midpoints are 1, 3, 5, 7, 9 (d) no

2.5 (a) Discrete (b) classes are 0–4, 5–9, . . . and the class relative frequencies are .167, .167, .167, .367, .133

2.7 (a) Classes are less than 4, 4 and under 7, 7 and under 10, . . . and the class frequencies are 3, 8, 5, 2, 2 (b) class width = 3, no

2.13 (b) The shape indicates that only one class has a low frequency and most of the data fell in the larger class intervals (c) no, the shape would not change

2.15 Classes are 5–9, 10–14, 15–19, . . .

2.19 (a) Classes are 0–3, 4–7, 8–11, 12–15, 16–19; frequencies are 12, 10, 7, 1, 1; cumulative relative frequencies are .39, .71, .94, .97, 1.00 (b) 29%

2.23 (b) Housing. Yes it is, by 17.6% (Old CPI) and 23.9% (New CPI).

2.25 (c) The suicide frequencies are somewhat uniformally distributed between ages 20 and 60 with small frequencies for ages less than 20.

2.31 (a) bar chart

2.33 (a) The outlier, 2.13, should be put into an open ended class (b) classes are .65 and under .75, .75 and under .85, . . .; frequencies are 4, 5, 4, 4, 1, 1, 1; cumulative relative frequencies are .20, .45, .65, .85, .90, .95, 1.00 (c) 85% of the astronauts have a response time less than 1.05 minutes.

2.35 (a) Classes are 5 and under 10, 10 and under 15, . . . (b) cumulative relative frequencies are .167, .445, .723, .834, .945, 1.0 (c) 7.5, 12.5, 17.5, 22.5, 27.5, 32.5

2.37 Classes are 35–39, 40–44, 45–49, 50–54, 55–59, 60–64, 65–69, 70–74, 75–79, 80–84

2.39 Classes are 41–50, 51–60, 61–70, 71–80, 81–90, 91–100; frequencies are 7, 7, 1, 3, 7, 5

2.41 (a) Classes are 1–5, 6–10, . . .; frequencies are 6, 2, 7, 7, 5, 3 (d) Six of the thirty people were able to memorize the material in five minutes or less.

2.43 (a) Classes are 15–19, 20–24, . . . (b) relative frequencies are .057, .286, .257, .171, .086, .057, .057, .029 (c) 60% of the total 35 weeks, there were less than 29 calls. During approximately 71% of the total 35 weeks, there were between 20 and 34 calls.

2.45 (a) Classes are less than 500,000, 500,00 and under 100,000, . . . (d) The distribution peaks between 500,000 and 1,499,999 and over 70% of the distribution is between 500,000 and 5,499,999.

2.47 (a) Classes are 150–169, 170–189, . . .; frequencies are 4, 9, 6, 3, 3; relative frequencies are .16, .36, .24, .12, .12

Chapter 3

3.1 mean = 93.13, median = 92, mode = 88

3.3 mean = 8.135, median = 6.35, mode = 8.6

3.5 approximately $25,000

3.7 (a) not overloaded (b) overloaded

3.9 mean = 74.35, median = 57.5, mode = 130

3.11 mean = 6.38, median = 6.3, mode = 6.3

3.13 (a) range = 87 (b) MAD = 24.09 (c) variance = 794.08 (d) $s = 28.18$ (e) $CV = 56.27$

3.15 (a) mean = 324.5 (b) range = 220.1 (c) MAD = 42.867 (d) variance = 5171.72 (e) $s = 71.915$

3.17 (a) For A: mean = 241.75, variance = 11495.48, $s = 107.22$. For B: mean = 70.75, variance = 1945.84, $s = 44.11$ (b) Yes (c) $CV(A) = 44.35$, $CV(B) = 62.35$

3.19 For A: mean = 43.2, $s = 33.37$. For B: mean = 36.6, $s = 24.68$. A is "better" and B is more consistent.

3.21 variance = 1742.13

3.23 approximately the 39th percentile

3.25 (a) 32 (b) 19 (c) 1st quartile = 10, 2nd quartile = 17, 3rd quartile = 27.5

3.27 (a) $Z = .78$ (b) $X = 63$ is .78 standard deviation to the right of the sample mean.

3.29 Mean is 37

3.31 $X = 13.25$

3.33 (a) Mean = 4.262, median = 3.93, $s = 1.096$ (b) $Sk = .909$. The data are skewed right.

3.35 48.5 to 96.5

3.37 Near zero

3.39 1 to 11

3.41 25 to 49

3.43 0 to 134.62

3.45 $s = \$.50$. 95% of the values will lie between $1.50 and $3.50

3.47 median = $26,833. The mean is $28,900.

3.49 median = 24.99

3.51 (a) Mean = 75.96, variance = 162.8049 (b) $s = 12.76$

3.53 Mean = 412.125, $s = 7.53$

3.55 Mean = 1760, $s = 2932$

3.57 Mean = .0013271, $s = .0001555$

3.59
1	4 1 8 9 7 8 6 7
2	1 8 6 1 7 2 6 0 5
3	5 8 0

3.61
1	
1	9
2	1 1 2 3 4 2 3 4 4 4
2	5 6 5 6 8
3	0 2
3	5 6

3.63
3	.6 .5 .8
4	.0 .2 .3 .5 .2 .3
5	.4 .3
6	.3 .0 .3 .1

3.65 $L = 18$, $H = 100$, Quartiles are 66, 80.5, 92

3.67 (a) Mean = 103.083, median = 102, no mode (b) range = 38 (c) s = 12.117

3.69 (a) Mean = 506 (b) s = 65.905 (c) Median = 507.5 (d) Sk = −.0683, yes (e) 95% between 374.19 and 637.81

3.71 (a) Mean 44.75 (b) s = 5.658 (c) Median = 45.772 (d) Sk = −.542

3.73 (a) Z = −4.0 (b) X = 45 is 4 standard deviations to the left of the mean

3.75 (a) Mean = 53.756, s = 2.78

50	.2	.8	.9
51	.4	.1	
52	.2	.8	.1
53	.8		
54	.3	.7	
55	.5	.9	
56			
57	.7	.6	
58			
59	.1		

3.77 (a) Ranges are 3.5, 90700, and 16 (b) Means are 2.758, 53208.333, and 8.25 (c) Medians are 2.85, 48000, and 7.5 (d) Standard deviations are 1.005, 32424.274, and 5.242 (e) Z scores are .44, −.5, and −.62

1	.8	.0	.3	
2	.8	.6	.7	.9
3	.8	.5	.2	.0
4	.5			

(g) Quartiles are 4.5, 7.5, and 11

Chapter 4

4.1 (a) Positive (b) No relationship (c) Positive (d) No relationship

4.3 Nearly a perfect negative linear relationship

4.5 Low positive relationship

4.7 Positive linear relationship

4.9 (a) Significant-correlated (b) Not significant-uncorrelated (c) Not significant-uncorrelated (d) Significant-correlated (e) Not significant-uncorrelated

4.11 (a) r = −.983 (b) Yes (c) Significant

4.13 (a) r = .697 (b) Yes (c) Significant

4.15 (b) r = −.8847 (c) There is a negative relationship (d) Significant

4.17 (a) Rank correlation = −.857 (b) Negative relationship (c) Significant

4.19 (a) Rank correlation = −.964 (b) Negative relationship (c) Significant.

4.21 (a) Rank correlation = .829 (b) Significant (c) r = .840, significant

4.23 (a) Rank correlation = .4286 (b) Positive relationship (c) Not significant (d) Yes

4.25 (a) There is a negative linear relationship (b) r = −.942 (c) Significant (d) Rank correlation = −.967 (e) The values are nearly the same.

4.27 (a) r = .950 (b) There is a significant linear relationship, but a cause-effect relationship cannot be concluded (c) Rank correlation = .943

4.29 (a) There is a negative linear relationship (b) r = −.949 (c) Significant

4.31 (a) Rank correlation = −.921 (b) Significant (c) Both are negative and close in value.

4.33 (a) r = .924 (b) r = .8845 (c) the correlation in part a is slightly higher.

4.35 (b) r = .983 (c) Significant (d) Positive relationship

4.37 r = .501. There is not a significant correlation between the dryness index and daily fire alarms.

Chapter 5

5.1 (a) Outcomes are {Red, Blue, Green} (b) Yes (c) 1/3

5.3 .75

5.5 5/4

5.7 (a) .16 (b) .38 (c) .82 (d) .62 (e) 0 (f) .18 (g) .03 (h) .36

5.9 (a) .09 (b) .42 (c) .73 (d) .04

5.11 (a) .048 (b) .559 (c) .158 (d) .034 (e) .111

5.13 (a) .40 (b) .62 (c) .14 (d) .24 (e) .484 (f) .389 (g) a) $P(W)$ b) $P(S)$ c) $P(U$ and $M)$ d) $P(A$ or $P)$ e) $P(W|S)$ f) $P(U|M)$ (h) (i) .16 (ii) .38 (iii) .30 (iv) .76 (v) .632 (vi) .75

5.15 (a) .4   (b) .96   (c) .15   (d) .10   (e) .25

5.17 .58

5.19 (a) 2   (b) 1   (c) 2   (d) 3   (e) 2   (f) 3   (g) 2

5.21 No

5.23 $P(A|B) = .4$. Events A and B are independent since $P(A \text{ and } B) = P(A)P(B)$.

5.25 (a) (i) .25 (ii) .25 (iii) .75 (iv) .25 (v) .75 (vi) .5 (vii) .5   (b) A and B are independent.

5.27 (b) .3   (c) .4   (d) .333   (e) (i) 0 (ii) .2 (iii) .1   (f) (i) 1 (ii) .6 (iii) .7

5.29 $P(H \text{ or } W) = .96$, $P(H \text{ and } W) = .15$

5.31 (a) (i) .36 (ii) .12   (b) (i) .358 (ii) .121

5.33 .167

5.35 $P(B) = .75$

5.37 (a) .024   (b) .08   (c) .92   (d) .48   (e) .336   (f) .976

5.39 Yes

5.41 .00148

5.43 (a) .216   (b) .432

5.45 26250

5.47 720

5.49 15504

5.51 90

5.53 24

5.55 60

5.57 (a) 1,073,741,824   (b) 1/1073741824

5.59 1/595

5.61 1/792

5.63 1/45

5.65 (a) .2167   (b) .231   (c) .7833   (d No

5.67 (a) .4666   (b) .643   (c) Events are dependent

5.69 (a) .281   (b) .641   (c) .719

5.71 (a) .203   (b) .726   (c) .071   (d) .394

5.73 (a) 90   (b) (i) .333 (ii) .5 (iii) .111

5.75 .075

5.77 No, since $P(A) + P(B)$ is greater than 1.

5.79 (a) .55   (b) .85   (c) 1   (d) .15   (e) 0   (f) .4706

5.81 847,660,528

5.83 (a) .21   (b) .476   (c) .79   (d) .714

5.85 $P(A \text{ and } B) = .15$, $P(A \text{ or } B) = .75$.

5.87 (a) .57   (b) .632

Chapter 6

6.1 The researcher would be interested in the number of ounces gained by the rats subjected to a special electro-mechanical stimulus. The variable would be continuous.

6.3 (a) discrete   (b) discrete   (c) continuous   (d) continuous   (e) discrete   (f) continuous

6.5

X	Y	P(X, Y)	X	Y	P(X, Y)
1	H	1/12	4	H	1/12
1	T	1/12	4	T	1/12
2	H	1/12	5	H	1/12
2	T	1/12	5	T	1/12
3	H	1/12	6	H	1/12
3	T	1/12	6	T	1/12

6.7 Each probability is .25, so $P(X = 3) = .25$.

6.9 Yes

6.11 $P(X = 0) = 9/49$, $P(X = 1) = 24/49$, $P(X = 2) = 16/49$

6.13 Yes

6.15 (a) $P(X = 2) = .20$ (b) .008

6.17 (a) .096   (b) .059   (c) .829

6.19 $P(\text{white}) = 1/6$, $P(\text{blue}) = 2/6$, $P(\text{black}) = 3/6$

6.21 Mean = .2, variance = .18

6.23 Mean = 3, variance = 3

6.25 Mean = 1.1428, variance = .6999

6.27 Mean = 3, variance = 2

6.29 (a) .667   (b) 4.5

6.31 (a) .714   (b) .286   (c) 6

6.33 (a) Yes   (b) Hispanic: mean = 2, variance = 2; Asian: mean = 2.5, variance = 1.25

6.35 (a) 9.5   (b) 3.452   (c) .417

6.37 .2963

6.39 (a) .049   (b) $p = .05$, $P(X \geq 3) = .98$ (c) approx. one

6.41 .098

6.43 (a) .094 (b) .094 (c) identical (d) .0137

6.45 (a) .157 (b) .502 (c) .057

6.47 Mean = 2.16, variance = 1.901

6.49 (a) .107 (b) .72 (c) .05

6.51 .8

6.53 .6353 using the binomial approximation

6.55 .9286

6.57 (a) .011 (b) .599 (c) variance = .454

6.59 Mean = 3, $P(X \geq 3) = .5768$

6.61 (a) Mean = 8, $P(X > 4) = .9004$ (b) 2.83

6.63 Mean = 5, $P(X > 6) = .2378$

6.65 (a) 2.89 (b) 2.99 (c) .7779, .7299

6.67 Mean = 3.5, variance = 18.85

6.69 (a) .029 (b) .237 (c) .117 (d) .085

6.71 .560

6.73 (a) .00345 (b) .999

6.75 (a) .168 (b) .773 (c) .397

6.77 .279

6.79 (a) .792 (b) .002 (c) .206

6.81 (a) .026 (b) .002 (c) No, they are not the same

6.83 (a) .035 (b) .956

6.85 (a) .265 (b) 2.235

6.87 X = number of males, $P(X = 3) = .4762$

6.89 (a) .3916 (b) .2308 (c) 3.2 (d) 9.8

6.91 (a) .5714 (b) .2857 (c) .1429

6.93 (a) .8407 (b) .4437 (c) .0021

Chapter 7

7.1 (a) 2 (b) 1.55 (c) .25 (d) .6925

7.3 (a) .524 (b) 23.15 (c) 1.819

7.5 (a) .273 (b) 12.5 (c) 10.083

7.7 $a = 126$, $b = 131$, $P(X > 128) = .6$

7.9 These parameters indicate where the curve is centered and how wide the curve is.

7.11 (a) .3413 (b) .0919 (c) .6826 (d) .0606

7.13 (a) .9439 (b) .0179 (c) .1986 (d) .9010

7.15 (a) $z = .35$ (b) $z = .67$ (c) $z = 1.57$
(d) $z = -1.62$

7.17 $z = .64$, $z = 1.57$

7.19 (a) .3632 (b) .2119 (c) .3842 (d) .1791

7.21 (a) .133 (b) .1423

7.23 (a) 2.37 (b) $Z = -.525$, $X = 64,250$ (c) .2514

7.25 .2877

7.27 (a) Mean = 16.45 (b) $s = 1.341$ (c) .3669
(d) .0314

7.29 (a) Z(Wechsler) = 1.5, Stanford-Binet score = 124 (b) (i) .2514 (ii) .2643 (c) .4972

7.31 8.62

7.33 (a) .2514 (b) .1056

7.35 .0054

7.37 (a) .506 (b) .503 (c) They are very close

7.39 .9612

7.41 .0694

7.43 (a) .387 (b) .4483 (c) approximately one
(d) .6700

7.45 .5438

7.47 .2997

7.49 (a) .6367 (b) .1446 (c) .9177

7.51 .223

7.53 .4541

7.55 (a) .2266 (b) .2957

7.57 (a) .0604 (b) .1515

7.59 $x = 7.025$

7.61 $x = 87.12$

7.63 .0764; .0436

7.65 .8264

7.67 .3998

7.69 (a) 865,000 (b) 144,337.567 (c) .63

7.71 (a) .0764 (b) .3594

7.73 .5889

7.75 (a) Mean = 1.0, standard deviation = .289 (b) .3

7.77 50%

7.79 (a) .6826   (b) .3811   (c) .3446   (d) .1587

7.81 .0294

Chapter 8

8.1 (c)

$\bar{X}$	Probability
1	1/36
1.5	2/36
2	3/36
2.5	4/36
3	5/36
3.5	6/36
4	5/36
4.5	4/36
5	3/36
5.5	2/36
6	1/36

8.3 (a) .2033   (b) .0019

8.5 (a) .017   (b) .1446   (c) .2974   (d) .7108

8.7 (a) .2033   (b) .99997   (c) approximately zero

8.9 without replacement, .0020; with replacement, .0031

8.11 .7960. We are assuming the amounts follow a normal distribution.

8.13 .0823

8.15 .0069

8.17 approximately zero

8.19 19.86 to 21.34

8.21 110.562 to 125.438

8.23 2.924 to 4.076

8.25 22.24 to 23.36

8.27 50.910 to 53.890

8.29 7.86 to 8.74

8.31 (a) 1.311   (b) 2.160   (c) 1.7341   (d) −1.325
      (e) −1.708   (f) approx. 1.282

8.33 (a) Zinc: 69.04 to 81.76; Magnesium: 2.49 to 3.27
      (b) Copper: 89.29 to 105.31; Iron: 58.11 to 102.89

8.35 (a) 62.25 to 73.75   (b) 63.21 to 72.79

8.37 302.37 to 317.63

8.39 (a) Mean = 30.907, $s = 15.345$ (b) 22.048 to 39.765   (c) the confidence interval is narrower

8.41 539.819 to 540.181

8.43 89

8.45 16

8.47 6.1033

8.49 4

8.51 303

8.53 39

8.55 $\bar{X}_s = 3.489$, $\sigma_{\bar{X}_s} = .2132$, CI is 3.1383 to 3.8397

8.57 $\bar{X}_c = 4.129$, $\sigma_{\bar{X}_c} = .118$, CI is 3.935 to 4.323

8.59 $\bar{X}_c = 3.511$, $\sigma_{\bar{X}_c} = .302$, CI is 2.919 to 4.103

8.61 (a) .2266   (b) .0329

8.63 (a) 33.43 to 39.07   (b) 2.675 to 3.065   (c) 2.25 to 2.71

8.65 59

8.67 (a) 117.7 to 122.7   (b) 13

8.69 $\bar{X}_s = 50.36$, $\sigma_{\bar{X}_s} = .728$, CI is 48.94 to 51.79

Chapter 9

9.1 (a) $k = 1.96$ (b) $k = 1.645$ (c) $k = 2.575$
     (d) $k = 2.33$

9.3 (a) False   (b) False

9.5 $Z^* = 4.65$, reject H_0

9.7 $Z^* = -3.486$, reject H_0

9.9 The 95% confidence interval does not contain 2.0, so reject H_0

9.11 $Z^* = 7.073$, reject H_0

9.13 $Z^* = -14.368$, reject H_0

9.15 $Z^* = 2.34$, reject H_0

9.17 (a) .9732   (b) 1   (c) .9131

9.19 (a) $Z^* = -10.541$, reject H_0

9.21 $Z^* = -7.83$, reject H_0

9.23 $Z^* = 6.62$, reject H_0

9.25 $Z^* = -2.67$, reject H_0

9.27 $Z^* = 4$, fail to reject H_0

9.29 (b) Mean = 3.16, $s = 2.251$ (c) $Z^* = 5.153$, fail to reject H_0

9.31 (a) Not significant (b) Significant (c) Not significant (d) Inconclusive

9.33 (a) .0174 (b) .05 (c) .0055 (d) .0644

9.35 $Z^* = -12.599$, reject H_0

9.39 (a) Mean = 4.8045, $s = 1.0322$
(b) $t^* = -.628$, fail to reject H_0 (c) Fail to reject H_0

9.41 (a) $.02 < p < .05$ (b) $.005 < p < .01$
(c) $p > .10$ (d) $.10 < p < .20$

9.43 (a) $t^* = -5.99$, reject H_0 (b) p value $< .01$

9.45 (a) $t^* = 2.75$, reject H_0
(b) $.005 < p$ value $< .01$

9.47 (a) 5.10 to 26.70, 3.37 to 15.63, -3.40 to 32.80
(b) (i) $t^* = -2.62$, fail to reject H_0 (ii) $t^* = -.05$, fail to reject H_0

9.49 $t^* = .409$, p value $> .10$, fail to reject H_0

9.51 chi-square = 30.795, p value $> .10$, fail to reject H_0

9.53 (a) 2.62 to 60.33 (b) 1.619 to 7.77 (c) fail to reject H_0 (d) p value $> .10$ (one-sided test)

9.55 $a = 9.39$, $b = 28.865$

9.57 chi-square = 15.04, p value $> .10$, fail to reject H_0

9.59 chi-square = 45.24, fail to reject H_0

9.61 $t^* = -2.018$, reject H_0

9.63 (a) Mean = 304.29, $s = 165.28$ (b) $t^* = .87$, fail to reject H_0 (c) p value $> .10$

9.65 $Z^* = -2.47$, reject H_0

9.67 (a) .0308 (b) .177 (c) $.02 < p < .05$
(d) $.1 < p < .2$

9.69 (a) $Z^* = 1.10$, p value = .1562 (b) fail to reject H_0 for all three significance levels

9.71 (a) 36.47 to 40.73 (b) $Z^* = -1.29$, fail to reject H_0 (c) p value = .0985

9.73 (a) 2.48 to 2.72 (b) reject H_0

9.75 (a) 65.941 to 637.290 (b) 43.583 to 1772.964
(c) chi-square = 5.407, p value $> .10$, fail to reject H_0

Chapter 10

10.1 Dependent samples matched by location

10.3 Dependent samples

10.5 Dependent samples

10.7 (a) Yes (b) No

10.9 (a) $Z^* = 1.77$, p value = .0384 (b) $Z^* = -.19$, p value = .8494 (c) $Z^* = -.06$, p value = .4761

10.11 7.485 to 8.171

10.13 (a) $Z^* = 5.62$, reject H_0 (b) p value is approximately zero, reject H_0

10.15 $Z^* = 7.942$, p value is approximately zero, reject H_0

10.17 54.275 to 125.725

10.19 $Z^* = -3.2$, reject H_0

10.21 $t^* = 1.53$, fail to reject H_0

10.23 (a) $t^* = 2.350$, reject H_0 (b) $t'^* = 3.139$, $df = 15$, reject H_0 (c) for (a), p value is between .02 and .05; for (b), p value is less than .005

10.25 .53 to 10.27

10.27 -48.364 to -1.636

10.29 (a) $t'^* = 3.555$, $df = 31$, reject H_0 (b) p value $< .005$, reject H_0

10.31 (a) Use t test that does not assume equal variances (b) Use t test that assumes equal population variances

10.35 $U = U_1 = 43$, fail to reject H_0

10.37 $U_1 = 9$, $U_2 = 16$, $U = \min(9, 16) = 9$, fail to reject H_0

10.39 $U_1 = 41$, $U_2 = 59$, $U = \min(41, 59) = 41$, fail to reject H_0

10.41 $U_2 = 28.5$, fail to reject H_0

10.43 .51 to 19.49

10.45 (a) $t_D^* = 1.239$, fail to reject H_0 (b) p value $> .2$ (c) Yes, these are dependent samples matched by location

10.47 $t_D^* = .2896$, fail to reject H_0

10.49 (a) $t_D^* = .638$, fail to reject H_0 (b) The samples are dependent

10.51 -110.28 to 1047.43

10.55 (a) yes (b) yes (c) yes (d) yes (e) yes (f) all of them (g) the samples are dependent

10.57 $T_+ = 1.5$, $T_- = 26.5$, table value $= 4$, reject H_0

10.59 $T_+ = 36$, $T_- = 0$, table value $= 0$, reject H_0

10.61 $T_+ = 0$, $T_- = 78$, table value $= 17$, reject H_0

10.63 .1273 to .3983

10.65 .511 to 12.753

10.67 Ratio of variances is .767

10.69 (a) $F^* = 1.056$, fail to reject H_0 (b) p value $> .10$ (c) .498 to 2.237

10.71 $Z^* = 2.51$, p value $= .006$, reject H_0

10.73 (a) $t^* = -4.87$, reject H_0 (b) p value $< .01$ (c) $t'^* = -.62$, p value $> .2$, fail to reject H_0

10.75 $t^* = 1.27$, fail to reject H_0

10.77 (a) $t^* = 3.49$, $df = 30$, p value $< .005$, reject H_0 (b) yes

10.79 (a) There is not a significant difference for all four courses (b) In all courses, except civil engineering, there is a significant difference (c) This appears to be accurate

10.81 $t^* = .803$, p value $> .10$, fail to reject H_0

10.83 (a) $-.068$ to 1.323 (b) fail to reject H_0 (c) The total sample size is $n = 21 + 12 = 33$.

10.85 (a) -5.998 to 1.998. Since this interval contains 0, fail to reject H_0 (b) The total sample size is $154 + 231 = 385$.

Chapter 11

11.1 (a) $n = 577$ (b) $n = 369$

11.3 (a) .483 to .557 (b) $n = .037$ (c) increase sample size or reduce confidence level

11.5 .363 to .748

11.7 (a) .3717 to .3983 (b) .1869 to .2131 (c) .0498 to .0702

11.9 (a) $\hat{p} = .126$ (b) .0545 to .1975 (c) $n = 812$

11.11 $n = 1168$

11.13 (a) $\hat{p} = .187$, confidence interval is .166 to .207 (b) $\hat{p} = .887$, confidence interval is .867 to .907 (c) $\hat{p} = .338$, confidence interval is .299 to .377

11.15 (a) $\hat{p} = .4797$, confidence interval is .4105 to .5489

11.17 (a) $\hat{p} = .2$, $Z^* = .403$, fail to reject H_0 (b) .6892

11.19 $\hat{p} = .12$, $Z^* = -2.123$, reject H_0

11.21 n must be at least 167

11.23 $Z^* = .3985$, fail to reject H_0

11.25 (a) .200 to .528 (b) $Z^* = .172$, p value $= .4325$, fail to reject H_0

11.27 $\hat{p} = .16$, $Z^* = -1.14$, p value $= .1271$, fail to reject H_0

11.29 (a) $\hat{p} = .71$, $Z^* = 1.59$, fail to reject H_0 (b) p value $= .0559$, fail to reject H_0

11.31 (b) $Z^* = -4.29$, reject H_0

11.33 The sign is changed

11.35 $-.097$ to .107

11.37 Total sample size is $332 + 332 = 664$

11.39 $Z^* = .98$, p value $= .1635$, fail to reject H_0

11.41 (a) Not all the respondents answered all the questions (b) (i) .5759 to .6681 (ii) .8881 to .9459 (iii) .2963 to .4377

11.43 (a) .087 to .491 (b) $E = .202$ (c) $n = 451$

11.45 (a) $Z^* = 2.18$, reject H_0 (b) $\hat{p} = .8$, confidence interval is .7462 to .8538

11.47 (a) .0127, .022, .0294 (b) 115, 340, 612

11.49 (a) .23 (b) .13 (c) 41

11.51 No

11.53 $\hat{p} = .4839$, $Z^* = 1.35$, fail to reject H_0

11.55 $Z^* = -.512$, fail to reject H_0

11.57 (a) The rate is declining (b) 6.73 (c) .0029 to .0271 (d) $\hat{p}_1 = .106$, $\hat{p}_2 = .028$, confidence interval is .0592 to .0968 (e) yes

11.59 .2499 to .4961

Chapter 12

12.1 (a) 6 (b) no

12.3 (a) 1.4784 to 3.2076 (b) $-.158$ to 1.586

12.5 (a) 3 levels and 8 replicates at each level (b) (i) 6.48 to 7.82 (ii) .2953 to 2.1797

(c)

Source	df	SS	MS	F
Factor	2	8.692	4.346	5.30
Error	21	17.224	.820	
Total	23	25.916		

(d) $.01 < p < .025$ (e) working environment
(f) $D = 1.13$, means 1 and 3, 2 and 3 are unequal

12.7 (a)

Source	df	SS	MS	F
Factor	2	54.111	27.056	12.817
Error	15	31.667	2.111	
Total	17	85.778		

(b) p value $< .01$ (c) Yes   (e) $D = 2.18$, means 1 and 3, 2 and 3 are unequal

12.9

Source	df	SS	MS	F
Factor	2	11.6667	5.8335	1.232
Error	27	127.8	4.7333	
Total	29	139.4667		

p value $> .10$, fail to reject H_0

12.11 (a)

Source	df	SS	MS	F
Factor	2	.009	.0045	0.03125
Error	12	1.728	.144	
Total	14	1.737		

(b) $F^* = .03125$, fail to reject H_0 (c) p value $> .10$, no   (e) a multiple comparisons analysis is not appropriate

12.13 The sum of squares are all divided by 10,000 but the resulting F value is unchanged.

12.15

Source	df	SS	MS	F
Factor	2	625.0	312.5	8.68
Error	18	647.6	36.0	
Total	20	1272.6		

$F^* = 8.68$, p value $< .01$, reject H_0

12.17 (a)

Source	df	SS	MS	F
Factor	2	1.2381	.61905	.074
Error	18	151.4286	8.4127	
Total	20	152.6667		

$F^* = .074$, p value $> .10$, fail to reject H_0
(b) Right eye: .6968 to 5.3032; left eye: 1.1254 to 5.7318; both eyes: 1.2682 to 5.8746

12.19 (a) Randomized block design   (b) blocks are schools, factor levels are Experimental Group and Control Group   (c) Two-way factorial design

12.25 (a) No, they are dependent   (b) Family status is the factor, school is the block   (c) There is none

(d)

Source	df	SS	MS	F
Factor	1	12	.12	1.765
Blocks	5	2.98	.596	
Error	5	.34	.068	
Total	11	3.44		

$F_1 = 1.765$, fail to reject H_0

12.27 $b = 5$, $F_1 = 4.7273$

12.29

Source	df	SS	MS	F
Factor	3	2759.454	919.818	14.984
Blocks	10	1259.727	125.973	
Error	30	1841.546	61.385	
Total	43	5860.727		

$F_1 = 14.984$, p value $< .01$, reject H_0

12.31

Source	df	SS	MS	F
Factor	3	293.1	97.7	15.5
Blocks	9	5160.2	573.355	90.955
Error	27	170.2	6.3037	
Total	39	5623.5		

Both F values are significant

12.33 (b)

Source	df	SS	MS	F
Factor	1	92.57	92.57	7.67
Blocks	6	16.43	2.73	0.23
Error	6	72.42	12.07	
Total	13	181.42		

$F_1 = 7.67$, p value is between .025 and .05, reject H_0 (c) .76 to 9.54

(d)

Source	df	SS	MS	F
Factor	1	16.07	16.07	1.84
Blocks	6	28.42	4.73	.54
Error	6	52.44	8.74	
Total	13	96.93		

12.35 (a)

Source	df	SS	MS	F
Factor	4	69.526	17.382	9.716
Blocks	1	16.641	16.641	9.302
Error	4	7.154	1.789	
Total	9	93.321		

$F_1 = 9.716$, reject H_0 (b) p value for F_1 is between .01 and .025, p value for F_2 is between .025 and .05 (c) $F_2 = 9.302$, reject H_0

12.37 (a)

Source	df	SS	MS	F
Keyboard (A)	2	242.67	121.335	68.242
Software (B)	2	8.22	4.11	2.312
Interaction	4	11.11	2.778	1.562
Error	18	32.0	1.778	
Total	26	294.0		

(b) $F_2 = 2.312$, fail to reject H_0 (c) $F_1 = 68.242$, reject H_0 (d) $F_3 = 1.562$, fail to reject H_0 (e) p value for $F_1 < .01$, p value for $F_2 > .10$, p value for $F_3 > .10$

12.39 (a) Yes (c) There must be more than one replicate per cell (d) Since the number of replicates per cell (r) must be at least two.

12.41 (a)

Source	df	SS	MS	F
Age (A)	1	275.234	275.234	4170.2
Chol. Level (B)	1	96.504	96.504	1462.2
Interaction	1	6.824	6.824	103.394
Error	8	.527	.066	
Total	11	379.089		

$F_1 = 4170.2$, p value $< .01$, reject H_0 (b) $F_2 = 1462.2$, p value $< .01$, reject H_0

12.43 $KW = 1.369$, fail to reject H_0

12.45 $KW = 6.167$, reject H_0

12.47 $KW = 2.577$, fail to reject H_0

12.49 $KW = 1.93$, fail to reject H_0

12.51

Source	df	SS	MS	F
Factor	3	3.3599	1.1200	1.2489
Error	32	28.6965	.8968	
Total	35	32.0564		

$F^* = 1.2489$, fail to reject H_0

12.53 lectures 1 and 2: $-.7867$ to 3.0723; lectures 2 and 3: -1.0723 to 2.7867; lectures 1 and 3: -3.9295 to .0705

12.55

Source	df	SS	MS	F
Factor	3	217.9	72.6333	706.55
Error	36	3.7	.1028	
Total	39	221.6		

$F^* = 706.55$, p value $< .01$, reject H_0

12.57

Source	df	SS	MS	F
Factor	9	190	21.111	2.5503
Blocks	2	250	125.0	15.1006
Error	18	149	8.2778	
Total	29	589		

12.59 $H = 2.64$, fail to reject H_0

12.61 $KW = 6.045$, fail to reject H_0

12.63 (a) No

(b)

Source	df	SS	MS	F
Factor	2	.0815	.041	4.8
Blocks	3	.096	.032	3.7
Error	6	.0515	.0086	
Total	11	.229		

$F_1 = 4.8$, fail to reject H_0 **(c)** p value for F_1 is between .05 and .10   **(d)** $F_2 = 3.7$, fail to reject H_0

12.65 (a)

Source	df	SS	MS	F
Factor	2	132.867	66.434	1.608
Blocks	9	858.3	95.367	2.308
Error	18	743.8	41.322	
Total	29	1734.967		

(b) $F_1 = 1.608$, fail to reject H_0 **(c)** $F_2 = 2.308$, fail to reject H_0

12.67 (a)

Source	df	SS	MS	F
Class (A)	2	780.778	390.389	5.93
Sex (B)	1	122.722	122.722	1.864
Interaction	2	84.778	42.389	.644
Error	12	790.0	65.833	
Total	17	1778.278		

(b) $F_1 = 5.93$, reject H_0 **(c)** $F_2 = 1.864$, fail to reject H_0 **(d)** $F_3 = .644$, fail to reject H_0 **(e)** p value for F_1 is between .01 and .025, p values for F_2 and F_3 are $>.10$

12.69 $F_1 = 2.375$, fail to reject H_0, no.

Chapter 13

13.1 **(a)** $Z^* = -.8386$, fail to reject H_0 **(b)** Chi-square value = .7033, fail to reject H_0 **(c)** The square of Z is equal to Chi-square.

13.5 **(a)** Chi-square value = 3.6213, fail to reject H_0 **(b)** p value $> .10$ **(c)** no

13.7 Pool the groups aspertaine, saccharin, and no preference. Chi-square value = 1.2, fail to reject H_0

13.9 **(a)** Chi-square value = .216, fail to reject H_0 **(b)** p value is greater than .1

13.11 $\hat{p} \cong .20$, Chi-square value = .4184, fail to reject H_0

13.13 **(a)** Mean = 506, standard deviation = 65.905, Chi-square value = .1609, fail to reject H_0 **(b)** p value is between .975 and .990

13.15 **(a)** Chi-square value = 182.976, reject H_0, p value $< .005$ **(b)** Chi-square value = .3244, p value is between .975 and .990, fail to reject H_0 **(c)** There is little evidence to support the existence of the subject's ESP.

13.17 Chi-square value = 387.788, p value $< .005$, reject H_0

13.19 **(a)** Chi-square value = 8.09, reject H_0 **(b)** the categories are dependent, p value is between .01 and .025   **(c)** at 5%, reject H_0; at 1%, fail to reject H_0

13.21 Chi-square value = 3.2511, reject H_0

13.23 Chi-square value = 6.06514, fail to reject H_0

13.25 **(a)** Chi-square value = 15.32, fail to reject H_0 **(b)** p value is between .05 and .10

13.27 Chi-square value = 8.8571, reject H_0

13.29 Chi-square value = 1.72, fail to reject H_0

13.31 **(a)** Chi-square value = 4.398, fail to reject H_0 **(b)** p value is between .10 and .90

13.33 **(a)** Chi-square value = 2.53, p value $> .10$, fail to reject H_0 **(b)** the X-rays are not useful

13.35 **(a)**

X	No. of Cities
0	2
1	5
2	6
3	3
4	2
5	1
6 or more	1
	20

(b) Categories are $\leqslant 1$, 2, $\geqslant 3$, Chi-square value = .078   **(c)** p value is between .95 and .975

13.37 **(a)** Categories are 0, 1, $\geqslant 2$, Chi-square value = .137, fail to reject H_0 **(b)** Yes   **(c)** $\hat{p} = 0.24$, Chi-square value = .842, fail to reject H_0 **(d)** .24

13.39 (a) Mean = 28.9, standard deviation = 21.66
(b) Chi-square value = 25.666, reject H_0 (c) p value < .005 (d) No

13.41 Chi-square value = 4, fail to reject H_0

13.43 (a) Chi-square value for Table 1 = 92.70, reject H_0; Chi-square value for Table 2 = 45.18, reject H_0 (b) Chi-square value = 136.88, reject H_0 (c) yes, since the df and critical values are the same

Chapter 14

14.1 (a) $r = -.943$, significant negative linear relationship (b) cov(X, Y) = -75.5 (c) $\hat{Y} = 82.81 - .2745X$ (d) $SSE = 25.446$

14.3 (a) Cov(X, Y) = 162.078 (b) $\hat{Y} = 33.83 + .286X$

14.7 (a) $\hat{Y} = .4543 + 4.0615X$ (b) $\hat{Y} = 19.5434$ (thousand)

14.9 (a) $r = .965$ (b) $\hat{Y} = -29.476 + .358X$

14.11 (a) $\hat{Y} = -3.872 + 5.819X$ (b) slope = 5.819, intercept = -3.872 (c) SSE = 129.535, $s^2 = 25.907$ (d) residuals are 6.652, -.240, -1.985, 4.682, -.748, -.822, -7.623

14.13 (a) $\hat{Y} = 1.7528 + .8998X$ (b) SSE = 6.4219, $s^2 = 2.1406$ (c) residuals are -.0122, -.1121, -1.7419, .0282, 1.8381 (d) yes (e) no

14.15 .3158

14.17 SSE = 155.94, $s = 2.6624$, all the residuals are within two standard deviations

14.19 Yes, the variance of the error component is not constant.

14.21 (a) $\hat{Y} = 27.421 + .61X$ (b) $t^* = 14.987$, reject H_0 (c) Yes (d) .5375 to .6824 (e) .04 (f) .4869 (g) p value < .005

14.23 (a) $t^* = 9.04$, p value < .01, reject H_0 (b) There is a positive relationship between X and Y (c) .0396 (d) .2112 to .5047

14.25 .7909 to 1.0155

14.27 (b) $\hat{Y} = 9.4629 + .000382X$ (c) $t^* = 6.307$, reject H_0 (d) p value < .005, the conclusion does not change at 5% and 1% (f) yes

14.29 (a) slope = -.964 (b) $t^* = -4.873$, reject H_0 (c) p value < .005 (d) -1.303 to -.6249

14.31 (a) $r = .942$ (b) $t^* = 6.875$, reject H_0 (c) p value < .01 (d) .887 (e) 88.7%

14.33 $r = .8371$, $t^* = 5.5172$, there is a linear relationship

14.35 (a) $r = -.92593$ (b) $t^* = -10.686$, reject H_0 (c) there is a negative relationship (d) p value < .005 (e) .8573 (f) .938

14.37 (a) No (b) $r = -.1225$, $t^* = -.4450$, fail to reject H_0

14.39 (a) $r = .449$ (b) $t^* = 3.30$, reject H_0 (c) .202 (d) 20.2%

14.41 (a) $\hat{Y} = 3.922 + .724X$ (b) $r = .824$ (c) .678 (d) SSE = 3.762, $s = .537$, $t^* = 5.988$ (e) 21.432 to 22.031 (f) 20.534 to 22.93

14.43 (a) $\hat{Y} = 69.085$ (b) 67.88 to 70.29 (c) 69.085 (d) 65.10 to 73.07

14.45 $\hat{Y} = -.0267 + 1.2233(.30) = .3403$, prediction interval is .2098 to .4708

14.47 The prediction interval is wider. Both intervals are the narrowest for $X = \bar{x}$

14.49 (a) $r = 0.855$, $t^* = 4.662$, reject H_0 (b) 73% (c) $\hat{Y} = 7.664$, confidence interval is 6.664 to 8.664 (d) 4.902 to 10.425

14.51 (a) $\hat{Y} = 108.3 - .575X$ (b) $\hat{Y} = 112.6 - 0.67X$ (c) -1 to -.33 (d) -1.025 to -.315

14.53 (a) regression line through the medians is $\hat{Y} = -27.155 + .335X$, least squares regression line is $\hat{Y} = -29.476 + .358X$ (b) medians: .16 to .44, least squares line: .2112 to .5047

14.55 (a) $\hat{Y} = 71 - 3.715(10.5) = 31.99$ (b) 1.75 to 6.5

14.57 (a)

	PDI	UAI	IDV	MAS
PDI	1.0	.76515	- .40284	.09149
UAI		1.0	- .73832	.36335
IDV			1.0	- .01749
MAS				1.0

(b) PDI and UAI, UAI and IDV($|r| > .514$)

14.59 (a) $\hat{Y} = 219.2880 - 1.2342X$ (b) $r = -.7253$, $t^* = -6.5794$, reject H_0 (c) SSE = 154.1407, confidence interval is -1.543 to -.926

14.61 (a) $\hat{Y} = -1.048 + .467X$ (b) SSE = 2.215 (c) 2.215 (d) .89 (e) 89%

14.63 SSE = 80.75, $s = 2.061$, one of the sample residuals lies outside two standard deviations, the histogram appears to be nearly symmetric

14.65 (a) $\hat{Y} = 2.82033 + .87090X$ (b) 73%   (c) $n = 20$ (d) (i) 152.407 (ii) 41.206 (iii) 111.201   (e) .60835 to 1.13344   (f) $t^* = 6.9694$, reject H_0 (g) The results indicate a linear relationship between X and Y, p value < .01

Chapter 15

15.1 (a) $\hat{Y} = 233.045$ (b) $\hat{Y} = 242.096$, net change = 9.051

15.3 (a) $\hat{Y} = 134.731 - 1.050X_1 + .048X_2 - .002X_3$ (b) SSE = 2174.24   (c) $s = 14.05$ (d) $\hat{Y} = 50.931$

15.5 (a) SSE = 4.4899   (b) 0.6414

15.7 6.6646

15.9 (a) $F^* = 196.5126$, reject H_0

15.11 (a) $t^* = 2.7678$, reject H_0 (b) $t^* = 6.496$, reject H_0

15.13 (a) $F^* = 9.57$, reject H_0 (b) $|t^*| = 3.13$, p value < .01, reject H_0 (c) for X_2, $|t^*| = 1.67$, fail to reject H_0; for X_3, $|t^*| = .69$, fail to reject H_0 (d) (i) -1.6523 to $-.4487$ (ii) $-.01535$ to .11185 (iii) $-.1048$ to .0094

15.15

Source	df	SS	MS	F
Regression	7	200	28.5714	10.3896
Residual	20	55	2.75	
Total	27	255		

15.17

Source	df	SS	MS	F
Regression	2	97.12	48.56	1.9994
Residual	17	412.88	24.287	
Total	19	510		

$F^* = 1.9994$, fail to reject H_0

15.19 (a) $\hat{Y} = -.50 + 2.40X_1 + 2.95X_2$ (b) 93.81%   (c) 20   (d) (i) 314.80 (ii) 19.50 (iii) 295.30 (iv) 128.39 (v) 3.59 (vi) 14.048 (vii) 2.110 (viii) 1.15

15.21 one

15.23 $F^* = 14.056$, X_2 and X_3 contribute

15.25 (a) $F^* = 55.52$, reject H_0 (b) p value < .01

15.27 $F^* = 2.6367$, the three variables do not contribute

15.29 (a) $R^2 = .64$

15.31 Presence of multicollinearity

15.33 Correlation matrix:

	X_1	X_2	X_3
X_1	1.0	.140	$-.854$
X_2	.140	1.0	.075
X_3	$-.854$	.075	1.0

The analysis indicates a high correlation between X_1 and X_3.

15.35 (a) .686   (b) .242   (c) Both variables should stay in the model.

15.39 (a) 3 dummy variables are needed   (b) Define X_1 = emission of nitrous and sulfur oxides, X_2 = water pollution index, X_3 = population density, $X_4 = 1$ if female, $X_5 = 1$ if urban, $X_6 = 1$ if suburban

15.41 (a) $\hat{Y} = 1.2011 + 1.828X_1 - .796X_2 - .756X_3 - .620X_4 - .327X_5$ (b) $\hat{Y} = .7926 + 2.5722X_1 + .0186X_2 - .0770X_3$ (c) .930   (d) .038   (e) .008   (f) no

15.43 (a) BDI   (b) $F^* = 15.489$, reject H_0 (c) yes

15.45 Stepwise regression can remove variables previously entered

15.47 Equal variance for error terms

15.49 (a) 37.68 to 63.00   (b) 16.90 to 83.79   (c) the confidence interval is narrower

15.51 The prediction interval is always wider than the corresponding confidence interval

15.53 (a) $\hat{Y} = 25.1434$, confidence interval is 22.9139 to 27.3729   (b) 18.7494 to 31.5374

15.55 (b) multicollinearity

15.57 (a) $F^* = 37.16$, reject H_0 (b) $F^* = 1.48$, X_2 does not contribute   (c) .0192 to .1465   (d) 1.4757 to 2.3519   (e) $\hat{Y} = 11.5657$

15.59 (a) $\hat{Y} = .10255 + .68808X_1 + .39844X_2$
(b) $R^2 = .504$ (c) $n = 20$ (d) (i) 170.7
(ii) 84.609 (iii) MSE $= 4.9770$ (e) $F^* = 8.649$,
reject H_0 (f) $t^* = 3.529$, reject H_0 (g) $t^* = 1.351$, fail to reject H_0 (h) 12.80105 to
23.98405   (i) 11.11815 to 25.6669

15.63 (a) (i) $\hat{Y} = 61.7 + 3.64X_1 - .436X_2 - .577X_3 + .289X_4$ (ii) $F^* = 5.88$
(iii) $t^* = -2.39$ (b) $F^* = 5.88$, reject H_0
(c) $|t^*| = 2.39$, reject H_0 (d) $\hat{Y} = 34.04$
(e) 20.22 to 47.85   (f) 28.22 to 39.86

15.65 (a) (i) $\hat{Y} = 32.199 - .035X_1 + .074X_2 + .2508X_3 + 8.931X_4$ (ii) 189.95 (iii) .4745
(iv) 2.26 (v) for X_1, $t^* = -.023$; for X_2,
$t^* = .038$; for X_3, $t^* = .476$; for X_4, $t^* = 2.307$
(b) $F^* = 2.26$, p value $> .10$, fail to reject H_0
(c) $t^* = .476$, fail to reject H_0 (d) for X_1,
$|t^*| = .023$, fail to reject H_0; for X_2, $t^* = .038$,
fail to reject H_0; for X_4, $t^* = 2.307$, reject H_0
(e) There is a very high correlation (in absolute
value) between X_1 and X_2 (f) 55.33 to 72.73
(g) 17.78 to 110.28

Index

Areas of the Standard Normal Distribution

The entries in this table are the probabilities that a standard normal random variable is between 0 and z (the shaded area).

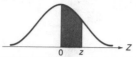

z	Second Decimal Place in z									
	0.00	0.01	0.02	0.03	0.04	0.05	0.06	0.07	0.08	0.09
0.0	0.0000	0.0040	0.0080	0.0120	0.0160	0.0199	0.0239	0.0279	0.0319	0.0359
0.1	0.0398	0.0438	0.0478	0.0517	0.0557	0.0596	0.0636	0.0675	0.0714	0.0753
0.2	0.0793	0.0832	0.0871	0.0910	0.0948	0.0987	0.1026	0.1064	0.1103	0.1141
0.3	0.1179	0.1217	0.1255	0.1293	0.1331	0.1368	0.1406	0.1443	0.1480	0.1517
0.4	0.1554	0.1591	0.1628	0.1664	0.1700	0.1736	0.1772	0.1808	0.1844	0.1879
0.5	0.1915	0.1950	0.1985	0.2019	0.2054	0.2088	0.2123	0.2157	0.2190	0.2224
0.6	0.2257	0.2291	0.2324	0.2357	0.2389	0.2422	0.2454	0.2486	0.2517	0.2549
0.7	0.2580	0.2611	0.2642	0.2673	0.2704	0.2734	0.2764	0.2794	0.2823	0.2852
0.8	0.2881	0.2910	0.2939	0.2967	0.2995	0.3023	0.3051	0.3078	0.3106	0.3133
0.9	0.3159	0.3186	0.3212	0.3238	0.3264	0.3289	0.3315	0.3340	0.3365	0.3389
1.0	0.3413	0.3438	0.3461	0.3485	0.3508	0.3531	0.3554	0.3577	0.3599	0.3621
1.1	0.3643	0.3665	0.3686	0.3708	0.3729	0.3749	0.3770	0.3790	0.3810	0.3830
1.2	0.3849	0.3869	0.3888	0.3907	0.3925	0.3944	0.3962	0.3980	0.3997	0.4015
1.3	0.4032	0.4049	0.4066	0.4082	0.4099	0.4115	0.4131	0.4147	0.4162	0.4177
1.4	0.4192	0.4207	0.4222	0.4236	0.4251	0.4265	0.4279	0.4292	0.4306	0.4319
1.5	0.4332	0.4345	0.4357	0.4370	0.4382	0.4394	0.4406	0.4418	0.4429	0.4441
1.6	0.4452	0.4463	0.4474	0.4484	0.4495	0.4505	0.4515	0.4525	0.4535	0.4545
1.7	0.4554	0.4564	0.4573	0.4582	0.4591	0.4599	0.4608	0.4616	0.4625	0.4633
1.8	0.4641	0.4649	0.4656	0.4664	0.4671	0.4678	0.4686	0.4693	0.4699	0.4706
1.9	0.4713	0.4719	0.4726	0.4732	0.4738	0.4744	0.4750	0.4756	0.4761	0.4767
2.0	0.4772	0.4778	0.4783	0.4788	0.4793	0.4798	0.4803	0.4808	0.4812	0.4817
2.1	0.4821	0.4826	0.4830	0.4834	0.4838	0.4842	0.4846	0.4850	0.4854	0.4857
2.2	0.4861	0.4864	0.4868	0.4871	0.4875	0.4878	0.4881	0.4884	0.4887	0.4890
2.3	0.4893	0.4896	0.4898	0.4901	0.4904	0.4906	0.4909	0.4911	0.4913	0.4916
2.4	0.4918	0.4920	0.4922	0.4925	0.4927	0.4929	0.4931	0.4932	0.4934	0.4936
2.5	0.4938	0.4940	0.4941	0.4943	0.4945	0.4946	0.4948	0.4949	0.4951	0.4952
2.6	0.4953	0.4955	0.4956	0.4957	0.4959	0.4960	0.4961	0.4962	0.4963	0.4964
2.7	0.4965	0.4966	0.4967	0.4968	0.4969	0.4970	0.4971	0.4972	0.4973	0.4974
2.8	0.4974	0.4975	0.4976	0.4977	0.4977	0.4978	0.4979	0.4979	0.4980	0.4981
2.9	0.4981	0.4982	0.4982	0.4983	0.4984	0.4984	0.4985	0.4985	0.4986	0.4986
3.0	0.4987	0.4987	0.4987	0.4988	0.4988	0.4989	0.4989	0.4989	0.4990	0.4990
3.1	0.4990	0.4991	0.4991	0.4991	0.4992	0.4992	0.4992	0.4992	0.4993	0.4993
3.2	0.4993	0.4993	0.4994	0.4994	0.4994	0.4994	0.4994	0.4995	0.4995	0.4995
3.3	0.4995	0.4995	0.4995	0.4996	0.4996	0.4996	0.4996	0.4996	0.4996	0.4997
3.4	0.4997	0.4997	0.4997	0.4997	0.4997	0.4997	0.4997	0.4997	0.4997	0.4998
3.5	0.4998									
4.0	0.49997									
4.5	0.499997									
5.0	0.4999997									

Reprinted with permission from *Standard Mathematical Tables*, 15th ed., © CRC Press, Inc., Boca Raton, FL.